GEOMETRY REVEALED

Selected works of Marcel Berger

RESEARCH

Le spectre d'une variété riemanniene (with Paul Gauduchon and Edmond Mazet), Springer, 1971

(Collaboratively under the pseudonym Arthur L. Besse) *Manifolds all of whose geodesics are closed*, Springer, 1978

(Collaboratively under the pseudonym Arthur L. Besse) *Einstein Manifolds*, Springer, 1978

PEDAGOGY AND POPULARIZATION

Geometry I, II. Springer, 1987, 2009

Differential Geometry: Manifolds, Curves and Surfaces (with Bernard Gostiaux), Springer, 1987

"Peut-on définir la géométrie aujourd'hui?", in *Results in Mathematics*, vol. 40, pp. 37–87, 2001

A Panoramic View of Riemannian Geometry, Springer, 2003

Cinq siècles de mathématiques en France, Éd. ADPF (Association pour la diffusion de la pensée française), 2005

Convexité dans le plan et dans l'espace: de la puissance et de la complexité d'une notion simple (with Pierre Damphousse), Ellipses, Paris, 2006

"Geometry in the 20th century", 2002, in *History of mathematics*, edited by V.L. Hansen and J.J. Gray, Encyclopedia of Life Support Systems (EOLSS), Developed under the auspices of UNESCO, Eolss Publishers, Oxford, UK

MARCEL BERGER

Geometry Revealed

A Jacob's Ladder to Modern Higher Geometry

 Springer

Author
Marcel Berger
Insitut des Hautes Études Scientifiques (IHES)
Bures-sur-Yvette
France

Translator
Lester J. Senechal
Professor Emeritus
Department of Mathematics
Mount Holyoke College
South Hadley, MA 10475
USA
lsenecha@mtholyoke.edu

Springer-Verlag thanks the original publishers of the figures for permission to reprint them in this book. We have made every effort to identify the copyright owners of all illustrations included in this book in order to obtain reprint permission. Some of our requests have however remained unanswered. We have inserted all sources and owners where known.

ISBN 978-3-662-50122-1   ISBN 978-3-540-70997-8 (eBook)
DOI 10.1007/978-3-540-70997-8
Springer Heidelberg Dordrecht London New York

Mathematics Subject Classification: 51-01, 01-01

Cover design: Wmx design

Cover illustration: Le Songe de Jacob (detail), Nicolas Dipre (D'Ypres), École d'Avignon, Musée du Petit-Palais, Avignon © bpk, Berlin, 2009

Printed on acid-free paper

Springer is part of Springer Science+Business Media (www.springer.com)

About the Author

Marcel Berger has played a unique role in the development of geometry in France. The important *exceptional holonomy* is due to him, and practically all the geometry that is dear to hearts of physicists is related to it. The elegance of his theorem on the 1/4 pinching continues to attract young mathematicians to Riemannian geometry.

A veritable school of geometry formed about him in the 1970s. His students, and his students' students (now about 90 in number), form the nucleus of geometry in France. He maintains contact with the association "Arthur Besse" where he and his students have written several books: the one on the spectra of Riemannian manifolds was for a while the bible of the subject. His book on closed geodesics turned out to be a scientific adventure story because important problems were solved during the writing process. The one on Einstein manifolds became a best-seller because it appeared at an opportune moment and popularized a little-known subject whose methods were primitive. It showed great flair, for today these manifolds constitute a common area of research for theoretical physicists and mathematicians of all specialties.

Last but not least, Marcel Berger recognized Mikhael Gromov's talent and convinced him to remain in Paris, which has been a determining factor for the development of geometry in France. Marcel Berger himself had likewise been the beneficiary of the support of André Lichnerowicz, a leading figure in differential geometry and relativistic mechanics, who gave Berger a thesis topic precisely on the holonomy groups that subsequently played an important role in his work.

As a teacher, Marcel Berger has been able to share his passion for geometry — from the outset that of elementary geometry in all its forms — with generations of students. His book *Geometry* remains unequaled as a survey, disclosing and modernizing all the various points of view that comprise elementary geometry, which of course is neither simple nor easy.

Marcel Berger was Professor at the Universities of Strasbourg (1953–1964), Nice (1964–1966), Paris (1966–1974), Director of Research at CNRS (1974–1985 and 1994–1996), Director of IHES (1985–1994). He was President of the French Mathematical Society 1979–1981.

Introduction

Numerous problems of geometry that are quite visual and can be presented in a very simple manner have one or more of the following properties in common:

- they remain unsolved, or have been solved only recently following great efforts;

- for being well understood — and eventually completely or partly solved — they require the creation of concepts and tools that vary in their degree of abstraction, which is in any case greater than what is required for stating the problem;

- the mathematical tools used in solving them were conceived for quite other purposes.

In this work we present a whole series of such problems while showing the necessity of abstract concepts and how they enter progressively into the solution. These are conceptual notions, each built "above" the preceding and permitting an increase in abstraction, represented metaphorically by Jacob's ladder with its rungs. This parable appears as a *leitmotiv* throughout this book.

We don't neglect mentioning problems that still remain open, an "openness" that may seem *a priori* astonishing, but less so once we understand the totality of the efforts and conceptual progress needed for solving similar problems. These classical problems are ever the object of vigorous research, all the while mathematical research is constantly suggesting new ones. And thus so-called elementary geometry is indeed very much alive and at the very heart of the work of numerous contemporary mathematicians.

This book pursues another goal: to show of course the unceasingly renewed vigor of the spirit of geometry, but also to offer readers the elements of a modern geometric culture. For mathematical instruction in our time presents a disquieting paradox. On the one hand, geometry is increasingly present in daily life; we live in a civilization of images. Virtual reality, robotic vision, aerial navigation and the conquest of space require more and more specialists: engineers, aerial controllers, navigators in space, etc. But at the same time geometry — in any case spatial geometry — is almost completely missing from the instructional programs of schools and universities. The small amount of three-dimensional geometry that nominally still belongs to those programs is in fact practically never treated.

It should seem obvious that geometry is of relatively large importance in the whole of mathematics, but the reality is quite different: the language of geometry, geometrical metaphors, have taken hold throughout as an expedient in modern mathematics. The most banal of these metaphors consists of calling a set of objects of the same type a "space" and the elements "points". This has spread to such an extent (and it's no paradox) that mathematicians throw off pictures of the space in order to warm to the objects created by them. Just think for example of the function spaces, whose introduction has thrown light on numerous problems in analysis.

"It may seem surprising that a simple change in language has brought such progress. The impact that it produces seems to come from what might be called a transfer of intuition," writes Jean Dieudonné in an article entitled *Domination universelle de la géometrie*. He adds: "in breaking out of its traditional boundaries, [geometry] has revealed its hidden powers, its flexibility and its extraordinary adaptation abilities, thus becoming one of the most used and universal tools in all branches of mathematics."

It also happens that a theory breathes new life by renewing contact with geometry in a rather unexpected way, so as Antaeus revived his strengths by contact with mother Earth: it's been the case, in recent history, with probability theory, with topology under the geometrization of Thurston and Gromov, and with that part of functional analysis which, under Alain Connes' influence has become noncommutative geometry.

For Alain Connes, "a geometer is a person with sufficient vision to be able to create sufficient mental images that permit treating varied mathematical problems." For "what is difficult and essential in mathematics is the creation of enough mental images to allow the brain to function."

An attempt at explanation was given by Michael Atiyah in his 2000 Fields Lecture *Mathematics in the twentieth century*: "Vision ... uses up something like 80 or 90 percent of the cortex of the brain. There are about 17 different centers in the brain, ... some parts are concerned with vertical, some parts with horizontal, some parts with color, perspective, finally some parts are concerned with meaning and interpretation. Understanding, and making sense of, the world that we see is a very important part of our evolution. Therefore spatial intuition or spatial perception is an enormously powerful tool, and that is why geometry is actually such a powerful part of mathematics — not only for things that are obviously geometrical, but even for things that are not."

There remains in the cortex a place for algebra, which Atiyah associates with time, with the succession of events, of operations. For the geometer, this one-dimensionality evokes the necessity of complying with the logical rules for proof, of metaphorically projecting our intuition onto a single axis. Geometers are often tempted to reject these rules when it is perceived that they bully the intuition, but know from experience that it's precisely these that lead to pushing our imagination beyond its limits so as to create completely new mathematical objects, as we attempt to show in this work.

As for the book's structure, we have on the one hand grouped these problems by their nature, affinity and similarity. For the most part the chapter can be read independently. Nevertheless, together with our book *Geometry* [B] for details beyond the conceptual, this work can serve for a course in geometry as seen from the cultural aspect. We can in fact perceive the various chapters as extensions of [B], illuminating it with some very recent results. We have therefore used [B] as a systematic reference for "elementary" geometry. Although biased, this choice is justified since only [B] treats all the notions used here; conversely, for each particular subject there are so many books that it is hopeless to give systematic references.

But more important is the fact that, in contrast to [B], the results studied are not, with rare exceptions, proved in detail. To lighten the reading, some definitions have been placed at the ends of chapters under the rubric XYZ. Only the crucial ideas and above all the abstract concepts introduced for attaining these results are elucidated.

In this respect we follow in the steps inaugurated by the absolutely remarkable book *Geometry and the Imagination* by Hilbert & Cohn-Vossen (original German, 1932, English translation, 1952), which filled a need for modern and easily accessible cultural geometry. It is this book that we hope to emulate, for it seems to us that a modern version is now much needed. This can only be at the price of a huge increase in size, given the exponential growth of results since the appearance in German in 1932 of Hilbert's course. In our preface to the 1996 republication (Hilbert & Cohn-Vossen, 1996) we emphasized that the work is not intended to be read from the first to the last page, but that we rather hoped that the reader would open it at random and page through it and plunge into this or that chapter with some pleasure depending on intuition and inclination. Will we likewise here be able to transmit our conviction that geometry is especially alive and that there are still innumerable ways to be explored and concepts to be created?

It is important to state that we have by no means covered all the directions of contemporary geometry. Thus we have made but little room for geometric probability, very little for combinatorics and none at all for some recent extensions of the notions of space and point. For this Cartier (1998) and Chap. 3 of Gromov (1999) may be consulted. A good reference for combinatorial geometry is Pach and Argawal (1995); we also find much that is well presented in Aigner and Ziegler (1998). A good idea of several new directions in mathematics and contemporary geometry can be obtained from the recent Carbone, Gromov and Prusinkiewisz (2000).

We should add that mathematics today is advancing extremely fast. We must therefore alert the reader — and especially the researcher — that we are certainly not completely up-to-date in all subjects treated.

This book began as a course given at the University of Pennsylvania in the fall of 1994, at SUNY Stony Brook in the winter of 1995 and at ETH Zürich in 1995–1996. We extend our gratitude to those mathematics departments for making it possible for us to offer courses which diverged very noticeably from traditional instruction.

For the editing we are greatly indebted to the Catherine and André Bellaïche, who bore the overall responsibility for the French edition for Editions Cassini and carried out important work in its realization. From the scientific point of view, practically all chapters have been painstakingly reviewed and edited by outside experts. These generous colleagues are: Patrick Popescu-Pampu for Chap. VII, Daniel Meyer for Chaps. VI and VIII, Pierre Arnoux and Sylvain Gallot for Chaps. XI and XII. Their criticisms and additions, sometimes very detailed, have been essential. André Bellaïche gave the entire text a final critical reading with special attention to Chaps. I and IX, where he rewrote several passages. Donal O'Shea provided many valuable comments on the early chapters during the translation process. To these friends I add with pleasure the name of Lester Senechal, who made the present English translation for Springer-Verlag with great dispatch, considering the compass of the book, and in the course of translation offered numerous remarks, corrections and criticisms important for the completion of this work.

Bibliography

[B] Berger, M. (1987, 2009) *Geometry I,II*. Berlin/Heidelberg/New York: Springer

Aigner, M., & Ziegler, G. (1998, 4th ed. 2010). *Proofs from the book*. Berlin/Heidelberg/New York: Springer

Atiyah, M. (2002). Mathematics in the 20th century. *The Bulletin of the London Mathematical Society, 34*, 1–15

Carbone, A., Gromov, M., & Prusinkiewisz, P. (2000). *Pattern formation in biology and dynamics*. Alghero: World Scientific

Cartier, P. (1998). La folle journée, de Grothendieck à Kontsevich. *Bulletin of the American Mathematical Society, 38*, 389–408

Dieudonné, J. (1980). *The universal domination of geometry*. Berkeley: International Congress of Mathematical Education IV

Dieudonné, J. (1981). *Domination universelle de la géométrie* (traduction du précédent). IREM de Paris-Nord

Gromov, M. (1999), *Metric structures for Riemannian and non-Riemannian spaces*. Basel: Birkhäuser

Hilbert, D., & Cohn-Vossen, S. (1932, 1996). *Anschauliche geometrie*. Berlin/Heidelberg/New York: Springer

Hilbert, D., & Cohn-Vossen, S. (1952). *Geometry and the imagination* (English translation). New York: Chelsea

Pach, J., & Argawal, P. (1995). *Combinatorial geometry*. New York: Wiley

Table of Contents

Chapter I
Points and lines in the plane

I.1. In which setting and in which plane are we working? And right away an utterly simple problem of Sylvester about the collinearity of points

We first work in the coordinate plane, which is familiar to everyone, with its *points* and *lines*. As is usual in the "elementary" geometry of school instruction, this has to do with Euclidean geometry, where there are distances (lengths), angles, circles, etc. This will also be the setting of the next chapter, but even in this first chapter we will see that we can already do many subtle and difficult things – and even find open questions – with only the so-called "affine plane". Affine geometry is a weaker structure than Euclidean geometry. Simply put: we won't be working with anything but points and lines; the mathematical definition is given in Sect. I.XYZ at the end of the chapter. Here we need only recall: two distinct points uniquely determine a line that contains them, along with a segment that joins them; two distinct lines intersect in a single point, with the sole exception of parallel lines. Regarding these, through each point exterior to a given line there passes a unique parallel to that line. Finally, there is a supplementary *affine notion*, more subtle than the merely set-theoretic ones of point and of line, which is the *affine invariant* attached to three collinear points: if a, b, c are collinear, there exists a real number (and one only) denoted $\frac{\overline{ab}}{\overline{ac}}$. It indicates a ratio (that can be negative, although negative numbers had been long forbidden in geometry, even by Poncelet and d'Alembert, until Chasles actually gave them rights of citizenship), a ratio obtainable by parameterizing the line considered, but which does not depend on this parameterization. We can thus speak about the midpoint of a segment, the third-way point, etc. See the necessary details in Sects. I.XYZ and I.3 below. The precise mathematical language is that of the *real affine plane*. If we adjoin a metric – which we permit ourselves occasionally, even in this chapter – we then speak of a *Euclidean plane*. An important remark about language: we can speak of "**the**", rather than "**an**", affine plane. For in fact any two affine planes are necessarily isomorphic, just as are two real vector spaces of dimension 2. The same remark applies to Euclidean spaces of any dimension.

But in this introductory chapter we will see that it is practically impossible to remain in the affine setting: to comprehend and unify certain things, by Sect. I.4 we will need to climb the ladder, know how "to go to infinity" and not interject the Euclidean plane but – more subtly – define the "projective plane". The degree of subtlety can be seen historically: projective geometry wasn't defined until Desargues in the 1650s and then only heuristically. The sound algebraic construction, following

M. Berger, *Geometry Revealed*, DOI 10.1007/978-3-540-70997-8_1,
© Springer-Verlag Berlin Heidelberg 2010

the synthetic attempts of Poncelet and Chasles in the years 1820–1840, was made
in the 1850s by the German school: Plücker, von Staudt and Grassmann, whereas
Euclid dates from 300 B.C.

◆

In 1893 Sylvester posed the following problem:

(I.1.1) *Let* E *be a finite set of points in the plane that has the following property: for
an arbitrary pair of distinct points of* E *there exists, on the line joining them,
a third point of* E. *Show that this is impossible, with the obvious exception of
the case where* E *consists of collinear points on a single line.*

Some readers may prefer the equivalent formulation:

(I.1.2) *If* E *is a finite set of points in the plane not composed of points belonging to
a single line, then there exists at least one line that contains only two of its
points.*

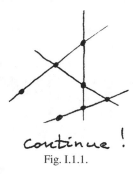

continue !

Fig. I.1.1.

As an exercise we can attempt to convince ourselves of Sylvester's conjecture
by making some sketches: we quickly see that we are forced into constructing sets
having an infinite number of points. But in spite of this easily won insight, it was
not until 1932 that there was a proof of this conjecture, found by Gallai. We owe
to Kelly in (Kelly, 1948) a proof that uses the following Euclidean argument: if
the points are not all collinear, there is a triple of non-collinear points a, b, c of
E forming a true triangle such that the distance from a to the line bc is minimum
among all such triples. We already have a contradiction if b and c are of the same
side of the altitude from a, for then the distance from b to ac, or else that of c to
ab, is less than that of a to bc. So b and c must be on opposite sides of the base
of the altitude. But there exists by hypothesis a third point d of E on bc, and we
are led to a new contradiction by considering either the triangle abd or the triangle
acd.

But this proof leaves us with a bad taste if we are at all purist: the problem
is strictly affine and we should be able to prove the conjecture in a purely affine
manner, without the aid of Euclidean geometry. The purely affine proof of Gallai

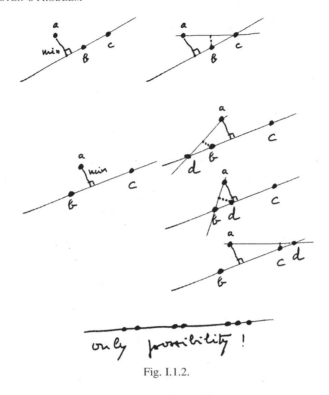

Fig. I.1.2.

is found on p. 181 of Coxeter (1989) . Courageous readers may attempt to find one of their own; but it is important to note that none of the proofs cited so far is *combinatorial* in the sense that in a combinatorial proof we compute the number of points on this or that line, how many lines joining two points of the set pass through a given point, etc., hoping to find relations that contradict the initial hypothesis. On the contrary, Gallai's proof uses the fact that a line in the plane divides it into two distinct *connected* regions; we can't pass from one to the other without intersecting the line. Apart from that, Gallai's proof doesn't introduce any new concept. Where then is Jacob's ladder? We will climb it in two different ways, but reluctant readers may skip immediately to the next section and the second problem of Sylvester.

The first way of ascending the ladder provides a conceptual and combinatorial proof of Sylvester's conjecture, due to Melchior in 1940; details can be found in Chaps. 8 and 10 of Aigner and Ziegler (1998).

We now use some tools whose motivation will be given subsequently: we extend (see Sects. I.7 and I.XYZ) the real affine plane under consideration to a real projective plane $\mathcal{P}$. There we consider not the finite set E of points satisfying Sylvester's condition, but its dual, i.e. the (necessarily finite) set of lines D dual to E under a duality of $\mathcal{P}$.

A duality consists of two mappings: the first associates with each point a of $\mathcal{P}$ a line denoted by a^*; the second associates with each line d a point denoted by d^*. The fundamental properties of a duality are the following:

– the mappings $a \to a^*$ and $d \to d^*$ are inverse to each other;

– if the line d passes through the point a, then the line a^* passes through the point d^*.

For example, corresponding to two points a and b lying on d, there are the two lines a^* and b^* intersecting in the point d^*. For a complete definition, see Sect. I.7.

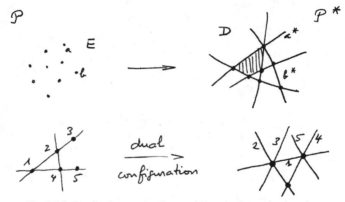

Fig. I.1.3. Duality between points and lines in the projective plane

For the configuration of points and lines provided by D we then obtain combinatorial relations between the two sequences of integers $\{p_r\}$ and $\{t_r\}$ defined as follows: p_r is the number of polygons of r sides that are found in the cellular decomposition of $\mathcal{P}$ that D defines, while t_r is the number of points that lie on r lines of D. We have the following relations, where f_0, f_1, f_2 denote the respective numbers of vertices, edges and polygons of the cellular decomposition: $f_0 = \sum t_r$, $f_2 = \sum p_r$, $f_1 = \sum r t_r = \frac{1}{2} \sum r p_r$. But algebraic topology (see the combinatorics of polyhedra in Sect. VIII.4) tells us that, for the surface $\mathcal{P}$, the Euler-Poincaré characteristic $f_0 - f_1 + f_2$ equals 1. To prove Sylvester's conjecture, we need to prove that $t_2 \geqslant 1$ (which implies that in the configuration defined by D there is one point lying on two lines and, in that defined by E, one line that contains only two points). Suppose to the contrary that we only have $t_r > 0$ when $r \geqslant 3$. The Euler-Poincaré formula yields on the one hand

$$\sum t_r + \sum p_r = 1 + \sum r t_r \geqslant 1 + 3 \sum t_r$$

and, on the other,

$$\sum t_r + \sum p_r = 1 + \frac{1}{2} \sum r p_r \geqslant 1 + \frac{3}{2} \sum p_r .$$

Upon multiplying the first relation by $\frac{1}{3}$, the second by $\frac{2}{3}$ and adding, we obtain a contradiction.

In Aigner and Ziegler (1998) or Aigner and Ziegler (2003) there is a variant of the proof by central projection, due to Steenrod, using graph theory and spherical geometry:

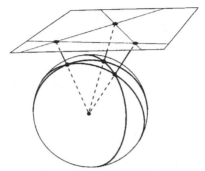

Fig. I.1.4. Aigner, Ziegler (1998) © G. M. Ziegler

♦

The second ascent entails introducing concepts of the *complex* affine plane and the *planar cubic*. We will see in Sect. V.14 that a generic cubic in the complex plane possesses nine distinct inflection points and, most importantly, that each two inflection points have the property that the line that joins them intersects the cubic once again in an inflection point. Sylvester's conjecture is thus false in the complex plane. This isn't so surprising, for we can't apply reasoning "a la Gallai" for the reason that a line in the complex plane only determines a single connected region. With some planar algebraic geometry, as in Sect. V.14, it is also easy to see that each complex planar configuration of nine points satisfying Sylvester's condition is equivalent to the one described above. However, there does exist an extension of Sylvester's result to complex affine geometry, necessarily of dimension higher than two, as will be seen in Sect. I.8, that requires a very high ascent on Jacob's ladder.

Finally, a result such as (I.1.2) will not completely satisfy a mathematical intellect, requiring as it does for the set E merely the existence of at least one line that contains only two of its points. A few sketches will convince readers that we might prove a stronger result, of a sort such as this: we will say that a line associated with a finite set of points is *ordinary* if it contains but two points of the set. We *denote* by $t(n)$ the minimum number of ordinary lines of a set E of n noncollinear points. Theorem (I.1.2) states that we always have $t(n) \geq 1$ for each integer n, but we might suppose that $t(n)$ may be rather large with increasing n. The question isn't yet settled. Here briefly is the present state of affairs; for more details and references see Problem F12 in Croft, Falconer and Guy (1991), Chap. 8 of Aigner and Ziegler (2003), and the second part of Pach and Agarwal (1995) — which is more conceptual — and also the Introduction, p. 679, of Vol. II of Hirzebruch (1997). The exact general value of $t(n)$ is unknown; the best we know presently is that we always have $t > [n/2]$ (*integer part of $n/2$*), which is due to Hansen, but we don't have an optimal answer. Moreover the proof of this result of Hansen doesn't at the moment seem to bring with it any new concept. Nevertheless, knowledge of the combinatorics of arrangements of lines in the real plane has recently increased considerably, see the reference Pach and Agarwal (1995). Finally, for the complex case, see Sect. I.8. For their aesthetic aspect and their naturalness, the configurations called *Sylvester-Gallai* remain much studied; see for example Bokowski and Richter-Gebert (1992).

♦

The name Erdös deserves special mention. Beyond his numerous results and his innumerable lectures, he was known first for having a rather long waiting line of researchers at the end of his lectures. Each in his turn would say: "Professor Erdös, I don't know how to settle this or that question". Almost invariably the response would be: "Here's how to do it. Write the article, we'll sign it jointly". Given then the innumerable articles written jointly with him, practically every mathematician of a certain age appears as a *connected component of Erdös* and even possesses an *Erdös number* defined thus: it's the minimum number of elements in a chain of several articles which ends with an article written jointly with Erdös. Your humble author didn't escape either; his Erdös number equals 3, via Aryeh Dvoretsky (who has seven articles jointly with Erdös — if we want to compute an *Erdös valence*) — and Eugenio Calabi. Another of Erdös's striking traits was his ease in making conjectures. For many of them he actually offered compensation (which he always paid) up to five thousand dollars, and he and his purse might thank the deity that he rarely deceived himself regarding their difficulty.

I.2. Another naive problem of Sylvester, this time on the geometric probabilities of four points

The second (still purely affine) problem of Sylvester from 1865 treats the arrangement of a quadruple of points in the affine plane: only two arrangements are possible (in the *generic* case, where three points are never collinear), either they form a *convex quadrilateral* or one of the points lies in the interior of the triangle formed by the other three. Then there is the natural question:

(I.2.1) *If four points are thrown randomly at the plane, what are the probabilities for obtaining one or the other of the possible configurations?*

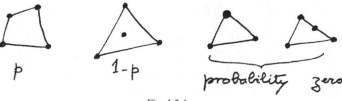

Fig. I.2.1.

There are really only two cases to consider; the degenerate ones have probability zero. But, for the question to make sense, i.e. in order to have a good notion of probability, we take as our target a planar domain D that is bounded and everywhere convex. The theoretical answer is then quite simple, the probability of obtaining four points such that one of them lies in the triangle formed by the other three is given by the triple integral

$$(\text{I.2.2}) \qquad \text{Sylv(D)} = \frac{4}{\text{Area}^4(\text{D})} \int_{\text{D}} \int_{\text{D}} \int_{\text{D}} \text{Area}(x_1, x_2, x_3) \, dx_1 dx_2 dx_3$$

where we integrate over all triples of points of D (i.e. over all the triangles contained in D) and where $\text{Area}(x_1, x_2, x_3)$ denotes the area of the triangle with vertices x_1, x_2, x_3. The proof is very simple: the probability that the first three points fall respectively in $x_1 + dx_1$, $x_2 + dx_2$, $x_3 + dx_3$ is $dx_1 dx_2 dx_3 / \text{Area}^3(\text{D})$. Knowing this, the probability that the fourth point is in the interior of the triangle formed by the first three is $\text{Area}(x_1, x_2, x_3) / Area(\text{D})$. From this we get the formula by observing that the event considered is the union of four mutually exclusive events of equal probability.

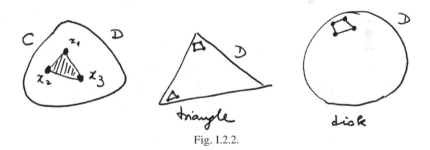

Fig. I.2.2.

The probability of having four points that form a convex quadrilateral is then simply equal to $1 - \text{Sylv(D)}$. The value of Sylv(D) depends on the "shape" of the domain D considered; we have $\text{Sylv(D)} = \frac{1}{3}$ for an arbitrary triangle and $\frac{35}{12\pi^2}$ for an arbitrary ellipse.

These results should give us much to think about. First, the value is the same for all triangles and for all ellipses. The reason is simple: in affine geometry, all triangles are "the same", all ellipses are "the same". We will return to all this amply in Sect. I.3, where we introduce notions that permit us to clarify what we mean by "the same". It will be noted that Sylvester's condition is purely affine. We now observe that, in Euclidean geometry, we clearly no longer have such equivalences for similarly shaped domains.

The two values above show in any case that the probability of having a quadrilateral is significantly lower for triangles than for ellipses. This is intuitive enough: when we take three points, each of which is close to a vertex of the triangle, there remains but little space for the fourth point outside the new triangle thus formed. In contrast, near the boundary of a round domain, we have more space. It is important to go further, since up to this point we know nothing about other domains. This problem was settled by Blaschke in 1917:

We always have $\quad \dfrac{35}{12\pi^2} \leqslant \mathrm{Sylv}(D) \leqslant \dfrac{1}{3} \quad$ *for any domain whatsoever.*

And surely our curiosity won't be completely satisfied until we know that Blaschke also showed that equality isn't attained for the lower and upper bounds except by triangles and ellipses, respectively: a nice characterization of triangles and ellipses! See Note I.4.5 of Santalo (1976) and Klee (1969) and Sect. 5.2 of Gruber and Wills (1993). We give here the two-fold idea of Blaschke. For the left inequality we use the *Steiner symmetrization* which we will encounter several times in Chap. VII (beginning in Sect. VII.5.A), but why not make quick use of it right off? It is described on the diagram: with each convex domain D and each linear direction Δ is associated the symmetrization $\mathrm{sym}_\Delta(D)$ of D for the direction Δ.

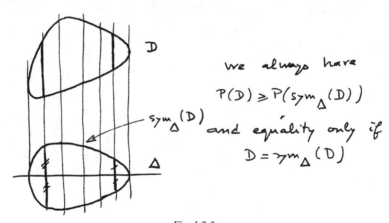

Fig. I.2.3.

Blaschke shows that each symmetrization can only diminish the integral (I.2.2). This is easy enough to perceive intuitively, for a few sketches quickly convince us that the symmetrization of a triangle interior to D often becomes a quadrilateral in $\mathrm{sym}_\Delta(D)$. Furthermore the diminution is strict provided that the convex set is not symmetric with respect to the direction considered. Knowing this, we effect some symmetrizations about well chosen lines (the directions alone matter), for example, by taking lines with inclinations that are irrational multiples of π.

For inequality in the reverse sense, Blaschke introduces the notion of *cosymmetrization*, which we haven't encountered anywhere else in the geometric literature.

It is easy to see that cosymmetrization, conversely, strictly increases the integral (I.2.2). We approximate D by polygons and thus obtains a reduction to the polygonal case. For the polygons for which one direction is orthogonal to a line

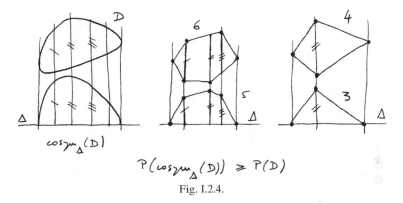

$$P(\cos\gamma_\Delta(D)) \geqslant P(D)$$

Fig. I.2.4.

joining two nonconsecutive sides, the cosymmetrization always has at least one vertex fewer than the initial polygon and we end up with a triangle, Q.E.D.

♦

Problem (I.2.1) seems well in hand, but in fact we have cheated a bit in requiring that the domain D be convex and bounded in order for the notion of probability to make sense. In truth it suffices for D to have *finite area*, which doesn't preclude "passage to infinity". Note that such a domain, extending to infinity, cannot — except for some very special cases — be convex and that Sylvester's second problem, cited frequently only for convex sets, continues to make sense for all sets of finite area. The four points may be in the domain, but the quadrilateral they determine may emerge from it, which actually needn't trouble us; it suffices to replace, in formula (I.2.2), Area(x_1, x_2, x_3) by Area(Triangle$(x_1, x_2, x_3) \cap D$). This more general non compact study was undertaken only very recently and is not yet well understood. Here is what we know, a recent reference being (Scheinerman and Wilf, 1994): on the one hand, the shape that yields the lower bound $p \equiv 1 - \text{Sylv}(D)$ over all D isn't known or even conjectured precisely. On the other hand this work provides a result that is amazing at first glance: even though we don't know the exact value of the optimal probability, it is possible to show that it coincides with another number, also unknown and extensively studied in combinatorial geometry, see Pach and Agarwal (1995), for it is related to planar realizations of the complete graph K_n with n vertices (a graph is complete when every pair of vertices is joined by an edge, and a result of Fany states that we only need use segments for joining vertices). Let $v(K_n)$ be the minimum number of points of intersection of the edges in an arbitrary planar realization of a complete graph K_n. By putting all the points on a circle, we get C_n^4 intersections. Readers may find a smaller number with other examples, but a classical result states that n^4 is the right order of magnitude. More precisely, there exists a positive real number v (finite) such that

$$\lim_{n \to \infty} \frac{v(K_n)}{C_n^4} = v.$$

The amazing result is that $v = p$.

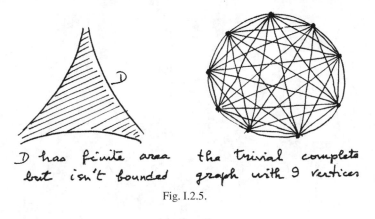

D has finite area   the trivial complete
but isn't bounded   graph with 9 vertices

Fig. I.2.5.

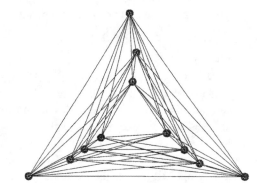

Fig. I.2.6. A very nice complete graph with 11 vertices, due to H. Jensen

For the proof, the connection between the two concepts is achieved thus: we choose n points in D at random and in a probabilistically independent manner. It is necessary to prove two opposite inequalities. In the one direction, we start with an optimal complete graph and surround each of its vertices with a small disk of radius ε. We can choose ε sufficiently small so that random points taken in this collection of disks yield another optimal graph. We then study Sylvester's probability, choosing for D the union of these disks; for ε sufficiently small we obtain the required inequality. Roughly speaking, $1 - \mathrm{Sylv(D)}$ is equal to the probability that the four points chosen at random form a convex set, i.e. the probability that among the three possible groupings of edges $ab - cd$, $ad - bc$, $ac - bd$, one of them gives rise to an intersection is $\frac{\nu(\mathrm{K}_n)}{\mathrm{C}_n^4}$. Thus we have roughly:

$$p = \min_\mathrm{D}(1 - \mathrm{Sylv(D)}) \leqslant \frac{\nu(\mathrm{K}_n)}{\mathrm{C}_n^4}.$$

More precisely, it is necessary to take into account the circumstance that the four points chosen may have the bad sense not to fall into four different small disks; but the asymptotic behavior of this bad case is in total of the order of $\mathrm{O}(\frac{1}{n})$ and thus goes to zero as n goes to infinity.

In the reverse direction, we start with any domain D in which we choose n points $\{p_i\}$ at random and probabilistically independently, and we assume that these n points are the vertices of a complete linear graph K_n. Then the number c of crossings of this graph is a random variable whose value is always at least $v(K_n)$. Moreover, consider the random variable

$$X = \sum_{a,b,c,d} \mathbf{1}_{\{p_a, p_b, p_c, p_d\}},$$

where the sum is taken over all quadruples of $\{1, \ldots, n\}$ and where $\mathbf{1}_{\{.,.,.,.\}}$ is a random indicator that equals 1 if the convex envelope of $\{p_a, p_b, p_c, p_d\}$ is a convex quadrilateral, and equals 0 otherwise. Since a random graph can't have more crossings than the mean, we have $v(K_n) \leqslant E(X)$ for the mathematical expectation of X. The desired result is obtained by letting n go to infinity.

The optimal shape of D isn't known, as already mentioned.

◆

We haven't yet finished with the problem (I.2.1), which violates the strict rules of the game: staying in the plane, in dimension two.

(I.2.2) *We randomly throw 5 points at a bounded region* D *of three dimensional space; what is the probability that they form a true polyhedron with five vertices? And the same problem with* $n + 2$ *points in the space of* n *dimensions.*

As before we compute the complementary probability to find the probability that the fifth point is in the interior of the tetrahedron formed by the four others. The formula is the strict generalization of that given above for an arbitrary dimension, which in dimension three will be:

$$\text{Sylv}(D) = \frac{5}{\text{Area}^5(D)} \int_D \int_D \int_D \int_D \text{Volume}(x_1, x_2, x_3, x_4) \, dx_1 dx_2 dx_3 dx_4,$$

where Volume(x_1, x_2, x_3, x_4) denotes the volume of the tetrahedron with vertices x_1, x_2, x_3, x_4. Here half the problems remain open at present; we only know what happens on one side of the conceivable inequalities. First, the value is known for *ellipsoids* (here again, there is but one ellipsoid in affine geometry, in which we continue to be situated, given the nature of the problem (I.2.2)); this is due to Klingman in 1969. In each dimension d he finds for the ellipsoid $\mathcal{E}^d$ (the binomial coefficients have their usual sense when d is odd and we use the gamma function to define the necessary factorials when d is even; the gamma function provides a factor $\sqrt{\pi}$):

$$\text{Sylv}(\mathcal{E}^d) = 2^{-d} \left(C_{d+1}^{(d+1)/2} \right)^{d+1} \Big/ \left(C_{(d+1)^2}^{(d+1)^2/2} \right).$$

In 1973, Grömer showed conversely that this value is attained only for ellipsoids. The value in question is thus a rational number when d is odd, and a rational multiple of π^{-d} when d is even. The method of proof is again the Steiner symmetrization, which is viable in all dimensions and which we will continue to encounter in

Sects. V.11 and VII.8. A recent reference is Sect. 5.2 of Gruber and Wills (1993) where there is a nice conceptual treatment.

On the other hand, for the maximum value, three problems remain open. Is it attained for tetrahedrons (in dimensions greater than three we say *simplex*)? Does it characterize the tetrahedrons? But above all, how can we calculate the above integral for tetrahedrons? Readers may find such ignorance surprising for so simple and ordinary a geometric object as the tetrahedron. In Sect. III.6 we will encounter two other unresolved problems on the volumes of tetrahedrons in the three dimensional sphere S^3. Readers may also try to see why Blaschke's cosymmetrization method doesn't work in dimension 3 or greater. We will encounter the P(D) in a remarkable way in Sect. VII.10. Many important results in this field have appeared quite recently; see a synthesis in Bárány (2008).

In dimension 3 or more we will not be satisfied with only an estimate of P(D) in the case of ellipsoids. The problem is to estimate (I.2.3),

$$P(D) = \frac{n+2}{\text{Area}^{n+2}(D)} \int_D \int_D \cdots \int_D \text{Volume}(x_1, x_2, \ldots, x_{n+1}) \, dx_1 dx_2 \ldots dx_{n+1},$$

as a function of invariants attached to the convex set D, where we are dealing with the volume of the simplex generated by the $n + 1$ points $x_1, x_2, \ldots, x_{n+1}$. We will find a partial answer in Sect. VII.10.F.

A final comment: we have just seen for the first time an interaction between geometry and probability. Historically the original problem is that of Buffon's needle; see the elementary exposition in Santalo (1976) and, for a contemporary treatment, Sect. 5.2 of Gruber and Wills (1993), already mentioned above. Recent directions in geometric research, in particular the Gromov's approach with mm-*spaces* (see Sect. I.XYZ), seem to indicate that the notion of *measure* − to which the notion of probability is equivalent − is every bit as important in geometry as that of distance, of metric. We will encounter other uses of geometric probability in Chaps. VII, XI and XII.

I.3. The essence of affine geometry and the fundamental theorem

We will attempt − as always without too much formalism − to enter further into a vision of the real affine plane. If we want to characterize affine geometry according to the philosophy of Klein at the turn of the twentieth century, it is necessary to study its automorphisms, by which we mean the bijections that map the affine plane onto itself and preserve its structure: lines, collinearity of points, intersections of lines, etc. In the modern definition given in Sect. I.XYZ these are the linear transformations combined with translations and thus the transformations that can be written, in arbitrary coordinates: $(x, y) \mapsto (ax + by + c, a'x + b'y + c')$, with the sole condition $ab' - a'b \neq 0$ for the six real numbers a, a', b, b', c, c'. Before returning to a purely geometric characterization of these automorphisms we will identify the affine invariants, that is to say the numbers, the situations, that are "respectable" and respected by all affine transformations. In any case, we must remember that affine transformations preserve lines (i.e. collinearity of points) and send parallel lines to parallel lines.

We begin with points. Two points do not give rise to any invariant since there always exists an affine transformation taking an arbitrary pair of points to another arbitrary pair; and it's the same for three points, which explains the fact noted above: all triangles in the affine plane are the same, are indistinguishable. This is further-more plausible — although this is not a proof — because a set of three points depends on exactly $3 \times 2 = 6$ parameters and the affine transformations also depend on the 6 parameters written above: a, b, c, a', b', c'. But if the three points considered are collinear we come upon the first affine invariant: for three collinear points a, b, c the real number denoted by $\frac{\overline{ac}}{\overline{ab}}$ is a characteristic invariant, i.e. it is preserved by every affine transformation, and two collinear triples a, b, c and a', b', c' are trans-formable into each other if and only if the corresponding invariants are equal. This invariant may be defined thus: $\frac{\overline{ac}}{\overline{ab}}$ is the value of the (unique) coordinate of the point c on the line defined by a, b, c in a coordinate system where a is the origin and b is the point with coordinate equal to 1.

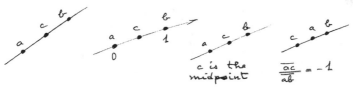

Fig. I.3.1.

As $\frac{\overline{ac}}{\overline{ab}}$ traverses the interval $[0, 1]$, the point c traverses the segment $[a, b]$ defined by a and b. The notion of *segment* is thus affine, as is that of *midpoint*: the midpoint of the segment $[a, b]$ is the point c such that $\frac{\overline{ac}}{\overline{ab}} = \frac{1}{2}$. Observe that this invariant is not Euclidean, but that if there is an additional Euclidean structure on our affine plane, then we may always compute it with the distances ab, ac (with assignment of the usual signs). Exercise: find conditions under which two sets of four (arbitrary) points can be transformed into one another.

Passing now to lines, all lines are first of all the same; then, two pairs of lines are indistinguishable, under the obvious condition that they are simultaneously inci-dent or simultaneously parallel. For three concurrent lines, there isn't any invari-ant: two triples of concurrent lines can always be taken into each other by an ap-propriate affine transformation. But it isn't the same for four concurrent lines D_i $(i = 1, 2, 3, 4)$: we can attach an invariant to them in a canonical manner, i.e. their *cross ratio* $[D_1, D_2, D_3, D_4]$. This is a characteristic invariant: we can define it in an affine manner. But we won't do this, for it is in fact a *projective* invariant as will be shown in an entirely natural and simple manner in Sect. I.6.

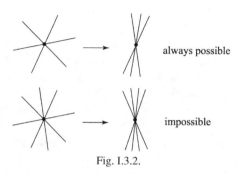

Fig. I.3.2.

Again, we can ask numerous questions on the subject of lines. Here is one of them: given two lines, three lines, or more, what is the number of possible *configurations*? For two or three, it's easy. For two: either they are concurrent or parallel. Difficulties begin with three and we encourage readers to sketch, to scribble: the lines may be concurrent or form a true triangle. But we must not forget the possibility of parallels, whence two other configurations: three parallel lines or two parallels and a third that intersects them. We see that for four and more, things become difficult; in particular we begin to get frustrated by the parallels. Here we find an additional incentive for projective geometry: parallelism doesn't exist!

♦

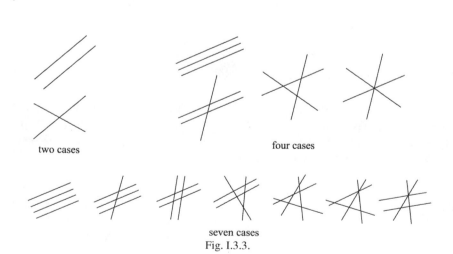

Fig. I.3.3.

The affine transformations map each line into a line, but if we want to completely capture the essence of affine geometry, e.g. by a purely axiomatic definition (without a vector space, etc.), we will want to be sure that there don't exist other transformations — beyond the affine ones defined above — of the affine plane to itself that transform each line into a line, i.e. that preserve the collinearity of points. We have a completely satisfactory answer to this question:

(I.3.1) *(Fundamental theorem of affine geometry) Each bijection of the affine plane to itself that takes lines into lines is an affine transformation.*

It is impossible to pass over the idea of the proof in silence, as much for its beauty and conceptual importance as for its allowing us to imagine what will happen in affine geometries over fields other than the reals — complex numbers, quaternions, etc. — that will be encountered in Sect. I.8. A detailed proof is found in 2.6 of [B]. We mention only this much: according to what has been said above we may suppose that our bijection f leaves three noncollinear points fixed, that we will use to define an origin and coordinates x, y; we then only need show that f is in fact the identity transformation. The fundamental remark is that parallel lines are transformed into parallels, since parallelism can be defined in a set-theoretic manner and f is bijective. Thus, in particular, parallelograms are transformed into parallelograms and it suffices to show that f acts identically on the first coordinate axis. To do this it will certainly be necessary to depart from this line, for any bijection of a line preserves that line, whether it acts identically or not. To define affine geometry we identify our line, the x axis, with the field $\mathbb{R}$ of real numbers. The figures below, based solely on parallelism, show that the restriction of f to $\mathbb{R}$ is an *automorphism*: $f(\lambda + \mu) = f(\lambda) + f(\mu)$ and $f(\lambda\mu) = f(\lambda)f(\mu)$.

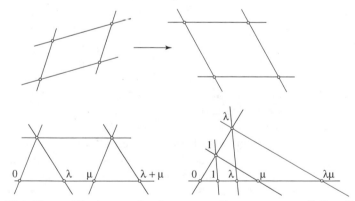

Fig. I.3.4. *Above*: a bijective mapping that preserves lines preserves parallelism. *Below*: construction of the abscissa points $\lambda + \mu$ and $\lambda\mu$ on the x axis

It is a classical exercise to show that only the identity is an automorphism of $\mathbb{R}$, but be careful not to use continuity, which has no reason to exist here; we have never required that f be continuous or even suggested that such a notion can make sense in the absence of distance!

We now pose a whole series of natural questions. First, the extension of (I.3.1) to all dimensions (> 2) is trivial; in contrast, (I.3.1) is false over the complex numbers, even adding continuity; see Sect. I.XYZ. A still more subversive question (a bit off the ladder) is to ponder the *local* and the *global*. But do we need the entire affine plane for our result? Certainly we do for the proof above, where parallelism is the key. But could we do without it? The answer is no. We will see definitively in Sect. I.5 that a bounded set in the affine plane admits plenty of other bijective transformations that preserve collinearity; these are the *projective transformations*. Thus for a deep knowledge of local affine geometry we need to climb at least one rung. In Sect. I.XYZ we will see that a good understanding of (I.3.1) in a general context and in good rapport with the axiomatics of the nineteenth century wasn't really achieved until 1950.

But back to the elaboration: what happens if we no longer require bijectivity or globality, or again if we study mappings between spaces of different dimensions? In the local but bijective case, readers will see, with the aid of passage to the infinite in the spirit of Sect. I.6, that the question is easily answered by reverting to the local affine case, but with full preservation of parallelism.

To finish our discussion of the essence of affine geometry, we pose two more questions. The first is that of incomplete duality: two distinct points determine a unique line, but in contrast two lines determine a point only if they are not parallel. Projective geometry will be the appropriate context (see Sects. I.5 and I.7) for having a *duality* without exception. A second question concerns topology: what is the topology of the set $\mathcal{D}$ of all the lines of the affine plane? What is its "shape"? The answer is that the topology of $\mathcal{D}$ is that of an open (no boundary) Möbius strip. We can convince ourselves with the sketches below. We puncture the plane at a fixed origin. With the exception of the lines that pass through the origin, the lines of the plane are associated in a one-to-one manner with the points of the punctured plane (take an auxiliary Euclidean structure and project the origin onto the line in question) and it only remains to "glue" (or sew) the punctured plane to the circle of lines that pass through the origin (caution! this is not the unit circle but is obtained by identifying antipodal points). The segments of the Möbius strip correspond to parallel lines. This operation, which consists of replacing a point by the set of lines that pass through it, is called the *blowing up* at the point; it is used in an essential way in algebraic geometry. More precisely, looking at the figure, we trace a disk about the blowing up point and replace it by a Möbius strip, while gluing the circle which bounds the disk to the circle bounding the Möbius strip. In this operation, the point is replaced by the median circle of the strip.

Analytically, the fact that the topology of the set of lines of the plane is not that of $\mathbb{R}^2$ is easily seen: it is not possible to obtain all lines with a single type of equation. For example, the two-parameter expression $y = ax + b$ allows the vertical lines with equation $x = c$ to escape. If we opt for the equation $ax + by + 1 = 0$, we lose the lines passing through the origin. We are thus forced to consider all the equations $ax + by + c = 0$; but then the triple (a, b, c) and the triple (ka, kb, kc), for $k \neq 0$, represent the same line. We are forced to pass to the quotient and to equivalence classes: this is precisely what we do in constructing projective geometry in Sect. I.5.

Fig. I.3.5. Correspondence between lines not passing through O and points of the punctured plane

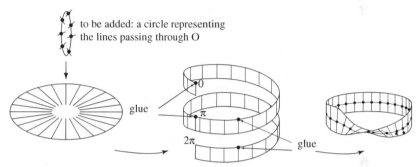

Fig. I.3.6. *Blowing up at the origin*. The half lines emanating from the origin O (and not containing the origin) are glued onto a *circle* of length π, two opposite half lines being glued to the same point of the circle. The punctured disk (or plane) thus becomes a Möbius strip

I.4. Three configurations of the affine plane and what has happened to them: Pappus, Desargues and Perles

We consider the three figures below, the first two are very old, the third dates from 1965. They seem innocent enough, but they are going to give rise, each in its turn, to very different phenomena. There are surely plenty of other plane affine configurations, but our choice has been dictated by the extensions for which the first two have given rise and the surprising consequences of the third.

Readers will be able to guess the significance of $(9_3, 9_3)$ and $(10_3, 10_3)$ and $(10_3, 10_3)$ or otherwise refer to Sect. I.XYZ or to Sect. I.9. The first configuration is that of *Pappus* (fourth century): given six points situated three apiece on each of two lines, then the three other points that can be derived from them, as indicated on the figure, are again collinear. In the second, *Desargues' theorem* (circa 1630), we have two triangles called *homological* (here abc and $a'b'c'$), which means that the lines joining corresponding vertices are concurrent. The conclusion is that the three points x, y, z indicated on the figure (the points of intersection of the homological sides) are again collinear. Finally, in the third, the conclusion is that the quotient of the affine invariants (see above) $\frac{\overline{12}}{\overline{13}} / \frac{\overline{42}}{\overline{43}}$ is forced, by the allignments drawn, to be equal to $\frac{1}{2}(3 - \sqrt{5})$.

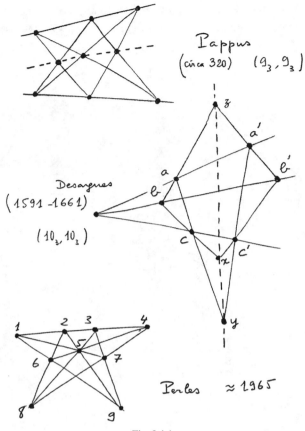

Fig. I.4.1.

◆

There are at least three things to mention regarding Pappus's theorem. The first, very briefly: when we have six points on two lines, we have a particular case of six points on a single *conic* because the pair of lines may be considered as a *degenerate* conic; see Chap. IV. In this more general case, the indicated collinearity still holds: this is the famous theorem of Pascal; see Sect. IV.2.

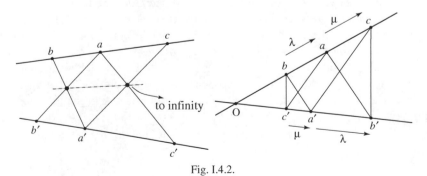

Fig. I.4.2.

We now speak about Pappus's proofs. The good proof, illuminating for the sequel, is one that uses projective geometry, considered amply in the next section.
Suppose that two of the three points of intersection constructed are "*at infinity*"; see
the figure. Then, as we will see, two pairs of lines that otherwise would intersect are
parallel. It is required to show that the third pair is also made up of parallels. We
pass from ac' to ca' by a *homothety* of ratio μ, and we pass from ba' to ab' by a
homothety of ratio λ (all these homotheties have center O). Thus we pass from b
to c by a homothety of ratio $\mu\lambda$ and from c' to b' by a homothety of ratio $\lambda\mu$. But
since $\lambda\mu = \mu\lambda$, the proof is complete.

Alerted by what has been said in regard to the fundamental theorem of affine
geometry, readers may ask what happens with the theorem for affine geometry over
the other fields and thus deduce that Pappus's theorem is true for *complex* affine
geometry, but not for *quaternion* affine geometry, since the quaternion field isn't
commutative. It is a consequence of an axiomatic study of affine geometry that the
commutativity of the underlying field can be characterized by the validity of the
configurations of Pappus. All this dates from the time indicated above in Sect. I.3;
see for example Artin (1957) or Baer (1952).

Recently Schwartz (1993) has given Pappus a second look. Here, very briefly, is
what it's about, see the original text for more details. The starting point is this naive
remark: to every pair of triples of collinear points, Pappus associates a third such
triple; we then have an operation on such triples. Whence two questions: what is
the algebraic nature of this operation? What happens if we iterate it a few or many
times, or even indefinitely? In Schwartz (1993) these two questions are resolved and
each is placed on an appropriate rung of the ladder; see also Berger (2005).

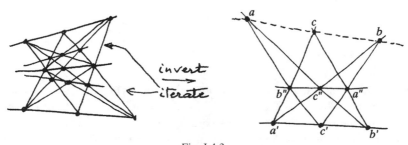

Fig. I.4.3.

The fundamental remark is that the operation "two triples produce a third" can be
inverted: we can go backwards. The reversal is illustrated by the figure on the right.
To study the iteration of this operation (after having composed two triples T and
T′ to obtain T″, we may compose T and T″, or T′ and T″, and so forth), Schwartz
introduced what he called "labeled boxes", consisting of two triples − in the box
labeled $((a,b,c),(a',b',c'))$, we have that $abb'a'$ is the box and c, c' are the points
labeled on the sides "above" and "below" − along with the transformations

$$\sigma : \big((a,b,c),(a',b',c')\big) \mapsto \big((a,b,c),(a'',b'',c'')\big)$$

and
$$\tau : \big((a,b,c),(a',b',c')\big) \mapsto \big((a',b',c'),(a'',b'',c'')\big).$$

It is easy to see that these two operations are related by only two conditions: $\sigma^2 =$ identity, $\tau^3 =$ identity. The group they generate is none other than the famous *modular group*, i.e. the group *denoted* $SL(2,\mathbb{Z})$, defined as the group of matrices $\begin{pmatrix} a & b \\ c & d \end{pmatrix}$ with integer entries and determinant $ad - bc = 1$. It is interesting to encounter in connection with Pappus this group that governs a good part of mathematics and is the most important after $\mathbb{R}$ and $\mathbb{C}$. We find it in number theory, complex analysis, Riemann surfaces and algebraic geometry, i.e. for elliptic curves; see Sect. V.14. We will encounter it again in connection with polygonal billiards in Chap. XI.

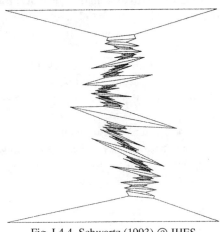

Fig. I.4.4. Schwartz (1993) © IHES

Now Schwartz has studied the figure obtained by applying the operations of this group to an initial box. It is drawn in Fig. I.4.4, to which we in fact need to add a whole complement (in order to go backwards), that turns out to be a Möbius strip (not drawn: this would be difficult). Schwartz shows that the discrete set of points marked by all the triples thus obtained can be extended by continuity to a closed continuous curve. If we start with one box that is an harmonic quadrilateral − and only in this case − all the marked points lie on a single line. In every other case the curve is *fractal*, but with an exceptional additional property: at each of its points, the line of support of the triple passing through this point intersects it in exactly one point, the one considered. This isn't the case for most fractals, where either there are plenty of lines that don't intersect the curve, e.g. the snowflake, or at the other extreme every line passing through this point intersects it amidst other points, e.g. a fractal curve that spirals. It seems that the only other known comparable example is that of the graph of Brownian motion in one dimension: at each of its points it behaves like the graph of the function $x \mapsto x^{1/2}$.

It is the moment to suggest that readers develop one or more purely affine proofs of Pappus's theorem, if only to appreciate projective geometry and in spite of the fact that they will need to climb a bit up the ladder.

We can also use projective geometry for a proof of the Desargues configuration by letting two of the collinear points go to infinity. We then only need use a homothety with center O. Thus the commutativity of $\mathbb{R}$ isn't needed, but the complete calculation will show readers that we use the *associativity* of $\mathbb{R}$: $\lambda(\mu\nu) = (\lambda\mu)\nu$ for all λ, μ, ν. This is important in the axiomatic theory of affine and projective spaces: we can replace the associativity of the object which must play the role of the underlying field by the requirement that Desargues' theorem hold. Interested readers will verify by calculation that to ascertain that two nonintersecting lines are parallel in an affine plane over an arbitrary field we need to use its associativity, it being understood that a line is a set defined by an equation $ax + by + c = 0$ and that two lines are parallel if and only if they are obtainable from each other by translation.

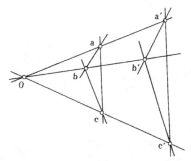

Fig. I.4.5. *Proof of Desargues theorem.* We can assume that x and y, the respective points of intersection of bc and $b'c'$ and of ac and $a'c'$ are at infinity, i.e. that bc and $b'c'$, and ac and $a'c'$, are parallel. It is then just a matter of showing that ab and $a'b'$ are parallel. But these hypotheses bring with them the existence of a homothety with center O that sends a, b, c to a', b', c', respectively. Hence the result

But there exists another proof that will subsequently appear less artificial. We embed the affine plane in the affine space of three dimensions and consider the figure obtained as the projection into dimension two of the figure below, where the three lines defining the projection between the two aren't coplanar. The result is then trivial: the three points x, y, z are collinear since they belong to the intersection of two planes, which is always a line.

The preceding explains why, in the axiomatic theory of affine or projective geometry, the situation in dimension two is completely different from the general case: affine or projective planes are hardly *categorical*. A typical example: there exists a

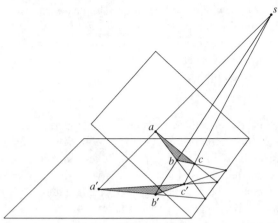

Fig. I.4.6. A figure necessarily drawn in the plane, but where we nonetheless see the perspective representation of a figure in space

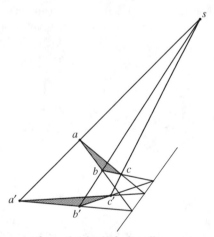

Fig. I.4.7. The same figure deprived of what allows us to see it "in space"

quasi-field, the *Cayley octonions*, denoted by $\mathbb{C}a$, where there is no longer associativity; see Sect. I.XYZ. Although a projective plane, denoted by $\mathbb{C}a\mathrm{P}^2$, can be well defined over $\mathbb{C}a$, we can never define $\mathbb{C}a\mathrm{P}^n$ for any $n \geq 3$; see Besse (1978). The reason for this is precisely that Desargues' theorem would be valid there according to the above figure; but we know that this would imply the associativity of the algebraic object, the Cayley octonions. We mention in passing that $\mathbb{C}a\mathrm{P}^2$ is for us one of the most beautiful of all geometric objects and that we could call it the *panda* of geometry. But in spite of this exceptional beauty, it is difficult to construct and extremely few authors construct it in detail; an exception can be found in 3.G of Besse (1978).

Finally, for a **dynamic** study of Desargues' configuration like that for Pappus, and by the same author, see Schwartz (1998). For another approach to iterations of geometric theorems, see Smith (2000), also cited at the end of Sect. II.1.

The philosophy of Perles's example is as follows: the configuration can never be realized in the *rational* affine plane, i.e. the subset of the affine plane made up of all points whose two coordinates are rational numbers in a given coordinate system (modulo which we always have isomorphic objects); the reason is simply that $\sqrt{5}$ is irrational.

The existence of irrational affine configurations was known before Perles, see for example the notion of *accessible point* on p.126 of Coxeter (1964). For computer enthusiasts this means that such configurations are not, *in an exact sense*, visible on the screen. On the other hand we can inject the irrationals *in a formal way*, especially a number such as $\sqrt{5}$, which can be defined for example by the equation $x^2 - 5 = 0$. But the precise Perles configuration has a much deeper interest: it allowed him to show the existence of *polytopes* in dimension 8, that can never be realized with the same combinatoric and with vertices having rational (or, equivalently, integer) coordinates. We will return to this question amply in Sect. VIII.12.

I.5. The irresistible necessity of projective geometry and the construction of the projective plane

We have had reason to be unhappy on several occasions above: first, while Pappus's theorem — like Desargues' in the purely affine context — presents several variants because of possibilities of parallelism. We have an even simpler question, encountered at the end of Sect. I.3: into how many regions do two, three, four, etc. lines divide the plane?

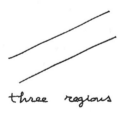

three regions   four regions

Fig. I.5.1.

Even though its formal definition in algebraic language may seem unproblematic, it requires a bit of time to begin to feel at ease with projective geometry and we thus beg readers to be patient and not to become discouraged. As further evidence of this difficulty it should suffice to remark that, even though introduced by Desargues

at the beginning of the seventeenth century, projective geometry wasn't firmly established until the second half of the nineteenth century. Desargues' naive definition
is as follows: the projective plane P* associated with the affine plane P extending
P is nothing other than P itself to which a line P_∞ of points at infinity is adjoined,
the elements of P_∞ (the line "at infinity" of P) being the set of directions of lines of
P: $P^* = P \cup P_\infty$. We then say that two distinct parallel lines intersect precisely at
the point at infinity that corresponds to their common direction. As for a line D of
P and the line at infinity, they intersect precisely at the point of P_∞ corresponding
to the direction of D. Finally, for lines joining two distinct points of P*: if one is in
P and the other in P_∞, the line joining them is the one that passes through the first
point with the direction given by the second; the line joining two points at infinity is
the line at infinity. Thus for two lines in P* — just as for two points — we can make
existence statements without exception, without fear of parallelism.

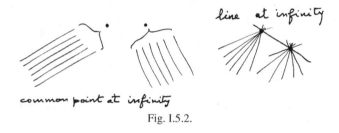

Fig. I.5.2.

But this construction is abstract. It demands an act of faith and furthermore
doesn't give us a basis for calculation, for which coordinates are needed. For finding
a concrete geometric construction of P*, we are inspired by the proof of Desargues's
theorem obtained by embedding P in a space Q of three dimensions and taking an
arbitrary point O of Q not in P. We have climbed a rung! With each point of P is
associated a unique line of Q which passes through O. We will call lines through
O "O-lines" for short and O the "origin" of Q; an O-plane of Q is likewise a plane
through the origin. Among the lines passing through O precisely those are missing
that are parallel to the plane P of Q; but we see that these are associated in a biunique fashion with the directions of the lines of P. We only need add that through
a point at infinity corresponding to a direction of a set of parallel lines of P there is
an O-line that has that direction. We thus define P* concretely as the set of **all** the
lines passing through the origin of Q. The lines of P* will be the **planes** (always
passing through the origin) of Q. The intersection axioms are now evident: two distinct O-lines uniquely determine an O-plane of Q; thus a line of P* and two distinct
O-planes Q intersect in a well determined O-line of Q. So now we no longer have
any exception or particular case, just as we have wanted.

Even though it isn't really necessary (and has no significance for projective
geometry over an arbitrary field), we can make this construction of P* still more

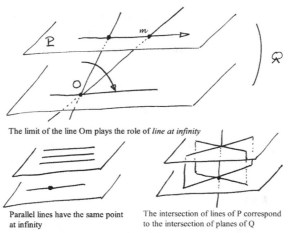

The limit of the line Om plays the role of *line at infinity*

Parallel lines have the same point   The intersection of lines of P correspond
at infinity   to the intersection of planes of Q

Fig. I.5.3.

plausible as follows: if a point m regresses to infinity along a line D of P, the line OD tends toward the directed line parallel to D.

Historically this construction of projective space simply reflects the need that painters have for representing a portion of space in a picture. The point O above is nothing other than the eye of the painter (the observation point) and the plane P the picture (the picture plane). The "empirical" rules for geometric constructions employed in the arts are consequences of projective geometry.

We can now **calculate** in P* since we have the vectorial calculus in Q at our disposal: the points of P* are none other than those of Q **within multiplication by a scalar.** Let us quickly see how things work. For an arbitrary coordinate system (x, y, z) in Q, the points of P* will thus be triples of reals, not all zero, modulo an equivalence relation: the triple (x, y, z) is equivalent to the triple (kx, ky, kz) for all nonzero real k, a triple of *homogeneous coordinates* for the same point. Most important is the case where the coordinate system is such that the plane P is defined in Q by the equation $z = 1$. Then the points of P have for homogeneous coordinates the triples (x, y, z) with $z \neq 0$: the point (x, y) of P will have homogeneous coordinates $(x, y, 1)$ and all associated triples. Conversely, the triple (x, y, z) associated with $(x/z, y/z, 1)$ will be a triple of homogeneous coordinates of the point $(x/z, y/z)$ of P. Thus the points of the line of the equation $ax + by + c = 0$ satisfy, in homogeneous coordinates, the equation $ax + by + cz = 0$. The passage from the first equation to the second is called *homogenization*.

In contrast, the points at infinity are those of type $(x, y, 0)$, and the point at infinity of a line satisfies the homogeneous equation of this line.

It is convenient to use the notation $(x : y : z)$ to represent the set of all triples of homogeneous coordinates that can be obtained from the triple (x, y, z) by scalar multiplication. We then have $(x : y : z) = (x' : y' : z')$ if and only if there is a nonzero scalar k such that $x' = kx$, $y' = ky$, $z' = kz$, i.e. $(x : y : z)$ and $(x' : y' : z')$ represent the same point of P*.

As an example of significance for us in the spirit of Sect. I.1, see the expression of Hesse's configuration in homogeneous coordinates in Sect. I.8. On the other hand, what we **see** globally is the projective plane, a quite different story that we will touch on later. The human mind doesn't like objects obtained through an equivalence relation that can't be embedded in any ordinary space.

This introduction of projective spaces may seem a bit artificial, but is in fact an essential tool for many problems where we have to consider things "within a scalar". We will see examples of this in II.6 and IV.7 for the space of all circles, or that of all spheres or of all conics.

An additional property of projective spaces is that they are **compact**, which is essential for certain problems; they are truly "round" (there are no longer points at infinity, they have been tamed): everything is "at a finite distance".

To respond to a whole array of natural questions we now need to study projective geometry (planar here, but see Sect. I.XYZ) from the points of view of geometry, algebra (group of transformations) and topology (topology of the projective plane). This study must be done for the structure itself, initially independent of its being an **extension** of affine geometry. But of course we will want to know subsequently how to *return* to the affine plane. A (**the**) projective plane $\mathcal{P}$ is defined a priori as the set of vectorial lines (one-dimensional subspaces) of a (the) real vector space **P** of dimension 3, the lines of this projective plane being the vectorial planes (two-dimensional subspaces). For the algebraist this will be the quotient of $\mathbf{P}\backslash 0$ modulo the equivalence: $v \equiv v'$ if there exists a real k such that $v' = kv$.

What are the good transformations of $\mathcal{P}$? In the spirit of (I.3.1) it is now easy for us to find biunique transformations of an affine plane that preserve lines, but only locally: simply consider the figure below and the projection starting at the origin of the space of three dimensions Q, where we have embedded two copies P and P′ of the affine plane.

We shouldn't fail to mention that we have a injective transformation from all of P onto all of P′, with just one line of P and one line of P′ removed. Note that, in the projective coordinates of P and P′ obtained starting with systems of ordinary coordinates (x, y, z) and (x', y', z') in Q in which P and P′ are respectively the planes $z = 1$ and $z' = 1$, these transformations are expressed in a linear fashion. These are the transformations we need to apply if we want to assemble different aerial photos in order to compose a single map. Whence the following definition: the projective transformations of $\mathcal{P}$ are the (invertible) linear mappings of Q applied to (vectorial) lines. So much for geometry, but for the algebraist we consider linear transformations within a scalar. For example, in a coordinate system, we deal with all 3×3 invertible matrices modulo the multiplication of all their terms by a single nonzero scalar. More conceptually: the group of projective transformations of Q is the quotient of the linear group of Q by nonzero multiples of the identity.

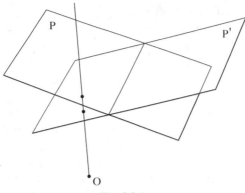

Fig. I.5.4.

In homogeneous coordinates a projective transformation will always have the form:

$$x' = ax + by + cz, \quad y' = dx + ey + fz, \quad z' = gx + hy + iz$$

or else, in affine coordinates:

$$x' = \frac{ax + by + c}{gx + hy + i}, \quad y' = \frac{dx + ey + f}{gx + hy + i}.$$

From this we deduce many things, in particular the important possibility of finding, for each quadruple of noncollinear points, coordinates written as

$$(1,0,0), \quad (0,1,0), \quad (0,0,1), \quad (1,1,1),$$

which is called a *projective frame*.

For the transformations of a projective line, see the following section. The preceding shows that a perspective (i.e. a central projection) of one line onto another is a homography (defined on the next page) and (see below) preserves the cross ratio: see Fig. 1.6.2.

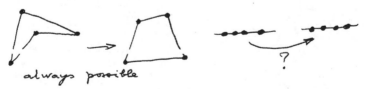

Fig. I.5.5.

♦

Now in the spirit of Sect. I.3 it's a rather easy exercise to show that, given two quadruples of non collinear points of $\mathcal{P}$, there exists a unique projective transformation taking one into the other. In the axiomatic theories this result is difficult, but essential; it is thus called the *second fundamental theorem of projective geometry*; see, in addition to Artin (1957) and Baer (1952), the classic (Veblen and Young, 1910–1918).

For us, in the vectorial context the result proceeds from the following fact (left to readers): for each quadruple a, b, c, d of noncollinear points, we can find a system of homogeneous coordinates such that $a = (1, 0, 0)$, $b = (0, 1, 0)$, $c = (0, 0, 1)$, $d = (1, 1, 1)$. The theorem then follows at once.

But now we have to answer the question for four collinear points. For collinear triples, we have of course transitivity. But for four, by definition of the projective plane, we need to know what happens for four vectorial lines of a vectorial plane, a question that we left open in Sect. I.3. In fact this opens an abyss under our very feet: we have completely forgotten to speak of the **projective line**! Otherwise expressed: what are the lines of $\mathcal{P}$? What is their geometry, assuming they have one?

I.6. Intermezzo: the projective line and the cross ratio

A (**the**) projective line is thus the set of lines of a vectorial plane, a set that we will denote by $\mathbb{R}\mathrm{P}^1$, in agreement with Sect. I.XYZ. The topologist is quickly satisfied here; the figure below shows that this set is in bijection with a (**the**) circle. This isn't astonishing, the construction of the projective line consists of completing the affine line by appending a single point ∞ at infinity; everything then closes up in a circle. It is well to emphasize that for the line there aren't two points at infinity, but one only. As in the affine plane it matters little in which sense we pass to infinity; we end up with the same point. This compactification of the line into a circle by a point at infinity is a particular case of a more general construction. Readers should be aware that there exist other types of compactification; we encounter some of these in Sect. II.3.

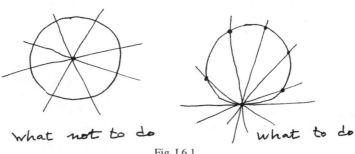

Fig. I.6.1.

♦.

Algebraically, a projective line D, in particular $\mathbb{R}\mathrm{P}^1$, can always be written as the set of pairs (x, y) of real numbers, not all zero, within equivalence: (kx, ky) is equivalent to (x, y) for each nonzero real k. As in the case of triples, we can use the notation $(x : y)$ to designate the pairs taken within multiplication by a scalar.

We recover the affine line as the set of pairs where y is nonzero: $(x, y) \equiv (x/y, 1)$, which provides an embedding $t \mapsto (t, 1)$ of the affine line into the projective line. The projective transformations, called *homographies* of the projective line, are the mappings $(x, y) \mapsto (ax + by, cx + dy)$. Interpreted for the affine line, these are the mappings $t \mapsto \frac{at+b}{ct+d}$ which thus extend onto the projective line by $-\frac{d}{c} \mapsto \infty$ and $\infty \mapsto \frac{a}{c}$, consistent with the notion of limit, to comfort us once more if need be. The projective group of the line — the group of projective transformations, homographies — has three parameters, permitting us to uniquely map each triple of points into a given triple. It clearly does not preserve, when restricted to the affine line, the invariant $\frac{\overline{ac}}{\overline{ab}}$ encountered in Sect. I.3. On the other hand, there does exist an invariant for four (distinct) points $\{m_i\}$ $i = 1, 2, 3, 4$, called the *cross ratio* of these four points and *denoted* $[m_i] = [m_1, m_2, m_3, m_4]$. Two quadruples of points are in projective correspondence if and only if their cross ratios are equal. In an arbitrary coordinate system, for points $m_i = (t_i, 1)$, this cross ratio equals: $\frac{t_3-t_1}{t_3-t_2} / \frac{t_4-t_1}{t_4-t_2}$. After a moment's reflection its value is no longer surprising, it being the quotient of two affine invariants associated in a natural way with the quadruple considered. On a projective line, likewise for four distinct points, it is necessary that the cross ratio accept the value ∞, e.g. for all x we have $x = [0, \infty, x, 1]$. We note that the fact that the mapping $m \mapsto [a, b, m, c,]$ establishes a bijection between a projective line and $\mathbb{R}P^1$ is equivalent to the fact that we are able to take (a, b, c) as the "projective frame", and is important for our being able to speak of "harmonic conjugation". Note that the cross ratio can be defined on an affine line; its invariance carries over by the fact that it is preserved by the point projection of the figure below, which furthermore allows it to be calculated for a quadruple of concurrent lines in the affine plane:

$$[[ap], [aq], [ar], [as]] = [p, q, r, s]_D;$$

see Fig. I.6.2. We have the interpretation: the lines passing through the point a form a projective line, which answers finally the question posed in Sect. I.3.

The above formula plays an essential role in certain geometric constructions. The case where the cross ratio equals -1 is particularly important; we say then that

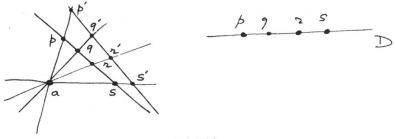

Fig. I.6.2.

the four points are in *harmonic division*. A purely geometric construction is given in Sect. I.7; note its systematic usage in Sect. IV.4.

The cross ratio is not invariant when we permute the points considered, but its behavior is simple and most interesting; see 6.3 of [B] for a detailed study. Direct calculation shows that $[b, a, c, d] = [a, b, c, d]^{-1}$ and $[a, b, c, d] + [a, c, b, d] = 1$, which allows us to calculate what happens for all the other permutations. But keep in mind for later (see Sect. V.14) that the simplest cross ratio λ which is invariant for all permutations of the four points is $\frac{(\lambda^2 - \lambda + 1)^3}{\lambda^2 (\lambda - 1)^2}$. We find in pp. 43–51 (Darboux, 1917) the calculation providing this invariance of λ for the four roots of an equation of fourth degree, as a function of the coefficients of this equation.

The real projective line and its group of transformations is not an object that has been artificially concocted by geometers for their exclusive enjoyment. First of all, the "homographic" functions $t \mapsto \frac{at+b}{ct+d}$ are encountered everywhere; they are quotients of affine functions and are very important in the complex case. An important physical application of the notion of projective line is found in the theory of centered systems in optics. Lenses, mirrors, etc., are arranged in some way on a line that is their common axis; zoom lenses of the most sophisticated variety are of this type. Then the correspondence between a point of the axis and its image is always a homography. To convince ourselves of this it suffices to study the case of a mirror or of a single lens; we succeed since the homographies form a group. Readers will surely remember the following formula from school:

$$\frac{1}{x} + \frac{1}{x'} = \frac{1}{f},$$

where x is the abscissa of the object, x' that of its image, and f the focal length, positive or negative. Note that in optics infinity is essential; here it provides the *focal point* (image of the point at infinity) and the *focal objective* (reciprocal image of the point at infinity).

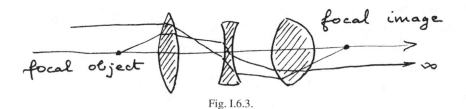

Fig. I.6.3.

The homographies, real (as here) or complex, are of primary importance in geometry; we will see this very soon in Sects. II.3 and II.4. Their classification — by their fixed points among other ways, especially as *involutions*, i.e. the homographies whose square equals the identity — is fundamental, but we will have but little to do with it (at the end of Sect. IV.6); see Chap. 6 of [B]. Finally, note that the complex projective line, in its role as a topological object, is nothing other than the sphere S^2, for it is obtained by appending a point to the complex line $\mathbb{C}$ (which is the real plane $\mathbb{R}^2$): $\mathbb{C}^* = \mathbb{C} \cup \infty = S^2$; we will see this again in II.4.

I.7. Return to the projective plane: continuation and conclusion

We haven't yet finished with the projective plane. We first note that the cross ratio allows us to recognize when two quintuples of points are projectively equivalent; compare with Sect. I.3 for the affine case and its invariant. We now study in depth the relation between affine geometry and projective geometry, if only to make rigorous the proofs of the theorems of Pappus and Desargues that were outlined in Sect. I.3.

Starting with P, we constructed P*, which contains the line at infinity. The essential things is that in $P^* = P \cup P_\infty$ and above all in any projective plane $\mathscr{P}$ whatever we can choose a line D and *make it* the line at infinity of the complement $\mathscr{P}\backslash D$ of D in $\mathscr{P}$. That is to say, in the construction of Fig. I.5.3, we replace the plane $z = 0$ by the plane defined by the origin and the desired line taken in the plane $z = 1$. The affine space so defined is "the" plane parallel to this new plane. For example the affine invariant $\frac{\overline{ac}}{\overline{ab}}$ of three collinear points on a line F equals the cross ratio $[c, b, a, \infty_F]$, where ∞_F denotes the point at infinity of the line F, i.e. $\infty_F = D \cap F$. To say for instance that c is the midpoint of ab is equivalent to saying that c, b, a, ∞_F form a harmonic division. In summary, we can accomplish this: in an arbitrary affine plane, completed to form a projective plane, we can alter the line at infinity, i.e. stay in the projective plane with all its advantages, all its properties, but decide on "*new parallels*". We can also speak of the "*transfer to infinity of one or more collinear points*": we let the new line at infinity pass through these points (or we can use a projective transformation sending the given line to the line at infinity). All this is certainly a rung up Jacob's ladder, where we can manage things a bit better. Fundamental is the fact that the cross ratio is conserved under these "transfers".

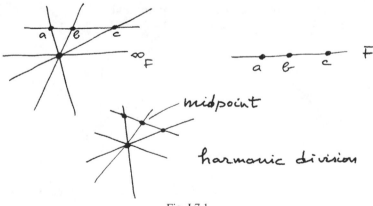

Fig. I.7.1.

The preceding technique was used to prove Pappus and Desargues in Sect. I.4. We now use it to demonstrate the classical property of the configuration of the complete quadrilateral:

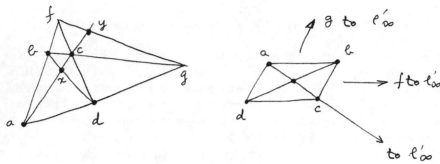

Fig. I.7.2.

In this figure the four points a, c, x, y are in harmonic division. To see this, it suffices to transfer the two points f and g to infinity. Then a, b, c, d becomes a parallelogram and our result simply translates the fact that the diagonals of a parallelogram intersect at their midpoints. Despite its simplicity, the configuration $(6_2, 4_3)$ of the complete quadrilateral may be seen as a geometric rendering of the fact that the solution of an equation of fourth degree may be reduced to that of a third degree equation. Indeed, this configuration associates in a canonical way a triple of points with a quadruple (find this in the figure). See Sect. I.XYZ for an entirely projective proof.

We now attack the question of *duality*, used in Sect. I.1 and imperfect in the affine context: there points and lines played similar, but not identical, roles. Furthermore, the space of all lines had a topology different from that of the points (the affine plane), among other reasons because we could not find a good one-to-one correspondence between these two sets (see Sect. I.3).

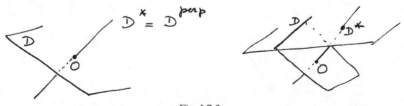

Fig. I.7.3.

In $\mathcal{P}$ the duality is perfect with regard to the line joining two points and to the intersection of two lines. However, we would like to obtain a one-to-one correspondence between $\mathcal{P}$ and the set $\mathcal{D}$ of all its lines: but this is utterly simple, since $\mathcal{P}$ is the set of lines through the origin of a vector space Q of dimension three and $\mathcal{D}$

the set of vectorial planes: we only need put a Euclidean structure on Q. With the directed line D we associate the perpendicular plane denoted by D^{perp}. There is a single defect: this bijection depends on the Euclidean structure chosen. The algebraist might prefer an alternative, but equivalent, presentation: let us choose some representation in projective coordinates, i.e. a projective frame of Q. Then the desired bijection consists of associating with the point (a, b, c) the line with equation $ax + by + cz = 0$. As for "modern" algebraists, they will observe that if $\mathcal{P}$ is the projective space associated with Q, then $\mathcal{D}$ is identified naturally with the vector space Q^* of Q. But, just as there doesn't exist a natural isomorphism between Q and Q^*, there doesn't exist a natural canonical isomorphism between $\mathcal{P}$ and $\mathcal{D}$.

For more on geometric dualities, the *correlations*, see 14.8.12 of [B] or p.260 of Frenkel (1973) for the general case, and Sect. I.8 below and Sect. IV.4 for the very particular case of Möbius tetrahedra. Duality will be unavoidable in a large part of Chap. VII. Furthermore this duality is completely geometric: given two points, the point of intersection of their two image lines has for an image precisely the line that joins the initial two points. This allows us to systematically obtain twice as many theorems, or to relate a desired theorem to another, perhaps simpler, theorem. In the sequel we will encounter examples in various contexts; see Sects. IV.4 and VIII.8 (conics, Pascal and Brianchon, inscribability of polyhedra). Right away readers can look for the duals of the theorems of Pappus and Desargues (see Sect. IV.4 as needed). Note that the mapping $D \to D^*$ of the left part of Fig. I.7.3 is imperfect: the origin doesn't have a dual; it is in fact the line at infinity.

◆

Attentive readers will not have missed noticing that, even though $\mathcal{P}$ and $\mathcal{D}$ are now in good bijection and have the same topology, this doesn't at all divulge the nature of the topology of $\mathcal{P}$. The first thing to observe is that $\mathcal{P}$ is **compact**, and this is also true for all the more general projective geometries of Sect. I.XYZ. In fact, if $\mathcal{P}$ is the set of points of $Q \backslash 0$ considered within multiplication by a nonzero scalar, this is also the set of points of the unit sphere of Q (a Euclidean structure is chosen for Q) modulo multiplication by ± 1, in other words the set obtained by identifying antipodal points of this sphere.

Nonetheless, the "shape" of $\mathcal{P}$ is not simple, and for an essential reason: if $\mathcal{P}$ is clearly a surface, moreover compact, it can nonetheless never be realized as a surface that is embedded in three dimensional space, this since $\mathcal{P}$ isn't *orientable*. The trick that was used for the projective line in Sect. I.6 doesn't work anymore. For

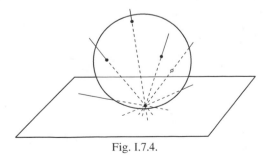

Fig. I.7.4.

on the oriented sphere, which must replace the oriented circle (compare Fig. I.6.1 with Fig. I.7.4), it is necessary to append all the points at infinity, and not just a single point; and in order to do that, cause the intervention of a "blowing up" (see Sect. I.3 and Fig. I.3.6).

A better way of understanding the topology of $\mathcal{P}$ is to see that not only can we obtain $\mathcal{P}$ by identifying antipodal points of the sphere, but that we can also be content to let this identification operate just on a hemisphere (boundary included): we need then only identify antipodal points of the equator. We can still choose to keep a band about the equator, it still being required that we identify antipodal points in this band. We obtain in this way a Möbius strip and $\mathcal{P}$ then appears as the union of a Möbius strip and a spherical cap, i.e. a disk sewn together without ambiguity.

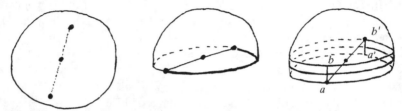

Fig. I.7.5. *Ways of seeing the projective plane. At left*: identify antipodal points. *Middle,* identify antipodal points of the equator. *At right,* identify antipodal points of a band (which comes down to preserving only the middle line of the band, while identifying ab and $b'a'$; pay attention to the direction of travel)

It is because $\mathcal{P}$ contains a Möbius strip that it is not an orientable surface; and it is for this reason that $\mathcal{P}$ is not embeddable in $\mathbb{R}^3$: if fact, to embed $\mathcal{P}$ in $\mathbb{R}^3$ would be to define a transformation of $\mathcal{P}$ into $\mathbb{R}^3$ that is continuous and injective; the geometer says *without double point* or *without self-intersection*. In view of compactness such a transformation would automatically realize a homeomorphism of $\mathcal{P}$ onto its image. However, a result from topology states that each compact surface in $\mathbb{R}^3$ without boundary possesses an interior and an exterior and is, for this reason, orientable.

A more direct proof of the fact that $\mathcal{P}$ is not embeddable in three dimensional space amounts to observing that a Möbius strip cannot be glued, following a common boundary, to a topological disk without the band and the disk intersecting. The boundary of the Möbius strip is indeed a circle (moreover unknotted), but a circle intertwined with the strip:

Fig. I.7.6.

We can nonetheless realize $\mathscr{P}$ as a surface in $\mathbb{R}^3$, but this surface must intersect itself. A first realization effectively consists, starting with a hemisphere, of identifying the antipodal points of the equator while cutting the boundary of the hemisphere into four arcs and gluing opposite arcs in an appropriate way:

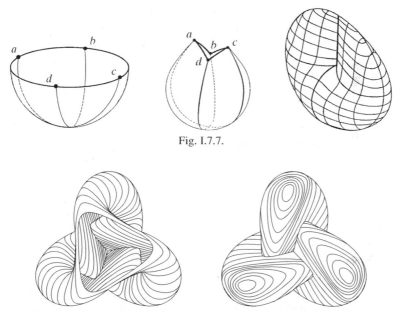

Fig. I.7.7.

Fig. I.7.8. Two views of Boy's surface. *At left*, a window has been opened to see the triple point

The figure obtained bears the name *cross-cap*. It possesses one ugly singularity at its vertex and a second at the other extremity of the curve of self-intersection, whereas *Boy's surface*, another realization of $\mathscr{P}$, only presents transverse self-intersections analogous to the intersection of two planes, with a single triple point. (What is called a singularity of a surface depends on its mode of definition. If, as here, the surface in $\mathbb{R}^3$ under consideration is presented as the image of another surface, the self-intersections are not considered singularities, provided that the sheets that intersect are regular open images of the plane. The singularities correspond to pinchings of the surface. These notions generalize those of a double point and of a cusp of a parameterized curve. Technically, a point is regular if at this point the rank of the differential of the parameterizing transformation is equal to 2; it is singular otherwise.)

Returning once again to the difficulty of visualizing the real projective plane, let us quote Léon Brunschvicg: "*what we see is in space, but we don't see the space.*" and "*space and its source in experience are attained in reason.*" On the other hand, we assert that there exists a very beautiful algebraic embedding of $\mathscr{P}$ in $\mathbb{R}^6$ given by the mapping $(x,\ y,\ z) \mapsto (x^2,\ y^2,\ z^2,\ \sqrt{2}yz,\ \sqrt{2}zx,\ \sqrt{2}xy)$. More precisely, let us consider the sphere S^2 as the set of (x, y, z) with $x^2 + y^2 + z^2 = 1$. As we have just seen, each point (x, y, z) of this sphere has an image in $\mathcal{P}$; in contrast a point of $\mathcal{P}$ originates from exactly two points of S^2, for the antipodes (x, y, z) and $(-x, -y, -z)$ define the same point of $\mathcal{P}$. We say that this transformation $S^2 \to \mathcal{P}$ is a "covering by two sheets". But once having chosen coordinates, we have a Euclidean structure on the space $\mathbb{R}^3$ and of course induce a metric − a notion of distance − on the sphere; but also a metric on $\mathcal{P}$ by defining the distance between two points of $\mathcal{P}$, regarded as lines of the space, to be equal to the angle (between 0 and $\pi/2$) that these two lines form. Now the transformation of $S^2 \to \mathbb{R}^6$ defined above also defines a transformation $\mathcal{P} \to \mathbb{R}^6$, according to the sign rule. This last transformation is injective, in contrast to the mapping of the sphere. The surface image, both of the sphere and the projective plane, is called the *Veronese surface*. It is found in many other situations, for example in the theorem on five conics of Chasles in Sect. IV.9. It's a justifiably superb geometric object, for numerous reasons. First, the embedding is isometric; it preserves distances in $\mathcal{P}$ and in $\mathbb{R}^6$. Moreover, it is *equivariant*, which is to say that the isometries of $\mathcal{P}$ are realized via this embedding, as induced by the global isometries of $\mathbb{R}^6$ (or S^5). Better still, it is algebraic in nature. Finally, it is *rigid*, for Kuiper has shown that, modulo an affine transformation of $\mathbb{R}^6$, it is the only surface whose topology is that of $\mathcal{P}$ and such that, for each direction of $\mathbb{R}^6$, it has only three perpendicular tangent planes in that direction. Knowledgeable readers are aware that there must always be at least three, according to Morse theory; for this theory see 4.2.24 of Berger and Gostiaux (1988), Bott (1988) and Bott and Tu (1986). The fact that there aren't more characterizes it. We can caricature this by saying that the Veronese surface is an extremely susceptible creature: the merest touch causes it to have more tangent planes in all directions; it becomes **dented**. For all this and for a second characterization by the condition that each tangent plane cuts it into exactly two pieces, see Kuiper (1984).

Readers who like the intrinsic can replace the transformation given in coordinates by the one given by $y \mapsto (x \cdot y)^2$ of $\mathbb{R}^3$ into the space of dimension 6 of quadratic forms on $\mathbb{R}^3$. Here we arrive at the forms of rank 1; see also Sect. IV.9. The Veronese surface exists over all fields and in all dimensions; we will encounter it in Sect. IV.9 and in studying Borsuk's conjecture in Sect. II.0. The transformation that defines the Veronese surface is easily extended to all projective spaces of arbitrary dimension and over any field. Moreover we can work with polynomials of arbitrary degree, not just of degree 2.

A third realization of $\mathcal{P}$ is Steiner's Roman surface. Like Boy's surface, it possesses a triple point and three lines of self-intersection, which are segments; but it also has singularities − pinchings − at the extremities of the lines of self-intersection, thus six in total; in return it possesses a double infinity of ellipses; see toward the end of Sect. II.6. It's the only surface, except for the ellipsoids, with this property. The fact that this surface has lots of ellipses arises simply from its being a projection in three dimensions of the Veronese surface (it's the image of the transformation $(x, y, z) \mapsto (yz, zx, xy)$ of S^2 into $\mathbb{R}^3$). The term "Roman" comes from the fact that Steiner discovered it when he was in Rome in 1844. We will encounter the Veronese surface in Sects. II.0 and V.9.

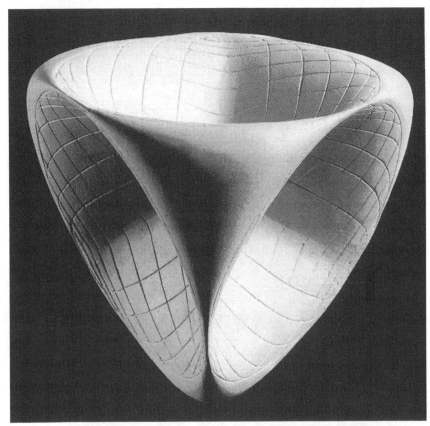

Fig. I.7.9. Steiner's Roman surface. Fischer (1986a) © G. Fischer

Note that the complement in $\mathcal{P}$ of a projective line is *connected*, in contrast to the affine case, infinity serving as the connection bond. The same thing is true for the median line of the Möbius strip. How do we see that we have the same phenomenon? By considering, in the projective plane, a band containing a given line D. The band situated between two lines parallel to D won't do, since it contains only a single point at infinity, but the region contained between the two branches of a hyperbola (situated on both sides of D) contains a whole segment of points at infinity, and it clearly has the topology of a Möbius strip, since it is obtained by identifying, in a rectangle, two opposite sides traversed in opposite senses.

This fact explains why a curve located on one side of its asymptote, when it tends toward infinity, always returns from infinity in the opposite direction. The curve is in fact tangent to its asymptote at its point at infinity, and if the contact is ordinary the curve does not cross its tangent. In particular this is the case for the point at

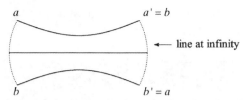

Fig. I.7.10. A neighborhood of a *line* in the projective plane has the topology of a Möbius strip

infinity of a hyperbola: the homography or homogeneous coordinate transformation, $(x, y, z) \mapsto (z, y, x)$, or $(x, y) \mapsto (1/x, y/x)$ in affine coordinates, transforms the hyperbola with equation $xy = z^2$ ($xy = 1$ in affine coordinates) into the parabola with equation $zy = x^2$ ($y = x^2$ in affine coordinates), tangent to the x axis at the origin. The curve does not cross its asymptote; it's the plane that makes a half turn like a Möbius strip.

As for the connectivity property indicated above, it explains the well-known trick of cutting a paper Möbius strip along its center curve and then continuing to cut along new median curves. It is left to readers to carry out the necessary experiments.

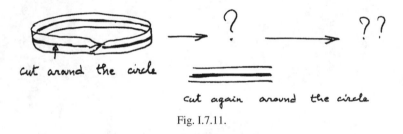

Fig. I.7.11.

We have seen that there exists a canonical metric structure on $\mathcal{P} = \mathbb{R}P^2$, derivative from that for the sphere: the distance between the two points $p, q \in \mathbb{R}P^2$ is the angle, between 0 and $\pi/2$, between the two lines of $\mathbb{R}^3$ which give rise to p and q. This geometry is called *elliptic*; it must be seen as a generalization of Euclidean geometry, for any two projective lines intersect in a single point (which is not the case for the great circles of spherical geometry). Here there are never any parallels, whereas in hyperbolic geometry in contrast there is an infinity of parallels for any given line. For more details on elliptic geometry, see Chap. 19 of [B]. But here is an example to which we should pay attention. It has to do with studying the cases of the equality of triangles: are two triangles for which two sides are the same equal, in particular are their three angles the same? An initial remark: two points in $\mathcal{P}$ are joined by a unique shortest path (projection of an arc of a circle onto the sphere) if their distance apart is less than $\pi/2$, otherwise there are exactly two shortest paths; but we will only discuss triangles with distances between vertices all less than $\pi/2$.

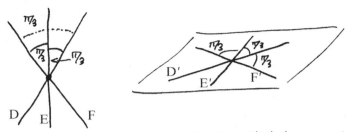

Fig. I.7.12. The *three lines* (D, E, F) and the *three lines* (D′, E′, F′) form *equilateral triangles* of side $\pi/3$ in $\mathcal{P}$. But these *triangles* are completely different: in the second the three angles equal π, in the first the three angles equal A, where $\cos A = \frac{1}{3}$. (apply the fundamental formula of spherical trigonometry)

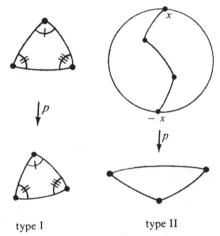

type I   type II

Fig. I.7.13. The two types of triangles in projective space, distinguished by their rise

We return to the equality of triangles, the exemplary case being, viewed as sketched in $\mathbb{R}^3$, that of a trihedron with three equal angles of $\pi/3$ and of a degenerate trihedron formed by three lines of a plane which makes among them the equal angles $\pi/3$. What can happen here, seeing that everything goes well in spherical geometry? To understand this, it is natural to go back to the sphere; but just one point of $\mathcal{P}$ provides two different points (antipodes) of the sphere. A curve of $\mathcal{P}$, here a side of a triangle, once a vertex has been lifted onto S^2, is lifted without ambiguity into S^2 because the projection of the sphere onto the projective plane is bijective and bicontinuous when restricted to a sufficiently small open set of S^2, typically the open hemisphere (spherical cap of aperture π) centered at a given point. Continuing in like manner for the two remaining sides we obtain a curve formed by three arcs of circles in S^2, but for which the terminal point is either the chosen point of departure or its antipode. These two cases are exactly those of the two triangles in $\pi/3$ considered. There are thus two types, I and II, of triangles in $\mathcal{P}$, but note that the type II will only be encountered if the sum of the sides is greater or equal to π. Readers will easily show, by deftly applying the case of equal spherical angles, that the equal angle case holds if, besides the equality of the respective sides, the two triangles considered are of the same type.

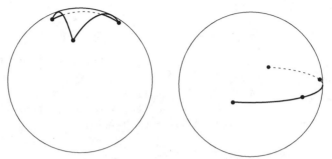

Fig. I.7.14. Lifting into S^2 of the two "exemplary" triangles of Fig. I.7.13

With the canonical metric structure of $\mathcal{P}$, the associated duality of Sect. I.7 is expressed thus: the projective line that is dual to a point p is made up of points of $\mathcal{P}$ located at a distance $\pi/2$ from p. For this associated geometry, perpendicular bisectors, etc., see [B]. It is interesting to note that elliptic geometry, which furnishes a trivial counterexample to Euclid's parallel postulate, was not known until well after hyperbolic geometry was discovered. This is due, among other reasons, to the difficulty of "seeing" the real projective plane.

Finally, we indicate why the only transformations that preserve lines are the projective transformations: the proof is achieved by fixing any line at infinity and applying the fundamental theorem of affine geometry to the complement. This result is often called the *second fundamental theorem of projective geometry*; see Sect. I.XYZ for the "first fundamental theorem".

I.8. The complex case and, better still, Sylvester in the complex case: Serre's conjecture

In Sect. I.1 we briefly alluded to affine geometry over the field of complex numbers, and even over the quaternions. The definition of the affine plane (over the reals), in which we have worked until now (see Sect. I.XYZ), extends trivially to the case of an arbitrary base field, not only to number fields, in particular the complex numbers $\mathbb{C}$ (commutative) and the quaternions $\mathbb{H}$ (non commutative), but also to all other fields, in particular the finite fields, the most simple among them being the field of two elements $\mathbb{Z}_2 = \mathbb{Z}/2\mathbb{Z}$. But there is no reason at all to restrict ourselves to dimension 2, since everything is constructed using only the algebraic theory of vector spaces (if necessary, refer to Sect. I.XYZ). Beyond what will be said in this section, see p. 9 of the introduction of Orlik and Terao (1992). For example, there is no strict notion of cross ratio in the non commutative case, typically over the quaternions, but only of conjugate classes, which take its place.

Let us repeat that Sylvester's theorem of Sect. I.1 is false over the complex numbers, the simplest numerical example is written in projective coordinates for the complex projective plane $\mathbb{C}P^2$; specifically, the nine points with projective coordinates

$$(0,1,-1), \quad (1,0,-1), \quad (1,-1,0),$$
$$(0,1,-\omega), \quad (1,0,-\omega), \quad (1,-\omega,0),$$
$$(0,1,-\omega^2), (1,0,-\omega^2), (1,-\omega^2,0).$$

Here ω denotes a cubic root of unity other than 1, e.g. $(-1 + i\sqrt{3})/2$. These nine points are the inflection points of the cubic (projective) equation $x^3 + y^3 + z^3 - 3axyz = 0$ $(a \neq 0)$. We verify by hand, without having need of the theory of planar cubics, that on each line joining two of these nine points there is always a third. Algebraically the condition for the collinearity of three points is translated by the fact that the determinant of their nine coordinates is zero. This configuration $(9_4, 12_3)$ is called *Hesse's configuration*. Readers who like simple calculations but hate projective coordinates will be able, with the aid of an appropriate projective transformation, to write the coordinates for a system of nine points of this type in the complex affine plane. There are too many zeros in the three possible places in the above array to allow us to proceed solely by division of the same coordinate. We will have observed that collineation here is complex, but we can regard the condition by taking a complex vector space as a real vector of twice the dimension; then Sylvester's condition will be that we always have a third point on the complex line generated by two given points; but of course this complex line is in effect a real **plane**.

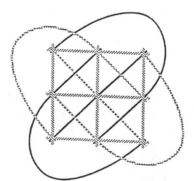

Fig. I.8.1. Hesse's configuration

In 1966 Jean-Pierre Serre announced the following conjecture: *"Let there be given, in a complex affine space of arbitrary dimension, a finite system of points satisfying Sylvester's condition: show then that this set of points is necessarily contained in a (complex) plane V"*. We emphasize that the proof, if there is one, cannot be purely combinatorial; and that if Sylvester was wrong in the complex case, it's because a line D does not separate the complex plane $\mathbb{C}^2$ into two regions: $\mathbb{C}^2 \setminus D$ is connected, we can "circle about D". Readers will have noted that the complex plane is of real dimension 4 and that lines are of real dimension 2 and thus "surfaces".

The answer is positive and is found in Kelly (1986); or see p. 802 of Hirzebruch (1987). But it is extremely difficult and requires climbing many rungs up the ladder. We indicate the principal steps, each of which constitutes at least one rung. We can place ourselves in dimension 3, for if the result holds there it will be true in any dimension, just as Sylvester's result for the plane immediately implies the same result in any dimension. Let therefore $\mathcal{P}$ be a system in $\mathbb{C}P^3$ satisfying Sylvester's condition. From a well chosen point in $\mathbb{C}P^3$, take the projection of the initial configuration onto a complex projective plane $\mathbb{C}P^2$; the projection is also clearly a Sylvester configuration. We construct all lines that join pairs of its points and let t_r denote the number of such lines that contain exactly r points. Let n be the total number of points; we have $t_n = 0$ (unless all the points are collinear) and, on the other hand, $t_2 = 0$. With this set of lines we create, as in Hirzebruch (1987b), an abstract algebraic surface in roughly the following way: we turn a certain integral numbers of times about the lines to obtain a covering of $\mathbb{C}P^2$, as is done in $\mathbb{C}$ about the origin with the transformation $z \mapsto z^k$. Finally, we must desingularize the places where the lines intersect each other. The miracle is that this surface possesses two invariants, its Chern classes $c_1(\mathcal{S})$ and $c_2(\mathcal{S})$. Since 1977 it has been known that these satisfy the inequality $(c_1(\mathcal{S}))^2 \leqslant c_2(\mathcal{S})$, a relation that is very difficult to prove. This was first accomplished in 1977 as a corollary to a theorem of Yau. This theorem, together with others, earned Yau the Fields medal; he showed that on certain algebraic surfaces Kähler metrics can be found for which the curvature is very special (so called Einstein varieties). When the Chern classes c_1 and c_2 are calculated there by precisely the same integral formulas obtained by Chern for "his" classes, using the curvature, the particularity of the curvature implies the required inequality. Yau's theorem is an ultra difficult global theorem about partial differential equations (PDE's), conjectured by Calabi in 1954 and which had defied all the experts in PDE's during those 23 years.

This inequality between c_1 and c_2 forces, for our configuration, some combinatorial relations that contradict, finally, the finiteness of p. These can be found in Hirzebruch; they bring with them the consequence that, with the notations of Sect. I.1, the two conditions $t_2 = t_n = 0$ imply $t_3 \neq 0$. We then fabricate a kind of translation that engenders an infinite number of points in the configuration, which is the contradiction that was sought. We recall that Sylvester's condition alone does not imply any such nontrivial combinatorial condition. Thus the conjecture of Serre has been settled, and even though its statement is elementary, it came only at the cost of a sizable portion of the modern theory of algebraic surfaces.

For those who are curious and who like open problems, we can ask what happens with the question of Sylvester-Serre in the quaternion spaces . To our knowledge the question has never been studied. The conclusion might be, by analogy with the real and complex cases, that the set of points must ultimately be in a quaternion subspace of quaternion dimension equal to 4.

Readers will perhaps not be surprised that the **trinity** of the three fields $\{\mathbb{R}, \mathbb{C}, \mathbb{H}\}$ is encountered in other parts of geometry and elsewhere in mathematics.

For a delightful presentation of a synthesis of this prevalence of $\{\mathbb{R}, \mathbb{C}, \mathbb{H}\}$, as well as for other trinities (e.g. the three kinds of conics to be seen in Sect. IV.1), read Arnold (1999) and Arnold (2000).

I.9. Three configurations of space (of three dimensions): Reye, Möbius and Schläfli

We will see in Sect. I.XYZ that a systematic study of *configurations* is not our objective. On the other hand, just as in Sect. I.3, we present here some configurations of the real space of dimension 3 that necessitate, for their understanding or even for showing their existence, a climb rather far up the ladder. In the space of three dimensions, a configuration of *type* (p_q, r_s) is a set of p points and of r planes such that each point belongs to exactly q planes and each plane contains s points.

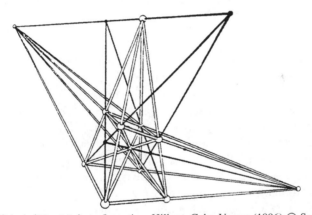

Fig. I.9.1. A $(12_4, 16_3)$ configuration. Hilbert, Cohn-Vossen (1996) © Springer

The easiest is *Reye's configuration*; it is of type $(12_6, 12_6)$ and its existence is elementary. Figure I.9.2 gives it with a cube and some points at infinity in the direction of the arrows: we only need verify that this actually works. But this configuration arises more naturally when we consider four spheres in space; the configuration is the one formed by their centers of homothety (see Chap. II). For three circles in the plane, we find a complete quadrilateral. In space our four spheres have, two at a time, two centers of homothety, i.e. 12 altogether for the 6 pairs of spheres. The sought-for planes are more difficult to locate, except for the four planes formed by the centers of triples of spheres. We leave this to readers. A reference for this configuration and for that of Schläfli is Hilbert and Cohn-Vossen (1952); see also below.

The Möbius tetrahedra are pairs of tetrahedra in real three dimensional space such that each vertex of either one belongs to a face of the other, giving the configuration $(8_4, 8_4)$. They can be found "by hand" easily enough, mainly by sending points to

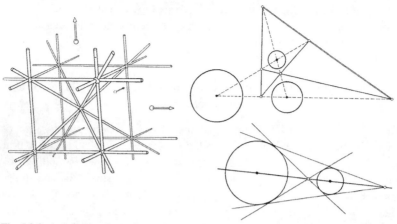

Fig. I.9.2. *At left*, Reye's configuration. *At right*, the configuration formed by the centers of homothety of *three circles* (below, recall the properties of the centers of homothety of *two circles*). Hilbert, Cohn-Vossen (1996) © Springer

infinity. But why can't we do the same thing for triangles? It is easy to convince ourselves of this impossibility "by hand", but that would not really get to the heart of the matter. We must climb the ladder, which we do in recalling the end of Sect. I.6, which turns out to be profitable in all dimensions. There we constructed a bijection between the points and lines of the plane, which with each triangle then associates a new triangle; but we can't require that this bijection have the property that the line associated with a point pass through that point. However, this is possible in a projective space (projective, so as to avoid exceptions) Q of three dimensions: there exists a bijective transformation f of Q onto the set Q^* of all its planes such that $p \in f(p)$ for each point p of Q. Of course we require further that the properties of collinearity and intersection be preserved. Thus any tetrahedron defines a pair of Möbius type by adding to it the tetrahedral image under f. And here is the "parachuting" of such an f: we set up projective coordinates in some way (any choice will do) and define $f((a, b, c, d))$ to be the plane with equation $-bx + ay - dz + ct = 0$. The membership condition is obviously satisfied and, things being linear, all properties of intersection, collinearity, etc. are preserved as we would like.

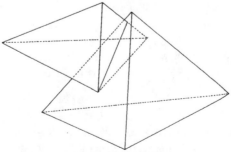

Fig. I.9.3. Möbius tetrahedra. I [B] Géométrie. Nathan (1977, 1990) réimp. Cassini. (2009) © Nathan Édition

In fact, we have indeed climbed the ladder somewhat. First, we see that we will be able do the same thing for every space of uneven dimension, the number of projective coordinates being even in this case. But above all the theory of all this was accomplished geometrically and very laboriously at the end of the nineteenth century. Now, algebraically, linear and multilinear algebra permit us to resolve completely all the questions that can be posed, and this in arbitrary dimension and over an arbitrary field. The essential problem is knowing what are the bijections between a space and its dual (the space of its hyperplanes) that preserve intersections and collinearity. The answer is that there are two possible types, ones that we have encountered: the type given by a Euclidean structure, i.e. a quadratic form (corresponding to a *symmetric* bilinear form) of maximum rank, and the type called *symplectic*, i.e. given by an antisymmetric bilinear form of maximum rank, it being understood that a symmetric or antisymmetric bilinear form of maximum rank on a vector space E defines a bijection of E onto its dual E^* and, by passage to the quotient, a bijection of the projective space $\mathcal{P}(E)$ onto the space of its hyperplanes. In the second type, we have the Möbius property that each point belongs to its dual; and it is trivial that these structures cannot exist in even dimensions (odd dimension for the vector space). Since the quadratic forms of maximum rank — at least in the real case — are categorical, as are the symplectic forms of maximum rank, we now know everything. The proof of this fundamental result is Exercise 14.8.12 of [B], a more complete reference is Frenkel (1973).

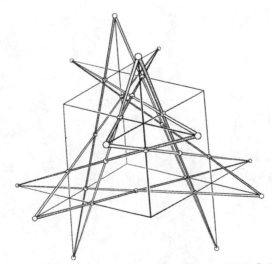

Fig. I.9.4. Schläfli's double six. Hilbert, Cohn-Vossen (1996) © Springer

♦

Schläfli's *double six* is given in the above figure. It's a configuration of type $(30_2, 12_5)$, but the notation here indicates 30 points and 12 lines (no planes). The proof of its existence isn't elementary; it makes essential use of algebraic geometry, specifically that a cubic surface (of degree 3) in three dimensional space and without singularities contains exactly 27 lines in the complex case, and in the real case 27, 15, 7 or 3 lines.

An "elementary" exposition is contained in §25 of Hilbert and Cohn-Vossen (1952). The point of departure is to take four lines in space in general position. There exist two lines that intersect them all, which is seen using what is found in Sect. IV.10: the set comprised by the lines intersecting three given lines in general position is a quadric surface, so that the desired lines, based on four lines, are obtained by looking for the points of intersection of the fourth line with the quadric surface defined by the first three. In general there are two such points, our four lines being "generic". The construction of the double six begins thus: we start with a line D_1, then construct at random five lines E_2, E_3, E_4, E_5, E_6 intersecting D_1. Let D_6 be the next line, other than D_1, intersecting the four lines E_2, E_3, E_4, E_5. Then

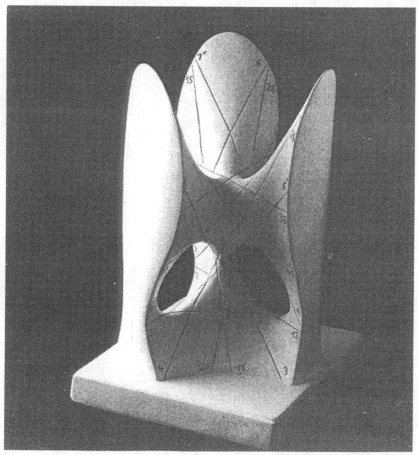

Fig. I.9.5. Clebsch's diagonal surface. Fischer (1986a) © G. Fischer

D_2, which in addition to intersecting E_3, E_4, E_5, E_6, likewise intersects D_3, D_4, D_5. It remains to define E_1 as the line likewise intersecting D_2, D_3, D_4, D_5. Now, we have numerous other intersections that form, by construction, a configuration of type $(30_2, 12_5)$. But the proof is difficult and, as already indicated, requires the use of algebraic geometry; see the book cited above for an heuristic proof.

We will encounter this configuration in Sect. V.16. There indeed exist real cubic surfaces containing 27 lines, all without singularities, e.g. Clebsch's diagonal surface. The theoretic considerations are found in Fischer (1986b), they depend on Fig. I.9.5.

I.10. Arrangements of hyperplanes

We have seen that there is a difficulty in studying the finite sets of lines in the real affine plane, because of the possibility of parallels. Nonetheless, the question of the **maximum** number of connected components of the complement of n lines was resolved before the twentieth century, and in fact by Schläfli for the complement of n hyperplanes in arbitrary dimension. In order to have the maximum number, it is necessary to be in the generic case; what happens when more than two lines intersect, and for the parallels, dates from 1889. This does not give a classification; such a study leads in fact to problems in topology, algebraic geometry, convexity, combinatorics, number theory, (arithmetic) analysis and geometry without any obvious relations. We will not speak further here of the real case; we refer readers to its introduction in Orlik and Terao (1992) or pp. 679–706 of volume II (Hirzebruch, 1987b), not to forget the commentaries of pp. 802–804.

The complex case has an even greater richness, and provides much pleasure when connections are found between seemingly different things. The book cited above (Orlik and Terao, 1992) is entirely devoted to it; it's a subject that ascends quite high on the ladder and has a very recent development. We can't speak about it in detail, or even partially. We have seen a first approach in Sect. I.7 above. In a second approach, we first remark that, if the complement of a hyperplane in a complex space is connected, contrary to the real case, in return it isn't *simply connected*: if we "make the tour" of a hyperplane, we obtain a loop (closed curve) that isn't contractible to a point. More generally, with a finite set of hyperplanes we associate the *fundamental group* of its complement, i.e. the group generated by the closed curves through a fixed point (in the complement, of course) that "turns" about the hyperplanes of the arrangement under consideration. We obtain this group by considering two closed curves to be equivalent if they are deformable each into the other, in a continuous fashion, without of course intersecting the figure about which they are turning. The nature of this group and more generally the topology of this complementary set are much studied nowadays and yield surprising relationships.

The simplest case is the topology of the complement of two intersecting lines of $\mathbb{C}^2$: it is easily seen that this set can be continuously deformed onto the product of two circles, i.e. a torus. In particular, the fundamental group of this complement is that of the torus, thus $\mathbb{Z}^2$.

I. XYZ

For the mathematical objects described above and for proofs, a general and complete reference is [B], to which we add Hilbert and Cohn-Vossen (1952) for the configurations.

Before mathematicians, curious and practical minds would ask how to make lines straight, planes "flat" — all this both in drawing instruments and in industrial practice of a more or less high degree of precision; and finally for the highest degree of precision, that of metrology. We will give some brief indications about this problem in Sect. II.XYZ.

The real affine plane is defined as follows: we consider a real vector space P of dimension 2, or, what amounts to the same thing — modulo isomorphism — the set $\mathbb{R}^2$ of pairs (x, y) of real numbers. The *points* are thus the elements of $\mathbb{R}^2$; the *lines* are all the subsets that are translations of a vectorial line of $\mathbb{R}^2$. We might say that our affine plane is a real vectorial plane **for which the origin has been forgotten**; there is no longer an origin, no special point. The properties enjoyed by lines and planes, and their relationships, are direct consequences of the axioms for vector spaces. This definition will not satisfy very abstract minds; there remains a mental trace of an origin. In [B] there is a more axiomatic presentation. Here we will retain our point of departure: with each pair (p, q) of points of the affine space there is associated the vector ("free" in the old language) that may be *denoted* $q - p$, or often also $\overrightarrow{pq}$, which belongs to the vector space, giving rise to the affine space under consideration.

An important element, which serves as the foundation for the more conceptual construction mentioned above (i.e. not favoring an origin, or even the ghost of one), is the **barycentric calculus**. Two points p, q cannot be added in an intrinsic manner (we can only subtract them, but the result is a vector and not a point of the space considered. On the other hand, the midpoint of the two points can be written $\frac{p+q}{2}$, and more generally we can divide a segment by a given ratio; see the affine invariant introduced in Sect. I.3. What matters in $\frac{p+q}{2}$ is that it can be written $\frac{1}{2}p + \frac{1}{2}q$. More generally we can define in an intrinsic and practically trivial fashion any finite sum $\sum \lambda_i p_i$ such that the sum of the coefficients (called barycentric) satisfies $\sum \lambda_i = 1$. Two facts are essential: the associativity of this operation and the uniqueness of the coefficients for the sums for three points forming a true triangle. This furnishes "purely affine" coordinates for the plane, coordinates called *barycentric*; for all the details see Chap. 3 of [B]. Briefly and in any dimension, since no changes are needed: if $\{p_i\}$ $(i = 1, \ldots, d + 1)$ denotes a set of $d + 1$ points in real affine space of dimension d forming a *simplex* — which means that none of them belongs to the hyperplane defined by the remaining d — then for each point x of the space there exists a unique expression (a "barycentric sum of the $\{p_i\}$") $x = \sum \lambda_i(x) p_i$

such that $\sum \lambda_i(x) = 1$. A simplex is: on a line, two distinct points; in the plane, a true triangle; and in three dimensional space a tetrahedron. The **full** simplex for a system of points is the set of all points for which the barycentric coordinates of the points of the system are nonnegative.

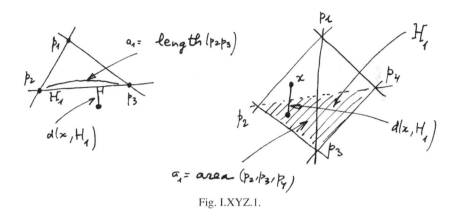

Fig. I.XYZ.1.

These coordinates $\{\lambda_i\}$ are sometimes essential; see the *spheres of the "combles"* in Sect. II.1. They are also much used in the study of *convexity*; see Chap. VII. It can in fact be shown that if, in addition, the space is Euclidean, V is the total volume of the simplex, a_1 the volume (a distance in the plane, an area in space) of the face H_1 opposite the vertex p_1 of the simplex under consideration (i.e. formed by the d points $p_2, \ldots, p_{d+1}$), then the absolute value $|\lambda_1(x)|$ is given by the formula $|\lambda_1(x)| = \frac{a_1}{dV} \cdot \text{distance}(x, H_i)$.

◆

But in fact, instead of $\mathbb{R}^2$, we can consider an arbitrary field K and an arbitrary dimension n: the *affine space of dimension n over* K is simply the vector space K^n, whose elements are points; the affine subspaces of dimension k are nothing other than translations of vector subspaces of dimension k of K^n. In dimension 1 we speak of *lines*, in dimension 2 of *planes* and in dimension $n - 1$ of *hyperplanes*.

In every affine space we have *homotheties*; the vectorial homotheties are the transformations $x \mapsto \lambda x$, where λ is an arbitrary nonzero element of the base field. The affine homotheties are obtained by translating these last. More precisely: the homothety with center a and ratio λ is the transformation $m \mapsto a + \lambda(m - a)$.

A frequent case is where the field is **finite** (frequently encountered in combinatorial geometry) and thus necessarily commutative. The most important field, after the field of real numbers, is the field $\mathbb{C}$ of complex numbers, to which we may add the (noncommutative) field $\mathbb{H}$ of quaternions; see 8.1 and the index of [B]. But also, in the spirit of the present book, see their appearance in polyhedra in Coxeter (1974) and, for the associated geometric groups, the important (Du Val, 1964).

Quaternion geometry is less developed, the non-commutativity of the field is a significant complication. We find that Pappus's theorem is false, nor is there a strict notion of **cross ratio**; the cross ratio is simply a conjugate class in $\mathbb{H}$ under the inner automorphisms. The three fields $\mathbb{R}$, $\mathbb{C}$, $\mathbb{H}$ form a trinity, for which there are many associated trinities. It can lead to discovery by analogy. For such a mindset and a list of trilogies, readers should look at the engaging text (Arnold, 2000). We take from it the Table I.XYZ.1 below, which is discussed in the text just mentioned.

	$\mathbb{R}$	$\mathbb{C}$	$\mathbb{H}$
(1)			
(2)	Morse theory handles attachment	Picard-Lefschetz theory Dehn twist	?
(3)	$\pi_0(\mathbb{R} \setminus 0) = \mathbb{Z}_2$	$\pi_1(\mathbb{C} \setminus 0) = \mathbb{Z}$	$\pi_3(\mathbb{H} \setminus 0) = \mathbb{Z}$?
(4)	$\mathbb{R}P^n$	$\mathbb{C}P^n$	$\mathbb{H}P^n$
(5)	$\mathbb{R}P^1 = S^1$	$\mathbb{C}P^1 = S^2$	$\mathbb{H}P^1 = S^4$
(6)	$\mathbb{R}P^1/\mathrm{Aut}\mathbb{R} = S^1$	$\mathbb{C}P^2/\mathrm{Aut}\mathbb{C} = S^4$	$\frac{\mathbb{H}P^4/\mathrm{Aut}\mathbb{H}}{\mathrm{Conj}} = S^{13}$
(7)	Quadratic forms	Hermitian forms	Hyperhermitian forms
(8)	Von Neumann – Wigner eigenvalues repulsion	Quantum Hall effect and Berry phase	?
(9)	Möbius S^0 bundle $S^1 \to S^1$	Hopf S^1 bundle $S^3 \to S^2$	Hopf S^3 bundle $S^7 \to S^4$
(10)	Monodromy of a covering	Curvature of a connection	Hypercurvature of a hyperconnection?
(11)	w	c	p
(12)	O, SO	U, SU	$Sp, ?$
(13)	Tetrahedron	Octahedron	Icosahedron
(14)	$(4, 4, 6)$	$(6, 8, 12)$	$(12, 20, 30)$
(15)	$x^2 + y^3 + z^4$	$x^2 + y^3 + yz^3$	$x^2 + y^3 + z^5$
(16)	$x^3 + y^3 + z^3$	$x^2 + y^4 + z^4$	$x^2 + y^3 + z^6$
(17)	$(\pi/3, \pi/3, \pi/3)$	$(\pi/2, \pi/4, \pi/4)$	$(\pi/2, \pi/3, \pi/6)$
(18)	A_3 o—o—o	B_3 o⇒o—o	H_3 o—⁵o—o—o
(19)	$2(1 + 3 + 3 + 5) = 24$	$2(1 + 5 + 7 + 11) = 48$	$2(1 + 11 + 19 + 29) = 120$
(20)	$(2, 4, 4, 6)$	$(2, 6, 8, 12)$	$(2, 12, 20, 30)$
(21)	D_4	F_4 o—o⇒o—o	H_4 o—⁵o—o—o
(22)	E_6	E_7	E_8
(23)	$\mathbb{C}[t]$	$\mathbb{C}[t, t^{-1}]$	$\mathbb{C}[t, t^{-1}, (1-t)^{-1}]$
(24)	Numbers	Trigonometric numbers	Elliptic numbers
(25)	H	K	Ell

Table I.XYZ.1.

In the case of the *Cayley octaves*, also called *octonions* and denoted $\mathbb{C}a$, we have to do with an object that is almost a field, but which lacks associativity. The octonions form a vector space of dimension 8 over $\mathbb{R}$. They are defined with a basis consisting of the identity element 1 and seven other generators $\{e_i\}$ ($i = 1, ..., 7$) for which the multiplication table is furnished by the triangles inscribed in a heptagon as in Fig. I.XYZ.2.

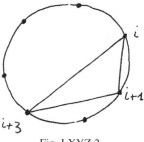

Fig. I.XYZ.2.

More precisely, we consider all triples of the form $\{e_i, e_{i+1}, e_{i+3}\}$ for i going from 1 to 7 and where the additions in the subscripts are computed modulo 7. Each of these triples obeys the same laws as the triple $\{i, j, k\}$ from the definition of the quaternions: $e_i e_{i+1} = e_{i+3}$, $e_{i+1} e_i = -e_{i+3}$. Even though this law is not ultimately associative, we can nonetheless define a reasonable geometry over $\mathbb{C}a$, not just in dimension 2, but also in dimension 3 and beyond. For references on this topic, which is especially subtle in the case of the construction of the "panda" encountered in Sect. I.4, the projective plane over $\mathbb{C}a$, see 4.8.3 of [B] and the references given there. See above all the entirety of Chap. XIV of (Porteous, 1969). A recent reference on octonions is Baez (2002).

We must pay strict attention to the tricks intuition can play, as soon as we are working over the complex field instead of over the reals. A plane over the complex numbers has real dimension 4. A typical example is this: what is the topology of the complement of a point in $\mathbb{C}$? It is that of a cylinder, which may be contracted to a circle, whereas in $\mathbb{R}$ it amounts to two intervals, which contract to two points; in each instance the complement in $\mathbb{R}$ is not connected, which is no longer the case in $\mathbb{C}$. And for the complement of two intersecting lines in the plane $\mathbb{C}^2$? In the case of the real plane, we find four connected components. But here the complement is connected; if we contract it onto the intersection of $\mathbb{C}^2$ with the unit sphere S^3 in $\mathbb{R}^4$, it turns out, according to Sect. II.5, that the complement has the same topology as the complement in S^3 of two orthogonal circles, i.e. of a torus T^2. Another motivation for geometry over the complex numbers is the recent notion of *quantum computers*; these are objects that truly operate in $\mathbb{C}$.

Another caveat is that, when we work in **higher dimensions**, our intuition loses practically all its rights, all its effectiveness. We will see numerous manifestations of this in Chap. VII. But high dimensions are absolutely necessary in numerous mathematical studies, both theoretical and applied, especially since about 1950. Loading an image on a computer requires working in spaces of, say, dimension 10 000. Think also about credit cards, of cryptography, of biology and the error correcting codes that are studied in Chap. X. All of linear programming involves a potentially large number of "consumers", and hence enormous dimensions.

Another need is that of functional analysis, which leads to the study of convex sets in very high dimensions; see Chap. VII. Our intuition will be put to the test there, starting with the volume of spheres of radius 1, for which the volume tends ultra-fast to zero as the dimension becomes large.

Still another need is that of physics, where statistical mechanics treats sets of particles with numbers of order 10^{23}, which thus lives in spaces having such dimensions.

Also interesting is the remark due to Pierre Cartier that, in fact, we scarcely comprehend more than dimension 1. The reason isn't merely that our brain works in a linear fashion, sequentially, for in fact (as far as is known today) it works rather like a parallel computer. It seems that this is simply for the crude reason "*that it's simpler*". For in fact dimension 2 isn't yet well understood; pattern recognition, for example, is in its infancy. Dimension 3, however indispensable for all the objects of space, harbors numerous open problems, as we shall see. Some people think that our difficulties in dimension 3 are due, at least in part, to the fact that the group of rotations in space is not commutative. But all these reflections on the profound nature of mathematics and the functioning of our brain are still in limbo. To paraphrase Pierre Cartier, in (Cartier, 1991): "The very subjective, if not to say blurred, character of art criticism is without doubt due to this characteristic. In our epoch of intensive use of computers and of computer-aided creativity, there is a regrettable gap." Making a leap from dimension 1 to dimensions 2 and 3 is one of the present obstacles to progress in mathematics. Similar obstacles have led to the creation of non-Euclidean geometry and to the discovery of Gödel's theorem.

The fundamental theory of affine geometry, seen for the real plane in Sect. I.3, is proved by the same technique for all dimensions and all fields. But there are two very important modifications that we give here; for more, see 2.6 of [B]. We proceed first to the obvious exclusions (the miracle is that there aren't others). First the case of dimension 1, where the collinearity condition is vacuous: every bijection preserves lines, since there is but one line here! But things get more complicated according to the qualities of the field K considered. It is necessary at first to completely exclude the field $\mathbb{Z}_2$ of two elements, for then the affine line contains only two points and the collinearity condition is again vacuous. But above all the proof of I.3.1 goes through thanks to the fact that only the identity is an automorphism of the real number field.

It's not entirely the same for other fields, e.g. the field of complex numbers admits the automorphism that transforms z into its complex conjugate $\bar{z}$: conjugation preserves sums and products. A *semi-affine* transformation of an affine space into a field K is, by definition, modulo a translation, a *semi-linear* transformation (more generally we can speak of a semi-affine transformation of one affine space into another), i.e. a transformation f such that $f(\lambda x + \mu y) = \sigma(\lambda) f(x) + \sigma(\mu) f(y)$, where $\sigma : K \to K$ is any automorphism of the field K. We point out that $\mathbb{C}$ admits lots of other automorphisms besides the identity and the conjugation $z \mapsto \bar{z}$, but if we further require continuity, then there remain only those two. The fields of characteristic k different from 0 admit the celebrated *Frobenius automorphism* $x \mapsto x^k$, whose importance in mathematics should not be underestimated, e.g. it enters Frobenius, Georg into Deligne's proofs of Ramanujan's conjectures; see Sect. III.3 (and Sect. III.6). On the other hand, the quaternion automorphisms are easy to classify; see 8.12.11 of [B]. And so these "refined" mappings are never so frightening as they can be in the case of the complex numbers.

We now define a projective space of dimension n over the field K as the set of "vectorial lines" (one-dimensional subspaces) of a vector space of dimension $n + 1$ over K. All these spaces are in fact the same, so we can speak of *the projective space of dimension n over* K; we *denote* it by KP^n. Algebraically, it's the quotient of $K^{n+1} \setminus 0$ modulo the equivalence relation such that $w \equiv v$ if and only if there exists $k \in K$ (necessarily nonzero) for which $w = kv$. The (projective) *lines, planes, hyperplanes* of KP^n correspond to vector subspaces of K^{n+1} of respective dimensions 2, 3 and n. The case of the projective line is the object of study of Chap. 6 of [B].

We introduced the projective spaces essentially for the purpose of completing affine geometry so that the intersection theorems could be presented without exceptional cases in their statements; but apart from the miracles mentioned below, the idea of considering the elements of a vector space "within a scalar factor" is very natural, if not to say indispensable. We will see two examples in this book, the first is the space of circles and spheres in Sect. II.6; the second is that of conics and quadrics (see Sect. IV.7). It isn't possible to really thoroughly understand circles, spheres, conics and quadrics without introducing the projective space formed **by their equations**. This is very much in the spirit of Jacob's ladder. Here now, very briefly, are the essential properties of projective geometries.

The complex projective spaces are among the most important objects of algebraic geometry.

The projective transformations are (bijective) linear mappings of $K^{n+1} \setminus 0$ **retracted** onto KP^n; they are also called *homographies*. They obviously form a group, called the *projective group*, denoted by GP(n ; K), whose structure is that of the quotient of the linear group GL(n ; K) by the group of multiples of the identity. In terms of coordinates, it is the multiplicative group of the $(n + 1) \times (n + 1)$ matrices with elements in K, modulo the group of multiples of the identity matrix. The projective

spaces over finite fields are encountered in combinatorial geometry. Here is a typical example in the spirit of Sylvester's problem of Sect. I.1: whatever its dimension,
the projective space $\mathbb{Z}_2 P^n$ over the field $\mathbb{Z}_2$ of two elements is a finite set such that,
for each pair of points, there is a third point on the line that joins them. In fact, all
projective lines over $\mathbb{Z}_2$ have three elements: two affine and their point "at infinity".

The homographies of the projective line are classified and studied in detail in
Chap. 6 of [B]. The involutions (homographies which when squared yield the identity) play a particularly important role. If a homography has two distinct fixed points
a and b, then the cross ratio $[a, b, m, f(m)]$ is constant; conversely, the relation
$[a, b, m, f(m)] = k$ defines a homography, an involution when $k = -1$.

As indicated above in Sect. I.7, there is a *second* fundamental theorem of projective
geometry valid in any dimension and over any field; but as in the affine case it is
necessary at first to make the obvious exclusion of dimension 1. But subsequently
there will be no need to exclude the field of two elements; for a projective line, in
contrast to an affine line, always contains at least three points. The conclusion is that
we find only semi-projective transformations, i.e. projective transformations modi-
fied if need be by an automorphism of the field; see as needed 5.5.8 of [B] and the
references mentioned there. The *first fundamental theorem of projective geometry*
states that the transformations in projective dimension n are sufficiently abundant
in order to transform two arbitrary $(n + 2)$-tuples of points into one another; see as
necessary 4.5.10 of [B]. The result is trivial using linear algebra, although it bears a
ponderous name, which comes from the fact that it is difficult to prove and is much
more concealed if we pursue the *axiomatic theory of projective geometry*. Here we
have parachuted the projective spaces with linear algebra, while axiomatic projective theory constructs them (more or less completely) with axioms bearing on points,
lines, etc., and properties required for their various intersections, properties that are
trivial in linear algebra. Readers interested in the axiomatic theory can consult the
two basic books that exist Artin (1957) and Baer (1952). The theory remains difficult and, in our opinion, of mediocre elegance for the case of projective *planes*,
since there is no Desargues' theorem − seen in Sect. I.4 − at our disposal; readers
may be able to form an opinion with a cultural text such as (Lorimer, 1983).

In complex geometry it is very important to know that, while topologically the
real projective line is the circle S^1, the complex projective line is the sphere S^2;
see the end of Sect. I.6. Note that the real plane − as a real object − is projectified
into $\mathbb{R}P^2$ by appending a line at infinity, topologically a circle. But the same real
plane, when seen as a complex line, is projectified with a single point at infinity,
which yields the sphere S^2 topologically. Thus we have two quite different com-
pactifications of $\mathbb{R}^2$. In Sect. II.3 we will see a third, also utterly essential, where
compactification provides the topology of a disk; the interior is $\mathbb{R}^2$ but the compactification adds the boundary circle. For more on the topology of projective spaces,
which are elements of constructions essential in algebraic topology, there are some
references in Chap. 4 of [B]. We mention only that $\mathbb{R}P^n$ is orientable for all odd

n, nonorientable otherwise. Let us add here, however, that the complex projective spaces are the generating elements of the cobordism ring in algebraic topology. The theory of cobordism, which dates from the 1950s, is a classification of compact differential manifolds; see Husemoller (1975). As for homographies $z \mapsto \frac{az+b}{cz+d}$, which dominate a large part of mathematics, we will encounter them in Sects. II.3 and II.4. For a modern proof of the fundamental theorem of geometry, affine or projective, see Faure (2002).

♦

A *configuration* of an affine or projective plane is simply a specification of a set of points and the lines joining those points. Such a configuration evidently does not have any interest unless we require particular properties. We say that a configuration τ is of type (p_q, r_s), where p, q, r, s are integers, if there are p points and r lines, such that for each point of τ there pass exactly q lines of τ and if every line of τ contains exactly s points of τ. An accounting implies the relation $pq = rs$. The proof of the existence of a configuration of a given type may be trivial, but also more or less difficult.

The complete quadrilateral is a configuration $(4_3, 6_2)$; existence is trivial but nevertheless interesting to interpret in Euclidean geometry as that formed by the six centers of homothety of three circles; this has served us above for the Reye configuration. In Sect. I.7 we saw its very useful property of harmonic conjugation, proved by "transfer to infinity". Here we give a purely projective proof "in situ".

The complete quadrilateral is thus the figure formed by four lines in general position (the sides) and their six points of intersection (the vertices). These points of intersection are joined pairwise by three diagonals. The property of harmonic conjugation mentioned above is the following: two vertices a and b not located on the same side (and thus located on one of the diagonals) having been chosen, and i being the point of intersection of the two other diagonals, the sides issuing from a divide the segment bi harmonically. This means that, if we let x and y denote the points of intersection of the sides passing through a with the line bi, we have $(b, i, x, y) = -1$.

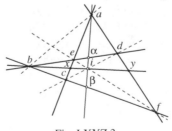

Fig. I.XYZ.3.

In Sect. I.7, this property was proved by transferring the points a and b to infinity, and thus in making a parallelogram out of the quadrilateral. The "classical" projective proof is based on the fact that the *perspectivity,* or *central projection,* of one line onto another is a homography and preserves the cross ratio. By considering the perspectivity with center a, we obtain $[b, \alpha, e, d] = [b, \beta, c, f]$. From the perspectivity with center i, we obtain $[b, \alpha, e, d] = [b, \beta, f, c]$. Thus $[b, \alpha, e, d] = [b, \alpha, e, d]^{-1}$, i.e. $[b, \alpha, e, d] = -1$, the value $+1$ being excluded. We have finally $[b, i, x, y] = [b, \alpha, e, d]$.

With the configuration of the complete quadrilateral is associated the notion of *polar of a point with respect to two lines.* We say that two *points* m and n are *conjugate with respect to the two lines* D *and* D' if they are conjugate harmonics with respect to the points of intersection with D and D' of the line that joins them. Otherwise expressed, denoting these points of intersection by p and p', we have $[m, n, p, p'] = -1$ on the line mn. The set of conjugates with respect to two lines D and D' of a point a not located on D or D' is a line passing through the point of intersection of D and D', called the *polar of the point a in relation to the lines* D *and* D'. To see this, it suffices to take D = be and D' = bd in Fig. I.XYZ.3. The point i is conjugate to the point a with respect to D and D', and the property of preservation of the cross ratio by perspectivity implies that a point of the plane is conjugate to a if and only if it is collinear with b and i. The polar of a is thus the line bi.

The point b represents a special case: we agree that it belongs to the polar and agree once for all that if a and b are two distinct points of a line, then $[a, b, b, b] = -1$, a convention that is justified by passage to the limit.

♦

In Sect. I.3, Pappus's theorem established the existence of configurations of type $(9_3, 9_3)$; Desargues' theorem proves the existence of configurations of type $(10_3, 10_3)$. The Hesse configuration, in the complex case, encountered as a counterexample to Sylvester, is of type $(9_4, 12_3)$ and establishing its existence was not trivial. Problems concerning the existence and classification of configurations have been very fashionable for some time. Personally we find the subject less agreeable when it is made systematic; see the figures on pp. 109–111 of Hilbert and Cohn-Vossen (1952). In contrast, a particular configuration can be fascinating, even critical, when it is related to geometric properties that are more or less profound. We have seen this above; we now mention some more examples.

The projective spaces over finite fields furnish important configurations because of their combinatorial consequences, the simplest example being the projective plane over the field $\mathbb{Z}_2$ of two elements; it contains 7 points and 7 lines and in its totality thus forms a configuration of type $(7_3, 7_3)$. The figure below includes some real (and continuous) lines (and a circle); they have no reality in $\mathbb{Z}_2 P^2$, they might actually be any curves; the fact that there is a curve through three points merely conveys the information that they are on the same projective line.

Readers will be able to show that such a configuration is impossible in real geometry; this is why we have drawn, in desperation, a circle in the above figure. What's important are the combinatorics of the figure, in the sense that it associates, by incidence relations, three objects of the second type with every object of the first type. The projective spaces over finite fields have abundant application: to geometry in

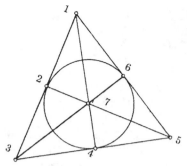

Fig. I.XYZ.4. The projective plane over $\mathbb{Z}_2$

the spirit of this chapter ("finite geometries"), to *combinatorics* and to error detecting codes. We treat these codes in Sect. X.4. As for combinatorics, they are already present in Sylvester's problem in Sect. I.1. Combinatorics is a very recent discipline and is establishing a presence in more and more mathematical domains. For more on projective spaces over finite fields, see Chap. 9 of Lidl and Niederreiter (1983) and the entire book (Hirschfeld, 1979). For more on combinatorial geometry, see all of Pach and Agarwal (1995); for a view of some very nice particular cases, see the corresponding part of Aigner and Ziegler (1998). The title *Proofs from the Book*, indicates this: it's a work containing a goodly number of proofs that are thought to be "perfect". Their collection is what Erdös called "The Book", saying: *"you don't necessarily have to believe in God but, as a mathematician, you have to believe in The Book"*; see its introduction.

Regarding problems on geometric probability, of which the first historical case is that of Buffon's needle, various references are: first the classic and easy (Santalo, 1976), then the more difficult and recent Part 5 of Gruber and Wills (1993). Finally the mm-spaces of Gromov, which are the subject of Chap. $3\frac{1}{2}$ of Gromov (1999a), are presented in an informal way in Berger (2000a); and, for a somewhat different perspective, see Talagrand (1995).

Bibliography

[B] Berger, M. (1987, 2009). *Geometry I,II*. Berlin/Heidelberg/New York: Springer
[BG] Berger, M., & Gostiaux, B. (1988). *Differential geometry: Manifolds, curves and surfaces*. Berlin/Heidelberg/New York: Springer
Aigner, M., & Ziegler, G. (1998). *Proofs from THE BOOK*. Berlin/Heidelberg/New York: Springer
Aigner, M., & Ziegler, G. (2003). *Raisonnements divins*. Berlin/Heidelberg/New York: Springer
Arnold, V. (1999). Symplectization, complexification and mathematical trinities. In E. Bierstone, B. Khesin, A. Khovanskii, & J. E. Marsden (Eds.), *The Arnoldfest* (pp. 23–28). Providence, RI: American Mathematical Society
Arnold, V. (2000). Polymathematics: Is mathematics a single science or a set of arts? In V. Arnold, M. Atiyah, P. Lax, & B. Mazur (Eds.), *Mathematics: Frontiers and perspectives* (pp. 403–416). Providence, RI: American Mathematical Society
Artin, E. (1957). *Geometric algebra*. New York: Interscience
Baer, R. (1952). *Linear algebra and projective geometry*. New York: Academic Press

Baez, J. (2002). The octonions. *Bulletin of the American Mathematical Society, 39*(2), 145–205

Bárány, I. (2008). Random points and lattice points in convex bodies. *Bulletin of the American Mathematical Society, 45*, 339–365

Berger, M. (2000a). Encounter with a geometer I, II. *Notices of the American Mathematical Society, 47*(2), *47*(3), 183–194, 326–340

Berger, M. (2005). Dynamiser la géométrie élémentaire: introduction à des travaux de Richard Schwartz. Atti della Accademia Nazionale dei Lincei. Classe di Scienze Fisiche, Matematiche e Naturali. Rendiconti Lincei. Serie IX. Matematica e Applicazioni. Accad. Naz. Lincei, Rome, Ser. 25, 127–153

Besse, A. (1978). *Manifolds all of whose geodesics are closed.* Berlin/Heidelberg/New York: Springer

Bokowski, J., & Richter-Gebert, J. (1992). A new Sylvester-Gallai configuration representing the 13-point projective plane in Υ^4. *Journal of Combinatorial Theory B, 54*, 161–165

Bott, R. (1988). Morse theory indomitable. *Publications mathématiques de l'Institut des hautes études scientifiques, 68*, 99–114

Bott, R., & Tu, L. (1986). *Differential forms in algebraic topology.* Berlin/Heidelberg/New York: Springer

Cartier, P. (1991, octobre). "Le calcul des structures à deux ou trois dimensions est un défi pour les mathématiciens". *Pour la Science, 168*, 8–10

Coxeter, H. (1964). *Projective geometry.* New York: Blaisdell

Coxeter, H. (1974). *Regular complex polytopes.* Cambridge: Cambridge University Press

Coxeter, H. S. M. (1989). *Introduction to geometry.* New York: Wiley

Croft, H., Falconer, K., & Guy, R. (1991). *Unsolved problems in geometry.* Berlin/Heidelberg/New York: Springer

Darboux, G. (1917). *Principes de géométrie analytique.* Paris: Gauthier-Villars

Du Val, P. (1964). *Homographies, quaternions and rotations.* Oxford: Oxford University Press

Faure, C.-A. (2002). An elementary proof of the fundamental theorem of projective geometry. *Geometriae Dedicata, 90*, 145–151

Fischer, G. (1986a). *Mathematische Modelle* [Mathematical models]. Braunschweig: Vieweg

Fischer, G. (1986b). *Mathematical models: Commentary.* Braunschweig: Vieweg

Frenkel, J. (1973). *Géométrie pour l'élève professeur.* Paris: Hermann

Gromov, M. (1999). *Metric structures for Riemannian and non-Riemannian manifolds.* In J. Lafontaine & P. Pansu (Eds.). Basel: Birkhäuser

Gruber, P., & Wills, J. (Ed.). (1993). *Handbook of convex geometry.* Amsterdam: North-Holland

Hilbert, D., & Cohn-Vossen, S. (1952). *Geometry and the imagination.* New York: Chelsea

Hilbert, D., & Cohn-Vossen, S. (1996). *Anschauliche Geometrie.* Berlin/Heidelberg/New York: Springer

Hirschfeld, J. (1979). *Projective geometry over finite fields.* Oxford: Clarendon Press

Hirzebruch, F. (1987a). *Selecta.* Berlin/Heidelberg/New York: Springer

Hirzebruch, F. (1987b). *Collected papers.* Berlin/Heidelberg/New York: Springer

Husemoller, D. (1975). *Fibre bundles.* Berlin/Heidelberg/New York: Springer

Kelly, L. (1948). The neglected synthetic approach. *The American Mathematical Monthly, 55*, 24–26. (Kelly's solution of Sylvester's problem can be found at the end of an article by H.S.M. Coxeter in the same issue.)

Kelly, L. (1986). A resolution of the Sylvester-Gallai problem of J.-P. Serre. *Discrete & Computational Geometry, 1*, 101–104

Klee, V. (1969). What is the expected volume of a simplex whose vertices are chosen at random from a given convex body? *The American Mathematical Monthly, 76*, 286–288

Kuiper, N. (1984). Geometry in total absolute curvature theory. In W. Jäger, J. Moser, & R. Remmert (Eds.), *Perspectives in mathematics: Anniversary of oberwolfach* (pp. 377–393). Basel: Birkhäuser

Lidl, R., & Niederreiter, H. (1983). *Finite fields.* Cambridge: Cambridge University Press

Lorimer, P. (1983). Some of the finite projective planes. *The Mathematical Intelligencer, 5*, 41–50

Orlik, P., & Terao, H. (1992). *Arrangements of hyperplanes.* Berlin/Heidelberg/New York: Springer

Pach, J., & Agarwal, P. (1995). *Combinatorial geometry*. New York: Wiley

Porteous, I. (1969). *Topological geometry*. London: Van Nostrand-Reinhold

Santalo, L. (1976). *Integral geometry and geometric probability*. New York: Addison-Wesley

Scheinerman, E., & Wilf, H. (1994). The rectilinear crossing number of a complete graph and Sylvester's "four point problem" of geometric probability. *The American Mathematical Monthly, 101*, 939–943

Schwartz, R. (1993). Pappus's theorem and the modular group. *Publications mathématiques de lílnstitut des hautes études scientifiques, 78*, 187–206

Schwartz, R. (2001). Desargues theorem, dynamics and hyperplane arrangements. *Geometriae Dedicata, 87*, 261–283

Smith, A. (2000). Infinite regular sequences of hexagons. *Experimental Mathematics, 9*, 397–406

Talagrand, M. (1995). Concentration of measure and isoperimetric inequalities in product spaces. *Publications mathématiques de lílnstitut des hautes études scientifiques, 81*, 73–205

Veblen, O., & Young, J. (1910–1918). *Projective geometry*. Boston, MA: Ginn and Co.

Chapter II
Circles and spheres

II.1. Introduction and Borsuk's conjecture

If the first chapter was essentially about affine and projective geometry, we now want to enter the Euclidean realm, i.e. we will now have a metric at our disposal, a notion of distance between points, with subsidiary notions such as *circles* and *spheres*. The basic reference for circles and spheres, completely authoritative at the time of its publication, is Coolidge (1916). We have made a critical selection from the enormity of classical results; see the very beginning of Sect. II.2. But of course above all we have chosen to talk about recent results, all the more if they require a climb up the ladder.

Borsuk's conjecture. In the spirit of this book and before touching on problems leading us to configurations that are natural but rather sophisticated, we must speak about Borsuk's conjecture. Its statement is trivial, except that it deals with arbitrary dimension. It is one of the simplest assertions in all of Euclidean geometry, for it doesn't involve anything but distance. Here it is formulated as a question, where $\mathbb{E}^d$ denotes d-dimensional Euclidean space without reference to any particular coordinatization (in contrast, $\mathbb{R}^d$ is d-dimensional Euclidean space with canonical coordinates):

(II.1.1) *In the Euclidean space $\mathbb{E}^d$, can we decompose any bounded part* E *into* $d + 1$ *parts of diameter strictly less than the diameter of* E?

Fig. II.1.1.

The figures above seem to show that the problem is indeed trivial. Recall that the *diameter* of a bounded set E of an arbitrary metric space is the supremum of the distance between points of E: $\mathrm{diam}(E) = \sup\{d(x, y) : x, y \in E\}$. The conjecture is certainly true for $d = 1$, and it is also evident that $d + 1$ is necessary; that d doesn't suffice is left to the readers. We are going to describe in a bit of detail some

M. Berger, *Geometry Revealed*, DOI 10.1007/978-3-540-70997-8_2,
© Springer-Verlag Berlin Heidelberg 2010

of the history of this conjecture, whose various aspects seem enriching. For the algebraic topologist, we need first say that in Borsuk (1933) that — among other things — an important result was proved that had been conjectured by Ulam, specifically that each continuous mapping of the sphere S^d into $\mathbb{E}^d$ sends at least one pair of antipodal points onto the same point, which is intuitive enough for $d = 1$ or $d = 2$. At the end of the text, Borsuk conjectured (II.1.1), or rather he merely stated the problem, for he was much too good a mathematician to make a conjecture without knowing a lot more. In fact his theorem just quoted implicitly suggests that we can't ever decompose the sphere S^d into only d pieces of smaller diameter, whence the idea for the problem. Borsuk's proof of the quoted theorem was a bit complicated, but H. Hopf pointed out to him that it is really instantaneous thanks to degree theory; see Chap. 7 of [BG].

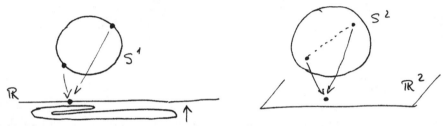

Fig. II.1.2. Borsuk's theorem for $d = 1$ and $d = 2$

Borsuk's conjecture didn't lack for dedicated investigators; but even for the plane, the matter isn't quite as trivial as it perhaps seems. For this and for the more recent history, the reference is 19.3 of Grünbaum (1993). We only mention that, for the plane, we can proceed thus: we inscribe E in a convex set E' of constant width (see the end of Sect. VI.9) equal to the diameter of E. This is easily done by taking the intersection of all the circular disks containing E. Next the continuity principle (turn around!) shows that there exists a regular hexagon circumscribed about this new body whose opposite sides are a distance apart equal to our diameter. We cut this hexagon into three (non regular) hexagons of diameter equal to $\frac{\sqrt{3}}{2}$, supposing diam(E) $= 1$; this cuts E a fortiori in the same way. We have thus succeeded, moreover with a gain of a factor always equal to at least $\frac{\sqrt{3}}{2}$.

For three dimensional space, it was necessary to await (Eggleston, 1957) for the first proof. No proof is really very simple. The best gain presently known is $0.9987...$; whereas $0.888... = \sqrt{(3 + \sqrt{3})/6}$ has been conjectured. The fact that it is much closer to 1 than for the plane allows us to predict a difficult prospect for dimension 4. Here again the proof consists of inscribing E in an "adapted" polytope, specifically and for the moment a regular octahedron whose parallel faces are a distance apart equal to the diameter in question and, after having trimmed it down at the vertices, subjecting it to some subtle dissections (see Fig. II.1.4). Later proofs of a combinatorial nature were found using finite point sets. The properties of their

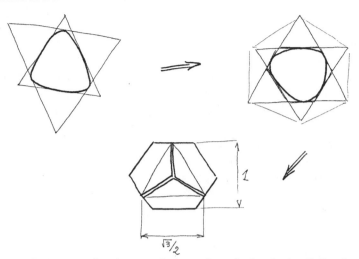

Fig. II.1.3. Circumscribe about our body *equilateral triangles* in all directions; by continuity at least two of these *triangles* are congruent. Moreover, as these two *triangles* have their parallel sides at a constant distance, their intersection must be a regular hexagon

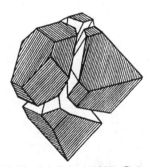

Fig. II.1.4. Boltyanski, Martini, Soltan (1997) © Springer and H. Martini

mutual distances are studied; see e.g. 13.15 of Pach and Agarwal (1995), Chap. 1 of Zong (1996) and Boltyanski, Martini, and Soltan (1997).

Meanwhile, for higher dimensions, it was necessary to make do with results for sets E having special properties. First, the solid balls in all dimensions are easily cut up – see the figure above – e.g. by inscription in a regular simplex. Next, every E having a center of symmetry is cut up using the following elementary observation: *when two points p, q of* E *attain the diameter of* E, *then* E *is contained between the two parallel hyperplanes that pass through p and q and are orthogonal to the line pq.* If there is a center of symmetry for the cut (and there must be, since E has a center of symmetry), this can't be anything other than the midpoint O of pq; and thus E is entirely contained in the ball whose center is O and whose boundary sphere passes through p, q. We now need only cut this ball as above or in some other way.

We owe to Hadwiger an extremely elegant proof of the fact that (II.1.1) is valid assuming E is a set whose boundary ∂E is a smooth hypersurface (a differentiable submanifold of dimension $d-1$ in $\mathbb{R}^d$). To do this we consider the Gauss-Rodrigues mapping (see as needed Sect. VI.6), i.e. the mapping that sends each point of ∂E to the point of the unit sphere S^{d-1} of $\mathbb{R}^d$ that defines the unit exterior normal at this point. We then cut S^{d-1} as above and consider the parts of E that are the inverse images of the pieces obtained. By the elementary observation above, all the points of such an inverse image are a distance apart strictly less than the diameter of E since they never result from antipodal points of S^{d-1}. As these parts are compact, their diameter is strictly less than that of E.

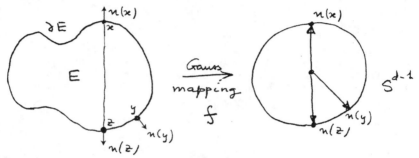

Fig. II.1.5. The Gauss mapping, discovered in fact also by Rodrigues, associates with each point of the smooth surface its unit normal vector

♦

Thus, if we have doubts about Borsuk's conjecture and seek a counterexample, it won't do to use either a set that is symmetric with respect to a point or a set whose boundary is a (smooth) hypersurface. The conjecture was effectively resolved negatively for the first time in Kahn and Kalai (1993). We introduce, for each dimension d, the *Borsuk number* Bors(d) as being the least integer such that each bounded subset E of $\mathbb{E}^d$ can be decomposed into Bors(d) subsets, each of diameter strictly less than that of E. Following Schramm, by techniques called *illumination of convex sets*, we know that Bors(d) $\leqslant (\sqrt{3}/2 + o(1))^{d/2}$ for d sufficiently large. It is a purely combinatoric result; for this and what follows see Chap. 15 of Aigner and Ziegler (1998) and Chap. 13 of Pach and Agarwal (1995). On the other hand, this apparently enormous bound isn't so ridiculous, since we now know that Bors(d) $\geqslant (1,25)^{\sqrt{d}}$ for d *sufficiently large*. This shows that the conjecture is false starting with 964 (use a pocket calculator), but refinements of the technique of Kahn and Kalai now permit a descent down to 561.

The counterexamples are of the following type: we start off from vertices of the discrete cube $\{-1, +1\}^n$ for which the first coordinate equals $+1$, and we construct a subset E of $\mathbb{R}^{n^2}$ whose n^2 coordinates are the products two at a time of those of the point in question. Some combinatorics on the mutual distances of points of E, along with some arithmetic, permit the conclusion. But the critical dimension, between 3 and 561, is not known at present.

Needless to say, the conjecture is open for dimension 4; and dimension 4 is the object of many other open problems, in particular the problems concerning the sphere S^3 that we will see in Sect. III.5.

♦

Here now is a thought, a reflection of Gromov (unpublished): a good candidate for refuting "Borsuk's conjecture", for "pure" geometry, is the Veronese manifold that generalizes the surface encountered in Sect. I.7 (see also Sect. IV.9), that is an embedding in $\mathbb{R}^6$ of in the real projective plane. The real projective plane is characteristically asymmetric since the central symmetry of the sphere has been destroyed precisely in forming the quotient by this symmetry. Note that the discrete version of the projective spaces is precisely the set of vertices of the cube $\{-1, +1\}^n$ for which the first coordinate is $+1$. This Veronese manifold is given in "coordinate free" fashion by the quotient mapping of the mapping $y \mapsto (x \cdot y)^2$ of S^{d-1} into the space $\mathbb{R}^{d(d+1)/2}$ of quadratic forms on $\mathbb{R}^d$ (y and $-y$ have the same image). In coordinates, this mapping is nothing other than $(\dots, x_i, \dots) \mapsto (\dots, x_i^2, \dots, \dots, x_i x_j, \dots)$. The result of Gromov is:

(II.1.2) *If* $\int_0^{\pi/2} \sin^{d-1} t \, dt / \int_0^{\pi/4} \sin^{d-1} t \, dt > d(d+1)/2$, *then the Veronese manifold in* $\mathbb{R}^{d(d+1)/2}$ *cannot be decomposed into* $\frac{d(d+2)}{2} + 1$ **connected** *pieces of smaller diameter.*

We thus obtain a counterexample, but under the restriction of cutting into connected subsets. In return, the dimension that refutes this (weak) conjecture is much lower, i.e. $d = 55$. Use a pocket calculator to find 10 starting with the inequality of (II.1.2), but 55 for $(10 \cdot 11)/2$.

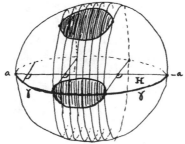

Fig. II.1.6. A symmetrization of the sphere S^{d-1}: apply Fubini's theorem; moreover the diameter can only diminish

The idea behind the proof of (II.1.2) is to think about Sect. VII.12.B, whose essence is that the caps of the spheres have a very small volume so long as we haven't reached the equator, i.e. the radius hasn't reached $\pi/2$. See as needed Sects. VII.6 and VII.12 for the volumes of spherical caps. Here we work in the real projective space $\mathbb{R}P^N$, which is realized isometrically by the Veronese manifold above. Let A be a **connected** portion of $\mathbb{R}P^N$ with diameter less than $\pi/2$. We can then lift it within S^N into A', which will be contained in a hemisphere. Now in S^N we have an *isodiametric* inequality which generalizes that of the penultimate subsection of Sect. VII.7:

> *Among all the domains of* S^N *contained in a hemisphere and of given volume, the smallest diameter is attained exactly by the spherical caps.*

The proof proceeds by the Steiner symmetrization (see Sect. VII.5) — for this method extends without difficulty onto the sphere — while paying attention to the convexity. It's for this reason that we need to be in a hemisphere from the outset. The diameter here being less than $\pi/2$, the greatest volume will be obtained by a cap of radius strictly smaller than $\pi/4$. We then finish with a calculation of volumes; the disjoint parts partitioning A' cannot be greater in number than the quotient of the total area of the hemisphere with that of a cap of radius $\pi/4$.

A recent text on Borsuk's conjecture is Raigorodskii (2004). For more on Borsuk's conjecture, as well as for other "strange phenomena" in geometry, see the book that is entirely devoted to them: Zong (1996), especially Sects. VII.11, VII.12, VII.13.D and the slicing conjecture in Sect. X.6.

II.2. A choice of circle configurations and a critical view of them

We present some figures formed by circle configurations (see Sect. II.XYZ and in particular "a scandal to repair") in *the Euclidean plane* $\mathbb{E}^2$. We will comment on them later, restricting our attention to those that arise from the idea underlying this work. Specifically, we study geometric situations that can be stated quite simply, but which have led, and possibly still lead, to more intensely conceptual developments, to rungs up Jacob's ladder. We comment on what exists for finding proofs (entire or partial) or for understanding them deeply or placing them in a more general context, etc. We hope that readers will take the trouble to contemplate these figures at length, to compare them and decide which are interesting and which, to the contrary, seem unaesthetic or otherwise unappealing. Chapter I was set in the **affine** context; we are now entering into the **metric** context. Readers destined to want to embrace everything, i.e. to climb the ladder in order to unify their vision, will find in Sect. II.XYZ how to connect the metric to the complex projective setting by projectifying, and then complexifying, the Euclidean plane.

For those who like constructions and practical matters, we should point out that circles are constructed with compasses, whereas **linear rulers** and also **graduated goniometers** (protractors) for **measuring** angles are always produced in a "physical" and thus approximate way; see Sect. II.3 below, Sect. V.7 and above all Sect. II.XYZ of this chapter for more on this.

And then there is the question of critical perspective: an historical example is that of "triangle geometry", a discipline which has seen a disproportionate flowering, but which to our knowledge has contributed absolutely nothing in the way of Jacob's ladders. The analysis (Davis, 1995) is very interesting, even if we don't completely share the conclusions. References on the subject can be found there of course, among them the classic (Lalesco, 1952).

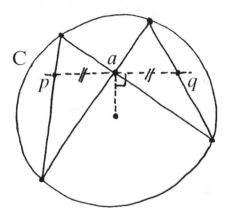

Fig. II.2.1. The butterfly theorem

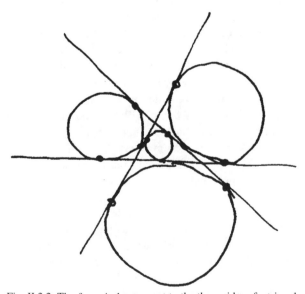

Fig. II.2.2. The four circles tangent to the three sides of a triangle

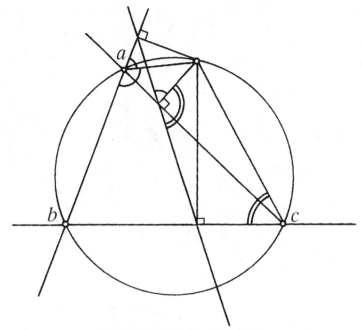

Fig. II.2.3. The Wallace-Simson line

♦

To our knowledge, the butterfly (Fig. II.3.1) has only limited interest: the correctness of the middle of this completely naive figure isn't easy to show and we leave
conviction to readers. In fact it's a trick problem: the result is formulated in a metric
(Euclidean) fashion, although it is really a theorem of projective geometry. In fact,
let us introduce the polar D of the point a with respect to the circle C and consider
the (harmonic) homology defined by the pair (a, D); it's an involutive projective
transformation (see Sect. I.7). By construction, this homology preserves C and D
and thus interchanges the two points p and q. But on the line D it is a homography
that preserves the point at infinity, thus an affine transformation. Now, it preserves
the midpoint a of pq: it's the metric symmetry of D with center a, and thus a is
clearly also the midpoint of pq. Apart from this, to our knowledge, the butterfly
theorem doesn't yield any movement on Jacob's ladder.

♦

The fact (Fig. II.3.2) that there exist four circles tangent to the three sides of any
triangle in the Euclidean plane is not profound; we use the interior and exterior bisectors of the triangle (which gives a new way of obtaining a complete quadrilateral,
cf. Sect. I.7). Nonetheless, we want to mention this figure, for its generalization to
dimension 3 (and higher) sets a trap, pointed out to the author by his school mathematics teacher, the late Jean Itard, when he was 16; a trap into which practically all
the mathematicians to whom the author has posed the problem have fallen, at least

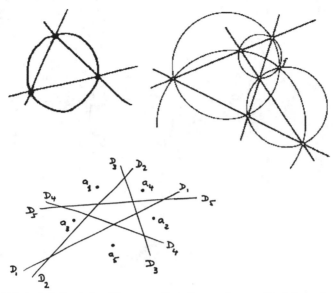

Fig. II.2.4. An infinite chain of theorems: *three lines*: a circumscribed circle; *four lines*: four coincident circles; *five lines*: (what is the result?); next, go to infinity

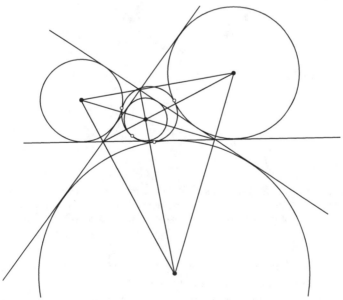

Fig. II.2.5. Feuerbach's result

when they were not given much time to reflect. The solution in every dimension does not require that we climb very high; it suffices to use *barycentric coordinates* (introduced in Sect. I.XYZ) and considerations of volume and area, all of which

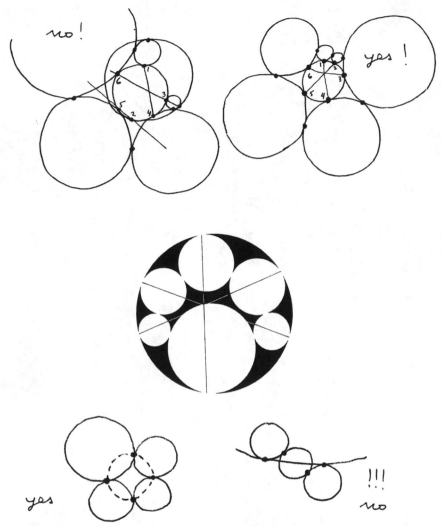

Fig. II.2.6. The seven circles theorem and that of five circles. Above and below, the
conditions for the tangency of the circles are satisfied, but the lines are not coincident;
or the four cocyclic points, and "one time in two"

can be found in detail in Sect. 10.6.8 of [B]; but we are going to give the essentials,
for the result is fascinating and the proof trivial, once barycentric coordinates have
been introduced. One of the appeals of this very elementary geometry problem is
that, once the dimension exceeds 2, the number of spheres tangent to the faces of
a simplex depends on the initial figure; for example, in dimension 3 this number
depends on the areas of the faces of the tetrahedron considered; it is equal to 8 typically, but decreases to 5 for the regular tetrahedron. It is equal to 2^d in general. In
dimension three, as far as pure geometry is concerned, it's a matter of looking at the

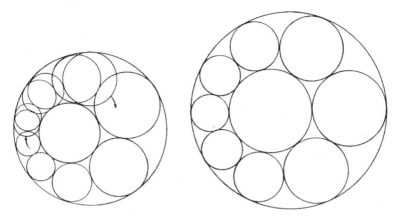

Fig. II.2.7. Steiner's alternative. Things either close up again no matter what the initial circle, or else they never close. I [B] Géométrie. Nathan (1977, 1990) réimp. Cassini (2009) © Nathan Édition

Fig. II.2.8. Eight circles tangent to three given circles. I [B] Géométrie. Nathan (1977, 1990) réimp. Cassini (2009) © Nathan Édition

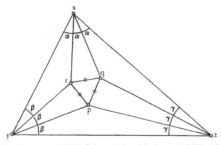

Fig. II.2.9. Morley's theorem. I [B] Géométrie. Nathan (1977, 1990) réimp. Cassini (2009) © Nathan Édition

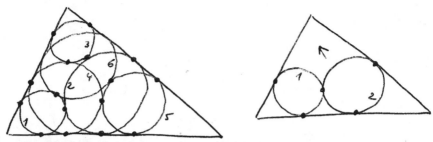

Fig. II.2.10. Six circle theorem: whatever the initial circle, the operation closes up at
the end of the six circles, or two rounds

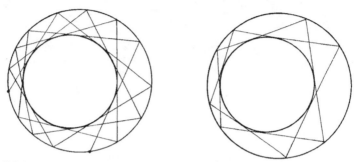

Fig. II.2.11. Poncelet's great theorem for circles. I [B] Géométrie. Nathan (1977, 1990)
réimp. Cassini (2009) © Nathan Édition

combles (for the meaning of the French word "comble" (literally "attic" in English,
see Fig. II.2.12 below). It's also a nice example for showing that regularity isn't always the gauge of optimality; we encounter this *symmetry breaking* in other places:
Sects. III.3 and X.6. Here now are the details for ordinary space: we evidently use
barycentric coordinates $\{\lambda_i\}$ $(i = 1, 2, 3, 4)$ which yield the tetrahedron, with what
was said in Sect. I.XYZ.

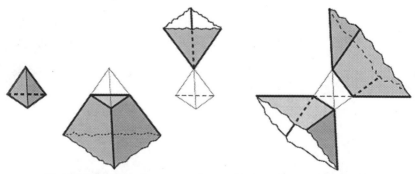

Fig. II.2.12. The interior, a truncation, a trihedron, two opposed combles

Now, if the a_i denote the areas of the faces associated with the tetrahedron, V its
volume and r the radius of a sphere tangent to the four faces, we must have

$$\frac{|\lambda_1|}{a_1} = \frac{|\lambda_2|}{a_2} = \frac{|\lambda_3|}{a_3} = \frac{|\lambda_4|}{a_4} = \frac{r}{3V}.$$

The sphere on the interior is that of radius $r = \frac{3V}{a_1+a_2+a_3+a_4}$, those of the *truncations* have for radii $r = \frac{3V}{a_2+a_3+a_4-a_1}$, etc. and thus always exist. In contrast, for two negative signs (which correspond to a pair of opposed *combles*, only one of the two numbers $a_1 + a_2 - a_3 - a_4$ and $a_3 + a_4 - a_1 - a_2$ is positive; moreover it may be zero for certain tetrahedra. Whence eight spheres in the general ("generic" is now the more common term) case. In particular, only five for a tetrahedron all of whose faces have equal areas (we leave to readers the study of these tetrahedra), and otherwise 6 or 7 according to the case. For a study, repulsive because of its quixotic refusal to use barycentric coordinates, see Rouché and de Comberousse (1912, v. II, p.653).

◆

Wallace's theorem (Fig. II.2.3), long attributed to Simson, states that the three projections of a point onto the three sides of a triangle are collinear if and only if the point in question belongs to the circumscribed circle of the triangle. It prepared the way for numerous developments, for example finding when two points of the circle have orthogonal Wallace-Simson lines; or to find the envelope curve of these lines of Simson when the point traverses the circle; for all of this see the references given in 10.9.7 of [B]. However, the solution isn't so very simple; for example, a brutal calculation using coordinates will discourage the majority of readers, even professionals. The solution "in depth" requires only a very small climb up the ladder, specifically the angular property of plane circles (called, in old French books, *"l'arc capable"*):

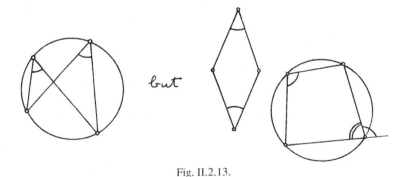

Fig. II.2.13.

However, it should be carefully noted that, for ordinary angles (between 0 and 180 degrees), there are two cases for the figures. We escape this difficulty with the notion of oriented angles for lines, but it is ultimately rather hard to explain well, see 8.7 of [B]. But we won't linger here; for, to our knowledge, no path has been cleared for any very interesting developments.

However, regarding the theory of oriented angles, we mention an application of a rather rare type, specifically the chain of theorems (Fig. II.2.4) associated with finite sets of lines in the Euclidean plane. Three lines determine a triangle, and we have its well defined circumscribed circle. Four lines determine four triangles, and we can ask what happens regarding their four corresponding circumscribed circles: the answer, far beyond what we might hope, is that they have a point in common, are coincident. Next, let there be five lines: with four at a time we associate the point of the preceding result. Then the five points belong to a single circle. It is left to readers to formulate the sequel and to prove all that, which is not hard with the angular property of circles; if there is difficulty, they can read the classic works or Section 10.9.8 of [B], but above all (Coolidge, 1916). It is fascinating that, for the case of five lines, a weak version of this very old result was presented at the ICM (International Congress of Mathematicians) in Beijing in 2002 as a difficult and little known problem, with the consequence that Alain Connes chose to give his own proof!

In contrast the three figures (Figs. II.2.5, II.2.6, II.2.7) that follow, associated with the names of Feuerbach, Steiner and (to the best of our knowledge) with an unknown person, require a new tool: *inversion*. This not only yields proofs of the three results that are rather obvious from the figures, but proofs that are very simple, that would be inextricable with direct calculations using coordinates. Furthermore, as we will see amply in the sequel, the inversions of the Euclidean plane are fundamental elements, the generators of the *conformal group*, and ultimately of *hyperbolic geometry*. We encounter the importance of hyperbolic geometry in numerous domains of mathematics: algebra, number theory, complex analysis, dynamical systems and theoretical physics.

We now give the assertions underlying the three figures in question. Feuerbach's theorem (Fig. II.2.5) asserts that the *Euler's circle* of a triangle (also called the nine point circle), i.e. the one that passes through the midpoints of the three sides of the triangle, but also through the bases of the three altitudes and through three other points (for readers to discover!). This Euler's circle is tangent to each of the four circles of Fig. II.2.2. We leave it to readers to show this with the aid of an appropriate inversion.

The seven circles theorem (Fig. II.2.6) can be proved using only inversion, even though the proof is more sophisticated than for Feuerbach's. However, we should beware that **one time out of two** holds. What does this mean? That one more condition on tangency of the circles is needed, that can only be defined by orienting each of the circles considered (by putting an arrow on them) and observing, at the point of contact, whether or not they agree. A conceptual proof thus requires placing oneself in the geometry of *oriented circles*. This geometry exists, but we don't

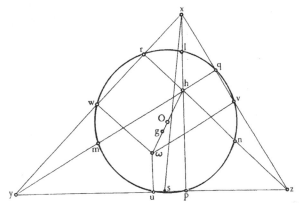

Fig. II.2.14. The nine point circle. I [B] Géométrie. Nathan (1977, 1990) réimp.
Cassini (2009) © Nathan Édition

know any good reference; what is said in Klein (1926–1949, §27), is too skeletal to permit finding the proofs that we need here. We know of no more modern exposition of them, all the elements are those given right at the end of Sect. II.XYZ. The circles are defined there (in the space of all circles) as points of a real projective space, thus via a quotient of an appropriate space by the real line $\mathbb{R}$. To apportion these circles as oriented circles, it suffices to take the quotient by only the real positive half-line $\mathbb{R}^*_+$. See also the article on "conformal geometry" in the Encyclopedic Dictionary (1985).

For an elementary proof (but of course via inversion) of this theorem and of the six circles theorem (Fig. II.2.10), see Evelyn, Money-Coutts, and Tyrelle (1975), but this "by hand". The author doesn't know how to attribute the seven circles theorem. The six circles theorem is astonishingly difficult to prove and is attributed to Money-Coutts; see in Tabachnikov (2000) several avatars, for example the figure below:

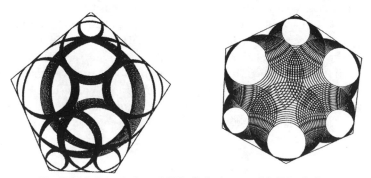

Fig. II.2.15. Tabachnikov (2000) © Springer and S. Tabachnikov

The five circles theorem states — see Fig. II.2.6 — that when four circles are tangent two at a time in succession, then the four points of contact are concyclic.

As with seven circles, it is in fact also true only one time in two, as readers will verify. To prove it (in the favorable cases) we perform an inversion about one of the points of contact, which transforms two of the circles into parallel lines; and then the result follows from completely elementary angular properties of triangles and parallel lines. An amusing application (but related to the result of Rivin on the possible inscribability of polyhedra in spheres, which will be seen in Sect. VIII.8) is that, if a quadrilateral (skew) has sides tangent to a sphere, then the four points of contact lie on the same ("small", as opposed to "great") circle of this sphere (proof left to readers).

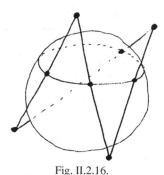

Fig. II.2.16.

The Steiner Alternative (Fig. II.2.7) concerns a pair of non intersecting circles; it asserts that the chain of circles (a priori infinite), starting out with an *initial* circle as in the figure, **either closes on itself** (but then this remains true for every other initial circle), or else never closes (and again, its then the same whatever the initial circle). Stated otherwise: the problem either has an infinite number of solutions (in fact a *continuous* infinity) or none. This depends only on the initial pair of circles: it's an alternative of the *all or nothing* sort. We will see the proof in detail, using inversion, in the following section. Moreover, this proof permits us to answer practically every question suggested by the result, e.g. what is the necessary and sufficient relationship between the radii and the distance between the centers of the circles considered so that there will be closure for a given integer n?

Figure II.2.8 doesn't give rise, as best we know, to conceptual developments. In contrast, it is quite different for the search for conics tangent to five given conics, a problem that was a source of inspiration for more than a century, until at least 1952. We will treat this question, the most difficult in the realm of conics, amply in Sect. IV.9.

Morley's theorem (Fig. II.2.9, strictly speaking without circles, but nonetheless of a Euclidean nature because of its angles) states that the trisectors of the angles of an arbitrary triangle always intersect in such a way a to form an equilateral triangle. It is celebrated as being, without any doubt, the simplest result of plane Euclidean geometry to state (not even circles appear in its statement) for which the proof is invisible; a good calculator in trigonometry will however get off without too much grief, but a non artificial geometric proof is still lacking. There wasn't any conceptual proof, although that of Chap. IV from the second part of Lebesgue (1950) climbs the ladder a little, but remains very computational. It is not until Connes (1999) that we find a very conceptual and even very natural proof, since it is based on the group of isometries of the Euclidean plane and extends to the case of geometries over more general base fields.

The six circles theorem (Fig. II.2.10) asserts the closure at the end of six steps in a chain of circles in a triangle, starting with an **arbitrary** tangent circle to two of the sides; the next circle must be tangent to two following sides and to the initial circle, etc. We don't know of any conceptual proof; see the proof in Evelyn et al. (1975) or a *dynamic* proof can be constructed by studying the effect on the other bisectors of a movement of the center of the initial circle along the first bisector. It doesn't seem that there is any connection with the theorem of closure of six spheres; see below in Sect. II.6.

The great Poncelet theorem (for circles) is an alternative, like that of Steiner. It states that for a pair of circles $\mathcal{C}$, $\mathcal{C}'$ and a given integer n, either there exists no polygon of n sides whose vertices are situated on $\mathcal{C}$ and whose sides are tangent to $\mathcal{C}'$, or else there exists an infinity of such polygons. More precisely, we can choose as initial vertex an arbitrary point of $\mathcal{C}$ (with obvious restrictions on arcs, if the circles intersect), but there are always as many such polygons as real numbers (the power of the continuum). In fact, Poncelet proved this theorem on two circles in order to generalize it by projective transformation to an arbitrary pair of conics. His proof wasn't completely conceptual, but in Sect. IV.8 there is a conceptual proof for circles, and a still more profound proof for the conics in general — even over an arbitrary field — with some references. This problem, as we will see, has haunted very many mathematicians, who have sought to comprehend it increasingly better. If there is a need, readers can gain respect for this result by attempting to prove it for themselves just for triangles ($n = 3$); in this case it's an old result of Chapple from 1746, "Ponceletified" by Mackay only in 1887. By hand the case $n = 4$ will be found easier by an expert in "elementary" conics; the case $n = 6$ can also be treated by hand. The general case is of a whole other order of difficulty. See Sect. IV.8 for the relation between the integer n and the radii of the two circles, where we will also see that Poncelet's theorem is still very much alive.

For configurations of regular hexagons, and their iteration (in the spirit of Sect. I.4), see the quite recent (Smith, 2000), typical of what new and profound things can always be done in very elementary geometry, such as the use of computers for divining new results, a trait that is common to the articles published by *Experimental Mathematics*. But see also Gutkin (2003), which studies the problem of finding triangles whose three vertices are situated on three concentric circles and which satisfy an extremal condition.

◆

See Sect. II.XYZ for some complementary objects relating to polarity with respect to a circle: pencils of circles. We describe them in detail, for it seems to us that this has been made necessary by the de-geometrization of mathematics instruction. For the configurations of spheres, and of circles *in space*, look to what accompanies the next topic.

II.3. A solitary inversion and what can be done with it

As an introduction we ask ourselves, in the spirit of the fundamental theorem of affine geometry seen in Sect. I.2: what are the bijective transformations of the Euclidean plane that preserve circles? For sure the *isometries* of the Euclidean plane, which preserve the metric, are included. These are: translations, rotations with arbitrary center and angle and the orientation reversing isometries, specifically the symmetry translations (glide-reflections); this is the place to remark, if needed, that many people believe that only the symmetries with respect to a line preserve distance while changing orientation. Then there are the homotheties, which of course change the radii of circles while preserving them as a whole. We define *similitudes* to be the transformations which are compositions of an arbitrary isometry and an arbitrary homothety; they form a group. It isn't difficult to deduce from the fundamental theorem of affine geometry that the only **bijections** of the Euclidean plane that preserve circles are the similitudes; see Sect. II.XYZ and, if needed, Chap. 9 of [B], where other characterizations of these similitudes can be found.

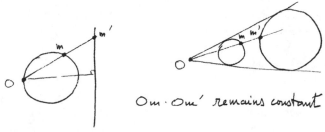

Fig. II.3.1. Inversion: the product $Om \cdot Om'$ must remain constant

Now comes the miracle, discovered by Descartes but little clarified and exploited by him, a miracle that has flourished in exuberant fashion only since nineteenth century: by removing just **one point** from the plane, we obtain the existence of transformations, other than the similitudes, that preserve the family formed of the line and circles of the Euclidean plane; these are the *inversions*. The inversion $I_{O,k}$ with *center* O and *power* k (an arbitrary nonzero real) is, by definition, the bijection defined on the plane with the point O removed, given by $m \mapsto m'$ where m' is the point situated on the line Om such that the product of the distances Om and Om' is equal to k; if k is positive, we take m and m' on the same side of O; if it is negative they are taken on opposite sides. Two inversions with the same center differ only by a homothety. It is practically trivial that $I_{O,k}$ permutes the circles not containing O among themselves and replaces lines (not containing O) with circles containing O, and finally conserves the lines passing through O (from which of course O has been removed). Moreover, just like the similitudes, inversion (we say inversion for short when it isn't necessary to specify center and power) *conserves angles*, and this in a larger sense, specifically not only angles between circles, between lines, between circles and lines, but also for *differentiable curves*; (see Chap. V). The angle between differentiable curves that meet in a point is, by definition, the angle formed by their tangent lines. However, there can be a change of sign if we use oriented angles. Inversion also preserves osculating circles, see Sect. V.6.

According to whether k is negative or positive, an inversion has respectively either no fixed point or a whole circle of fixed points, of radius $\sqrt{k}$, called the *circle of inversion*. In the $k > 0$ case, the interior of the associated disk is transformed to the exterior and this permits, according to a classical jest, catching a lion in the desert without difficulty or risk. Note finally that an inversion changes orientation,

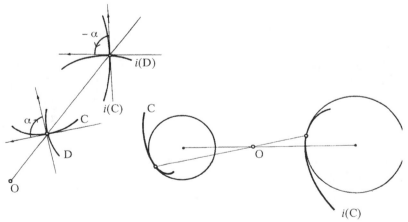

Fig. II.3.2. Inversion preserves angles and osculating circles. I [B] Géométrie. Nathan
(1977, 1990) réimp. Cassini (2009) © Nathan Édition

and that it is *involutive*, i.e. its square is the identity transformation so that iteration
leads nowhere.

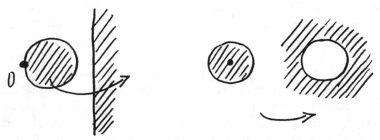

Fig. II.3.3. How to catch a lion in the desert

Inversions can be realized mechanically, with systems of articulated rods. In
the two figures below the O points (centers of inversion) are fixed, and the inversion points for the two are m and m'. To invert a line, we could demonstrate its
property with Ptolemy's theorem (see in Sect. II.XYZ the "scandal to be repaired").
Historically, since the appearance of steam engines, these inverters have had some
importance, having to do with describing straight lines by articulated systems; such
a system eluded Cayley in spite of his desire for one. It seems that these *inverters*
have only exceptionally been applied in practice, such as for industrial systems or
for tracing perfect lines in metrology; see the beginning of Sect. II.2 and above all
Sect. II.XYZ of this chapter. Incidentally, Peaucellier's inverter was used to ven-
tilate the Crystal Palace in London. We will encounter this inverter at the end of
Sect. IV.8. For the rest, in order to trace lines in practice, rulers (rather good, or
sliders, according to what is required) are used, and not inverters: see Sect. II.XYZ
for these considerations, which are rather little known.

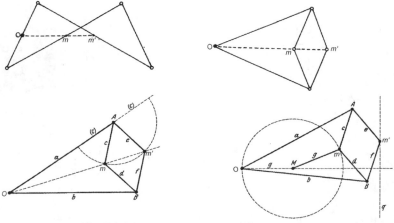

Fig. II.3.4. Peaucellier's inverter and Hart's inverter

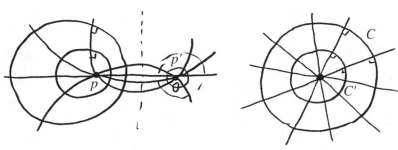

Fig. II.3.5.

We now prove Steiner's alternative for a pair $\mathcal{C}$, $\mathcal{C}'$ of circles. It is left to readers to create the figure below, i.e. to see that there exist two points p and p' such that the circles passing through them are orthogonal (intersect at right angles) to $\mathcal{C}$ and $\mathcal{C}'$. It then follows that an inversion with center p or p' transforms the pair $\mathcal{C}$, $\mathcal{C}'$ into a pair of concentric circles, the common center being the image under this inversion of the second point, since all lines passing through this point cut the new pair at a right angle, which is characteristic of the center of a circle. Now Steiner's alternative is trivial for a pair of concentric circles; the profound reason is that these circles are conserved by the group of rotations for which the center is their common center.

Interested readers can make some calculations for finding the condition linking the integer n with the number of circles in the chain and the triple formed by the two radii and the distance between their centers; it suffices to know how inversion transforms the radii of the circles. Finally, they can prove Feuerbach's theorem, then the seven circles theorem (figures in Sect. II.2). In applying inversion, the problem is to properly choose an appropriate center (on the other hand the power, in general, matters little) in obtaining the simplest figure. If the center of inversion is the point of contact between two tangent circles, these latter are transformed to two parallel lines. In Sect. II.6 a typical and spectacular example of this technique in dimension three establishes the theorem known as that of six spheres.

A naive question: why not use the same technique to show Poncelet's great theorem, i.e. find a transformation (perhaps defined only in a portion of the plane) preserving lines and transforming a pair of arbitrary circles into a pair of concentric circles? If we want to preserve lines, we know (according to Sect. I.XYZ) that it is necessary to choose a projective transformation. But it is easy to see that there exists no projective transformation that takes nonconcentric circles into concentric circles; in fact, two concentric circles, when they have been complexified and projectified, become two bitangent conics at their cyclic points; see II.XYZ and Sects. IV.8. This is why the great theorem of Poncelet is of a whole other order of difficulty; see Sects. IV.8 and see there also, at the end, a result of Emch that encompasses the Steiner and Poncelet theorems.

II.4. How do we compose inversions? First solution: the conformal group on the disk and the geometry of the hyperbolic plane

And here we have the great desire — the great and good temptation! — to compose inversions with different centers (the product of two inversions with the same center is a homothety with the same center), and this for various reasons. For example, we could hope to obtain results that are stronger than in the three examples above; but also to form a group, whether just because we like groups or, following Klein, ca. 1900 we know that the study of a geometry is very strongly tied to the structure of its groups of isomorphisms. So the idea is to compose inversions, of arbitrary number, hoping all the while that things will terminate at the end of finitely many steps; for the inversions would then just generate a group of *finite dimension*, as do plane isometries having symmetries about lines as generators. We would then have, in the case of inversions, *generators* that are easy to treat. Note that an inversion is a type of symmetry, not with respect to a line, but with respect to a circle. Finally, the great hope: the group obtained will be the one that preserves circles and also the one consisting of all transformations that preserve angles — all this in a broader sense to be defined, because only the similitudes accomplish it in the strict sense of bijections of the Euclidean plane, as seen at the beginning of this section.

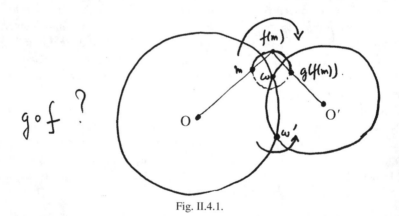

Fig. II.4.1.

Suppose we compose two inversions with respect to circles as in the above figure; with a bit of sketching we see that we obtain a type of rotation about the two points of intersection ω and ω' of the two circles, which is rather nice. On the other hand it will now be necessary to remove not only two points from the plane, specifically the two centers, but also their images and inverse images sufficient for this transformation being well defined on the set considered. Surely our composition preserves the set of lines and circles. It also preserves angles and orientation, but as we continue, we must remove more and more points, and this is not satisfying. The definitive and global solution will come momentarily in Sect. II.5, but we start with an intermediate solution, which will be essential for the sequel.

The idea is to find a domain of the plane and lots of inversions that preserve it, without having to remove any points. We will then be able to compose as many times as we wish. The right domain is simple; it's the interior of a circle, any circle whatsoever.

Fig. II.4.2. Inversions that preserve the area of a circle

Each point on the exterior of the circle is the center of a well defined inversion that preserves the circle entirely; we can thus compose any number of such inversions without any problem. However, when do we stop? We can see it geometrically, but it's really hopelessly complicated. It suffices to write the formula giving the inversion about the pole outside the disk to see that — once our Euclidean plane is identified with $\mathbb{C}$, taking the center of the disk as origin — it can be written in the form of the homography $z \mapsto e^{i\theta} \frac{z-z_0}{1-\overline{z_0}z}$ composed with complex conjugation $z \mapsto \overline{z}$ (with the sole condition $|z_0| < 1$, θ being any real number). We only need remark that the transformations $z \mapsto e^{i\theta} \frac{z-z_0}{1-\overline{z_0}z}$ of this type form a group, and a larger group if we add their compositions with conjugation $z \mapsto \overline{z}$. This group is called the *conformal group of the disk*, but we will now give it, with some explanation, the name *Möbius group of the circle* S^1 or *for dimension* 1 and denote it by Möb(S^1). Without the conjugation, we obtain the connected component of the identity element. In both cases we have exactly **three real parameters**. Be aware that, to generate it geometrically, we need to add to the inversions that give rise to it the inversions with center infinity, i.e. the symmetries with respect to the origin of the disk (take a limit in order to convince yourself). The group Möb(S^1) contains all rotations about the origin, the symmetries about lines through the center, the compositions of two inversions whose circles intersect at a point interior to the disk; these are *generalized rotations*, oblique in some way. This group plays an essential role in the study of holomorphic functions.

We show, as a particular case of more general results to come, that this group coincides with that of transformations of the disk that preserve circles (more precisely the part intersected by the disk), or alternatively (differentiable) transformations of the disk which preserve angles. Finally, we now justify now the appearance of the circle S^1 in the description. In fact, the homographies above − or the inversions − evidently preserve the circle S^1, for it is the boundary of the unit disk. Moreover the action on the circle determines, and is determined by, what happens on the interior. These mappings of S^1 into itself cannot be characterized as *conformal*, i.e. (true or infinitesimal) angle preserving, because we are in dimension 1; but it will be different in Sect. II.5, which follows. The geometry invariant under this group is that of the *hyperbolic plane*; see below in this section and in Sect. II.XYZ, where we will have a metric and angles.

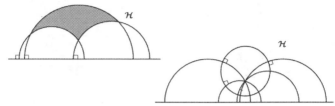

Fig. II.4.3. Poincaré's half-plane: the lines are the semi-circles orthogonal to the boundary; Euclidean circles are also hyperbolic circles

A very useful model of the geometry of the hyperbolic plane − see for example Sect. IX.3 − is the *Poincaré half-plane* $\mathcal{H}$. To obtain this model it suffices simply to perform an inversion whose center is on the boundary circle of the preceding disk. The interior of the disk thus becomes an open half-plane, whose boundary is taken as the real axis in $\mathbb{R}^2 = \mathbb{C}$ and the hyperbolic group becomes the group of mappings $\mathcal{H} \to \mathcal{H}$ written $z \mapsto \frac{az+b}{cz+d}$ where a, b, c, d are real and $ad - bc = \pm 1$. The connected component containing the identity element is isomorphic to $SL(2; \mathbb{R})$. In this model the lines are semi-circles that are orthogonal to the real axis and the vertical half-lines. In this model it is the easiest to prove the fundamental formula for the area of a triangle T in the hyperbolic plane as a function of its angles α, β, γ (compare with formula (III.1.2) for the spherical case):

(II.4.1) $\text{Area(T)} = \pi - (\alpha + \beta + \gamma).$

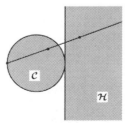

Fig. II.4.4. Models of the hyperbolic plane: Poincaré half-plane and Poincaré's disk correspond under inversion

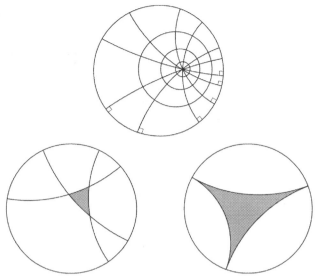

Fig. II.4.5. Circles, triangle, ideal triangle in Poincaré's disk

A remarkable consequence, essential for certain applications, is that the area of a triangle is always less than π and equals π only for the *ideal* triangles, i.e. those whose three vertices are at infinity.

We now have all that is necessary in order to introduce the *hyperbolic plane*; see the historical remark at the end of this section. We will not dwell on hyperbolic geometry (also not in higher dimensions), despite its essential importance in many branches of mathematics and the fact the it was "the solution" to Euclid's fifth postulate. Expositions can be found in many works on geometry and other subjects. We note only that hyperbolic geometry is represented by **five** very different models, which, although we won't show it, of course yield completely isomorphic geometries. Many authors develop just one or two models, explaining that these are the best or simplest. This is an illusion: on the contrary, it is the case that one or another model is best adapted to the particular problem being studied. In Chap. 19 of [B] we gave the four models that emerge from "elementary" geometry. They will be presented briefly in Sect. II.XYZ . The fifth is not given in [B], for it uses Riemannian geometry: hyperbolic geometry is the Riemannian structure induced on a semi-hyperboloid in $\mathbb{R}^{n+1}$ by the quadratic form $q = -x_1^2 - \cdots - x_n^2 + x_{n+1}^2$.

In this hyperbolic geometry that we want to define, as in Euclidean geometry, there are angles but above all a metric. The angles are those induced locally by the Euclidean structure on the disk. They are indeed conserved by the group Möb(S^1) since its elements are holomorphic functions of a complex variable (or antiholomorphic if $\bar{z}$ appears). This property is in fact nothing other than the definition of *holomorphic* functions. These are functions U $\to$ $\mathbb{C}$ (where U is an open subset of

$\mathbb{C}$, identified with $\mathbb{E}^2$; see Sect. II.XYZ) for which the (complex) derivative $f'(z)$ exists. When $f'(z) \neq 0$ they are then automatically local similitudes, since multiplication by a nonzero complex number $f'(z)$ is a similitude with ratio equal to the modulus of $f'(z)$ and angle equal to its argument. We need to find a metric invariant under this group.

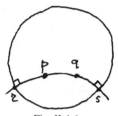

Fig. II.4.6.

Now is the moment to recall that the homographies have an invariant, i.e. the cross ratio encountered in Sect. I.6. Given two points p, q in the interior of the disk, consider the unique circle through p, q that is orthogonal to the unit circle bounding the disk. It intersects the unit circle at two points r, s, whence we have four points p, q, r, s and we define the (hyperbolic) distance δ of p to q by $\delta(p, q) = |\log([p, q, r, s])|$, where $[p, q, r, s]$ is the cross ratio of the four complex numbers p, q, r, s on the affine line $\mathbb{C}$. We can then prove the basic formula from which all others can be derived, i.e. that if a triangle has sides a, b, c and angle α at a, then

(II.4.2) $\operatorname{ch} a = \operatorname{ch} b \cdot \operatorname{ch} c - \operatorname{sh} a \cdot \operatorname{sh} b \cdot \cos \alpha$,

where ch and sh denote hyperbolic cosine and sine. And we can verify without difficulty that in this hyperbolic geometry Hyp^2, the *lines* are the arcs of circles (or segments of lines passing through the origin) orthogonal to the unit circle, i.e. that these are the shortest paths from one point to another. Finally, the group with which we began is the group of isometries of this metric. For more on plane hyperbolic geometry (and in all dimensions: Hyp^n), see Sect. II.XYZ and the references mentioned there.

Finally, in the spirit of Sect. I.7 in particular, the geometry *at infinity* of the hyperbolic plane is that of the circle; each of its points is a point at infinity, as opposed to the complex affine line, which has but a single point at infinity (which will play an essential role in the following section), or the real affine plane whose points at infinity are associated with directions of lines, and not as here with directions of vectors (or oriented lines).

Historical remark: hyperbolic geometry had to wait a long time before being solidly defined by Riemann in 1854. Apart from philosophical reasons, there was the mathematical fact shown by Hilbert (see Sect. VI.7) that this geometry cannot be completely realized by a surface in three-dimensional space. The surface that Beltrami proposed had a double disadvantage: it wasn't complete and it wasn't simply connected — it had to be developed by unrolling it; see Sect. VI.7 for all this.

II.5. Second solution: the conformal group of the sphere, first seen algebraically, then geometrically, with inversions in dimension 3 (and three-dimensional hyperbolic geometry). Historical appearance of the first fractals

We can find the unconditional algebraic solution for the composition of inversions if we remember the preceding section. We continue to identify $\mathbb{E}^2$ with the complex line $\mathbb{C}$. Then the inversion through the origin and the power 1 is nothing other than $z \mapsto \frac{1}{\bar{z}}$; thus each inversion of the plane has the form $z \mapsto \frac{a\bar{z}+b}{c\bar{z}+d}$. But we know how to complete $\mathbb{C}$ with a single point at infinity: $\mathbb{C} \cup \infty$, which we write as the sphere S^2. As for $z \mapsto \frac{az+b}{cz+d}$, we have seen in Sect. I.6 that the right extension is given by $\infty \mapsto \frac{a}{c}$ and $-\frac{d}{c} \mapsto \infty$, and likewise with $\bar{z}$.

Written in homogeneous coordinates (see Sect. I.5), the inversions become quasilinear: $(z,t) \mapsto (a\bar{z} + bt, c\bar{z} + dt)$. Finally, the composition of an arbitrary number of inversions belongs to the group $GP(1;\mathbb{C}) \cup GP(1;\mathbb{C})^*$, which comes in two pieces. The first of these is the projective group of the complex line, denoted $GP(1;\mathbb{C})$; the second piece can be denoted $GP(1;\mathbb{C})^*$, having the form of elements of $GP(1;\mathbb{C})$ composed with complex conjugation. We *denote it* by $\text{Möb}(S^2)$ and call it the *Möbius group in dimension* 2 or else the *conformal group of* S^2. It has *six* real parameters (we can divide numerator and denominator of the fraction by an arbitrary complex number) and two connected components. The component containing the identity element consists of orientation preserving transformations. It can be shown (see 18.10 of [B]) that this group coincides with the group of those differentiable transformations of S^2 that preserve infinitesimal (local) angles, i.e. those for which the derivative is a similitude. But the algebraic construction above, albeit sound, doesn't tell us much about the *geometric* nature of the elements of $\text{Möb}(S^2)$.

The geometric trick here is to introduce stereographic projection, which is the geometric realization of the completion $S^2 = \mathbb{C} \cup \infty$ of $\mathbb{C}$. This doesn't take us much up the ladder, but is a new use of three dimensions for resolving questions posed in dimension 2.

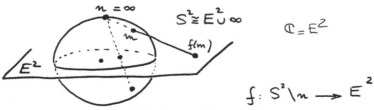

Fig. II.5.1. Stereographic projection

Stereographic projection — see the figure — puts the points of the Euclidean plane in bijection (one-to one-correspondence) with the points of the sphere **minus the north pole** n. Thus, if we identify the Euclidean plane with the complex line $\mathbb{C}$, there is a bijection between the points of the sphere **in its entirety** and the points of the complex projective line $\mathbb{C}P^1 = \mathbb{C} \cup \infty$, the north pole now corresponding to the point ∞ at infinity of $\mathbb{C}P^1$. First of all, the definition of inversion extends to Euclidean spaces of **arbitrary** dimension without any change. These inversions preserve the set of hyperplanes and spheres (we don't say hypersphere) and hence circles in the space. They also preserve angles between them as well as the angles between tangents to curves at their points of intersection. Stereographic projection is nothing other than the restriction to the sphere of inversion through the north pole. It thus preserves circles (the "small" circles of the sphere), except those that pass through the same pole, which yield lines of the plane (Fig. II.5.2); and it preserves angles. The fact that it preserves circles explains why it has been used in astronomy for so long, for if our sphere is thought of as the celestial vault, the stars there all trace out circles; we can thus represent their trajectories by circles on a plane map. For us, its interest is to furnish a geometric compactification that preserves the angles of the Euclidean plane.

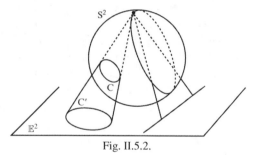

Fig. II.5.2.

We now proceed as in Sect. 11.4 and consider the inversions of three dimensional space, where we effect stereographic projection and restrict ourselves to inversions of the space whose center is exterior to the solid ball defined by the sphere and which preserve this ball (Fig. II.5.3): they preserve simultaneously the interior of the ball and also the sphere. We can thus compose them as many times as we wish.

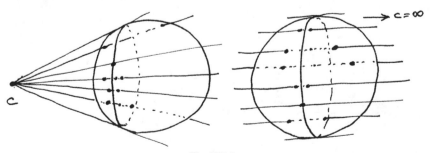

Fig. II.5.3.

Now it an entertaining but nontrivial exercise to see that, transported by stereographic projection in the plane, the effect of these inversions on the sphere becomes an inversion of $\mathbb{C} \cup \infty$, always with the convention that the north pole corresponds to the point at infinity ∞ of the plane. Proceeding in the reverse sense, we can compose plane inversions as many times as we want. It can be shown, as usual, that this group, taken on S^2, coincides with the group of differentiable transformations of S^2 that are conformal, i.e. preserve angles. This group has already been denoted Möb(S^2). We will see that it is canonically isomorphic to that of hyperbolic geometry of dimension 3.

♦

Here now is an historical example proceeding from the composition *ad infinitum* of planar inversions; it's the first historical appearance of *fractals*. We consider a family of mutually tangent circles and the inversions for which the invariant circles are the circles of the family. Next we consider the effect of these inversions on the set of these circles, and we repeat *ad infinitum*. What is the "curve" formed by the points of contact of all these circles? This doesn't have to do with a curve in the

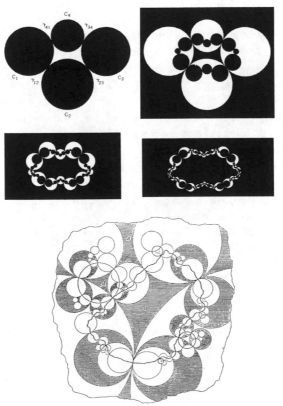

Fig. II.5.4. Mandelbrot (1982) © Mandelbrot

strict sense (like those of Chap. V), but in a broader sense, since we have an infinite, but denumerable, number of points. The answer is a fractal — see Fig. II.5.4 — is a "curve" that has a tangent nowhere (we except the case where the initial circles are all orthogonal to the same circle, in which case the curve is that circle itself).

This fractal, studied by Klein at the end of the nineteenth century, is doubly interesting: on the one hand, it is a natural example of a fractal object appearing where smooth object was expected. But above all it arises naturally in studying certain discrete subgroups of Möb(S^2), called *Kleinian*, which occur in the analysis of functions of a complex variable. A recent book is entirely devoted to them: Maskit (1988). In fact, the group here is that generated by the (finite) set of inversions for which the circle of inversion is one of the given circles. These groups, defined by circles, are called *Schottky groups*. They play a natural and essential role in various problems; they are used in Sects. II.8 and II.9 for the study of certain pencils of circles.

We now define hyperbolic geometry of dimension 3, denoted by Hyp^3, and its group of transformations, as the space constituted by the open ball above, acted on by Möb(S^2). Distance is defined as for the hyperbolic plane in the preceding section. The trouble is that we can't read the group algebraically in the real coordinates of dimension 3. The solution requires climbing the ladder into dimension 4, utilizing the linear projective model of hyperbolic geometry. Furthermore, it's no harder in any higher dimension. Let us say here only that the geometry studied, viewed on S^n, is in fact the conformal geometry of the Euclidean space of dimension n, but **compactified** with a single point at infinity (compactification without regard to real projective geometry of dimension n, that of $\mathbb{R}P^n$). For the study of transformations in Möb(S^2) **viewed** on the sphere, see Sect. III.2.

To conclude with the transformation of angles, we are right to ask whether there might exist still more marvelous transformations than inversions (and their compositions) that preserve angles. We have just seen that the answer is no, but we utilized the fact that the transformations considered are bijective. Defined on the whole Euclidean space, these are the similitudes; on the compactified space in S^n, these are the inversions and their compositions. But let us suppose that we are concerned only with local geometry; then, in the case of the Euclidean plane viewed as $\mathbb{C}$, each holomorphic function defined on an open set is conformal except possibly at points where the derivative is zero: it preserves angles. We thus have an enormous class (not depending at all on a finite number of parameters) of local transformations that preserve angles (but not circles). For bijections of the whole plane, the theorems cited earlier on characterization of similitudes and inversions are the generalizations of Liouville's theorem, which states that the only holomorphic and bijective transformations of $\mathbb{C}$ are the similitudes, and those of $\mathbb{C} \cup \infty$ are the functions of the form $z \mapsto \frac{az+b}{cz+d}$. What happens in dimension 3? The question was settled by the very same Liouville:

> *Each transformation of an open subset (no matter how small) of the Euclidean space of dimension n $\geqslant$ 3 that preserves angles is necessarily the restriction to this open set of an element of* Möb(S^n).

This is a good example of passage *from local to global*; we also speak of *rigidity*. However beautiful Liouville's proof may be, it remains in part artificial and complicated as it comes down in the end to showing that the transformation in question transforms spheres into spheres. Moreover, as was typical of the time, it was formulated in only three dimensions. Liouville's result has attracted numerous mathematicians, who have given varied proofs, all rather complicated. The first direct proof is due to Nevanlinna and is relatively simple: it consists solely in applying three times the condition that the derivative of the mapping considered is a similitude at each point. Such a calculation is almost impossible when expressed in coordinates, i.e. with partial derivatives. Before Nevanlinna (1960) all the differential calculus on the surface of the planet was expressed in coordinates. In 1960 Nevanlinna wrote a whole little text to show that the modern language of linear algebra permitted intrinsic expression of derivatives, a language introduced for the first time in great profusion in Dieudonné (1960) and which then spread like lightning. Nevanlinna, as a great mathematician, apologized for the simplicity of his article, but added that for a new notation to be really interesting it must have applications. He then gave the application to Liouville's theorem, see 9.5.4.6 of [B]. In conclusion, beginning with dimension 3 conformal geometry is reduced to that controlled by Möb(S^n). For the proofs and more, see Sect. 9.4 of [B]. A recent proof is presented in Frances (2003).

Calculation of volumes of tetrahedra in hyperbolic geometry of dimension 3 is both useful and very difficult. For these reasons it is much studied. There is no magic formula, but only calculations in particular cases, which are closely and fascinatingly connected with number theory, topology and analytic function theory. See, for example, Vinberg (1993) and Cho and Kim (1999) and the references mentioned therein.

In Langevin and Walczak (2008) there is an interesting geometric characterization of holomorphic functions with the aid of sphere pencils.

II.6. Inversion in space: the sextuple and its generalization thanks to the sphere of dimension 3

Inversion in space has applications just a spectacular as those of Steiner's alternative. We deal first with that of the *sextuple*: consider three spheres in three dimensional Euclidean space (**our** space), which are mutually tangent and exterior. We easily convince ourselves that there exists a continuous infinity (in one real parameter) of spheres that are tangent simultaneously to these three spheres. Now, as in Steiner's alternative, start with one such sphere tangent to three given spheres and construct the next sphere of the chain, requiring that it be tangent to four initial spheres. Continue progressively in this manner, choosing a sphere tangent to the three initial spheres and the preceding sphere of the chain. Drawings are difficult to produce because the spheres pass alternatively above and below the three initial spheres, which makes it hard to see what is happening. Do we have an alternative a la Steiner? The solution is immediate if we effect an inversion whose center is one of the points of contact between two of the three spheres considered. Then the figure image is made up of two parallel planes and a tangent sphere "sandwiched" between the two. But any sphere tangent to these three is itself sandwiched, thus of the same radius as the third, and is tangent furthermore, so the center is on a circle: and things then close up always at the end of **six** times exactly. There is no alternative here but only automatic closure as in the theorem of the six circles in Sect. II.2.

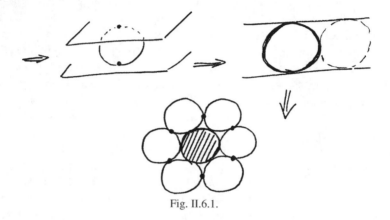

Fig. II.6.1.

The generalized sextuple is scarcely more difficult to define and besides isn't a sextuple in general. We start with an arbitrary torus of revolution and associate with it two continuous families of spheres, the first that of spheres tangent to the torus on the interior along its meridians, the second that of exterior spheres tangent along its parallels, a family to which we need to adjoin for continuity the two planes tangent to the torus along the extreme parallels. We proceed as before, starting with a given interior sphere and constructing a chain of interior spheres such that each is tangent to the preceding and ask whether the resulting chain is finite. This is a generalization of the problem of the sextuple inasmuch as we begin with three mutually exterior spheres (not necessarily tangent) instead of starting with a torus. With an inversion we get reduction to the case where the spheres are of equal radius, and the family of spheres tangent to the three spheres coincides with the second family associated with the torus that they define.

Since it's a torus of revolution, things are as in Sect. II.3: either the chain never closes or it closes no matter how the initial sphere is chosen. We let p denote the associated rational number, since things are in general starred. More precisely, we say that the chain is of type $p = \frac{a}{b}$, where a, b are relatively prime integers, if it includes a spheres and closes up after making b circuits about the center. Now an inversion shows that we have exactly the same dichotomy for the exterior spheres; if there is closure there will be an associated rational number q.

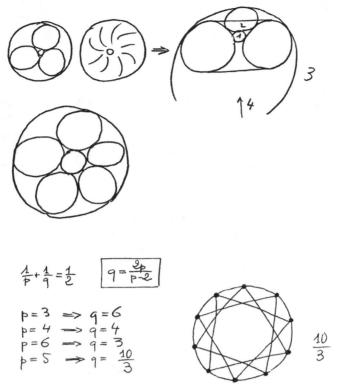

Fig. II.6.2. The generalized sextuple: five spheres in a torus, how many are there on the outside?

The elegant theorem of the *generalized sextuple*, discovered by Steiner, is as follows:

> *Either the torus admits no chain of type p (whatever the initial spheres), in which case the same is true for the exterior chains: they never close; or the torus admits an interior chain of type p (and thus infinitely many, with any initial sphere), but then it admits exterior chains (with any initial sphere) of the same type q, where q is given by the relation $\frac{1}{p} + \frac{1}{q} = \frac{1}{2}$, i.e. $q = \frac{2p}{p-2}$. The first pairs are $(3, 6)$, $(4, 4)$, $(6, 3)$, $(5, \frac{10}{3})$, etc.*

This theorem was considered by Steiner to be one of the most beautiful in all of geometry. The case $(3, 6)$ is that of the sextuple above; we leave it to readers to prove the case $(4, 4)$ by hand with the inversion utilized for the sextuple. As for the symmetry of the relation between p and q, it can't be a surprise: we can, with an inversion, interchange the interior and the exterior of a torus. Now for the integers > 3, beginning with 5, we can't get away with an inversion in dimension 3, we must "climb" into dimension 4 and study the geometry of the sphere S^3 (three-dimensional, but situated in $\mathbb{E}^4$); then everything becomes transparent, as we will see right away. We can also say that we lift the situation studied in dimension 3, by a stereographic projection onto S^3, into the space of dimension 4. Readers who wish to stay in dimension 3 can try a proof of their own and thus appreciate the force of Jacob's vision.

◆

The sphere S^3 is a principal object of modern geometry; it is central to topology and, like S^1, admits a group structure − the group of quaternions of norm equal to 1, see 8.9 of [B]. Our main interest here is that S^3 is a covering of just two sheets of the group SO(3) of rotations of $\mathbb{E}^3$. For the intrinsic geometry of a sphere we refer to the next chapter. We will see in Sect. III.5 that we are far from knowing everything about S^3. To prove the generalized sextuple, we begin by identifying the Euclidean space $\mathbb{E}^4$ of dimension 4 with the complex plane $\mathbb{C}^2$ of pairs (z, z') of complex numbers. Then S^3 consists of the pairs such that $z\bar{z} + z'\overline{z'} = 1$ (or $x^2 + y^2 + s^2 + t^2 = 1$ if $z = x + iy$ and $z' = s + it$). We have two distinguished circles C and C′ on S^3 that are associated with the pairs of type $(z, 0)$ and $(0, z')$, respectively. These belong to two orthogonal planes of $\mathbb{E}^4$. Since these planes generate $\mathbb{E}^4$, we see that S^3 is the union of circular arcs that terminate on either C or C′ and which have length equal to $\pi/2$; these circular arcs are automatically orthogonal to C and C′ at the points that define them; we will denote the set of all of them by $\mathcal{A}$.

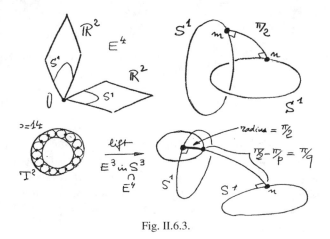

Fig. II.6.3.

◆

We now have everything we need to solve the problem of the generalized sextuple posed earlier. We easily see that stereographic projection transforms a torus of revolution, if the axis of revolution is the axis of stereographic projection, into the set $\mathcal{T}$ of points of S^3 consisting of points situated on all the arcs of $\mathcal{A}$ at a given distance d from their initial point in C. Then this torus $\mathcal{T}$ clearly contains a chain of inscribed spheres of type the rational number p if this distance d is equal to π/p. Now the spheres exterior to $\mathcal{T}$ correspond to the torus $\mathcal{T}$, but seen as a torus for which the interior is the complement in S^3 of the interior of $\mathcal{T}$ complementary to $\mathcal{T}$ — a torus now regarded as formed by the points situated on the arcs of $\mathcal{A}$ which are located this time at a distance $\pi/2 - \pi/p$ from the initial point, but on the circle C'. But by the same reasoning as above this complementary torus contains a chain of type the rational number q if and only if its radius equals π/q. Now by construction its radius equals $\pi/2 - \pi/p$, Q.E.D.

♦

We can't talk about S^3 without mentioning the *Clifford parallels* and the *Hopf fibration*. The latter is essential in algebraic topology, where it is the archetype of certain geometric fibrations. It will also be utilized at the end of the following section. A coordinate-free, purely geometrical, construction can be given; see Chap. 18 of [B]. Here we will proceed as in most of the texts which deal with this matter: we parachute using a description in terms of complex numbers. It suffices to remark that the one-parameter group formed by multiplication by complex numbers λ of modulus 1 acts on $\mathbb{C}^2$ by the mapping $(z, z') \mapsto (\lambda z, \lambda z')$, operations which are isometries and thus preserve distances on S^3. The orbits, i.e. trajectories, are circles;

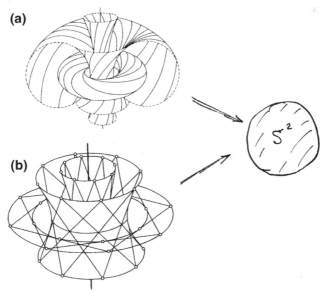

Fig. II.6.4. *Two models of fibration by Clifford parallels.* In the model at the *top* we obtain the sphere S^3 by compactifying the space $\mathbb{R}^3$ by a single point at infinity (and then the vertical axis becomes a circle). The linear model is more complicated to explain, and the figure is imperfect at the level of the horizontal plane. It has in fact to do with a Hopf fibration of the projective space $\mathbb{R}P^3$. The space $\mathbb{R}^3$ of the figure must now be compactified in $\mathbb{R}P^3$ by adjoining the plane at infinity (the lines however all become topological circles); we remark that the completion of a hyperboloid of one sheet is a (topological) torus. We have of course to remove from the hyperboloids all generators of one of the two families. **(a)** Penrose (1978) © Springer; **(b)** Hilbert, Cohn-Vossen (1996) © Springer

and these circles have the property that, taken two at a time, each point of one is at a constant distance from the other (which explains the term *parallel*). These properties of S^3 are often presented with the aid of quaternions, even though this is not at all essential; again see Chap. 18 of [B].

The lower figure, based on quadric surfaces, is a **linear** model of the one above (see Sect. IV.10). It is also very important to find the structure of this family of circles in their entirety, insofar as there is one. The answer is that they naturally form the sphere S^2. We have all that is needed to explain this, since the quotient obtained by the equivalence provided by $(z, z') \equiv (\lambda z, \lambda z')$ is nothing other than the complex projective line $\mathbb{C}P^1$ (see Sect. I.XYZ); we have simply to remark that $S^3 \subset \mathbb{C}^2$ doesn't omit any complex line of $\mathbb{C}^2$. The mapping of $S^3 \rightarrow S^2$ thus obtained, whose fibers are all circles, is called the *Hopf fibration*. It plays a considerable role in algebraic topology, as do its generalizations for the spheres S^{2n+1} and S^{4n+1} obtained from the projective spaces $\mathbb{C}P^n$ and $\mathbb{H}P^n$ in all dimensions (in the latter case, the fibers are the quaternions of modulus 1, namely the spheres S^3).

The *Clifford tori* are tori formed by points at a given distance from a great circle of S^3. These are the inverse images in the Hopf fibration of a "small" circle of S^2. Certain authors reserve the name Clifford torus for the torus equidistant to two orthogonal circles of S^3: it is conjectured, see Sect. III.5, that these are the only surfaces in S^3 that are minimal and of toroidal type.

II.7. Higher up the ladder: the global geometry of circles and spheres

As things stand, the associated geometry Möb(S^n) remains insufficient for giving account of the immensity of theorems on circles and spheres, results that have accumulated over centuries. We need to climb just one more rung up the ladder for a concept that is quite natural: we seek to endow the space of all spheres in the Euclidean space of dimension n with an appropriate structure. We present this structure very briefly, then give a few examples of configurations that it allows us to treat conceptually. We stop there; for more "depth" see Chap. 20 of [B]. However, we must mention Coolidge (1916, 1999) above all for all the classic results on the subject. The reason that we stop is that, in the spirit of Jacob's ladder, the tool that follows allows us to treat all possible problems in this branch of the subject; to our knowledge it isn't necessary to climb higher. We use the language and notations set forth at the end of Sect. II.XYZ.

The first configuration (yet another six circles theorem) − for which we owe the interpretation to Adrien Douady − is classic and concerns circles in the plane. It states that, in the figure of four circles below (Fig. II.7.2): if the four points a, b, c, d are on a circle, then so are the four other points e, f, g, h on a circle. This can be proved "by hand" with the classic notions of *power with respect to a circle* and of *radical axis*, etc.; see Sect. II.XYZ.

However, the conceptual proof is meteoric: in the space $\mathcal{C}$ of circles constructed in Sect. II.XYZ, which is a projective space of dimension 3, the set of circles passing

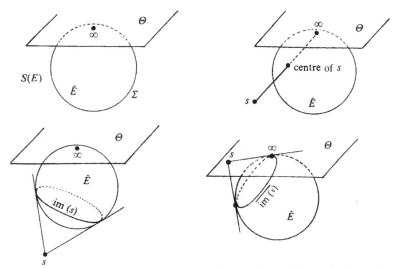

Fig. II.7.1. The space of circles in the plane, Im(s) denotes the set of point circles (radius zero) that constitute the points of a circle s; the point circles form a sphere. [B] Géométrie. Nathan (1977, 1990) réimp. Cassini (2009) © Nathan Édition

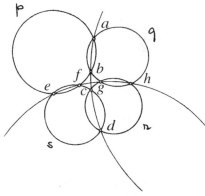

Fig. II.7.2. Yet another six circles theorem. [B] Géométrie. Nathan (1977, 1990) réimp. Cassini (2009) © Nathan Édition

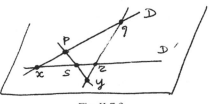

Fig. II.7.3.

through two given points form a projective line of $\mathcal{C}$. Then, if in $\mathcal{C}$ our four circles are denoted p, q, r, s, the hypothesis of the theorem is that the lines D and D′ defined by the pairs (p, q) and (r, s) respectively have a common point x. But these two lines are in the same plane and thus the lines defined by (p, s) and (q, r) intersect in a point y of $\mathcal{C}$, which is the desired circle! Interested readers will subsequently see a third circle, but then will be traumatized by the fact that the desired points of intersection don't exist. This reflects the fact, see for example Chap. 20 of [B], that the lines of $\mathcal{C}$ don't correspond just to circles having two points in common, but also to a *pencil of circles* of type "having no point in common":

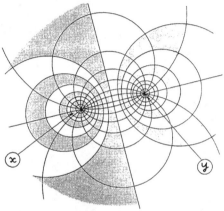

Fig. II.7.4. A pencil of circles without a common point is a line in the space of circles

The second configuration is in the space of spheres $\mathcal{S}$ of the Euclidean space $\mathbb{E}^3$. It figures in many French books on "classical" geometry from the first half of the twentieth century, in the sections called "anallagmatic geometry". We don't cite them, but they use 50 pages at a time for proving results that we will develop much more briefly, this length being due either to an outdated desire for geometric purity (they even speak occasionally of "higher geometry", with a fragrance of romanticism) turned toward the past or to conceal composition of inversions in the group Möb(S²). Yet we will see from a particular case that all problems of this sort are resolved very rapidly, and above all "in depth", as soon as the geometry of $\mathcal{S}$ is utilized. Moreover the concepts work well, without additional effort, in arbitrary dimension. We have allowed ourselves this little polemic because the geometry of $\mathcal{S}$ was widely known outside France by the end of the nineteenth century, especially in Germany by Felix Klein.

Here is the configuration we need, the *paratactic ring*. We begin with the case of the ring of orthogonal circles, easily treated by inversion, that. A circle C of $\mathbb{E}^3$ and its axis, the line D, have the property that each sphere passing through C intersects D at a right angle; but also each plane passing through D intersects C at a right angle. Now a general inversion transforms the union of D and C into a pair of circles such that each sphere passing through the one cuts the other at a right angle (this of course applying to both).

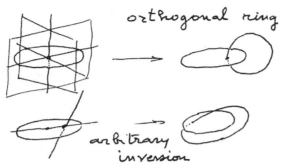

Fig. II.7.5. An orthogonal ring and a paratactic ring

But the thing is true for any angle: for each angle α there exist pairs of circles such that each sphere passing through the one cuts the other by the angle α; the same thing for the other circle and the same angle α. These pairs of circles are called the *paratactic ring for the angle* α. By hand, and even with inversion, it is rather difficult and long to establish their properties, still much longer to show that, within an element of the conformal group, there is but one paratactic ring of angle α; this takes a good fifty pages, for example in Hadamard (1911). If one of the circles is a line, it is called a focal line of the other circle; this designation resulting because, in projecting onto a plane perpendicular to the line, the line must yield a point which is a focus of the elliptical projection of the circle (see Section IV). Things are much simpler in the present case and are left as an exercise. But this is not the good conceptual approach.

To prove the existence and uniqueness of paratactic rings with the language of the space of spheres $\mathcal{S}$, we must first know that in $\mathcal{S}$ the circles are represented by lines (made up of points corresponding to the spheres that contain the circle). Therefore we look for pairs of lines such that each point of the one is at a constant distance to the other (the distance being the same for both) given by the angle α. But we have found such lines in Sect. II.6, the Clifford parallels. We only need still find the sphere, "Euclidean of dimension 3", in $\mathcal{S}$. In fact, what we find are the projective spaces endowed with their elliptic structure (quotient geometry that of the sphere). To do this, it suffices to take any imaginary sphere whatsoever; considered as a point in $\mathcal{S}$ it has a dual hyperplane with respect to the basic quadric of $\mathcal{S}$. This dual hyperplane is thus endowed with the restriction of the basic quadratic form, which here is well defined and positive if an imaginary sphere has been chosen from the start. This is then certainly an elliptic projective space; see more details as needed in Sect. II.XYZ and 20.5.4 of [B]. To establish that such rings are essentially unique, it suffices to utilize the strong transitivity of the group Möb(S^3). The case where a circle is a line is of course covered, but it is not so easy to prove by hand.

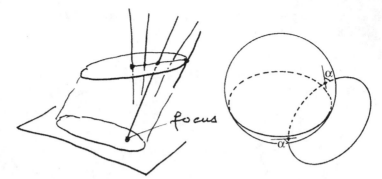

Fig. II.7.6. The focal lines of a circle and the paratactic rings

◆

Now I can't resist speaking about the result in the geometry of ordinary space which made the greatest impression on my life (along with, in truth, Poncelet's theorem on conics: see Sect. IV.8). We can see it by regarding what is given by Clifford parallelism in the sphere S^3 under stereographic projection to ordinary space and concerns the so-called *Villarceau circles*; see 18.9 of [B]. We consider a torus of revolution and we cut it with the tangent planes on the interior: the astonishing thing is that the section always decomposes into two circles. So we find, on the tori of revolution, **four** continuous families of circles, the parallels, the meridians and the two Villarceau families. Moreover, these "exotic" circles are the loxodromes of the torus: they cut the meridians by a constant angle, the double of the angle at their points of intersection.

I will soon cease being very personal, but at 16 was so astonished that I actually wanted to saw out a ring in wood to verify the theorem. It was in my father's terminal instructional material (Rouché and de Comberousse, 1912) for the baccalaureate degree where I found the theorem. The proof is completely elementary, but good spatial perception is needed. Then, in the preparatory class for the university, I learned the clear and profound proof not unlike the one here. The torus contains the umbilical (see Sect. II.XYZ) as double line, but a tangent plane cuts it in a curve that has two double points (at the points of contact). Whence, finally, for this curve section, there are four double points; but since it is of degree four, it is decomposed into two conics (see Sect. V.14), which are thus circles.

This proof is more than quick, and nonetheless it's just this way that it was taught us. In fact you will understand all that follows: first projectify the space where our torus is considered, then complexify it. Then the complexified torus is that to which the above reasoning is applied. The two circles obtained are in fact complex circles, they have the topology of a sphere S^2. We subsequently keep only the real part of the figure. This going and coming from the real to the complex is short-circuited in entire books, or entire parts of (now old) textbooks, of so-called modern geometry. Things go well because a torus is defined by an algebraic equation (a polynomial of degree four), and complexifying and projectifying is automatic, we keep the same

Fig. II.7.7. See other nice photos in Berger (2002). Berger (2002) © Pour la Sciénce, Éditions Belin

equation, but throw in homogeneous quadruples; then we allow complex homogeneous quadruples, and there we have the torus with which we can reason as we have just done.

To my knowledge, some historical points remain open: who really first discovered these circles, known in at least one particular case to the stone masons of the middle ages (see the Fig. II.7.7)? Who discovered the loxodromic property? Moreover, the view in $\mathcal{S}$ shows that an arbitrary pair of Villarceau circles of the same family form a paratactic ring for the angle α of loxodromy. The topologist will not have failed to note that the Villarceau circles, thus all the paratactic rings, are formed from *enlaced* circles. If it is surprising that four distinct families of circles

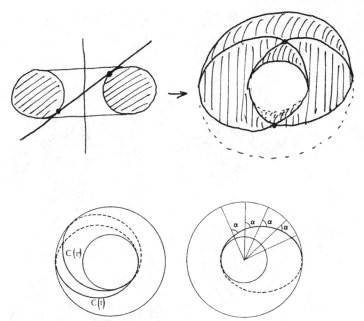

Fig. II.7.8. Readers will convince themselves in their own ways of the reality of the
Villarceau circles

can exist on a single surface as simple as the torus (its coordinate equation is only of
degree 4); it can be seen in 20.7 of [B] that there even exist surfaces, always of degree four, called *cyclides*, that contain six distinct families of circles. A special class
is formed by the Dupin cyclides, which are nothing other than the inverses of tori of
revolution; besides [B], see also 11.21 of [BG]. Yvon Villarceau, who has a street
in Paris named after him, was an astronomer. The oldest reference that we know is
Yvon Villarceau (1948), where the circles are noted in about ten lines and without
any explanation!

Readers will ask if there exist other surfaces that contain as many or more one-
parameter families of circles. This question was completely settled in
Darboux (1880). The answer is that ten families of circles is the maximum, and
that ten is only attained by the circles discussed earlier. However, Darboux's text is
ambiguous, for as described in Darboux (1917) (see also [B]), the real cyclides can't
actually have **ten** families of circles, for that they must be complexified, as it seems
to us. This sort of ambiguity is frequent in the texts of the nineteenth century and
has often fed quarrels, the one between Poncelet and Cauchy being most celebrated.
In Darboux (1880) the problem studied is that of surfaces that contain lots of conics.
The answer is that the maximum number of parameters for continuous families of
conics is ten (only six in the real domain), except for the ellipsoids and Steiner's
Roman surface, already encountered in Sect. I.7, for which we have an infinity of
conics dependent on two parameters.

II.8. Hexagonal packings of circles and conformal representation

What has preceded is beautiful geometry, with some ascent of the ladder, but was essentially known and explained in depth prior to the twentieth century. The results of this section and the one that follows are different both with respect to topicality and depth. The matter begins with the drawing exercise that follows, which we advise readers to carry out for themselves with paper and pencil, especially if they are good drawers, for it is very instructive. Start with a system of seven circles, as in Fig. II.8.1, consisting of a central circle surrounded by six circles tangent to it and the corresponding tangent lines, and continue the figure by first adjoining a ring of twelve circles, with the condition that each of the first six circles around the central circle will be surrounded by six circles, each tangent to it and to two of the other six. Such a packing is called *hexagonal*. For the moment we ask nothing of boundary circles regarding the number of circles they touch. The numbers of the generations is indicated in the figure. It's a matter of continuing indefinitely, if possible, so as to fill the entire plane.

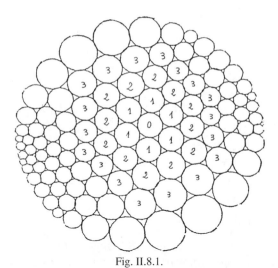

Fig. II.8.1.

Fig. II.8.2.

However well we draw, it will be confirmed that this is impossible. Very quickly a catastrophe occurs: we can no longer continue. If this happens with the figure on the left of Fig. II.8.3, it's because the large initial circle has many more than six neighbors! But nonetheless we know of an hexagonal packing, of the whole plane, called the *regular hexagonal packing*. Where is the contradiction?

There isn't any; it simply happens that in the other drawings the six imposed circles of the first *generation* didn't have equal radii. Precisely, we introduce the *default* def(C) of a circle surrounded by six circles with radii r_i $(i = 1, 2, 3, 4, 5, 6)$:

$$\mathrm{def(C)} = \sup_{i \neq j} \left| 1 - \frac{r_i}{r_j} \right|.$$

We speak later of packings having n generations: see Fig. II.8.1 above.

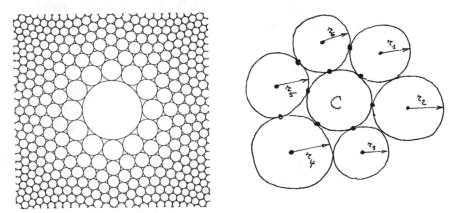

Fig. II.8.3. Pach, Agarwal (1995) © John Wiley & Sons, Inc

In 1987 Rodin and Sullivan showed that:

(II.8.1) *There exists a sequence $\{u_n\}$ indexed by the positive integers n, such that u_n tends to zero as n tends toward infinity, and which has the property that if a circle C is the starting place (zero-th generation) of an hexagonal packing having n generations, then def(C) $< u_n$.*

Therefore if we fill the plane the number of generations is infinite and thus def(C) = 0; in particular the circles about C have the same radius, necessarily equal to the radius of C itself, and thus step by step all the radii are equal, Q.E.D. Thus:

(II.8.2) *The only hexagonal packings of circles that fill the whole plane are those with equal radii.*

We can thus say that such a packing is unique within similitude. We are dealing here with an extremely difficult result. The amateur can try proving it "by hand", but this author has wracked his brains over it during very many hours of waiting in airports. Before speaking of motivation, applications and proof, here is a bit of history, with some additional remarks for interested readers. We mention first that the result of Rodin and Sullivan immediately unleashed a veritable avalanche of work: an up-to-date bibliography on the subject would contain well over a hundred titles. We also study generalizations of the Rodin-Sullivan result: packings of circles in the plane, hexagonal and also more general, under the condition that they be of *finite valence* (the valence is the number of circles that touch a given circle); and indeed plane packings of still more general objects. We will see another motivation, apart from the geometric beauty, below.

Never had a result as strong as that of Rodin-Sullivan been anticipated. Fejes Tóth had merely conjectured in 1977 that an hexagonal packing of the whole plane would have one of the radii either tending toward zero or toward infinity as the number of generations increases. In 1984 Barany, Füredi and Pach proved the conjecture of Fejes Tóth, and even a bit more. But that is far from Rodin and Sullivan; and it furthermore tells us nothing about the sequence $\{u_n\}$, apart from the fact that it tends toward zero. The first thing to do is to define the optimal u_n, the only interesting values. That is to say, we define, for each integer n, the number s_n as equal to the largest default def(C) possible so that an initial circle C can admit a packing having n generations. Readers will easily calculate the default of the initial circle in the packing of n generations obtained by an inversion through a center as close as possible to the packing of equal radii; they will thus prove that $s_n \geqslant \frac{4}{n}$. But what is clearly interesting is finding an upper bound for s_n. We have almost complete answers at present, due to Doyle, He and Rodin:

(II.8.3) *On the one hand there exists a constant K such that $s_n \leqslant \frac{K}{n}$ for each n; on the other hand,* $\lim_{n \to \infty} n s_n = 2 \sqrt[3]{2} \Gamma^2(\frac{1}{3}) / \Gamma(\frac{2}{3})$, *where Γ denotes the gamma function.*

This number is not as mysterious as it might seem; it is in fact equal to 4/R, where R is the conformal radius of the regular hexagon; see just below for the definition. But we still don't yet know the best K possible in complete generality.

♦

For the proofs, the essential difficulty is that the algebraic relation (it certainly exists) that connects the six radii r_i of the circles about a given circle of radius r is unworkable. The seven circles theorem (see Sect. II.2) is in any case useless. Barany, Füredi and Pach proved the two inequalities

$$\sum_i \frac{1}{r_i} \geqslant \frac{6}{r} \quad \text{and} \quad \sum_i r_i^n \geqslant 6r^n \quad \text{(for each } n > 1\text{)}.$$

They concluded in a relatively elementary way with the left equality by reasoning on the graph of the centers, where we can — it is now classic — define a *Laplacian* (of discrete type) and speak of subharmonic functions. But there isn't any subharmonic function except for the constants *a la Liouville's*. Interested readers can consult Sect. 8 of Pach and Agarwal (1995).

The proof of Rodin and Sullivan has at first the appearance of a vicious circle; to prove (II.8.1), they used (II.8.2). In fact, to deduce (II.8.1) from (II.8.2) is elementary, proceeding by contradiction with the help of a pretty lemma illustrated by the figure below and passing to the limit:

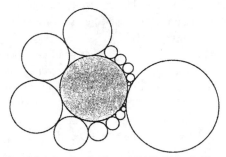

Fig. II.8.4. Rodin, Sullivan (1987) © D. Sullivan

(II.8.4) *(Ring lemma.) There exists a constant $r(n)$ (i.e. depending only on n) such that, if we are given a system of n circles tangent two at a time in succession and all tangent to the unit disk, then all these circles have radius greater than $r(n)$.*

The proof of (II.8.2) (uniqueness of hexagonal packings in the plane) uses the theory of the geometry at infinity − including some quite recent results − of the discrete groups studied by Klein, so-called *Kleinian groups*, which are discrete subgroups of $GL(1; \mathbb{C})$, encountered in Sect. II.5. They enter analysis in an essential way, since they are connected by definition to functions of a complex variable. The effect of these groups on S^2 is examined by studying their effect on the circles of S^2. We must see what happens when we iterate this or that transformation and so encounter a *geometrically dynamic* composition that we have already seen with Pappus's theorem in Sect. I.1. The Rodin-Sullivan proof thus utilizes recent and profound results on Kleinian groups. But this still does not explain the depth of matters in (II.8.2). The starting idea is common to much rather recent work in analysis and geometry and uses the notion of a *quasiconformal mapping*. Such a mapping, say from a portion of S^2 into itself (although this notion is in fact very general, applicable to every metric space), preserves *spheres* within a deformation factor that is universally bounded on the open set considered; in addition it is required that we have a homeomorphism. Precisely, there must exist a $k > 0$ such that the image under f of each sphere $S(p; r) = \{q : d(p, q) = r\}$ with center p and radius r satisfies the two conditions: $f(S(p; r))$ is contained in the ball which is the interior (a ball) of the sphere $S(f(p); kr)$ and $f(S(p; r))$ contains in its interior the sphere $S(f(p); k^{-1}r)$. Such a definition evidently has interest only if we go to infinity, i.e. if its domain of definition isn't compact. Now we consider an hexagonal packing $\mathcal{H}$

of the plane and the regular packing $\mathcal{H}'$ (with equal radii). It may seem natural to define a quasi-conformal mapping of the plane by conformally mapping the disks of $\mathcal{H}$ onto the corresponding disks of $\mathcal{H}'$, worrying subsequently about the interstices between the disks. But this approach is too naive: it is impossible to attain agreement at the points of contact, since a conformal mapping of one disk into another is determined once we know the images of three points of the boundary.

In fact, it's with the interstices that we must begin: there exists a unique conformal mapping that takes an interstice defined by three mutually tangent circles onto an interstice of the same type, taking the three vertices of the first interstice onto the three vertices of the second. To see this, it suffices to consider the circle passing through the three vertices of the interstice:

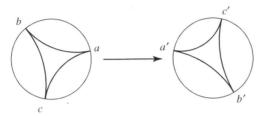

Fig. II.8.5.

Next, for the interiors of the circles of $\mathcal{H}$ we do the same thing for the image circles of the circles of $\mathcal{H}$ by the inversions for which the circle of inversion is in $\mathcal{H}$. Likewise for $\mathcal{H}'$. We thus create new interstices (shaded in Fig. II.8.6) which are made to correspond by new conformal mappings. By iterating this operation, we obtain on both sides a partial darkening of the plane by interstices, the images of the original interstices by the Schottky groups (see Sect. II.5) of $\mathcal{H}$ and $\mathcal{H}'$; the two darkenings correspond to each other by a mapping which can be shown to be quasiconformal. But what remains is easily seen to be of measure zero; see Sect. II.9. We can also interpret all this in S^2 by stereographic projection. But in this situation a fundamental result of the theory states that a quasiconformal mapping of a "very large" part of S^2 onto itself arises from a quasiconformal mapping on all S^2, thus an element of Möb(S^2). Hence $\mathcal{H}$ and $\mathcal{H}'$ are "the same", and thus the elements of $\mathcal{H}$ also have equal radii.

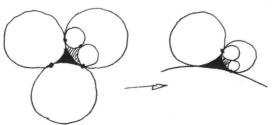

Fig. II.8.6. $H_1 3$ and $H_2 3$ are the images of H_1 and H_2 by inversion in the circle H_3

It seems that as yet there is no very simple and conceptual proof of the uniqueness of hexagonal packings of the plane. That of Schramm from 1991 provides a
rigidity for all packings of a rather general type. This result is topological, rather
long and hard to follow in detail. The essential property that allows conclusion in
the case of circles is that the difference of two circles, in contrast to convex sets generally, never has more than two connected components. This excludes for example
rectangles with different side ratios, which relates to the fact that the side ratio of a
rectangle is a conformal invariant.

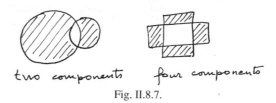

two components   four components

Fig. II.8.7.

So much for (II.8.2). But how do we prove (II.8.3)? For the upper bound in
$\frac{K}{n}$ we do exactly the same thing as in the proof of Rodin-Sullivan, except that we
abandon the finite packings $\mathcal{H}_n$ and $\mathcal{H}'_n$: $\mathcal{H}_n$ has equal radii and n generations (think
of a regular hexagon), whereas $\mathcal{H}'_n$ is arbitrary hexagonal and at least n generations.
We still have interstices, finite in number. We show that what remains once the
interstices are removed (which evidently is not of measure zero) has a very small
measure, i.e. less than $\frac{K'}{n^2}$ (where K' is a universal constant). Let us say that if a
point is thrown at random, the probability is very strong that it will fall in one of
the interstices. It remains then to show by a (reasonable) analysis that the mapping
constructed − which is conformal on a subset of very large measure − since it is
here defined on a compact set, has the global property of quasiconformity and is
explicitly evaluable as a function of n. This allows an estimation of the default of
·the initial circle in $\mathcal{H}'_n$.

Finally, to get the asymptotic estimate of s_n in (II.8.3), we apply the preceding
technique with $\mathcal{H}_n$ and $\mathcal{H}'_n$ to an arbitrary packing, but whose combinatoric is exactly the canonical one with n generations, and thus the interior of a regular hexagon;
see Fig. II.8.1. We therefore have the mappings $g_n : \mathcal{H}_n \to \mathcal{H}'_n$. We show that they
have a limit and that this limit is a conformal representation of the regular hexagon
on the disk.

This provides us with a connection to the motivations: where could Rodin and Sullivan have gotten the idea for their result? It in fact did not appear as a result in the
theory of Kleinian groups, but rather their motivation lay in a conjecture of Thurston
from 1985 concerning the geometric construction − an algorithm − for the conformal representation of a domain in the plane. We have seen in Sect. II.5 that there
exist plenty of local conformal mappings: every holomorphic function defined on an

arbitrary open subset U of $\mathbb{C}$ is *conformal*, i.e. conserves angles; computationally its complex derivative (real differential) is a similitude. The problem of *conformal representation* is of considerable mathematical importance. We are given an open subset U of $\mathbb{C}$ and seek a conformal transformation (a biunique conformal function) which maps it onto the open unit disk D. We would have thus "parametrized" the open set U with the disk, which is trivial, but not if conformality is required. We have what is called a *conformal representation* of U. We could thus reduce the study of this or that problem on U to a problem on the unit disk. Apart from its pure mathematical interest, conformal transformations are encountered in applications. There is the case of the study of heat dispersion and that of subsonic flow about an airfoil. We will now describe in detail an application to electricity. We fix a point O of U and consider a long cylinder in space having section U, representing an electric conductor or electrostatic, with a central wire represented by O. A voltage of 10 000 V is applied to the exterior, the boundary of the cylinder, and 0 V to the central conductor. We want to find how the voltage is distributed on the interior, between the central conductor and the exterior surface. In other words we seek the *equipotentials* of the problem.

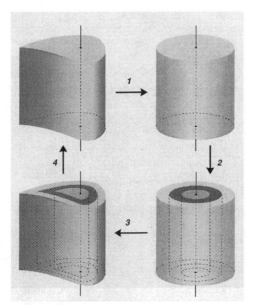

Fig. II.8.8. Conformal mappings are well-known for their use in determining electrical equipotentials and flow about an airfoil. To calculate the equipotentials in the case of the cylindrical surface indicated above on the left, we deform the problem by a conformal mapping so as to obtain a circle about its center for which the equipotentials are known: these are concentric circles. In effecting the inverse mapping, we obtain the equipotentials for the original problem under study. Pour la Science, no 176, June (1992), Marcel Berger, Les placements de cercles © Pour la Sciénce, Éditions Belin

It's easy to see that the answer is given by the level curves of the conformal mapping taking the pair (U, O) onto the pair (D, 0). We deduce from this a physical demonstration of the existence and uniqueness of the conformal representation of U. This result was announced by Riemann — in a more general setting, by the way — with a completely insufficient proof, in terms of convergence of mappings minimizing an appropriate functional. Still other mathematicians thought that the theorem did not require proof, given the physical interpretation. It was not until toward the beginning of the twentieth century that Koebe gave the first complete proof. As for uniqueness, it is easy if a single image point is fixed, for example the center. Then we have uniqueness, within transformations of the conformal group for the disk, i.e. the group Möb(S^1) encountered in Sect. II.4. This allows the imposition of three boundary points; see the figure below. We mustn't forget that the necessary condition is evidently that the topology of U be that of the disk, which is easily seen to be equivalent to requiring that U be simply connected. Further on we will see what happens in the general case.

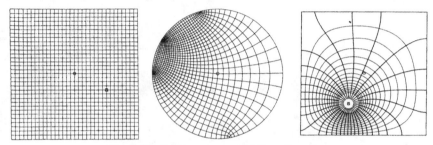

Fig. II.8.9. Conformal representation of a square

The problem of Thurston is to obtain a geometric realization for the conformal transformation. With a succession of improvements over a long period of time, we have extremely rapid algorithms at our disposal; see the references in the introduction of Pommerenke (1992). But these have nothing to do with geometry. Here is Thurston's strategy: we fill U with regular hexagonal lattices with a rather small common radius r. We see (in terms of the hyperbolic polyhedra in Sect. VIII.8) that there always exists an "image" in the disk of such a packing by an hexagonal packing (except at the boundary) and having the same combinatoric as the initial packing of the lattice, this combinatoric being the one imposed by the behavior at the boundary.

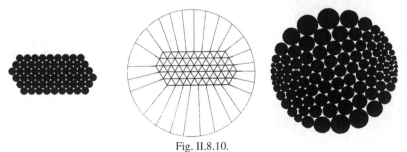

Fig. II.8.10.

With each r we associate an $f_r : U \to D$ by requiring first that centers of circles in U be sent to centers of the corresponding circles in D. We then fill the triangles thus obtained by affine mappings (which exist and are unique, as seen in Sect. I.2). Thurston conjectured that, if r tends to zero, the mappings f_r have a limit mapping and that this is conformal. The proof of this result was the goal of the article of Rodin-Sullivan, and (II.8.1) gives a preview: at the center of circle C, the default in conformity is carried over by the default def(C) of this circle. But we know that this default tends to zero with r, Q.E.D. We can now also say that it is constant by (II.8.2), namely 4/R, where the so-called *conformal radius* R of the regular hexagon is the similitude ratio (ratio of conformity) at the origin of the conformal mapping (unique within a rotation) that sends the regular hexagon onto the unit disk while preserving the centers.

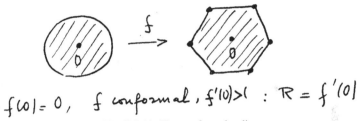

Fig. II.8.11. The conformal radius

What happens next is very interesting. The various authors mentioned earlier who attacked the generalization of (II.8.1) had begun to sense a connection between circle packings and problems of conformal representation. We point out this important fact: that although the numerous results obtained so far indisputably demonstrate such a connection, a deep and truly conceptual result remains to be found. We return to the case of conformal representation, this time for domains U that are not necessarily simply connected, i.e. U is punctured by (simply connected) compact sets on its interior, e.g. points, etc. The *Kreisnormierungsproblem* is to generalize the existence of classic conformal representation by showing that, for any U, there always exists a conformal representation onto the unit disk D that is *punctured* in holes that are either points or disks. So the boundary will be completely *traversed*. Thanks to the intuition forged by the study of packings, the specialists knew how to show this for a disk punctured in no more than a *denumerable* number of holes. However, a point of history: when Schramm recounted this result at a conference, Sachs — also a specialist in conformal representation — pointed out that, at least for a finite number of holes, the proof had already been obtained by Koebe by 1936, but only in the simplicial case, i.e. triangulations (in fact decompositions into polygons) where each vertex belongs to only three polygons, or the dual situation where the polygons are all triangles; see also Sect. VII.9 and Ziegler (1995) for references. Koebe utilized a passage to the limit, easy to see with hexagonal packings. It remains to be discovered today whether or not the *Kreisnormierungsproblem* has a solution in the case of a *non denumerable* number of holes.

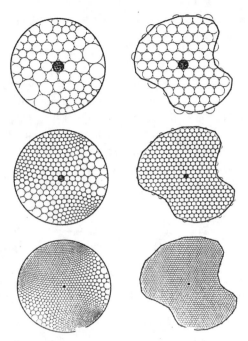

Fig. II.8.12. In these geometric examples we can **see** the conformal representation

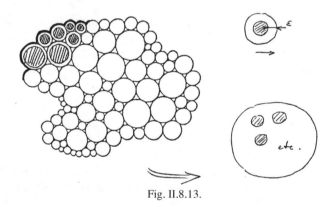

Fig. II.8.13.

For all the preceding, readers can consult the general exposition (Berger, 1993) and the references therein, to which we need to append some more recent ones, e.g. Doyle, He, and Rodin (1994) and the references mentioned there, as well as Mathéus (1999). A recent experimental reference is Bobenko and Hoffmann (2001). Recent expositions are Stephenson (2003) and the book entirely dedicated to it (Stephenson, 2005).

This whole section might seem to readers like a an aesthetically pleasing game of pure geometry, but Thurston's Thurston technique for realizing conformal representation with circle packings has just recently been applied to construct planar representations of the brain which may give better account of all the connections within its complete internal structure; see the site www.math.fsu.edu/~mhurdal of M. Hurdal.

II.9. Circles of Apollonius

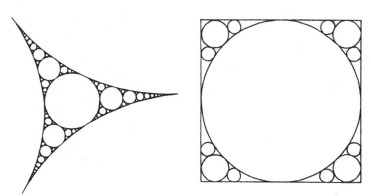

Fig. II.9.1. The circles of Apollonius and the Apollonian packing of a square

The problem is to know what happens when we fill the interstices formed by three mutually tangent circles **ad infinitum**. That is, we trace the circle tangent on the interior to the three given circle, then we do the same with the three new interstices obtained, and we iterate indefinitely. Here is a new intrusion of *iteration* into geometry, in other words a certain dynamic feature. The first problem lies in knowing the area (the measure) of the surface that remains at the end of these infinitely many operations. Readers can try to show that this final area is, might be suspected, *zero*, which can be done quite nicely by hand. It is interesting to note that the first reference to the proof seems to date from 1943; for this whole section readers can consult (Apéry, 1982). We remark that we are in the situation inverse to that of the preceding section: here it's the interstices "that survive", for which the set itself is of measure zero. For those who know the construction of Lebesgue measure with the aid of squares, the preceding discussion shows that we can just as well use circular disks to define this measure. Such disks are sufficient to fill a square "completely", in the sense of the measure.

Now there is a conceptual ascent into the theory of Kleinian groups that explains − or puts in a good context − the following result on Apollonian circles. Here briefly is the essence of things: the set of disks for removal by the discrete (Kleinian) group are defined thus: Let A, B, C, D be the four circles of the Fig. II.9.2. There exist two Möbius transformations f, g that are well defined by the conditions: first they are transformations called *parabolic*, i.e. conjugates of translations. We can obtain them all − since translation is the product of two symmetries about parallel lines − by composing two inversions whose circles are tangent to each other (two parallel lines, in conformal geometry, are tangent at the point at infinity). Second, we require that:

$$f(A) = A, \quad f(B) = B, \quad f(C) = D, \quad g(C) = C, \quad g(D) = D, \quad g(A) = B.$$

We can draw things on the sphere, thanks to stereographic projection:

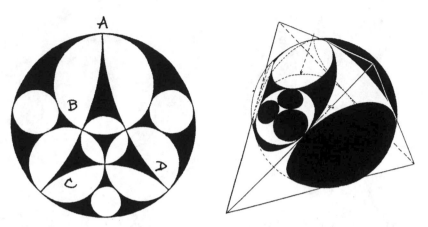

Fig. II.9.2. We view things on the sphere by stereographic projection. Apéry (1982)
© Societé Mathématiques de France and F. Apéry

We see for example that f slides the circles in the space included between A and B, beginning by sending C onto D. Finally, the group Γ generated by f and g (which is a free group by the way) is exactly the one which provides, beginning with the three initial circles, all the disks of the Apollonian Circles.

Now we need to interpret Γ as a subgroup of the group of hyperbolic space $\mathcal{H}^3$ of three dimensions. Its geometry is tied to that of its *fundamental domain* (also called a *Dirichlet* or *Voronoi domain*); we encounter this type of domain in Sects. III.3 and IX.3. This domain $\Delta(p)$ (associated with a point p, but they are all isomorphic by the definition itself) is the set of points that are closer to p than all the other points $\gamma(p)$, where $\gamma(p)$ runs through all the elements of Γ other than the identity. In hyperbolic geometry of dimension 3, the points at an equal distance from two distinct points always constitute, as in Euclidean geometry, a plane.

The boundary of $\Delta(p)$ is thus made up of pieces of such planes. We see that this domain is a tetrahedron of $\mathcal{H}^3$. But, as a discrete subgroup Σ more general than $\text{Möb}(S^2) = \text{Isom}(\mathcal{H}^3)$, this boundary may be made up of an infinite number of planes. We say that Σ is *geometrically finite* in the case where the boundary of the fundamental domain is made up of a finite number of planes. Then the theorem on the Apollonian Circles is a very, very special case of a general theorem of Ahlfors from 1966: *for each geometrically finite Kleinian group, its limit set on S^2 is of measure zero*. The limit set of Σ is associated with it just as the Apollonian circles were associated with the group Γ. That is what puts the Apollonian Circles in the right conceptual context.

A question: The Apollonian circles give a packing, within measure zero, of the sphere S^2 by disks: is this the optimal way? This question seems open, but we can nonetheless think that a *fractal dimension* provides part of the answer. For fractals see the classic references: among others, after the historic (Mandelbrot, 1977–1882), the three books Feder (1988), Falconer (1990) and Cromwell (1997). We consider here, along with the Apollonian Circles, the subset of $\mathcal{C}$ made up of **circles** (not disks) — an infinite (but still denumerable) set of curves. The "finite" curves have a positive 1-*dimensional* measure but are of measure zero with respect to 2-dimensional measure. Now, what is it for the infinite set of curves $\mathcal{C}$? It's a set of fractal type obtained by an iteration process repeated infinitely many times, as in a *snow crystal*, which is the best known. For each part of S^2, and of a metric space more generally, we define its *Hausdorff dimension*. For a nonempty open subset of S^2, it is always equal to 2; for each (differentiable) curve of S^2, it is equal to 1. This dimension measures in some way the *density*, the complexity, the packing efficiency, etc. To evaluate this efficiency, we evaluate the total "α-dimensional" measure of the balls required to cover the object considered as the maximum radius tends to zero. Precisely, we seek the limit inferior, for r tending toward zero, of the sums $\sum (\mathrm{radius}(B_i))^\alpha$, where the balls B_i cover our set and are of radius $\leqslant r$. It is shown that there is at most one value of α for which this limit is neither infinite nor zero. This real number α is, by definition, the Hausdorff dimension of the set. In considering the value of this limit for our set and its subsets, we have furthermore a corresponding measure for this dimension: the *Hausdorff measure*. Note that in simple cases we can be content with finite coverings by balls of equal radius and replace the above sum by $N(r) \cdot r^\alpha$, where $N(r)$ designates the number of balls of radius r necessary for covering the set; but there are situations where denumerable coverings are required, and then the balls can not all have the same radius.

For the Apollonian Circles $\mathcal{C}$, we are obviously between 1 and 2 and — as for the snow crystal — we therefore expect, in one dimension, a Hausdorff dimension between 0 and 1. For the snow crystal, the answer is classic: its Hausdorff dimension equals $\frac{\log 4}{\log 3}$.

The matter is infinitely more difficult for the Apollonian circles; in fact the exact value of its Hausdorff dimension is **unknown**. The profound reason why we know it for snow crystals is that these are *self-similar* fractals, i.e. defined by an iterative procedure involving similitudes, for which we know the ratios for the areas and lengths when we iterate to infinity. In contrast, for the Apollonian circles, it is inversions that are used and that must be iterated, and they transform neither areas nor lengths uniformly. For computer enthusiasts, calculations can aid in formulating a result and they may believe that they need only program the calculations on the right computer. But it is essential to realize that Hausdorff dimension, which requires an infinite process in order to obtain an exact limit, is by its very nature inaccessible to computer calculation to the extent that such calculation is "explicit" and nonmathematical: the computer knows only rational numbers for which the number of significant digits is bounded (certainly the bound depends on the computer, but it is always **finite**). Nonetheless we owe to Sullivan, in a very subtle study, a method for calculating the Hausdorff dimension of $\mathcal{C}$ theoretically, which furnishes the crude estimate of 1.3. The title of the text (Sullivan, 1984) shows how fully up the ladder of our subject it is situated.

II. XYZ

The practical measurement of lengths (distances). Before presenting any mathematical theory, we have the pleasure of giving — for inquisitive minds — some information on the way in which lengths and angles are measured from practical and theoretical points of view. On the one hand because there is mathematics behind the practice, but on the other because, as it seems to us, the practice is most often unknown to professional mathematicians, teachers and researchers. There are roughly three levels of precision to be distinguished: to start that of rulers, the protractors of schoolchildren or graduated straightedges for "ordinary" drawing. Then that of industry and scientific laboratories: in this case the degrees of precision demanded can vary enormously according to the objects to be made. For angles, the scholarly word is *goniometer*. Today the machining of car engines and reactor rotors is performed with increasing precision, but such precision is also needed in scientific laboratories. The exemplary case is that of telescope mirrors. The final level is that of metrology, for which the laboratories furnish on the one hand theoretical results so that measures on the planet might be, as much as possible, the same in different places, and on the other hand instruments utilized by industry. There are objects called *calipers*, both for distances or for angles.

In order to produce straightedges (see also Sect. V.7): defective, approximate rulers are rubbed (with an appropriate abrasive) one against the other, dealing with three of them (with two only we obtain circles); imperfections are further repaired by interferometry or, recently, by laser — imperfections that are then corrected "by hand", with or without a polishing tool, and more and more today with CAD (computer aided design).

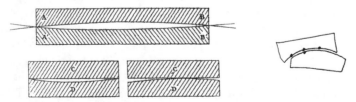

Fig. II.XYZ.1. The curve of contact necessarily being of constant curvature, it's a piece of a circle. With three pieces we therefore end up, in theory, with portions of perfect lines. Bouasse (1917). With kind permission from Delagrave Éditions, public domain

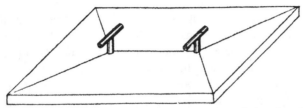

Fig. II.XYZ.2. A "marble" Bouasse (1917). With kind permission from Delagrave
Éditions, public domain

As for planes, we will see in Sect. VI.6 why, when three "defective" planes
are rubbed each against the others, three perfect planes are obtained in the limit.
In practice two things are done. First, proceeding as just indicated in the case of
straightedges, finishing with corrections by hand (since time is limited and reality
isn't theory). In industry, such planes are called *marbles*. To make other planes (or,
as is said, to *prepare* a surface), a "bad plane to be prepared" is rubbed with a
dye against the reference marble and corrected by abrading the colored portions by
hand, repeating until everything finally is just about evenly colored. In metrology,
the "perfect" marble is of a transparent material, for example of the ceramic glass
ZERODUR; it is applied to the future plane and then imperfections are repaired
by interferometry (classic or laser), always with subsequent correction "by hand".
Of course in all these cases the operations are iterated until a satisfactory result is
obtained. Ceramics are employed, in that they are materials for which the coefficient
of thermal expansion is extremely low.

For all these fabrications, the inquisitive mind will pose the question about the
abrasion paradox: "how can we wear down steel with a felt grinder?" The answer
is given on p. 46 of (Bouasse, 1917 # 1728): when two different bodies rub against
each other with interposition of an abrasive, it isn't the harder material that wears
less quickly, but the one that better retains the abrasive. The grains of abrasive, of
emery, find lodging in the felt, the paper, or the cloth. Everything happens as if the
steel were rubbing against a block of emery.

This provides lines, we dare say sufficient for the purposes of affine geometry.
But now we need to measure lengths. The first thing to do is define a unit of length
common to all nations. Progress in this definition has been historically almost simultaneous with that in graduation, the measurement of lengths as a function of
a unit length. Finally we have witnessed a spectacular turn of events. The world
community of metrologists, having made better and better measurements of the
speed of light with rather sophisticated meter standards and using optical means
of measurement based on classical interferometry, then based on fringes for well
determined wave lengths of radiation corresponding to transitions between energy
levels of given atoms — this community has reversed everything in a double fashion
and ended up defining the unit of length (always called *meter*) as follows. In 1963
the meter was equal to $1, 650, 763.73$ wave lengths, in a vacuum, of radiation corresponding to the transition between the levels $2p_{10}$ and $5d_5$ of the krypton 86 atom.
Since 1983: the meter is the length of the path traversed by light in a vacuum during

$1/299,792,458$ seconds. Technically what has made these successively new definitions possible is the appearance of lasers in 1958 and their stabilization in frequency
at molecular transitions. The second (unit of time) is defined in atomic fashion (the
actual precision in the measure of time is known). There remains today the mass,
which still hasn't been defined in an atomic fashion and for which the kilo remains
the mass of an object that is an iridium alloy of platinum and kept at Sèvres (which
still keeps the meter standard, but now only as an historic souvenir). But it is clear
that the atomic definition of unit mass is much studied, but nevertheless still poses
difficulties for physicists. As possible references for the history of the definition of
length, readers can consult (Bouasse, 1917) (whose humor and virulent attacks are
to be taken with caution and forewarning), but above all Bouchareine (1996) for
interferometry, and Giacomo (1995).

Let's return to the practice of the measurement of lengths. The crudest procedure
is with a **graduated** ruler, which always remains difficult, more difficult still as
we seek not only equal graduation (that would be trivial enough) but we need first
to have an accurate basic length, say a meter, and then to divide it. The problem
of the good length encompasses first of all that of expansion, see above. Metals
have the advantage of not rotting, but in return they expand with heat considerably.
This is why the old standard was kept precisely at $4°$ centigrade. Today, just as
for lengths, we have the same for angles (*angular calibration*): the ceramic glass
called ZERODUR mentioned above is used. It is also used now for the mirrors of
large telescopes. Primitive graduations are made by division and by repetition for
verification.

For refining measurement beyond that, a practical device is the *vernier*, whose
theory is closely related to the continued fractions, which are discussed in Sect. IX.1.
This trick permits measurement of a tenth of a millimeter, and even of a twentieth,
by interpolation on a ruler graduated in millimeters. The *micrometer*, which was
(and is) used in average precision industry, permits measurement to better than a
tenth of a millimeter, it uses a screw thread (and the theory of continued fractions,
at the lowest level to be sure). We will come back to the screw thread in a moment.

For the theory of the vernier and that of the micrometer (which is always an
entry point of continued fractions), see again Bouasse (1917). In industry, distances
are now easily measured to the order of a micron. To go to a higher degree of precision in the measurement of length, optics and interferometry are used, which had
fundamentally been done for the first precise measurement of the speed of light by
Michelson in 1881, then still more precisely by Michelson and Morley in 1887. Now
we have come so far that the laser is an everyday object, to the extent that the tracks
of high speed trains are corrected to the ideal curve with the aid of lasers; they have
come to be used likewise for correcting errors in linearity and planarity. The best
performances in the domain of dimensional metrology are obtained by a laser system with fringe counting. The wave length of a laser in a vacuum can be stabilized to
within a few multiples of 10^{-9}, and the measurement of air pressure and temperature
permit determination of wave length in air to several 10^{-7}. These devices can display a tenth of a micron by interpolation on the interior of a fringe. The wave length
exploited is that of the transition of neon (633 nm) and the scrolling of a fringe cor-

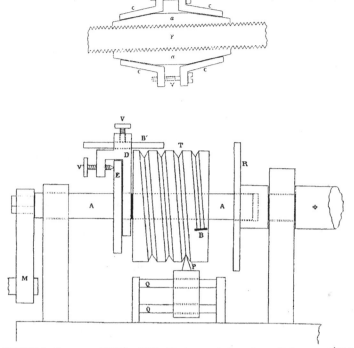

Fig. II.XYZ.3. Bouasse (1917). With kind permission from Delagrave Éditions, public domain

responding to displacement of a half wave length, say $0.316\,\mu$m. Screw threads are used, their role however being limited to that of moving the displacements, which are read either by a scale engraved on rulers with the aid of a microscope or by numerical rulers on which an optical reader counts lines that scroll, with an inter-

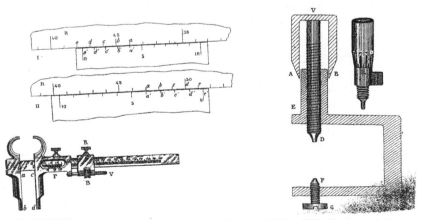

Fig. II.XYZ.4. A caliper and a micrometer. Bouasse (1917). With kind permission from Delagrave Éditions, public domain

polation system between lines. Such a reader can handle about a tenth of a micron, thus a precision of 10^{-7}.

It is important to remark that here we are concerned with the measurement of **absolute** lengths. In contrast, in measuring **variations** of lengths, precisions of a whole other of magnitude are now attained, totally astonishing; specifically, of order 10^{-21}, it is hoped that gravitational waves will soon be detected in this way. Their existence is one of the great problems of contemporary physics, for these waves are predicted by the general theory of relativity. This is astonishing for the following reason: the diameter of the atomic nucleus is of the order of 10^{-9} μm, the diameter of the atom (with its electron cloud) of order 10^{-3} μm only. Well, presently, measurements of 30 m are made with this relative error of 10^{-21}, or thus an absolute error of order $3 . 10^{-14}$ μm.

It seems that there is a contradiction here, since the atomic dimensions are much larger. The explanation is that, if the surface mirror that is used in the experiments is almost perfectly plane or spherical, it will produce some compensations, thus an averaging effect. As for detecting gravitational waves, it's not precision that is lacking presently, but the time for constructing tubes sufficiently accurate and three kilometers long. All these experiments require excellent parabolic mirrors, whose construction is being refined constantly.

The screw thread is a much prized device in dimensional metrology. Making perfect screw threads is just as simple as making perfect spheres (cf. Sect. VII.13), for here again there aren't three but only two objects to adjust one against the other. It suffices to take a long bolt, a long screw thread and a nut − approximations − and to rub them carefully; see e.g. Fig. II.XYZ.3. When all goes well, when they have been rubbed well and when the red powder used leaves a uniform trace on the screw thread, that's what matters for a "perfect" helix. This may be seen in two ways, either that we know that the helices are the only curves of $\mathbb{E}^3$ of constant curvature and torsion (see 8.6 of [BG] for the curves of $\mathbb{E}^3$, about which we have practically no time here to speak explicitly and readily acknowledge that they will need to be broached in Chap. V); or again that the only one-parameter subgroups of the group of motions of $\mathbb{E}^3$ are the helical displacements. Having a perfect helix, we have a correspondence − an exact proportionality − between lengths and angles that had been sought. Nowadays interferometry is used. We point out that, historically, the measurement of the speed of light by Michelson succeeded with a very high precision because, being a very, very, good experimenter, he had made very good sliders and an extremely precise helix. For all this history and more, see the reference Bouasse (1917) which still remains up-to-date for the majority of fabrications of objects with strong geometric constraints. For surface fabrication, and for that of spheres and planes, see VI.6. For the fabrication of balls see Sects. VI.8 and VII.13.C.

The problem of angle measurement is simultaneously easier theoretically but more difficult practically. For what is easy, first there is no problem in tracing circles with a compass. Then, about its center, we are certain, even if the circle itself is badly drawn, to turn 360 degrees. It's not like having to define a unit of length such as the meter. Of course we need to take care to turn steadily while remaining in a plane, which is one of the major difficulties in practical astronomy. Besides, dilation has practically no effect on angles. Nevertheless more and more the ceramic glass ZERODUR is being used for angular calibration. Now the problem is to divide our circle into equal parts to obtain the best possible goniometers. The method is incredibly primitive, but there isn't any other. We divide approximately and correct until reaching the point where, by repeating such and such an angle, the division is visually good. We obtain in this way goniometers with engravings to the order of several seconds. To measure some angle with precision, to be verified to some standard, it has been discovered rather recently that the best thing to do is to make the dozen measurements obtained in turning the object twelve times through 30 degrees and to take the mean. The precision of this mean is in theory several tenths of a *second* (a degree is divided into 60 min and the minute into 60s). Better results are obtained in this way than by the classical least squares method applied to repeated single measurements. If things go well, it's because we have turned through exactly 360 degrees (see however the caveat above). There is also the fact that an error in the center of the goniometer is compensated by the symmetry about the center.

For the rest, the measure of angles today in the best metrological laboratories on the planet don't achieve maximum precision of an order less than that of the second, and again it is necessary to make comparisons between such laboratories. As for measurements made by astronomers and every other profession requiring precision angles, things are much more difficult because of the variation in axes of rotation, etc. We end up at best with an accuracy of order 10^{-6}.

Let us return to "pure" mathematics with an eye only for the formalization of geometry. Recall at first, if necessary, that a *metric space* is a set E endowed with a function d with nonnegative real values that associates with each pair of points x, y of E their *distance* $d(x, y)$, satisfying only three axioms: the *symmetry axiom* $d(x, y) = d(y, x)$ for all x, y of E, the *triangle inequality* $d(x, z) \leq d(x, y) + d(y, z)$ for all x, y, z in E and finally the *separation axiom*: $d(x, y) = 0$ if and only if $x = y$. It's often this last that is most difficult to show when defining a metric space.

Thanks to linear (and multilinear) algebra we nowadays define a *Euclidean vector space of dimension n* simply as a vector space of dimension n endowed with, among other things, a positive definite quadratic form q. We then define the *norm* of a vector v as the square root of the value we get upon applying the quadratic form of the definition, denoted by $\|v\| = \sqrt{q(v, v)}$, the *scalar product* being denoted by $\langle v, w \rangle$ or $v \cdot w$. The *angle* α (between 0 and π) between two nonzero vectors v and w is defined by $\cos \alpha = v \cdot w / \|v\| \cdot \|w\|$. The Euclidean space itself is the associated affine space whose subspaces (lines, planes, hyperplanes) are defined. We have in addition a *metric d* defined by $d(p, q) = \|q - p\|$. We note the *Euclidean heredity*:

all the subspaces (of whatever dimension) of a Euclidean space are automatically endowed with a canonical Euclidean structure, inherited trivially. All the Euclidean spaces of the same dimension n being isomorphic to one another, we often speak of *the* Euclidean space of dimension n and use the notation $\mathbb{E}^n$ (or rigorously $\mathbb{R}^n$).

It is often of interest to identify the Euclidean plane $\mathbb{E}^2$ with the field $\mathbb{C}$ of complex numbers; to do this we choose an orthonormal basis $\{0, v, w\}$ (a point and two orthogonal unitary vectors). Then the complex number associated with the point with coordinates (x, y) in this basis is the complex number $z = x + iy$. The (direct, i.e. orientation preserving) similitudes with center O are represented by the operation $z \mapsto kz$ (k an arbitrary nonzero complex number), and if we want to change the orientation (to find the *inverse similitudes*) it is necessary to consider the $z \mapsto k\bar{z}$. In particular, multiplication by i is nothing other than rotation about the origin (or vectorial) through an angle $\frac{\pi}{2}$.

Poles and polars with respect to a circle. Let C be the circle with center O and radius R. We say that two points of the plane m and m' are *conjugate* with respect to C if

$$\overrightarrow{Om} \cdot \overrightarrow{Om'} = R^2.$$

The points of the circle C are those that are conjugate to themselves. If m isn't the center of C, the set of points conjugate to m is a line perpendicular to Om, situated a distance $R^2/d(O, m)$ from O. We call it the line *polar* to m, or simply the polar of m, denoted m^*. If m is interior to C, then m^* doesn't intersect C, if m is exterior, then m^* intersects C, and if m is a point of C, then m^* is the tangent to C at the point m. These instructions suffice to give a construction of the polar of a point m exterior to C: it's the line that joins the points of contact of the tangents leading to m.

The mapping $m \mapsto m^*$ possesses all the properties of a duality if we agree to eliminate the point O from the plane $\mathscr{P}$ and the lines passing through O from the set of lines of $\mathscr{P}$. First of all, it's a bijection. In particular, for a line d not passing through O, the unique point d^* for which d is the polar is called the *pole* of the

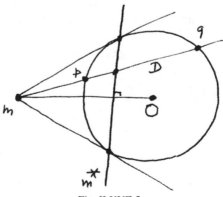

Fig. II.XYZ.5.

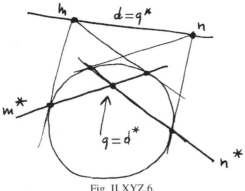

Fig. II.XYZ.6.

line. Next, it transforms collinear points into intersecting lines. Moreover, when the point m describes a line d (see the figure below), its polar m^* describes, with a single exception, the pencil of lines passing through d^*, denoted again d^*. This correspondence between d and d^* is homographic, i.e. it preserves the cross ratio.

A second way of defining the conjugate points will lead to another definition of the polar: two points m and m' of the plane are conjugate with respect to C if the points of intersection p and q of the line form with m and m' an harmonic division: $[m, m', p, q] = -1$. The points p and q can be complex or identical (the case where mm' is tangent to C and $m = m'$ is the point of contact).

A rather simple calculation allows us to show the equivalence with the first definition, but the deep reason is as follows: if we are restricted to a line D on which an abscissa x is chosen, the relation $\overrightarrow{Om} \cdot \overrightarrow{Om'} = R^2$ takes on the (symmetric) form

$$axx' + b(x + x') + c = 0.$$

We solve for x' as a homographic function of x. This homography is an involution that admits the abscissas of the points p and q as fixed points (since they are solutions of the equation $ax^2 + 2bx = c = 0$). By expressing that they preserve the cross ration we obtain the relation $[m, m', p, q] = [m', m, p, q]$, from which we deduce $[m, m', p, q] = -1$; see Sect. I.6.

To conclude, we indicate two other ways of defining conjugation with respect to a circle, valid in all circumstances, and which avoid, for those who care, recourse to the scalar product and to complex numbers.

- Two points m and m' are conjugate with respect to the circle C if the circle with diameter mm' is orthogonal to C.

- In the figure below, the points m and m' are conjugate with respect to C.

To see this, we complete the figure (see Fig. II.XYZ.8): the point α is conjugate to the point m with respect to the pair of lines D and D' (figure of the complete quadrilateral). It is thus the conjugate harmonic of m with respect to the points of intersection of the line $m\alpha$ with D and D', which are also the points of intersection

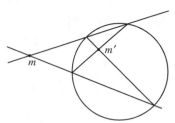

Fig. II.XYZ.7. Construction of a conjugate point (we can begin either with m, or with m')

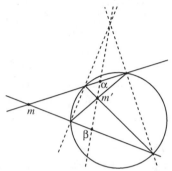

Fig. II.XYZ.8. Justification of the construction of Fig. II.XYZ.7

of this line with C. In other words, α is the conjugate of m with respect to the circle C, and it similarly for β, which shows that the polar of m with respect to the circle C is the line $\alpha\beta$, and in particular that this polar passes through m'.

It remains to dispose of the exceptions in the pole-polar correspondence. This is done by passing to the projective plane and by giving the definition of conjugation in homogeneous coordinates.

The equation of the circle is given in homogeneous coordinates by setting a quadratic form equal to zero:

$$Q(x, y, z) = x^2 + y^2 - R^2 z^2 = 0$$

if the coordinates are well chosen, but the left hand side of the equation remains a quadratic form in any coordinate system. In this case, the points $(x : y : z)$ and $(x' : y' : z')$ are conjugate with respect to C if we have

$$xx' + yy' - R^2 zz' = 0.$$

For points at a finite distance, we verify immediately that this relation is equivalent to the relation $\overrightarrow{Om} \cdot \overrightarrow{Om'} = R^2$. In general two points m and m' of the projective plane are conjugate with respect to C if the corresponding triples $(x : y : z)$ and $(x' : y' : z')$ are *orthogonal* (we also say *conjugate*) with respect to the bilinear form B associated with Q. Frequently B is called the *polar* form of Q, and in coordinates it is obtained by *polarizing* the quadratic form: a square term z^2 gives zz' and a rectangular term yz gives $\frac{1}{2}(yz'+y'z)$. The coincidence of vocabularies is evidently not a coincidence.

We can now associate a polar with the point O: it's the line at infinity, and the poles of the lines passing through O: these are the points at infinity situated in the perpendicular direction. We thus have a perfect duality between $\mathscr{P}$ and the set $\mathscr{P}^*$ of lines of $\mathscr{P}$.

It is clear that, in the form that we have just given it, the theory applies to an arbitrary conic; see Chap. IV. For the mapping $m \mapsto m^*$ to be bijective, it is clearly necessary that the quadratic form Q be nondegenerate, but the mapping is interesting even when Q is degenerate.

In the case where Q is only of rank 2, i.e. where Q is the product of two equations of distinct lines D and D', the polar of a point m with respect to the conic is well defined if m is distinct from the point a of intersection of D and D', and it is a line passing through a: i.e. the locus of the points m' of the plane such that $[m, m', p, q] = -1$, where p and q denote the points of intersection of the line mm' with D and D'. We thus recover the classic notion of polar with respect to two lines.

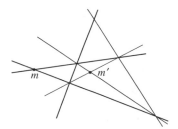

Fig. II.XYZ.9. Conjugate points with respect to two lines

Harmonic homology. Homology comes up in the proof of the butterfly theorem; see Fig. II.2.1. Let D be a line of the projective plane $\mathscr{P}$, and a a point not belonging to D. The harmonic homology with center a and axis D is the mapping h that associates with a point m of the plane distinct from a and not belonging to D the harmonic conjugate of m with respect to the point of intersection of the line am with D, in other words the unique point m' collinear with a and m such that $[a, p, m, m'] = -1$, where p is the point of intersection of the line am with D. According to the usual conventions regarding the cross ration, we have $[a, m, m, m] = -1$, $[a, a, a, m] = -1$ on the line am, provided that m is a distinct point of a, which allows extending h to D and to the point a: h leaves the point a and all the points of D invariant. An extension by continuity will have the same effect.

If $\mathscr{P}$ denotes the projectified affine plane, and if a is a point at infinity, the harmonic homology with center a and axis D is the (oblique) symmetry with respect to D, parallel to the direction of a. If a is at finite distance and if D is the line at infinity, the harmonic homology is the symmetry with center a. These examples show, without need for an additional calculation, that the general harmonic homology is a homography.

Pencils of circles. Let C be a circle in the plane with center O and radius R; moreover let m be a point of the plane and D be a line passing through m and intersecting C at the points p and q. Then the product $\overline{mp} \cdot \overline{mq}$ is independent of the line D chosen. The value of this product is called the *power of the point m with respect to the circle* C, and is denoted by $p_C(m)$. If m is exterior to the circle, then $p_C(m)$ is the square of the length of the tangents leading from m to C (which is called the *tangential distance* of m to the circle). In general we have $p_C(m) = Om^2 - R^2$, and $p_C(m) = 0$ if and only if m is a point of the circle.

In coordinates, $p_C(m)$ is nothing other than the first term of the equation of the circle, provided that this term is normalized so as to begin with $x^2 + y^2$.

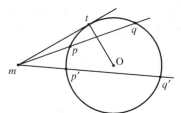

Fig. II.XYZ.10. Power of a point with respect to a circle: $\overline{mp} \cdot \overline{mq} = \overline{mp'} \cdot \overline{mq'} = mt^2 = Om^2 - R^2$

We call the line defined by the equation $p_C(m) = p_{C'}(m)$ the *radical axis* of the two circles C and C'. If the circles intersect in two distinct points a and b, their radical axis is the line ab; if the circles don't intersect, the radical axis is the line that passes through the centers of their common tangents; if the circles are tangent, the radical axis is the common tangent.

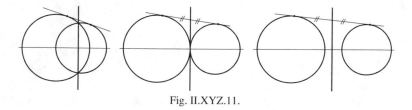

Fig. II.XYZ.11.

The *pencil of circles* defined by two circles C and C' is the set of circles given by the equation

$$\lambda p_C(m) + \lambda' p_{C'}(m) = 0.$$

For $\lambda = -\lambda'$, we obtain not a circle but a line: the radical axis. We can normalize the above equation by writing $p_C(m) + \alpha p_{C'}(m) = 0$, but we then lack the circle C', for which we must stipulate that it is obtained for $\alpha = \infty$. We can also normalize it by writing $\beta p_C(m) + (1 - \beta) p_{C'}(m) = 0$, but we must stipulate in this case that the radical axis is obtained for $\beta = \infty$. For β finite, the first member of the equation has for higher degree terms $x^2 + y^2$ and thus represents the power of m with respect to the circle $C(\beta)$ with parameter β. Two properties are thus evident: the points of the radical axis have the same power (or the same tangential distance) with respect to all the circles of the pencil; the ratio of powers of a point of a given circle of the pencil of circles with basis C and C′ is independent of the point chosen.

When the circles C and C′ intersect, the pencil defined by C and C′ is the set of circles passing through the points of intersection a and b of C and C′ (see the figures taken from [B]). When C and C′ don't intersect, the pencil corresponds to two circles of radius zero situated on the line of the centers. When C and C′ are tangent at a point a, the pencil is the set of circles tangent at a to the common tangent of C and C′. We note that in all these cases, the radical axis of two circles of the pencil is the same as that of C and C′.

We can now forget the two base circles C and C′ and give the classification of the pencils:

- *pencil with two base points:* the set of circles passing through two given distinct points a and b;

- *pencil with two limit points:* the set of circles defined by an equation of the form $mp/mq = $ const., where p and q are two given points;

- *tangent pencil:* the set of circles tangent to a given line D at a given point a.

Completing the plane by a line at infinity and introducing the cyclic point may clarify the situation to a certain extent: a pencil of generic conics is defined as the set of conics passing through four given points a, b, c, d; it entails three degenerate conics: the pairs of lines ab and cd, ac and bd, ad and bc.

The pencil with base points a and b is simply the pencil of conics passing through the four points a, b, I, J. Of the three degenerate conics that it contains, two are complex, but the third is the union of the radical axis and the line at infinity, which is a degenerate circle, as it contains the line at infinity.

In the case of a pencil with limit points p and q, the points common to all the circles of the pencil aren't interesting, but the three degenerate circles are clearly visible: first that formed by the radical axis and the line at infinity, then the circles p and q of radius zero, which can be considered as being the union of lines of slope i and $-i$ (isotropic lines).

Here is another way of (partially) clarifying the situation: the space $\mathcal{C}$ of circles is a real projective space of dimension 3 (see toward the end of XYZ), and a circle pencil $\mathcal{F}$ is a line in the space of circles, whereas the set of circles of radius zero is a quadratic in this space. For the circle with equation $x^2 + y^2 - 2ax - 2by + c = 0$, we have $R^2 = a^2 + b^2 - c$; it's a quadratic form if we pass to homogeneous coordinates. There are thus three possible situations: $\mathcal{F}$ doesn't intersect this quadric, $\mathcal{F}$ intersects this quadric in two points, $\mathcal{F}$ is tangent to this quadric.

A necessary and sufficient condition for two circles C and C′ with centers O and O′ and radii R and R′ to be orthogonal is that $R^2 + R'^2 - OO'^2 = 0$. What is remarkable is that the quantity $OO'^2 - R^2 - R'^2$, within a factor of 2, is the polar form associated with the quadratic R^2, and which is expressed in bilinear fashion with respect to the equations of two circles. In fact, if these are written respectively

$$x^2 + y^2 - 2ax - 2by + c = 0$$
$$x^2 + y^2 - 2a'x - 2b'y + c' = 0$$

we have
$$OO'^2 - R^2 - R'^2 = 2aa' + 2bb' - c - c'.$$

We can call the quantity $2aa' + 2bb' - c - c'$ the *scalar product* of the circles C and C'; and two circles are orthogonal if they are, in the space of circles, conjugate with respect to the quadric of circles of radius zero. Further, the set of circles orthogonal to two given circles is a line in the space of circles, i.e. a pencil of circles, and each pencil can be defined in this way. In particular, the set of circles orthogonal to the circles of a pencil is a pencil − the pencil with limit points a and b when the given pencil is the pencil with base points a and b, and conversely.

This is particularly interesting when we complete the plane with a single point at infinity, in the context of conformal geometry. First, the radical axis becomes a circle like the others, provided that we adjoin the point at infinity. But above all, the fact that a pencil can be defined as the set of circles orthogonal to two given circles

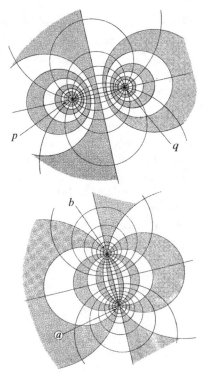

Fig. II.XYZ.12. Pencil of limit points (alternation of the *white* and the *gray*) and its orthogonal pencil. Pencil of base points and its orthogonal pencil

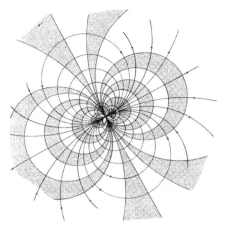

Fig. II.XYZ.13. Tangent pencil

provides an immediate proof of the fact that the image by an inversion of the pencil of circles is again a circle pencil — which isn't at all evident from the calculation in the case where this is necessary, i.e. in the case of limit point pencil.

A scandal to repair. In all this we have neither defined a *circle* nor, a fortiori, a *sphere*. Recall that a sphere of radius r and center O in a Euclidean space is the set $S(0; r)$ formed of points situated a distance r from the point O, and which we call a *circle* in the case of a plane. Let us speak primarily of the plane and of circles and the problem of having tools for studying different figures, like those of Sect. II.2. We first remark that three noncollinear points, forming a true triangle, have a well defined, and unique, circumscribed circle. Therefore: four points being given, when are they on a circle? We saw in Sect. II.2 a condition that is necessary and sufficient, based on angles. But we are right to expect a purely metric condition that allows intervention of the six mutual distances that separate the four points. In fact, the notion of angle is not primary, and we have seen that it is necessary to be very sophisticated if we want to include at the same time the case of the convex quadrilaterals and ones that cross.

Now such a condition is purely metric, non angular, and almost never taught in any school or university curriculum. That is a scandal, to be repaired, even though there is a dazzling excuse: it's practically impossible to prove the overwhelming majority of theorems on circles using this relation. Here it is, however; it's called *Ptolemy's theorem* and states that, for the four points a, b, c, d to be on a circle, it is necessary and sufficient that one of the four relations $ab \cdot cd \pm ac \cdot db \pm ad \cdot bc = 0$ between the six distances in question be satisfied. We utilize this expression so as not to have to specify whether the quadrilateral is convex, in which case the relation is very pretty: "the product of the diagonals is equal to the sum of the products of the opposite sides". A perverse but miraculous proof uses angles; for a purely metric proof and its extension to arbitrary dimensions (five point on a sphere in ordinary space, etc.), see 9.7 of [B] or the great classic (Blumenthal, 1970). All the relations are written in the language of determinants. If we apply Ptolemy's theorem when two points are the ends of a diameter, we find the relation that gives the sine of the sum of two angles; in fact, Ptolemy used his theorem for constructing tables of the sine function (or the equivalent at his time: tables of "chords".: $\operatorname{crd} \alpha = 2 \sin \alpha/2$)

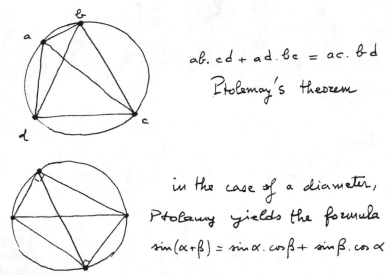

$$ab.\,cd + ad.\,bc = ac.\,bd$$

Ptolemay's theorem

in the case of a diameter,

Ptolemy yields the formula

$$\sin(\alpha+\beta) = \sin\alpha.\cos\beta + \sin\beta.\cos\alpha$$

Fig. II.XYZ.14. Ptolemy's theorem: $ab\bar{c}d + ad \cdot bc = ac \cdot bd$. In particular, in the case of a diameter, we find the formula $\sin(\alpha + \beta) = \sin \alpha \cos \beta + \sin \beta \cos \alpha$ by using the basic fact: in a triangle the ratio of a side to the sine of the opposite angle is always equal to the diameter of the circumscribed circle

It's also nice to know that the six distances of four (arbitrary) points aren't independent, as can be seen in the references and as one might discover from a plane figure, for when five distances are given there are just two possibilities for the sixth, since three sides determine a triangle.

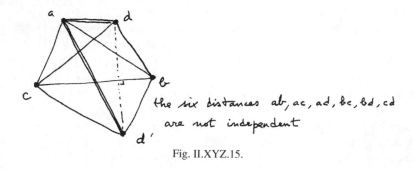

the six distances ab, ac, ad, bc, bd, cd are not independent

Fig. II.XYZ.15.

It can be shown that there exists a universal relation between the six distances of each quadruple of points in the Euclidean plane but this is, contrary to that of

Ptolemy, of second degree in each distance; see once again 9.7 of [B]. This relation is almost *unwritable* by hand, without the language of determinants. Interested readers might ask if there exist other geometries (of given dimension n) where the $(n + 2)(n + 1)/2$ mutual distances of each $(n + 2)$-tuple of points are always connected by a universal relation. In fact, there aren't any such geometries other than Euclidean, spherical and hyperbolic; see Berger (1981). These Euclidean relations don't only serve the interests of geometers, but are used in particular in celestial mechanics; see Albouy and Chenciner (1998).

♦

The *isometries* of a Euclidean space are the mappings that preserve distances. In the case of the plane we have seen that these are the rotations about a point, the translations and the symmetry-translations, specifically compositions of symmetry about a line and a translation in the direction of this line. Every isometry is a product of at most three symmetries with respect to lines, symmetries that thus form a pleasing (and useful) set of generators. The classification of isometries of Euclidean spaces of arbitrary dimension is given in [B]; Theorem 9.3.1 is taken from Frenkel (1973), where it seems to appear for the first time in general and precise form. The symmetries with respect to hyperplanes remain generators, a much more general result; see Chap. 13 of [B] and the references mentioned there. Note the separation of isometries into two classes, according to whether or not they preserve *orientation*; see Sect. 2.7.2 of [B] for this notion.

The similarity transformations (similitudes or similarities for short) of Euclidean spaces are the transformations that multiply distances by a fixed ratio, called the *similitude ratio* or *aspect ratio* of the similarity. They differ from isometries, in the main, only by the simplest similarities, specifically the *homotheties* (see Sect. I.XYZ). A homothety of center O and ratio the real number λ is the mapping that, when the affine space is vectorialized about the point O, is written $v \mapsto \lambda v$. Among the bijections, the similarities are characterized, in arbitrary dimension, by the following equivalent conditions: they preserve spheres, preserve angles, preserve right angles, see 9.5 and 9.6 of [B]. The initial idea is to arrive at using the fundamental theorem of affine geometry (of Sect. I.3).

♦

In Sect. II.2 we promised readers that we would "retrieve" Euclidean geometry with the help of projective geometry; here, briefly, is how we can achieve a good bit of this, which we explain in the planar case. We have seen in Sect. I.7 how to retrieve the affine invariant $\frac{ac}{ab}$ of three points arranged on a line F: its value is nothing other than the cross ration $[c, b, a, \infty_F]$, where ∞_F of course denotes the point at infinity of the line F, i.e. $\infty_F = D \cap F$ when the affine plane considered has been projectified. We can calculate ratios of distances, so it thus now suffices to know how to calculate angles in order to control all of Euclidean geometry. For that, we need to do two things: first, complexify our (Euclidean) real affine $\mathbb{R}^2$ in the complex affine plane $\mathbb{C}^2$; we thus embed $\mathbb{R}^2$ trivially in $\mathbb{C}^2$. We could do the complexification more intrinsically, but then much more abstractly; but this is too complicated for the goal pursued; see Chap. 7 of [B]. Then we projectify $\mathbb{C}^2$ in $\mathbb{C}P^2$. The essential idea is that the Euclidean structure (to within a change of scale) of $\mathbb{R}^2$ is *coded* in the pair $\{I, J\}$ of points of $\mathbb{C}P^2$, called *cyclic points* of $\mathbb{R}^2$, and which are, in **complex** projective coordinates, the two points at infinity $\{(0, 1, i), (0, 1, -i)\}$ (here i is the square root of -1, that which in real coordinates is written $(0, 1)$) in $\mathbb{C}$. Be aware that the real dimension of $\mathbb{C}^2$, like that of $\mathbb{C}P^2$, is equal to 4.

An important remark: we cannot choose any one of the cyclic points in a canonical fashion; it necessary to consider the *pair*, to the extent that we only know the Euclidean structure of the plane considered, identified in fact with $\mathbb{R}^2$ by an "unknown" isometry; thus we can't say that we know the point $(0, 1, i)$, for we know only the quadratic form and not its expression in coordinates. However, interested readers can show that we can specify the cyclic points in a *pair* $\{I, J\}$ if the Euclidean plane is, in addition, *oriented*; see 8.8.7 and 8.5.1 of [B] for this and what follows.

Briefly, the notion of cross ratio for 4 points of $\mathbb{C}$ (or of $S^2 = \mathbb{C}P^1$) furnishes a nice characterization of the cocyclicity of points: they are cocyclic (or collinear) if and only their cross ratio is **real**.

Now we have hope of retrieving Euclidean geometry, i.e. the quadratic form that defines it, for this form is $x^2 + y^2$. Now, in $\mathbb{C}^2$, the equation $x^2 + y^2 = 0$ defines a conic (see Chap. IV), here degenerated into the pair of lines $x \pm iy = 0$. We can in an equivalent fashion replace consideration of cyclic points by that of these two lines, called *isotropic lines*; we denote the pair of them by $\{I^*, J^*\}$. Their points at infinity are exactly the cyclic points. This accomplished, all that remains is to give the formula, due to Laguerre:

$$\text{angle}(D, D') = \frac{1}{2}\left|\log\left([\infty_D, \infty_{D'}, I, J]\right)\right| = \frac{1}{2}\left|\log\left([D, D', I^*, J^*]\right)\right|,$$

where the angle(D, D') of two lines D, D' passing through the origin of $\mathbb{R}^2$ is expressed with the help of the cross ratio, with a choice either of four points ∞_D, $\infty_{D'}$, I, J of the line at infinity of $\mathbb{C}P^2$, or of four lines D, D', I^*, J^* passing through the origin. The young Laguerre was still a student in the preparatory class when he concocted this formula, not being really satisfied with his professor's course. The proof today is banal, it utilizes only the fact that the complex exponential e^{it} equals $\cos t + i \cdot \sin t$. We are thus finished with the problem. We will see in Sect. IV.7 how the whole "elementary" theory of the Euclidean conics can be treated with the considerations above.

In the geometry of space, the role of the cyclic points is played by the *umbilical*. In Euclidean projective space, it's the curve with homogeneous equations $x^2 + y^2 + z^2 = 0, t = 0$. It collects the cyclic points of all the planes of space (two parallel planes have the same cyclic points) and, for a non degenerate quadric to be a sphere, it is necessary and sufficient that it contain the umbilical. A torus of revolution contains the umbilical as a double line (curve of self intersection); we have used this observation apropos Vilarceau circles.

The umbilical plays a large role in the modern theory of stereoscopic vision and its applications to computer vision (see the work of Olivier Faugeras on *epipolar geometry*).

♦

To now make clear the nature of *hyperbolic geometry* of dimension n, *denoted* Hyp^n, we follow Chap. 19 of [B], i.e. the four models of hyperbolic geometry, to which we adjoin a fifth, the Riemannian model, that was not presented in [B] since it requires the notion of Riemannian manifold. Other references are mentioned in [B], but after the appearance of numerous works on geometry, we remain very fond of Benedetti and Petronio (1992) and Ratcliffe (1994), since both go very high on the ladder.

The most efficient model for exhibiting *hyperbolic geometry in any dimension* n is the projective model (also called *linear* or *Klein's model*), where the price for rapidity that must be paid is knowledge of the notion of projective space, always difficult to visualize, since it requires an increase by one in dimension. We are placed in $\mathbb{R}^{n+1}$, where the starting point is no longer a Euclidean structure but that defined by the quadratic form $q = -x_1^2 - \cdots - x_n^2 + x_{n+1}^2$. For what follows, carefully examine the figures below. The basic geometric object is the cone $Q = q^{-1}(0)$; its interior consists of points where q is negative. But we now need to consider it in the projective space $\mathbb{R}P^n$ associated with $\mathbb{R}^{n+1}$. We find two things: the projections of the points where q is negative, the set of which forms an open ball $\mathcal{B}_n$ of dimension n, and the projection of the cone Q. We can, for example, obtain this ball (a disk when $n = 3$) as the interior of the cone cut by the hyperplane $x_{n+1} = 1$. For what follows − the space of spheres − it is well to remark that in $\mathbb{R}P^n$ the projection of the cone is a projective quadric (see Chap. IV) for which the topology is that of a sphere S^{n-1}, since it is the boundary of the ball $\mathcal{B}_n$.

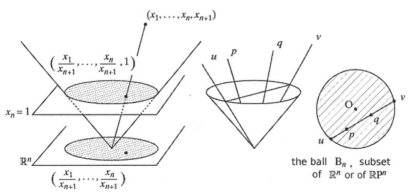

Fig. II.XYZ.16. II [B] Géométrie. Nathan (1977, 1990) réimp.
Cassini (2009) © Nathan Édition

now remains to define the metric $d(p, q)$ on Hyp^n: it is equal to the quantity $\frac{1}{2}|\log([p, q, u, v])|$, where $[p, q, u, v]$ denotes the cross ratio of the four points considered in the projective space $\mathbb{R}P^n$. Here p and q are seen in the projective and $\{u, v\}$ is by definition the pair of points of the boundary of $\mathcal{B}_n$ where it is cut by the projective line joining p with q. We might prefer to see the cross ratio as that of four lines D, D′, U, V passing through the origin associated with the four points considered. With the quadratic form q and its associated polar form P (i.e. the analog of the scalar product in the Euclidean case), we can calculate $d(p,q)$ by the formula utilizing the hyperbolic cosine:

$$\operatorname{ch}(d(p,q)) = \frac{P(\xi,\eta)}{\sqrt{q(\xi)q(\eta)}},$$

where ξ and η are any two representatives of p and q in $\mathbb{R}^{n+1}$. In Hyp^n, by definition, the subspaces (lines, planes, hyperplanes) are those of $\mathbb{R}P^n$ **restricted** to Hyp^n. The lines are effectively those that attain minimum length among curves joining two points.

By construction, all these linear mappings of $\mathbb{R}^{n+1}$ that preserve q, lowered into $\mathbb{R}P^n$, are isometries of Hyp^n, and it is easy to see that there aren't any others. We *denote* this group Isom(Hyp^n). Here it appears in linear form in each dimension, more precisely in linear projective form. But if interpreted on the cone, then the projective quadric S^{n-1} is the Möbius group Möb(S^{n-1}). This is the projective version, let us say linear, of hyperbolic geometry. We can say it is linear because the lines, the subspaces, are the same as in the affine geometry of the unit ball embedded in $\mathbb{R}^n$. But of course the angles are no longer the same as in the canonical Euclidean geometry of $\mathbb{R}^n$.

The presentation we saw in Sects. II.4 and II.5, in contrast to the preceding, does not conserve lines but does conserve angles. The lines are pieces of circles orthogonal to the boundary. It is rightly called the *conformal model* or the *Poincaré model*. We can pass from the linear model to the conformal model by the inverse of a stereographic projection followed by an orthogonal Euclidean projection, as can be seen in the figure below:

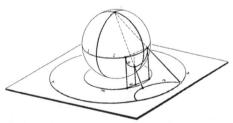

Fig. II.XYZ.17. Correspondence between the Poincaré's disk model and the linear model Hilbert, Cohn-Vossen (1996) © Springer

The *Poincaré's half-space* model consists simply in transforming this conformal model by an inversion through a center situated on the spherical boundary of the conformal model. The interior of the ball is thus transformed into an open half-space. This model allows very simple calculations for certain problems, in particular in the dimension 2 case; see Sect. II.4. The lines are the semicircles of the plane that are orthogonal to the boundary line of the half-space.

The quickest and simplest definition is that of Hyp^n as a Riemannian model. Its drawback is being nonelementary, i.e. it requires knowledge of the basics of

Riemannian geometry dealt with in Chap. III; also see Berger (2003) or any book on Riemannian geometry. It consists of saying that the underlying set is the upper sheet of the hyperboloid (see Sect. IV.10) defined by the equation $q = 1$, i.e. $x_{n+1} = \sqrt{1 + x_1^2 + \cdots + x_n^2}$. It's a hypersurface of $\mathbb{R}^{n+1}$ for which the tangent space, translated to the origin, never intersects the cone Q and thus on which the restriction of the quadratic form q is positive definite. By definition, this makes the sheet of the hyperboloid into a Riemannian manifold. A Riemannian geometry in fact consists of putting a Euclidean geometry (infinitesimal thus) on each tangent space. Its Riemannian geometry is nothing other than that of Hyp^n, which can be seen, for example, from the group of isometries being the same. For more on hyperbolic geometry, see the references mentioned.

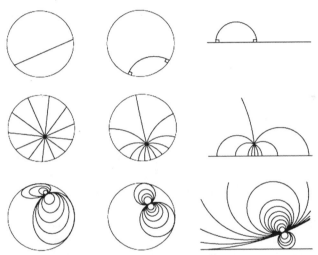

Fig. II.XYZ.18. *Three versions of hyperbolic geometry.* From *left* to *right*: Klein's linear model, Poincaré's disk model, Poincaré's half-plane model. From *above* to *below*: a line, a pencil of lines passing through a point, a family of circles passing through a point and tangent at that point to a given fixed direction. We note the *horicycles*, circles for which the center has been moved to infinity: they are represented by circles tangent to the horizon in the two Poincaré models and by ellipses that are surosculating to the horizon in the Klein model

♦

In Sect. II.7, we encountered the space $\mathcal{C}$ of circles of the Euclidean plane and the space $\mathcal{S}$ of spheres of the Euclidean space of dimension 3. As a reference for what follows, in modern language, we know only Chap. 19 of [B]. To define them correctly (things are the same in each dimension) we will thus construct the space $\mathcal{S}^n$ of spheres of the space $\mathbb{E}^n$ (we might say hypersphere beginning with $n = 4$, as we say hyperplane, but the term sphere is presently most common). The idea is to describe a sphere in $\mathbb{R}^{n+1}$ by its equation, which may always be written in the

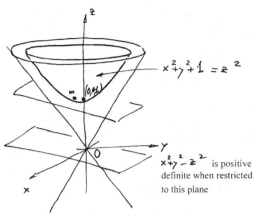

Fig. II.XYZ.19. The quadratic form $x^2 + y^2 - z^2$ restricted to the plane is clearly positive definite

form $x_1^2 + \cdots + x_n^2 + b_1 x_1 + \cdots + b_n x_n + c = 0$. But if we wish to calculate, to add, we can't do it, the form changes. The idea is to write the form slightly more generally, specifically $a(x_1^2 + \cdots + x_n^2) + b_1 x_1 + \cdots + b_n x_n + c = 0$; but then two proportional expressions give the same sphere. So that this doesn't happen, we pass to projective space! Here it is thus a $\mathbb{R}P^{n+1}$ resulting from $\mathbb{R}^{n+2}$, defined by the $n + 2$ coordinates $(a, b_1, \ldots, b_n, c)$. In taking all $\mathbb{R}^{n+2}$ we have appended to the "true" spheres the object of the equation $1 = 0$, which is therefore empty but that we interpret as **the** *sphere of radius zero at infinity*, which we like, for we have seen that it interposed too the objects where $a = 0$, which are the planes, and we know that they must be introduced at the same time as spheres in conformal geometry. To have a geometry, and hopefully a distance, on $\mathcal{S}^n = \mathbb{R}P^{n+1}$, the idea is to introduce a quadratic form, which is in the disguise of the *square of the radius*. We define ρ on $\mathbb{R}^{n+2}$ above by $\rho((a, b_1, \ldots, b_n, c)) = \frac{1}{4}(b_1^2 + \cdots + b_n^2 - 4ac)$. But, within a trivial change of coordinates, this quadratic form r is the same as the quadratic form q of hyperbolic geometry (we say that they are of signature $(1, n + 1)$, by having in diagonal form a single positive square and $n + 1$ negative squares). We let $\mathcal{S}^n$ denote the projective space $\mathbb{R}P^{n+1}$ endowed with the quadratic form ρ. The group of this geometry, called *generalized spheres*, denoted here Möb(S^n), is thus the same as Isom(Hyp^n) seen above. We now see why this answers practically all the questions about spheres, see [B] for the details.

We can say that the geometry of the sphere is an extension of hyperbolic geometry. First, there is the hyperbolic geometry in $\mathcal{S}^n$; these are spheres for which the square of the radius is **négative**. We can say that the radius is a pure imaginary, i.e. that ρ is negative. The spheres of radius zero are the points of the space to which one adds the point at infinity; the totality makes up S^n. The true spheres are those for which ρ is positive, but there are also the hyperplanes on which ρ is positive. We can see geometrically where the hyperplanes in $\mathcal{S}^n$ are: these are the points of the hyperplane tangent to $\mathcal{S}^n$ at its point at infinity. The space of the true spheres is thus difficult to "see", for its topology is that of $\mathbb{R}P^{n+1}$ from which we have removed an open ball; it's what remains when we have removed the hyperbolic geometry, a sort of complement of this geometry. As a quadric, not yet projectified in $\mathbb{R}^{n+2}$, it's that of the hyperboloid of *one* sheet with equation $\rho = 1$ (see Sect. IV.10). It is immediate, for example, that the (generalized) circles are represented, in $\mathcal{S}^3$, by the projective lines (made up of the set of spheres that contain that circle). We do the same with all the k-spheres and the projective subspaces of all dimensions. Finally, the angle of two spheres plays a role analogous to that of the hyperbolic metric; if R is the polar form of ρ, then we define the *angle* α of two spheres s and s' as the real number in $[0, \pi/2]$ such that $\cos \alpha = \frac{|R(s,s')|}{\sqrt{\rho(s)\rho(s')}}$. This angle doesn't alway exist, but when it does, i.e. if in fact the spheres intersect, it is of course the angle they make in the ordinary sense.

The geometry of oriented spheres consists simply — instead of considering $\mathcal{S}^n = \mathbb{R}P^{n+1}$, where $\mathbb{R}P^{n+1}$ is the set of vectorial lines of $\mathbb{R}^{n+2}$ — of taking the space $(\mathcal{S}^n)^*$ of *half-lines*, i.e. the quotient of $\mathbb{R}^{n+2}$ by the equivalence relation $v \equiv w$ if $w = kv$, with k a positive real number. We thus obtain a space for which the topology is that of the sphere S^{n+1}. See an alternative presentation in the article on conformal geometry in the Encyclopedic Dictionary (1985).

Bibliography

[B] Berger, M. (1987, 2009). *Geometry I,II*. Berlin/Heidelberg/New York: Springer

[BG] Berger, M., & Gostiaux, B. (1987). *Differential geometry: Manifolds, curves and surfaces*. Berlin/Heidelberg/New York: Springer

Aigner, M., & Ziegler, G. (1998). *Proofs from THE BOOK*. Berlin/Heidelberg/New York: Springer

Albouy, A., & Chenciner, A. (1998). Le problème des n corps et les distances mutuelles. *Inventiones Mathematicae, 131*, 151–184

Apéry, F. (1982). La baderne d'Apollonius. *Gazette de la Société math. France, n° 19* (juin), 57–86

Benedetti, R., & Petronio, C. (1992). *Lectures on hyperbolic geometry*. Berlin/Heidelberg/New York: Springer

Berger, M. (1981). Une caractérisation purement métrique des variétés riemanniennes à courbure constante. In P. Butzer & F. Fehér (Eds.), *E. B. Christoffel* (pp. 480–492). Basel: Birkhäuser

Berger, M. (1993). Les paquets de cercles. In C. E. Tricerri (Ed.), *Differential geometry and topology*. Singapore: World Scientific

Berger, M. (2002). *Les cercles de Villarceau*. Pour la Science, n°292 (février 2002), 90–91

Berger, M. (2003). *A panoramic view of Riemannian geometry*. Berlin/Heidelberg/New York: Springer

Blumenthal, L. (1970). *Theory and applications of distance geometry*. New York: Chelsea

Bobenko, A. & Hoffmann, T. (2001). Conformally symmetric circle packings : a generalization of Doyle's spirals, *Experimental Mathematics, 10*, 141–150

Boltyanski, V., Martini, H., & Soltan, P. (1997). *Excursions into combinatorial geometry*. Berlin/Heidelberg/New York: Springer

Borsuk, K. (1933). Drei Sätze über die dreidimensionale euklidische Sphäre. *Fundamenta Mathematicae, 20*, 177–190

Bouasse, H. (1917). *Appareils de mesure*. Delagrave, reprinted by Blanchard en 1986

Bouchareine, P. (1996). Charles Fabry métrologue. *Annales de physique, 21*, 589–600

Cho, Y., & Kim, H. (1999). On the volume formula for hyperbolic polyhedra. *Discrete & Computational Geometry, 22*, 347–366

Connes, A. (1999). Le théorème de Morley. *Publ. Math. Inst. Hautes Études Sci., numéro special des 40 ans*

Coolidge, J. (1916, 1999). *A treatise on the circle and the sphere.* Oxford/New York: Clarendon Press/AMS Chelsea

Cromwell, P. (1997). *Polyhedra.* Cambridge: Cambridge University Press

Darboux, G. (1880). Sur le contact des coniques et des surfaces. *Comptes Rendus, AcadÈmie des sciences de Paris, 91,* 969–971

Darboux, G. (1917). *Principes de géométrie analytique.* Paris: Gauthier-Villars

Davis, P. (1995). The rise, fall and possible transfiguration of triangle Geometry: a mini-history. *The American Mathematical Monthly, 102,* 204–214

Dieudonné, J. (1960). *Foundations of modern analysis.* New York: Academic Press

Doyle, P., He, Z.-X., & Rodin, B. (1994). The asymptotic value of the circle packing constant. *Discrete & Computational Geometry, 12,* 105–116

Eggleston, H. (1957). Covering a three-dimensional set with sets of smaller diameter. *Journal of the London Mathematical Society, Second Series, 30,* 11–24

Encyclopedic Dictionary. (1985). *Encyclopedic dictionary of mathematics* (3rd ed.). Cambridge: M.I.T

Evelyn, C., Money-Coutts, G., & Tyrelle, J. (1975). *Le théorème des sept cercles.* Paris: CEDIC

Falconer, K. (1990). *Fractal geometry.* New York: John Wiley

Faugeras, O. (1993). *Three-dimensional computer vision: A geometric viewpoint.* Cambridge: MIT Press

Feder, J. (1988). *Fractals.* New York: Plenum Press

Frances, C. (2003). Une preuve du théorème de Liouville en géométrie conforme dans le cas analytique. *Líenseignement mathématique, 49,* 95–100

Frenkel, J. (1973). *Géométrie pour l'élève professeur.* Paris: Hermann

Giacomo, P. (1995). Du platine à la lumière. *Bulletin BNM, 102,* 5–14

Grünbaum, B. (1993). *Convex Polytopes.* Berlin/Heidelberg/New York: Springer

Gutkin, E. (2003). Extremal triangles for a triple of concentric circles. *Preprint IHES,* M03-46

Hadamard, J. (1911). *Leçons de géométrie élémentaire.* Paris: Armand Colin, reprint Jacques Gabay, 2004

Hilbert, D., & Cohn-Vossen, S. (1996), *Anschauliche Geometrie.* Berlin/Heidelberg/New York: Springer

Kahn, J., & Kalai, G. (1993). A counterexample to Borsuk's conjecture. *Bulletin of the American Mathematical Society, 29,* 60–62

Klein, F. (1926–1949). *Vorlesungen über höhere Geometrie, 3ème édition, bearbeitet und herausgegeben von W. Blaschke.* Berlin/Heidelberg/New York: Springer, 1926, reprint Chelsea 1949

Lalesco, T. (1952). *La géométrie du triangle.* Vuibert, reprint Jacques Gabay, 1987

Langevin, R. & Walczak, P. (2008). Holomorphic maps and pencils of circles, *American Mathematical Monthly, 115*(8), 690–700

Lebesgue, H. (1950). *Leçons sur les constructions géométriques.* Paris: Gauthier-Villars, reprint Jacques Gabay, 1987

Mandelbrot, B. (1977). *Fractals: Form, chance and dimension.* New York: Freeman

Mandelbrot, B. B. (1982). *The fractal geometry of nature.* New York: W.H. Freeman and Company

Maskit, B. (1988). *Kleinean groups.* Berlin/Heidelberg/New York: Springer

Mathéus, F. (1999). Empilements de cercles et discrétisation quasiconforme: comportement asymptotique des rayons. *Discrete & Computational Geometry, 22,* 41–61

Nevanlinna, R. (1960). On differentiable mappings. In *Analytic functions* (pp. 3–9). Princeton, NJ: Princeton University Press

Pach, J., & Agarwal, P. (1995). *Combinatorial geometry.* New York: Wiley

Penrose, R. (1978). Geometry of the universe. In L. A. Steen (Ed.), *Mathematics today - twelve informal essays.* Berlin/Heidelberg/New York: Springer

Pommerenke, C. (1992). *Boundary behavior of conformal maps.* Berlin/Heidelberg/New York: Springer

Raigorodskii, A. (2004). The Borsuk partition problem: The seventieth anniversary. *The Mathematical Intelligencer, 26,* 4–12

Ratcliffe, J. (1994). *Foundations of hyperbolic manifolds.* Berlin/Heidelberg/New York: Springer

Rodin, B., & Sullivan, D. (1987). The convergence of circle packings to the Riemann mapping. *Journal of Differential Geometry, 26,* 349–360

Rouché, E. & de Comberousse, C. (1912). *Traité de Géométrie* (two volumes). Paris: Gauthier-Villars, reprint Jacques Gabay, 2004

Smith, A. (2000). Infinite regular sequences of hexagons. *Experimental Mathematics, 9,* 397–406

Stephenson, K. (2003). Circle packing: A mathematical tale. *Notices of the American Mathematical Society, 50,* 1376–1388

Stephenson, K. (2005). *Introduction to circle packing.* Cambridge: Cambridge University Press

Sullivan, D. (1984). Entropy, Hausdorff measures old and new, and limit sets of geometrically finite groups. *Acta Mathematica, 153,* 259–278

Tabachnikov, S. (2000). Going in circles: variations on the Money-Coutts theorem, *Geometriae Dedicata, 80,* 201–209

Vinberg, E. (1993). Volumes of non-Euclidean polyhedra. *Russian Mathematical Surveys, 48,* 15–48

Yvon Villarceau, A. (1948). Note concernant un troisième système de sections circulaires qu'admet le tore circulaire ordinaire, *C. R. Acad. Sciences Paris, 27,* 246

Ziegler, D. (1995). *Lectures on Polytopes,* Graduate Texts in Mathematics, New York: Springer Verlag

Zong, C. (1996). *Strange phenomena in convex and discrete geometry.* Berlin/Heidelberg/New York: Springer

Chapter III

The sphere by itself: can we distribute points on it evenly?

III.1. The metric of the sphere and spherical trigonometry

As we shall see throughout this chapter, the geometry of the "ordinary sphere" S^2 — two dimensional in a space of three dimensions — harbors many pitfalls. It's much more subtle than we might think, given the nice roundness and all the symmetries of the object. Its geometry is indeed not made easier — at least for certain questions — by its being round, *compact*, and bounded, in contrast to the Euclidean plane. Sect. III.3 will be the most representative in this regard; but, much simpler and more fundamental, we encounter the "impossible" problem of *maps* of Earth, which we will scarcely mention, except in Sect. III.3; see also 18.1 of [B]. One of the reasons for the difficulties the sphere poses is that its group of isometries is not at all commutative, whereas the Euclidean plane admits a commutative group of translations. But this group of rotations about the origin of $\mathbb{E}^3$ — the *orthogonal group* $O(3)$ — is crucial for our lives. Here is the place to quote (Gromov, 1988b):

"$O(3)$ *pervades all the essential properties of the physical world. But we remain intellectually blind to this symmetry, even if we encounter it frequently and use it in everyday life, for instance when we experience or engender mechanical movements, such as walking. This is due in part to the non commutativity of* $O(3)$, *which is difficult to grasp.*"

Since spherical geometry is not much treated in the various curricula, we permit ourselves at the outset some very elementary recollections, which are treated in detail in Chap. 18 of [B]. Typically we will deal with the sphere S^2 of radius 1 centered at the origin of the Euclidean space $\mathbb{E}^3$, but all spherical geometries are the same within a change of scale. The interesting distance between two points p and q of S^2 isn't their distance in $\mathbb{E}^3$ but that on S^2, i.e. the length of a shortest path on S^2 that joins p to q. We are dealing here with the metric that is called *intrinsic* or *internal*, as opposed to the *induced* metric that is the distance between these points in $\mathbb{E}^3$ and evidently without much interest for Earth's inhabitants: digging tunnels is expensive. It's classical that this path is unique and that is the arc of a great circle that joins p to q, which can be proved rigorously by a symmetry argument, or here with the fundamental formula (III.1.1) below. Note that in saying **the** arc of a great circle we necessarily exclude *antipodes*, for which all the great circles starting from the one arrive at the other after a distance π. We let $d(p,q)$ denote this *distance*, which varies from 0 to π. It is given in terms of the scalar product by the formula:

M. Berger, *Geometry Revealed*, DOI 10.1007/978-3-540-70997-8_3,
© Springer-Verlag Berlin Heidelberg 2010

$\cos(d(p,q)) = p \cdot q$. But it may also be regarded as the angle through which we see the points p and q from the origin O. We can say, in an entirely equivalent manner, that we are dealing here with seeing the sphere as a set of half-lines emanating from the origin, which is crucial in descriptive astronomy: the sphere is the celestial vault and there is an essential need of the formulas that will follow. As astronomy is very old, these formulas date from several centuries ago, but we will read them on S^2, not as would astronomers.

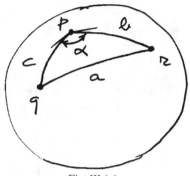

Fig. III.1.1.

We remain on S^2. It is clear that when we have three points p, q, r of S^2 we may speak — in analogy to Euclidean geometry — of the *spherical triangle* $\{p,q,r\}$. Now if we know the distances of $d(p,q)$ and $d(p,r)$, as well as the angle between the sides that emanate from p, we sense in advance that the distance of $d(q,r)$ is determined, thus theoretically calculable. The formula that provides this distance is called the *first fundamental formula of spherical trigonometry*. Let us call a, b, c the lengths (distances) of the sides of $\{p,q,r\}$, and α, β, γ its angles (between 0 and π). Then we always have:

(III.1.1) $$\cos a = \cos b \cos c + \sin b \sin c \cos \alpha.$$

The proof is a direct application of the definitions and of the scalar product, whose neglect in school mathematics explains why this formula isn't better known to students. It was prohibited in the French lycées until about 1950, which explains the incredible contortions in treating the inequality of days and nights, which was part of the required program. From it we can deduce — cautioning that the value of the sine does not determine the angle unambiguously — all the formulary given in detail in [B]. The practical importance is considerable; this or analogous formulas provide the solution to the following problem relating to three directions (or three half-lines) in space. Such a configuration is associated, as is the corresponding spherical triangle, with six numbers: the three angles between each pair of lines, and also the angles between the planes determined by pairs of these lines. This play with formulas allows us to calculate all these elements as a function of only three of them, but with some precaution. Think about the "dubious" case of equality of triangles. Readers will be able to investigate some practical applications: to astronomy, computation of the length of days in the course of the year, positions of artificial satellites, geodesy, petroleum research, GPS, etc.

◆

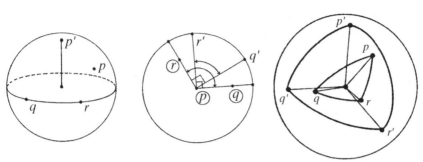

Fig. III.1.2. Construction of the spherical triangle dual to $\{p,q,r\}$. In the order: construction of p', construction of q' and r' in the plane orthogonal to the ray Op (ⓟ, ⓠ, ⓡ represent the projections onto the plane of points p, q, r), the triangle $\{p,q,r\}$ and the dual triangle. II [B] Géométrie. Nathan (1977, 1990) réimp. Cassini (2009) © Nathan Édition

It is important to mention the *duality* that exists for spherical triangles, that simplifies the derivation of the formulas. This duality associates with the triangle $\{p,q,r\}$, by taking appropriate perpendiculars, a (the) dual triangle $\{p',q',r'\}$ whose elements (denoted using apostrophes) are simply the angles complementary to π, but exchanging the roles of sides and angles:

$$a + \alpha' = b + \beta' = c + \gamma' = a' + \alpha = b' + \beta = c' + \gamma = \pi.$$

The triangle dual to $\{p',q',r'\}$ is the original triangle $\{p,q,r\}$. This duality is the spherical version of that given in Sect. I.7 for the real projective plane $\mathcal{P}$. In the language of half-lines, i.e. of *trihedra*, it is expressed thus: with the trihedron $\{D, E, F\}$ we associate the trihedron $\{D', E', F'\}$, called *supplementary*, defined by the conditions: F' is perpendicular to D and E and is located on the same side of the plane defined by D and E as F, and likewise for the two other half-lines. An historical note on "school" mathematics, interesting but doubly depressing: in French instruction the duality of supplementary trihedra was included in the program for the terminal classes at the lycées. Until 1950 the proof of duality (involutivity) was done geometrically, since use of the scalar product was strictly forbidden. Now with the notion of "same side" involutivity was rather obscure to prove, even in the very good work (Hadamard, 1911). It took several pages, with complicated reasoning about the poles associated with a great circle, etc. Such reasoning is always subtle; we will see this clearly in Sect. III.4 for the problem of "the thirteenth sphere". Now this involutivity is trivial with the use of the scalar product: if $\{x, y, z\}$ and $\{x', y', z'\}$ are nonzero vectors representing our trihedra of our spherical triangles, the six conditions defining $\{x', y', z'\}$ are:

$$x' \cdot y = x' \cdot z = 0, \quad y' \cdot x = y' \cdot z = 0, \quad z' \cdot x = z' \cdot y = 0$$
$$x' \cdot x > 0, \quad y' \cdot y > 0, \quad z' \cdot z > 0.$$

The set in its totality is clearly symmetric. The author learned this algebraic formulation from his classmate André Aragnol when he was preparing for his "agrégation" (French competition for specially privileged teaching positions at the secondary level). We ask ourselves why so many schoolchildren and their teachers need suffer because of a perverse sort of love for *pure geometry*. We have seen this phenomenon above regarding the fundamental formula, we will encounter it again regarding the duality with respect to a circle in Sect. IV.2. Avoiding use of coordinates and − still worse − the scalar product, is an obsession that has long haunted "pure" geometers. Now the scalar product figured in Grassmann's work from 1840, but in writing that was far too prophetic; it wasn't really popularized until Gibbs in 1860. But resistance to Gibbs raged from the disciples of Hamilton, who wanted to keep the scalar product exclusively for quaternions (thus in dimension 4) and prepared the way for a long and bitter correspondence, which is today hilarious: see Gibbs (1961).

The second point, sadder perhaps, is that this discussion is hardly relevant today, since geometry, especially that of space, has all but completely disappeared from the mathematics curriculum of schools and universities − this while space visualization is becoming more and more necessary for practical applications such as three dimensional animation and robotics.

So, with the fundamental formula and the duality, plus some calculations, astronomers have all the necessary formulas; see the formulary 18.6.13 of [B]. They were much more developed until the introduction of computers; the calculations were done by hand using logarithmic tables. So that they could be done as quickly as possible, it was necessary to transform as much as possible all those operations bearing on trigonometric properties, sums in particular, into products or quotients.

The sphere possesses a canonical measure that is unique within a scalar factor if we require invariance under O(3); see Sect. III.6. This is what physicists call a *solid angle*, in visualizing half-lines emanating from the origin. The total area of the sphere equals 4π. The spectacular theorem that goes back to Thomas Harriott in 1603 but that was published only later by Albert Girard (1595–1632) states:

(III.1.2) *The area of any spherical triangle with angles α, β, γ is equal to*
$$\alpha + \beta + \gamma - \pi.$$

We can compare this with formula (II.4.1) from hyperbolic geometry. If the three angles are commensurable with π (are rational multiples of π), then the surface area of such a triangle will also be commensurable with π. In Sect. III.7 we will see that the analogous question in higher dimensions remains open. The unpublished proof of Harriott amounts to looking at the classic figure below. By calculation alone things are much more difficult and do not give a hint of the final simplicity of the formula. To make use of the figure, it suffices to know that the area of a *lune* on a sphere formed by planes through a diameter of angle α equals 2α, which is evident from the proportion to the total area 4π of the sphere (calculated in Sect. VII.6.B).

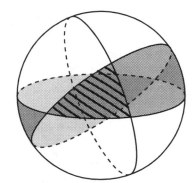

Fig. III.1.3. Harriott (1560–1627), Girard (1595–1632)

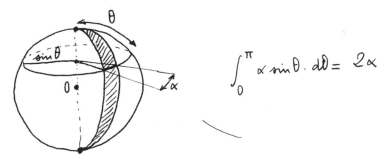

Fig. III.1.4. A spherical lune with angle α

◆

Here we briefly mention the *isoperimetric problem on the sphere* (we will encounter isoperimetric problems extensively in this book, in particular those on convex sets in Chap. VII). First, the isoperimetric inequality is true for triangles in S^2: among all the triangles of S^2 the maximal area (measure of the surface) is attained for the equilateral triangles and for them only – a direct consequence of the formula that gives this area σ as a function of the lengths a, b, c of its three sides:

$$\tan \frac{\sigma}{4} = \sqrt{\sin p \sin(p-a) \sin(p-b) \sin(p-c)}$$

where $p = \frac{a+b+c}{2}$. This formula for the area may be compared to Heron's, given in Sect. VIII.3. We will see a comparable formula in Sect. III.5 for the sphere of dimension 3. At the other extreme of the subsets of S^2, we have the general domains for which an area and a *perimeter* can be defined. As in the Euclidean plane (see Sect. V.11) the general isoperimetric inequality is true on the sphere: among the domains having a given perimeter, it is the *disks*, and they alone, that have maximum surface area. The disks of S^2 are the closed balls for the canonical metric, the $B(p ; r) = \{q : d(p, q) \leqslant r\}$ of center p and radius r. In the literature the term *spherical cap* is frequently used. We encounter the canonical measure of the sphere in Sect. III.6.

Fig. III.1.5. The spherical caps attain, for a given area, the minimum perimeter

Although a little outside of our general theme because of the setting in more abstract geometries, we mention the problem of finding geometries for which we know how to calculate the distance between two points, starting with a universal formula in which nothing appears except for the distances of the two points to another point (fixed, so to speak) and the angle subtended by the two points from the fixed point. These geometries are all known, and in fact they are nothing other than Euclidean geometry (where the formula is $a^2 = b^2 + c^2 - 2bc \cos \alpha$), spherical geometry and hyperbolic geometry (see Sect. II.XYZ). In fact, if we want notions of distance and angle in the context of Riemannian geometry, then the spaces sought are those for which the local curvature is constant. We can then show that there are only the three geometries mentioned; see any work on Riemannian geometry, for example Berger (2003). For more general spaces we don't know much about the problem; one of the difficulties is knowing which are the most general metric spaces for which we can define a reasonable notion of angle.

◆

Another problem, concerning *geometric optics*, was posed in 1927 by Blaschke: do there exist other metric geometries on the topological sphere for which each point has a well determined antipodal point and on which each point can be joined to its antipode by a shortest path, starting in any direction from the initial point? It's the problem of constructing a perfect optical instrument that is completely *stigmatic*. In practice the corresponding instrument is realized in certain animals: it's Maxwell's fish eye. In fact only the canonical metric of the sphere is everywhere stigmatic, but this has only been known since 1963 and is due to Leon Green; see

Deschamps (1982) for the systematic construction of partially stigmatic objects, as well as for references.

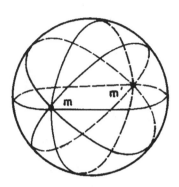 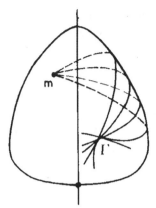

Fig. III.1.6. Only the sphere is a completely perfect lens. But there exist other objects that are stigmatic on just a portion of their surface. Deschamps (1982) Gauthier-Villars.
© Elsevier

III.2. The Möbius group: applications

In Sect. II.5 we introduced a group on the ordinary sphere S^2 that is larger than the group of isometries (which has three parameters), i.e. the Möbius group $\text{Möb}(S^2)$, which has *six* parameters. It's the group of its conformal transformations, generated by restricting to the sphere those inversions of space that preserve the sphere. We can study the geometric nature of all these transformations, interesting for being more general than isometries, but all the while preserving angles. The conformal type mappings most removed from isometries are those that lift the homotheties of $\mathbb{R}^2$ centered at the origin onto S^2 by stereographic projection. We see what they do: they preserve the north and south poles; they preserve meridians while moving individual points (to a variable extent that depends on the latitude) toward the north pole if the homothety ratio is greater than 1. For a very large homothety ratio, points are moved radically toward the north pole. If from the outset the sphere is considered as a distribution of masses, there will be more and more mass shifted toward the north pole as the homothety ratio increases. Matters are still more graphic if we (think dynamically!) iterate such a transformation f. In the limit all the mass is going to be concentrated at the north pole; all other points, the south pole in particular, are left without mass. By composing such a transformation with a rotation about the north-south axis, we obtain what are called *loxodromic* transformations. For example, we can simultaneously compose the group of rotations with the one-parameter group of homotheties of variable ratio and look at the trajectory of a point: this will be a curve, one that cuts all the meridians at a constant angle. These are then, for navigators, the curves that describe the movement of a boat for which the

heading is constant; they are called *loxodromies*; and on marine maps they appear as lines. These curves are liftings of logarithmic spirals by stereographic projection, written in polar coordinates $(\rho, \theta) : \rho(\theta) = e^{k\theta} \cdot \theta$. Indeed, stereograhic projection preserves angles and the logarithmic spirals are exactly the curves that make a constant angle with the radial lines.

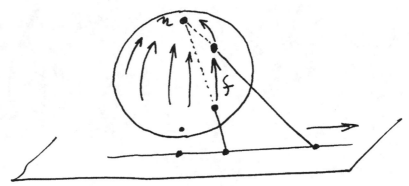

Fig. III.2.1. Action of a homothety of the plane lifted stereographically onto the sphere

For interested readers, we mention two applications of such transformations to geometry. The first concerns rather general metrics g, called *Riemannian*, on the projective plane $\mathcal{P}$. Here two constants are associated with g: the first is the total measure, the area Area(g); the second is its *systole* Sys(g), defined as the shortest length, relative to g, of the curves of $\mathcal{P}$ that are topologically equivalent to projective lines (in fact they are curves that are not contractable to a point, for $\mathcal{P}$ is not simply connected). Thanks to the Möbius group, we can show that we have an inequality of global isoperimetric type: Area$(g) \geqslant \frac{2}{\pi}$ Sys$^2(g)$ with equality only for the elliptic metric, thus canonical. The proof uses the fact that the Möbius group is very large. For such isosystolic inequalities, and for references, see e.g. Berger (1998) or Berger (2003).

Let us return to the sphere S^2 and endow it too with a Riemannian metric g. The theorem on conformal representation encountered at the end of Sect. II.8 implies here that we can always assume that $g = f \cdot d$, where f is a numerical function on the sphere (we say that g is *consistent* with the canonical metric d). We can thus regard the pair $(S^2, g = f \cdot d)$ as a variable mass distribution on the sphere, the *density* being f. If we consider this object as **vibrating**, it will have frequencies: it will be musical. The lowest frequency is the one of reference, call it $\Lambda(g)$. Then we have an *isomusical* inequality (characteristic of the standard sphere) between this frequency $\Lambda(g)$ and the total area Area(g), namely $\Lambda(g)$ Area$(g) \leqslant 8\pi$ (equality holding only for the standard metric). The proof consists of showing that the Möbius group allows conformal transformation of the density f such that the center of gravity in the space of this massive sphere is the origin O of the space. Intuitively, this is how we proceed: we have seen above that we can move all the mass toward the north pole, but likewise toward any point of the sphere. This shows, by a covering type argument, that the initial center of gravity, subjected to all these transformations, must surely "travel through the origin" at least one time. The proof is completed with a classical minimal property for the function that gives the vibration of least frequency. For references and more, see Berger (1998) or Berger (2003).

III.3. Mission impossible: to uniformly distribute points on the sphere S^2: ozone, electrons, enemy dictators, golf balls, virology, physics of condensed matter

The problem that arises in very many practical applications is to **equally** place points, more or less large in number, on a sphere. We can always assume we are dealing with S^2. It is first a matter of esthetics. A bit more serious is the case of golf balls, to which we will return; but in fact the problem is much more important. We can think about the heart of a spherical nuclear reactor: how should we distribute the combustible bars?. In architecture, it has to do with constructing a sphere (or almost a whole sphere) as a *geodesic dome*, i.e. with finding an appropriate triangulation; an example is the Géode in the Villette Park in Paris. But it also has to do, in physics, with studying the distribution of electrons in atoms and with studying various molecular structures. It arises in molecular chemistry, in solid state physics, in virology and the physics of condensed matter. Finally there is the document scanner and medical imaging.

Readers may reproach us for spending too much time on a single problem. Other than the motivations above, we mention that the problem relates to the seventh in the list of 18 mathematical problems that Smale has proposed for the twenty-first Century (Smale, 1998). Smale in fact encountered this problem in a totally different context, different at least in appearance: that of discovering good algorithms for finding the roots of a polynomial. The connection between the sphere S^2 and such algorithms comes from this: a good quantity for considering the complexity of these algorithms is a pair of complex numbers, or rather their ratio, or still more precisely such pairs modulo their product with an arbitrary complex number. But we know that this is nothing other than the complex projective line; in Sect. I.XYZ we correctly identified it with the sphere S^2.

Uniform distribution of points on the sphere is not just a study of *configurations* to be studied or discovered. More important, we will encounter this problem in each discipline where distribution of points on a sphere is involved: information receptors at the given points, numerous enough for the information to be good, but not too numerous for economic reasons. A typical case is that of ozone: if we want to know the total quantity of ozone in the atmosphere, it is necessary to measure its density at different points and take the mean of these values. It is clear that these points must be well distributed, so as not to bias this or that portion and so as "not to neglect any". The points considered may be various receptors, or various analyzers: sounding balloons, but also satellites.

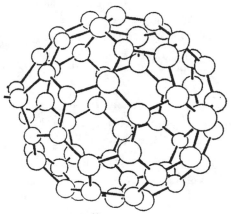

Fig. III.3.1. The carbon molecule C^{60}: a fullerene discovered rather recently and named in honor of Buckminster Fuller, architect and inventor of the geodesic dome. Spherical fullerenes are also called "buckyballs"

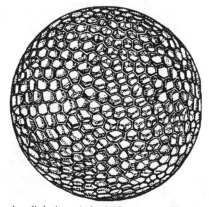

Fig. III.3.2. Hexagonal radiolarian: *Aulonia hexagona*. The skeleton of this animal can't consist only of hexagons (we observe pentagons)

For the mathematician, the problem presents itself when it is required to approximate the **integral** of a function on the sphere. We know that, for ordinary integrals, the interval of $\mathbb{R}$ on which the function is to be integrated gets partitioned, ideally into N equal intervals; then N is made to tend toward infinity.

These matters are also rather easy on the circle, which can easily be partitioned into equal arcs of whatever number, but also for integrals over the plane and over

Fig. III.3.3. The Géode in the Villette Park

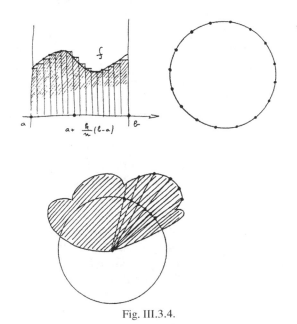

Fig. III.3.4.

space: we take points that form a *lattice* (see Chap. IX), etc. For the plane, the best functioning lattice is "obviously" regular hexagonal, an old friend encountered in Sect. II.8.

On the sphere we would like − if the points are denoted by x_i ($i = 1, \ldots, N$) and for any reasonable function f (say continuous) − that the sum $\frac{1}{N} \sum_i f(x_i)$ approximate well the total mass $\int_{S^2} f(x)\, dx$ of the function, or rather its *mean value* $\frac{1}{4\pi} \int_{S^2} f(x)\,dx$, where dx denotes the canonical measure on the sphere. What is asked of the approximation evidently varies according to the problem. We might suspect that there could be several sorts of criteria. We are going to review several of them, but the conclusion will be rather negative: it is very difficult to distribute points well on the sphere. It is ultimately impossible, except for some small values of N, that don't suffice for computing integrals; or on the contrary − as has just recently been discovered − it seems that such partitions *are* possible, but require more than 10 000 points. Before studying the different approaches to this problem and concluding that it is finally and in a certain sense impossible, some heuristic words for getting a sense of its difficulty.

A first observation is that the sphere, even a tiny piece of it, can never be isometric to a piece of the Euclidean plane, i.e. that a **perfect map** does not exist. It is possible for a map to preserve angles, which is almost always done for reasons of navigational bearings, but not distances; there will always be some distortion. Otherwise we would only need to "place" a plane (or a part of the plane) on the sphere and we could mark out regular hexagonal lattices, with increasingly smaller scale. We might say, as has already been noted, that an essential reason for this impossibility regarding distances is the non commutativity of the group of isometries of the sphere, whereas the affine spaces possess the commutative translation group. Another observation: we might think of using points that are the vertices of regular polyhedra. Now, contrary to the case of the circle where there are regular polygons with arbitrarily many sides, there do not exist in space regular polyhedra with a large number of vertices, not even semi-regular polyhedra (which would suffice); see Sect. VIII.4. Still worse, we will see that the cube and the dodecahedron do not yield optimal lattices.

We barely touch on the related problem of scanning a solid. Theoretically it has to do with reconstructing the density of a ball starting with all plane projections, in other words with classic x-rays taken in all directions. In practice, only a finite (but perhaps rather large) number of projections are used and there is a need for effective algorithms for distributing these "evenly"; see Helgason (1980).

Historically, we can't but be surprised by the fact that mathematical work on the subject is so incredibly recent. But above all, if we except the "foundations" by Habicht and van der Waerden (1951), the first edition of Fejes Tóth (1972) in 1953, and several others, e.g. Stolarski (1973) and Delsarte, Goethals and Seidel (1977), the subject didn't really begin to be studied until 1990. We can convince ourselves of this with the bibliography of basic references in Conway and Sloane (1993), to which it is necessary to add the brief and informal synthesis (Saff and Kuijlaars, 1997) that we largely follow, but above all the recent (Hardin and Saff, 2004) which is, with its technical references, the new global reference text. This is corroborated by the nonexistence of books treating numerical integration on the sphere, whereas the works treating numerical integration on the line or the plane are legion. We also cite the book (Melissen, 1997), very detailed for the "elementary" problems, for example for configurations with few points, and which has a lot of pictures. We are now going to study the different approaches to the problem. See also how to use the theorem of Dvoretsky (cf. Sect. VII.12) for evenly distributing points on spheres of arbitrary dimension in Wagner (1993), or the end of 7.3 of Giannopoulos and Milman (2001).

The Tammes problem. The formulation that is historically first and the simplest for geometry consists of being given an integer N and deciding that "the best configuration" for these N points is the one furnished by "the" solution of the problem of the *enemy dictators*, also known as the *Tammes problem*. It is studied in the synthesis (Saff and Kuijlaars, 1997), but the first treatment historically is in a set of works of van der Waerden and his collaborators; all the references are in van der Waerden (1952). The case of small values of N is treated in great detail in Melissen (1997), whereas Croft, Falconer and Guy (1991) is no longer sufficiently up to date. The "best partition", by whatever criterion, is of course understood modulo an isometry of the sphere.

Tammes was a Dutch botanist who in 1930 published an article on the partition of tiny holes on the surface of pollen grains. The criterion for the desired configuration $\{x_i\}$ $(i = 1, \ldots, N)$ is to require that each of the points be as far as possible from the others, i.e. to maximize the minimum of the distances $d(x_i, x_j)$ $(i \neq j)$. It amounts to the same thing as seeking the largest possible radius r for N disjoint disks of radius r placed on the sphere. The points sought should be as far as possible from one another in order to maximize their territory, whence the language of the dictators. The territory of the dictator p will of course be a Dirichlet (or Voronoi) domain about p, a notion we will encounter in Chaps. IX and X, i.e. the set of points of the sphere that are closer to p than to any of the other dictators (this defines the realm of the dictator). For reasons of compactness, there always exist such optimal configurations: consider the product of the sphere with itself N times and the (continuous) function equal to the smallest of these mutual distances. The solution is furnished by the point (or points) of the product space that gives the function its maximum value. We will say at first what happens for small values of N, this to give a feeling for the difficulty of the problem, also because it is very geometric and a source of pitfalls!

Two points get placed at antipodes, three are placed on an equator, the vertices of an equilateral triangle; four points are vertices of a regular tetrahedron, the proof being left to readers. We thus confirm that there is essentially one optimal solution for the values 2, 3, 4, and that these are configurations that have the greatest symmetry possible, which conforms well to the intuition mentioned above. It can be shown next that, for 6 points, there is only one configuration possible, the vertices of a regular octahedron. Up until now our intuition has worked well.

Having gotten started, it is natural to think that we can do better for 5 points than for 6. We remove a point from the regular octahedron and thus have room to move the 5 remaining points, distancing them from each other. The answer is **no**, you can displace them in such a way as to find a different configuration that gives the same

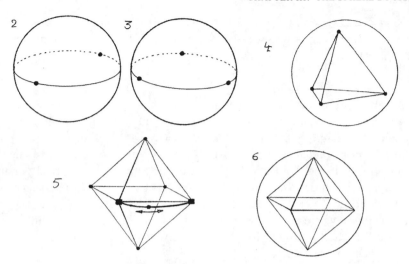

Fig. III.3.5. The optimal configurations for 2, 3, 4, 5 and 6 points. For 5 points there
is no uniqueness, and the configurations are not very aesthetic!

value of the optimal distance as do 6 points, which equals $90° = \pi/2$ incidentally
(not less!). Readers can search for this second configuration.

We pass to the case of 8 and distribute the 8 points on the sphere at the vertices
of a cube. Can we do better? Yes, we turn one of the faces, which preserves the
distances between vertices in two faces, the one that we turn and the one opposite,
but strictly increases all the other distances. Whence the idea, if the turned face
and its opposite are horizontal and we lower the top face and raise the other, while
keeping it a square, then we increase the sides of these squares; and if we move them
just enough the distances between points of the resulting (non-regular) octahedron
will be equal, and greater than they were initially (equaling $70° \, 32'$, by way). Finally,
Schütte and van der Waerden showed that for 8 points the optimal configuration can
be obtained precisely by the above method; it consists of what is called a *square
antiprism* i.e. two equal squares with the same axis, which we take to be vertical,
but one of the squares is turned by 45° in relation to the other. Subsequently we
adjust their distance apart so that the sides joining the vertices of these opposing
squares have the same length as the sides of the squares themselves.

Such a figure admits symmetries (find all of them!), but fewer than the cube.
Thus our intuition wasn't well founded; the solution of the Tammes problem for the
values 4, 6, 8, 12, 20, isn't always given by regular polyhedra. It is for 4 and 6, but
not for 8. Readers familiar with the notion of *symmetry breaking*, very important
in physics, know that it isn't always configurations having maximum symmetry that
provide optimal solutions. Nature loves symmetry only tepidly! The simplest known
symmetry breaking concerns the square: we seek to connect its four vertices, e.g.
four villages, by a set of paths for which the total length is as small as possible.
The solution is given in the figure below. It does not have all the symmetries of the
square and so there are two solutions.

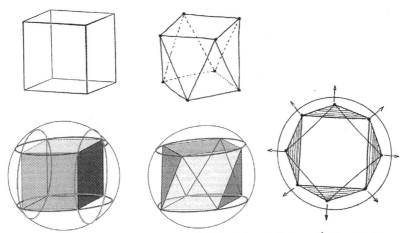

Fig. III.3.6. Square antiprism. Berger (1992) © Pour la Sciénce, Éditions Belin

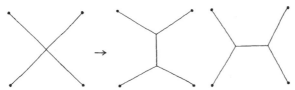

Fig. III.3.7. A very classical symmetry breaking

Alain Connes likes to give an example of symmetry breaking where at a circular restaurant table the server puts the bread, or anything else, symmetrically between the guests, for whom there are no firm left-right conventions. The first person who takes his bread breaks the symmetry!

For 7 points of the sphere the configuration isn't very pretty, which isn't surprising. We will encounter symmetry breaking again in Sect. VII.11.C. In physics, for the states of a system admitting symmetries, there was first Curie's principle that affirmed that each solution admitted all these symmetries. A false principle! What *is* true is that it's the set of solutions that admits these symmetries. A physics text on symmetry breaking is Brézin (2001).

Where are we at present? To the knowledge of the author, the **only** numbers for which the optimal value are known include all N through 12 but only one additional value, i.e. $N = 24$. For $N = 12$ we get the regular icosahedron, the unique solution, but we also see the phenomenon already encountered for 5 and 6. I.e. for $N = 11$ we can't do better than the regular icosahedron minus a point, and this is not the only configuration! For 20, it is known that we can do much better than with the regular dodecahedron, i.e. $47° 26'$ as against only $41° 49'$ for the regular dodecahedron. Readers will easily see that an improvement is possible; proceed as with the cube. For $N = 16$ some alloys are found in nature where *Friauf's configuration* is encountered, and we might think that it seeks to solve the Tammes problem, but it seems that no one knows if it is the best possible. For this subject,

Fig. III.3.8. Drawings by Geoffroy Wagon

see *Nature*, vol. 352, July 1991. In *Nature*, vol. 351, May 1991, there is still better: for the analogous (dual) problem of **covering** the sphere most economically with a number of identical spherical caps of the same radius (that of the sphere), for the cases of 16 and 20 caps, objects are found in nature (cages of clathrine — clathrine is a protein) which have done better than the conjectures of mathematicians. In the case 16, the dual polyhedron, with 16 vertices, is exactly Friauf's configuration.

We should also say that we know that the semi-regular polyhedra, like the soccer ball which furnishes a configuration with 60 points, don't in general yield the best value. The case $N = 24$ is thus special. Moreover, the only configuration giving the optimal angle value ($43°\,41'$) is the one that is called the *snub cube*. This semi-regular polyhedron is a most interesting object, astonishingly little known. Together with the snub dodecahedron it is the only semi-regular polyhedron that doesn't admit an orientation-changing symmetry, thus in particular a mirror symmetry, not even simply a center of symmetry! Returning to the general problem, the proof for 24

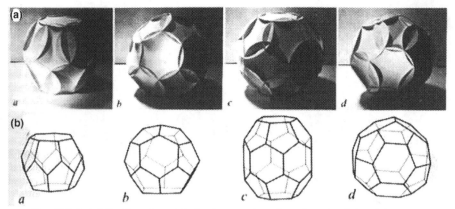

Fig. III.3.9. **(a)** Cardboard models for the covering of a sphere by disks of the same radius: *a* and *c* are the optimal solutions conjectured by mathematicians for the cases of 16 and 20 discs; *b* and *d* are the actual solutions observed by Tarnai in the study of cages formed by the molecule of clathrine. **(b)** Wire polyhedra with equal sides corresponding to modules from A above: *a*, the "heptagonal drum", formed from two special heptagons surrounded by two rings of seven pentagons; *b*, polyhedral cage formed from 12 pentagons and 4 hexagons placed tetrahedrally; *c*, the "hexagonal barrel", polyhedral cage formed from two rings of 6 pentagons, a ring of 6 hexagons and 2 special hexagons; *d*, the "tennis ball", polyhedral cage formed from two concave bands of 4 hexagons, separated by a belt of 12 pentagons placed in space like the seam of a tennis ball. Stewart (1991). © Nature Publishing Group, a division of Macmillan Publishers Limited

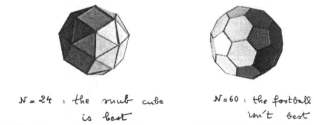

Fig. III.3.10. The snub cube and the soccer ball. Holden (1971) © Columbia University Press

is very long and is not conceptual, but of course makes extensive and subtle use of spherical trigonometry; see Robinson (1961).

It's the moment for reassuring − or perhaps of disquieting − computer enthusiasts. It's necessary to realize that devising a computer program for solving the Tammes problem (obtaining the best bound and configuration) is extremely difficult. We will return to this later and say here only that it's an optimization problem having to do with a number of variables equal to $3N$. Many have devised programs, by different methods, but all those programs require enormous computational resources for which the value of N is important. In all these numerical studies, generally treating other criteria in addition to the distribution of points on the sphere, there is a major difficulty due to the existence of a number of local minima that is exponential in N.

◆

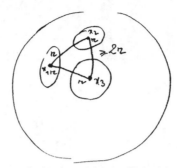

Fig. III.3.11. How to obtain a bound for the $d(N)$ of the Tammes problem

As in numerous other problems that we encounter, where exact solutions either don't exist or are only theoretical, we must be content with asymptotic estimates. It's definitely not easy either to obtain these estimates or to find configurations that satisfy them. Here in the Tammes problem the task is to study the *optimal value* $d(N)$, i.e. the upper bound of the minima $\inf\{d(x_i, x_j) : (i \neq j)\}$, ranging here over all possible configurations $\{x_i\}$ $(i = 1, \ldots, N)$ of N points on the sphere. For this is exactly the requirement that each of the points be as far as possible from the others. Even not knowing $d(N)$ exactly, we desire a best possible approximation. This was accomplished in Habicht and van der Waerden (1951), where we find the double estimate:

$$\left(\frac{8\pi}{\sqrt{3}}\right)^{1/2} \frac{1}{\sqrt{N}} - C\frac{1}{\sqrt{N^3}} \leq d(N) \leq \left(\frac{8\pi}{\sqrt{3}}\right)^{1/2} \frac{1}{\sqrt{N}}$$

which is thus very satisfying, apart from the value of the constant C, which is not well known. In fact, the circumstance that the inequality is double with the same factor $\frac{1}{\sqrt{N}}$ shows that things necessarily have this precise asymptotic order. The inequality on the right is not difficult to obtain; at best the spherical caps of an optimal configuration touch each other three at a time (the boundary circles are tangent to each other). Three such centers form the vertices of a spherical triangle with sides of length equal to the optimal distance $d(N)$ that is sought. Thus the triangle is known on the sphere and we can calculate its surface area as an explicit function of $d(N)$ by the formulas of spherical trigonometry. At best all these triangles cover the sphere without holes; they are N in number and the total surface area of the sphere equals 4π. All calculations done, we find that d is bounded above by the value given by the equation

$$\cos d(N) = \cos e/(1 - \cos e) \quad \text{where} \quad e = N\pi/3(N - 2).$$

We have derived the bound on the right.

To get the inequality on the left, it is necessary to explicitly construct configurations with a rather large $d(N)$. This geometry problem has been encountered by makers of *golf balls*, which generally contain hundreds of dimples. For those who don't know, it is interesting to recall that originally these balls were smooth and that one day a player, lacking new balls, recognized that the very worn balls went farther than the newer smooth balls. We know that the dimples (originally just scratches) have the effect of reducing the *drag*, i.e. the force caused by turbulence behind the ball; the dimples allow diminution of separation from the boundary layer.

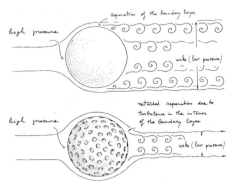

Fig. III.3.12. How air flows around a smooth golf ball and around a rough ball. Walker (1979) © J. Walker

The patent filed by the Slazenger, for which the principle coincides with the construction of Schütte and van der Waerden, was natural enough. We can't impose a regular hexagonal lattice on the sphere, as we have seen; but we can do something similar on the interior of a spherical equilateral triangle on the sphere: we take its center — which gives us four triangles — and we iterate the operation as many times as we wish. Note that we start with an equilateral triangle, but the triangles that make up the successive subdivisions are not equilateral on the sphere. We can also project pieces of a planar hexagonal lattice onto the sphere. It remains to find a nice partition of the sphere into equilateral triangles, as numerous as possible: we see below that 12 is the maximum, but as we know the regular icosahedron, we proceed thus: we place twelve points on the ball at the vertices of a regular icosahedron, which thus partitions the sphere into twelve equilateral spherical triangles. Now the faces of the icosahedron are planar equilateral triangles, so we can fill them with as many triangles as we wish in hexagonal packing as explained above; these are the integers of the form $n(n + 1)/2$, there being in total $N = 10n(n + 1)/2$ points.

Proceeding as we have seems like the most natural way to distribute a significant number of points on the sphere. But now a brief commercial digression. If you obtain a certain model of Slazenger ball and mark the points with a felt-tip pen where a dimple touches only 5 others, you will see that they are located at the vertices of a regular icosahedron. Now we don't advise you to make and market such balls — Slazenger has patented that; but if this game interests you, obtain some other models

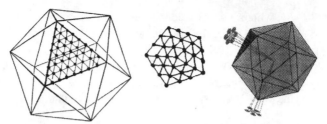

Fig. III.3.13. How to obtain the left hand bound for the $d(N)$ of the Tammes problem, also how to make golf balls *à la Slazenger*. Berger (1992) © Pour la Sciénce, Éditions Belin

Fig. III.3.14.

of balls. Some of them will be of a completely different type, about which we won't speak. But you will find balls where the 12 exceptional points are drawn so that at least one whose position has been shifted, and thus the Slazenger patent has been "gotten 'round"!

Let us see why 12 is an inevitable minimum. The reason for it is *Euler's formula* for convex polyhedra (see Sect. VIII.4), which says that $s - a + f = 2$, where s, a, f are the respective numbers of vertices, edges, faces of the polyhedron considered. But if we regard them closely and accept that some vertices belong to but 5 triangles (each spherical cap touches only 5 others with some exceptions, hopefully small in number, and everywhere else there will be six), then Euler's formula shows that this works with exactly 12 exceptions. In fact, since each face is a triangle and each edge determines two faces, we have $f = 2a/3$. Let s_5 and s_6 be the respective number of vertices where 5 or 6 triangles meet. Then, since each edge produces two vertices, we must have $2a = 5s_5 + 6s_6$. All together we have (the s_6 disappears, thus we can place as many as we wish, as we have seen!) $\frac{s_5}{6} = 2$, where $s_5 = 12$. Readers should investigate the exceptional pentagons in Fig. III.3.2 showing radiolaria and Fig. III.3.3 showing the Villette Géode. Typically in Fig. III.3.15 are shown both the points (the triangulation is apparent) and the tiling of the sphere; the dual obtained from the Dirichlet–Voronoi domains, composed of hexagons and 12 pentagons. We can apply Euler's formula to it with, of course, the same conclusion.

Let us return to one of our motivations for regular distribution of points on the sphere, i.e. for knowing how to approximate the integral or the mean value of a function defined on the sphere, while knowing its value only at certain points, more and more numerous if we want to approach better and better. The method of van der Waerden–Slazenger isn't good for integrating arbitrary functions on the sphere, not only because the twelve points distort things close to them, but also because the triangles that partition the faces of the icosahedron have more or less unequal areas. It seems thus that we are at an impasse. We are going to see where we are at the present moment, which we will do a bit in caricature – for it is difficult to get our bearings in the sea of studies that regularly appear on the subject, given its practical importance – returning once more to the references already given.

The energy viewpoint of physicists. The physicist who reads this, or someone whose thinking has been partly influenced by physics, will likely be exasperated. For such a person it is so simple, so trivial, to find a good distribution for N points by a physical principle: take N points $\{x_i\}$ and put them anywhere on the sphere, thinking of them as equally and positively charged particles; they repel each other mutually according to Coulomb's law (i.e. in $\frac{1}{r^2}$, where r is the distance between them) and thus they are certainly going to uniformly distribute themselves ultimately, i.e. they will minimize the *total energy* $\sum_{i \neq j} \frac{1}{|x_i - x_j|}$. We could also consider small magnets, likewise signed with the same magnetic force. We consider here the Euclidean distances $|x_i - x_j|$, but since they are intrinsically connected to distances on the sphere, it amounts practically to the same thing. Finally, your physicist will say, you can make your calculations on a computer, for you have only to minimize an explicit function of a set of points, which is much simpler than maximizing $\inf\{d(x_i, x_j) : (i \neq j)\}$ as above in the Tammes problem. But other energies can also be introduced, e.g. all those of the form $\sum_{i \neq j} \frac{1}{|x_i - x_j|^s}$, where s is a positive real number. We remark that, as s tends toward infinity, with N fixed, we tend toward the Tammes problem, since the energy is dominated by the terms of small distances. Finally, we might think of maximizing the product $\prod_{i \neq j} |x_i - x_j|$ of mutual distances, which amounts to minimizing the energy given by $\sum_{i \neq j} \log \frac{1}{|x_i - x_j|}$. This should be mentioned since this question was introduced in Shub and Smale (1993), where the problem is to study the complexity of the algorithmic solution of polynomial equations!; see the end of the present section.

What will we then obtain as a consequence of these energies? We recall first the double algorithmic difficulty of the problem on computers. In the first place, the number of variables is large, i.e. $3N$ if N is the number of implanted points. More seriously, if we are to have any prospects, is the fact that the energy functionals considered have all the messiness of local minima. These minima are moreover very near each other; we must find the true minimum from among them. A simple example: if you take 12 points at random on the sphere and apply a dynamic program that follows a descending energy path, the limiting configuration obtained won't always be a regular icosahedron.

This having been said, recently we have seen emerge results of computer calculations for the minima of energies $\sum_{i \neq j} \frac{1}{|x_i - x_j|^s}$ for various s and for numbers of points of the order of several hundreds at most. There are several crucial remarks to make in view of the configurations obtained, like that below in Fig. III.3.15 for 122 electrons. First, astonishingly, they practically all resemble each other, whatever the exponent s in the definition of the energy: they are of Schütte-van der Waerden - Slazenger type, i.e. partitioning the sphere into hexagons, with twelve exceptional pentagons. Secondly, none of these configurations is good for integrating arbitrary functions on the sphere. This is not surprising; see the remarks already made previously on the configurations of Schütte-van der Waerden-Slazenger type. We will subsequently see that it is possible to be very effective, but at the price of only integrating polynomials whose degree is bounded by a given constant or of working with much larger numbers of points.

Fig. III.3.15. 122 electrons in equilibrium and their Dirichlet–Voronoi domains. Saff, Kuijlaars (1997) © Springer and E. Saff

There is also the problem of estimating the minimum energy obtained, at least in asymptotic fashion, as a function of the number of points considered. We will return to this later, at the very end of the present section.

Integrating only polynomials. The impossible mission is realizable in the case of polynomials of degree with a fixed bound, presently of low degree. Since (Delsarte et al., 1977) we know how to construct sets of points $\{x_i\}$ that give the exact value

$\frac{1}{N} \sum_i f(x_i) = \frac{1}{4\pi} \int_{S^2} f(x) \, dx$ for no matter which polynomial f (polynomial in three variables of $\mathbb{R}^3$) of degree less than a given integer. We find the present state of affairs in Saff and Kuijlaars (1997). We know, for example, how to construct 94 points which work for degrees up to 13 − this requires large-scale computer calculations. For higher degrees, other than explicit construction, we ask for the minimum number of points necessary for a degree bounded by k; it is conjectured that it is of order k^2.

◆

Other attempts. We find in (Saff and Kuijlaars, 1997) a couple more ideas, one by partitioning the sphere into domains of the same surface area, the other with spirals; see Fig. III.3.16 below. The idea with spirals is very natural for some: place the points, from the north pole to the south pole, deftly arranged on a slicing of the sphere into pieces obtained by cutting across equally spaced planes orthogonal to the north–south axis. This yields an explicit formula and some sketches like those below.

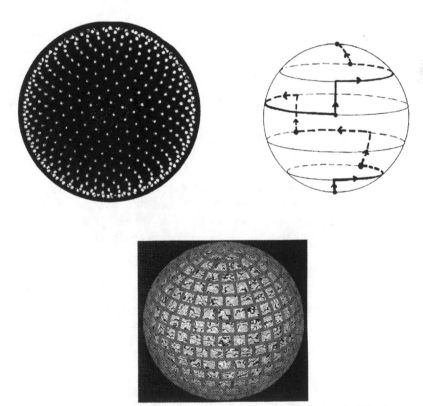

Fig. III.3.16. 400 parts of equal area; 700 points on a generalized spiral. Saff, Kuijlaars (1997) © Springer and E. Saff

♦

Uniformly distributed and optimal (?) infinite subdivisions. For those who don't fear the infinite, a good subdivision can be defined for an infinite sequence of points $\{x_i\}_{i=1,2,...}$ by means of an integral. We say that the sequence is *asymptotically uniformly distributed* if for each function f on the sphere we have:

$$\lim_{N\to\infty} \left| \frac{1}{N} \sum_{i=1,...,N} f(x_i) - \frac{1}{4\pi} \int_{S^2} f(x)\,dx \right| = 0.$$

On the other hand, the geometer will prefer to count points in the spherical caps A, that being more visual. We have a good subdivision if for any cap A, the average number of points it contains tends to a limit, i.e. towards the mean area of the cap. In a formula:

$$\lim_{N\to\infty} \frac{\#\{1 \leq i \leq N \ : \ x_i \in A\}}{N} = \frac{\text{Area(A)}}{4\pi}.$$

The analyst won't have any difficulty seeing the equivalence of these two definitions, for we can recover each continuous function with the help of characteristic functions of caps (the characteristic function of a set has value 1 on the set and value 0 on the complement). On the other hand, we repeat here that what matters in practice for having a manageable number of points isn't that there is a limit, but that we have an estimate of the speed of convergence, i.e. that we have integrals of the form

$$\left| \frac{1}{N} \sum_{i=1,...,N} f(x_i) - \frac{1}{4\pi} \int_{S^2} f(x)\,dx \right| < \text{"something in N"}$$

and the same for caps. In Lubotsky, Phillips, and Sarnak (1987) we find a rather revolutionary method, but it answers the question in an optimal manner only with two disadvantages that we will examine after first having seen how the sequences of points are constructed. We give the simplest case, but the other constructions of the authors are completely analogous, for the method is very general. For more information, beyond the original paper, see Colin de Verdière (1989) and the book Sarnak (1990); see also the very similar Lubotzky (1994). Readers will perceive a rather lofty ascent of Jacob's ladder.

$$\cos \alpha = -\frac{3}{5}$$

Fig. III.3.17.

We fix three rectangular coordinate axes D, E, F in space and perform rotations about each of these axes through the angle 126° 52′, more precisely the angle α such that $\cos \alpha = -3/5$. The number $-3/5$ is less mysterious than it seems; we also find it when we attempt to construct discrete subfields of the quaternions fields (recall that quaternions provide an algebraic description of the rotations of $\mathbb{E}^3$, see 8.9 of [B]). We then take an arbitrary point x on the sphere and submit it to these rotations, composed in all possible ways, a certain number of times. We arrange the elements of the group G generated by these rotations in an infinite sequence $\{\gamma_i\}$, the indexing being made for example by the length of the words required for writing an element. The sequences of points considered by these authors are the various **orbits** of the group G acting on a point x of S^2. With each point x we thus associate a sequence $\{\gamma_i(x)\}$ and we can consider, for each function f on the sphere, the absolute value of the difference

$$\left| \frac{1}{N} \sum_{i=1,\dots,N} f(\gamma_i(x)) - \frac{1}{4\pi} \int_{S^2} f(x)\,dx \right|.$$

It isn't every difference that the authors majorize (and here is the first difficulty), but the mean in x of these differences with respect to the norm L^2. The exact result is:

$$\left(\int_{S^2} \left(\frac{1}{N} \sum_{i=1,\dots,N} f(\gamma_i(x)) - \frac{1}{4\pi} \int_{S^2} f(x)\,dx \right)^2 dx \right)^{1/2} = \|f\|_2 \frac{\log N}{\sqrt{N}}$$

for every numerical function, where $\|f\|_2 = (\int_{S^2} |f|^2)^{1/2}$. That is not to say that the orbits are all good for all functions f; it says only that, for each given function f, the orbits $\{\gamma_i(x)\}$ will be good for this given function f and for the majority of initial points x. The authors have another result, also optimal, which says this: for each function f and each point x we have

$$\left| \frac{1}{N} \sum_{i=1,\dots,N} f(\gamma_i(x)) - \frac{1}{4\pi} \int_{S^2} f(x)\,dx \right| \leq C(f) \frac{(\log N)^{3/2}}{N^{1/2}},$$

where the constant $C(f)$ depends only on the Lipschitz constant of f and on the integral (Sobolev) norm $(\int_{S^2} |df|^2)^{1/2}$. Here matters are thus optimal only when the function is well-controlled; this is the second of the difficulties we mention.

From this result, with the help of a good grasp of spherical harmonics, the authors deduce, for any initial point x, that its orbit is uniformly distributed, moreover with the following estimate in N, valid for each spherical cap A:

$$\left| \frac{\#\{1 \leq i \leq N \,:\, \gamma_i(x) \in A\}}{N} - \frac{\text{Area}(A)}{4\pi} \right| = O(\log N / N^{1/3}).$$

Here are several pictures of orbits, taken from Lubotsky et al. (1987). They have been zoomed in on as indicated.

The pictures help convince us that the subdivisions are indeed uniform; also that they have a completely different aspect from what we saw in Slazenger or other

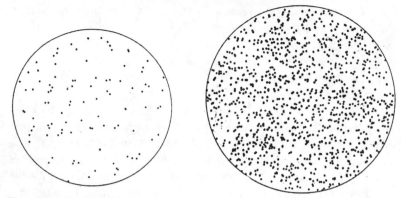

Fig. III.3.18. Number of points: 23 437. *At left*: area equal to 0.00444 of the total area.
At right: area equal to 0.05082 of the total area

figures given previously; this is why we have to distrust our intuition. However, if we know that the "O" for the functions is optimal for the distribution in the caps, as it was in the restricted sense seen earlier, the power $N^{1/3}$ isn't optimal; we can rightfully hope one day to have something in $N^{1/2}$.

Most extraordinary in Lubotsky et al. (1987) are the mathematical tools necessary for **proving** that things go well, better than the best that can be done, in a sense that we will define. They have shown that, if we proceed with a discrete group, there will be a limit for the optimal efficacy of the good distribution with respect to the L^2 norms above, and that this limit is obtained by their procedure. The idea is to analyze the action of such a group G on the sphere by regarding its effect on the various spaces of spherical harmonics (see Sect. III.XYZ). And this finally brings in number theory. In particular, in the end they use Deligne's proof of the Weil conjectures. These conjectures of André Weil, formulated in 1949, concern arithmetic, i.e. number theory. They are very abstract and involve a mixture of the geometry of curves and arithmetic. The curves in question will be encountered in connection with the error correcting codes in Sect. X.7. These conjectures were the object of longing of very many mathematicians. It was in 1973 that Deligne succeeded in proving them, using a whole arsenal of very abstract mathematics amassed by Grothendieck from 1960 to 1970. The truth of these conjectures implies in particular that of a conjecture of Ramanujan, which is very easy to state. It has to with knowing the number of ways an integer n can be decomposed as the sum of 24 squares of integers x_i $(i = 1, \ldots, 24) : n = x_1^2 + \cdots + x_{24}^2$. Geometers prefer to view these numbers as points of the lattice $\mathbb{Z}^{24}$ in $\mathbb{R}^{24}$ which are situated on the sphere in $\mathbb{R}^{24}$ with center O and radius n. Here again we are interested in the asymptotic behavior of the number $r(n)$ when n is large. In fact there exist theoretical formulas for calculating $r(n)$, but these are not explicit. It is known that the first order of magnitude is $\frac{16}{691}\sigma_{11}(n)$, where $\sigma_{11}(n)$ is equal to the sum of the eleventh powers of all the divisors of n. Between 1910 and 1920, Ramanujan conjectured that the growth in n isn't very strong, precisely that we have $r(n) = \frac{16}{691}\sigma_{11}(n) - \frac{128\,259}{691}\tau(n)$ with, for $\tau(n)$, the precise inclusion $-2n^{11/2} < \tau(n) < 2n^{11/2}$.

We encounter the dimension 24 at least two times: i.e. in the next section, then again with the error correcting codes of Sect. X.7. Readers who likes unexpected relationships will want to know that the enumerations relative to sums of squares are encountered naturally in problems of frequency of vibrations in balls; for this see Sect. IX.2.

The technique of the authors makes essential use of *modular forms*. These are series that are essential in number theory, but the book (Sarnak, 1990) already mentioned shows their many applications, including the one above to points on the sphere and another to the uniqueness of measure on the sphere; see Sect. III.6 below. Finally, there is an application to combinatorial geometry, i.e. to the problem of bipartite graphs. Let us repeat that we do practically no combinatorial geometry in this work, despite its growing importance, but we mention it several times; for this subject see the book Pach and Agarwal (1995). The problem is of practical importance: it deals with connecting two networks of n persons by cables (or any other theoretically comparable means), and this in an economic way. To join persons two at a time requires n^2 cables, which is prohibitive when n is large. It is necessary to proceed otherwise. Without entering into the details, let us say that the theoretical existence of optimal graphs has been established (based on a computational calculus), but that the first explicit construction giving optimal results can be found in Lubotsky, Phillips and Sarnak (1988); see the book Lubotzky (1994). Discrete quaternions are used here too and above all the Ramanujan conjectures.

The radically different approach of Bourgain and Lindenstrauss (1993) merits special mention; it uses the notion of *zonotope*, which we won't have time to discuss in Chap. VII, even though this is an important type in convexity theory. Refer to the end of Giannopoulos and Milman (2001). It is a topic once again full of mysteries; see the index of Gruber and Wills (1993).

We find there a striking illustration of the true nature of mathematics: the structures and the concepts that it creates are paradoxically the most natural and, at the same time, the most abstract. This is why its constructions can be applied suddenly to totally unexpected contexts. Not only in mathematics, but in all of science. The platonists don't lack for evidence to bolster their arguments. As for us, our view is that the nature of mathematics is halfway between Platonism and reductionism, but perhaps such a compromise is not very courageous? For this category of questions, see Berger (2001).

In spite of its seductiveness for the mathematician the preceding method is not practically applicable, for the results are not optimal in a probabilistic sense. Just recently double progress has been made, in both theory and practice, this with the physicist's approach to minimizing the potential energy. We return to the basic reference (Hardin and Saff, 2004) for more details.

We begin with some notation. We denote by ω_N any set of N different points of the sphere:

$$\omega_N = \{\text{N different points } x_1, \ldots, x_N \text{ of } S^2\}$$

and define the associated *s-energy*:

$$E_s(\omega_N) = \sum_{i \neq j} \frac{1}{|x_i - x_j|^s}.$$

For $s = 1$ this is the Coulomb–Thompson potential, for $s = \infty$ these are the enemy dictators. And for $s = 0$ we get

$$E_0(\omega_N) = \sum_{i \neq j} \log \frac{1}{|x_i - x_j|},$$

or simply the product (for maximization)

$$\prod_{i \neq j} |x_i - x_j|.$$

We recall (see above) the relation between S^2 and the solution of polynomial equations $P = 0$. In Newton's method what matters is the pair $(P(z_0), P'(z_0))$, but only within a scalar multiple: now the $(z : z')$ defined within a scalar is $\mathbb{C}P^1 = S^2$.

The two basic questions are:

(Q_1) *If* $E_s(N)$ denotes *the minimal energy*

$$E_s(N) = \inf\{E_s(\omega_N) : \omega_N \subset S^2\},$$

how are the configurations for minimal energy distributed for large values of N *and how can we construct them?*

(Q_2) *What is the asymptotic of the minimal energy as a function of* N *and* s*?*

Presently we find in Kuijlaars and Saff (1998) that the basic answer is known for Q_2 and that, for all the s-energies when $0 < s < 2$, we have

$$E_s(N) \approx_{N \to \infty} C(s) \cdot N^2.$$

For $s = 2$, the order is in $N \cdot \log N$, for $s > 2$ it is in $N^{1+s/2}$. The constants $C(s)$ are explicit, for $0 < s < 2$. The proof uses analysis, i.e. potential theory and spherical harmonics (see Sect. III.XYZ). But in addition the configurations for minimal energy are asymptotically uniformly distributed. Attention: this says nothing about the question Q_1. As for the case $s = 0$ of Smale, the conjecture is that an ω_N can be found in polynomial time (by algorithms to be precisely constructed) such that $E_0(\omega_N) \leqslant E_0(N) + C \cdot \log N$. The technique of spirals and that of constant areas, seen previously and cited in Rakhmanov, Saff, and Zhou (1994) and Saff and Kuijlaars (1997), go in this direction but haven't yet provided everything that is desired. They are however good up to $N = 12\,000$ with $C = 114$.

Now Q_1 is a totally different story, among other problems that rely on the power of computers. But even with this approach the supplementary difficulty of finding the minimum and the corresponding configurations is considerable, for there is a number exponential in N of *local* minima that are extremely close to the true minimum. Apart from the references given above, for the experimental point of view see Katanforoush and Shahshahani (2003).

However two types of recent work, working with the N in excess of 1 000 points, contradicts the intuition, which may induce configurations for several hundreds of points from configurations seen above. In those previous configurations, we found twelve pentagons (based on the regular icosahedron), and all the other cells were hexagons. Here on the contrary we see heptagons appear, which are akin to pentagons; they are arranged along the edges of the icosahedron, forming what are called *cicatrices*. Cicatrices are found starting with $N = 300$.

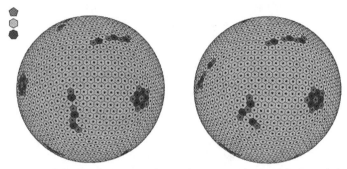

Fig. III.3.19. Almost optimal configurations for 1 600 points and for $s = 1$ (*at left*) and $s = 4$ (*at right*). Hardin, Saff (2004) © Ed. Saff

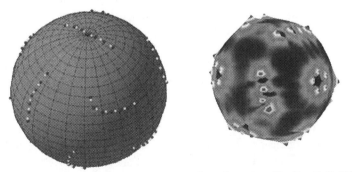

Fig. III.3.20. A physics experiment on condensed matter. Hardin, Saff (2004) © Ed. Saff

In the physics of condensed matter, there is no mention of cicatrices, but rather of "defects" and rather of "disclinations". In Busch et al., *Science* 2003 physical experiments are done with beads of polystyrene (one micron in diameter) attached to drops of water immersed in an oily mixture; "cicatrices" are also found.

Now this has to do with evaluating the uniformity of the above distributions. For the energy method, the method of point counting in the caps is replaced by the evaluation of point energies. These are the

$$E_s(x_i) = \sum_{j=1...N; j \neq i} |x_i - x_j|^{-s}.$$

Figure III.3.20 represents the point energies for the figure with $s = 1$; the hexagons are the dominant sea, whereas the pentagonal points have higher energy and the heptagonal a lower energy.

The conclusion of experts is that we will need to wait until configurations with values of N of order 10 000 or more are obtained in order to have really good distributions. Another big problem is to find, for the $E_s(N)$, at least the second term for their asymptotic evaluation. Thus the mission is likely possible, but tens of thousand of points are needed.

III.4. The kissing number of S^2, alias the hard problem of the thirteenth sphere

The *kissing number* has to do with knowing how many solid and disjoint balls, all of radius 1, can be arranged in space so as to touch the unit ball. Two balls tangent to S^2 will be exterior to each other precisely if their two points of contact p, q satisfy $d(p,q) \geq \pi/3$. The problem here is thus a sort of partial inverse to that of the enemy dictators from the preceding section: we seek the maximum number of points that can be put on the sphere whose distances from one another are at least equal to a given number, but whose value here is fixed at $\pi/3$. Or again, to stack spherical caps of radius $\pi/6$. This optimal number is called the *kissing number* or sometimes the *contact number*. We can see why it is called *Newton's number* on p. 849 (Sect. 5.3, Chap. 3.3) of Gruber and Wills (1993), but ask whether it wasn't for the sake of propriety that the authors abandoned now classic terminology. Interest in this number is evident when we seek to stack spheres in space in a densest possible way, a problem we treat amply in Sect. X.1. But in each case the process begins around a first sphere. The problem can be posed in all dimensions; we will see some of this in Sect. III.5. For the case of the plane, this number is clearly equal to six, for on the circle S^1 things are trivial, because they are somehow *linear*. In the space of three dimensions we can easily distribute 12 such balls, or 12 points, on the unit sphere. Impinging a bit on Chap. X, we can find them by regarding a bunch of canon balls or seeing how grocers stack round fruit. The geometer, worshiper of symmetries, may prefer the much more regular configuration furnished by the vertices of a regular icosahedron, conditional on verifying that the distance between vertices (to be calculated) is just enough greater than $\pi/3$, i.e. 63° 44′. We can thus still move the spheres, and enough to be able to permute them among each other (always without interpenetration) in an adept way; it is the *planetarium* of Figure 1.7 of Conway and Sloane (1999); see below Fig. III.4.1 below. We thus arrive at stacking configurations, even though these configurations have a gridlocked appearance:

it is surely thus for the spheres one at a time, but not if we move them all simultaneously in order to free up a bit of space and arrive at the regular icosahedron; see the details of the proof on p. 29 of Conway and Sloane (1999). We will return to this exasperating, but unavoidable, situation in Sect. III.7 and above all in Sect. X.2.

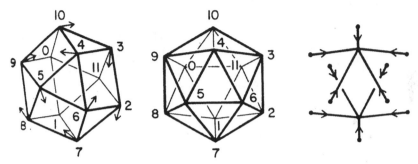

Fig. III.4.1. How to permute the planets. Conway, Sloane (1999) © Springer

But the problem we started with was the *maximum* number: can we add a thirteenth ball? An evident motivation being to obtain better stackings than those that are known. The problem was the object of an animated controversy between Newton and Gregory around 1694. Gregory assured Newton that it was possible to place a thirteenth sphere by an argument from the calculus of areas: the total area of the sphere is 4π while the area of a spherical cap of radius $\pi/6$ equals $2\pi(1 - \frac{1}{\sqrt{3}})$, thus the quotient $4\pi/2\pi(1 - \frac{1}{\sqrt{3}}) = 14.9280...$ is not only greater than 13, but also 14. Newton was too good a mathematician to believe Gregory: a measure argument isn't sufficient for an essentially metric problem. There then came several purported proofs of the impossibility of a thirteenth sphere, but all contained errors; spherical geometry is treacherous enough, as we will see again in connection with Cauchy's theorem in Sect. VIII.6. The first presumably correct proof that at most twelve spheres are possible is in Schütte and van der Waerden (1953). It is long and in German; it isn't clear to this author whether anyone has really read this proof in its entirety, and whether it is complete. There is, let us say, an epistemological-sociological reason for this and what follows. Publishing a complete and careful proof won't bring glory to its author, for the result is "evident" or "well known".

Here are some references: first (Leech, 1956), an extremely short text which was long thought to be the obligatory reference. Our colleague Marjorie Senechal wanted to achieve clarity and organized a seminar to complete Leech, but after some time the participants abandoned. A sad part of the history, but which may be consoling to those who believed themselves infallible and didn't recognize the traps of spherical geometry. In the first edition of Aigner and Ziegler (1998) the proof given rested on a lemma that was clearly false. In spite of their efforts the authors weren't able to set up a complete (correct) proof, and in the second edition of this remarkable book the theorem of the thirteenth sphere no longer figured. Subsequently this author has received several new purported proofs, but admits to not having verified any in detail. Here are a few references on the subject: Pfender and Ziegler (2004), Casselman (2004), Anstreicher (2004), Krauth and Loebl (2004).

Why is this so difficult? The initial idea is so simple: we triangulate the sphere with the given points as vertices. By hypothesis, all these triangles have three sides of length $\geq \pi/3$. The naivety consists then of believing that these triangles have an area greater than that of an equilateral spherical triangle of side $\pi/3$. Then 13 or more triangles would have a sum of areas more than the total area of the sphere. But it is grossly false for triangles that are too flattened, since the area can be very, very small, indeed nil. It is thus necessary to choose a triangulation, or a tiling, by spherical polygons for which the angles aren't too small, and to thus obtain a reduction of their areas. Matters are simple enough for the elimination of 14 vertices, but for 13 it is much more difficult. We refer to the cited texts to see the various attacks.

It seems that there can't exist a truly conceptual proof; the reason is that the twelve spheres aren't rigid — contrary to the case of the plane (6 disks, trivial), which is a rigid situation. It is interesting to know that the attempts at conceptual proofs, i.e. using linear programming and spherical harmonics, don't allow the conclusion in dimension 3 (14 spheres on S^2) but "easily" allow the conclusion in dimension 8 (with 240 spheres on S^7) and in dimension 24 (with 196 560 spheres on S^{23}) for the proof shows moreover that the situation is rigid, unique (within isometry of course), i.e. we can't move anything at all; see Sect. X.2 in this regard.

III.5. Four open problems for the sphere S^3

Two open problems for S^3 concern the volume of its tetrahedra, i.e. the figure formed by four points not in the same plane, its *faces* being spherical triangles. The total volume of S^3 is equal to $2\pi^2$ (see Sect. VII.6.B). In contrast to the case of dimension 2, where we have two very simple formulas for calculating the area σ of a triangle: $\sigma = \alpha + \beta + \gamma - \pi$ as a function of the angles, and $\tan\frac{\sigma}{4} = \sqrt{\sin p \sin(p-a) \sin(p-b) \sin(p-c)}$ as a function of the sides, for the tetrahedra of S^3 there does not exist any simple formula. Here are two problems on them without solution up to the present.

The first is that of the isoperimetric inequality: to find the tetrahedra of S^3 for which the volume is fixed and the area (i.e. the sum of the areas of its four faces) is *minimal*. The problem is solved for Euclidean space $\mathbb{E}^3$: the minimum is attained by regular (equal sides) tetrahedra, and for these only. We will see in Sect. VIII.7 what the solution is more generally for the isometric inequalities of polyhedra. In fact, the same conclusion is valid in hyperbolic space, but more difficult to show: Fejes Tóth (1963). On the other hand, the problem is open in S^3. We merely note for enthusiasts for algorithms and complexity that this problem is subject to **Tarski's principle**: each problem that can be written as the solution of a finite number of equalities and (real) polynomial inequalities may be entered into a computer; but this isn't sufficient for our problem. In Sect. X.5 we will encounter other problems of this type, involving the combinatorics of polyhedra in Sects. VIII.10 and VIII.12 and also involving Hermite's constant.

The second problem begins just as simply: the formula $\sigma = \alpha + \beta + \gamma - \pi$ implies that, if a triangle of S^2 has angles that are rational multiples of π, then its surface area too is rational to π, or again rational relative to the total area of the sphere, which equals 4π. Intuitively, this is not unreasonable: if we think about angles of the form π/n, where n is an integer, we then think of being able to **tile** the sphere with copies of our triangle, and thus the result is evident. In the case of rationals more general than the $1/n$, we might think of an argument by a covering the sphere. The same question thus in dimension 3: if a tetrahedron of S^3 has each of its dihedral angles (six in number) rational to π, is its volume also rational to $2\pi^2$, the volume of S^3? We must be very clear at the outset that these aren't the *solid angles* at the four vertices, as in Sect. III.3, that are interesting here, but rather the dihedral angles, for it is these that control the geometry of the tetrahedron and enter also, by the way, into problems on tilings. Unfortunately, even if the dihedral angles have the form π/n, n an integer, and even if tiling proceeds well about the edges, the catastrophes occur at the vertices; we will encounter this difficulty in Sect. XI.11.

We must definitely mention the surprising fact: this question of rationality has come up recently, and in an essential way, in the theory of the *secondary characteristic classes* of Chern, Cheeger and Simons. There is an historical explanation of this ascent of the ladder in Simons (1998). It can be seen there that this rationality hypothesis is, underneath its simple appearance, actually extremely profound – and still open.

The kissing number of S^3 is now known to be exactly equal to 24; this is completely recent; see Musin (2004). The proof is based on the technique explained in Sect. III.7 below.

Lawson's conjecture concerns the minimal surfaces in S^3: certainly the surfaces which are equators are minimal, but are they the only minimal surfaces of the topological type of the sphere? The answer is affirmative, a result that we might now call classic. The general idea is the same as that of Hopf in Sect. VI.8: to exploit the complex structure of the sphere. But then, in a topology that is just a bit more complicated, we can ask for the tori that are minimally embedded in S^3. We know them, they are the Clifford tori, i.e. those introduced in Sect. II.6; here we use the name only for tori made up of points equidistant from two orthogonal circles of S^3. Lawson's conjecture is that there aren't any others that are minimal. For a generalization, due to Calabi, from the study of minimal spheres in spheres of higher dimension, see the informal text (Berger, 1993b) and the references mentioned there, in particular Michelsohn (1987). See also the nice figure by Herman Gluck: Fig. VII.6.3.

The *Poincaré conjecture* does not fit the spirit of this book very well; we mention it to allow our ignorance of dimension 3 to be perceived once again. Moreover, it is trivial to state, once the notions of manifold and simply connected are known. For the notion of abstract manifold, readers can consult any work on differential geometry, e.g. Berger and Gostiaux (1987). In fact, the conjecture is stated in the same way for manifolds that are only topological. *Simply connected* means that every *closed curve* is contractible to a point; see also the end of Sect. V.XYZ. The circle is not simply connected, but in any dimension greater than 1 a sphere is always simply connected, although we must pay special attention to the proof when the closed curve isn't differentiable. Poincaré's conjecture maintains that every compact and simply connected manifold of dimension 3 is in fact the topological sphere S^3. It forms, with "Fermat's theorem" and the Riemann hypothesis on the zeros of the ζ function, the set of three great mathematical conjectures. Now that Fermat's theorem has been proved, there remain no more than two "very great" conjectures. As for seniority, the (merely "great"?) conjecture of Kepler was older, but we will see in Sect. X.2 that it has just very recently been proved.

Perelman has announced a proof of the Poincaré conjecture. To learn more, read the end of Sect. X.3 for the evidence, as well as Milnor (2003), Anderson (2005), Morgan (2005). There thus remains today only the Riemann hypothesis.

♦

It should also be mentioned that the sphere S^3 is an important object for comprehending the geometry of singularities of an algebraic curve in the complex plane (end of Chap. V): we cut the curve by small spheres centered at the singularity to be studied; these spheres are in a space of real dimension 4, thus spheres S^3. This study is perfectly motivated and presented in the book (Brieskorn & Knörrer, 1986).

III.6. A problem of Banach–Ruziewicz: the uniqueness of canonical measure

It can be shown easily enough by analysis (by performing a convolution to make things regular in the sense of measure, i.e. an integral of translation with an infinitely differentiable function) that the canonical measure on the sphere is the only measure (within a scalar multiple) that is invariant under the group O(3) of isometries. For, since O(3) is transitive on the sphere, if the measure is given by a function in terms of canonical measure, this function must be constant. Recall that a Lebesgue measure must be additive for **denumerable** unions of disjoint open sets; in 1920 Ruziewicz posed the question of knowing what happens if we only require the measure considered be additive for **finite** unions: does there exist, under this weaker hypothesis, a noncanonical measure invariant under the group O(3)?

Since Banach in 1921 we know that such measures exist abundantly (which is clearly appalling) on the circle S^1, even if they are very pathological. For their construction, it seems best to consult the book (Sarnak, 1990), since we have seen that there he also treats the problem of Sect. III.3; but there is also Lubotzky (1994), which is quite similar. We also find out what happens in higher dimensions. The answer is that, starting with dimension 2, there is no longer another invariant measure that is both finitely additive and canonical, but the proofs are all very involved and very abstract. Uniqueness itself is not so surprising, for the group $O(3)$ on S^2, then $O(n+1)$ on S^n, is very active, much more than just transitive; the subgroup – called an *isotropy group* – that leaves a point fixed, and which is $O(n)$, is a good evaluator of this "abundance". What *is* surprising: the difficulty of all these proofs. Numerous mathematicians have attacked this very natural problem. Margulis in 1980 and Dennis Sullivan in 1981 showed the desired uniqueness by using that $O(n)$, starting with $n = 4$, is a really large group; the conceptual tool was the notion of a group having Kazdan's *property* (T), a powerful tool invented in 1967. The whole book (de la Harpe & Valette, 1989) is dedicated to property (T), and its Chap. 7 to the present problem. But for S^2 the group $O(3)$ isn't all that big, the isotropy group is just the circle S^1 (doubled, to be precise), which is commutative. In 1984 Drinfeld proved the conjecture for S^3 and S^4, but this necessitated not only a quite conceptual use of automorphic forms, but above all the truth of the Ramanujan conjectures established by Deligne (see above in Sect. III.3), and we might well ask what number theory has to do with all this. The proof given in Sarnak (1990) is new and the first that runs through all dimensions at once. But it too uses Deligne's solution of the Ramanujan conjectures. The basic tool is the construction on the sphere of appropriate sets, a bit like in Sect. III.3.

III.7. A conceptual approach for the kissing number in arbitrary dimension

The spheres of higher dimension (volumes, isoperimeters, etc.) will be amply studied in Chap. VII on convex sets. However, we treat here the problem of the "kissing number" (number of contacts) in arbitrary dimension, as much for reasons of equilibrium between chapters as for giving an interesting conceptual ascent. The notion of the *kissing number* $\tau(n)$ is defined in all dimensions, in addition to the dimension 2 which was treated in Sect. III.4. The kissing number in dimension n, for the sphere S^{n-1}, is the maximum number of solid spheres of unit radius that can touch the unit ball of the sphere S^{n-1} without interpenetration. It is thus the maximum number of points that we can put on S^{n-1} such that the distances between pairs is never less than $\pi/3$.

Which conceptual approach can we expect for this problem? If we ask physicists, they will say right off: "*why not use spherical harmonics?*" We will speak about this briefly in Sect. III.XYZ. The fundamental idea is that the spherical harmonics are the basic functions on the sphere; every other function on the sphere can be expressed as a convergent series of them, just as any periodic function on the line (i.e. on the circle S^1) possesses a Fourier series composed of sines and cosines. It is thus natural to treat a problem (of distances or otherwise) on the sphere by using these harmonics, all the more since the first spherical harmonics are precisely the cosines of distances to a point, in fact restrictions to the sphere of linear functions. We give this method of attack in some detail, first because it is conceptual, but above all because it yields the optimal answer in dimensions 2, 8 and 24 (and better; see below).

This approach, and the results it furnishes, date from Odlysko and Sloane (1979); it also figures in Chap. 13 of Conway and Sloane (1993). Here are the essential stages. We need to adeptly find a suitable polynomial $f(t)$, for $t \in [-1, 1]$, that satisfies the hypotheses to come. We should remark that here in fact t represents the cosine of distance on the sphere, which goes from 1 to -1 as the angle goes from 0 to π. Then we have the result that is key to all:

(III.7.1) *Suppose that the polynomial $f(t)$ satisfies first $f(t) < 0$ for $t \in [-1, 1/2]$ and then that, when expressed as a sum $f = \sum_i f_i Z_i^{(n)}$ of zonal spherical harmonics, its coefficients satisfy $f(0) > 0$ and $f_i \geq 0$ for positive i. Then the kissing number $\tau(n)$ for dimension n satisfies the inequality $\tau(n) < \frac{f(1)}{f(0)}$.*

If we make the change of variable $t = \cos d$, we see that the underlying idea is to find a function f^* of the distance on the sphere that is negative for those pairs of points for which the distance is less than $\pi/3$, points that don't count in the enumeration with the help of the function f^*. These are the points that have no right to exist as points of contact with S^2 of the exterior balls that touch it without interpenetration. Here is the proof: we let $A(t)$ denote the distribution equal to the sum, in the sense of distributions, which equals $u(t) \cdot \frac{\delta(t)}{N}$, where $\delta(t)$ is the Dirac distribution in t and $u(t)$ is the number of pairs (x, y) such that $\cos(d(x, y)) = t$. In the calculations that follow, the integrals will be ultimately extended in the sense of distributions. Then the kissing condition of (Sect. III.7.1) says that f^* is negative or zero for all pairs for which the distance is greater than or equal to $\pi/3$, that is for a distribution of "kissing" type, i.e. that $A(t) = 0$ for $t \in {]}1/2, 1[$. And of course $A(1) = N$, the total number of points of the distribution considered. Then: $\int_{-1}^{1} A(t)\, dt = N^2$. Next: $\int_{-1}^{1} A(t)\, dt \leq A(1) f(1)$ from the preceding; but $\int_{-1}^{1} A(t)\, dt = \int_{-1}^{1} \sum_i f_i Z_i^{(n)}\, dt = f_0 \int_{-1}^{1} A(t)\, dt + \sum_{i>0} f_i A(t) Z_i^{(n)}(t)$ and a property of the spherical harmonics is that of having a *positive kernel*, so that for each $i > 0$ and for each distribution of points $\{x_k\}$ we always have $\sum_{k,h} Z_i^{(n)}(x_k . x_h) \geq 0$. We thus have

$$f(1)\mathrm{N} \geq f_0 \int_{-1}^{1} A(t)\, dt \geq \mathrm{N}^2 f_0 , \quad \text{QED.}$$

Now we need to find effective polynomials $f(t)$. Here is what has been discovered up to the present. For the dimensions 8 and 24 we take the polynomials $f(t) = \frac{320}{3}(t + 1)(t + \frac{1}{2})^2 t^2 (t - \frac{1}{2})$ and $f(t) = \frac{1490944}{15}(t + 1)(t + \frac{1}{2})^2(t + \frac{1}{4})^2 t^2 (t - \frac{1}{4})^2 (t - \frac{1}{2})$ respectively. These polynomials are excellent, for they provide the exact value of $\tau(8)$ and $\tau(24)$. In fact, they yield the inequalities $\tau(8) \leq 240$ and $\tau(24) \leq 196560$. But we will see in Chap. X that these numbers are attained with the aid of spheres centered at points of the lattice *denoted* E_8 and Λ_{24} (the Leech lattice). We thus finally have $\tau(8) = 240$ and $\tau(24) = 196560$.

It is very important to know the following: when we follow the path to equality in the proof, we can show that there are as many relations satisfied in the spherical harmonics as arise from the points of contact and that the positions of these points are completely determined within an isometry of the space, i.e. that the configuration of spheres is given completely: it is a **rigid**. Rigid here means that any two configurations of spheres can be derived one from each other by an orthogonal transformation. The details are given in Chap. 14 of Conway and Sloane (1993); it is seen there that the rigidity is still stronger. For the case of the plane and $\tau(2) = 6$, we leave it to readers to see if an effective polynomial can be found following this method so as to determine the contact number 6, even though geometrically matters are trivial.

The drama here is that this method — seemingly perfect theoretically and in practice actually so in dimensions 2, 8, 24 — goes so badly wrong for the other dimensions that it doesn't even furnish, at least for the polynomials $f(t)$ currently known, more than the inequality $\tau(3) \leqslant 13$, whereas we have seen that $\tau(3) = 12$ and $\tau(4) = 24$.

III. XYZ

The *spherical harmonics* provide a perfect Fourier analysis on spheres, even though they don't allow, as we have seen, for the solution of all problems there. They are found in practically all the books on mathematical physics, for example Courant and Hilbert (1953), but also in Berger, Gauduchon, and Mazet (1971), as well as in more specialized books such as Andrews, Askey, and Roy (1999) or Takeuchi (1994). A starting point for finding them is to think of "vibrations of the sphere" or "water waves on the sphere". On the circle, viewed as the periodic line, the functions that give Fourier series — $\sin(nt)$ and $\cos(nt)$ (n any integer) — are in fact solutions of the equation $f'' + \lambda f = 0$, which yields simultaneously the eigenfunctions and eigenvalues (and thus the frequencies): $\lambda = n^2$. The thing that plays the role of the second derivative on the sphere S^{n-1} is the *Laplacian* Δ. We get an infinite sequence of eigenvalues and the associated eigenfunctions, and every reasonable function on the sphere can be expressed as a convergent series in terms of them. These functions are very simple, being the restrictions to the sphere of harmonic polynomials f on the space $\mathbb{R}^n$, i.e. of polynomials f such that $\sum_i \frac{\partial^2 f}{\partial x_i^2} = 0$. The first are thus the linear functions, and their restriction to the sphere are, when normalized, exactly the functions $f(q) = \cos(d(p, q))$ for each point p. Then come the quadratic forms with trace zero, etc.

In fact, we can generate, by linear combinations, all the spherical harmonics using only *zonal functions*. For each nonnegative integer i there exists a unique polynomial $Z_i^{(n)}$ of degree i such that, for each point p of the sphere, the function $Z_i^{(n)}(\cos d(p, \cdot))$ is a spherical harmonic. The complete sequence of these functions for dimension n is *denoted* $\{Z_i^{(n)}\}$.

◆

There exists an *isoperimetric inequality* for the sphere S^2, which says that among all domains of given area, the spherical caps (and only these) attain the minimal perimeter. The proof is the same as for domains of arbitrary surfaces with the powerful technique of GMT (geometric measure theory); see Sect. VI.11. We won't speak about this further.

Bibliography

[B] Berger, M. (1987, 2009) *Geometry I,II.* Berlin/Heidelberg/New York: Springer

[BG] Berger, M., & Gostiaux, B. (1987). *Differential Geometry: Manifolds, Curves and Surfaces.* Berlin/Heidelberg/New York: Springer

Aigner, M., & Ziegler, G. (1998). *Proofs from the book.* Berlin/Heidelberg/New York: Springer

Andrews, G., Askey, R., & Roy, R. (1999). *Special functions.* Cambridge: Cambridge University Press

Anstreicher, K. (2004). The thirteen spheres: A new proof. *Discrete & Computational Geometry, 31,* 613–626

Berger, M. (1992). Les placements de cercles. *Pour la Science, 176,* 72–79

Berger, M. (1993a). Encounter with a geometer: Eugenio Calabi. In P. de Bartolomeis, F. Tricerri, & E. Vesentini (Ed.), *Conference in honour of Eugenio Calabi, Manifolds and geometry (Pisa)* (pp. 20–60). Cambridge: Cambridge University Press

Berger, M. (1993b). Les paquets de cercles. In C. E. Tricerri (Eds.), *Differential geometry and topology.* Alghero: World Scientific

Berger, M. (1998). Riemannian geometry during the second half of the century. *Jahrbericht der Deutsch. Math.-Verein. (DMV), 100,* 45–208

Berger, M. (2001). Peut-on définir la géométrie aujourd'hui? *Results in Mathematics, 40,* 37–87

Berger, M. (2003). *A Panoramic introduction to Riemannian geometry.* Berlin/Heidelberg/New York: Springer

Berger, M., Gauduchon, P., & Mazet, E. (1971). *Le spectre d'une variété riemannienne.* Berlin/Heidelberg/New York: Springer

Berger, M., & Gostiaux, B. (1987). *Géométrie differentielle: variétés, courbes et surfaces.* Paris: Presses Universitaires de France

Bourgain, J., & Lindenstrauss, J. (1993). Approximating the sphere by a Minkowski sum of segments with equal length. *Discrete & Computational Geometry, 9,* 131–144

Brieskorn, E., & Knörrer, H. (1986). *Plane algebraic curves.* Boston: Birkhäuser

Casselman, B. (2004). The difficulties of kissing in three dimensions. *Notices of the American Mathematical Society, 51,* 884–885

Colin de Verdière, Y. (1989). Distribution de points sur une sphère (a la Lubotzky, Phillips and Sarnak). In Séminaire Bourbaki, 1988–1989. In Astéerisque, 177–178, 83–93

Conway, J., & Sloane, N. (1999). *Sphere packings, lattices and groups* (3rd ed.). Berlin/Heidelberg/New York: Springer

Courant, R., & Hilbert, D. (1953). *Methods of mathematical physics.* New York: Wiley

Croft, H., Falconer, K., & Guy, R. (1991). *Unsolved problems in geometry.* Berlin/Heidelberg/New York: Springer

de la Harpe, P., & Valette, A. (1989). *La propriété (T) de Kazdhan pour les groupes localement compacts,* Astérisque 175, Société mathématique de France

Delsarte, P., Goethals, J., & Seidel, J. (1977). Spherical codes and designs. *Geometriae Dedicata, 6,* 363–388

Deschamps, A. (1982). Variétés riemannienne stigmatiques. *Journal de Mathématiques Pures et Appliquée, 4,* 381–400

Fejes Tóth, L. (1963). On the isoperimetric property of the regular hyperbolic tetrahedra. *Mag yar Tud. Akad. matematikai Kutato Intez. Közl, 8*, 53–57

Fejes Tóth, L. (1972). *Lagerungen in der Ebene, auf der Kugel und im Raum* (2nd ed.). Berlin/Heidelberg/New York: Springer

Giannopoulos, A., & Milman, V. (2001). Euclidean structure in finite dimensional normed spaces. In W. B. Johnson & J. Lindenstrauss (Eds.). *Handbook of the geometry of banach spaces* (Vol. 1). New York: Kluwer

Gibbs, J. (1961). *The scientific papers of J. Williard Gibbs* (Vol. 2). New York: Dover

Gromov, M. (1988b). Possible trends in mathematics in the coming decades. *Notices of the American Mathematical Society, 45*, 846–847

Gruber, P., & Wills, J. (Eds.). (1993). *Handbook of convex geometry*. Amsterdam: North-Holland

Habicht, W., & van der Waerden, B. (1951). Lagerungen von Punkten auf der Kugel. *Mathematische Annalen, 128*, 223–234

Hadamard, J. (1911). *Le'cons de géométrie élémentaire*. Paris: Armand Colin, reprint Jacques Gabay, 2004

Hardin, D., & Saff, E. (2004). Discretizing manifolds via minimum energy points. *Notices of the American Mathematical Society, 51*(10), 1186–1195

Helgason, S. (1980). *The radon transform*. Boston: Birkhäuser

Katanforoush, A., & Shahshahani, M. (2003). Distributing points on the sphere, I. *Experimental Mathematics, 12*, 199–210

Krauth, W., & Loebl, M. (2006). *Jamming and geometric representations of graphs*. http://www.emis.de/journals/EJC/Volume_13/PDF/v13i1r56.pdf

Kuijlaars, A., & Saff, E. (1998). Asymptotics for minimal discrete energy on the sphere. *Transactions of the American Mathematical Society, 350*, 523–538

Leech, J. (1956). The problem of the thirteenth sphere. *The Mathematical gazette, 40*, 22–23

Lubotsky, A., Phillips, R., & Sarnak, P. (1987). Hecke operators and distributing points on the sphere, I, II. *Communications on Pure and Applied Mathematics, 39*(40), S149–S186

Lubotsky, A., Phillips, R., & Sarnak, P. (1988). Ramanujan graphs. *Combinatorica, 8*, 261–277

Lubotzky, L. (1994). *Discrete groups, expanding graphs and invariant measures*. Boston: Birkhäuser

Melissen, H. (1997). *Packing and Covering With Circles*. Ph.D. Thesis, Universiteit Utrecht

Michelsohn, M. L. (1987). Surfaces minimales dans les sphères. *Astérisque, Théorie des variétés minimales et applications*, Astérisque 154–155. Société mathématique de France, 131–150

Morgan, F. (2005). Kepler's conjecture and Hales proof – a book review. *Notices of the American Mathematical Society, 52*, 44–47

Odlysko, A., & Sloane, N. (1979). New bounds on the number of unit spheres that can touch a unit sphere in n dimensions. *Journal of Combinatorial Theory. Series A, 26*, 210–214

Pach, J., & Agarwal, P. (1995). *Combinatorial geometry*. New York: Wiley

Pfender, F., & Ziegler, G. (2004). Kissing numbers, sphere packings and some unexpected proofs. *Notices of the American Mathematical Society, 51*, 873–883

Rakhmanov, E., Saff, E., & Zhou, Y. M. (1994). Minimal discrete energy on the sphere. *Mathematical Research Letters, 1*, 647–662

Robinson, R. (1961). Arrangement of 24 points on a sphere. *Mathematische Annalen, 144*, 17–48

Saff, E., & Kuijlaars, A. (1997). Distributing many points on a sphere. *The Mathematical Intelligencer, 19*, 5–11

Sarnak, P. (1990). *Some applications of modular forms*. Cambridge: Cambridge University Press

Schütte, K., & van der Waerden, B. (1953). Das problem der dreizehn Kugeln. *Mathematische Annalen, 125*, 325–334

Shub, M., & Smale, S. (1993). Complexity of Bezout's theorem III: Condition number and packing. *Journal of Complexity, 9*, 4–14

Simons, J. (1998, June). Interview with Jim Simons. The Emissary, MSRI Berkeley, 1–7

Smale, S. (1998). Mathematical problems for the next century. *Mathematical Intelligencer, 20*(2), 11–27

Stewart, I. (1991). Circularly covering clathrin. *Nature, 351*, 103

Stolarski, K. (1973). Sum of distances between points on a sphere. *Proceedings of the American Mathematical Society, 41*, 575–582

Takeuchi, M. (1994). *Modern spherical functions*. Providence: American Mathematical Society

van der Waerden, B. (1952). Punkte auf der Kugel. Drei Zusätze. *Mathematische Annalen, 125*, 213–222

Wagner, G. (1993). On a new method for constructing good point sets on spheres. *Discrete & Computational Geometry, 9*, 111–129

Walker, J. (1979a). More on Boomerangs, including their connection with a climped golf ball. *Scientific American, 240*,

Walker, J. (1979b). The amateur scientist, column: More on boomerangs, including their connection with the dimpled golf ball. *Scientific American, 1979*, 134–139

Chapter IV
Conics and quadrics

IV.1. Motivations, a definition parachuted from the ladder, and why

Even though there isn't to our knowledge any important open problem concerning the conics — for quadrics it's a different story — we are going to stay with them for a long time, but talk about the quadrics only very briefly. We hope, however, that the chapter will please many readers. More knowledgeable — but not necessarily omniscient — readers may skip all the beginning material and just look at Sects. IV.8 and IV.9. Here are our motivations: we have already stated how much the teaching of geometry, however useful it is nowadays, has almost completely disappeared from instruction, whether in middle or upper schools, or in the university. If a few circles remain, the other conic sections are gone, even though they are an integral part of many things in our everyday lives. Here are a few examples, to which readers may append their own.

The summit of the Eiffel tower describes an ellipse (of diameter about a meter in high wind), but also that of every analogous object (the Montparnasse tower, your skyscraper of choice). Even though the amplitude may very, it is never zero. More important is the fact that when you look at a circle, you see it from most viewpoints as an ellipse. In every drawing, in every painting, circles generally appear as ellipses. The shadow cast on a wall by a circular lamp shade is an hyperbola (or in reality one piece of an hyperbola — the Greeks incidentally didn't always work with both pieces), the trajectories of thrown objects (military projectiles or otherwise) are approximately parabolas. The trajectories of the planets of the solar system are approximately ellipses, while those of comets may be any of the three types: ellipse, parabola, hyperbola. Architects frequently use conics, but also sometimes quadrics. For example, when a roof must cover a non-rectangular trapezoid, they construct it in the form of an hyperbolic paraboloid. We also know that ellipses are found in physics (e.g. harmonic oscillators), parabolas and hyperbolas in optics (large and small mirrors of telescopes and automobile headlights), even though in fact paraboloids and hyperboloids of revolution are involved.

But once again triality, the triptych (ellipse, parabola, hyperbola) invades the language and, more profoundly, in mathematics it is the structure and the behavior of the objects concerned. The ellipse and the hyperbola are classic figures of style, whereas the word "parabola" is used with different meanings in different contexts. Gromov likes to compare the triad (elliptic, parabolic, hyperbolic) with the three physical states of matter: solid, liquid and gaseous.

M. Berger, *Geometry Revealed*, DOI 10.1007/978-3-540-70997-8_4,
© Springer-Verlag Berlin Heidelberg 2010

Here now is a picture of these respective realms. We note here the double label "parabolic, subelliptic". In fact, in numerous cases, the intermediate realm — left over from the elliptic and the hyperbolic cases — is not well defined, not well understood, even completely unknown. It's typical in the case of discrete groups. Gromov again likes to compare finite groups to ellipses, the hyperbolic groups (a notion he introduced in 1987 and which dominates the present theory of discrete groups) "to themselves". Like the gases, the hyperbolic groups are expansive: they spread out everywhere. In contrast — typically — there still does not exist today a good definition of what a parabolic group might be. And in one sense there exist/be: what remains between finite groups and hyperbolic groups will require a (likely very subtle) sub-classification. We also know that in physics the solids and liquids are rather badly understood. The case of partial differential equations in analysis is exemplary: elliptic equations (very rigid), parabolic equations (e.g. the heat equation), hyperbolic equations (e.g. the wave equation). For the Riemannian manifolds, the case of positive curvature and the more subtle one of strictly negative curvature correspond well to the terms ellipse and hyperbola, whereas to the contrary the parabolic case is not well defined, for it cannot be solely the case of the manifolds with nonpositive curvature nor that of the manifolds with nonnegative curvature. Last but not least: elliptic geometry is that of the sphere or that of the real projective spaces; hyperbolic geometry is that of the hyperbolic spaces of Chap. II. The name "elliptic curve" that we encounter in various places does not enter the realm of the triptych, since for example in the complex domain they are tori; the name is explained solely by the fact that "elliptic" functions — required for computing the arc length of an ellipse, hence their name — are the very functions that are needed for parameterizing the planar cubic curves; see Sect. V.14.

In Strichartz (1987) there is a rather elaborate table (requiring more than a little theoretical knowledge) of three-fold partitions in several mathematical domains. For the three-fold aspect, e.g. the trinity of fields $\mathbb{R}$, $\mathbb{C}$, $\mathbb{H}$, we have already seen the references Arnold (1999, 2000) at the end of Sect. I.8.

We will now persecute our too impatient readers a bit for wanting to know "what is a conic really"? Here is the right answer:

> *A conic is a curve of second degree in the complex projective plane, i.e. the set of the $(x : y : z)$ satisfying an equation $P(x, y, z) = 0$, where P is a polynomial of second degree in the three variables x, y, z that is not identically zero. (The notation $(x : y : z)$ was introduced in I.5 and denotes triples of complex numbers that are not all zero, taken modulo multiplication by a scalar.)*

Readers may ignore this definition **for the moment**; it will appear again in Sect. IV.7.

In this chapter we pursue three goals more or less simultaneously:

(i) To state the essential properties of the conics and the most beautiful and profound concerning them;

(ii) to explain why and how mathematicians, in studying (i) in the course of history, have been naturally and necessarily led to the doubly abstract but final definition given in the above statement. Doubly abstract in fact, by the introduction of the projective plane — with its points at infinity — and by the introduction of complex numbers — with their "imaginary" aspect. It's thus a climb of three rungs up the ladder: the first — already accomplished by Descartes — replaces metric geometry by an equation (here of second degree), the second projectifies the real affine plane, the third complexifies the real projective plane into a complex projective plane, and all this just for the good definition of a conic. For the two theorems we have chosen as being principal, we need to climb several rungs higher still.

(iii) to say a little about quadrics.

The conics and many of their often subtle properties were known to the Greeks. An ideal historical reference is, among others, Coolidge (1968) and Van der Waerden (1954–1975), Vol. I, p. 247; but for a complete inventory of what was known in that epoch, the chapters dedicated to the conics in the Encyclopedia of the Mathematical Sciences (Dingeldey, 1911) (which is an elaborated translation of the original work in German) is unavoidable, even though it only gives the statements or even just references. This is why, as it seems to us, few recent works are as detailed and complete — for a relatively elementary exposition – as the collection formed by the five chapters 13–17 of [B], to which we can turn for more details about what is treated here. The other references are not always easy to find in libraries, and Dingeldey is of an incredible density.

IV.2. Before Descartes: the real Euclidean conics. Definition and some classical properties

We give at the outset the historical definition, and an etymological translation, of the conics. These are all the plane curves that can be obtained as sections of cones of revolution in Euclidean space. Here are some drawings in space:

Which yield in the plane:

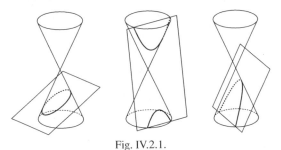

Fig. IV.2.1.

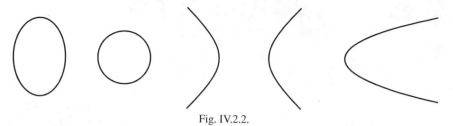

Fig. IV.2.2.

In the Euclidean plane $\mathbb{E}^2$, the simplest curves other than lines are the circles (simpler than lines, by the way, to trace mechanically). We can't stay with circles, for the images of circles under affine/projective transformations are generally not circles. In $\mathbb{E}^2$ there are four types of conics (later we will add the qualification "proper" or "non-degenerate"), specifically:

− circles, the set of which *is denoted by* $\mathcal{C}$;

− ellipses, the set of which *is denoted by* $\mathcal{E}$;

− parabolas, the set of which *is denoted by* $\mathcal{P}$;

− hyperbolas, the set of which *is denoted by* $\mathcal{H}$.

The union of these preceding sets is *denoted by* $\mathcal{K} = \mathcal{C} \cup \mathcal{E} \cup \mathcal{P} \cup \mathcal{H}$. The Greeks knew more or less (often a half hyperbola only), that the union of these four types was exactly all the sections by planes not passing through the apexes of cones of revolution in the Euclidean space $\mathbb{E}^3$, as shown in the above figures.

So − by leaving the plane and going into space − the lines, a circle and a plane suffice for finding all the conics. Note that for the ellipses $\mathcal{E}$ there are two still simpler ways of finding them: by projecting a circle onto a plane or by cutting a cylinder of revolution by a plane not parallel to the axis of the cylinder.

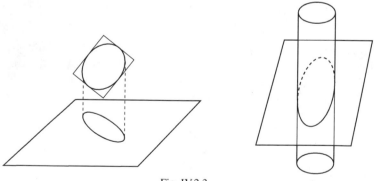

Fig. IV.2.3.

♦

Of course our preference is to find the elements of $\mathcal{K}$ without leaving $\mathbb{E}^2$. Here are the methods:

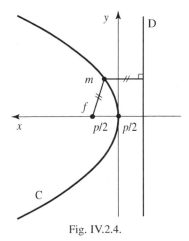

Fig. IV.2.4.

(i) In $\mathbb{E}^2$ let a point f and a line D not containing f be given. The curve P $=$ $\{m \in \mathbb{E}^2 : d(m, f) = d(m, \mathrm{D})\}$ formed of points at equal distance from D and f is a parabola; and conversely each parabola can be obtained in this way. We say that f is the *focus* of P and that D is its *directrix*.

(ii) Same given f, D as in (i) but in addition a real $e \in \,]0, 1[$. Then the curve E $= \{m \in \mathbb{E}^2 : d(m, f) = e \cdot d(m, \mathrm{D})\}$ is an ellipse in $\mathbb{E}^2$, and reciprocally every ellipse can be obtained in this way. We say that f is a *focus* of E and that D is *the associated directrix* for f; in fact there exists (this needs to be shown) another pair (f', D') yielding the same ellipse (for the same e), for ellipses always have a center of symmetry. We say that f and f' are *the* foci of E. The number e is called the *eccentricity*.

(iii) Same data as in (ii) but here $e > 1$. We then obtain all hyperbolas. Same definitions of foci and directrices.

(iv) Let f, f' be two distinct points of $\mathbb{E}^2$ and a a nonzero real such that $2a > d(f, f')$. Then the set E $= \{m \in \mathbb{E}^2 : d(m, f) + d(m, f') = 2a\}$ is an ellipse whose foci are f and f'. Moreover, every ellipse can be obtained in this way.

(v) Same data as in (iv), but here $2a < d(f, f')$. Then H $= \{m \in \mathbb{E}^2 : |d(m, f) - d(m, f')| = 2a\}$ is an hyperbola, with foci f, f'. Moreover, every hyperbola can be obtained in this way. Note that $\{m \in \mathbb{E}^2 : d(m, f) - d(m, f') = 2a\}$ is only a demi-hyperbola.

The Greeks knew these propositions; the definitions (i), (ii), (iii) are referred to as *monofocal* and (iv), (v) as *bifocal*. The property of equal angles (see the figures) for the tangents with the rays emanating from the foci explains the optical properties of conics and the uses indicated below in the section on quadrics. It remains to learn how to show the equivalence of these definitions; we will return to this problem in

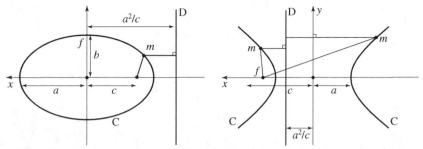

Fig. IV.2.5. Here are shown, for the ellipse and for the hyperbola, the two foci but only one directrix

Sect. III.3. The bifocal definition is generalized thus: gardeners use a rope that goes around two stakes to trace ellipses. But Graves' theorem (see Fig. IV.2.11) says that if a rope goes around an ellipse instead of two points, then the pointer again describes an ellipse (with the same foci as the initial ellipse).

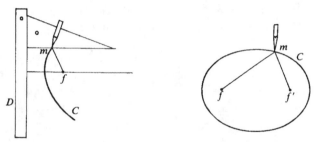

Fig. IV.2.6. The schoolboy's parabola. The gardener's ellipse. II [B] Géométrie. Nathan (1977, 1990) réimp. Cassini (2009) © Nathan Édition

◆

Readers must begin to be somewhat unnerved by this separation of $\mathcal{K}$ into the four parts $\mathcal{C}$, $\mathcal{E}$, $\mathcal{P}$, $\mathcal{H}$ and must feel a desire for unification (without leaving $\mathbb{E}^2$). Actually, two partial unifications are immediate. The first is that $\mathcal{P} \cup \mathcal{E} \cup \mathcal{H}$ can be obtained with the same monofocal definition by letting e take on all positive real values: $\mathcal{P}$ for $e = 1$, $\mathcal{E}$ for $e < 1$ and $\mathcal{H}$ for $e > 1$. But here the circles escape hopelessly: it is in fact the case that the "directrix" of the center $f = 0$ is at infinity, or again that the two foci are merged. This can be demonstrated in an appropriate sense by limits. We begin to sense here the necessity of infinity, i.e. of the projective plane. A second unification allows putting together $\mathcal{C}$, $\mathcal{E}$, $\mathcal{H}$ by the same bifocal definition. To obtain $\mathcal{P}$, it is necessary to transfer the focus f' to infinity.

The Greeks knew – they even had the notion of coordinates but in the complicated language which for them played the role of our modern algebraic language (see Coolidge, 1968) – that the circles, ellipses, hyperbolas and parabolas might be defined by equations of the form

$$\frac{x^2}{a^2} \pm \frac{y^2}{b^2} = 1$$

for

$$\mathscr{C} \cup \mathscr{E} \cup \mathscr{H}$$

and

$$y^2 = 2px$$

for $\mathscr{P}$. The equation of ellipses shows easily that these are the harmonic oscillators of physics: $x = a \cos \omega t$, $y = b \sin \omega t$. For the movement of the planets and comets it suffices to see (and this is easy) that, in polar coordinates, if e is the eccentricity, then every conic can be written

$$\rho = \frac{p}{1 + e \cos \theta} \qquad (p > 0)$$

(the circle is obtained for $e = 0$). We then need to know that when Newton's law of universal attraction is expressed in polar coordinates, the second derivative of $\frac{1}{\rho}$ satisfies the relation

$$\left(\frac{1}{\rho}\right)'' + \frac{1}{\rho} = \text{const.}$$

The calculation takes several lines. We need at first to write everything as a function of time: $\rho(t)$, $\theta(t)$, next apply the law of universal attraction (or gravitation), then eliminate t. We use the fact that, as the acceleration is central, the quantity $\rho^2 \frac{d\theta}{dt}$ remains constant. The law in $1/r^2$ had been presented before Newton, but the above calculation had only succeeded with him and for its time was an absolute *tour de force* of differential calculus. We note that, in the same epoch but before Newton, Hooke had constructed a laboratory machine for verifying if this law indeed furnished "elliptoid" curves and also if it was susceptible to numerical integrations. For the details of this history and the very violent relations between Hooke and Newton, see Arnold (1990), to be read with some caution because the author has an inclination toward the spectacular to the detriment of a completely well documented history, even though he surely is one of the great mathematicians of our time.

The angular properties of tangents are indicated in the figures below. They explain the use of conics in optics, automobile headlights and numerous lamps and — for large distances — telescopes. In a large telescope, the principal mirror is a paraboloid of revolution, but the small mirror which reflects everything to the eye of the astronomer is a piece of an hyperboloid of revolution (see Figs. IV.10.1 and IV.10.2). An acoustic property has been attributed to ellipses, permitting persons situated at the foci to communicate so that no one else can hear (old usage by the ticket punchers of the Paris Metro, older still for confessing lepers). This is true only in small part. The true explanation of such a property, for example that of the circular wall of the Temple of Heaven in Beijing, is the fact that the vibrations of higher frequencies in the interior of a flat space are concentrated near the boundary. In more learned language, the Bessel functions — the harmonics of the planar disk — of a somewhat elevated order are very, very flat close to the center and don't really manifest themselves except very close to the boundary.

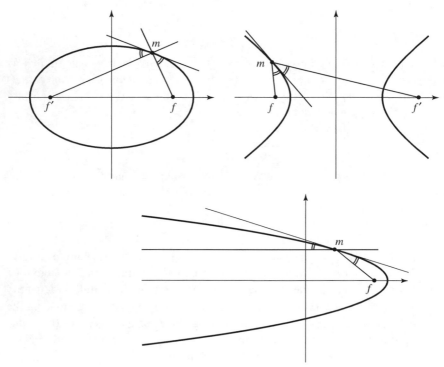

Fig. IV.2.7. In these figures, we see the angular property of the bisectrix for the tangents

♦

A very important figure, encountered in numerous contexts, is that of *homofocal conics*. Analytically the family is written

$$(*) \qquad \frac{x^2}{a^2 + \lambda} + \frac{y^2}{b^2 + \lambda} = 1 \qquad (a > b).$$

We obtain ellipses for $\lambda \in \,]-b^2, \infty[$ and hyperbolas for $]-a^2, -b^2[$. In mathematical physics, this family provides a coordinate system called *elliptic*, by associating with a point the pair $\{\lambda, \mu\}$ furnished by the ellipse and the hyperbola that pass through this point. This is useful for example in studying the proper frequencies (or "eigenfrequencies") of a vibrating elliptic plate; see Morse and Feshbach (1953). The hyperbolas give interference curves when the two foci emit waves. This was already used for radio tracking for some time before GPS. In studying dynamic geometry in Sect. XI.9 we will encounter the homofocal conics while examining

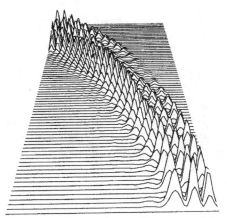

Fig. IV.2.8. Distribution of intensity for the vibrations of the disk associated with the 40-th Bessel functions. McDonald, Kaufman (1988) © American Physical Society

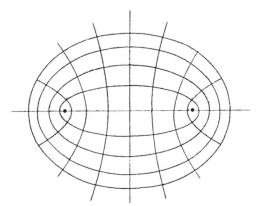

Fig. IV.2.9. Homofocal conics

the trajectories of a billiard ball (or of light) on the interior of an ellipse and we will see revealed a completely atypical phenomenon.

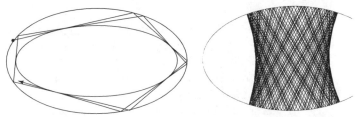

Fig. IV.2.10. Trajectories of a billiard ball in an ellipse (reflections following Descartes' law), programmed by John Hubbard

This property of the billiard ball is equivalent to the fact that the tangent at the point describing the second ellipse is a bisectrix of the two tangents issuing from that point to the interior ellipse. It thus implies the following spectacular consequence: if we trace, on the interior of an ellipse, the trajectory of a light ray (or a billiard ball), this trajectory will always remain tangent to a fixed ellipse (or an hyperbola if we start at a point on the segment joining the foci).

This is to say, in the language of dynamics, that these trajectories will never be everywhere dense since a whole portion of the elliptic domain eludes them. Whereas, for a convex set taken at random, P. Gruber has just shown everywhere density (for almost all the trajectories): see Sect. XI.10.C. We can furthermore say that the ellipses have a *continuous family* of *caustic curves* that fill their whole interior. The problem remains open, at the present moment, as to whether only ellipses have this property; a recent reference is Section 2.4 of Tabachnikov (1995), where a variety of more or less definitive forms that this conjecture (called Birkhoff's) might take are presented. See also Sect. II.4 and, for what happens if the trajectory closes, Sect. IV.8 below.

A final property is Graves' theorem mentioned earlier: if we stretch a cord about an ellipse, the pointer will describe a homofocal ellipse. The analog for quadrics has been awaited for a very long time and remains very difficult; see the result of Staude in Sect. IV.10.

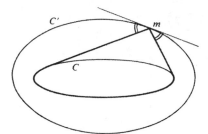

Fig. IV.2.11. Graves' theorem

Here are some remarks, geometric proofs and rather easy calculations that permit us to obtain quite rapidly most of the equivalences between the different definitions of conics.

To obtain the monofocal properties we take coordinates (x, y), where the directrix D is parallel to the $0y$ axis and the focus f is the origin. Then the relation $x^2 + y^2 = e^2(y - c)^2$ becomes, after a change of variable $x \to x + d$, one of the equations $\frac{x^2}{a^2} + \frac{y^2}{b^2} = 1$ that we will obtain below.

For the bifocal definitions, the calculations requires a small trick. If we take $f = (c, 0)$, $f' = (-c, 0)$ and attempt to get rid of the radicals in the equation of definition $\sqrt{(x - c)^2 + y^2} + \sqrt{(x + c)^2 + y^2)} = 2a$, we may well risk going in a metaphorical circle. The starting point consists in calculating mf and mf' while noticing that, if $m = (x, y)$, then

$$mf'^2 - mf^2 = 4cx = (mf' + mf)(mf' - mf) = 2a(mf' - mf).$$

Thus

$$mf' - mf = \frac{2cx}{a}, \quad mf' = a + \frac{cx}{a}, \quad mf = a - \frac{cx}{a},$$

whence

$$mf'^2 + mf^2 = 2x^2 + 2y^2 + 2c^2 = 2a^2 + 2\frac{c^2 x^2}{a^2},$$

and whence the equation

$$\frac{x^2}{a^2} + \frac{y^2}{a^2 - c^2} - 1 = 0.$$

The above calculation — there is an analogous one for the hyperbola — furnishes just as rapidly, in polar coordinates, the equation $\rho = \frac{p}{1+e\cos\theta}$. This then clearly shows what was said above concerning planetary motion; for, $\frac{1}{\rho}$ being linear in $\cos\theta$, we have that $\frac{1}{\rho} + (\frac{1}{\rho})''$ is constant.

Finally a very simple trick: if we start with a focus and a directrix, and if we can show (left to readers, but it isn't obvious) that there is a center of symmetry, then there is a second focus and associated directrix, parallel to the first. Thus $mf + mf' = e(d(m, D) + d(m, D'))$ equals the constant distance between D and D'.

The very brief calculations above have shown very quickly the equivalence of the monofocal and bifocal definitions. The author has never understood, in the country of Descartes, why for so many years and until at least 1950, degree candidates in mathematics had to know how to prove in a purely geometric way (no coordinates — they were forbidden) the equivalence of the monofocal and bifocal definitions. One of the topics of the preparation classes for the "agrégation" (examination for privileged secondary school teaching positions) — at this time there was no choice of subject — was "equivalence of the definitions of a conic". Incredulous readers can consult the book (Lebesgue, 1942) which provided instruction in preparation for the "agrégation" and gives several rather elegant ways of showing this geometric equivalence. Of course Lebesgue didn't have the courage to stop there: after 40 pages of submission to the establishment we find lots of things that we have encountered or will encounter: the focal lines of circles (see Sect. II.7), Poncelet's theorem (see Sect. IV.7), Villarceau's circles (see Sect. II.7), Morley's theorem (see II.1) and Poncelet's polygons of Section 8. In fact, one reason was an attachment to the romance of "pure", i.e. coordinate-free, geometry. Many books speak about this; see Berger (2005) in regard to Poncelet.

We will also want to obtain and unify these metric properties in a still more conceptual way, in light of Cayley's philosophy heralded in Sects. II.1 and II.XYZ: each

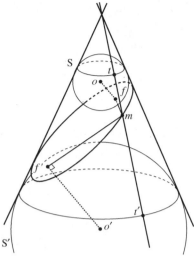

Fig. IV.2.12. Bifocal definition of the ellipse according to Quetelet and Dandelin:
$mf = mt$ (tangential distance to S), $mf' = mt', mf + mf' = mt + mt' = tt' =$ const

geometry arises from complex projective space. In Sect. IV.7 we indicate how this is done; it's also the philosophy ascribed to Plücker.

It wasn't until the nineteenth century that an elementary and elegant and purely spatially geometric way was found for showing that the plane sections of cones satisfy the mono- and bifocal definitions without using the equations of the conics. This was done by two Belgians, Quetelet and Dandelin. The idea is that a cone of revolution, in which a sphere S has been inscribed tangent along a circle in the plane P, can be defined as the locus of points for which the *tangential distance* (the tangential distance of a point to a sphere is the common length of the tangents from the point to the sphere) to S is proportional to the distance to P, the ratio being $\sin \alpha$, if α is the angle at the vertex of the cone. The proof is left to readers as a drawing exercise. They should pay close attention to the circumstance that, even if it is obvious that the points of the cone satisfy this property of distance, it is a little more subtle to see that they are the only ones that do; for the bifocal definition makes use of the fact that the cone is the locus of points for which the sum or difference of the tangential distances of two spheres inscribed in them is a constant. See Figs. IV.2.12 and IV.2.13 below and consult 17.3 of [B] for more details as needed.

The Greeks knew other things that are more difficult, for example that the locus of points for which the distances d, d', d'' to three given lines D, D', D'' satisfy $d^2 = kd'd''$ (k real positive) is always a conic. But they never succeeded in resolving the "locus for four lines", i.e. to prove that the set of points that satisfy $dd' = kd''d'''$, for the distances d, d', d'', d''' to four given lines D, D', D'', D''', is a conic. No more than did their successors until Descartes, whose solution to the problem will appear in the next section; see pp. 20, 21 of Coolidge (1968).

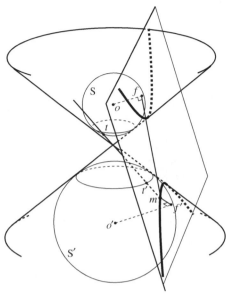

Fig. IV.2.13. Bifocal definition of the ellipse according to Quetelet and Dandelin:
$$|mf - mf'| = |mt - mt'| = tt' = \text{const}$$

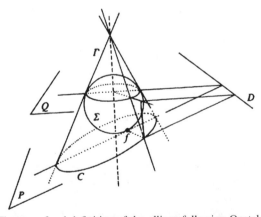

Fig. IV.2.14. The monofocal definition of the ellipse following Quetelet et Dandelin:
$mf = mt$, $mt/mh = \text{const.}$ for all points of the cone, $mh/md = \text{const.}$ for all
points of the plane P, thus $mf/md = \text{const}$

◆

The Greeks knew plenty more things, in particular the *diametral properties*: when a
conic is cut by parallel lines, the midpoints of the segments so obtained are collinear
on a line called a *diameter*. We see however that we only get segments, and almost
never an entire line, and would thus like a general context. This property is purely

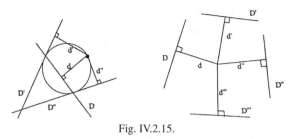

Fig. IV.2.15.

affine, but we might also be interested, in the Euclidean case, in the lengths of the diameters associated with an ellipse. Appolonius knew this result.

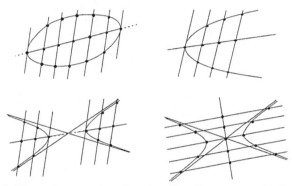

Fig. IV.2.16. Diametric properties of conics. The diameters are the lines passing through the centers for the ellipses and hyperbolas, the lines parallel to the asymptotic direction for a parabola

Let us put matters differently: in the affine plane an ellipse is symmetric with respect to all directions of the plane, as are the circles for the Euclidean symmetries. The diametral properties can thus be seen very quickly analytically, whereas in "pure" geometry we see them for ellipses by transferring to them the diametral property of circles by an affine transformation of the plane. For the hyperbola, readers can try it for themselves.

Do there exist other curves having this property? The answer is no. It is easy enough to put together a proof for the conics, but for the general quadric (in any dimension) it is true, but difficult to prove; see 15.6.9 of [B]. The proof uses the so-called John-Loewner ellipsoid that we will explain in Sect. VII.10.C.

Pascal's theorem is exceptional in the history of the conics by its novelty and its proof not being instantaneous: *the three pairs of opposite sides of an arbitrary hexagon inscribed in a conic intersect in three points that are collinear.* Pascal's theorem is never easy to show, nor is it so very difficult. It poses a problem for historians: Pascal states that it reduces to the case of circles thanks to the "Greek

definition" — but really mostly created by Desargues, of whom Pascal was a disci-
ple, see Berger (2005) — since each conic, being the section of a cone of revolution,
is thus obtained by the spatial projection of a circle. Such a projection preserves
lines and their intersection properties. But Pascal never said how he proved the theorem for a circle. Readers can try it for themselves. Here is such a proof, but it is
not at all purely Euclidean. We easily see that we can, by a projective transformation preserving a circle, ensure that ab and de on the one hand, bc and ef on the
other, are parallel. Then (by an "angular" exercise — use for example Fig. II.2.13)
cd and fa are parallel! For details on the history of Pascal's theorem see p. 33
of Coolidge (1968). Note finally that if the conic is degenerate in two lines, then
Pascal's theorem becomes that of Pappus; see Sect. I.4.

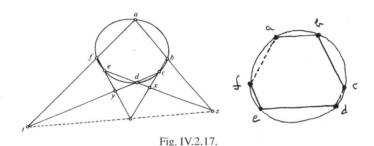

Fig. IV.2.17.

We can use Pascal's result to construct, point by point with a ruler, the conic
that passes through five points of the plane that are in general position. Why does a
well defined conic pass through *five* points of the plane? We will see the answer in
Sect. IV.7, where a proof is again furnished by Pascal's theorem. This existence is
of first order importance in positional astronomy, for to calculate the trajectory of
a planet, it is assumed that it is a conic (at least for an astronomically brief time);
it then suffices to know just five of its positions to determine all the elements of its
orbit. In practice, matters are admittedly more complicated and, moreover, computational resources have become considerably better over the centuries, from tables of
logarithms (astronomy was the only profession to use six place tables, as opposed
to the five places of normal tables) to computers.

◆

In the nineteenth century the duality — *polarity* — associated with a conic came to be
well understood. Here we extract some essential facts from what was given in detail
in Sect. II.XYZ. We begin with a circle and its visual properties. At each point m
of the plane other than the center O of the circle C (whose radius we take to be 1),
we associate its *polar* line m^*, which is the line that is perpendicular to the line Om
and is at a distance $d(O, m*)$ to O such that $Om \cdot d(O, m*) = 1$.

We need to show that this is really a duality. Each line not passing through O
originates from a unique point called its *pole*. All the lines that pass through m have
a pole located on m^* and thus the pole of the line joining two points m and n is

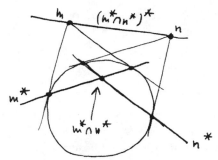

Fig. IV.2.18. Duality interchanges the point of intersection of the polars with the line passing through the poles

nothing other than the point of intersection $m^* \cap n^*$ of the polars of m and n. None of this would get us very far if we didn't know that, if a line D passing through m is such that it cuts the circle in two points p and q, then the four points m, $m^* \cap D$, p, q are in harmonic ratio: $[m, m^* \cap D, p, q] = -1$. This furnishes a geometric construction of the polar with the aid of a complete quadrilateral.

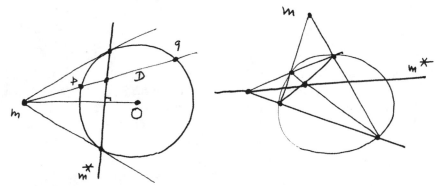

Fig. IV.2.19. *At left*: polar of a point exterior to a circle. *At right*: how to construct the polar of an arbitrary point with the help of the harmonic property of the complete quadrilateral

Before 1950 the pitiable students of the terminal class of the French lycées had to know how to prove all that "by hand with pure geometry". An embellishment had been invented between the two world wars: for two points m and n to be such that $n \in m^*$ (we say that m and n are *conjugate with respect to* C) it is necessary and sufficient that the circle of diameter mn be orthogonal to the circle C. All this was torture due to the prohibition of the scalar product. In fact, let us suppose that the plane is vectorized about O and that the circle has radius 1. Then all the above properties amount to the bilinearity of the scalar product: m, n are conjugate if and only if their scalar product equals 1; the polar of m is the line of the equation $m \cdot n = 1$, etc. We will have noted several complications: it's necessary to remove the center of the circle and the lines that pass through it (projectifying is the remedy for that); then all the lines passing through m don't intersect C (we need to complexify to remedy that). For all this see Sects. IV.4 and IV.6. We have already stigmatized this prohibition in Sects. III.1 and IV.4, but we are sad to have to note that, as recently as 1999 students in the graduating class could (or had to) mention the scalar product but were forbidden to use its bilinearity!

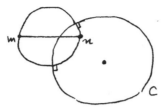

Fig. IV.2.20. The points m and n are conjugate with respect to the circle C

♦

Like the circle, in the affine context any conic C furnishes a duality between points and lines of the plane: at a point p of the plane, we associate the line (called the *polar* of p) containing the harmonic conjugates of p with respect to the sectional segment mn (if it exists). We need to show that these conjugates are all collinear. We see in the preceding array of figures all the exceptions that need be made: the center of an ellipse yields nothing except the conclusion that the polar is the line at infinity. We have a good polarity (duality): the polars D and F of two points p and q have a point of intersection, if it exists, for which the polar is nothing other than the line pq (with some exceptions). Each line (with the exasperating exceptions) is the polar of a well determined point, called its *pole*. And there are all the exceptions, including the intersection at infinity for parallels: it thus becomes intolerable to stay in the affine context: we need to projectify. For example, the diameters will be the polars of the points at infinity. A celebrated example of the use of the duality is Brianchon's theorem: *the diagonals of a hexagon circumscribed about a conic all pass through a common point*. This result is nothing other than the dual of Pascal's theorem.

Fig. IV.2.21. Brianchon's theorem

All of the preceding is more or less difficult geometrically; we will see in Sect. IV.4 that algebraically and in the projective plane it borders on being child's play.

IV.3. The coming of Descartes and the birth of algebraic geometry

Descartes' basic idea is that what is common to the elements of $\mathcal{K} = \mathcal{C} \cup \mathcal{E} \cup \mathcal{H} \cup \mathcal{P}$ is their capability of being defined by an equation of second degree: $\frac{x^2}{a^2} \pm \frac{y^2}{b^2} - 1 = 0$ or $y^2 = 2px$. We remark that an equation of second degree in x, y

$$(*) \qquad\qquad ax^2 + 2b''xy + a'y^2 + 2b'x + 2by + a'' = 0$$

remains one in any affine system of coordinates of the plane $\mathbb{E}^2$. We of course restrict ourselves to equations that are truly of the second degree, i.e. a, b'', a' aren't all zero. Descartes showed that each equation of type $(*)$ can, in an appropriate system of orthogonal coordinates for $\mathbb{E}^2$, be written

$$ax^2 + a'y^2 + a'' = 0 \quad \text{or} \quad ax^2 + 2by = 0.$$

In modern language, this is simply the theorem of linear algebra that permits the diagonalization of real quadratic forms with respect to a positive definite quadratic form, the latter defining the Euclidean structure (and translations of axes). Thus every equation $(*)$ represents an element of $\mathcal{K}$ or else one of the "aberrant" cases that follow (interpreted for the two reduced forms):

- the empty set, if a, a', a'' are all nonzero and of the same sign

- one point only, if a, a' are nonzero and of the same sign, and $a'' = 0$;

- two distinct lines, if a, a' are of different signs, and $a'' = 0$;

- a single line (called in the sequel a *double line* or *two merged lines*), if a is nonzero and a', a' zero.

This leads us to realize (a classic phenomenon) that this unification demands a price — in fact a double price — that need be payed. If we exclude the single point and empty set cases, it is still required to include pairs of lines and the double lines. We will see in Sect. IV.9 that this is necessary for certain problems. We might note, by the way, that the two cases are obtainable geometrically, the first in (iii) of Sect. IV.2 for $f \in D$, the second in case (i) and $f \in D$ again. If we want to remove the aberrant cases of the empty set and a single point (these also can be obtained geometrically by taking the constant $2a$ to be negative or zero and $f = f'$) in (iv) of Sect. IV.2, it is necessary to introduce the complex numbers (see Sect. IV.7). There is a second reason for complexificaton, coming from the connection between the geometric object "conic set of points in the plane" and its equation. We see in $(*)$ that there are six parameters, whereas the conics obviously depend on five only. This has simply to do with the fact that we get the same conic if the six coefficients of $(*)$ are multiplied by the same nonzero real. In spite of that the correspondence isn't totally satisfactory, for in the case of the empty set we have much more choice than just "within multiplication by a scalar"; in effect we have a continuous infinite choice.

In modern language, that of quadratic forms, the equation (∗) is written $Q(m) + L(m) + c = 0$, where Q is a quadratic form, L a linear form and c a constant. This somewhat cumbersome expression will be proved in the projective context of the following section.

As must be the case for truly innovative concepts, Descartes' discovery was not just a new interpretation for the conics, but brought with it a whole harvest of new results. We mention three of them: the first is that every affine (or projective) transformation, when applied to a conic, yields a conic. The second is that every plane section of a cone of second degree (of revolution or not) is a conic. The third is the solution of the problem of the "locus for four lines", mentioned at the end of Sect. IV.2. In fact, the distance of a point to a line is, in arbitrary coordinates, the absolute value of a linear form $|px + qy + z|$. Thus the locus of four lines is defined by an equation of the form

$$|px + qy + z||p'x + q'y + z'| = k|p''x + q''y + z''||p'''x + q'''y + z'''|.$$

Whence we have **two** conics (in the broad sense above, the one being possibly empty, etc.). We make two pleasant observations. First, Descartes wrote that this locus, that had vainly been sought by his predecessors, is **a single** conic, even though in fact there are two of them (because of having to get rid of absolute values). This circumstance gave rise to a bitter discussion between Descartes and Roberval; in fact Descartes always took great care with signs, for he knew that the sign of $px + qy + z$ depends on which side of the line $px + qy + z = 0$ the point (x, y) is located. Next, even for the circle, readers will see that it is difficult to show in an elementary way this property of the points m of a circle that passes through four points; for evidently the four intersection points (if they exist) $D \cap D''$, $D \cap D'''$, $D' \cap D''$, $D' \cap D'''$ always satisfy $dd' = kd''d'''$ (for any k). Besides the circle, there is an hyperbola, which in the case of the square degenerates into the pair of lines formed by the diagonals of the square (when $k = 1$).

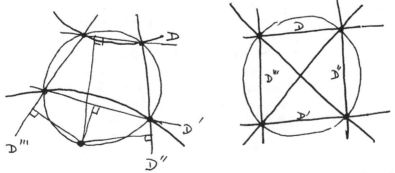

Fig. IV.3.1. *At left*: the "locus of four lines". *At right*: for $k = 1$ and the four sides of a square, the locus of four lines consists of the circumscribed circle for the square and the hyperbola degenerated into the two diagonals of the square

Much more important was that Descartes' study was for him a particular case of consideration of curves with equation $f(x, y) = 0$, where f is a polynomial in x and y, recognized as being the cradle of algebraic geometry (see p.74 of Coolidge 1968), of which we will see a very little bit in Sect. V.13. For the essence, vision and history the reference Dieudonné (1985), despite its density, stands out most strongly.

IV.4. Real projective theory of conics; duality

By unifying the efforts of Sect. I.5 and Sect. IV.3, we now have all the elements needed for handling what follows. We place ourselves in $\mathbb{R}P^2$. Recall that, in non-intrinsic fashion, $\mathbb{R}P^2$ is the set of triples $(x : y : z)$ of reals, not all zero, considered modulo multiplication by a nonzero scalar (see Sect. I.5). We know from Chap. I that the real plane $\mathbb{E}^2$ has a natural completion in $\mathbb{R}P^2$, in which it is embedded by $(x, y) \mapsto (x : y : 1)$. A conic with equation

$$(*) \qquad\qquad ax^2 + 2b''xy + a'y^2 + 2b'x + 2by + a'' = 0$$

is thus extended naturally to the conic

$$(**) \qquad\qquad ax^2 + a'y^2 + a''z^2 + 2byz + 2b'zx + 2b''xy = 0$$

of $\mathbb{R}P^2$, which is often called the *homogenization* of the equation $(*)$. But then $(**)$ represents a (general) *quadratic form* on $\mathbb{R}^3$ and we will thus very naturally call *conic* each portion of $\mathbb{R}P^2$ of the form $p(Q^{-1}(0))$, where p designates the canonical projection $p : \mathbb{R}^3 \setminus 0 \to \mathbb{R}P^2$, where Q is any quadratic form. Things depend on the sign and rank of Q. If Q is only of rank 1, we find a double line (a projective line); if Q is of rank 2, we find two distinct lines or a point, according to the sign of Q. In these two cases we say that the conic is *degenerate*. Finally, if Q is of maximum rank 3, we say that the conic is *proper*: we get the empty set if Q is of constant sign (i.e. positive definite or negative definite, in the language of quadratic forms), and otherwise **a single** conic. Here is some more about hyperbola, parabola, ellipse: all have the topology of a circle and even possess an *interior* (in $\mathbb{R}P^2$), which corresponds to the convex pieces associated with the affine plane (one piece for the ellipses and parabolas, two pieces for the hyperbolas). Modulo an homography (a projective and bijective transformation), there is but a **single** proper conic. Note that the projective lines in $\mathbb{R}P^2$ are also topological circles; but they don't have interior, for a line of the real projective plane does not divide it into two regions.

Surely readers will understand how to revert to the affine plane: we trace the conic C and the line at infinity D. If $C \cap D$ consists of two distinct points, a hyperbola remains in the affine plane; if $C \cap D$ reduces to a single point, we find a parabola; and finally we have an ellipse if $C \cap D$ is empty.

The majority of the affine and projective results on the conics stem from the double fact in the following chain: first we define a bijection $f : C \to D$ between the points

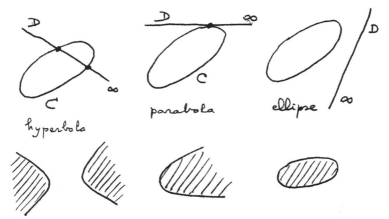

Fig. IV.4.1. A projective conic always has an interior

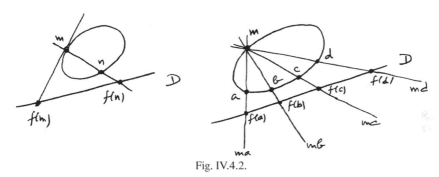

Fig. IV.4.2.

of a proper projective conic and some projective line D (which isn't tangent to it) by the following figure. We fix $m \in C$ and the bijection is $f : n \mapsto f(n) = mn \cap D \in D$. When $m = n$ the line mm is, by convention, the tangent to C at m.

The essential fact is that this provides a parameterization of C by the points of D (the simplest typical case is $(s{:}t) \mapsto (s^2{:}st{:}t^2)$, which parameterizes the conic $xz = y^2$), but above all the two parameterizations of the type $f : D \to C$, $g : E \to C$ (where E is a second projective line) are always such that the bijection $g^{-1} f : D \to E$ is an homography. A principal consequence of what precedes is that each invariant of the projective line carries over to the conics. In particular, with four points a, b, c, d of a conic C we can associate their *cross ratio, denoted* $[a, b, c, d]_C$. It is defined, for any parameterization $f : D \to C$ whatever, as $[a, b, c, d]_C = [f(a), f(b), f(c), f(d)]_D$. Note that we thus have, by definition, $[a, b, c, d]_C = [ma, mb, mc, md]$, where the last cross ratio is that of four lines of $\mathbb{R}P^2$ passing through m, an arbitrary point of C. Thus we can define a conic as the locus of points m such that $[ma, mb, mc, md] = k$, where a, b, c, d are four points in the plane such that no three of them are collinear and k is a nonzero real. When the four points are fixed and k varies, we obtain the whole family of nondegenerate conics passing through the four points (a *pencil*, see Sect. IV.6).

With the exception of the most elegant theorem of all on conics (Poncelet polygons, see Sect. IV.8) we can derive from the preceding practically all the properties of conics (real or, later, complex), the projective properties as well as the affine properties, and also even the metric properties of Sect. IV.2, this by complexifying and using the cyclic points introduced in Sect. II.XYZ. We restrict ourselves here to some important classical examples and some commentary.

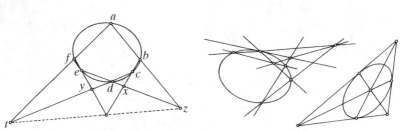

Fig. IV.4.3. Pascal's theorem. *At left*, general case; *middle*, case where two of the six vertices are merged; *at right*, case where the six vertices are reduced to three, each counted twice

We first of all prove Pascal's theorem stated in Sect. IV.2. The proof is rapid: if we set $x = bc \cap ed$, $y = cd \cap ef$, $z = ab \cap de$, $t = af \cap dc$, we have

$$[z, x, d, e] = [ba, bc, bd, be] = [a, c, d, e]_C = [fa, fc, fd, fe] = [t, c, d, y]$$

(see Fig. IV.4.3) so that zt, xc, ey are coincident. Note also that Pascal's theorem remains true when two vertices such as a and b are merged if we define aa (initially undefined) to be the tangent to the conic at the point a.

We may regard Pascal's theorem as the generalization of Pappus's theorem, encountered in Sect. I.4: it has then to do with a conic degenerated into two distinct lines.

The conics can be generated as the intersection of two lines that turn: if m, n are two points of $\mathbb{R}P^2$ and f an homography between the pencil of lines passing through m and that of lines passing through n, then the point $D \cap f(D)$ describes a conic as D turns about m (passing through m and n, proper if $f(mn) \neq mn$). By duality we deduce that if m and n describe two lines D and E of $\mathbb{R}P^2$ while staying in projective correspondence, then mn envelops (that is to say, makes up the family of tangents to) a conic tangent to D and E. Otherwise expressed, if f is an homography (projective correspondence) between two lines D and E of $\mathbb{R}P^2$, then the lines $xf(x)$, as x traverses D, envelop a conic tangent to D and E. For example, if two points traverse two given lines of the plane $\mathbb{R}^2$ with constant speeds, then the line that joins them envelops a parabola.

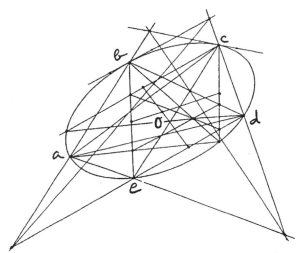

Fig. IV.4.4. Construction of the center of the conic passing through a, b, c, d, e. II [B]
Géométrie. Nathan (1977, 1990) réimp. Cassini (2009) © Nathan Édition

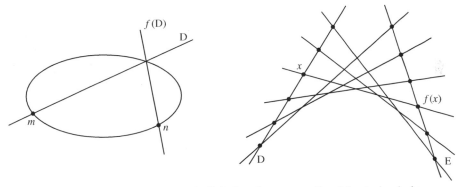

Fig. IV.4.5. When two lines D and f(D) describe two pencils while staying in homographic correspondence, their point of intersection describes a conic. When two
points x, $f(x)$ traverse, while staying in homographic correspondence, the respective
lines D and E, then the line $x f(x)$ that joins them envelops a conic

The above figure (right side) appears to "be in space". There is a good reason for
that, which will be explained in Sect. IV.10 with the aid of *ruled* quadrics. See Fig.
IV.10.4, where we see the projection − the apparent contour − of a ruled quadric.

◆

We can easily prove Poncelet's theorem for triangles (Figs. II.2.11 and IV.4.6 below): we take two conics C and C′ and seek a triangle inscribed in the one and circumscribed about the other. The answer is surprising: either there isn't any, or there
are just as many as there are points on C (we have however encountered the same
phenomenon for circles that are tangent in succession to two circles; see Fig. II.2.7).
The proof is trivial; we have: $[b, d, e, c] = [b, b', c', c]_C = [d', b', c', e']$. We will

see in Sect. IV.8 that the result is true for polygons with an arbitrary number of sides, but vastly more difficult to prove.

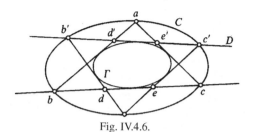

Fig. IV.4.6.

The duality (polarity) with respect to the conic $C = p(Q^{-1}(0))$ simply translates the fact that a quadratic form arises, by definition, from a symmetric bilinear form P, i.e. that $Q(x) = P(x, x)$ for each x. Then the polar line m^* of a point $m = p(u)$ will be the projective line resulting from projection onto $\mathbb{R}P^2$ of the set $\{v : P(u, v) = 0\}$ of $\mathbb{R}^3$. The word *duality* is sacred in mathematics; we find it in Sects. I.1, I.7, IV.4 and a large portion of Chap. VII, where it is unavoidable. As for figures, see those given for circles; there is no fundamental change. All the properties of points and their polars arise simply from the fact the P is a symmetric bilinear form. Don't forget that, if $m \in C$, then its polar line m^* is nothing other than the tangent to C at m; in particular, $m \in m^*$. In fact, the points of C are characterized as those that belong to their polar. All the properties stated in Sect. IV.2, diametral and otherwise, are now trivial. The polar of a focus is the associated directrix. If $P(u, v) = 0$ we also say that the associated points are *conjugate* with respect to the conic: if the line that joins them cuts the conic in two real points, we then have four points in harmonic ratio.

There exists another duality, no longer between points and line of $\mathbb{R}P^2$, but between lines (resp. points) of $\mathbb{R}P^2$ and points (resp. lines) of $(\mathbb{R}P^2)^*$, that depends neither on a basic conic nor on a choice of coordinates (in what follows, we could replace $\mathbb{R}P^2 = P(\mathbb{R}^3)$ and $(\mathbb{R}P^2)^* = P((\mathbb{R}^3)^*)$ by $P(Q)$ and $P(Q^*)$, where Q is a real vector space of dimension 3 and Q^* the dual).

With each line d of $\mathbb{R}P^2$ is associated canonically a point d^* of $(\mathbb{R}P^2)^*$. Geometrically: if d arises from the canonical projection p of a subspace of dimension 2 of $\mathbb{R}^3$, then d^* arises from the line of $(\mathbb{R}^3)^*$ that is orthogonal to this subspace. In coordinates: for each system of projective coordinates in $\mathbb{R}P^2$, there is a unique system of projective coordinates in $(\mathbb{R}P^2)^*$ such that the point d^* corresponding to the line d with equation $ux + vy + cz = 0$ has coordinates $(u : v : w)$.

This duality extends to conics in the following manner. When d ranges over the set of tangents to the conic C of $\mathbb{R}P^2$, the corresponding point d^* describes a conic C* in $(\mathbb{R}P^2)^*$, called the *dual conic*. This is easily proved by using the matrix representation of the equation for conics (for the details, see Sect. IV.XYZ or 14.6 of [B]).

Regarded as a family of lines of $\mathbb{R}P^2$, the conic C* envelops C. This is why we say that it is a *tangent conic*.

We haven't mentioned the duality between points of $\mathbb{R}P^2$ and lines of $(\mathbb{R}P^2)^*$. It is easily derived from the fact that the lines of $(\mathbb{R}P^2)^*$ are called pencils of lines in another context and thus define a point of $\mathbb{R}P^2$. We then show that as q ranges over the set of tangents to C*, the corresponding point q^* describes the conic C in $\mathbb{R}P^2$. In other words, we have $(C^*)^* = C$. Apparently then there is no difference between point conics and tangential conics. But pay attention that this is only true in the case of proper conics. It is thus that if the first case of degeneracy of a conic is a pair of distinct *lines*, for a tangential conic this degeneracy will become the set of lines passing through one or the other of two distinct *points*. This fact will be fundamental in Sects. IV.8 and IV.9; we will see there that the good notion that needs to be used for a conic is neither a set of points nor a set of lines, but rather the union of its set of points and its set of tangents. It's a good rung up the ladder. By way of an example, in considering C* we can define, for a proper conic C, the cross ratio of four tangents to C and show that it is equal to the cross ratio of the four points of contact.

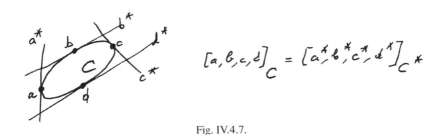

$$[a,b,c,d]_C = [a^*,b^*,c^*,d^*]_{C^*}$$

Fig. IV.4.7.

More generally, if C and C' are two proper conics, we may ask for the set of polars m^* of the points m of C, but with respect to C'. The answer is that these polars envelop a third conic C'', called the *dual conic* (or *polar* of C *with respect to* C'. In fact, the preceding case of C*, when $\mathbb{R}P^2$ is identified with $(\mathbb{R}P^2)^*$ by the mapping $(a : b : c) \mapsto (ax + by + cz = 0)$, corresponds to the case where the conic C' is that with equation $x^2 + y^2 + z^2 = 0$, which as a set is empty and exists only in an imaginary fashion − a good reason for complexifying everything in Sect. IV.6.

IV.5. Klein's philosophy comes quite naturally

If a conic C and a projective line D are thus put in biunique correspondence, we can legitimately ask ourselves: what, mapped on C, are the homographies (see Sect. I.6) of the projective line D? These are interesting transformations; they form a group, and as this group (as is easily seen) is independent of the parameterizations used, we call it the *group of the conic* C. This should inspire us: we have succeeded in attaching, in canonical fashion, a group to a conic. We say no more, for three reasons: first a minor reason, which is that it is not our purpose in this book to treat all the properties of conics; for that see 16.3 of [B]. But beyond that there are two fundamental reasons. First, this presentation isn't the right one − it's better to say that this group is that formed by set theoretic restrictions to C of the set of homographies of the projective plane that preserve the set of points of C. We need to show (which is easy) that these two groups coincide. But, if our proper conic is defined by the quadratic form Q, then this group is clearly the group consisting of transformations of $\mathbb{R}P^2$ that arise from bijective linear mappings of $\mathbb{E}^3$ that preserve Q. Already it can be said that the duality with respect to a conic is completely described, **coded**, by the quadratic form Q; but now we must dare to say that the whole geometry of the conic is coded by the group of (linear) automorphisms of Q (only take care to quotient by multiples of the identity to be in the projective space).

Now all the quadratic forms that we have to consider are isomorphic to $Q = x^2 + y^2 - z^2$. The linear group that preserves this form is *denoted* in general $GL(2, 1; \mathbb{R})$; here we have its projective version $PGL(2, 1; \mathbb{R})$. But in Sect. II.XYZ we already encountered quadratic forms of this type when we defined the hyperbolic geometries. In particular, here it's the case of the hyperbolic plane and thus the group of a conic is isomorphic to the group of the hyperbolic plane, which we have denoted $\text{Möb}(S^1)$ and called the Möbius group of the circle (see Sect. II.4). The action on the circle corresponds here to the actions on the conic $Q = 0$. The essential remark of Felix Klein from 1872 (when he was but 23) is that a geometry is characterized by its group of automorphisms. In particular, we don't have to redo a theory every time: if we know hyperbolic geometry well, we automatically know the geometry of a conic well. Of course we must make up a dictionary, but subsequently the translation is automatic. Here is an element of it: the transformations that preserve the unit disk of $\mathbb{E}^2$ were generated by the inversions centered at the exterior of that disk. For the conic C, the dictionary translation of an inversion is what is called a *homology of a conic*. A general *homology* of a projective space is the transformation associated with a pair $\{p, H\}$ composed of a point and a hyperplane not containing that point (see *harmonic homology* in Chap. II). At each point m distinct from p we associate the point of the line $f(m)$ which is such that the four points $p, pm \cap H, m, f(m)$ are in harmonic ratio. If we have a conic, it will be preserved by all the homologies of the form $\{p, p^*\}$ where p^* is the polar of p with respect to C. The fixed points of this homology are exactly the base point p and all the points (individually) of H. A consequence, for example, is that the group of a conic is generated by its homologies.

There exist numerous geometric analogies of this sort; it's what had driven Felix Klein in his famous Erlanger Program in 1872 (see Klein, 1974, in French) to reduce in some way any sort of geometry whatever to the theory of groups, considering only those results interesting that are an "automatic translation" of a theorem of this or that "geometry". We will amply see in our work that geometry in a still

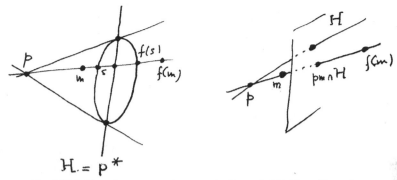

Fig. IV.5.1. Homology conserving a conic. General definition of homology

broader sense remains a fascinating workplace. But let us illustrate Klein's principle with situations encountered in the course of Chaps. II and IV. The projective group $PGL(3, 1; \mathbb{R})$ of automorphisms of the quadratic form $x^2 + y^2 + z^2 - t^2$ appear there three times: the first time in the hyperbolic geometry of dimension 3; then in the group $\text{Möb}(S^2)$ of the sphere; or again the group of all the inversions of the Euclidean plane, made sufficient by adjoining a point at infinity; but it's also the group of the space of circles in the plane (see Sect. II.6); finally, in the projective space $\mathbb{R}P^3$, the proper quadric with equation $x^2 + y^2 + z^2 - t^2 = 0$ will have an associated group. All four geometries are, however, isomorphic. Readers will be able to compose a dictionary, most likely incomplete, and use it to prove the things that interest them.

This systematic association of a group with a geometry explains why the theory of Lie groups has been an essential subject for mathematics from the end of the nineteenth Century to the present day. It also explains the importance accorded the groups that preserve a nondegenerate quadratic form. It is why, in [B], the study of conics and quadrics is preceded by a chapter dedicated only to quadratic forms. A large part of the geometry of these forms, and thus of geometric objects that have this as their automorphism group, results from two essential theorems associated with the names of Witt and Cartan-Dieudonné. The latter theorem states that the group $O(Q)$ of a quadratic form in a space of dimension n is generated by hyperplane symmetries; better yet, each of its elements is the product of at most n hyperplane symmetries ([B], 13.7.12).

Let us see what this theorem says in some particular instances. For Euclidean geometry, each isometry (preserving an origin) is the product of at most two symmetries with respect to lines. For the conformal group $\text{Möb}(S^1)$, each isometry of the hyperbolic plane is the product of at most 3 inversions (preserving the disk); but for a conic, this will be the product of at most three homologies. For a quadric of $\mathbb{R}P^3$, it will be a product of at most 4 homologies; for the conformal plane it will be a product of at most four inversions and of five for the conformal group of three-dimensional space. All these results of products in precise number are helpful in the geometries studied, but their proof in the context of Klein is reduced to that of a single theorem, that of Cartan-Dieudonné. The refusal to see that we have an explanation that is systematic and quick and that it is absolutely necessary to make use of it, and to want at all costs to do things by pure geometry, leads to the sort of aberrations already mentioned in Sects. III.1 and IV.2. For the conformal group, for example, we can refer to the long Note L of Hadamard (1911), already mentioned in Sect. II.7. It's the moment to quote Yuri Manin: "*a good proof is a proof which make us wiser*".

And yet it turns out the theory of Klein with its groups doesn't exhaust the possible geometries. If the examples above seemed to indicate that it does, it's because the geometries considered have "big" transformation groups, whether the linear group for projective geometry or the entire group preserving a quadratic form. These groups act in a very overdetermined fashion; in particular they act transitively. Recall that the action of a group G on a space X is called *transitive* if, given $p, q \in X$, there exists $g \in G$ that takes p into q, whereupon we say that X is a *homogeneous space* of G. We then have $X = G/H$, where H is the isotropy group of a given point of X, i.e. the group leave this point invariant (if we change points we obtain an isomorphic subgroup). The affine and projective spaces, spheres, the hyperbolic plane, etc. are all homogeneous spaces of their natural symmetry groups (for spheres this is the larger isometry group, or Möbius group). On the other hand, there exist on the sphere S^3 and on $\mathbb{R}P^3$ — which are groups in their own right — geometries invariant only under the group, but with isotropy group reduced to the identity. Within a change of scale, these geometries depend on two parameters and can be very different, in particular from the canonical geometries on S^3 and on $\mathbb{R}P^3$ (for which the isotropy group is O(3) and thus has six parameters). The much greater portion of our book will study topics "beyond Klein" — outside the Erlanger Program — where the groups are most often nonexistent.

IV.6. Playing with two conics, necessitating once again complexification

We have seen that modulo a homography there exists but one proper conic. What should we say of an attempt to classify pairs of conics? Let C and C' be two proper conics of the (let's say projective) plane: how do they intersect? We are interested only in those cases where there are points in common. To study them we solve simultaneously the two equations of C and C' thus: we can always parametrize C as in Sect. IV.4 by $(s^2 : st : t^2)$ and substitute these coordinates into the quadratic form which is the equation of C'. The result is thus a polynomial P, homogeneous and of degree 4 in $(s : t)$. We call a pair $(s : t)$ a root if it is a zero for P and we are interested for the moment only in the case where all four roots are real (note that this implies that if there is, for example, just a single root, then it must be or order 4). If they are distinct, we obtain below the anticipated figure; but we have in fact five possibilities for these four roots and their multiplicities: $\{1, 1, 1, 1\}, \{2, 1, 1\}, \{2, 2\},$ $\{3, 1\}, \{4\}$. What is the geometry of these situations? It is clearly seen in the figures; but why would we want to draw all these conics anyway, instead of just two? The reason — amply justified in various contexts, of which we will see several — is that for studying the pair of conics $\{C, C'\}$ there is great interest in considering the one parameter family of conics whose equations are all the linear combinations of the Q and Q' of C and C'. Specifically, the conics $p(\lambda Q + \mu Q')$ for the pairs of reals $\{\lambda : \mu\}$ are the ones drawn below. Such a family is called a (linear) *pencil of conics*. All of these conics intersect two at a time in the same way, with the exception of those that are degenerate; see below.

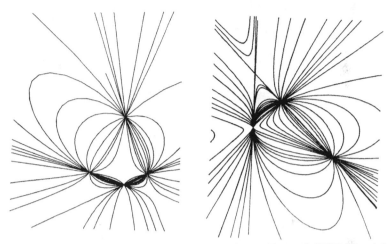

Fig. IV.6.1. *At right*: the type $\{1, 1, 1, 1\}$; *at right*: the type $\{2, 1, 1\}$. II [B] Géométrie. Nathan (1977, 1990) réimp. Cassini (2009) © Nathan Édition

The case of double roots is easy to interpret geometrically; it signifies exclusively that the conics are tangent at the points in question. But the two cases $\{3, 1\}$ and $\{4\}$ are more subtle. In fact, these have been omitted in numerous textbooks that we will be so kind not to name. In the case $\{3, 1\}$ we say that the conics are *osculating* at the point of order 3, and in the case $\{4\}$, we call them *superosculating*. In the language of differential calculus, this is to say that at the point in question the two conics have contact of order 3 (resp. 4) — for mere tangency, the order would be 2. Geometrically, if we endow the affine plane with a Euclidean structure, being osculating is equivalent to the radii of curvature at the point being equal. For superosculation, it is necessary to use projective transformations; see 16.5 of [B] for all this. We extract figures as shown below, where we can read osculation and superosculation with collinearities.

The next figure represents the trajectories of a small ball on the interior of a circle situated in a vertical plane, when these balls are launched from the lowest point (some readers may prefer the language of the simple pendulum, but with a string and not a rigid rod). The trajectory always begins with the circle on which the ball rests; then, at a point which depends on the initial velocity (if the speed is not too large), the ball leaves the circle and describes a parabola in free fall (in the figure we have subsequently forgotten the circle, but if the ball bounces on it, it's a source of many amusing problems to know what happens at moments of successive rebounds; see Problem 17.9.2 of [B]. Where the ball leaves the cir-

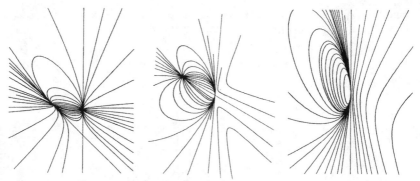

Fig. IV.6.2. The types $\{2, 2\}$, $\{3, 1\}$ and $\{4\}$. II [B] Géométrie. Nathan (1977, 1990)
réimp. Cassini (2009) © Nathan Édition

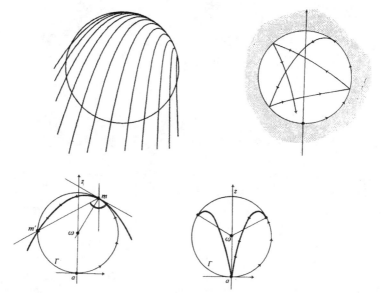

Fig. IV.6.3. Different trajectories of a ball launched in a circle from the lowest point,
with different initial speeds

cle is precisely where the circle and the parabola osculate; because the accelerations (which are of second order) coincide at this point, there will be third order contact.

In algebraic language, osculation and superosculation are easy to describe: C and C′ will be of type $\{3, 1\}$ if a quadratic form Q′ defining C′ is of the form $Q' = k \cdot Q + D \cdot D'$, where D and D′ are two linear forms defining two lines. For superosculation, this will be $Q' = k \cdot Q + D \cdot D$, where D is an equation of the tangent at the point in question.

We can do unusual things with pencils of conics, but we will opt to stay with our guiding point of view, giving special mention to difficult results that require more abstract concepts or lead us to open problems. For more than what we can say briefly, see [B] or Coolidge (1968). But the next item will serve us well in the sequel: the most important thing is first to look for degenerate conics in the pencil of the pair $\{C, C'\}$. They can be seen in the figure: there are three pairs of lines, thus three such conics. Algebraically, these will be (within a scalar) the pairs $\{\lambda, \mu\}$ such that the quadratic form $\lambda Q + \mu Q'$ isn't of maximum rank 3. But that is to say that the determinant of $\lambda Q + \mu Q'$ is zero (the determinant being zero is an invariant, even though the determinant itself isn't). We thus find a homogeneous equation in $\{\lambda : \mu\}$ of third degree and the figure shows that, in the case $\{1, 1, 1, 1\}$, it has three distinct roots (guess what it is in the other cases in the figures). In the case of four distinct points, we have a nice picture:

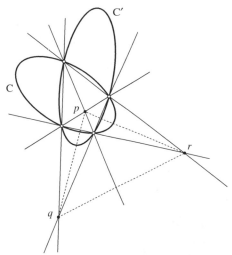

Fig. IV.6.4. *Pencil with distinct base points.* The three degenerate conics of the pencil and the autopolar triangle

An important and even useful remark is that the figure shows the existence of an *autopolar triangle* simultaneously with respect to the two conics, i.e. a triangle such that the polar of each vertex is the side opposite the vertex. Notice that this autopolar triangle is common to each of the conics of the pencil generated by these two conics. The figure shows it in the case $\{1, 1, 1, 1\}$, where the triangle is p, q, r. Algebraically it's easy; the geometric proof is the construction of the polar of a point with the complete quadrilateral.

We can't but mention, because of its elegance, the fact that the cross ration of four points common to one of the conics, is equal to the cross ratio of the four tangents to the other conic.

We can also investigate how the conics of a pencil intersect a fixed line D. The answer is that these pairs of points are in involution, i.e. that there exists an involutive homography of the line that sends the one onto the other; see Sect. I.XYZ. In particular there exist two conics of the pencil that are tangent to D (two at most in the real case). All this can be seen by replacing, in the equation of the conic, the coordinates by a (linear) parametric representation of the line; the pair $\{\lambda, \mu\}$ defines a conic tangent to D when this equation has a double root, which is a condition of second degree. We see right away what is the right interpretation of a pencil of conics.

Finally, we can ask about the morphology of the pencils of (real) conics that aren't of the types studied above, i.e. where the points of intersection aren't all real. Such a classification can be found in Levy (1964).

IV.7. Complex projective conics and the space of all conics

We have seen in several places the necessity of complexifying; we thus now define a conic by the definition that was parachuted in Sect. IV.1:

> A conic *is a curve in the complex projective plane, i.e. the set of* $(x:y:z)$ *that satisfy an equation* $P(x, y, z) = 0$, *where P is a homogeneous polynomial of degree 2 in the three variables x, y, z (and not identically zero).*

But over the complex numbers, there is but a single invariant for quadratic forms, their *rank*. Thus there aren't more, modulo a projective transformation, than three types of conics, specifically those defined by the equations: $x^2 + y^2 + z^2 = 0$, $x^2 + y^2 = 0$, $x^2 = 0$. i.e. the *proper* conics, the conics that degenerate into two lines, and those that degenerate into one *double* line. Of course, this has to do with complex projective lines. Here we may seek to "see" things, but $\mathbb{C}P^2$ is actually an object of real dimension equal to 4, which furthermore has a complicated topology. Intrinsically, a conic is a surface (in the real sense). We generalize all of what was said in Sect. IV.4; for example, each proper conic is in bijection with a line and we have a notion of cross ratio for four points of a conic. Since a projective line has the topology of the sphere S^2, a proper conic has thus the topology of a sphere. We always have parameterizations of second degree, such as for example $(s:t) \mapsto (s^2 : st : t^2)$. But all this is not the most important issue: we always work with essentially geometric language, but visualizing isn't always necessary or possible; we must think by analogy. All this, by the way, in fact depends on ones psychology. Here then, in cascade, are two essential facts.

The first is that the set of points $p(Q^{-1}(0))$ in $\mathbb{C}P^2$ **determines** the quadratic form Q within a scalar multiple. Of course, two proportional quadratic forms define the same conic, but the converse is totally false in the real case, where all the positive definite forms correspond to the empty set. The proof is practically trivial for conics. Its essence is that an equation of second degree (with coefficients in $\mathbb{C}$) is determined within a scalar multiple by the values of its roots. This easy uniqueness is in fact a particular case of the very general theorem that is basic for algebraic geometry, namely, the "*Nullstellensatz*"; see Sect. V.13.

But "within a scalar multiple" makes us think of projective spaces. Now, the set of all quadratic forms Q on the vector space $\mathbb{C}^3$ in fact forms a vector space over $\mathbb{C}$ under addition and scalar multiplication; its dimension is 6, since there are exactly six coefficients: $Q(x, y, z) = ax^2 + a'y^2 + a''z^2 + 2byz + 2b'zx + 2b''xy$. Finally, the preceding can be reformulated as follows: the set of all the conics is in bijection with the complex projective space of dimension 5; we use the *notation* $\mathcal{CCP} = \mathbb{C}P^5$ (*space of complex projective conics*). We will see in Sect. IV.9 that this structure is unavoidable for certain problems, and we can already put it in place, install it on the present rung of Jacob's ladder. For example, the condition that a conic pass through a given point is linear, but the set of conics passing through a given point in $\mathcal{CCP}$ is a projective hyperplane. It's hardly astonishing that there is one and only one conic that passes through five given points (in general position). Pay attention, though, that such hyperplanes are special in $\mathcal{CCP}$, if only because they depend on two parameters (since they are points of $\mathbb{C}P^2$) and therefore the hyperplanes depend in all generality on 5 parameters. The geometric nature of the conics of a general hyperplane of $\mathcal{CCP}$ is described in 14.5.4 of [B]; see Sect. IV.9 below for the condition of "being tangent to a given conic".

♦

Now a pencil of conics, if written $p(\lambda Q + \mu Q')$, will thus be nothing other than a projective line of $\mathcal{CCP}$. If for example we seek degenerate conics, we can say that they are the points where this line intersects the *subset* $\mathcal{CD}$ of $\mathcal{CCP}$ formed by the degenerate conics. As the determinant (within a scalar multiple) of a quadratic form is of third degree (in its coefficients, for example), in the language of algebraic geometry we have that $\mathcal{CD}$ is a hypersurface of third degree of $\mathcal{CCP}$, and thus a line intersects it in three (generally distinct) points; we will see a good bit more in Sect. IV.9.

The classification of the pencils in $\mathcal{CCP}$ is identical to that given in Sect. IV.6: morphologically there are exactly five types, but clearly the five figures given below are but a nice support for the mind, and indeed also an aid for certain reasoning.

We must take care not to fall into the following trap: the classification of *pairs* of conics is finer than that of pencil types. In the case of four common distinct points a, b, c, d a pair $\{C, C'\}$ is known modulo a projective transformation if the two cross ratios $[a, b, c, d]_C$ and $[a, b, c, d]_{C'}$ are given (quite an old result). To prove it, we remember from Sect. I.XYZ that the group of homographies is transitive on quadruples of points. But since a conic passing through four points is exactly described as the locus of a point such that the pencil of lines joining that point to four given points has a given cross ratio, we're done. The author has not failed to fall into the trap of false classification by type, but was saved in time by Jacques Tits. The assertion 16.5.1 of [B] is true, for it is addressed to the set of conics of a pencil and not to the pair that generate it. There is moreover a good reason — encountered in the next section — for two pairs of conics not to be projectively isomorphic: it's the existence or nonexistence, for each pair, of polygons of N sides (N fixed) inscribed in the one and circumscribed about the other.

In the other direction − for the types {3, 1}, {4} − there isn't any invariant and
two pairs of the same type are always projectively "the same". For the types {2, 1, 1}
and {2, 2}, there is a single invariant, which is not immediately identified from C
and C′. On the other hand, this invariant − on the projective line that defines in
$\mathcal{CCP}$ the lattice constructed with {C, C′} − is nothing other than the cross ratio of
points represented by C, C′ and the points represented by *two* degenerate conics of
the pencil. We can thus find more easily the geometric definition of this invariant.
When C and C′ are transformed projectively into two circles of the Euclidean plane
$\mathbb{E}^2$, then in the case of type {2, 1, 1} we obtain two tangent circles and the invariant
sought is the ratio of their radii; whereas for the type {2, 2} we obtain two concentric
circles and the sought-for invariant is the square of the ratio of their radii!

◆

As promised in Sects. II.1, II.XYZ and IV.2 we now explain the principle that permits recovery of all the metric geometry of the Euclidean conics, which is the philosophy of Plücker . Recall that (see Sect. II.XYZ) we can embed the Euclidean plane
$\mathbb{E}^2$ in $\mathbb{C}P^2$ by first projectifying first and then complexifying. There is the canonically attached pair {I, J} of points of $\mathbb{C}P^2$, called the *cyclic points* of $\mathbb{E}^2$ (even though
they are in $\mathbb{C}P^2$). We also note that a conic C in $\mathbb{E}^2$ can be embedded canonically
in a complex projective conic C** in $\mathbb{C}P^2$. With this notation, we interpret first the
circles: they are exactly the conics passing through the cyclic points, i.e. C is a circle if and only if {I, J} ⊂ C**. The center of the circle is the pole of the line at
infinity; two circles will thus be concentric if and only if they are bitangent in [I, J].
But above all the angular property of the points of a circle (seen in Sect. II.1) result
similarly from Laguerre's formula:

$$\text{angle}(D, D') = \frac{1}{2} |\log([\infty_D, \infty'_D, I, J])|$$

and the constancy property of the cross ratio seen in Section 4:

$$[a, b, c, d]_C = [ma, mb, mc, md].$$

The drawings below are just an absurdity that can be helpful in certain arguments.

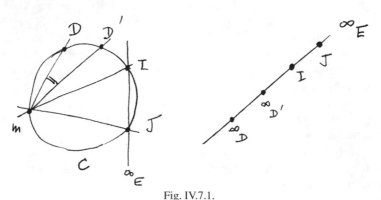

Fig. IV.7.1.

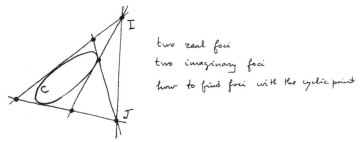

two real foci

two imaginary foci

how to find foci with the cyclic point

Fig. IV.7.2.

For general conics, we fist need to find the foci: these will be the points (again an "absurd" figure) f such that the pair of tangents to C^{**} emanating from f contain the pair $\{I, J\}$. There are thus four foci, but only one pair "is real". For an *a priori* object in $\mathbb{C}P^2$ *to be real* means that it is in fact in $\mathbb{R}P^2$. It can be seen in 17.4 and 17.5 of [B] how to recover all these metric properties of C by regarding $C^{**} \subset \mathbb{C}P^2$ and by utilizing the projective properties and the duality for conics. The directrix of a focus is its polar (if necessary, restricted to the Euclidean plane).

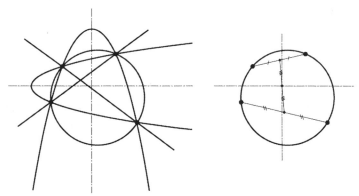

Fig. IV.7.3. *At left*, pencil of conics passing through 4 cocyclic points. *At right*, construction of the intersection of the axes of the parabolas of the pencil; the directions are obtained by considering one of the degenerate conics of the pencil, formed by two lines

With this sort of consideration using cocyclic points — and using appropriate pencils — we obtain some nice Euclidean theorems, such as: through four cocyclic points of a Euclidean plane $\mathbb{E}^2$ there pass two parabolas with orthogonal axes; moreover, these axes pass through the center of mass of the four points (see [B], 17.5 for the details and for other results). More generally, the points of intersection of two conics are cocyclic if and only if these conics have parallel axes of symmetry: this is seen by regarding how the points where these conics intersect the line at infinity of $\mathbb{C}P^2$ are situated with respect to cocyclic points. We stop here; the preceding shows how the great majority of classic results on conics can be obtained in a unified way with the three rungs of the ladder: projectify, complexify and finally introduce the projective space of all conics. In return, we are going to dwell further on two theorems, simultaneously for their elegance, their difficulty, their historical importance and the fact that they are going to force us to climb still higher on the ladder.

IV.8. The most beautiful theorem on conics: the Poncelet polygons

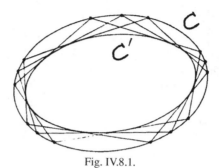

Fig. IV.8.1.

We consider first the case of two real affine conics C and C' such that C' is entirely interior to C. The result of Poncelet is of the type "all or nothing", like Steiner's porism in Sect. II.2:

> *For each integer* N $\geq$ 3 *either there exists no polygon with* N *sides inscribed in* C *and circumscribed about* C', *or else there exists an infinity of them; furthermore, we can take any point of* C *as initial vertex.*

It is natural to believe to the contrary that only certain points, chosen carefully, could be the vertices of such a polygon. This is what happens for pairs of "ordinary" curves: first it never works for all N; next if it works at all it's only for a finite number of polygons, but not with the right to choose an arbitrary point of C for a vertex. We have already encountered such a theorem — Steiner's — in Sect. II.3. But we remarked in Sect. IV.1 that there does not in general exist a projective transformation which takes the pair $\{C, C'\}$ into a pair of concentric circles; if such were the case, we'd be through, for such a transformation preserves lines and properties of tangency. And now we know, after the preceding section, the profound reason: two concentric circles are bitangent, while a general pair $\{C, C'\}$ of conics intersect in four points (more precisely here, the complexified-projectified pair $\{C^{**}, C'^{**}\}$).

This theorem of Poncelet for polygons has a double attraction, which is why numerous great mathematicians have been interested in it, right up to the present — as we shall see — and this interest is likely to persevere. The first attraction comes from the elegance of the result, but the second is linked to the first, i.e. the difficulty of its proof (in contrast to Steiner's porism, which is proved by an inversion). Next, there is the challenge of finding, among the equations of conics, the condition as a function of the integer N for the existence of Poncelet polygons with N sides. Since we too have been seduced by it, and for a longish time, we are going to dwell a while on this topic. Readers will have noticed that there is no reason to limit ourselves to the real case and that all the complex conics, affine and projective, are also amenable to such a result, likewise even the conics over arbitrary commutative fields, since the notions of conics, lines and tangency remain valid (for tangency, the purely algebraic language of double root of an equation applies). As for references beyond those that will be mentioned here and there in the sequel, the basic one is Bos et al. (1987). Just a bit of history however: above we proved the result for triangles, a result which for circles dates to William Chapple in 1746 — this for the statement of the "all or nothing" type — but the proof is inadequate; see more in Bos et al. (1987), and likewise for the case of quadrilaterals. This last case is easily treated with polarity and pencils of conics.

The interest in the Poncelet polygons appears, for example, as a typical application in treatises on elliptic functions: Appell and Lacour (1922) mention them only briefly, whereas (Halphen, 1886, 1888), in Chap. X of Volume II, treats the problem in great detail, including the formulas of Cayley that we will see further on.

But in fact history connects elliptic functions in a more essential fashion with Poncelet's theorem, i.e. that Jacobi by 1828 had confirmed a direct link between it and the addition and associativity formulas for elliptic functions; see Sect. IV.5 of Bos, Kers, Oort, and Raven (1987), also Hrasko (1999) and Hrasko (2000).

◆

Here first is a proof in the real case — a Euclidean proof in fact. We show first, without much difficulty, that a pair of conics as above can be transformed projectively into a pair of circles, one interior to the other. We can do it by hand with the geometry of $\mathbb{E}^3$ and conic sections, but also, by the way, in $\mathbb{C}P^2$ using the cyclic points. It remains then to prove the theorem for two circles C and C′. To our knowledge there doesn't exist any relatively short proof that uses only what has been said so far about the projective theory — even the complex version — of conics and pencils. The best proof we know has been taken from Shen (1998), but we know that it was already in the air (in this explicit form) for several years, and in essence known already by Jacobi. To each point $m \in$ C we attach its *tangential distance* L(m) to the circle C′. Then we endow the circle C with a **measure** different from the canonical Euclidean measure ds, but which is locally proportional to it, specifically $\frac{ds}{L}$. We denote by F the mapping C $\rightarrow$ C of the figure (it needs to be decided in which sense to traverse C). The basic lemma is that F **preserves the measure** $\frac{ds}{L}$ (if we want to avoid the specialized language of "measure", we can use only the notion of integral). If s designates a parameterization by the arc length of C (see Sect. V.6), then for each interval $[a, b]$ we have

$$\int_a^b \frac{ds}{L(s)} = \int_{F(a)}^{F(b)} \frac{ds}{L(s)}.$$

The verification requires just a little differential calculus, based on two essential facts: on the one hand, the line joining $m \in$ C to its image F(m) cuts the circle C′ at equal angles; on the other hand, the two tangents issuing from a point of a circle are always of equal length. In particular, if we reparameterize C with a parameter t

with the aid of this new measure, the mapping F becomes a translation $t \mapsto t + k$. But if we *denote* by Λ the new total measure of C, we have a polygon of N sides in the pair $\{C, C'\}$ if and only if $Nk = \Lambda$. Now this condition is independent of the point (of parameter t) chosen initially, Q.E.D.

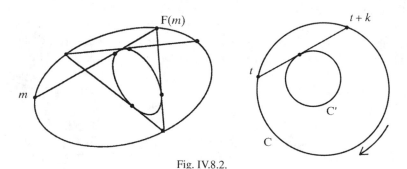

Fig. IV.8.2.

Note the ascent of the ladder with the notion of measure. In fact we could do much better, as Poncelet himself did, and we will be able for example to know what happens with the sides of the star polygons associated with a family of Poncelet polygons, or indeed with the diagonals when n is even! Poncelet in fact proved the following lemma, which we call the *general lemma* in the sequel:

> Let C″ be an arbitrary circle (interior to C) and belonging to the pencil of circles defined by the pair of circles $\{C, C'\}$. We traverse all these circles in the same sense, and with each $m \in C$ we associate the point n where the tangent issuing from m to C′ encounters C again; then from n we follow the tangent to C″ which cuts C again at p. Then the line mp, when m traverses C, envelops a circle C‴ belonging to the pencil considered.

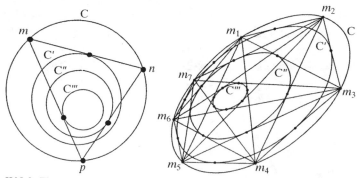

Fig. IV.8.3. The general lemma. *At left*: case of circles. *At right*: case of ellipses, applied to various diagonals of the polygons. II [B] Géométrie. Nathan (1977, 1990) réimp. Cassini (2009) © Nathan Édition

This lemma implies everything we want. First, the result on the polygons of the pair $\{C, C'\}$: for if we apply the result several times in succession, we see that if we start off from a point of C, at the end of N times we have a line mn that envelops a circle. But if the polygon closes once, this circle will then have N points in common with C (in fact 2N because things get counted twice); it is thus C itself. But by applying again the iterated lemma to a Poncelet polygon, we see that the star polygons of arbitrary type $\{N, k\}$ (N vertices, k turns) remain continuously circumscribed about a circle of the pencil in question. In the case where N is even, the diagonals thus pass through a fixed point. Needless to say that all this carries over to pairs, and to pencils, of real conics.

As for the proof of the lemma, it suffices to know this: the tangential distances from a point of C to two circles C' and C'' of a pencil containing C are in a constant ratio. To see this, it's simplest to remark that the square of the tangential distance of m to C' is equal to $Q'(m)$, the value for m of the quadratic form Q' that defines the circle C' (Q' must be normalized to start with by $x^2 + y^2$). Now membership in a pencil is a linear condition, and if m traverses C, then $Q(m) = 0$ if Q is the normalized quadratic form defining C. Then if the tangential distances $L(m)$ and $L'(m)$ are proportional, the associated measures $\frac{ds}{L}$ and $\frac{ds}{L'}$, when normalized, will coincide. So, with the right parameterization, the chord mn will correspond to a translation of the parameter, and each translation generates an envelope which is a circle of the pencil (proceed in the reverse sense and by uniqueness).

◆

All this seems to tell us nothing about the case of complex conics. In fact our whole presentation is not geometrical, but purely algebraic and, above all, the operations performed are all algebraic (construct a tangent, intersect a line). We could appeal to the *extension principle for algebraic identities*. Briefly: the algebraic entities (relations, identities) true over the real field $\mathbb{R}$ remain true over the complex field $\mathbb{C}$; see Bourbaki (1981), Chap. IV, § 2, Section 3, Theorem 2. There is here again an ascent of the ladder. In contrast, in the nineteenth century, Poncelet and many others appealed to the *continuity principle*, but that was more philosophically than solidly founded. Whether with regard to Poncelet's theorem or to other results, due to Chasles and many other geometers including Plücker and Steiner, the quarrel over the continuity principle raged, sometimes rather violently, throughout a good part of the nineteenth century; see the references and citations in 7.0 of [B]. In any case, we can now consider Poncelet's theorem as being achieved — in its more general form over the complex numbers — and in consequence return, as desired, to the real case for circles or, more generally, intersecting conics, e.g. an ellipse and a homofocal hyperbola.

The preceding does not answer the question about Poncelet's theorem for the conics over an arbitrary (commutative) field, finite for example; nor the question of an explicit, and much wanted, condition in N on the equations of conics. In 16.6 of [B] there is an elementary proof using only the projective geometry of conics over an arbitrary commutative field, and thus valid over all commutative fields; but it is rather long.

If we restrict ourselves to the "all or nothing" theorem — as it is rightly called — for a pair of conics, then there exists an ultra-rapid algebraic proof based on the so-called theory of *algebraic correspondences*. This method consists of parameterizing C to the second degree with, say, one parameter: let $m(t)$ be the point of C with parameter t and note that the condition for the line $m(t) \cdot m(t')$ to be tangent to C′ is of second degree simultaneously in t and in t'. More generally, if we construct the broken line starting from $m(t)$ that has N sides successively tangent to C′, the final point $m(s)$ is such that the relation between t and s is of second degree in t and in s. The difficult part of the proof is that this iterate $Q \circ Q \circ \cdots \circ Q$ of the original correspondence Q is again of second degree. To say that the broken line closes up is to imply that this relation (of fourth degree then in t) is satisfied for a certain t; but it is also for $N - 1$ ($N \geqslant 3$) values, to which we need to adjoin the points of $C \cap C'$. The fact that this relation is algebraic (polynomial) of type $\{2, 2\}$ and that it is zero for more than four values implies then that it reduces identically to zero. It is thus satisfied for all t.

♦

Here now is the proof of Poncelet's theorem which goes furthest toward getting to the bottom of things; we give a completely modern version; but, in essence, it is that of the theory of elliptic functions, known since Jacobi in the nineteenth century. We partly use the equivalent language of elliptic curves, specifically of cubics (see Sect. V.14). The fact is that these have a group structure and that the elliptic functions provide them with parameterizations (not at all algebraic). Here is the most rapid formulation there is, laid out here for the complex domain. To the extent that the theory of elliptic functions remains valid over other fields, what follows is more generally applicable. We begin with two conics C and C′ for which the four points of intersection are distinct. The subtle idea which follows consists in doing what is necessary for avoiding the double event that from a point of C two tangents depart, and that one tangent to C′ cuts C in two points (this had been avoided in the case of real circles by orienting the circle of departure).

We introduce the algebraic set that is the product $C \times C'^*$, where C'^* denotes the dual conic of C′ (i.e. the set of its tangents). In this product the *ad hoc* object is the subset $\Gamma = \{(m, D) : m \in D\} \subset C \times C'^*$ that is formed of pairs of a point of C and of a tangent to C'^* such that $m \in D$. What is this object? The fact that the tangents to C′ at the points of $C \cap C'$ are distinct implies that Γ is a curve without singularities in $C \times C'^*$. The first coordinate of the mapping $\Gamma \rightarrow C$ is a covering of two sheets, except at the four points of $C \cap C'$, which are of simple ramification of order 2. The elementary geometry of Riemann surfaces shows that the topology of Γ is that of the torus T^2 (said to be of genus 1). It's thus what is called an *elliptic curve* (complex, on $\mathbb{C}$) and it is classic that it possesses a group structure (once an identity element is chosen anywhere on the curve; this is amply treated in Sect. V.14). The group is written in additive form.

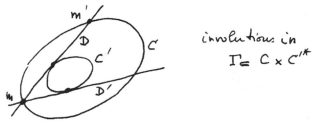

Fig. IV.8.4.

Now Fig. IV.8.4 defines two involutions $\gamma, \delta : \Gamma \to \Gamma$ on Γ, specifically $\gamma(m, \text{D}) = (m', \text{D})$ and $\delta(m, \text{D}) = (m, \text{D}')$. The construction of a polygon of Poncelet type starting from an (m, D) consists of iterating the composite operation $\delta \circ \gamma$. The polygon departing from the pair (m, D) closes on itself at the end of N times if and only if $(\delta \circ \gamma)^{\text{N}}(m, \text{D}) = (m, \text{D})$. But the elementary theory of elliptic curves assures that an involution having at least one fixed point is of the form $x \mapsto -x + a$ (for an appropriate a), thus the product $\delta\gamma$ will be a translation $x \mapsto x + k$. Thus the polygon closes if $\text{N}k = 0$. But — oh miracle! — this condition is intrinsic, it does not depend on the point of departure, Q.E.D.

$\blacklozenge$

There remains the problem of finding the condition leading to the equations of C and C′ and the integer N which will tell if the pair {C, C′} admits a Poncelet polygon with N sides (or vertices). It's Cayley who gave the answer in (Cayley, 1861); here we explain his solution following the exposition of Griffiths and Harris (1978a), where by the way the presentation just given figures too. We need to find a representation of the elliptic curve Γ that is explicit as a function of the equations of C and C′ and where we can also make explicit the operation $\delta \circ \gamma$. The ingenious *coup* of Cayley is first to find a cubic associated with a pencil of conics and the connection with its group law. In fact, Cayley introduced two cubics. The first, not really unique (but this is not important), consists specifically of a point p being fixed in the plane, taken somewhere on the initial conic C; the locus of the points of contact of the tangents issuing from this point to the conics of the pencil considered is a cubic (without double point). Let us say it has to do with Cayley's "concrete" cubic, and the formidable connection between this cubic and Poncelet's theorem on triangles, which we have called in the case of a circle pencil the *general lemma*, is going to be precisely this. We recall the general lemma:

> Let a pencil of conics, generated by the two conics C and C′, be given. If a triangle inscribed in C has its first side tangent to C′ and its second side tangent to another conic C″ of our pencil, then the third side is tangent to a third conic C‴ of the same pencil.

The connection discovered by Cayley is that, among the six points of contact of the tangents issuing from p to these three conics C′, C″, C‴, at least three are collinear, which gives us the group law of this cubic (see Sect. V.14). And all at once

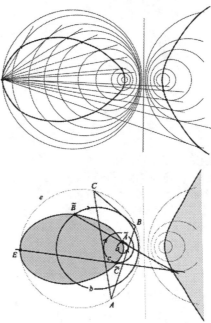

Fig. IV.8.5. In the figure we see two drawings of the first cubic of Cayley in the case
of a pencil of circles. In the first of these we see the cubic and the pencil. In the second we see the triangle inscribed in the first circle, then the three collinear points corresponding to a tangent to three circles, issuing from the point E which defines the
cubic, for the three circles of the pencil which are tangent to the three sides of the
inscribed triangle ABC considered

there appears the notion of a point of order N on this cubic when we have a polygon
with N sides inscribed in C and circumscribed about C′. To see this it suffices to
decompose the polygon into triangles, but this does not easily allow the calculations
for finding the condition between N and the equations of C and C′.

To find it Cayley introduces a second cubic − an abstract cubic − specifically the elliptic curve (Riemann surface) Σ defined by the algebraic function
$y = \sqrt{\det(x \cdot Q + Q')}$, i.e. the curve $y^2 = \det(x \cdot Q + Q')$ of $\mathbb{C}P^2$. Here Q and
Q′ are the quadratic forms defining C and C′, and the determinant can be taken (for
example) in the canonical basis of $\mathbb{C}^3$. Note that $\det(x \cdot Q + Q') = 0$ possesses three
distinct roots $\{x_i\}$ ($i = 1, 2, 3$) corresponding to the three degenerate conics of the
pencil defined by C and C′; also we denote by $\{m_i\}$ ($i = 1, 2, 3, 4$) the four points
of C ∩ C′. Finally, let C_x be the conic of the equation $x \cdot Q + Q' = 0$ (with x
different from the x_i). Now we define a mapping $\Sigma \to \Gamma$ thus: with an x (different
from the x_i) we associate the point m of C different from the m_1, where the tangent
at m_1 to the conic C_x intersects the conic C. We then choose any point of Γ which
projects onto m. Then, since we are on coverings (ramified, to be sure), we extend
without difficulty the mapping $x \mapsto \tau$ analytically to a mapping $\Sigma \to \Gamma$, complete
since the t_i correspond correctly to the m_i and $x = \infty$ to the point m_0. We verify
finally that we have a good isomorphism between elliptic curves.

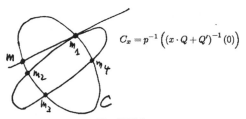

$$C_x = p^{-1}\left((x \cdot Q + Q')^{-1}(0)\right)$$

Fig. IV.8.6.

On Σ we thus see, if we take the point $x = \infty$ as identity element of the group law, that the operation $\delta \circ \gamma$ corresponds to the translation τ which goes from $x = \infty$ to $x = 0$. In this universal description, we must thus be able to see whether $N\tau = 0$ in the single expression for $\det(x \cdot Q + Q')$. It's a problem on elliptic curves and we resolve it with a theorem of Abel; it's reduced to finding the points of hyperinflection of the so-called normal elliptic curve traced in $\mathbb{C}P^{N-1}$ with the aid of derivatives up to order $N - 1$ in the parameterization by x. We have thus to write that a determinant of order N is zero. The corresponding derivatives are given by the development in a series $\det(x \cdot Q + Q') = a_0 + a_1 \cdot x + a_2 \cdot x^2 + a_3 \cdot x^3 + \cdots$ and the conditions obtained by Cayley are thus the vanishing of a determinant:

$$\begin{vmatrix} a_2 & \cdots & a_{m+1} \\ \vdots & \vdots & \vdots \\ a_{m+1} & \cdots & a_{2m} \end{vmatrix} = 0 \quad \text{if } n = 2m + 1$$

and

$$\begin{vmatrix} a_3 & \cdots & a_{m+1} \\ \vdots & \vdots & \vdots \\ a_{m+1} & \cdots & a_{2m} \end{vmatrix} = 0 \quad \text{if } n = 2m.$$

Readers will find it instructive to calculate several of these conditions for the case of two circles. Let R and r be their respective radii and d the distance between their centers. By 1827 Steiner had given the conditions

$n = 3 \; R^2 - a^2 = 2rR$
$n = 4 \; (R^2 - a^2)^2 = 2r^2(R^2 + a^2)$
$n = 5 \; r(R - a) = (R + a)[(R - r + a)(R - r - a)]^{1/2}$
$\qquad\qquad\quad + (R + a)[(R - r + a)2R]^{1/2}$
$n = 6 \; 3(R^2 - a^2)^4 = 4r^2(R^2 + a^2)(R^2 - a^2)^2 + 16r^4a^2R^2$
$n = 8 \; 8r^2(R^2 - a^2)^2 - r^2(R^2 + a^2)$
$\qquad\qquad\quad \times \{(R^2 + a^2)[(R^2 - a^2)^4 + 4r^4a^2R^2] - 8r^2a^2R^2(R^2 - a^2)^2\}$
$\qquad\quad = [(R^2 - a^2)^4 - 4r^4a^2R^2]^2 .$

For $n = 5, 6, 8$, Steiner gave no explanation! We therefor don't know how he obtained these formulas. Moreover his formula is false for $n = 8$, as may be seen in Pécaut (2000) which is a novel and elementary treatment of Poncelet's theorem (and where the correct formula for $n = 8$ and others for higher degree can be found).

But all this says nothing — for fanciers of explicit formulas — about how to calculate the coefficients a_i above. On p. 609, Vol. II of Halphen (1886, 1888) there are double recursion formulas of the "continued fraction" type (see Sect. IX.1.A for this theory), which seem to us very complicated compared to those in the pair $\{\alpha, \gamma\}$ that are to come below (and which are already complicated enough).

The preceding easily permits us to recover the general lemma of Poncelet. In fact it suffices to know how to compare the elliptic curves $\Gamma \subset C \cap C'^*$ and $\Gamma' \subset C \cap C'^*$ corresponding to the pairs $\{C, C'\}$ and $\{C, C''\}$, where the third conic C'' belongs to the pencil defined by C and C'. But in fact we have just seen that Γ is isomorphic to the elliptic curve which is the Riemann surface Σ of the algebraic function $\sqrt{\det(x \cdot C + C')}$ (and thus Γ' is isomorphic to the surface of $\sqrt{\det(x \cdot C + C'')}$). But, as $C'' = a \cdot C + C'$, we see that there is a change of parameter x to be made, whence the general lemma of Poncelet by reasoning as above.

There is also something else we might desire (mathematicians being insatiable): a pencil based on C being given along with an integer N, to know how many conics there are in the pencil that furnish Poncelet polygons with N sides. This problem is treated in Barth and Michel (1993). It is in fact a problem in number theory since an elliptic curve is isomorphic to the quotient of $\mathbb{R}^2$ by an appropriate lattice Λ: we need to count how many of the elements v of $\mathbb{R}^2$ are such that $Nv = 0$ (modulo Λ), but while paying attention to counting only the "primitive" (or "primary") elements, i.e. to eliminate those of the form kv which correspond to a polygon traversed several times.

Readers may have sensed — felt — a small scandal. If the formulas of Cayley are a technical *tour de force* — typical of the nineteenth century and in the spirit of "analytic geometry" — it should perhaps be asked why we shouldn't look for the existence, for C and C', of polygons with N sides inscribed in C and circumscribed about C' as being among the invariants that classify such pairs of conics under the group of homographies, in particular the pair of complex numbers $\{\alpha = [a, b, c, d]_C, \gamma = [a, b, c, d]_{C'}$; see Sect. IV.7. We are thus here more in the style of "geometry" than "analytic geometry". In Halphen (1886, 1888), Vol. II, p. 377, can be found the formulas — with this exact notation in $\{\alpha, \gamma\}$ — that give, for each N, the relation imposed between α and γ, again where these formulas are obtained through recurrence relations (building on the two preceding terms like the archetypical Fibonacci sequence defined by the relation $u_N = u_{N-1} + u_{N-2}$). It's difficult not to give the relation for $N = 4$, for it is $\alpha = \gamma^2$, which we can prove in a completely elementary fashion. For $N = 3$, it is already a lot less amusing: $\alpha^2 - 2\gamma(2\gamma^2 - 3\gamma + 2)\alpha + \gamma^4 = 0$. For this case and the general case, we need essentially the theory of elliptic functions. We define first two numbers x and y by:

$$x = -\frac{[\alpha^2 - 2\gamma(2\gamma^2 - 3\gamma + 2)\alpha + \gamma^4]^3}{2^8 \alpha^2 (\alpha - 1)^2 \gamma^4 (\gamma - 1)^4}$$

and

$$y = -\frac{(\gamma^2 - \alpha^2)(\gamma^2 - 2\gamma + \alpha)(\gamma^2 - 2\alpha\gamma + \alpha)}{2^3 \alpha(\alpha - 1)2\gamma^2(\gamma - 1)^2}.$$

Then the conditions are: $x = y$ for N = 5, $y = x + y^2$ for N = 6, $(y - x)x = y^3$ for N = 7 and $(y - x)(2x - y) - xy^2 = 0$ for N = 8. The relations between x and y are defined step by step by a recurrence relation. The true space of pairs of conics is not the set of $\{\alpha, \beta\} \subset \mathbb{C}^2$, but precisely the sets $\{0, 1, \infty, \alpha, \beta\}$ of five different points of $\mathbb{C}P^1 = S^2$, a set that plays an important role in physics.

♦

We may well ask whether there exist generalizations, variants in other contexts, for example in the space of three dimensions, for the polygons **zigzaging** between the generators (see Sect. IV.10) of two quadrics or with certain circles, etc. In Barth and Bauer (1996) there is not only a very long list of theorems of "Poncelet" type,

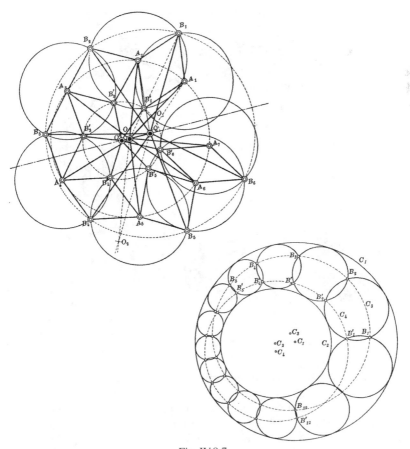

Fig. IV.8.7.

but a very general scheme. In particular, there is a modern treatment of the result of Emch (1901), which encompasses both Steiner's theorem and that of Poncelet. This result is the one of the figure below; to deduce from it the result of Poncelet, take a circle reduced to a point; then, by an inversion, all the circles passing through that point become lines and form the desired polygons. In Emch (1900) there is a mechanical realization of all that − the manuals for producing these mechanisms with an Erector or Meccano kit or something similar. This fabrication is done with parallelograms (deformable) linked to one another (see the figure). These articulated systems will be encountered briefly in Sects. V.16 and VIII.3. For an elementary treatment of Poncelet and zigzags, see Hrasko (2000) and Pécaut (2005).

The Poncelet polygons continue to fascinate mathematicians; they are found in foldings of polygons; see Sect. VII.4 and Benoist and Hulin (2004). See also Barth and Michel (1993). But Schwartz, (2001) (unpublished) seems to us to be an aston-ishing revival. In Berger (2005) there is an explanation for the magnificent figure of Schwartz below:

Fig. IV.8.8.

IV.9. The most difficult theorem on the conics: the 3264 conics of Chasles

The problem is to find the number of tangent conics to five given conics. We won't repeat again "in general position" after this. The notion can be made precise, but that is not our mission here; we leave it to readers to absorb it and to do more if they are curious. We have seen in Sect. II.1 that there are as many as eight circles

Fig. IV.9.1.

tangent to three given circles, but this may be interpreted in $\mathbb{C}P^2$ as the search for conics passing through the two cyclic points (see Sect. II.XYZ) and tangent to three conics that have these two cyclic points in common. From the viewpoint of conics, this is thus not at all a general result.

This problem of five conics has a long and fascinating history. It seems to be the only problem on conics that resisted beyond 1900; in particular it's an implicit part of the 15th Hilbert problem. Moreover, it has been one of the principal motivators in the development of the fundamentals of algebraic geometry, for to be well understood it requires a solid foundation of *intersection theory*. We will find in Sect. V.13 what needs to be understood and the importance of such a theory in numerous mathematical domains other than algebraic geometry itself. For this reason — and for the beauty of the problem in its own right — we are going to describe a little of its history. For more details, see Kleiman, 1980) and the introduction of Ronga, Tognoli, and Vust (1997).

We place ourselves exclusively in the complex projective domain, i.e. in $\mathbb{C}P^2$ — see at the end for the real case — but we can surely work with other base fields. In 1848 Steiner (always the same one) announced that there exist 7776 conics tangent to five given conics (in general position). It is in some way the most difficult problem that bears on the conics, and it is normal to impose five conditions to determine one conic, since the space they form is of dimension five. We have seen that five points determine a unique conic; and that four points and one line determine two of them, those that pass through these points and are tangent to the line. The reason is that, in the space $\mathcal{CCP}$ of all the conics, to pass through a point is a linear condition, having to do with a hyperplane of $\mathcal{CCP}$. To be tangent to a line is an equation of second degree: to see this we substitute for the values of (x, y, z) into the equation $Q(x, y, z) = ax^2 + a'y^2 + a''z^2 + 2byz + 2b'zx + 2b''xy$ of the unknown conic by a linear parametric representation in (s, t) of the line. We obtain an equation of second degree in the six coefficients of C. The tangency between the line and C requires that this equation have a double root, which yields a condition of second degree in these coefficients. When the line is given, the set of conics that are tangent to it constitutes a hypersurface S of degree two in $\mathcal{CCP}$, what is called a quadric of $\mathcal{CCP}$ (or better here, a *hyperquadric*; see Sect. IV.10). Finally, the conic we seek will be formed by the points of intersection of four hyperplanes and this quadric S.

Without appealing to general theorems, we will be intersecting S with the line of intersection of the four hyperplanes, which in general yields two distinct points.

Steiner generalized this general vision as follows: it has first to do with knowing which set C^T contains the conics of $\mathcal{C}\mathcal{C}\mathcal{P}$ that are tangent to a given conic C. We can always write C in the form $xz - y^2 = 0$, and thus parameterize it by (s^2, st, t^2). We substitute these values in $Q(x, y, z) = ax^2 + a'y^2 + a''z^2 + 2byz + 2b'zx + 2b''xy$ and find thus an equation in (s, t) of degree 4. There will be tangency if this equation has a double root, which is classically expressed by its *discriminant*, which is a polynomial in the coefficients of degree $4 \cdot 3 = 12$. In fact, the true degree is only 6. Here's why: it suffices to remark that our two conics with equations Q and Q' will be tangent if the equation $\det(Q + \lambda Q') = 0$, of third degree in λ – which furnishes degenerate conics of the pencil that they determine – possesses a double root (look at Fig. IV.6.1). This time the discriminant of this equation is of degree $3 \cdot 2 = 6$ (in the coefficients of the equation) and we have only to verify that this discriminant also remains of degree 6 with respect to the coefficients of Q (make the calculation with $Q' = x^2 + y^2 + z^2$). For example, there will be, in general, six conics passing through four points and tangent to a given conic:

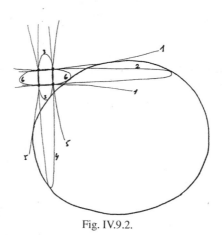

Fig. IV.9.2.

The tangent conics to five given conics C_i ($i = 1, ..., 5$) will thus be the points of intersection of 5 hypersurfaces C_i^T of degree 6 of $\mathcal{C}\mathcal{C}\mathcal{P}$. The general theorem of Bézout (see any fairly recent book on algebraic geometry) states that there are $6 \cdot 6 \cdot 6 \cdot 6 \cdot 6 = 7776$ points common to these five hypersurfaces.

In 1859 de Jonquières criticized the reasoning of Steiner and found the right number: 3264. But he didn't publish his result, for the stature of Steiner's geometry in the epoch intimidated everyone. Chasles didn't believe the reasoning of Jonquières and published his own proof in 1864. But a complete proof needed to wait for the XXth Century. The fact of a hypersurface of degree 6 had been made rigorous by Bischoff in 1866, but it was the number 7776 that caused the problem. In fact, the following objection was raised to Steiner's reasoning: Steiner's

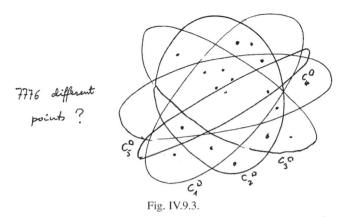

Fig. IV.9.3.

proof led to finding $2^5 = 32$ conics tangent to five given lines, whereas it is well known that there exists but one of them, thanks to duality (see Sect. IV.4). Where is the contradiction? The reason wasn't found until 1864, by Cremona: every conic that is a double line must be considered, in the algebraic sense, as tangent to whatever conic; thus all the hypersurfaces of $\mathcal{CCP}$ of this type contain the submanifold $\mathcal{V}$ of $\mathcal{CCP}$ formed of double lines, i.e. all quadratic forms of rank equal to 1. It is interesting to remark that it is a submanifold isomorphic to the set of lines of $\mathbb{C}P^2$, itself isomorphic to $\mathbb{C}P^2$. The embedding of $\mathbb{C}P^2$ in $\mathcal{CCP}$ by the quadratic forms of rank 1 is nothing other than the complex Veronese surface $\mathcal{V}$, whose real form was encountered in Sects. I.7 and II.0. We have here the same mapping $(x, y, z) \mapsto (x^2, y^2, z^2, \sqrt{2}yz, \sqrt{2}zx, \sqrt{2}xy)$.

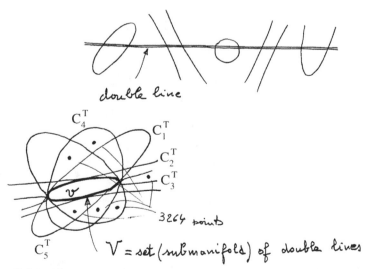

Fig. IV.9.4. *Above*: a double line is always tangent to any conic; *below*: a more or less symbolic sketch of the five C_i^T and of $\mathcal{V}$ in $\mathcal{CCP}$

Bézout's theorem is thus here too crude; we need in some way to know how to **deduct** the points of intersection for the fact that the five C_i^T contain all $\mathcal{V}$, by removing $\mathcal{V}$. We thus need a very general theory of intersection. But this did not acquire its definitive form until about 1960; it is the object of the book (Fulton, 1984). But from Cremona's time until today numerous mathematicians studied the problem, with varying rigor but with good ideas. Chasles, following de Jonquières, was thus the first to innovate, beginning in 1864, and to find the right number: 3264. It remained only to put his ideas into rigorous form, which was attacked by numerous geometers. In reading (Kleiman, 1980) it seems difficult to say who first obtained a truly complete proof; we mention Severi and van der Waerden in the first half of the twentieth century. We find two modern proofs in Griffiths and Harris (1978b) and we will explain one of these, whose idea is that of Chasles. In contrast, a proof based on the rigorous theory of intersection remains too costly, for it isn't easy to show that the Veronese surface counts, or "has a multiplicity", for $7776 - 3264 = 4512$.

The other method, described in Griffiths and Harris (1978b), is classic in intersection theory. It consists of "blowing-up" $\mathcal{V}$ and counting the number of times it then intersects itself; see in Sect. V.13 for a very tiny bit of this and for the Bezout theorem for curves. The idea − to in some way get rid of double lines − consists of considering *complete conics*, i.e. that a conic must simultaneously be the set of its points in $\mathbb{C}P^2$ and the set of its tangents in the dual $(\mathbb{C}P^2)^*$, thus reasoning in $\mathbb{C}P^2 \times (\mathbb{C}P^2)^*$. Then the complete conics can degenerate in double fashion: either into two lines (point perspective) or into two points (tangential viewpoint), having then to do with the set of lines that pass through one or the other of two distinct points.

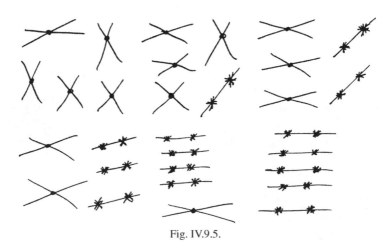

Fig. IV.9.5.

Now we do this: we degenerate in this double way the five conics C_i into the couples $\{(D_i, D_{i'}), (p_i, p_{i'})\}$ of a pair of lines and of a pair of points, there being six possible cases. The technical work of making rigorous what Chasles allowed himself to call the "continuity principle" amounts to showing two things during this quintuple degeneracy of the five pieces $C_i^T \setminus \mathcal{V}$. On the one hand the pieces tend toward the set of (true) conics that are either tangent to one of the lines or contain one of the points; on the other hand these five pieces intersect each other continually transversally, which shows that the number of their points of intersection is constant. There is nothing left but to calculate the number of the five double pairs $\{(D_i, D_{i'}), (p_i, p_{i'})\}$. There will always be 2^5 choices, but six possible types: containing 5 points, containing 4 points and being tangent to one line, containing 3 points and being tangent to two lines, and the three dual cases remaining. The first furnishes a conic, the second two conics (see above regarding pencils of conics) but with five choices; for the third type it can be shown rigorously that there are 4 conics but with $C_5^2 = 10$ choices. For the three remaining choices the numbers are the same by duality. The total number is thus exactly $2^5(1 + 2\cdot5 + 10\cdot4 + 10\cdot4 + 2\cdot5 + 1) = 3264$.

To finish we can ask what happens in the real case: certainly 3264 is a maximum, but it is difficult to conceive that there can exist such a number of conics tangent to five given conics in the real projective (or affine) plane. Now such configurations really *can* be found, the idea being to begin with five well-chosen degenerate conics, then to show that we can deform them continuously in such a way that they have **real** persistence for the a priori complex solutions. This is very recent: Ronga, Tognoli, and Vust (1998); there was previously a proof by Fulton from 1980, but unpublished. It's the occasion to point out that real algebraic geometry has been for a long time the poor relative of the complex version, but that there has recently been a role reversal; see the book Bochnak, Coste, and Coste-Roy (1998). The configuration of Ronga-Tognoli-Vust consists of five hyperbolas very close to their asymptotes, said asymptotes being pairs of lines very near the lines supporting the sides of a regular pentagon:

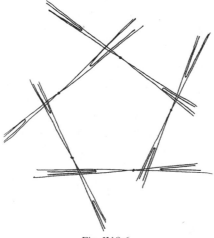

Fig. IV.9.6.

In the spirit of the following section, we can surely also seek to find how many quadrics (in three-space) are tangent to 9 given quadrics (the space of quadrics is of dimension $4 \cdot 5/2 = 10$). Armed nowadays with a complete theory of intersection in any dimension, 666 841 088 have been found, but it can also be done in higher dimensions. For all that see Sect. IV.10.4 of Fulton (1984).

IV.10. The quadrics

We consider, without going into detail — for lack of space — the real or complex quadrics: affine, Euclidean or projective. We deal first with the study, in the spaces $\mathbb{E}^3$, $\mathbb{R}^3$, $\mathbb{R}\mathrm{P}^3$, $\mathbb{C}\mathrm{P}^3$, of the surfaces defined by equations of second degree. We consider only *proper (or nondegenerate)* quadrics. Secondly, we can consider the same objects, always defined by equations of second degree (a quadratic form), in spaces of arbitrary dimension. Some use the word *hyperquadric* when the dimension exceeds three. Here is a caricatural selection of questions and answers on quadrics; for more, consult Chaps. 14 and 15 of [B].

In the Euclidean space $\mathbb{E}^3$, quadrics are encountered naturally as the simplest surfaces after planes and spheres.

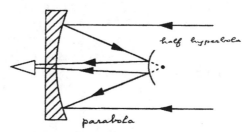

Fig. IV.10.1. Mirrors of a telescope; the two mirrors are: a piece of the paraboloid of revolution for the principal mirror and a piece of the hyperboloid of revolution for the small mirror

For example it is known that the Earth is, to a first approximation, an ellipsoid of revolution; that solid bodies have an ellipsoid of inertia and that their movement can be interpreted geometrically by rolling an ellipsoid of fixed center on a plane, see Sect. VII.13.C. The paraboloids of revolution are the mirrors of telescopes (and the small mirror is a piece of an hyperboloid of revolution), whereas the hyperbolic paraboloids are fundamental in architecture, just as are the hyperboloids (of revolution) by the way, for they are ruled surfaces, i.e. composed of lines; see Sect. VI.1. The ellipsoids of revolution have two foci; this property is utilized in the lamps of dentists and operating theaters, called by the pretty name *scialytic*. Until a few year ago, the headlights of automobiles were also paraboloids of revolution; nowadays they are very sophisticated surfaces, composed of several pieces, but paraboloids, ellipsoids and hyperboloids of revolution remain essential elements. For the possible detection of gravitational waves, an essential problem of general relativity, there is presently being constructed a telescopic mirror whose focal length is three kilometers, which is an extremely elaborate technical problem. Note then here that what is reflected on these surfaces is not light but electromagnetic waves (see the parabolas for picking up satellites). See in [B] several photographs of varied architecture where quadrics are used systematically.

The ellipsoids play an essential role in algorithmics and optimization – see Grötschel, Lovasz, and Schrijver, 1988 – as well as in the theory of normed spaces; see for example Pisier (1989). But we will in fact encounter ellipsoids amply in Chap. VII, in particular in Sect. VII.6.B. There exists a whole array of characteristic properties of ellipsoids; see Gruber and Wills (1993). But we will also see in Sect. V.3 that geometry on an ellipsoid isn't yet completely well understood. As for the mystery of *rolling stones*, i.e. the fact that the pebbles on a beach are roughly ellipsoidal, this will be partly elucidated in Sect. VII.13.C.

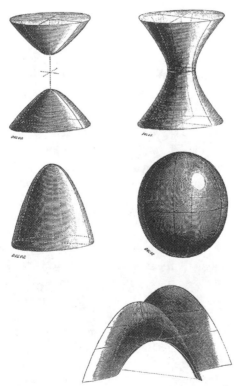

Fig. IV.10.2. The five types of proper quadrics, discovered and named by Monge (see for example Berger, 2005): hyperboloid of two sheets, hyperboloid of one sheet, elliptic paraboloid, ellipsoid and hyperbolic paraboloid. Rouché, de Comberousse (1912) © Elsevier

The proper quadrics in $\mathbb{E}^3$ are:

I - *The ellipsoids*

$$\frac{x^2}{a^2} + \frac{y^2}{b^2} + \frac{z^2}{c^2} - 1 = 0$$

(of revolution if $a = b$).

II - *The elliptic paraboloids*

$$\frac{x^2}{a^2} + \frac{y^2}{b^2} - 2z = 0$$

(of revolution if $a = b$).

III - *The hyperbolic paraboloids*

$$\frac{x^2}{a^2} - \frac{y^2}{b^2} - 2z = 0.$$

IV - *The hyperboloids of two sheets*

$$\frac{x^2}{a^2} + \frac{y^2}{b^2} - \frac{z^2}{c^2} + 1 = 0$$

(of revolution if $a = b$).

V - *The hyperboloids of one sheet*

$$\frac{x^2}{a^2} + \frac{y^2}{b^2} - \frac{z^2}{c^2} - 1 = 0$$

(of revolution if $a = b$).

The affine classification thus contains but five types, the Euclidean classification consists only in appending the values of a, b, c figuring in the above equations. The projective classification in $\mathbb{R}P^3$ has room for two types only: the topology of the first type is that of a sphere; the topology of the second that of the torus $\mathbb{T}^2$. Another essential matter distinguishes them. The quadrics of the second type contain lines, projective or affine: there are two families of them, $\mathcal{F}$ and $\mathcal{F}'$ for each quadric; their lines are called *generators* of these ruled quadrics. In the projective context, these are thus topologically circles and we note that those of the same family don't intersect one another but are enlaced, whereas two generators of different families intersect in a unique point. The first type is formed of convex quadrics and the tangent plane intersects the quadric in one point only; for the second type the tangent plane intersects the quadric precisely in two secant lines at the point of contact, one from each family. The global topological configuration of these two families $\mathcal{F}$ and $\mathcal{F}'$ is the same as that of Villarceau circles, encountered in Sect. II.7; inquisitive readers will ask if we can pass from one situation to the other by an appropriate transformation. Analytically, the generators are found thus: the two types correspond to quadratic forms with signatures $(3, 1)$ and $(2, 2)$ (since $(4, 0)$ and $(0, 4)$ yield the empty set). In the case $x^2 + y^2 - z^2 - t^2 = 0$, the generators are given by the pairs of equations:

$$x - z = k(y + t) \quad \text{and} \quad x + z = -k^{-1}(y + t)$$

and

$$x - z = k(y + t) \quad \text{and} \quad x + z = -k^{-1}(y - t).$$

This immediately furnishes an affine (and projective) way of "tracing" quadrics. We take three quadrics D, D′, D″ in the space, no two in the same plane. Then the set of lines of $\mathbb{R}^3$ or $\mathbb{R}P^3$ that simultaneously intersect D, D′ and D″ form a ruled quadric; and they are all so obtained. This property was used to construct the double six of Schläfli in Sect. I.9. The following is easily shown: whatever the four lines $D_i \in \mathcal{F}$ and the line $E \in \mathcal{F}'$, the cross ratio $[D_1 \cap E, D_2 \cap E, D_3 \cap E, D_4 \cap E]$ of four points depends only on the D_i and not on E.

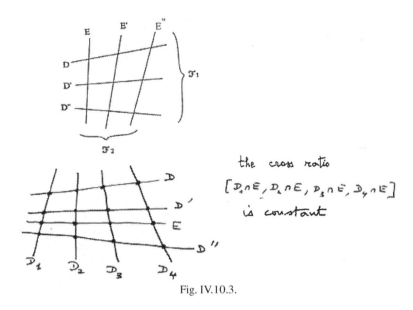

the cross ratio

$$[D_1 \cap E, D_2 \cap E, D_3 \cap E, D_4 \cap E]$$

is constant

Fig. IV.10.3.

Whence also this: let D and D′ be two lines not in the same plane and $f : D \to$ D′ any homography. Then the line $m \cdot f(m)$ describes a ruled quadric. Or again this: let D, D′ be two lines not in the same plane, and f a homography between the planes passing through D and those passing through D′. Then the line $P \cap f(P)$ describes a ruled quadric. A particular affine case is this: if two points traverse two lines D, D′ not in the same plane with constant speeds, then the line that joins them describes a hyperbolic paraboloid, the shape of many building roofs with non rectangular walls. This is also the explanation of Fig. IV.4.5; we need only remark that the apparent contour of a hyperbolic paraboloid is a parabola. If architects and builders love hyperbolic paraboloids of one sheet, it's because they have **two** series of lines, so that concrete can be doubly reinforced, etc.

The metric definitions analogous to those of Sect. IV.2 exist but are not very inspiring; see a quite detailed study in Coolidge (1968), §1 of Chap. XI. To find a quadric

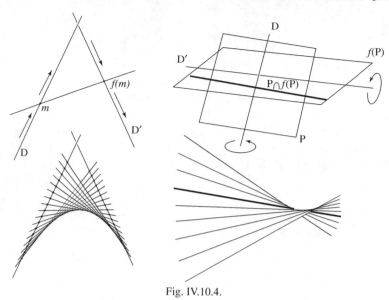

Fig. IV.10.4.

of $\mathbb{E}^3$ with three unequal axes (i.e. not of revolution) we need to assume a point f, a line D, a planar direction P and a constant e. Then we obtain practically all the quadrics of $\mathbb{E}^3$ as the set of points m satisfying $d(m, f) = e \cdot d(m, p(m))$ where $p(m)$ is the point of D obtained by cutting D by the plane passing through m and parallel to P. The generation analogous to bifocal generation and even its extension of that of Graves, discussed in Sect. IV.2, is due to Staude and only about 1850!; see below. The ruled quadrics themselves can be obtained much more simply as the set of points for which the ratio of the distances to two given lines is constant. Hilbert is credited with saying that Staude's result on the generation of the quadrics with string was one of the great mathematical results of the nineteenth century, but the basis for this opinion isn't clear. In fact, if we go to the bottom of things, it has to do with hyperelliptic functions and their additive properties; see below for references on this subject.

◆

The complex quadrics are not so very interesting in $\mathbb{C}P^3$. Pay close attention to what these complex surfaces are: objects in four real dimensions. There is but one type of proper quadric and this in arbitrary dimension, since the quadratic forms over the complex numbers of maximum rank are all isomorphic to $\sum_{i=1}^{n+1} z_i^2$. The topology of the complex quadrics is interesting for algebraic topologists – it's that of a product $S^2 \times S^2$ of two spheres, but in higher dimensions the discovery of the topology of hyperquadrics commenced only with Cartan (1932).

In dimension 2, **the** complex quadric possesses generators with properties entirely identical to the real case; the equations are the same. The properties announced earlier about generation and the cross ratio remain valid. In higher dimension, generating lines are generalized by projective subspaces, but things depend on the

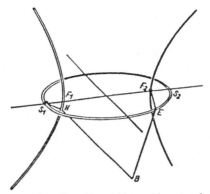

Fig. IV.10.5. Staude generation of an ellipsoid by a string stretched between an hyperbola and an ellipse; see also Fig. IV.10.9 Hilbert, Cohn-Vossen (1996) © Springer

parity of the dimension. The fundamental result is Witt's theorem on quadratic forms, valid over an arbitrary field. It goes back to 1936; see 13.7.1 of [B]. This theorem also regulates the major portion of questions concerning the group of isometries of quadratic forms, thus a goodly part of the geometry of quadrics. To show the mental state of geometers not so long ago, we complexify the sphere S^2. It thus has generators, and it is thus that the geometers of the nineteenth century spoke profusely of the "generators of the spheres"; see for example Darboux (1917), also Dieudonné (1985). But the first was the romantic Poncelet; see Berger (2005). To dive into Duporcq (1938) was a rule for the author in his final year at the lycée in 1944, but Koszul informed him at Strasbourg in 1957 that − although this book is surely nice to read − it is regrettably devoid of any serious foundation for algebraic geometry.

The whole theory of duality seen for the conics generalizes immediately to quadrics, because this duality is the geometric interpretation that a quadratic form is, by definition, the diagonal of a bilinear symmetric form, this both in the real and in the complex domain. In the complex case, i.e. in $\mathbb{C}P^{n+1}$, we have again a *Nullstellensatz* in each dimension, and in particular we can define the space of all the quadrics (of dimension n) of $\mathbb{C}P^{n+1}$. It's a complex projective space of dimension $n(n+3)/2$. For the "ordinary" quadrics, it is thus of dimension 9. Equipped now with modern techniques, we won't fail to calculate the number of quadrics tangent to 9 quadrics in general position; see the end of the preceding section.

Now we have but little space left for speaking both of properties of quadrics − such as the intersection of two or three quadrics, the geometry on a given quadric, etc. − and the entities that extend the quadrics to spaces of any dimension n. The second part of this program can be found in Chaps. 13,14, and 15 of [B].

The intersection of quadrics leads to very subtle questions. An important starting point is that these intersections are curves that are parameterized naturally with elliptic functions. Here are four references storied in time: Appell and Lacour (1922), Halphen (1886, 1888), Donagi (1980), and Tjurin (1975).

To conclude we choose to speak of a very important configuration of the Euclidean space $\mathbb{E}^3$, called *homofocal quadrics*, whose interest beyond its intrinsic beauty is, as we will see, that it overlaps quite varied problems that have no immediate relation to quadrics. Here such a family has one real parameter λ and is made up of quadrics with equations

$$(**) \qquad \frac{x^2}{a^2 + \lambda} + \frac{y^2}{b^2 + \lambda} + \frac{z^2}{c^2 + \lambda} - 1 = 0,$$

that generalize Equation $(*)$ of Sect. IV.2. The case where a, b, c are distinct is the most interesting; we thus suppose from now on that $a > b > c$.

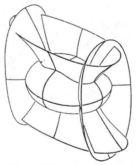

Fig. IV.10.6. Hilbert, Cohn-Vossen (1996) © Springer

According as $\lambda \in \,]-\infty, c^2[, \,]c^2, b^2[, \,]b^2, a^2[$, we obtain respectively: an ellipsoid, a hyperboloid of two sheets, a hyperboloid of one sheet. The figure formed by these three one-parameter families enjoys properties that are many and remarkable. Here is a sampling:

First, through each point of space located outside the union of the three coordinate planes there passes one and only one quadric of each sort, whence the mapping $(x, y, z) \mapsto (\lambda, \mu, \nu)$. If we restrict ourselves, for example, to the region $x > 0$, $y > 0$, $z > 0$, then this mapping is bijective and provides a parameterization by the so-called *homofocals* of each of the 8 quadrants of space. These coordinates allow us to study questions rather systematically, whether they relate to the whole collection or to a fixed quadric such as the ellipsoid E such that $\frac{x^2}{a^2} + \frac{y^2}{b^2} + \frac{z^2}{c^2} - 1 = 0$, for example for studying vibrations on or interior to this ellipsoid; see e.g. Morse and Feshbach (1953).

An elementary property is that the three quadrics, at a point that they all contain, have mutually orthogonal tangent planes. The proof can be hard if the calculations are made in (x, y, z) coordinates and without some finesse.

Recall (see Chap. XII) that the geodesics of a surface (or more generally of a reasonable metric space) are curves that provide the shortest paths from one point to another. But the geodesics defined by this local minimum property remain interesting when they are prolonged indefinitely, which is always possible if there is compactness.

If we want to study the geodesics on an ellipsoid E, we will then set $\lambda = 0$ in the triple (λ, μ, v) and the surface of the ellipsoid will be parameterized by (μ, v). These are called *elliptic coordinates*. The geodesics of E are defined by a very simple family of differential equations in $\mu(t)$, $v(t)$, with the additional condition of speed equal to 1, dependent on one parameter. This permits us to study them almost completely. A fairly recent text is Knörrer (1980); see also the references at the end of Sect. 10.4.9.5 of [BG]. Here are some properties, but also look in Sect. VI.3. With one fixed value of the parameter mentioned above is associated a hyperboloid H (of one sheet or two depending on these parameter values) which is such that the geodesics having this parameter oscillate between the two curves of intersection of E with H. They may, in $\mathbb{E}^3$, be defined geometrically by the condition that their tangents are the lines of $\mathbb{E}^3$ constantly tangent to H. But we will see in Sect. VI.3 the unfinished history of some recalcitrant ellipsoids.

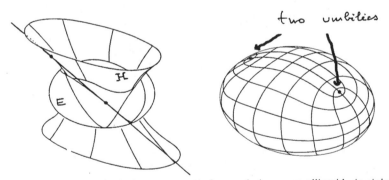

Fig. IV.10.7. *At left*, the tangency property for geodesics on an ellipsoid. *At right*, a figure discovered by Monge, that of lines of curvature of the ellipsoid Hilbert, Cohn-Vossen (1996) © Springer

The above figures correspond to the case where H has one sheet; as an exercise readers can sketch the case where H has two sheets. The intermediate case is where H is reduced to the hyperbola H* (see Fig. IV.10.9)

$$\frac{x^2}{a^2 - b^2} + \frac{z^2}{b^2 - c^2} - 1 = 0$$

of the plane $y = 0$. The intersection $H^* \cap E$ is composed of 4 points a, a', b, b'. They
are called the *umbilics* of E, for they are the points where the surface E in $\mathbb{E}^3$ has
two equal radii of principal curvature; see Chap. VI or Berger and Gostiaux (1987)
for the vocabulary for what follows. But here they have the following extraordinary
focalization property, which is nothing other than the limit of the oscillation property
mentioned above: each geodesic issuing from a (resp. b) passes through a' (resp. b').
See more of this in Chaps. VI and XII. We don't know if the geometry of ellipsoids
of dimension greater than two − for example what generalizes the focalization of
umbilics − has been studied; see however Joets and Ribotta (1999) for the focal
sheets of these ellipsoids in higher dimensions; see also the end of Sect. VI.3.

The curves of the intersections $E \cap H$ are all *lines of curvature* of E (and,
moreover, of H also), i.e. the curves for which the tangent always has a direction of principal curvature (see the references given above). It involves a definition that is valid for every surface. With elliptic coordinates, we can easily show
that all the lines of curvature are defined as the loci of points for which the sum
of the intrinsic distances (see Sect. VI.1) to the two umbilics is constant (compare this with the bifocal properties of conics in Sect. IV.2). Note that this is now
the difference which becomes constant if we change an umbilic into its antipodal!
We find this nice figure, due to Monge, that will arouse our enthusiasm again in
Sect. VI.3.

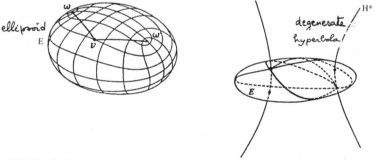

Fig. IV.10.8. The lines of curvature of an ellipsoid can be defined, just as for ordinary
ellipses in the plane, by a bifocal condition where the two foci are the umbilics

Just as the bifocal definition of conics can be extended by Graves' theorem
(Sect. IV.2), so perhaps can the above property $d(a, v) + d(b, v) = $ constant be
extended to give initial results of Graves type for a given quadric. But they especially long resisted the ardor of geometers, in particular in attempting to generate an
entire quadric with a single stretched string. The case $d(f, x) + d(f', x) = $ constant
in $\mathbb{E}^3$ yields in fact only ellipsoids of revolution. It was necessary to wait for Staude
− toward the end of nineteenth century − to find this generation. The simplest is
that of Fig. IV.10.5.

The hyperbola is

$$\frac{x^2}{a^2 - b^2} - \frac{z^2}{b^2 - c^2} - 1 = 0, \; y = 0$$

and the ellipse is

$$\frac{x^2}{a^2} + \frac{z^2}{b^2} - 1 = 0, \; z = 0.$$

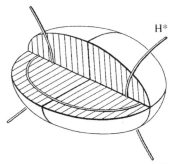

Fig. IV.10.9. Hilbert, Cohn-Vossen (1996) © Springer

But Staude extended this result "a la Graves": the string is stretched between an ellipsoid and a hyperboloid of one sheet which is homofocal with it: the points which tighten the string describe an ellipsoid. For a proof of Staude's results, see any of the following: p. 450 of Salmon (1874) or Chap. XIII, §3 of Coolidge (1968), or p. 179 of Coolidge (1940–1963), or again Staude (1904–1992).

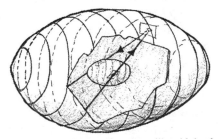

Fig. IV.10.10. Playing billiards on the interior of an ellipsoid that has been cut open so as to see what happens (drawing by Geoffroy Wagon)

A property of homofocal ellipsoids, that generalizes the one for ellipses what was mentioned at the end of Sect. IV.2, is again a property of light caustics: a ray of light propagates on the interior of the ellipsoid E, reflecting on the interior surface with the symmetry which is imposed with respect to the tangent plane at the point of reflection, remains always tangent to an ellipsoid E', homofocal and interior to E.

This if it stays on the interior; if not, it will remain tangent to one of the hyperboloids of the family. A result, by the way, that is purely local and of little difficulty is that: the only Euclidean surfaces that admit a caustic surface are pieces of quadrics; cf. Berger (1995).

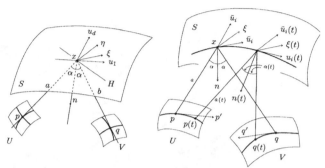

Fig. IV.10.11. Represented on the right is what happens along a line of curvature of S
Berger (1995) © Société Mathématiques de France

The proof of the "causticity" of the quadrics − in fact valid in arbitrary dimension − is accomplished very nicely with duality, complexification and the use of what generalizes the cyclic points of the plane (see Sect. II.XYZ) to $\mathbb{E}^3$, specifically the conic Ω in the plane at infinity (called *umbilical*) defined by the cone $x^2 + y^2 + z^2 = 0$. For example the spheres are the quadrics for which the complexification contains Ω. This classic proof is found for example in Douady (1982).

Recently an extraordinary and unexpected relation was discovered among the geodesics of the ellipsoid (more precisely the obvious extension of the homofocal quadrics to $\mathbb{E}^n$) and the solutions "a la Peter Lax" of the KdV (Korteweg-de Vries) equation, solutions called *solitons*. The KdV partial differential equation describes the movement of water in a channel. A connection is made by isospectral deformations of matrices. Several other problems are directly connected to it: that of the movement of a point that moves on a sphere under the influence of a quadratic potential, and that due to H. Knörrer which generalizes the fact that the intersection of two quadrics of $\mathbb{C}P^3$ is parameterized by an elliptic function and which involves the manifold formed of subspaces of dimension $n - 1$ of the intersection of two quadrics of $\mathbb{C}P^{2n+1}$. For all the preceding consult (Knörrer, 1980; Moser, 1980) and the quite recent (Audin, 1995).

IV.XYZ

Conics over arbitrary fields. Conics and quadrics too can clearly be defined, in any dimension, over commutative fields other than the reals or the complexes. When the field is no longer commutative − in the case of the quaternions $\mathbb{H}$ for example − the

notion of quadratic form becomes more difficult. A very little bit of the study of
quadrics over $\mathbb{H}$ can be found in Porteous (1969).

Over finite fields, we are led to problems in number theory and combinatorics.
Geometric intuition must then be handled with great caution. Here are some examples that illustrate this remark. It can be shown (but is not obvious) that over a finite
field no conic is ever empty. If the field has and even number q of elements, then a
conic has at most $q + 1$ points (because it must be isomorphic to a projective line)
and all the tangents pass through the same point! But if q is odd, then through every
point not belonging to the conic there are either two tangents to the conic or else no
tangent to the conic passes through the point; see Lidl and Niederreiter (1983).

Matrix representation of the equation for conics and duality. We represent the
point $m = (x : y : z)$ of $\mathbb{R}P^2$ by the column matrix $X = \begin{pmatrix} x \\ y \\ z \end{pmatrix}$ (defined within a
constant multiple). We also represent the line with equation $ux + vy + wz = 0$ by
the column matrix $U = \begin{pmatrix} u \\ v \\ w \end{pmatrix}$, in such a way that the equation of the line is written
$^tUX = 0$. We can then consider U as the column matrix with the coordinates of d
in the dual projective plane $(\mathbb{R}P^2)^*$.

Now let C be a conic in $\mathbb{R}P^2$, with equation

$$F(x, y, z) = ax^2 + a'y^2 + a''z^2 + 2byz + 2b'zx + 2b''xy = 0.$$

This equation can be rewritten

$$^tXAX = 0,$$

where A is the symmetric matrix

$$A = \begin{pmatrix} a & b'' & b' \\ b'' & a' & b \\ b' & b & a'' \end{pmatrix}.$$

We assume in what follows that A is regular, i.e. of rank 3.

(i) Two points X and Y are *conjugate* with respect to the conic if and only if

$$^tXAY = 0.$$

We prove the case where X and Y are distinct. The line joining these two points is
parameterized by $(s : t) \mapsto sX + tY$, or $\lambda \mapsto X + \lambda Y$ if we allow the value ∞
for λ. The points of intersection P and Q of the line with C are given by the roots λ'
and λ'' of the equation

$$^tXAX + 2\lambda \cdot {}^tXAY + \lambda^2 \cdot {}^tYAY = 0.$$

We have $[X, Y, P, Q] = [0, \infty, \lambda', \lambda''] = \lambda'/\lambda''$. This cross ration is equal to -1 if
$\lambda' + \lambda''$, the sum of the two roots, is zero, i.e. if $^tXAY = 0$.

(ii) It follows that the polar of the point X is the line represented by $U = AX$,
and that the pole of the line U is $X = A^{-1}U$.

(iii) A pointX belongs to C if it is conjugate to itself, or again if it belongs to its polar. In the latter case, the polar X is the tangent to C at X. We in fact show that the coefficients of the equation for the tangent to a curve with homogeneous equation

$F(x, y, z) = 0$ are given by the column matrix $\begin{pmatrix} F'_x \\ F'_y \\ F'_z \end{pmatrix}$, which here yields 2AX.

(iv) Consequently a line U is tangent to C if and only if we have

$$^t U A^{-1} U = 0.$$

This is the equation of the dual conic C* (in the dual projective plane).

Involutions. An *involution* of $\mathbb{C}P^1$ is a homography $f : \mathbb{C}P^1 \to \mathbb{C}P^1$ such that $f^2 = \mathrm{Id}$, $f \neq \mathrm{Id}$ (we treat only the complex case — readers can carry out the adaptation to the real case). The homography

$$f(x) = \frac{ax + b}{cx + d}$$

$(ad - bc \neq 0)$ is an involution if and only if $a = -d$. The fixed points of the involution are the roots of the second degree equation $cx^2 - 2ax - b = 0$. By hypothesis, $a^2 + bc \neq 0$. An involution therefore has two distinct fixed points (if $c = 0$, one of them is ∞).

If we denote the fixed points by α and β, and if x is distinct from these, then we have $[x, f(x), \alpha, \beta] = [f(x), x, \alpha, \beta] = [x, f(x), \alpha, \beta]^{-1}$, whence the relation

$$[x, f(x), \alpha, \beta] = -1,$$

since a cross ratio can't equal 1 for distinct points. From this relation, symmetric incidentally in x and $f(x)$, we can extract $f(x)$ as a homographic function of x, which shows that an involution is determined by the specification of its two fixed points.

Finally, we frequently define a mapping $f : \mathbb{C}P^1 \to \mathbb{C}P^1$ such that $f^2 = \mathrm{Id}$, $f \neq \mathrm{Id}$, but without being able (or even wanting) to compute $f(x)$ explicitly. Once we know that $f(x)$ is a rational function of x, we are assured of having an homography (we don't need the condition $f^2 = \mathrm{Id}$; the fact that f is injective suffices.

Equations of second degree and involutions. The solutions of an equation of second degree that is linearly dependent on a parameter, e.g.

(E_λ) $\qquad\qquad (a + \lambda a')x^2 + (b + \lambda b')x + c + \lambda c' = 0$

are "in involution". In modern language this means that there exists an involutive homography f such that

$$x'' = f(x'), \quad x' = f(x'')$$

where $x' = x'(\lambda)$ and $x'' = x''(\lambda)$ denote the two roots of E_λ. In applications, the parameter λ and the unknown x are taken in $\mathbb{C}P^1$, but we avoid using homogeneous coordinates: if the coefficient of x^2 vanishes, one of the roots is infinite, and if $\lambda = \infty$, the equation is simply $a'x^2 + b'x + c' = 0$.

The proof is very simple. This sort of computation incidentally is the has been the delight of the fifth form (eleventh grade) students for scarcely more than a generation.

We assume that the E_λ *don't have a common root.* The three coefficients of the equation are necessarily related by a relation of the type

$$A(a + \lambda a') + B(b + \lambda b') + C(c + \lambda c') = 0.$$

In view of the properties of the sum and product of the roots of a second degree equation, we then have

$$A - B(x' + x'') + Cx'x'' = 0,$$

whence we find

$$x' = \frac{Bx'' - A}{Cx'' - B}, \quad x'' = \frac{Bx' - A}{Cx' - B}.$$

The fixed points of f are the roots α and β of the equation $A - 2Bx + Cx^2 = 0$.

According to the preceding, they are distinct and we have $[x', x'', \alpha, \beta] = -1$.

A fixed point of f is also a double root of E_λ. Since there are two distinct fixed points, the equation E_λ has a double root for exactly two values of λ.

Application 1. (Desargues-Sturm Theorem.) If the line D doesn't intersect the base of the pencil, the conics of a linear pencil determine an involution on D. Two distinct conics of the pencil are tangent to D, the points of contact being the fixed points of the involution.

Application 2. (Frégier's Theorem.) Let C be a proper conic and p a point not belonging to C. The mapping that takes each point m of C to the point where the line pm intersects C is an involution. Every involution of a conic has this form.

The forward implication is immediate in view of what has already been said. For the converse, we indicate that, an involution of C being given, we obtain p as the intersection of the tangents at the fixed points of the involution.

Bibliography

[B]Berger, M. (1987, 2009). *Geometry I,II*. Berlin/Heidelberg/New York: Springer

[BG]Berger, M., & Gostiaux, B. (1987). *Differential geometry: manifolds, curves and surfaces*. Berlin/Heidelberg/New York: Springer

Appell, P., & Lacour, E. (1922). *Fonctions elliptiques*. Paris: Gauthier-Villars

Arnold, V. (1990). Huyghens and barrow, Newton and Hooke. Basel: Birkhäuser

Arnold, V. (1999). Symplectization, complexification and mathematical trinities. In E. Bierstone, B. Khesin, A. Khovanskii, & J. E. Marsden (Eds.), *The Arnoldfest* (pp. 23–28). Providence, RI: American Mathematical Society

Arnold, V. (2000). Polymathematics: Is mathematics a single science or a set of arts? In V. Arnold, M. Atiyah, P. Lax, & B. Mazur (Eds.), *Mathematics: Frontiers and Perspectives* (pp. 403–416). Providence, RI: American Mathematical Society

Audin, M. (1995). Topologie des systèmes de Moser en dimension quatre. In H. Hofer, C. Taubes, A. Weinstein, & E. Zehnder (Eds.), *The Floer Memorial Volume* (pp. 109–122). Basel: Birkhäuser

Barth, W., & Bauer, T. (1996). Poncelet theorems. *Expositiones Mathematicae, 14*, 125–144

Barth, W., & Michel, J. (1993). Modular curves and Poncelet polygons. *Mathematische Annalen, 295*, 25–49

Benoist, Y., & Hulin, D. (2004). Itération de pliages de quadrilatères. *Inventiones Mathematicae, 157*, 147–194

Berger, M. (1995). Seules les quadriques admettent des caustiques. *Bulletin de la Société Mathématique de France, 123*, 107–116

Berger, M. (2005). Dynamiser la géométrie élémentaire: introduction à des travaux de Richard Schwartz. *Atti della Accademia Nazionale dei Lincei. Classe di*, Ser. 25, 127–153

Bochnak, J., Coste, M., & Coste-Roy, M.-F. (1998). *Real algebraic geometry*. Berlin/Heidelberg/New York: Springer

Bos, H., Kers, C., Oort, F., & Raven, D. (1987). Poncelet's closure theorem. *Expositiones Mathematicae, 5*, 289–364

Bourbaki, N. (1981). Algèbre, chapitre IV. Masson

Cartan, E. (1932). Sur les propriétés topologiques de la quadrique complexe. *Public. math. Uni. Belgrade, ou OEuvres complètes, 1-2, 1227-1246*, 55–74

Cayley, A. (1861). On the Porism of the in-and-circumscribed polygon. *Philosophical Transactions of the Royal Society of London, CLI*, 225–239

Coolidge, J. (1940, 1963). A history of geometrical methods. Oxford: Oxford University Press, Dover reprint

Coolidge, J. (1968). *A history of the conic sections and quadric surfaces* (1st ed. 1945). New York: Chelsea, Dover reprint

Darboux, G. (1917). *Principes de géométrie analytique*. Paris: Gauthier-Villars

Dieudonné, J. (1985). *History of algebraic geometry*. Monterey, CA: Wadsworth

Dingeldey, F. (1911). Coniques et systèmes de coniques. In *Encyclopédie des sciences mathématiques pures et appliqués*. Paris et Leipzig: Gauthier-Villars and Teubner

Donagi, R. (1980). Group law on the intersection of two quadrics. *Annales scientifiques Ecole norm. sup., 7*

Douady, R. (1982). Applications du théorème des tores invariants. *Thèse Paris VII*

Duporcq, E. (1938). *Premiers principes de géométrie moderne*. Paris: Gauthier-Villars

Emch, A. (1900). Illustration of the elliptic integral of the first kind by a certain link work. *Annals of Mathematics, 1*, 81–92

Emch, A. (1901). An application of elliptic functions to Peaucellier link-work (inversor). *Annals of Mathematics, 2*, 60–63

Fulton, W. (1984). *Intersection theory* (2nd ed. 1998). Berlin/Heidelberg/New York: Springer

Greenberg, M. (1974). *Euclidean and non-Euclidean geometries*. New York: Freeman

Griffiths, P., & Harris, J. (1978a). On Cayley's explicit solution to Poncelet's porism. *L'enseignement math., 24*, 31–40

Griffiths, P., & Harris, J. (1978b). *Principles of algebraic geometry*. New York: John Wiley

Grötschel, M., Lovasz, L., & Schrijver, A. (1988). *Geometric algorithms and combinatorial optimization*. Berlin/Heidelberg/New York: Springer

Gruber, P., & Wills, J. (Ed.). (1993). *Handbook of convex geometry*. Amsterdam: North-Holland

Hadamard, J. (1911, 1988). *Leçons de géométrie élémentaire* (Reprint Jacques Gabay). Paris: Armand Colin

Halphen, G. (1886, 1888). *Traité des fonctions elliptiques, I, II*. New York: Gauthier-Villars

Hilbert, D., & Cohn-Vossen, S. (1999). *Geometry and the imagination*. Providence, RI: American Mathematical Society

Hrasko, A. (1999). Letter to A. Shen. *The Mathematical Intelligencer, 21*(3), 50

Hrasko, A. (2000). Poncelet-type problems, an elementary approach. *Elemente der Mathematik, 55*, 1–18

Joets, A., & Ribotta, R. (1999). Caustique de la surface ellipsoïdale à trois dimensions. *Experimental Mathematics, 8*, 57–62

Kleiman, S. (1980). *Chasles's enumerative theory of conics: An historical introduction, in studies in algebraic geometry* (pp. 117–138). Washington, DC: The Mathematical Association of America

Klein, C.F. (1872). Erlangen Program. http://www.xs4all.nl/~jemebius/ErlangerProgramm.htm#Introduction

Knörrer, H. (1980). Geodesics on the ellipsoid. *Inventiones Mathematicae, 59,* 119–143

Lebesgue, H. (1942, 1987). *Les coniques* (Reprint Jacques Gabay). New York: Gauthier-Villars

Levy, H. (1964). *Projective and related geometries.* New York: McGraw Hill

Lidl, R., & Niederreiter, H. (1983). *Finite fields.* Cambridge,UK: Cambridge University Press

McDonald, S., & Kaufmann, N. (1998). Wave chaos in the stadium: Statistical properties of short-wave solutions of the Helmholtz equation. *Physical Review A, 37*(8), 3067–3086

Morse, P.M., & Feshbach, H. (1953). *Methods of theoretical physics.* New York: McGraw-Hill

Moser, J. (1980). *Geometry of quadrics and spectral theory* (pp. 147–188). In The Chern symposium. Berlin/Heidelberg/New York: Springer

Pécaut, F. (2005). Équivalence du grand théorème de Poncelet pour deux cercles et du théorème des zigzags. *Quadratures, 58,* 13–18. EDP Sciences, Les Ulis

Pisier, G. (1989). *The volume of convex bodies and Banach space geometry.* Cambridge, UK: Cambridge University Press

Porteous, I. (1969). *Topological geometry.* London: Van Nostrand-Reinhold

Ronga, F., Tognoli, A., & Vust, T. (1997). The number of conics tangent to five given conics: The real case. *Revista Matemática Complutense, 10,* 391–421

Rouché, E., & de Comberousse, C. (1912). *Traité de géométrie* (2 vols.). New York: Gauthier-Villars

Salmon, G. (1874). *A treatise on the analytic geometry of three dimensions* (Reprint Chelsea). Dublin: Hodges

Schwartz, R. (2007). The Poncelet grid. *Advances in Geometry, 7,* 157–175

Shen, A. (1998). Mathematical entertainments. *The Mathematical Intelligencer, 20,* 31

Staude, O. (1904, 1992). III 22. Quadriques. *Encyclopédie des Sciences mathématiques.* J. Molk, Teubner, trad. Gauthier-Villars, reprint Jacques Gabay, III, 1–162

Stricchartz, R. (1987). Realms of mathematics: Elliptic, hyperbolic, parabolic, sub-elliptic. *The Mathematical Intelligencer, 9*(3), 56–64

Tabachnikov, S. (1995). *Billiards.* Paris: Société mathématique de France

Tjurin, A. (1975). On intersection of quadrics. *Russian Mathematical Surveys, 30,* 51–105

Van der Waerden, B. (1954–1975). *Science awakening I.* Groningen: Noordhoff

Chapter V
Plane curves

V.1. Plain curves and the person in the street: the Jordan curve theorem, the *turning tangent theorem* and the isoperimetric inequality

Here are three examples of "facts" — but rather more of "results", of "theorems" — that illustrate the difference between the mathematician and the "person in the street". They are excellent for explaining the nature of mathematics to a nonprofessional. The first is the *Jordan curve theorem*:

(V.1.1) *In the plane we trace a closed curve that doesn't intersect itself. Then it has an interior, a region that it encompasses.*

Fig. V.1.1. The interior of a simple closed curve

The instinctive reaction is that it's obvious! — but the mathematician on the contrary requires a **proof**; but first — and this is also important — a **definition**, formalized from the notion of interior. It's about the same for the *turning tangent theorem* known in German as Hopf's *Umlaufsatz*:

(V.1.2) *We trace a smooth closed curve that doesn't intersect itself and look at how much the tangent to the curve turns as we make a complete circuit: it turns 2π in all.*

The second result is surely already a little more mathematical in its statement, but it remains nonetheless very intuitive. It was, by the way, long considered evident by practically all mathematicians. Another typical example of a result that was initially considered obvious, then was the object of more or less rigorous proofs before being truly completely proved, is what figures in the beginning of Sect. V.11, that is to say the *isoperimetric inequality*:

M. Berger, *Geometry Revealed*, DOI 10.1007/978-3-540-70997-8_5,
© Springer-Verlag Berlin Heidelberg 2010

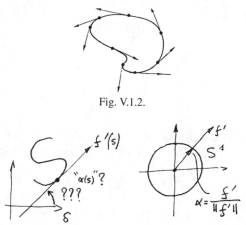

Fig. V.1.2.

Fig. V.1.3. To be able to specify what it means *to turn*, a mapping α from the curve to the circle is introduced, while the geometric velocity is normalized as a vector of length 1. If f' is this velocity, we set $\alpha = \frac{f'}{\|f'\|}$. If an axis is chosen, we can identify α, an element of S^1, with the angle that f' makes with this axis

(V.1.3) *Among all plane curves of a given length, the circle is the one which encompasses the largest area possible.*

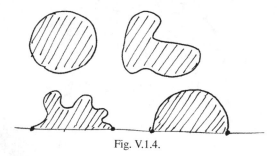

Fig. V.1.4.

This last result is the one that in folklore is the most ancient; this sort of thing was known to the Greeks. Moreover, its practical importance is considerable, for example for plane silaging problems. By symmetry, the same result can be deduced for a semi-circle among all curves of given length which are attached to a fixed line at its extremities.

Now, to convince our ordinary person that things are not so evident as they appear, we need to show more complex drawings and appeal to the imagination.

In such drawings it becomes more difficult to know what is the actual interior and still more difficult to believe that this time the tangent ends up turning only 2π, i.e. that all its contortions have been finally cancelled out to leave room in the end for **only** a single turn. From the conceptual point of view and that of Jacob's ladder, it is much the same with our two first "obvious" theorems. The isoperimetric inequality will be the subject of Sect. V.11 and V.12.

Fig. V.1.5. In the two figures it can be seen that the notion of *interior* isn't so very clear!

First and foremost we need to define *curve* and, in addition, "closed and not self-intersecting". If we read the definitions given in Sect. V.2 or in V.XYZ, then we don't at all see anymore what the thing called the interior is supposed to be. In Sect. V.4 we sketch a proof of the Jordan curve theorem that is relatively elementary when the curve considered is differentiable, but requires in any case the introduction of the concept of *index* of a point with respect to a curve; but the result is in fact still true when the curve is only required to be continuous. A classic proof consists of identifying $\mathbb{R}^2$ with $\mathbb{C}$ (see as needed Sect. I.XYZ) and applying Cauchy's theorem of the theory of functions of a complex variable, the one that figures in many works as a geometric application of this theorem, for example in Dieudonné (1960). Jordan's proof appeared in the second edition of his famous *Cours d'Analyse* in 1893. This work had a very important global impact, the more profound reason being that the proof was done by *algebraic topology*, of which it is one of the very first applications; a very lucid exposition can be found for example in Greenberg (1967). The theorem that is proved is more precise than affirming only that there is an interior. It says this: the complement in $\mathbb{R}^2$ of a *simple* (the technical term for "not self-intersecting") closed curve is composed exactly of two connected components, one of which is bounded (the interior), the other unbounded (the exterior). This approach moreover furnishes de facto the extension to arbitrary dimension.

The proof is carried out on the sphere S^2, on which a simple closed curve C is considered. The result is then that the complement of C in S^2 has exactly two connected components. Now, taking the north pole on the interior of one of them, the associated stereographic projection furnishes two connected components in the plane, one of which is bounded and the other, originating from the north pole, unbounded.

♦

For the *turning tangent theorem* or *Umlaufsatz*, there is at first the problem of correctly defining what is meant by "how much the tangent turns". This is accomplished by putting in place what is needed in order to say what is meant by "keeping track by continuity of the angle formed by the tangent line with a fixed direction" when the curve is traversed. To do this, an orientation of the plane and a direction of reference δ are fixed. At a point of parameter s of the curve, parameterized by

arc length; see Sect. V.4. The problem is that the angle $\alpha(s)$ of this tangent with δ is
an element of the unit circle S^1 and not a real number. This mapping is continuous
if the curve is continuously differentiable (see Sect. V.XYZ). But the real line rolls
onto the circle S^1 by the mapping $t \mapsto (\cos t, \sin t)$ and no more is needed than
to **lift** the mapping $s \mapsto \alpha(s)$ to a mapping $s \mapsto \alpha^*(s)$ in $\mathbb{R}$. This lifting is done
in small pieces, subsequently joined (or "glued" by continuity. We could say that
"we have pursued one of the determinations of $\alpha(s)$ within $2k\pi$ by continuity". If
now the curve has length L, the mapping α^* takes $[0, L]$ into $\mathbb{R}$, and to obtain the
theorem we need to prove the equality $\alpha^*(L) = \alpha^*(0) + 2\pi$. All this will take form
in Sect. V.5.

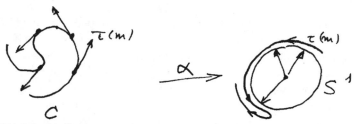

Fig. V.1.6. In this figure, we suppose that the velocity $f' = \tau$ is always equal to a unit
vector, giving a (partial) example where this mapping *reverts onto itself*

This theorem was long considered obvious, even by numerous "professional"
mathematicians, Euler first, but also Rolle, Watson (see Watson, 1916) and Riemann.
In numerous rather old works that treat curves, the *turning tangent theorem* was considered "settled", or "evident". Mathematicians often use the term "folklore" in reference to such a result. What to our knowledge are the first two complete proofs appeared in the same issue of the journal *Compositio mathematica*: Ostrowski (1935)
and Hopf (1935). Ostrowski proceeded by approximating the curve by polygons, the
resulting proof being elementary. Hopf gave a direct version of it, without approximation by polygons, but his proof is intricate; we will illuminate its essential ideas
in Sect. V.5. We know of no conceptual proof that is truly simple, even at the cost
of a rather steep ascent of Jacob's ladder.

◆

Historical remark. Question relating to Jacob's ladder: in spite of its astonishing simplicity and in contrast to other objects of this book such as e.g. conics or surfaces,
the notions of curve and plane curve in particular have waited until just recently
for mathematicians to invent the concepts needed for progressing in their study,
as is indicated by the dates given in the present chapter. Our book is organized
mainly under the aegis of differentiable curves, and we single out notions which
arise from affine geometry, let us say in $\mathbb{R}^2$ and those that require a presence in the
Euclidean plane $\mathbb{E}^2$, without completely omitting the projective spaces, treated very
briefly in Sect. V.9. In the three examples seen above, the first (Jordan) is affine, the

second (*turning tangent*) is half Euclidean, whereas the third (isoperimetric inequality) makes sense only in $\mathbb{E}^2$. There will also be an essential difference between the *simple* closed curves (without self intersection) and arbitrary closed curves. The following section will clarify what is meant by that. Our presentation will be brief; we will refer to Chaps. VIII and X of [BG] for an exposition that was rather complete at the time of its publication in 1987. There exist plenty of books on differential geometry that treat curves (and surfaces), but there are few that explain in good detail the notions of geometric curves and of kinematic curves; we will see why. We will have need of the language of differentiable mappings − of which we will sketch the definitions and properties in Sect. V.XYZ − and also notions of *genericity*, of robustness; see for example Sects. V.2 and V.10. Beyond the three results that we have used in the way of introduction, we will give other global results. But certain among them, typically the four vertex theorem and its generalizations, have been the object, since about 1990, of a revolution due to Arnold, that replaced these results in a variety of very general conceptual contexts with vast generalizations, that is the very justification for introducing a new concept. This revolution has unleashed publication of very numerous studies, for which we will be able to give useful introductory concepts, as well as some applications. The theory itself is presently in full flight.

In spite of this prolificness, in order to stay within the spirit of this book we will ultimately devote but few sections to algebraic curves. In fact we will see that early on, starting with Descartes and Newton, algebraic curves were the object of numerous and profound studies, to the extent that we can say that by the end of the nineteenth century almost everything was known about them, it being stipulated that this has to do with plane curves over the real or complex fields. We will see however an exemplary exception, the subject of a problem of vigorous current research, i.e. the topology of real algebraic curves. In contrast, for the curves that are defined over finite fields, the entire area of number theory comes into play and is always in full swing, the quite recent proof of Fermat's theorem being a foremost example.

We now risk a few explanations of this historical difference between algebraic curves and merely differentiable (smooth) curves. In the algebraic context, the curves of given degree form a space of finite dimension (5 for the conics; see Sect. IV.7) − a nice projective space, but metrizable. A whole collection of algebraic tools is needed. On the other hand, the merely smooth curves form a space of infinite dimension, essentially a space of functions, but the classic tools of analysis can't be directly adapted to their study, as we shall see; we have to be much more creative: we must use algebraic topology and recent notions of differentiable manifolds. Note also that algebraic curves appear as truly mathematical objects, whereas the smooth curves are more objects of kinematics. But − contrary to our own discipline − kinematics is not much concerned with global problems. It is typical that, until very recently, the abundant instructional material implicitly considered every curve to be algebraic; see a critical analysis for the case of duality in Berger (1972). Also, for those interested in pedagogy and history, Dombrowski (1999) is a remarkable book; we will mention it several times.

Important remark. Of course, the notion of curve, both geometric and kinematic, exists in three dimensional space as well, and just as well in arbitrary dimension. However, the literature is not so abundant and important results are rare. We have chosen not to speak about these matters, as few as they are, so as not to have to expand the present book. But, although progressing slowly, it is an area that is very much alive. We give a recent reference, that will allow readers to review previous contributions to the extent they want, i.e. Uribe-Vargas (2004b).

V.2. What is a curve? Geometric curves and kinematic curves

For the person in the street a *curve* is something that is drawn with a pencil on a sheet of paper, or else the trajectory of a car, etc. For secondary school students, it may also be a graphic representation of a function, and how they were made to suffer, until very recently calculators assumed much of the task. What connection is there between these two very different definitions?

(V.2.1) *A geometric curve is a subset* C *of* $\mathbb{R}^2$ *which is locally — with well chosen axes — the graph of a numerical function of class* C^1.

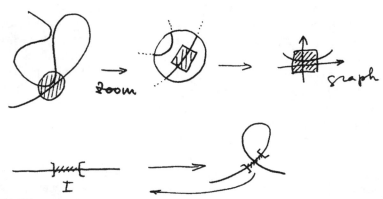

Fig. V.2.1. In these figures, we have zoomed in twice on a point, so as to be able to realize that, with suitable axes, what we see of the curve becomes a graph. See also Figs. V.2.2 and V.2.3

This may necessitate us occasionally having to zoom in considerably on particular points of the curve. Note also that by zooming in more and more, what we see of the curve resembles more and more a straight line segment along the tangent at this point (this because the function that comes into play is differentiable). Moreover, an equivalent definition consists precisely in saying that, about each point of the curve, we can find a small piece of the plane that is diffeomorphic to a piece of plane where the corresponding piece of curve is an open segment of a line. The essential lemma is:

(V.2.2) *Let $f : I \to \mathbb{R}^2$ be a mapping of class C^1 defined on an open interval $I \subset \mathbb{R}$ and $t \in I$ such that $f'(t) \neq 0$. Then there exists $\varepsilon > 0$ such that the image of $]t - \varepsilon, t + \varepsilon[$ is a geometric curve in $\mathbb{R}^2$.*

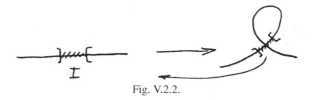

Fig. V.2.2.

We thus see appear here the notion of **tracing** a curve. Which brings us then to the definition:

(V.2.3) *A kinematic curve (or parameterized curve) is that given by a pair (I, f) formed by an open interval I of $\mathbb{R}$ and a mapping $f : I \to \mathbb{R}^2$ of class C^1 such that $f'(t) \neq 0$ for each $t \in I$.*

(If $f'(t) \neq 0$, we say that the point is not *stationary* for the parameter value t.) In analytic language, we write $x = f_1(t)$, $y = f_2(t)$, with the condition that $f_1'(t)$ and $f_2'(t)$ are nowhere simultaneously zero. In formal language, we call such an object an *immersion* in $\mathbb{R}^2$ of an open interval of $\mathbb{R}$.

Important note on methodology. The definition (V.2.3) may be considered too restrictive because there is in fact no reason not to study the curves where the vector $f'(t)$ is allowed to be zero at certain points. Even though kinematic motions — trajectories — may actually have this more general nature, it seems that there do not exist any profound or geometrically spectacular results relating to it that pertain to Jacob's ladder.

Numerous questions now arise, turning mainly about the question: Proposition (V.2.2) furnishes a **local** equivalence between two categories of objects: geometric curves and kinematic curves. What is this relationship globally? An essential difference is that a kinematic curve can have multiple points (i.e. *self-intersections*), but not so a geometric curve. But this remark is quite far from answering the question completely from a formal point of view. That is done in [BG], but for us what is said in the next section will suffice. The essential caveats are perhaps more important than the presentation itself. They concern the natural connection between geometric and kinematic curves, specifically the entirely natural consideration of the *image* $f(I)$ of a kinematic curve.

There are three caveats. The first is the classic pitfall: it is not because the immersion $f : I \to \mathbb{R}^2$ is injective that the image is actually a geometric curve: in the figure below the point of "virtual crossing" is the necessary counter example: in any neighborhood of this point there will be three parts, the normal part plus one coming from $+\infty$ and another from $-\infty$. In formal language, this phenomenon is due to the fact that the mapping considered is not *proper*: a mapping is proper when the inverse image of every compact set is again a compact set in the source set for the mapping.

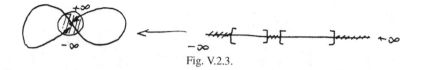

Fig. V.2.3.

The second caveat is that the image, even if it is a compact set, does not determine the immersion *even within reparameterization*. The two pairs of figures below show this well. We note for Sect. V.10 that this is due to the existence of points where the image is tangent to itself, i.e. the same point with the same tangent line is obtained for different values of the parameter t, but the curve does not actually cross itself.

We now explain *reparameterization*. It is clear that if $g : I \to I$ is a bijection of class C^1, then the images of f and of $f \circ g$ will be identical. This leads to an inevitable definition if we want to study kinematic curves and images, namely to consider curves obtained by reparameterization as above as equivalent, and "pass to the quotient"; see [BG] for the presentation. But it is important to remark that there are two classes of diffeomorphisms of an interval, those with positive derivative and those with negative derivative: we are dealing with *oriented* curves.

The third caveat is that the image of a general kinematic curve can be pathological. Some examples appear in Fig. V.6.7; but readers will be able to imagine worse. Conclusions: first, it is useless to seek a *classification* of these curves by their images. Next, for getting results, we need to limit ourselves to categories of (kinematic) curves having additional properties. These will be the *generic curves* of Sect. V.10, specifically those which have at worst double points and which are never self-tangent. In particular they don't have any triple points. These two exclusions are natural when thinking of *robust properties*, i.e. those that persist under small defor-

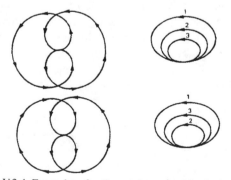

Fig. V.2.4. Examples of paths not determined by the image

mations. Triple points and self-tangencies are eliminated right away; see the figures of Sect. V.10. In Fig. V.2.3 it is the points of self-tangency that allow a change in direction for traversing the curve. Moreover, here the image of curve, as a plane set, determines the curve within orientation.

V.3. The classification of geometric curves and the degree of mappings of the circle onto itself

The classification of geometric curves is really quite simple, especially as regards the statement of the result, but the complete proof requires some care. We always assume that they are *connected*; then there are but **two** possible topological types — the *line* and the *circle*. In the first case, the geometric curve C may be obtained as the image of a kinematic curve, necessarily injective. In the second case, C can be obtained as the image of a *periodic kinematic curve*, that is an $f : \mathbb{R} \to \mathbb{R}^2$ for which there exists T > 0 such that $f(t+T) = f(t)$ for each real t. It is necessary that we have injectivity on the interval $[0, T]$. We always stipulate that T is in fact the smallest period (to preclude the curve being described redundantly); and we also pay attention that the images of periodic kinematic curves are not geometric curves in general. In fact, the preceding permits us now to use alternative — and somewhat simpler — language for the two types of curves that we are going to consider.

(V.3.1) *By a simple closed curve is meant a geometric curve (or one of its periodic parameterizations) that is of circle type (thus compact). By a closed curve is meant a periodic kinematic curve.*

In practice, (kinematic) closed curves are considered within a change of parameter, i.e. the change has to be global and always with nonzero derivative. In more formal language, this means that we form the quotient with the group of *diffeomorphisms* of the circle, in the same spirit as reparameterization above. The curve is called *oriented* when we only form the quotient with diffeomorphisms with positive derivative. In Sect. V.6 an essential point will be that if the curve is traced in the Euclidean plane, it can be reparameterized by arc length.

The above classification of geometric curves into only two topological categories, circles and lines, isn't difficult in any astonishing way, as seems quite evident; there is, however, something to prove. This classification results from the more general one of *differential manifolds* of dimension 1. Briefly these are objects which are unions of intervals of $\mathbb{R}$, pieces to be fastened together using local diffeomorphisms, which for curves means specifically a change in parameterization with derivative always nonzero. These parameterizations are those furnished by definition (V.2.2). We briefly follow 3.4 of [BG]: the basic idea is that if I and J are two such intervals and if $I \cap J$ is connected, they can be joined in a new interval $I \cup J$. This is continued as far as possible until a complete interval is obtained, i.e. $\mathbb{R}$ topologically, this since intersections of intervals are connected. There remains the case where two intervals I, J are such that $I \cap J$ has two connected components; but then the union is topologically a circle. To carry out the preceding, we reparameterize the intervals so that the joins are all translations (changes in orientation are easily excluded), which is possible since a local diffeomorphism of R with positive gradient is always conjugate to a translation precisely because a real function with strictly positive derivative on an open interval always has an inverse function. These considerations by themselves show that we construct, in the case of a circle, a periodic parameterization.

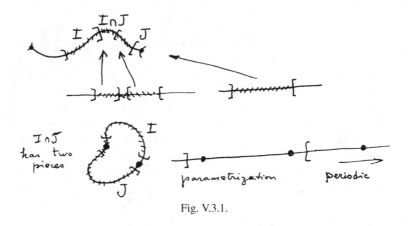

Fig. V.3.1.

♦

Here now is the first concept (rung of Jacob's ladder) necessary for advancing in our understanding of plane curves. We consider an arbitrary continuous mapping $f : S^1 \to S^1$ of the circle S^1 into itself. We orient the circle and consider the mapping (universal covering of the circle) $\exp: \mathbb{R} \to S^1$ given by $t \mapsto e^{it} = (\cos t, \sin t)$. The basic fact is that f is lifted to a mapping $f^* : \mathbb{R} \to \mathbb{R}$ that

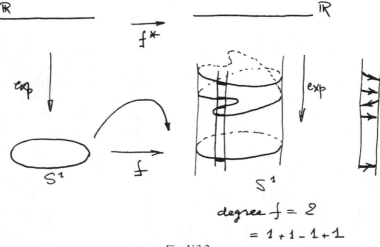

Fig. V.3.2.

commutes with the two exponentials: $\exp \circ f^* = f \circ \exp$. This comes from f being continuous and exp being locally a diffeomorphism. Now the quantity $\frac{f^*(2\pi)-f^*(0)}{2\pi}$ is an integer if $\exp(f^*(2\pi)) = \exp(f^*(0))$ and is called the *degree* of $f : S^1 \to S^1$. Heuristically it represents "the number of times that the circle is covered (globally) by the mapping f". We can calculate it at a single point if f is well-behaved about this point; we count the inverse images of this point under f with coefficient equal to $+1$ or to -1 according as to whether the orientation is preserved or changed. In particular the degree is always equal to zero if the circle isn't covered completely by the image $f(S^1)$. An essential property of degree is that it is invariant under homotopy of mappings, i.e. under continuous deformations; see Sect. V.XYZ. Since the degree is an integer, this isn't very surprising. See 7.3 of [BG] for the details.

V.4. The Jordan theorem

Here is the proof of Jordan's theorem for simple closed curves of class C^2, with the aid of the notion of index of a point with respect to a closed plane curve C (oriented and just continuous for the moment). For each point m of the complement $\mathbb{R}^2 \setminus C$ we define the *index* of m with respect to C as being equal to the degree of the mapping $f : S^1 \to S^1$ defined thus: we choose a periodic parameterization $s(t)$ on $[0, 2\pi]$ of C and define f by $f(t) = \frac{s(t)-m}{\|s(t)-m\|} \in S^1$. The invariance of the degree under homotopy shows that this index is constant on each connected component of the complement $\mathbb{R}^2 \setminus C$. Moreover, if m is sufficiently far from C then the degree is equal to zero, since the image $f(S^1)$ doesn't cover all S^1 (i.e. not all the directions).

We next need to find how the index behaves as the curve is traversed. A figure allows us to guess that it changes from $+1$ (or -1) because very near the curve we have nearly π on the one side and $-\pi$ on the other; see [BG] for the full argument, which uses the differentiability of C.

Now $\mathbb{R}^2 \setminus C$ has but two connected components; it is here that we need to use of the properties that C is simple and closed. We see this by taking a tubular neighborhood T of C; then $T \setminus C$ clearly has two connected components, and by continuity

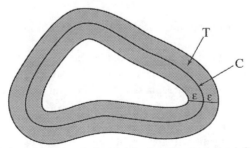

Fig. V.4.1. A tubular neighborhood of a (simple closed) curve is, by definition, obtained by moving at each point of the curve along the normal to the curve (i.e. along the line perpendicular to the tangent at that point) a given distance ε in both directions. It is easily seen that the simplicity of the curve indicates that the neighborhood is just like that of the figure if ε is chosen sufficiently small

each connected component of $\mathbb{R}^2 \setminus C$ intersects just one of these components. Thus we have finally shown:

> *If C is a simple closed curve, then $\mathbb{R}^2 \setminus C$ has exactly two connected components, the one (the exterior of C) made up of points whose index with respect to C is equal to zero, the other (the interior of C) made up of points whose index with respect to C is equal to +1. We could also characterize them by the fact that one is bounded and the other isn't.*

What was said in Sect. V.3 for calculating the degree of a mapping $S^1 \to S^1$ provides the following criterion given in the figure:

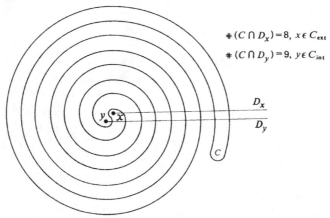

$\# (C \cap D_x) = 8, \; x \in C_{ext}$

$\# (C \cap D_y) = 9, \; y \in C_{int}$

D_x

D_y

C

Fig. V.4.2. In this figure, in order to find whether a given point is interior of exterior to the curve, a half-line is traced emanating from the point which always cuts the given curve transversally (for such a line to exist, reasonable conditions are needed for the curve, which readers can find for themselves). Then the point is on the interior if the number of points of intersection is odd, on the exterior if it is even. [BG] Berger, Gostiaux (1987) © Presses Universitaires de France

V.5. The turning tangent theorem and global convexity

In this section we consider **simple** closed curves of class C^1 and parameterize them periodically by $g : [0, 2\pi] \to \mathbb{R}$, where the derivative $g'(t)$ is continuous and nonzero. We take any Euclidean structure and thus have a natural mapping $f : S^1 \to S^1$ defined by $f(s) = \frac{g'(s)}{\|g'(s)\|}$. The *turning tangent theorem* states that the degree of f is equal to ± 1, the sign being dependent on the orientation chosen. That is, globally the tangent will turn exactly 2π when we traverse the curve a single time. This theorem, indeed visually evident when the curve is strictly convex, is a lot less so if it "turns some in each sense", and in any case the proof is tricky. Here is an outline, the details are in [BG] and also in the excellent book on curves, (Dombrowski, 1999), in portions of 1.3.

The process consists of finding a deformation f which will clearly be a mapping of degree 1. But we have seen in Sect. V.3 that the degree is invariant under (continuous) deformation. The deformation depends on a parameter k; it can be seen clearly in the figure for small k. We replace the tangents by the segments associated with the arcs corresponding to parameters distant from $k = t - s$, i.e. precisely $\frac{g(t)-g(s)}{\|g(t)-g(s)\|}$, which we define on a domain formed by the s, t with $0 \leqslant s \leqslant t \leqslant 2\pi$. At the end of the deformation we find the $\frac{g(2\pi)-g(s)}{\|g(2\pi)-g(s)\|}$, for which the set constitutes all the unit vectors with origin the point $g(0) = g(2\pi)$, which turns thus through 2π as we traverse the curve. But we need to pay attention to some details in order to have a good deformation that is continuous over the entire domain of definition of s and t right up to the boundary.

After having chosen the lowest point of the curve as the origin, supposing that g moves from there to the right, we define the deformation f_k as the path traversed by the vector

$$\frac{g(t) - g(s)}{\|g(t) - g(s)\|}, \quad \text{ou} \quad \frac{g'(t)}{\|g'(t)\|} \quad \text{si } s = t$$

when s, t increase from 0 to 2π in the following way:

– we let t increase to $t = k$ while s remains at 0;
– then s and t increase equally, i.e. $s = t - k$ until $t = 2\pi$;
– then t remains equal to 2π as s increases to the same number.

We can describe this path by introducing two functions $s_k(u)$ and $t_k(u)$ of the same parameter u, ranging over the interval $[0, 2\pi + k]$. The fact that the interval changes with k isn't important. The essential thing is that f_k doesn't pass through the origin (because g is simple and g' is always $\neq 0$), and because the extremity of f_k coincides with its origin $g'(0)$. The index of f_k is thus conserved under the deformation.

Now f_0 is the path $t \mapsto g'(t)$ ($0 \leqslant t \leqslant 2\pi$), and $f_{2\pi}$ consists of the path $t \mapsto \frac{g(t)-g(0)}{\|g(t)-g(0)\|}$ ($0 \leqslant t \leqslant 2\pi$), which makes a *half turn* about 0, from 0 to π, and of the path $s \mapsto \frac{g(2\pi)-g(s)}{\|g(2\pi)-g(s)\|}$ ($0 \leqslant s \leqslant 2\pi$), which makes a second *half turn* about 0, from π to 2π. The index of f_0 is thus, like that of $f_{2\pi}$, equal to 1.

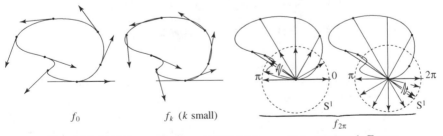

f_0 f_k $(k$ small$)$ $f_{2\pi}$

Fig. V.5.1. [BG] Berger, Gostiaux (1987) © Presses Universitaires de France

♦

Here is a first application of the *turning tangent theorem*, for which it is indispensable: "local convexity versus global convexity". We consider a closed curve C of
class C^1 and one of its points m where it does not intersect itself. We say that C is
locally convex at m if there exists a neighborhood U of m such that in U the image
of C is entirely in one of the half-planes determined by the tangent to C at m. A
closed curve can be everywhere locally convex without having to be the boundary
of a plane convex set; see Chap. VII.

In contrast it seems evident that the curve must be globally convex if, in addition,
it is simple. Such is actually the case and is an immediate consequence of the *turning
tangent theorem*:

> *Every locally convex simple closed curve is globally convex, i.e. it is the
> boundary of a convex and compact subset of the plane.*

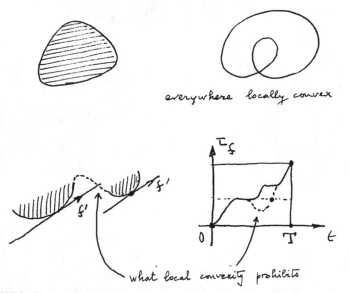

Fig. V.5.2. The (non simple) closed curve of the figure, although locally convex,
is not convex globally; moreover its index equals 2. In the graph of τ_f (argument
of f'), the *dotted* portion is impossible because of the local convexity. [BG]
Berger, Gostiaux (1987) © Presses Universitaires de France

It suffices to show that the curve is entirely on one side of each tangent, for it is
classical in the theory of convexity (see Chap. VII) that the intersection of all these
half-planes will yield a convex set. Let m thus be a point where there are points
on both sides of the tangent T at m; by compactness there are points of tangency
parallel to T on both sides. Of the three parallel tangents thus obtained, two have
definitely the same sense. We then look at the graph of the lifting into $\mathbb{R}$ of $\frac{g'}{\|g'\|}$,
which is denoted by τ_f in the caption for Fig. V.5.2. The local convexity implies that
the graph is that of a function that is nondecreasing on the whole interval $[0, 2\pi]$.

The turning tangent theorem states that the final value in 2π of this graph equals exactly 2π. Thus our tangents that are parallel with the same direction correspond to a single value in $[0, 2\pi]$ and the graph will be horizontal (constant) between the two tangents; thus the curve must be linear over the interval and the tangents must be identical, which is the desired contradiction.

V.6. Euclidean invariants: length (theorem of the peripheral boulevard) and curvature (scalar and algebraic): Winding number

We are now essentially in the Euclidean plane $\mathbb{E}^2$ and we consider a kinematic curve $f : I \to \mathbb{E}^2$, of class at least C^1. We can thus speak of velocity for f, precisely the *velocity vector* at t of f is $f'(t)$ (recall that we are considering only kinematic curves for which $f'(t) \neq 0$). The *(scalar) velocity* of f at t is by definition $\|f'(t)\|$. Also of importance is the *distance traveled*, or the arc length, of the curve f from t to t', defined as the integral $\int_t^{t'} \|f'(\tau)\| \, d\tau$. We justify this abstract definition by showing, without difficulty, that this length is equal to the limit superior of the length (perimeter) of all polygons with vertices $\{f(t), f(t_1), \ldots, f(t_n), f(t')\}$ $(t < t_1 < \cdots < t_n < t')$ inscribed in the curve between the values t and t'.

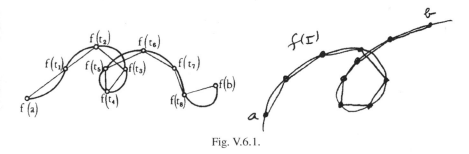

Fig. V.6.1.

The essential thing — realizable since the velocity is never zero — is that we can reparameterize the curve by the *arc length*, starting from a fixed origin $m_0 = f(t_0)$ on the curve. In fact we define the function s by $s(t) = \int_{t_0}^t \|f'(\tau)\| \, d\tau$, and then the new parameterization of f by g is such that $g(s(t)) = f(t)$. This time, by construction, the scalar velocity is constantly equal to 1, since $\|g'(s)\| = 1$ for all s. We say that the curve is *parameterized by arc length*, and we may always suppose that this is the case for every kinematic curve, by an appropriate reparameterization.

♦

Here is a first result, perhaps rather evident, although perhaps not all drivers will be persuaded — but which also requires the turning tangent theorem of Sect. V.5 — i.e. that the *peripheral boulevard theorem*. The problem is to know whether there is a significant difference in distance traveled in making a circuit of the town in question by the interior lane of the periphery rather than by the exterior lane. It seems clear that the interior trajectory is always shorter for convex peripheries, but is it really shorter than the exterior trajectory if there are lots of contortions? To make the calculation quickly and correctly, we identify $\mathbb{E}^2$ with $\mathbb{C}$ as explained at the end of Sect. II.XYZ; in particular, multiplication by i is the rotation by $\frac{\pi}{2}$. Let g be a reparameterization by arc length of the median curve (of class C^2) of the periphery. We write the unit tangent g' in complex form, i.e. in the form $g' = e^{i\tau}$, and let ε be the lane width, so that the parameterization of the interior will be $g + i\varepsilon e^{i\tau}$. The velocity is $(g + i\varepsilon e^{i\tau})' = g'(1 - \varepsilon\tau')$ and the scalar velocity is $1 - \varepsilon\tau'$. The total length of the interior is thus $\int_0^L (1 - \varepsilon\tau') \, ds = L - \varepsilon(\tau(L) - \tau(0))$. But the turning tangent theorem assures us that, precisely: $\tau(L) - \tau(0) = 2\pi$, whence the interior distance sought is $L - 2\pi\varepsilon$. For the exterior it will thus be $L + 2\pi\varepsilon$, and the total increase is always $4\pi\varepsilon$. It is thus constant and ultimately quite minor, regardless of the various twists and turns of the peripheral.

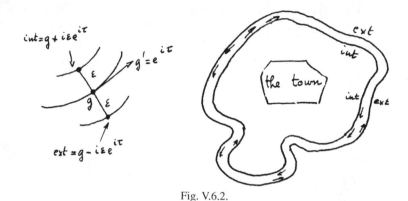

Fig. V.6.2.

♦

We are now necessarily in class C^2. The expression $g' = e^{i\tau}$ signifies in fact that we have chosen an axis so as to be able to speak of the *angle* of the tangent to the curve with this axis. And the quantity τ' encountered above is thus the *velocity at which the tangent turns* for the kinematic curve considered. A sketch shows that, the larger this velocity is in absolute value, the more the curve is **curved**. For example for a circle of radius R this velocity will be $\frac{1}{R}$. We thus define the *scalar curvature* of the curve under consideration to be equal to $|\tau'|$, and *denote* it by K. If we have oriented the plane, we give τ' a sign; this will be the *algebraic curvature*, denoted by k; and for the scalar curvature: $K = \|g''\|$ for the second derivative of g.

In the science of kinematics it is useful to know how to calculate K for an arbitrary parameterization f. We find

$$K(t) = \frac{f'(t) \wedge f''(t)}{\|f'(t)\|^{3/2}},$$

where $\wedge$ denotes the exterior product, which may also be calculated as a determinant in orthonormal coordinates (this determinant depends on the orientation of the plane, but here we can take the absolute value). In thinking of the case of the circle, where K is not zero, we say that $R = K^{-1}$ is the *radius of curvature* of the curve considered

Fig. V.6.3. [BG] Berger, Gostiaux (1987) © Presses Universitaires de France

at $f(t)$. In kinematics there is also a need to decompose the acceleration vector of the curve f into two components, one along the tangent, the other along the *normal* to the curve (the normal is by definition the perpendicular to the tangent). But if we choose a unit vector there in a way that is connected intrinsically to the curve, this will only be possible when the acceleration vector is not parallel to the tangent. We thus define a standard normal $\{\tau, \nu\}$ at $f(t)$ by taking $\tau = \frac{f'(t)}{\|f'(t)\|}$ and the unit vector ν located on the normal on the same side as $f''(t)$. Thus the said formula of *intrinsic components of the acceleration* is written: $f''(t) = v' \cdot \tau + \frac{v^2}{R} \cdot \nu$ (here v' is the derivative with respect to the time t, of the scalar velocity). The *centrifugal force* is $\frac{v^2}{R}$; and we see that it is indeed large if the velocity is large, but also if curvature is large (i.e. if the radius of curvature is small).

We see that the curve *turns* toward the side of the acceleration in the following sense: at a point m where the curvature isn't zero, we can define a half-plane attached to the curve at that point, denoted **convex**$_m$C in the figure above. This half-plane is bounded by the tangent to the curve and contains the acceleration vector. If we change the parameterization the new acceleration vector will remain in this

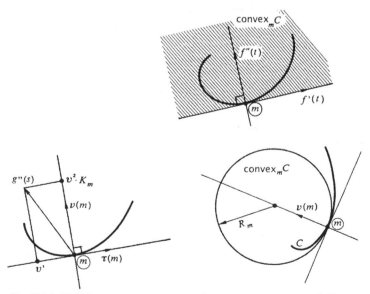

Fig. V.6.4. [BG] Berger, Gostiaux (1987) © Presses Universitaires de France

"convexity half-plane" for our curve. The term *convexity* is justified because, locally, the curve is situated in the half plane and moreover is the graph of a convex function.

The consequences of this formula in practical life are legion, whether we suffer from travel sickness, are marking out roads or are laying train tracks. The problem is to find the markings that connect the various "turns"; it essential that the curvature change gently when the velocity is large. As a straight line has zero curvature and a circle a constant nonzero curvature, if is thus necessary to **connect** the circles to lines by means of appropriate curves. Our cherished conics are not at all good in this regard, for their curvature is never zero. Numerous curves have been proposed, and the matter is still not settled, all the more so in that the markings have to conform to imperatives of economy, of visibility, etc. For railroads (Alias, 1984), can be consulted; we mention, moreover, for those interested in technical matters, that two successive opposing curves need to be connected by a piece of straight line long enough to permit the train to regain its equilibrium.

♦

The tangent gives a first approximation to the picture — to the image — of the curve; the tangent has contact of order 2 with the curve. The curvature, when it is nonzero, furnishes an approximation for the next order. In Euclidean geometry, we think of "circles"; the circle that has contact of order 3 with the curve is called the *osculating circle* (from the Latin *osculari*, to kiss, to embrace, terminology due to Lagrange). It is defined when the curvature is nonzero, has radius $R = K^{-1}$ and is thus the circle centered on the normal at a point situated a distance R from the point considered. Its center is called the *center of curvature* of the curve. It can be found by considering the envelope curve for the family of normals to the curve; the family that envelops a curve is called the *development of the curve*. We will scarcely speak of developments — there is not the space — but say here only that the point of contact of the normal with the development is in fact the center of curvature.

In particular, at a point of nonzero curvature, the curve is locally *strictly convex*: it is entirely on one side of its tangent — the side of the acceleration — with a single point of contact.

How is the curve situated locally with respect to its osculating circle? Things go well if, in parameterizing by arc length, the derivative K' of the curvature is nonzero (a condition on the third derivative), in which case the curve traverses its osculating circle in the neighborhood of the point considered. We recommend trying to draw a curve and one of its osculating circles, which is very difficult to do in such a way

Fig. V.6.5.

that the circle is really osculating — to draw it so that it isn't, at least locally, either on the interior or the exterior. It is more difficult still to draw the family of all the osculating circles, because this shows (it's a good exercise) that, in the neighborhood of a point where $K' \neq 0$, the osculating circles are mutually nonintersecting. In the figure below, which was provided us by Étienne Ghys, it will be observed that the curve itself seems to have been drawn, but in fact it isn't; it's an optical illusion. For more on this topic, see the reference given in Sect. V.10 regarding human vision. This figure is also a very good — entirely natural — example of a continuous vector field in the plane for which the integral curves aren't unique: we consider the vector field formed by all unit vectors tangent to osculating circles to a given oriented curve; then the trajectories passing through an arbitrary point outside the curve are unique, but at a point of the curve we can either follow the circle or else the curve while remaining tangent to the vectors of the field the whole time. This figure is also a good example for the theory of envelopes: it's a classical "result" that the point of contact of a curve of a family with its envelope is the limit of the point of intersection of two neighboring curves of the family. This monstrosity is spelled out in many books, along with others concerning duality. Such a result is true in certain special circumstances, but certainly not in general since the osculating circles don't intersect at all! It's a result that is true for envelopes of lines. For more on this topic, see Berger (1972).

It is important to point out that there does not exist any classification of the *local form* for curves of class C^∞, even if the velocity remains nonzero. The tangent can be traversed an infinite number of times in the neighborhood of such a point. And if

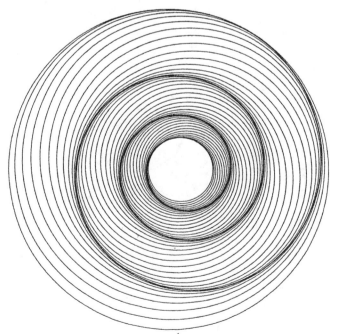

Fig. V.6.6. © Étienne Ghys

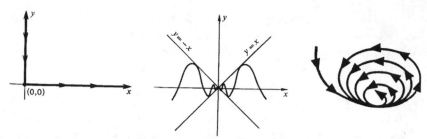

Fig. V.6.7. Possible shapes of a curve of C^∞ at a point where the velocity vanishes, and also where it doesn't vanish. [BG] Berger, Gostiaux (1987) © Presses Universitaires de France

the velocity becomes zero, there may very well not be a tangent. It's left to readers to imagine really appalling situations.

♦

We now consider closed curves C of class C^2 in the affine plane $\mathbb{R}^2$, so that readers may want to review the degree theory of Sect. V.3 and its use in Sect. V.6. We choose a Euclidean structure, an orientation and any parameterization $f : [0, 2\pi] \to \mathbb{E}^2$, so that the mapping $\frac{f'}{\|f'\|} : S^1 \to S^1$ has a degree, called the *winding number* of f. It's an integer denoted $wind(f)$, and because of its invariance under deformation (Sect. V.3), depends only on C and orientation but not on specific parameterizations, whence it is an invariant $wind(C)$ attached to each closed plane curve.

Now the calculation of Sect. V.6 remains unchanged and thus, with some Euclidean structure, we can calculate the winding of a closed curve of length L by means of the algebraic curvature k, using the formula: $wind(C) = \int_0^L k(s)\, ds$. For the scalar curvature $K = |k|$ we will thus always have $\int_0^L K(s)\, ds \geq |wind(C)|$. Readers can study the case of equality. We will return to this winding number very

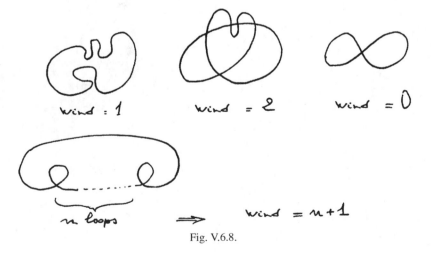

Fig. V.6.8.

extensively in Sect. V.10, in particular to its connection with the number of points, called (*double points*), where the curve intersects itself.

For a very recent "DNA" inequality, bearing on the integral of the curvature, see Tabachnikov (2001).

V.7. The algebraic curvature is a characteristic invariant: manufacture of rulers, control by the curvature

Scalar and algebraic curvature are Euclidean invariants of kinematic curves: they are preserved under isometry (but pay attention to orientation in the case of algebraic curvature). We are interested in the converse: is a curve characterized, within isometry, by its curvature? For example, are circles (resp. lines) the only curves of nonzero (resp. zero) constant curvature? We suppose throughout that the curve is parameterized by arc length. Then for the curvature zero case we have $g'' = 0$, so that g is a vector function of the form $g = t \cdot x + y$: and the image is thus linear (in fact, a piece — dependent on the interval of definition — of a line).

Instead of treating the case of circles by hand, we proceed in an oriented plane as follows:

> *If two kinematic curves g and h, defined on the same interval, are such that their algebraic curvatures are equal for each s: $k(g(s)) = k(h(s))$, then there exists a an orientation-preserving isometry $\mathcal{I}$ such that $\mathcal{I}(g(s)) = h(s)$ for each s.*

We can thus say that the function $k(s)$ is the *intrinsic equation* of the curve. We place ourselves in $\mathbb{C}$ with the language of Sect. V.6: in coordinates with $g = (x, y)$, we have (since the velocity a unit vector) $x'(s) = \cos(\tau(s))$ and $y'(s) = \sin(\tau(s))$. But $k(s) = \tau'(s)$, so that $\tau(s) = \int_0^s k(t)\,dt + C$ and we have $x(s) = \int_0^s \cos(\tau(t))\,dt + a$, $y(s) = \int_0^s \sin(\tau(t))\,dt + b$. The curve g is thus determined within a translation and a rotation. In particular the circles are clearly the only curves with nonzero constant curvature. Observe also that the scalar curvature K, which only yields $|\tau'|$, doesn't suffice: if K becomes zero in places we can't draw the conclusion. Take for example the graph of $x \mapsto x^3$ and reverse one of the two pieces about 0.

Fig. V.7.1. [BG] Berger, Gostiaux (1987) © Presses Universitaires de France

♦

We mention an application (little known to students in schools today, so it seems, but which the author learned from his teacher during his final year) of the fact that the only curves of constant curvature are lines and circles. How are really straight rulers − not graduated rulers, but straightedges, called in some instances "runners" by technicians − manufactured? One solution would be to use one of the two of the inverters mentioned in Sect. II.3, that of Peaucellier or that of Hart. It may seem astonishing that still today the practical procedure is as follows: three approximate straightedges are taken and rubbed against each other, two at a time. When a rubbing between two curves is perfect, i.e. if there is sliding of one curve on the other, it's necessary that the curvature of each be constant. Thus three circles (or lines) are obtained. But circles are impossible two at a time: the convexities are opposed for each pair. The only possibility is that they all three be lines. Those interested in groups and more abstract language will have noticed that it amounts to the same thing to say that the only one-parameter subgroups of the group of planar Euclidean displacements are those of rotations about a fixed point and those of parallel translations. In practice a product is obtained that can be checked for its adequacy. This is done by interferometry, where recently lasers have been used. More precisely, defects are measured and corrected by polishing (possibly with the aid of CAD); the result is checked again, etc., until sufficient precision is obtained. We are repeating here in part what was said in Sect. II.XYZ.

Even though the laser provides perfect lines of light, it is only used at present for estimating − very effectively, to be sure − **deviations** (of a lot of things, for example in installing railroad tracks). We should know that if Michelson was the first to be able to estimate the velocity of light with a precision sufficient for distinguishing c from $c \pm v$, where v is the velocity of the earth in its orbit, it was because he was an excellent experimenter, constructor of slides, of rulers that were finely graduated; see Sect. II.XYZ.

For the fabrication of surfaces, plane or spherical, see Sect. VI.6. We have already treated, in Sect. II.XYZ, the manufacture of graduated rulers and length measurement, and likewise that of goniometers for the measure of angles. For surfaces, fabrication of spheres, of planes, see Sect. VI.6; for the fabrication of balls, see Sect. VII.13.C.

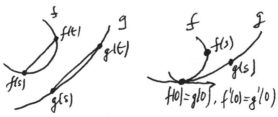

Fig. V.7.2.

♦

The fact that knowledge of the curvature of a plane curve as a function of arc length determines the curve leads us to anticipate that more can be done. Typically: if we have control over the curvature, can we hope to control the curve? Here are two results, both completely intuitive, concerning two curves where the one is always more curved than the other:

(1) (Schur, 1920). *If for each s we have $k(f(s)) \geqslant k(g(s))$, then for sufficiently small s and t we have, for the length of the chords: $d(f(s), f(t)) \leqslant d(g(s), g(t))$.*

(2) *Under the same conditions, if f and g have a common origin at $s = 0$ and are tangent at this point, and if moreover the algebraic curvature of g is positive, then for s sufficiently small the image of f is contained in the interior of g.*

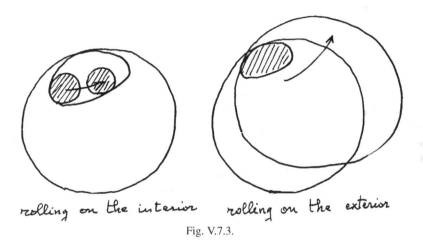

rolling on the interior   rolling on the exterior

Fig. V.7.3.

Once the curves are no longer simple, the restriction to small parameters is indispensable. Readers should be wary of these two results which, although not difficult, aren't entirely immediate either. For the second, the simplest thing is to use Euler's equation for curves, which will be given in Sect. V.8. We note a corollary for mechanics: let C be a simple closed curve, and m (resp. n) a point where the curvature is maximum (resp. minimum). Then we can roll the osculating circle at m around the interior of C, and we can roll the image of the curve around the interior of the osculating circle at n.

V.8. The four vertex theorem and its converse; an application to physics

The curvature function of a planar simple closed curve (of class C^2), as a function of arc length, gives rise to a continuous function on the circle S^1. Is it an arbitrary function? Otherwise expressed, can we always find a simple closed curve for which the curvature as a function of arc length is the given arbitrary continuous function k on S^1? Certainly not, for if we look at the formulas given in the preceding section and if L is the length of the curve, it is necessary that the function τ defined by $\tau(s) = \int_0^s k(t)\,dt$ satisfy the two conditions $x(L) - x(0) = \int_0^L \cos(\tau(s))\,ds = 0$ and $y(L) - y(0) = \int_0^L \sin(\tau(s))\,ds = 0$, equalities that have no reason for existing in general. But we might gain hope in being less demanding, i.e. in no longer requiring that curvature be given as a function of are length, but only as an arbitrary parameterization of the circle. Formally we seek an embedding (not necessarily parameterized by arc length) $f : S^1 \to \mathbb{E}^2$ such that the curvature of the image $f(S^1)$ at the point $f(s)$ is equal to $k(s)$. The mathematical problem consists thus of reparameterizing the circle, i.e, finding a diffeomorphism $g : S^1 \to S^1$ such that the function $f = k \circ g$ satisfies the two zero integral conditions above. Now this is not always possible.

We will call a point of a curve where the curvature is zero a *vertex*. The nomenclature comes from the fact that this is what happens at the four vertices of an ellipse, but in general a vertex may be a local maximum, a local minimum or more generally a more complicated point. Now in Mukhopadhyaya (1909) it was shown that: *a convex simple closed curve must have at least four vertices*, whereas a continuous function on the circle can very well have only a single maximum and a single minimum, the derivative at all other points being nonzero. We are going first give the ideas of two proofs of this result, called the *four vertex theorem*, the one because it's very intuitive (while most other proofs are completely obscure), the other because it's a really good conceptual proof and because we will see lots of generalizations of it later. The result is true whether the curve is convex or not, but it was not until (Kneser, 1912) that the general case was obtained, although the proof remained obscure. The proof preferred by the pure geometer may well be that of Osserman (1985). A detailed history of the four vertex theorem can be found in Dombrowski (1999).

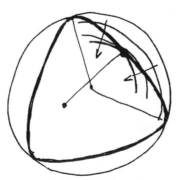

Fig. V.8.1.

Osserman begins by enclosing the curve in a circle of smallest radius R that contains it; this circle will thus be tangent to the curve at no fewer than two points. Moreover Osserman shows that if the circle is tangent at N points, then the curve has at least 2N vertices; the visually appealing idea that, between two points of contact, the curvature must have a minimum which is smaller than R^{-1} is easily realized by diminishing the radius of the circle until it is tangent; and since at points of contact the curvature is $\geq R^{-1}$, we are done. The proof can be adapted without difficulty to the nonconvex case.

♦

The second proof, more conceptual, is based on a remark about periodic functions on $\mathbb{R}$ – alias functions on the circle S^1 – say of period 2π. Moreover, the same idea will turn out to be basic in Sect. V.10 for a whole series of recent results (and surely for more yet to come). It seems to be in Blaschke (1916) that this conceptual proof appeared for the first time. However, now it will be necessary to suppose that the curve has nonzero curvature everywhere and thus is strictly convex, as we have seen above. With the notations of Sect. V.7, if $g' = e^{i\tau}$, we thus have $K = \tau' = \frac{d\tau}{ds} > 0$, which allows us to take τ as parameter; it varies from 0 to 2π. The formulas of Sect. V.7 thus become $\int_0^{2\pi} K^{-1}(\tau) \cos \tau \, d\tau = 0$ and $\int_0^{2\pi} K^{-1}(\tau) \sin \tau \, d\tau = 0$. In the language of Fourier series, this means that the function K^{-1} is orthogonal to the first two harmonics, i.e. to $\sin \tau$ and $\cos \tau$, which implies that K^{-1} must have at least two minima and two maxima. The general idea is that $f(\tau) = \sum_{k \geq 2}(a_k \cos k\tau + b_k \sin k\tau)$ (the derivative of K^{-1} is of this form) has at least as many zeros with change of sign as its first term, i.e. 4. The proof is as follows: the function $f(\tau)$ necessarily changes sign, since it has mean zero (except for the case $K^{-1} = $ const.). If $f(\tau)$ becomes zero and changes sign at a finite number of points $x_1, \ldots, x_N$, then N is even because of the periodicity. If $N = 2$, we can find a, b, c such that $a \cos \tau + b \sin \tau + c$ becomes zero at x_1 and x_2 and has the same sign as f at all

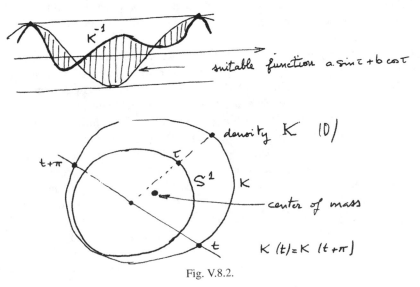

Fig. V.8.2.

other points, whence $\int_0^{2\pi} f(\tau)(a\cos\tau + b\sin\tau + c)d\tau > 0$, a contradiction. Thus $N \geqslant 4$.

Here are two small variations in the presentation: *Euler's equation* for convex curves consists of writing the curve as the envelope of its tangent, the latter being the line that makes the angle θ with a fixed axis and u denoting its distance to an origin situated on the interior of the curve; see Fig. V.8.3. It is then shown that the radius of curvature equals $u + u''$, where u'' denotes the second derivative of u with respect to the variable θ. We can then reread the preceding in two ways: by periodicity the integrals $\int_0^{2\pi}(u + u'')\cos\theta\,d\theta$ and $\int_0^{2\pi}(u + u'')\sin\theta\,d\theta$ are zero and we have thus at least two maxima and two minima; but we can also look for points where the derivative of the radius of curvature is zero, i.e. $u' + u''' = 0$. Again, we have both $\int_0^{2\pi}(u' + u''')\cos\theta\,d\theta$ and $\int_0^{2\pi}(u' + u''')\sin\theta\,d\theta$ equal to zero. But, just as functions orthogonal to the first two harmonics have at least four extrema, we see also that they have at least four zeros. We will encounter this point of view of zeros of functions orthogonal to harmonics in Sect. V.10.

More geometrically, and this will also serve for the sequel, we can consider K^{-1} as a mass distribution on the circle S^1, so that the fact that the two integrals above are zero says that the center of gravity of the total mass is at the origin. This is clearly impossible if K has only one maximum and one minimum, so consider a line passing through the origin such that $K(t) = K(t + \pi)$, which always exists. Then all the density is larger on one half of the circle and smaller on the other, thus the center of gravity is not on this line. The simplicity condition is essential; the *limaçon of Pascal* (of the father, not of Blaise) of the figure (for the curve with polar equation $\rho = 1 + 2\cos\theta$) has a curvature strictly increasing from 5/9 to 3 (more than a half) and is surely closed and convex.

In Sect. V.8 we will encounter, in a more general context, the use of orthogonality to the first harmonics in obtaining functions having multiple minima and/or points where they become zero.

Can we float logs when there is weightlessness? The four vertex theorem will answer the question. Here are the details of this anecdote (we will encounter logs again, but then with gravity, in Sect. VII.13.C). Two physicists sought out the author in 1991 with the following elementary problem: let C be a convex compact set in the Euclidean plane and $\alpha \in [0, \pi/2]$ a given angle: to find a line D which cuts C at an angle α at the two points of intersection. The physical motivation is simple: we consider a very long prismatic object in equilibrium in an interface situation between two liquids, everything under weightlessness. (A log floating on water is a simple interpretation; in fact, the physicists were interested in long molecules in problems of condensed matter.) The equilibrium positions correspond to the lines sought above, where α is given by the formula $\cos\alpha = \frac{\gamma_{13}\gamma_{23}}{\gamma_{12}}$, the γ_{ij} denoting the interfacial tensions of the system, 1 and 2 for the two liquids and 3 for the object itself.

This is a very simple problem; typically it, like many analogous problems, must be solved by the intermediate value theorem. The idea here is to consider the area function of a triangle formed by two tangents to C for which the angle at the vertex

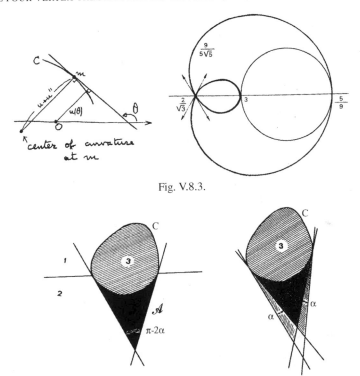

Fig. V.8.3.

Fig. V.8.4. In the figure on the *right*, the two shaded surfaces must be equal; this shows, in the limit, the desired condition on the angles. V.8.4. Raphaël, di Meglio, Berger, Calabi (1992) © EDP Science

equals $\pi - 2\alpha$ (area $\mathcal{A}$ represented in black in Fig. V.8.4). The derivative of the area will be zero if and only if the angles at the base are equal, and thus equal to α. We have thus found two positions of equilibrium, the maximum and the minimum of this function. Moreover, one at least is thus stable.

Unfortunately (but ultimately fortunately) for the author, these physicists knew the four vertex theorem, and thus in fact conjectured the existence of at least four equilibrium positions. This is true, but the proof is more difficult; we owe to Calabi at least four different solutions. Here is the one he prefers. We give it because, beyond its elegance, it uses the Euler equation introduced above for representing strictly convex plane curves. The curve is parameterized by the angle θ of the tangent with a fixed axis, but the curve is defined as the envelope of its tangent at the point of angle θ; the function sought is thus the distance $u(\theta)$.

We consider two tangents to the curve at angles $\theta + \alpha$ and $\theta - \alpha$: the triangle formed will be that sought if the two lines with angle θ passing through the points of contact with the two tangents coincide. Calabi calculates their distance $\Delta(\theta)$ and finds

$$\Delta(\theta) = -\big(\hat{u}'''(\theta) + \hat{u}'(\theta)\big) \int_{-\alpha}^{\alpha} (\cos \Phi - \cos \alpha)\, d\Phi.$$

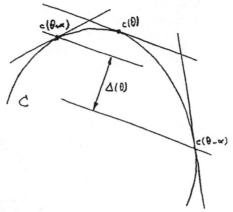

Fig. V.8.5.

Here the function $\hat{u}$ is the natural mean value

$$\hat{u}(\theta) = \frac{\int_{-\alpha}^{\alpha}(\cos\Phi - \cos\alpha)u(\theta - \Phi)\,d\Phi}{\int_{-\alpha}^{\alpha}(\cos\Phi - \cos\alpha)\,d\Phi}.$$

Let us caste a glance at the correspondence $u \mapsto \hat{u}$ between functions, which in fact yields a correspondence between curves given by their Euler equation. We

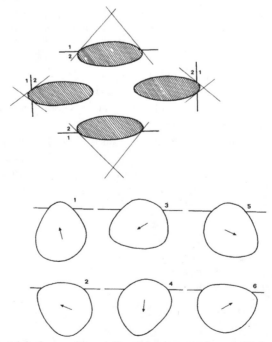

Fig. V.8.6. Certainly four positions (*ellipse*), but in general more. Raphaël, di Meglio, Berger, Calabi (1992) © EDP Science

have seen that the curvature equals $u + u''$; thus the zeros of Δ correspond exactly to the vertices of the curve of equation $\hat{u}$ and there are at least four of them, Q.E.D. More details can be found about this personal account in Berger (1993b), Raphaël, di Meglio, Berger, and Calabi (1992) and Raphaël (1992).

In dimension 3 the problem is of a totally different order of difficulty, because in general an arbitrary convex set never admits a plane section making a constant angle with the tangent plane; it is thus necessary that the separation surface is somewhat deformed about a plane (to be determined); see Raphaël and Williams (1993) for this problem in the physics of condensed matter, which remains essentially completely open.

◆

We must now return to the problem posed right at the beginning of this section: to realize an arbitrary (continuous) function on the circle as the curvature of some simple closed curve. The four vertex theorem shows that this is not possible without

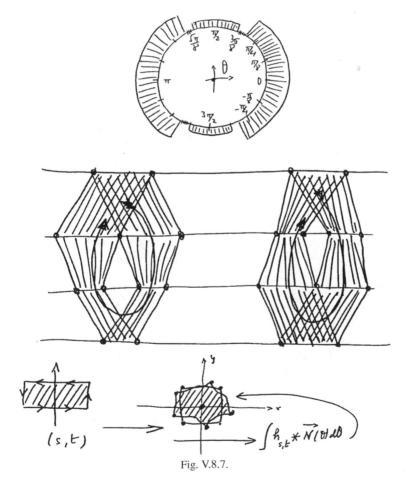

Fig. V.8.7.

some supplementary condition. But it turns out that: *each function that has at least two maxima and two minima can be effectively realized in this way.* This result awaited (Gluck, 1971), it still being required there that the function in question is never zero, i.e. that we work only in the realm of curves that are strictly convex. But we now know that the result is valid in all generality; see the very complete and historical exposition of De Turck, Gluck, Pomerleano, and Shea Vick (2007). Note that the analogous result in higher dimensions, for example for surfaces, is valid for every function; see Gluck (1972). For example there exist surfaces where the Gauss curvature (see Sect. VI.5) has only a single minimum and a single maximum. Returning to the case of plane curves, the proof of this converse theorem is difficult. The idea concerns sliding masses along the circle in such a way that the center of gravity is the origin. We begin with a distribution as in the figure; we effect two types of sliding and obtain thus four points as in the figure; they "encircle" the origin. Thus a homotopy in two parameters will give a ruled surface that will cover the origin at one point at least.

For a discrete presentation of the four vertex theorem with the aid of polygons, and more, see Ovsienko and Tabachnikov (2001).

V.9. Generalizations of the four vertex theorem: Arnold I

The four vertex theorem haunted many mathematicians seeking to know whether there are concepts hidden behind it. We will soon see such concepts appearing, but we first look at some intermediate results. The four vertex theorem is a Euclidean theorem; we can say it asserts that, among the osculating circles to the curve, at least four have contact with it of order higher than 3. But, subsequent to Chaps. I and II, we like purely **affine** geometry and also **projective** geometry, which are simultaneously somewhat richer than the Euclidean version, as we shall see. We first give the results and subsequently see how it is with the proofs.

We first see what replaces the four vertex theory in the purely affine case. In the affine plane, the conics replace circles. Just as the osculating circle at a point of a kinematic curve can be obtained as the limit of the circle passing through three points that tend toward the given point, so each (affine) kinematic curve — a conic being determined by five points — admits at each point a well-determined conic having a contact of order at least 5 with the given curve. What corresponds to the vertices in the Euclidean case are then the conics that have a contact of order 6 at least; they are called *sextactic*. The corresponding points are called *affine vertices*. We thus have Mukhopadhyaya's theorem (Mukhopadhyaya, 1909):

> *Each simple convex curve in an affine plane possesses at least six affine vertices.*

This theorem remains valid in the projective context, conics there always having the same citizenship rights as does the notion of convex curve; see Sect. IV.4 as needed. What is peculiar to the projective context is precisely the case of non convex curves, precisely the simple closed curves that aren't contractible and thus

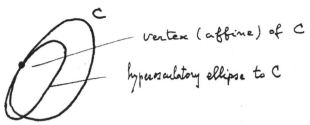

vertex (affine) of C

hyperosculatory ellipse to C

Fig. V.9.1.

are homotopic to a projective line. We could define an inflection point of a smooth curve as a point where the curve crosses the tangent, but the right definition consists of stipulating that the two first vector derivatives are collinear. The curve crosses its tangent in general, but not necessarily; see the graph of $x \mapsto x^4$. Then an old theorem, attributed to Möbius, is:

> Each simple closed curve in the projective plane that is homotopic to a projective line possesses at least three points of inflection.

For this theorem of Möbius and its generalizations, see p. 4 and Sect. 3.7 of Haupt and Künneth (1967). It should be mentioned that if a cubic in the affine plane has only two points of inflection, it is because the third is at infinity. We will encounter this result in Sect. V.14. Notice that a projective line has no inflection point, but in a more precise sense (consider the first two vector derivatives) all its points are inflection points. In contrast to the other results mentioned, Möbius's is by nature purely topological.

Mukhopadhyaya's is a generalization of what he had used in the Euclidean case of four vertices and remains ultimately rather mysterious. The idea is to see, in the Euclidean case, how some osculating circles are situated with respect to others (think of Fig. V.6.6). In the projective case, he likewise studied the various osculating conics to the curve: as in the Euclidean case, at a point of contact of order 5, but not more, the conic crosses the curve at the point considered. In Bol (1950) a more conceptual proof can be found, in part analogous to the proof of Blaschke given above. For a curve in the projective context we define a projective arc length and a projective curvature h (here these invariants depend on the third derivatives): Bol shows that for the reparameterized curve, and in arbitrary projective coordinates $\{x_i\}$ ($i = 0, 1, 2$), we have $\int h x_i . x_j = 0$. This leads to many conditions, from which we infer that h must become zero at least six times; this provides six projective (and affine) vertices.

It is here that we must mention that there exists an entire *affine* differential geometry, with notions of affine length and affine curvature. For example the sextactic points are those where the affine curvature is zero. The inflection points are, in contrast, the points where we can't really even define affine elements. For this geometry, amidst a broad literature, see Guggenheimer (1963), Buchin (1983) and Chap. 9 of the synthesis (Dillen and Verstraelen, 2000). This affine differential geometry indeed exists in all dimensions, for example for surfaces, but is then much more complicated. The essential reason, which may be inferred from what we will say just a little further on, is that the invariants associated with surfaces and arising from third derivatives are not the quadratic forms of the Euclidean theory of surfaces (see Sect. VI.6), but rather cubic forms. These are algebraic objects that are much more intricate to manipulate than quadratic forms.

We are indebted to Arnold for placing the six vertex theorem in a more general context, which permitted him to obtain a whole series of generalizations. All this has to do with the study of points called *extactic* of order $d(n) + 1 = \frac{n(n+3)}{2} + 1$. Here we consider the algebraic curves of degree n which have a contact of order $d(n) = \frac{n(n+3)}{2}$ with a projective curve (typically nonalgebraic). At each point there exists one of them, because we will see in Sect. V.13 that the space of all (plane) algebraic curves of degree n is of dimension $\frac{n(n+3)}{2}$. We then find 2 for the lines ($n = 1$), 5 for the conics and 9 for the cubics. The points where the contact is of order greater than $d(n) = \frac{n(n+3)}{2}$ are called n-*extactic* or, n being understood, simply *extactic*. Involved is an attempt to show that there are, with increasing n, more than the preceding would lead us to expect. On the other hand, we shall see that Arnold's context only works, at least presently, for curves that are sufficiently close to standard curves.

Here very briefly is the scheme. Just as the inflection points correspond to a zero second derivative trivially parallel to the first derivative, the sextactic points of zero projective curvature correspond to third derivatives, linear combinations of the first two derivatives. The n-extactic points correspond to n-th derivatives that are linear combinations of the preceding derivatives. Arnold interprets these as flattening points of the *Veronese curve with* associated with a plane curve C (we recall the Veronese surface from Sects. I.7 and II.1). We are working in the projective context; the Veronese mapping $\mathcal{V} : \mathbb{R}\mathrm{P}^2 \to \mathbb{R}\mathrm{P}^{d(n)}$ is, in homogeneous coordinates,

$$(x : y : z) \mapsto (x^n : y^n : z^n : x^{n-1}y : x^{n-1}z : \ldots)$$

and we define the n-th Veronese curve of C to be the curve $\mathcal{V}(C)$ of $\mathbb{R}\mathrm{P}^{d(n)}$. We see rather easily that the n-extactic points of C correspond to *flattening* points of $\mathcal{V}(C)$, which by definition are the points where the osculating hyperplane has contact of multiplicity at least $d + 1$ with $\mathcal{V}(C)$. A small technical annoyance: this correspondence isn't valid without a supplementary convexity condition, which helps to explain the statement of the theorem to follow.

This flattening condition is thus the vanishing of $\det(Y, Y', \ldots, Y^{(d(n)+1)})$, an (ordinary) differential equation of order $d(n) + 1$, where $Y : t \mapsto Y(t)$ is a parametric representation of $\mathcal{V}(C)$. It can be shown by long and difficult work that we are in a situation generalizing that of the second proof of the four vertex theorem, specifically that we have a periodic function for which the Fourier series is lacking a certain number of initial harmonics . We then know (from a theorem with the names Hurwitz-Kellogg-Sturm-Tabachnikov) that it has sufficiently many zeros: $\sum_{k \geq n}(a_k \cos k\tau + b_k \sin k\tau)$ has at least as many zeros as its first term, i.e. $2n$.

And so Arnold's theorem is:

> *Under an appropriate nearness condition for an algebraic curve, as technically required at present: for each n every plane curve admits at least $d(n) + 1$ extactic points.*

For example consider a plane cubic having an oval. Then each curve sufficiently close to this oval admits at least 10 points where the osculating cubic at this point will have contact of order higher than 9. We have been inspired by presentations at L. Guieu's seminar at Montpelier; a text by Arnold (among others, treating somewhat related subjects) is Arnold (1996).

A recent synthesis is given in the Thorbergsson and Umehara (1999), which in particular treats space curves; and see also the very recent Thorbergsson and Umehara (2002, 2004).

V.10. Toward a classification of closed curves: Whitney and Arnold II

We begin now to take into account results concerning general, i.e. not necessarily simple, closed curves. Here we deal with global results, the final goal being, if possible, a complete classification of these curves. We first remark that such a classification has only a topological sense; plane curves are considered equivalent when they have the "same shape", the "same appearance". We suppose everything to be C^{∞} for simplicity. Precisely, two curves are considered equivalent when they are obtainable one from the other by a diffeomorphism of the plane $\mathbb{R}^2$. The Jordan theorem implies that all simple closed curves are equivalent among themselves, and to a circle in the Euclidean setting. Note that there is an orientation problem, both for curves and for diffeomorphisms of the plane. We will not always specify the context in which we work, both for simplicity and because it isn't at all essential. The description below of more or less recent results has to be brief; we leave it to readers to consult the works cited for most of the definitions required. In the spirit of this book the idea is to show that, in order to progress in the theory of curves, we need to ascend several rungs up the ladder.

In all areas of mathematics, the ideal classifications are those where we dispose over a series of invariants whose equality yields the equivalence of the objects considered. Section V.6 provides us with a first direction for investigation: with each closed curve C is associated an **invariant**, its winding number $wind(C)$ (defined only within ± 1 if there aren't two orientations, one for the curve and one for the plane). By construction, and since it is an integer, it is invariant under diffeomorphisms of the plane. But it is also invariant under homotopies between curves (see the definition right at the end of Sect. V.XYZ). This invariant isn't sufficient for classifying curves modulo a plane diffeomorphism, as is shown by the figures:

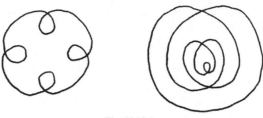

Fig. V.10.1.

However, Whitney rushed in immediately behind Hopf's breakthrough proof of the turning tangent theorem by proving, in Whitney (1937), that: *two plane closed curves are homotopic (in the class of closed curves) if and only if they have the same winding number)*. A *homotopy* between two curves (here it's a particular case of a very general and completely analogous definition) is a continuous deformation of the one into the other, with the condition that all the intermediate curves are again really curves of the type considered, in this instance immersions (and not just any, in which case all curves would be trivially homotopic among themselves). The proof consists of constructing a homotopy by integrating − at the vector velocity level − the homotopy between the two mappings of the given circle by the tangents (the vector derivatives). The name Grauenstein is sometimes attached to this result because Whitney had said that the theorem and its proof were due to him. In fact, the theorem was already present without proof in Boy (1903). We remark, for the sequel, that such deformation in general necessitates passing through intermediate curves possessing triple points and/or points of self-tangency.

Unfortunately − and inevitably in view of the above figure − numerous curves with the same winding number don't have the same shape, aren't so to speak **situated** the same way in the plane, the precise definition being that there don't exist diffeomorphisms of the (whole) plane sending the one onto the other. This is equivalent to saying that there doesn't exist a homotopy between the two where all intermediate curves are generic in the sense of Sect. V.10.1.

In order to progress and be able to hope for a reasonable classification, we appeal to the notion of **robustness** (that we will see in Sect. VI.10 and also in Sect. VII.13.D). The idea is that the really interesting geometric entities are those that we see; we can neglect those that have probability zero of materializing. If there is a measure, which isn't always possible, we can replace probability zero by measure zero. A more precise way of broaching the problem is to consider only those objects as interesting whose form is stable under small deformations, which is the origin of the word robustness. This idea is due to Whitney in the context of curves; but it is René Thom to whom we are indebted for finding the general context for the notion of genericity, of robustness, as well as the tool that is necessary and sufficient for treating it, that of *transversality*. For the planar closed curves, the right notion is that of *generic (closed) curve*:

(V.10.1) *We say that* C *is generic* (normal *for Whitney*) *if the only multiple points of the curve (points where the parameterization is no longer injective) are the double points with distinct tangents.*

Fig. V.10.2. Forbidden for a generic curve

    This definition is obviously robust; but it is also optimal, for Whitney showed in the article mentioned that each closed curve can be made generic by a small deformation. This is visually evident for eliminating self-tangencies and triple points, but for the general case it is necessary to use Whitney's structural results on differentiable functions. Throughout the article he succeeds in calculating the winding number of a generic oriented curve C, with the aid of its double points, from the formula that he proves:

$$wind(C) = \pm 1 + N^+(C) - N^-(C),$$

where $N^+(C)$ (resp. $N^-(C)$) is the number of positive (resp. negative) double points. A double point is called *positive* if, when we arrive there for the first time, we see the second branch of the curve cross before itself from left to right (*mutatis mutandis* for the *negative* case). The ± 1 can be determined by the orientations. Readers will have the courage to verify Whitney's formula in the two cases of Fig. V.10.1.

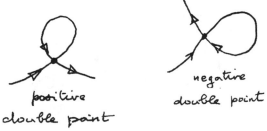

positive
double point

negative
double point

Fig. V.10.3.

    The proof consists of applying the *turning tangent theorem* to each loop furnished by the double points and then forming an appropriate sum. A nice corollary is:

(V.10.2) *We always have* wind(C) $\leq n + 1$, *where n denotes the number of double points.*

Whatever the beauty of Whitney's formula, it scarcely does more than scratch the surface of the desired classification for the shapes of curves. The two curves of Fig. V.10.1 have the same winding number and the same numbers N^+ and N^-, specifically $5, 4, 0$. If we exclude the result, rather isolated in its spirit, of Fabricius-Bjerre and Halpern mentioned right at the end of this section, no work on the classification problem appeared before 1994, the date of the revolution brought about by Arnold (1994b); see also Arnold (1994a), (1994c)). Numerous novel concepts and results are to be found there. We mention some of the most striking. First a **complete classification** of **minimal** generic curves: a generic curve is said to be *minimal* when its total number of double points $n(C)$ is minimum for a given winding number, say $n(C) = wind(C) - 1$, the case of zero winding number being excluded.

Fig. V.10.4. The standard (minimal) curves

To classify the minimal generic curves, we must introduce the combinatorial figure called the *Gauss diagram* of a curve. It amounts to marking values on a circle (representing a parameterization of the curve) that yield, in increasing order of their appearance on the curve, the double points. Two values are joined by a chord in the disk when they give the same double point:

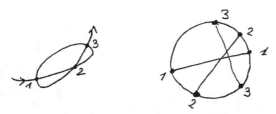

Fig. V.10.5.

It is easy to see that these chords never cross if the curve is minimal; they thus divide the disk into a number of connected components equal to $wind(C)$ and we make a *tree*, for which the vertices are precisely the components and whose edges correspond to the chords separating two of these components. Finally we root the tree, marking in black the connected component whose boundary chords correspond to the points located *exterior* to the curve. An entire book Fiedler (2001) is dedicated to Gauss diagrams and the study of space curves.

For curves a bit simpler than those of Fig. V.10.21, we obtain:

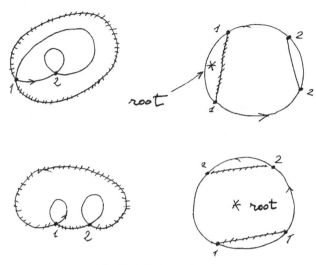

Fig. V.10.6. Gauss diagrams

Arnold's classification theorem states that, for each integer n, the equivalence classes (by diffeomorphisms of the plane, i.e. the forms of curves) of generic minimal curves with n double points are in bijection with the rooted trees with n vertices. The proof uses nothing conceptual beyond Whitney's formula. Shown below are the 9 classes of curves associated with the 9 possible trees for the minimal curves with four double points, as well as the case of two double points; readers should treat the case of three double points:

Readers can look at larger numbers of double points to see the type of growth of the number of classes. It is important to remark that Arnold's theorem is definitely in the spirit of introducing combinatorics into geometry; see for example Sects. II.1, then VIII.4 and VIII.10. In Biggs, Lloyd et al. (1995) can be found the exact number of rooted trees as a function of the integer n; in any case the growth is exponential in n.

Before continuing, we mention that the complete classification of shapes not necessarily minimal curves has not terminated; it is not even certain that this classification is at all inspiring. It can be found up to five double points in Arnold (1994b). The corresponding table might suggest to some readers that the notion of minimal curve is the right one; but from an aesthetic point of view the table might seem discouraging.

One of the obstacles to the classification is this: a Gauss diagram can always be associated with a generic curve, but in general the chords can now intersect and, furthermore, each diagram doesn't necessarily originate from just one curve.

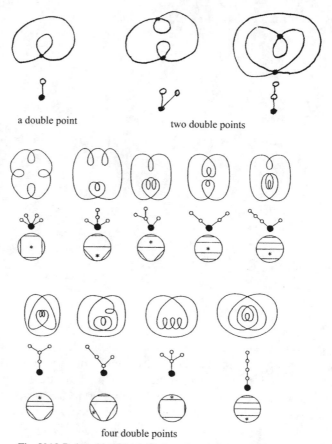

a double point   two double points

four double points

Fig. V.10.7. Arnold (1994a) © American Mathematical Society

We now look at some motivations for the sequel. Arnold was inspired by the desire to place the four vertex theorem in a more general context; he tried to compare it with results of what are called *symplectic* geometry and *contact* geometry We can only explain in a few words what the latter is about, referring to Arnold (1994a, 1994b) and their bibliographies, for it concerns a very general subject allowing treatment of numerous problems in mechanics; see below the knot associated with a plane curve. Moreover, Gromov's prediction is that "in the twenty-first century a good part of geometry will be symplectic". See also a whole series of references below. What now follows issues very largely from this philosophy.

Arnold's idea for finding new invariants, finer than the three introduced by Whitney, consists of putting ourselves high on the ladder and looking at *the space of all planar closed curves*: without seeking to give a very precise formulation − which would be quite ponderous since this space $\mathcal{C}$ is of infinite dimension − we consider the hypersurface $\mathcal{D}$ in $\mathcal{C}$, called the *discriminant*, which is formed by the curves which are not generic. The above figure is of course oversimplified. Classifying the

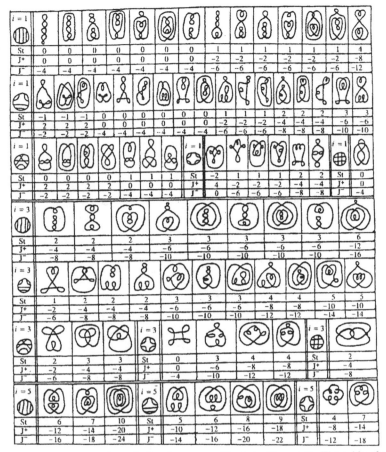

Fig. V.10.8. Classification of closed curves with 4 double points. The table also gives the Gauss diagrams and the values of the Arnold invariants. Arnold (1994a) © American Mathematical Society

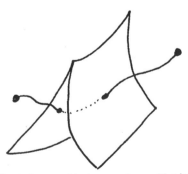

Fig. V.10.9. The discriminant and its intersections with a homotopy of curves

forms of curves amounts to classifying the connected components of $\mathcal{C} \setminus \mathcal{D} = \mathcal{G}$, which is the set of generic curves. An *invariant* is thus a function defined on $\mathcal{G}$ that is constant on the connected components. The idea of Arnold is to construct invariants while requiring them to jump by 1 or by 2 when a path connecting two generic curves of different classes cross the discriminant. We must also fix their initial value for the curve types, called *standard* by Arnold, which are shown in Fig. V.10.4 and which clearly are the most natural.

One of the important steps in proving that invariants can be so defined consists of showing that the discriminant is naturally *co-oriented*, which is to say that we can canonically choose *a side* of this hypersurface. In particular when a path goes through it, we will know if this occurs positively or negatively. For the passage through a triple point, this co-orientation results from the fact that there are two types of passage through a triple point, for the passage through a point of self-tangency, matters are simpler still:

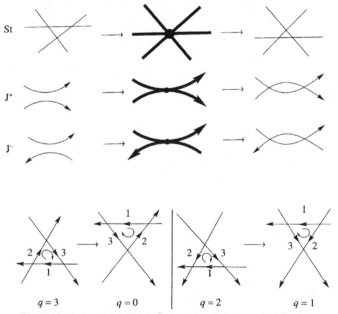

Fig. V.10.10. Arnold (1994a) © American Mathematical Society

This finally leads directly to **three** invariants, for we easily see that a path between two generic curves can be deformed into a path for which the nongeneric curves have only at most three types of singularities: the triple point, points of common tangents having the same sense and points of common tangents having the opposite sense. By examining what happens upon crossing the discriminant, it is shown that these three invariants in fact exist; they are denoted St, J^+, J^-. The invariant St jumps by 1 upon crossing $\mathcal{D}$, whereas J^+ jumps by 2 and J^- by -2.

The construction of these invariants itself indicates the minimum number of triple points and self-tangencies that are necessarily encountered when two generic curves are connected by a homotopy, whence the fundamental theorem:

> In a homotopy between two generic curves C and C', at least $|St(C) - St(C')|$ triple points, $\frac{1}{2}|J^+(C) - J^+(C')|$ points of positive self-tangency and $\frac{1}{2}|J^-(C) - J^-(C')|$ points of negative self-tangency must be encountered among the intermediate curves.

We see in the examples that these three invariants, whatever their interest, are not sufficient for classifying generic curves. There also remains the fact that Arnold doesn't have a general formula for calculating them, apart from the case of St for minimal curves. In fact he shows that — read on the associated rooted tree — the value of St is nothing other than the sum of the distances of all the vertices of the tree to its root. Thus we see that, for our two favorite curves (Fig. V.10.1), we have the values 4 and 10 indicated below. Consequently, passing from the one to the other requires at least six points of each of the three types:

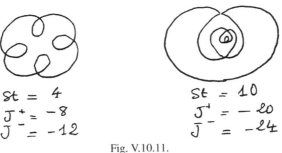

$$St = 4$$
$$J^+ = -8$$
$$J^- = -12$$

$$St = 10$$
$$J^+ = -20$$
$$J^- = -24$$

Fig. V.10.11.

This explains the notation St, which signifies *strangeness*. The curve on the left is the most natural curve with winding number 5, whereas that on the right is the least natural, the most strange. More generally, for minimal curves, the standard curve is the least strange (Fig. V.10.4 or V.10.11 on the left); the curve on the right in Fig. V.10.10 is the most strange. Readers can confirm that the latter curve is the most difficult to *draw*.

For minimal curves we indeed always have $St = -2J^+$. For not necessarily minimal generic curves, we find in Chmutov and Duzhin (1997) a series of formulas for calculating the three invariants of Arnold. For J^+ and J^- the formula involves the following: the curve separates the plane into connected components; with each of these we can associate an index. Take any interior point and see how many multiples of 2π the vector that joins it to a point of the curve turns when the entire curve is traversed; see Sect. V.4. At each double point four connected components are thus associated in this way; the formulas in Chmutov and Duzhin (1997) furnish our three invariants as a function of all these indices.

Here now are various complementary ideas.

We may be interested in closed curves traced on the **sphere**. The classification is in fact simpler, but even so not complete. We merely remark that the winding number only makes sense modulo 2, which means it only takes on the values 0 or 1. To see this, "make a portion of the curve pass through the north pole": the two curves below are in the same class on the sphere:

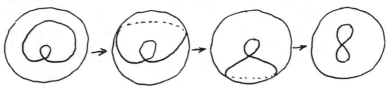

Fig. V.10.12. How to pass from winding number 2 to winding number 0

A spectacular theorem involves the tennis ball:

> *If a simple closed curve on the sphere separates it into two regions of equal area, then the curve possesses at least four inflection points.*

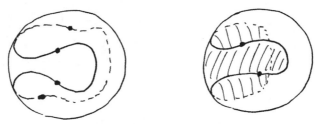

Fig. V.10.13. The tennis ball

The stated result can be found at the end of Arnold (1994a); it fits into the context of wave fronts as we will see right away; for curves on the sphere see also Arnold (1995). But in fact, as can be seen in Thorbergsson and Umehara (1999), this theorem was practically already in Segre (1968). A new approach to curves in space is to be found in Tabachnikov (2002) and the very recent Uribe-Vargas 2004a, 2004b).

The notion of *wave front* comes up in geometry when we seek to study curves parallel to a given curve: we transport points a constant distance t along the normals to the curve C, which yields the curves C_t. We can see in the drawings that these parallel curves are in general going to develop singularities, typically cusps; we want to know how many. In Arnold (1994a) there is a theorem on four cusps for sufficiently small deformations of a circle (attention: they can be counted twice). Contrary to Arnold's theory of three invariants, which only uses simple tools for the proofs, here we need to climb the ladder and place ourselves in an abstract three-dimensional space. Note also the use of a cancellation theorem in at least four values as for the four vertex theorem. We may note that the general context of wave allows us to effectively treat the problem — much more subtle than it might appear — of *envelopes of lines*. Here they are the normals to the initial curve.

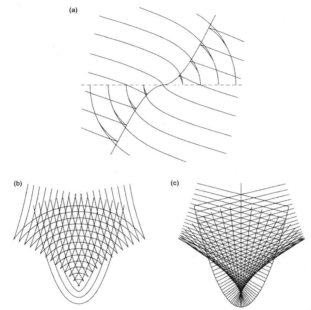

Fig. V.10.14. Evolutions of wave fronts: (**a**) wave fronts: parallel to $x + y^3 = 0$ in the neighborhood of the inflection point; (**b**) wave fronts: parallel to a parabola and (**c**) normal to a parabola. Porteous (1994) © Cambridge University Press

◆

We now treat in more detail a doubly interesting construction, for it is also connected to the study of vision, i.e. making a **knot** in space from a plane curve. To do this, we consider the three-dimensional space formed by lines in the plane which are marked at a point. This space $\mathcal{E}$ is thus the product of the plane with the circle of lines passing through a point. It has the topology of the interior of a torus: here the core of the torus and the parallel curves represent the lines passing through a given point; the plane is represented by the open sectional disks of the torus formed by planes perpendicular to the core.

With a curve C in the plane we associate the subset C* of $\mathcal{E}$ composed of directed lines that are the tangents to the curve and their point of contact. If the curve is generic, it will not have double tangents and thus the closed curve C* will not

Fig. V.10.15. *At left*, a (relatively) realistic drawing where above each point of the plane a topological circle (called the fiber above this point) is represented, composed of a set of lines passing through this point (a projective space of dimension 1!). *At right*, the view is, in contrast, somehow completely changed, another representation of the total space $\mathcal{E}$ has been drawn in which the plane, although infinite in reality, is represented by an open disk (which has the same topology), which varies along a circle S^1

have double points. This will be a simple closed curve of the space $\mathcal{E}$; but what is interesting is that in general it is *knotted*.

To go further, it is necessary to put a supplementary structure on $\mathcal{E}$ called the *contact structure*: the lifted curves of type C^* are not just any curves of $\mathcal{E}$. At each of their points they are tangent to a distinguished plane of $\mathcal{E}$; typically the parallels of the torus are not appropriate. This contact structure is precisely that given, at each point of $\mathcal{E}$, by a tangent plane that depends on the point when the point of $\mathcal{E}$ considered traverses the circular inverse image of a point of the plane; these circles are called the *fibers*. This plane always contains the tangent to the fiber, but turns somehow. This abstract space $\mathcal{E}$ is one whose consideration is indispensable in the study of wave fronts, a notion that was very briefly introduced above (Fig. V.10.14).

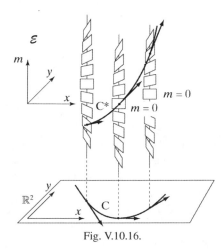

Fig. V.10.16.

A curve of $\mathcal{E}$ is a lifting of a plane curve if and only if it is tangent at each of its points to this family of tangent planes of $\mathcal{E}$. The verticals correspond to the point curves, the horizontals to the lines.

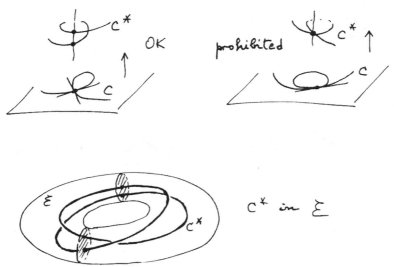

Fig. V.10.17. It is shown in the two figures above why there may be double points for C* if C has points of self-tangency, and why no double points of C* if the tangents to C at each one of its double points are different. The figure below illustrates clearly that C* is in general knotted

To return to the notion of contact structure, which is fundamental in mechanics, there is Appendix 4 of the great classic Arnold (1976), for which there is an English translation Arnold (1978) and which has seen numerous editions. There is a very close relationship between contact structures and symplectic structures, which are also essential in mechanics. Apart from Chap. VIII of Arnold, references are: McDuff (2000), McDuff and Salamon (1998), Hofer and Zehnder (1994).

To come back to our knot, it is thus special and is called *Legendrian*. A small miracle: such a knot possesses an invariant, introduced by Bennequin in 1983: it has been found that this invariant for C* coincides with the invariant St(C). However, in the final reckoning, the theory of plane curves is finer than that of knots, which is too bad, for we would have hoped that this new theory of plane curves would come to the support of knot theory. We can consult first and foremost the very popular Sossinski (1999), then for example Kaufman (1994) and Turaev (1994), the connections with mathematical physics being very strong and profound. For another very recent and very geometric approach, see A'Campo (2000b, 2000a).

The contact structure, defined above, that automatically (canonically) defines the plane, has appeared very recently — at least with precise mathematical formulation — as a part of the structure of the human brain associated with vision; a reference is Petitot and Tondut (1998). It seems that the cortex includes a stack of planes, each of which is a discretized representation of the contact structure of the plane that is realized by a particular neuronal connectivity (that is called a functional architecture). The stack of these implantations isn't three dimensional just because

$\mathcal{E}$ is, but − as is conjectured − only for redundancy of verification and improving the resolution. Furthermore, it has been shown by very sophisticated experimentation that, when one cell is activated, the neighboring small cells are pre-activated, but not all, only those that are close in both position and direction, i.e. that don't turn too much.

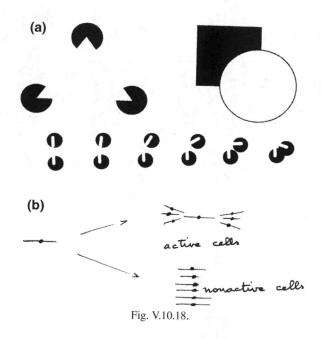

Fig. V.10.18.

That is to say, the pre-activation translates the integral curves of the contact structure. It is thus that the brain can, in receiving signals, distinguish curves − at least certain curves, typically the contours of objects − that play an essential role in human vision, also in the technology of image compression, presently much studied. We will encounter *contours* again, this time of surfaces, in Sect. VI.10.

An atypical result in the theory of closed curves is that of Fabricius-Bjerre (1962), which is rather similar to the Plücker formulas toward the end of Sect. V.14. It says this for generic curves, in a sense to be made precise: for a closed curve, four integers are introduced: the number D of double points, the number I of points of inflection, the number N^+) (resp. N^-) of double tangents where the normals have the same orientation (resp. opposite orientation): then $N^+ = N^- + D + \frac{1}{2}I$. The initial proof used "finite geometry"; see Sect. V.16. A more powerful proof, due to Halpern, uses the notion of index of vector fields; see V.9.8 of [BG].

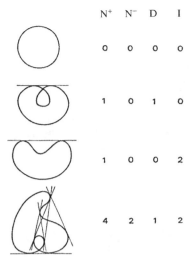

| | N⁺ | N⁻ | D | I |

Fig. V.10.19. [BG] Berger, Gostiaux (1987) © Presses Universitaires de France

V.11. Isoperimetric inequality: Steiner's attempts

The history of the isoperimetric inequality — announced in Sect. V.1 — is extraordinary and has continued to the present; it's a perfect example for illustrating Jacob's ladder, and not merely one of its rungs. Here is a statement that is both precise and complete:

(V.11.1) *For each simple closed curve, if the enclosed area is denoted by* A *and the length of the curve by* L, *then* $\frac{L^2}{A} \geq 4\pi$, *equality holding only if the curve is a circle.*

We can formulate its content equivalently by saying:

(V.11.2) *For all the compact domains of the Euclidean plane, if* A *is the area and* L *the length of the boundary, then* $\frac{L^2}{A} \geq 4\pi$, *equality holding only for circles.*

The current basic reference for practically all the isoperimetric inequalities is the remarkable (Burago and Zalgaller, 1988); we can also look at what there is in Berger (1990b) and Berger and Gostiaux (1987) for the proof with Stokes' formula. We also need to add Osserman (1978) and Osserman (1979) and we will add more recent references in the sequel. However, only Porter (1933) treats the historical aspect in much detail. The assertions Sect. V.11.1 were part of statements admitted without discussion, without requirement of proof, until Steiner devoted a portion of his energy toward attempting to prove it. Steiner is, with respect to "elementary" geometry, one of the greatest geometers of the nineteenth century, both for the variety of his work and for its depth; we will encounter his name several times. Yet he failed with regard to the isoperimetric inequality. Here are two of his attempts.

The first is ultra elementary and based on the property of diameters of the circle: they are viewed from some point of the circle at a *right* angle (equal to $\pi/2$). We comment on the key drawings for this proof (Fig. V.11.1). We take a curve Γ which realizes the minimum of $\frac{L^2}{A}$ and goes on to show by *reductio ad absurdum* that it is a circle. Take any point x on Γ and its antipode y which divides the length into two equal parts. By symmetry and the minimum property, we see that the area is also divided into two equal parts by the line xy. We then replace the domain by the union of the upper part and its symmetrical part. For each p of the upper arc (and the same for the lower) there must be a right angle between px and py, otherwise we can construct a domain with strictly greater area in rotating the two shaded pieces, end of argument.

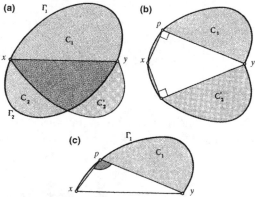

Fig. V.11.1. (a) If the area of C_1 is > than that of C_2, the domain obtained by replacing C_2 by C_2' (symmetric to C_1 with respect to xy) has the same perimeter as the given domain, and a strictly larger area. The significance of (b) and (c) is left to the reader

Weierstrass had great difficulty convincing Steiner that the above argument isn't a proof, for it lacks **the existence** of a curve minimizing the isoperimetric ratio $\frac{L^2}{A}$. To meet his objections, Steiner then proposed the now famous *Steiner symmetriza-tion*, one of the most beautiful and useful inventions in all geometry. It is also much used in analysis, along with its generalizations; see for example Bérard, Besson, and Gallot (1985). We encounter symmetrization operations several times: in Sect. I.2 for Steiner's problem, in Sect. II.1 for isodiametric inequalities in Borsuk's problem and for the MacBeath number in Sect. VII.9. The basic idea is that these symmetriza-tions diminish (or increase) various numerical invariants attached to the object that is symmetrized.

We describe it in the plane, but we will amply see in the sequel, in particular in Sect. VII.5.A, that not only can it be defined in all dimensions but also in hyper-bolic spaces and on spheres. It was used, for example, at the end of Sect. II.1. This symmetrization is most often used for domains which are a priori convex, but that isn't required here. We work with a plane domain D and we consider some line H

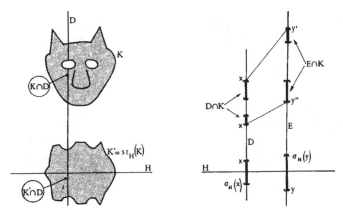

Fig. V.11.2. I [B] Géométrie. Nathan (1977, 1990) réimp. Cassini (2009) © Nathan Édition

(letter chosen to depict a hyperplane!): the *symmetrization of* D *with respect to* H, denoted $\sigma_H(D)$, is obtained as in the figure by intersecting D with lines orthogonal to H. The segments (or unions of segments) of these intersections are subsequently set symmetrically **astride** H, keeping the same total length. The set of these new segments is the $\sigma_H(D)$ in the figure; it has three essential properties: it is symmetric with respect to H, it has the same area as D and for the lengths of the boundaries we have $L(D) \geqslant L(\sigma_H(D))$; moreover there is equality for these lengths only if D was initially symmetric with respect to H. Thus Steiner constructed an operation that strictly decreases the isoperimetric ratio (choose a good direction!) as soon as we aren't dealing with a circle. Again Weierstrass had to convince him that something essential was lacking in his proof: the fact that $\frac{L^2}{A}$ had in addition to satisfy a universal inequality $\frac{L^2}{A} \geqslant k > 0$ for a certain k, or something like that — what analysts call an a priori *inequality*. Readers can try to prove such an inequality if they think they can do better than Steiner and Weierstrass. It is even asserted that Weierstrass said to Steiner: "take the sequence $\frac{1}{n}$: for each $\frac{1}{n}$ there is one that is strictly smaller, and nonetheless there isn't a positive lower bound". This well illustrates the point at which physical intuition makes it impossible to conceive that $\frac{L^2}{A}$ could be arbitrarily small. This anecdote can be found on p. 4 of Blaschke (1949), but above all in Porter (1933).

It wasn't until (Schwartz, 1884) that the first correct proof was published. In fact, Weierstrass was the first, in the 1870s, to give such a proof, but that was in his course at the University of Berlin. Schwarz thanks Weierstrass explicitly in his text. The "worst" thing was that this text had as its goal a proof of the isoperimetric inequality for surfaces in space, which is obviously much more difficult; in fact his text also proved the plane case. His proof also introduced for the first time the "circular" symmetrization, transforming an arbitrary surface into a surface of revolution. The proof was carried out in two steps: the first shows that the symmetrization improves the isoperimetric ratio, with characterization of the case of equality; then things are proved for surfaces of revolution. In both cases inequalities are proved for a suitable type of integral, for the product of two functions among others, which is the celebrated Schwarz inequality in current terminology.

We will see in Chap. VII what is necessary for finding a complete proof "a la Steiner", that is to say completely geometrically. Here we say only that this was done by Blaschke, who truly answered the objection of Weierstrass by showing with a compactness argument that when symmetrization is carried out with respect to more and more lines in varied directions, then a limit object is actually obtained. That it is a disk is the trivial part, for it is clear that only the circle is symmetric with respect to all directions.

The ratio $\frac{L^2}{A}$ is thus finally bounded, in an ideal fashion, **from below**. A really trivial drawing shows that there does not exist an upper bound: $\frac{L^2}{A}$ can be arbitrarily large. However, in the realm of affine geometry, there has been recent interest in the upper bound $\frac{L^2}{A}$ when renormalization by an affine transformation is allowed; what has been discovered will be treated in detail in VII.10.E.3. We must also mention the very beautiful formula obtained in Hurwitz (1902) for the strictly convex case, which gives an estimate of the *deficit*:

$$\frac{L^2}{4\pi} - A \leq \frac{1}{4\pi} \times (\text{area of the developable}).$$

The *developable* of a plane curve is the envelope curve of its normals. The figures below motivate it, but this isn't a proof! For ellipses that are more and more flattened the area of the developable is larger and larger:

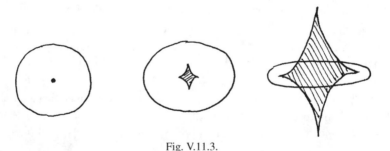

Fig. V.11.3.

V.12. The isoperimetric inequality: proofs on all rungs

Unlike the person in the street, readers armed with a certain mathematical culture will very quickly propose at least four possible attacks for the isoperimetric inequality.

(1) A simple closed curve being essentially a periodic object in one parameter (even if there are two coordinates), we must be able to get away with an analysis **a la Fourier.**

(2) The inequality in question is a relation between the area of a domain — thus its interior — and its boundary. But there exists a very general formula relating the two: the **Stokes** formula, which is an equality between an integral over the boundary and an interior integral, this for two objects derivable from one another in a universal way, the first being of our choice. We would thus certainly have to be able to prove the isoperimetric inequality with the aid of Stokes' formula by cleverly choosing the function to be integrated on the boundary.

(3) Since we know that the circle is optimum, shouldn't we be able to **deform** systematically (and in a continuous manner, as opposed symmetrization, which is a discrete operation) a curve by letting it become more and more a circle, in a natural way, which surely (?) decreases the isoperimetric ratio. We can even hope that the limit exists and is a circle.

(4) The final approach is the simplest: first show that there **exist** curves such that the isoperimetric ratio is minimum, then show that such a curve is necessarily a circle.

These four approaches have been successfully pursued but, apart from the first, are rather recent or even very recent; here briefly is what has transpired.

◆

The proof by Fourier analysis appeared in Hurwitz (1902). But it is necessary to be adroit and parameterize the two coordinates $\{x(\theta), y(\theta)\}$ of the curve C by the angle θ that the tangent makes with a fixed axis and not, as would be more natural, by the arc length. Subsequently, letting $K(\theta)$ be the radius of curvature, the calculation yields the equality:

$$\frac{L^2}{4\pi} - A = \frac{\pi}{2} \sum_{k \geqslant 2} \frac{a_k^2 + b_k^2}{k^2 - 1}$$

where the Fourier series is that of $K = a_0 + \sum_k (a_k \cos k\theta + b_k \sin k\theta)$. But $a_1 = b_1 = 0$ because the curve is closed.

Part of the interest in this proof is that the **deficit** $L^2 - 4\pi A$ in circularity, is measured by a sequence of geometric invariants. Here is some information on this natural problem: to go further than $\frac{L^2}{A} \geqslant 4\pi$ and interpret the deficit $L^2 - 4\pi A$ geometrically. This is an important characteristic of mathematicians: We always want to "go further". The invariants introduced by Hurwitz have not really been exploited, but in any case they form a discrete infinite sequence of invariants characterizing the curve within an isometry of the plane. This sequence is the series of ratios between the $(a_k^2 + b_k^2)$. Hurwitz obtained related results: the one on the area of the developable mentioned at the end of the preceding section and some that could lead to the four vertex theorem (Sect. V.8).

In Santalo (1976) can be found a proof that is not "a la Fourier", but uses integral geometry, i.e. geometric probabilities in disguise; see Sect. I.2. It also provides an explicit estimate of the deficit. For what can be done better presently regarding the deficit, see Osserman (1979) and 1.3.1. of Burago and Zalgaller (1988), especially the very beautiful Bonnesen inequality: $L^2 - 4\pi A \leqslant \pi^2(R - r)^2$, in which R (resp. r) denotes the smaller (resp. greater) radius of a circle containing C (resp. contained in C).

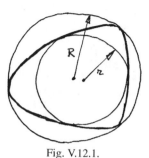

Fig. V.12.1.

We mention an open problem from this realm: in analogous fashion, prove the isoperimetric inequality for surfaces (or better for manifolds in all dimensions) with the aid of spherical harmonics — a very natural desire, since Fourier analysis on the sphere is given by the spherical harmonics. We will speak of this again in Sect. VI.11.

◆

It wasn't until (Schmidt, 1939) that we had a proof using Stokes' formula, but the proof is extremely difficult, although it was the first that worked in the not necessarily convex case. It would be desirable that this proof of Schmidt be cast in modern language. In Knöthe (1957) there is a very simple proof using Stokes, but it is concealed among much more general results in arbitrary dimension. It was explained to the author by Gromov in the train between Paris and Bonn and was featured for the first time in 9.3 of [BG], in [B], and 12.11.7 of Berger and Gostiaux (1987). We will encounter Knöthe's mapping in Sect. VII.8; see also 2.2 de (Giannopoulos and Milman, 2001). Stokes' formula for a compact domain D of $\mathbb{R}^2$ and its boundary C, says that $\int_D d\alpha = \int_C \alpha \, ds$, where α is an arbitrary differential 1-form and $d\alpha$ the 2-form which is the exterior derivative of α. The form α is not obvious to find; it is obtained by constructing a mapping of D into the unit disk D^* which is obtained thus: we normalizes the area of D to equal π, we then slice D and D^* by parallels in a fixed direction, so that the areas are in the same ratio, then divide the sectional intervals so that they have the same ratio. This mapping is interpreted as a vector field f on D and we write Stokes' formula

$$\int_D \text{divergence}(f) \cdot dx \cdot dy = \int_C \langle f \cdot \text{unit normal} \rangle \, ds$$

for the scalar product with the unit normal of the curve C, which is the *flux* of f. By construction the Jacobian of f, i.e. the determinant of the differential of f, is equal to 1, whereas the divergence of f is its trace. We complete the proof with Newton's inequality (VII.2.1). In return, this proof is subtle in the non convex case.

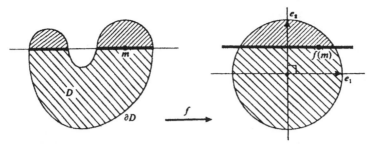

Fig. V.12.2. Knöthe's mapping. II [B] Géométrie. Nathan (1977, 1990) réimp. Cassini (2009) © Nathan Édition

This proof calls for several remarks. First there are problems of differentiability, then of convexity, problems that are circumvented by suitable approximations; these presently have not been detailed anywhere. But above all the proof applies without any modification in arbitrary dimension; it's this that is given in the references above. Finally the proof yields as a corollary the case of equality: only the circles (spheres) attain the minimum; but see Sect. VII.14 or Burago and Zalgaller (1988) for details concerning this uniqueness: the idea is that the domain D, beginning with dimension 3, must be sufficiently regular, or convex, to eliminate objects such as a sphere with hairs, since hairs change neither the area of the sphere nor the enclosed volume. Typically such hairs don't exist either on convex sets or on submanifolds.

The Steiner symmetrization, used for making an arbitrary curve circular, is discrete, harsh. Can the same result be obtained in a continuous fashion? We must then think of **curvature**: circles are the curves of constant curvature. The idea is then to deform a curve C by transporting along the normal at a point a distance proportional to the curvature at this point, thus (see the figure) tending to equalize differences in curvature. If the curve is not convex, we will need to carry things out with the appropriate sign; to have a continuous deformation, we define a one-parameter (the parameter can be thought of as the time t) family of curves C(t) such that, for each t, the deformation is that of the figure. We then need to write the C(t) precisely in the form C(s; t), where s is a suitable parameter for each C(t), and thus C(s; t) is a point of the plane dependent on two real parameters. We thus obtain a partial differential equation for the deformation that is of parabolic type in t. This equation thus admits, thanks to classical theorems, solutions for sufficiently small values of t: we start on the curve C(0) = C. In contrast, the existence for all t is a difficult study, begun in Gage and Hamilton (1986). For interested readers, the parabolic partial differential equation is:

(V.12.1)
$$\frac{\partial C}{\partial t} = \kappa(t, \phi) N(t, \phi)$$

where t is the time, ϕ the curve parameter C(t, ϕ) which evolves with time, κ the curvature and N the unit normal vector of C pointing toward the interior.

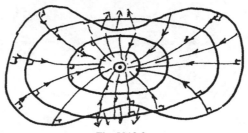

Fig. V.12.3.

Readers will surmise, correctly, that as t tends toward infinity, $C(t)$ converges toward a limit object, typically a point. This is not a circle as we had hoped. To gain satisfaction, it remains to show two things: the first is that, appropriately renormalized, the curves $C(t)$ really do converge to a circle; the second, for the goal we are pursuing, is to prove that the isoperimetric ratio $\frac{L^2(C(t))}{A(t)}$ is decreasing (nonincreasing, more precisely) in t. For such a renormalization, a sort of absolute zoom, there are several choices. For example, we might require that the area be constant, but holding the length constant is also a possibility. The area is the simplest to treat. Since we then have a circle in the limit, we will have proved the isoperimetric inequality as well; and we will not have failed to notice here the promised intrusion into modern geometry of the **dynamics** of evolutionary systems. All that precedes is true; the final point for the exact problem studied is in Gage (1986). However, interest in the preceding does not come down to a very conceptual and dynamic proof of the isoperimetric inequality for plane curves. The interesting point is that this technique may be used in more general contexts, for example for obtaining periodic geodesics on surfaces, a difficult problem that we will examine amply in Chap. XII. We mention also that this deformation of curves into circles comes up in physics in what happens when metal in fusion is poured on a plate at very high temperature. A generalization in the case of surfaces will be seen in Sect. VII.13.C. We mention this "descent" method again in Sect. VII.14 along with its very flamboyant recent application in proving the Poincaré conjecture.

There is also a mechanical context where curves are deformed, not proportional to the curvature at a point, but in an entirely different spirit. With each closed (here not necessarily simple) curve C is associated a sort of energy integral $\int_C k^2(s)\,ds$ of the square of the curvature. We place ourselves in the space of curves and consider the functional $F : C \mapsto \int_C k^2(s)\,ds$. The curves which render this functional critical, i.e. of derivative zero with respect to all deformations, are called *elastica*, which are the forms taken on by metal rope. To find them we take any rope and decrease the functional by following, in the space of closed curves, the gradient directions (those of steepest descent) of the functional F. For recent progress, see e.g. Angenent (1991).

Finally, in Sect. VII.13.C we will discuss at some length surface deformations analogous to those of Gage-Hamilton above for curves, for they bear on a problem of industrial physics and metrology; see also Angenent (1992).

◆

For curves, a proof like the one here is nowhere else to be found, for the result is almost obvious once we have acquired the profound results of geometric measure theory (GMT) which we now discuss. This is because the technique in question is applicable in the very general context of Riemannian manifolds and in other contexts too; see for example TOP.1.B and TOP.10.C of Berger (1999), or 7.1.2 and 14.7.2 of Berger (2003). But the basic idea is in Gromov (1980). It's the most natural possible idea for studying all minimization problems, in particular to prove "Steiner's postulate" and what made it so deficient: the need to show the existence of curves minimizing the ratio $\frac{L^2}{A}$. If we were able to make a suitable compact set out the set of all curves, the ratio in question being obviously continuous, there remains nothing more to say than that a continuous function attains a minimum somewhere on the compact set. In fact, let's for the moment assume existence and suppose that the minimizing curve is differentiable. Then a little differential calculus shows that if the derivative of the ratio $\frac{L^2(t)}{A(t)}$ at $t = 0$ is zero for every variation $C(t)$ of curves about $C(0) = C$, then C has constant curvature: it is thus a circle, Q.E.D. The computation is made as follows: we obtain the curves by transporting a variation $f(s)$ along the normal $n(s)$ for a parameterization of the curve by arc length; then (see the computation of the periphery in Sect. V.6):

$$\frac{\partial(A(C(t)))}{\partial t} = \int_0^L f(s) \cdot ds \quad \text{and} \quad \frac{\partial(L(C(t)))}{\partial t} = \int_0^L k(s) \cdot f(s) \cdot ds.$$

It's a classic trick of the calculus of variations that, if we want the second integral to be zero each time the first is, it is necessary that the curvature $k(s)$ be constant, and thus the minimizing curve is clearly a circle.

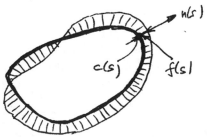

Fig. V.12.4. Variations of area and length

Compactness theorems for sets of curves were obtained in the wake of work on geometric measure theory (GMT) begun in 1960, then more generally for manifolds in a broad sense to be defined precisely and in such a way that numerous geometric minimization problems have reasonable objects as solutions. We will encounter several examples in Sect. VI.11 and, for the case of convex sets, in Chap. VII. In every case here, GMT assures that the minimizing curve is differentiable. We thus very

rapidly obtain a solution, modulo the complex and lengthy foundations of GMT itself, of the isoperimetric inequality for surfaces not only in space (see Sect. VI.11),
but also for all Riemannian manifolds — not to mention the very popular theme of
minimal surfaces and soap bubbles.

We may resolutely wish to ignore that the curves of constant curvature are circles, for both in higher dimensions and in the more general context of Riemannian
geometry the analogous problem can be resolved by the following *packing method,*
which is due to Gromov. We give this proof in detail, for it doesn't to our knowledge
appear anywhere else explicitly (in the Euclidean case, strictly speaking) except for
I.5.G of Berger (2003). Furthermore, we will see in Sect. VI.11 that it extends to
all dimensions. In contrast this technique is used in depth, and is unavoidable, in
the case of Riemannian manifolds with a Ricci curvature that is bounded below; see
Chapt. TOP. 1.A of Berger (1999) or Berger (2003). The idea is to fill the minimizing domain D by normals to the boundary 'curve C. In fact, we proceed in the reverse
direction: for each point m of D we look for a point of C which is at minimum distance; at least one such point q (called a *foot*) = *foot*(m) exists by compactness. The
fact that the distance is minimum assures that the segment mq is normal to C. The
whole domain is thus filled by the normals to C, and the property of the minimum
then proves that the distance mq is less than of equal to the radius of curvature k^{-1}
(which is, we recall, constant) of C at q. In the figure, k^{-1} is also denoted foc(s),
denoting the *focal value* and representing the distance at which the normal at $c(s)$
meets the envelope of normals (and thus the normals that are "infinitely close"). We
stop on the normal at the value of the cut cut(s), a distance beyond which the foot
of m is no longer $c(s)$. We have cut(s) $\leqslant$ foc(s) and computation of the double
integral shows that

$$A = \text{Area}(D) = \int_0^L \int_0^{\text{cut}(s)} (1 - kt)\, dt\, ds \leqslant \int_0^L \int_0^{k^{-1}} (1 - kt)\, dt\, ds = \frac{1}{2k}L,$$

whence the inequality $k^{-1}L \geqslant 2kA$. But we don't know what k is! The astute-
ness of Gromov was to notice that the quotient $\frac{L^2}{A}$ is also minimum, by writing that
its derivative is zero (looking at the above formulas and setting $f = 1$), we find
$k^{-1}L = 2A$. There is thus equality throughout in the above inequality, which im-

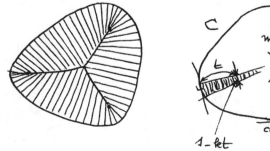

Fig. V.12.5.

plies that $\mathrm{foc}(s)$ = the constant k^{-1}, which is tantamount to saying that our packing is that of a circle.

Just because we have proposed four natural methods doesn't mean that there aren't others. We find no less than ten at the very beginning of Burago and Zalgaller (1988); or in this book look a bit in Sect. VII.14. See alsoRitore and Ros (2002).

V.13. Plane algebraic curves: generalities

Chapter IV gave us almost complete satisfaction for the simplest curves after the circles, i.e. the conics. The books on the subject of plane algebraic curves are numerous, we mention only Coolidge (1959), a very precious little bible, valid up to the date of its first publication in 1931; then a classic work Walker (1950) from the intervening period; finally the completely modern Fulton (1969) and the very recent Chenciner (2006); and a separate mention for the very pedagogical Brieskorn and Knörrer (1986). But the more general algebraic curves are in particular very well treated in the bible Griffiths and Harris (1978), which covers a good part of algebraic geometry. But given the importance, the naturalness and indeed the beauty of the subject, there exist a whole array of other references; for the history, see Dieudonné (1985).

We will broach but very few matters; our choice is dictated by two requirements: to be brief and to stay within the spirit of this book, elementary at least at the level of the problems that are posed. Moreover, we will above all else insist upon problems encountered elsewhere in the book, such as the 3264 conics of Chasles, the inflection points of cubics (Serre's conjecture), the Poncelet polygons. Which is to say: we decline to be interested, at least at the outset, in objects that are already very high on Jacob's ladder . We want to stay well below at the beginning of a problem, of a concept. Another reason is that, contrary to curves that are only differentiable, algebraic curves were very quickly and intensely studied (Descartes and Newton, for a start) and by the end of the nineteenth century numerous results, both local and global, were already known. The problems that have long remained open, that are indeed in some instances still unsettled, are rather those where algebraic curves encroach upon the theory of numbers, the exemplary case being that of Fermat's theorem and the use of elliptic curves. In comparison to those that are complex, the real algebraic curves have enjoyed a resurgence of interest, as have the real algebraic varieties of higher dimension, this while the natural objects of robotics are real algebraic. In Sect. V.15 we will see some interesting problems concerning real algebraic curves.

For us an algebraic curve will almost always be, to begin with, a plane algebraic curve, i.e. a set of points of $\mathbb{R}^2$ (or of $\mathbb{C}^2$) defined by an equation $P(x, y) = 0$, where P is a polynomial in the two variables, real or complex, x and y. We consider the complex case right away, since the study of conics has shown us that this is almost an absolute necessity. But then we know that we also need to extend things to the projective context. Then an algebraic curve will be the set of points of $\mathbb{C}P^2$ (or $\mathbb{R}P^2$) defined by $P(x, y, z) = 0$, where P is a *homogeneous* in x, y, z. We won't forego passing from the affine situation to the projective, for we have known since Chap. I how this is done. It's an important fact that the projective algebraic curves are always **compact** sets.

In all that precedes, the *degree* of an algebraic curve is the degree of the polynomial (the smallest degree; we need to avoid powers of the same polynomial) which defines it. In the **complex** case we have precisely a Nullstellensatz, which states that two irreducible polynomials that yield the same curve are proportional. We can thus consider, as for the conics, the space of all curves of a given degree n, a projective space of dimension $\frac{n(n+3)}{2}$. We thus find 2 for lines, 5 for conics, 9 for cubics, etc. To characterize the Nullstellensatz a bit more, we need pay attention to the case where the polynomial P of the definition can be written $P = QR$, in which case we say that the curve is *decomposable*. Otherwise it is called *irreducible*. In the sequel we always assume irreducibility without mentioning it explicitly.

We can — and should as well — study algebraic curves over any field, but especially over commutative fields. For intuitively clear reasons those over finite fields are connected to combinatorics, and also in an obvious way to the theory of numbers; see for example Chap. V of Silverman (1986) and all of Silverman and Tate (1992). But there are also the fields that are fundamental in arithmetic, specifically the field $\mathbb{Q}$ of rational numbers. In all these cases, we work either in the affine or in the projective context. The case of cubics is exemplary; see below.

Bézout and singularities. We treat essentially the complex case. The problem here is that of the intersection of two algebraic curves C and C' of respective degrees n and m, not to forget the trivial case of intersection with a line. Each line intersects a curve of degree n in at most n points, exactly n if we are in the complex case and count multiplicities (defined as the order of a root of a polynomial), since we obtain an ordinary equation of degree n. Two merged points of intersection correspond to a tangent. For the natural question of knowing whether only algebraic curves have this property of always being intersected by a line in a bounded finite number of points, see Sect. V.16.

We return to the general case: by eliminating y between the equations P and Q of C and C' by appropriate algebraic manipulations, we obtain an equation in x of degree equal to nm and we see that there are at most nm common points. But this is the trivial part of the matter; the case of conics in Sect. IV.6 already has shown us that the subtle problem is that of multiple roots, the common points "merged a certain number of times". It is necessary to do two things in the general case: first to define for a common point how many merged points must be counted; then to show that the sum of all these multiplicities is in fact equal to nm. The figure below is exemplary;

the two curves are cubics, $n = m = 3$. They have three common points, the one obvious point counts for 1; for the other two we can try to guess "by deformation". The drawings thus suggest counting 3 for the point of tangency but with crossing (think for example of a curve and its osculating circle, see Sect. V.6). For the more complicated point, we must deform the cusp and observe at how many points it intersects each branch of the double point of the first curve. The small sketches thus suggest counting this point for $3 + 2 = 5$. In total we thus find $1 + 3 + 5 = 9$.

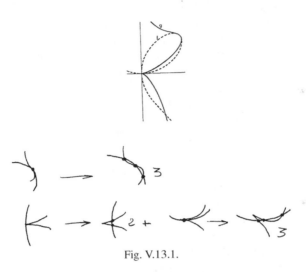

Fig. V.13.1.

For implementing the preceding, the deformation method is too complicated; besides, it would only be valid in the case of fields such as $\mathbb{R}$ or $\mathbb{C}$, not in the case of finite fields — whereas all that follows is completely general. In brief, we do this: at a point of a curve, we define the *branches* that pass through it, and with each of these branches we associate a development of a local parameterization (for example, for a curve with a simple cusp it is (t^2, t^3)); the technical name is *place*. At a common point, the order of the one curve with respect to the other is calculated by replacing the expansion of places of the one in the equation of the other. The local parameterizations can all be made with the help of development in fractional powers of t, called *Puiseux developments*. It remains to show: first, that at a common point the sum of these operations is symmetric in the two curves, in spite of an asymmetric definition; then to show that the total sum is indeed nm. These matters had been expedited with complete correctness by the end of the nineteenth century. The final result is called *Bézout's theorem*. Complete proofs are to be found in the references; the book by Walker is the most accessible for a first approach.

For the study of singularities and of global formulas, see below "Duality and Plücker's formulas" at the end of Sect. V.14.

We have emphasized Bézout in order to indicate the formidable difficulties to be dealt with in order to create an intersection theory for higher dimensions and to obtain, for N hypersurfaces of $\mathbb{C}P^N$, a number of "points" equal to the product of their degrees, of which we had need of in Sect. IV.9.

V.14. The cubics, their addition law and abstract elliptic curves

We are going spend some time on cubics, because it turns out that these curves play an essential role, not only for the Poncelet polygons encountered in Sect. IV.8, but in a large part of mathematics and are of first importance in *number theory* (recall that we say either *number theory* or *arithmetic* indiscriminately; these two names cover essentially the same areas of mathematics). Here are some reasons why we treat cubics in preference over other curves: in the first place, these are the simplest curves after the conics; but then in fact the cubics are richer than the conics, because (at least for those without singularities) they are the only ones that possess a group structure, which is Abelian. This wasn't the case for the (real or complex) conics: they admit a group that acts on them, but it is of but one parameter. The case of the Euclidean circle, seen as a group, is deceptive: it is not an intrinsically algebraic object. Besides, just as the conics and more generally, the unicursal algebraic curves, admit a parametric representation, the cubics admit parametric representation (of course not rational) with the aid of the so-called *elliptic* functions, which are the simplest functions after the sine and cosine — the term "elliptic" has its origin in their use for calculating the arc length of an ellipse. Like the sine and cosine functions, which are of dazzling importance in all mathematics, these elliptic functions turn up in many places; we will see this for the geodesics of ellipsoids of revolution in the following chapter and for the movement of a solid about its center of gravity in Sect. VII.13.C. Next, we can describe this group law geometrically. The cubics are a common thread between Abelian integrals (integrals of algebraic functions), geometry and number theory.

In summary, one of the riches of the elliptic cubic curves is that they are mathematical objects that are susceptible to three equivalent definitions, but definitions completely different in their nature; moreover, the connections between them are very subtle, not to mention the concepts that are needed for interposing them. Typically, in the case of the complex field, these three versions of the same object are:

(i) the singularity-free cubics of $\mathbb{C}P^2$;

(ii) the toruses of dimension 2 viewed as the quotient of $\mathbb{C}$ by a lattice (and the accompanying elliptic functions);

(iii) the abstract algebraic curves of genus 1.

♦

The Abelian group law of plane cubics. The *cubics* (we also say *elliptic curves* and we will see why) are by definition algebraic curves defined by a polynomial of degree 3, whatever the base field: $\mathbb{R}$, $\mathbb{C}$, $\mathbb{Q}$, a finite field, etc. Let us first look at this famous group structure, for which the interest is double, first while it is situated geometrically in the concrete case (curves of KP^2), but also abstractly when we regard abstract cubics as toruses (see below). In what follows, claims are made without justification; readers should refer to the works cited for the proofs.

We dispose at the outset of the cubics with a singularity. Once there is a double point or a cusp, we can parametrize the cubic by two rational functions of degree 3, the parameter being for example the slope of a line passing through the singular point. The algebraic curves thus parameterizable are called *rational* or *unicursal*, and are of much less interest that the fortunate ones that remain.

We begin with almost any commutative field. A cubic C always admits an inflection point p. This point is going to be the identity element of the group that we seek. If m and n are two points of C, the line mn joining them intersects C at a third point q. Then the line pq intersects C at a third point $f(m,n)$. We show (we need to do a little algebra and with some care, thus climbing at least one more rung of the ladder) that the law $(m,n) \mapsto f(m,n)$ is an Abelian group that we denote simply $m + n$, the point p which is the origin and identity element being thus the 0. The only non obvious axiom is the associativity of this law: we can prove it geometrically with the help of the lemma (which requires the careful algebra of which we spoke): if three lines intersect a cubic in the points a_i, b_i, c_i ($i = 1, 2, 3$) and if simultaneously the a_i and the b_i are collinear, then the c_i too are collinear. Here is the diagram:

In other words, three points x, y, z of C are collinear if and only if $x + y + z = 0$. All this has a precise sense, even with the points that get counted twice (as in the case of the conics, mm will be tangent to C at m). In particular the tangent to C at x intersects C at the point $-2x$, etc. There is also no question of reality here since we are dealing with equations of third degree which already have two roots, so that the third always exists.

One of the most utilized cases for graphics, for number theory and for error correcting codes is where the chosen point of inflection is at infinity and the curve is symmetric with respect to the x axis, because it can be shown that a singularity-free real cubic can always be written, after a suitable homography, in the form $y^2 = P(x) = x^3 + ax + b$. There are but two possible forms according as P has three real roots or only one. With this group law we can now resolve practically all geometric problems posed, in the complex case at least, for it can be shown (again a bit of algebra, to be done carefully) that the $3n$ points of intersection of $\{x_i\}$ with an algebraic curve of degree n obey the necessary and sufficient condition $\sum_i x_i = 0$ — which calls for a small celebration!

The points of inflection p are the triple points: $3p = 0$. There is one or three (at most) in the real case, but always 9 in the complex case; we saw in Sect. I.8 that they have very simple coordinates via the homogeneous equation $x^3 + y^3 + z^3 - 3axyz = 0$. And we can now explain the counterexample to Sylvester's situation in Sect. I.1: for if p and q are two points of inflection, then $3p = 0$ and $3q = 0$, and thus the third point r where pq intersects C is $-(p + q)$, and thus is again an inflection point because $3r = -3p - 3q = 0$.

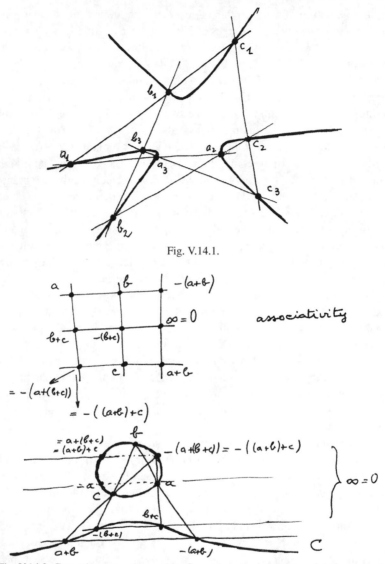

Fig. V.14.1.

Fig. V.14.2. Group law on a plane cubic. Case where the inflection point chosen as an identity element is at infinity

We now intersect C with a conic to obtain six points (with their orders of multiplicity). For example, at each point m of C there exists an osculatory conic, i.e. one which has with C at m five merged points (a conic is defined by five points). It intersects C at a sixth point n, which is the point $-5m$. If we think of the four vertex theorem (for the superosculating circles), or of the sextactic points encountered in Sect. V.9 for not necessarily algebraic curves, we then look for the points m such that $6m = 0$. But we know them, for to say that $6m = 0$ is to say that $2(3m) = 0$, thus that the tangent to m at C intersects it at $p = -2m$ and thus p is an inflection point since then $3p = -6m = 0$. The sextactic points are thus the points of contact of the curve with the tangents emanating from its points of inflection. We will see in the duality (at the end of the present section) that through each point of the plane, in the complex case, there pass *six* tangents to C. If the point is on the curve, there will be only four such tangents apart from the tangent to this point, which counts for two, and three only if it is a point of inflection. The sextactic points are thus $9 \cdot 3 = 27$ in number. In the real case with two connected components we find 9 of them (other than the inflection points themselves), 6 on the oval and 3 on the infinite branch (a projective line topologically). The sextactic conics corresponding to points of inflection are the tangents at these points, but considered as conics degenerated into double lines.

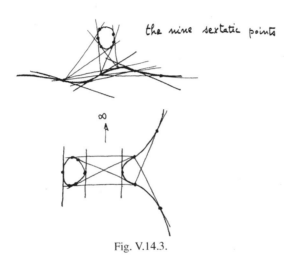

Fig. V.14.3.

We mention that all the points m for which we have $km = 0$ for some integer k play an important role in the cubics that come up in arithmetic; they are called *torsion points*. Over $\mathbb{Q}$ their existence depends on the type (see below) of the cubic.

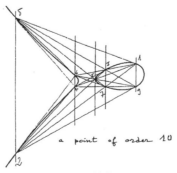

Fig. V.14.4.

◆

Projective classification of planar cubics. The projective-minded geometer will
take the viewpoint ii) still further: it can be shown that the cross ratio of four tangents
(when there are four, which occurs according to the type of cubic in the real case)
issuing from a point of C is independent of this point and is thus an invariant of
the curve (take inspiration from the figure below and see a proof in Walker 1950).
But above all it can be shown that this invariant is *characteristic*, i.e. two complex
cubics (without singularities) are projectively equivalent if and only if the cross ratio
is the same for both. The proofs can be very geometric and rather simple; see the
end of Walker (1950), whence a projective classification of cubics. In the case of
$y^2 = P(x)$ at three points on the x axis, the four tangents issuing from the point at
infinity on the y axis are thus the vertical tangents at the three points $P(x) = 0$ of
the x axis and the point at infinity. Thus the cross ratio is immediately calculable
with the third degree polynomial P.

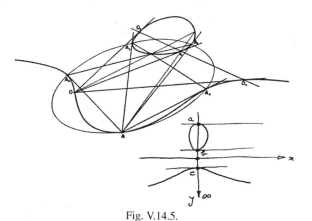

Fig. V.14.5.

To be more precise in the classification, we need to recall the effect of permutations on the cross ratio, for in our cubics the four tangents issuing from a point are
not ordered. The correct invariant is thus not just one of the cross ratios λ, but rather
the function $\frac{(\lambda^2-\lambda+1)^3}{\lambda^2(\lambda-1)^2}$; see Sect. I.6. We will soon see how to calculate it explicitly
from the equation of the curve.

We show that the cubics (without singularity) are never parameterizable by rational functions, as are the conics or the cubics with singularity. But we would like to
parameterize them using reasonable functions, generalizing the parameterization of
the circle by sine and cosine. Said circle can also, by the way, be written in rational
functions $(\frac{2t}{1+t^2}, \frac{1-t^2}{1+t^2})$. These functions exist, they are the elliptic functions; see
above. Just as sine and cosine originate from the integral $\int \frac{dx}{\sqrt{1-x^2}}$, the elliptic functions originate from first integrals that can't be calculated with rational functions,
or with sine and cosine, i.e. the integrals $\int \frac{dx}{\sqrt{x(x-1)(x-\lambda)}}$. We also see clearly that

we can parameterize the cubic with equation $y^2 = x(x - 1)(x - \lambda)$. These elliptic functions surely exist, but we need to study them in the complex domain, where they belong to the class of doubly periodic holomorphic functions (with points where they become infinite, called *poles* − referred to as "meromorphic functions").

Digression: plane lattices (alias the flat torus). The connection with the group law in the complex case is going to fall naturally under the point of view (ii) (see the beginning of the present section) with a question we should have asked a long time back, as we did for the conics in Sects. IV.4 and IV.7: what is the topology of a singularity-free conic? We stay in the projective setting. For the conics we have a circle in the real case and a sphere in the complex case. Here we have a torus in the complex case; in the real case we can have one circle or two (see the above figures). But in the real case, if there are two circles, in $\mathbb{R}\mathrm{P}^2$ the one is convex (has an interior) while the other is not, for it is homotopic to a projective line. In the complex case, we need to visualize the torus $\mathrm{T}^2 = \mathrm{S}^1 \times \mathrm{S}^1$ in $\mathbb{C}\mathrm{P}^2$. If the topology is always that of $\mathrm{S}^1 \times \mathrm{S}^1$ and if addition modulo the periods is always the same as the naive one on $\mathrm{S}^1 \times \mathrm{S}^1$, it turns out quite differently as soon as we compare cubics as objects of type (i), (ii) or (iii), i.e. as curves of the projective plane, complex tori or abstract algebraic curves. In each case the classifications obtained in these three cases coincide. In addition to some of the references already given, Serre (1970) is very dense and very precise.

Here we need to introduce some objects that we will encounter amply in the sequel (Sect. IX.4): the lattices known as the *flat tori*. It is necessary to understand thoroughly that the geometry of a torus that is embedded in $\mathbb{R}^3$ (see the geometry of surfaces in the following chapter) is never compatible with a group law; it only holds at most for one period for tori of revolution and rotations along the meridians; but the rotations along the parallels don't respect the geometry. Whether we want to or not, for these tori we have to climb a bit up the ladder. A *lattice* in a Euclidean plane is obtained by specifying two independent (thus necessarily nonzero) vectors (u, v) of $\mathbb{R}^2$, the lattice Λ strictly speaking being the subset of $\mathbb{R}^2$ composed of the points $ku + hv$ for all integers k, h. We thus define an *associated torus* as the object obtained by the equivalence relation on $\mathbb{R}^2 = \mathbb{C}$, obtained by identifying two points x and y when $y - x \in \Lambda$. The topology shows itself by identifying the opposite sides of the full parallelogram $\{xu + yv : x \in [0, 1], y \in [0, 1]\}$ (Fig. V.14.6).

When we parameterize a cubic with elliptic functions whose two periods are precisely u and v, topologically we obtain the torus $\mathbb{C}/\Lambda$; but above all the Abelian group law is the law of addition of vectors in $\mathbb{C}$ induced by the equivalence relation. The torsion points are then seen very clearly, the nine points of inflection, etc. Note here that the geometry is affine rather than a group geometry: contrary to the embedded case, where it is necessary that the identity element of the group is an inflection point, here we can take any point whatsoever as the identity element. We will look for collinearities in the figure, thanks to periodicity.

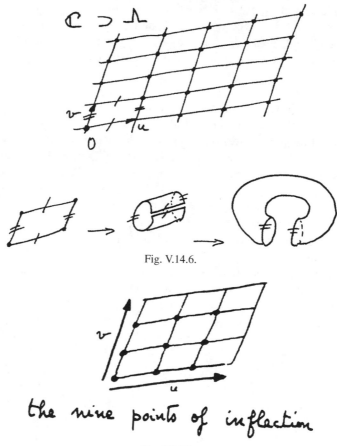

Fig. V.14.6.

the nine points of inflection

Fig. V.14.7.

It remains to connect the classification of cubics by the cross ratio invariant encountered above with that of pairs $\{u, v\}$ of associated periods. What matters is just the ratio $\frac{v}{u}$, where u and v are regarded as complex numbers. Note that, for the complex cubics that are simply the complexifications of real cubics, the ratio $\frac{v}{u}$ is real (and conversely). But in fact we can change the basis for a lattice, the set of these changes (conserving orientation) is the modular group $SL(2; \mathbb{Z})$, i.e. the group of matrices with integer coefficients and determinant one: $\left\{ \begin{pmatrix} a & b \\ c & d \end{pmatrix} : ad - bc = 1 \right\}$. This group joins $\mathbb{R}$ and $\mathbb{C}$ as being one of the most important groups in all of mathematics. Thus the good invariant for the complex cubics is the complex number $\frac{v}{u}$, but modulo $SL(2; \mathbb{Z})$, that is to say, in taking the quotient by the action $\begin{pmatrix} u \\ v \end{pmatrix} \mapsto \begin{pmatrix} au + bv \\ cu + dv \end{pmatrix}$, where the matrix $\begin{pmatrix} a & b \\ c & d \end{pmatrix}$ belongs to $SL(2; \mathbb{Z})$. In general this quotient of $\mathbb{C}$ by $SL(2; \mathbb{Z})$ is most frequently represented by the famous *modular domain*:

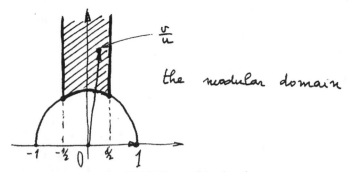

Fig. V.14.8. The modular domain

To have the true quotient, we need to identify the two vertical half-lines on the one hand, and the two half arcs of the circle on the other. The points on the vertical axis represent the rectangular lattices, those on the boundary of the lattice cells. The *square* lattice corresponds to the point $(0, 1) = i$ and the *hexagonal regular* lattice to the two lowest points. This gives the torus that is richest in symmetries. As for the right geometric structure to place on the modular domain, it's not the Euclidean (which is deceptive) that we see so clearly, but the locally hyperbolic metric induced by the Poincaré half-plane (see Sect. II.4), so that we have invariance under the **natural** group SL$(2; \mathbb{Z})$; see also Sect. IX.4.

We are thus within our rights to ask for the connection between this $\frac{v}{u}$ and the invariant coming out of the cross ratio, given the two classes of cubics. The connection is (unfortunately, but inevitably) transcendental, as we will see. But we will do more and will find the "explicit" link between (i) and (ii) (see the beginning of the present section), i.e. between the equation of a planar cubic and the lattice that it determines. In fact the connection is made from (ii) to (i) in the following way, which is an operation inverse to the calculation of the integrals $\int \frac{dx}{\sqrt{x(x-1)(x-\lambda)}}$ introduced above.

Passage from (ii) to (i). We prove all that follows; see the references given, to which it is interesting to add that of Serre. We begin with a fixed lattice Λ. For each integer k, we define the series $G_k(\Lambda) = \sum_{\gamma \in \Lambda'} \frac{1}{\gamma^{2k}}$, where the (double) sum is over all nonzero elements of Λ (*notation* Λ'). Then the abstract cubic defined by Λ has the equation:

$$y^2 = 4x^3 - g_2(\Lambda)x - g_3(\Lambda),$$

where we have put $g_2(\Lambda) = 60 \, G_2(\Lambda)$ and $g_3(\Lambda) = 140 \, G_3(\Lambda)$. The two so-called Weierstrass canonical elliptic functions $\{p_\Lambda(z), p'\Lambda(z)\}$ parameterizing this curve are given by the series:

$$p_\Lambda(z) = \frac{1}{z^2} + \sum_{\gamma \in \Lambda'} \left(\frac{1}{(z-\gamma)^2} - \frac{1}{\gamma^2} \right), \text{ and the derivative } p'_\Lambda(z) = \frac{\partial p_\Lambda(z)}{\partial z}.$$

The parameterization sought is $\{x = p_\Lambda(z), y = p'_\Lambda(z)\}$. Contrary to the case of the circle with sine and cosine, the Weierstrass functions cannot be everywhere defined in the usual sense, for they must have singularities (called *poles*) where they become infinite, but in a reasonable way, i.e. as rational functions, because a bounded holomorphic function (in particular one that is periodic) must reduce to a constant.

And here finally is the relation with the cross ratio of the four tangents issuing from four points that classifies the cubics. For the equation $y^2 = 4x^3 - g_2(\Lambda)x - g_3(\Lambda)$, we will forget its lattice Λ from time to time; we introduce its *discriminant* Δ, for which being nonzero guarantees three distinct roots, and being zero at least a double root, i.e. $\Delta = g_2^3 - 27g_3^2$. Thus the *modular invariant* $j(\Lambda)$:

$j = 1728 \frac{g_2^3}{\Delta}$. The terminology comes from the fact that $j(\Lambda)$ is invariant under $SL(2; \mathbb{Z})$ when we let $SL(2; \mathbb{Z})$ act on an expression $z = \frac{v}{u}$, where $\{u, v\}$ is a basis of Λ. With the associated notation $j(z)$ we thus have

$$j(z) = j\left(\frac{az + b}{cz + d}\right)$$

for each $\begin{pmatrix} a & b \\ c & d \end{pmatrix}$ in $SL(2; \mathbb{Z})$. This posed, for our cross ratio λ we thus have: $j = 2^8 \frac{(\lambda^2 - \lambda + 1)^3}{\lambda^2(\lambda - 1)^2}$. Moreover, the function $j(z)$ is holomorphic in the entire half-plane $\mathrm{Im}(z) > 0$ and defines, by passage to the quotient by $SL(2; \mathbb{Z})$, a bijection from the modular domain (Fig. V.14.8 above) onto $\mathbb{C}$. To go back from the equation $y^2 = 4x^3 - g_2(\Lambda)x - g_3(\Lambda)$ to Λ it is necessary to use the inverse function of j, which does not have a simple expression.

As information useful for the sequel: the invariants $G_k(\Lambda)$ introduced above, written in $G_k(z)$ (for $z = \frac{v}{u}$), are no longer invariant under $SL(2; \mathbb{Z})$, but still behave very well because they satisfy $G_k(z) = (cz + d)^{-2k} G_k\left(\frac{az + b}{cz + d}\right)$. These are called *weighted modular forms* $2k$ and are essential in arithmetic, on the one hand because it can shown that there aren't any others, and better yet because they are generated by G_2 and G_3. Precisely, the algebra M of all modular forms is the algebra $\mathbb{C}[G_2, G_3]$ of the polynomials in G_2 and G_3; the set M_k of those with weight $2k$ has for a basis the family of the $G_2^\alpha G_3^\beta$ with $2\alpha + 3\beta = k$. On the other hand they serve for computing the functions that count the number of points of given norm in a lattice of arbitrary dimension; we encounter this in Sect. X.8. Yet once again a succinct, but perfect, exposition can be found in Serre (1970).

Abstract elliptic curves (iii). Let us now recall the "higher" notion of a 1-dimensional manifold as compared to that of a plane curve, see Sect. V.3. In the setting of algebraic geometry (here complex for simplicity), we can also define in an abstract manner, independent of any embedding, the notion of *algebraic curve* (alge-

braic variety of dimension 1). The definition copies that of differentiable manifolds
— see Sect. XYZ — but here we require that the changes of chart be given by rational functions. Such a curve possesses an invariant, its *genus*. The abstract algebraic curves of genus 1 are none other than the elliptic curves which are obtained
as quotients $\mathbb{C}/\Lambda$. They can be embedded in $\mathbb{C}P^2$, but also in the $\mathbb{C}P^n$ of higher
dimension.

Return to the Poncelet polygons. It's this abstract presentation that provides everything readers want in connection with Poncelet's theorem on polygons and the
cubics in Sect. IV.8, and we now have practically everything that is needed. We
still need to show that the abstract curve defined by a point of the first conic and
a tangent issuing from this point to the second is really an elliptic curve. Now the
object obtained covers the first conic two times algebraically, with the exception of
four points of intersection with the second. But in these four points the covering
singularity is of the form $z \mapsto z^2$ in $\mathbb{C}$. Now there exists a very general formula
called Riemann-Hurwitz for calculating the genus of such an algebraic curve; see
2.1 of Griffiths and Harris (1978). It says this: consider a holomorphic mapping
$f : S \to S'$ between two algebraic curves. Apart from a finite number of ramification points, the covering is throughout a constant number n of sheets. At the
ramification points q the mapping is written $z \mapsto z^{v(q)}$ and we have:

$$\text{genus}(S) = 1 + n \cdot (\text{genus}(S') - 1) + \frac{1}{2}\sum_q (v(q) - 1).$$

Here we indeed find the genus 1, Q.E.D.

Then in Sect. IV.8 we used the classification of the involutions of an elliptic
curve, which states that all are of the form $x \mapsto -x + a$. This result is a particular
case of the theorem that states that the only holomorphic mappings of an elliptic
curve into itself are the group automorphisms, followed by a translation; see 2.6,
p. 326 of Griffiths and Harris (1978). An involution whose square is the identity is
thus necessarily clearly of the form $x \mapsto -x + a$.

Finally we use the expression of a cubic as $y^2 = P(x)$ for finding the explicit
Cayley conditions via the pencil of conics determined by our two conics.

The classification of elliptic curves over different fields isn't yet complete, but it has
known steady — even spectacular — progress, which among other things has led to
resolution of Fermat's theorem. Even though this is but a crude caricature, the idea
of showing that the equation $a^n + b^n = c^n$ has no nontrivial solution in integers for
$n \geq 3$ has consisted of studying the elliptic curve, nowadays called Frey's curve (but
for what it will be in the future, see Hellgouarch, 2000): $y^2 = x(x - a^n)(x + b^n)$
and to show that the Fermat relation is too strong: the elliptic curve in question must
finally have too numerous properties, ultimately contradictory.

♦

Duality and the Plücker formulas. What can we do for curves of degree > 3? We stay with the complex numbers for the better part of this subsection. There is no longer any group law by which we could completely describe, for the cubics, the intersections with the other algebraic curves. There is no longer anything so simple starting with degree four (at least for general curves − the unicursals are a trivial case to be treated separately). There exists a theorem, called *Riemann-Roch*, that best describes these various intersections, but it cannot be stated in just a few words; we refer the reader to Walker (1950) and Coolidge (1959) for "elementary", i.e. visual, expositions, at least to begin with, not requiring us to climb many rungs. However, we can't get to the heart of things by staying just with those works; we must ascend higher in order to progress.

We have already encountered the *genus* of algebraic curves. For a singularity-free curve of degree n the genus equals $\dfrac{(n-1)(n-2)}{2}$. For the visual person, and for the complex projective curves of $\mathbb{C}P^2$, which are thus compact oriented surfaces, it is absolutely necessary to know that our genus coincides with the topological genus that we will encounter in Sect. VI.1, specifically the *number of holes*. We should find zero holes for curves of degree one and two − lines and conics − which we have seen to be topological spheres; and even a torus, for the cubic. But for higher degrees and for singularity-free curves, we don't get all the integers, but only $0, 1, 3, 6, 10, \ldots$.

The topological appearance of all the compact surfaces of three-dimensional space is shown above. These figures represent the topology of complex algebraic curves as situated in the complex projective plane $\mathbb{C}P^2$, which is real four-dimensional! The spherical case is that of the (complex) conics; the toroidal case (one hole) is that of the cubics without double point.

In contrast, we can find all the genuses if singularities are permitted. The genus will soon be defined for abstract algebraic curves, but its connection with the surface with singularities of $\mathbb{C}P^2$ is more delicate; see Griffiths and Harris (1978). The curves of genus 0 are called *unicursal* or *rational*, which means that they are parameterizable by rational functions. The lines and the conics are such, as are cubics having one singularity; more generally, we can apply Plücker's formulas, which will bring out the fact that many singular points imply unicursality.

The notion of *duality* was essential in studying the conics; see Sect. IV.4. Here too an algebraic curve C (over any field) has a dual curve C*, i.e. the set of its tangents − taken in $(\mathbf{KP}^2)^*$, a set of lines of $\mathbf{KP}^2$ identified with $\mathbf{KP}^2$ itself − is the

Fig. V.14.9.

set of points of an algebraic curve; but, if C is of degree n, then C* is of degree $n(n-1)$. This degree is, by its construction, the number of tangents that can be placed on the curve starting with a point of the plane and is called the *class* of the curve. Here is how we show that it equals $n(n-1)$: the line $xX + yY + zZ = 0$ will be tangent to the curve $P(x, y, z) = 0$ if, when we write the equation (of degree n in X, Y, replacing x by its value) which gives the points of intersection, easily seen to be of degree $n-1$, has a double root. Moreover, we have a proper duality: $(C^*)^* = C$. But then this number $n(n-1)$ opens up an abyss under us, for when n is greater than 2 we get a weird degree for $(C^*)^*$. The explanation is that the class in fact depends on singularities. For example a singularity-free cubic will have a dual of order 6, but that dual will have an enormous number of singularities. Following a first attempt by Poncelet to remove this "apparent" contradiction, the complete answer would lie in the *Plücker formulas*, which had been rigorously proved by the end of the nineteenth century. Here are these formulas, which apply to the complex case, for curves having only double points, cusps of the type (t^2, t^3) and inflection points of the type $(t; t^3)$. We have two direct formulas:

$$m = n(n-1) - 2\delta - 3\kappa \quad \text{and} \quad \iota = 3n(n-2) - 6\delta - 8\kappa,$$

where n is the degree, m is the class, ι the number of points of inflection, δ the number of double points and κ the number of inflection points. In the duality, the double points correspond to double tangents, the inflection points to cusps, whence we get two other Plücker formulas by applying the first formulas to the dual curve:

$$n = m(m-1) - 2\tau - 3i \quad \text{and} \quad \kappa = 3m(m-2) - 6\tau - 8\iota,$$

where τ is the number of double tangents of the curve. Readers will not have forgotten to verify that this time there is no contradiction in the degree of $(C^*)^*$! The value $3n(n-2)$ for the "general" number of points of inflection is explained thus: with each curve $P(x, y, z) = 0$ is associated its *Hessian*, which is the determinant of the 6 second derivatives $\frac{\partial^2 P}{\partial x^2}$, etc. and is of degree $3n(n-2)$. Now the points of inflection are easily seen to be the points of intersection of the curve with its Hessian. The first formula shows the elementary result: an algebraic curve of degree n cannot have more than $\frac{n(n-1)}{2}$ singular points.

Thus, the correct invariant for algebraic curves, in the duality, is neither the degree nor the class, but exactly a sort of mixture of the two and is nothing other than the *genus* already encountered several times above, for its value is effectively both $\frac{(n-1)(n-2)}{2} - (\delta + \kappa)$ and $\frac{(m-1)(m-2)}{2} - (\tau + \iota)$.

Just because we have Plücker's formulas doesn't mean that we know all the actual possibilities for the numbers that arise; not all combinations of integers satisfying the four Plücker formulas are possible; see p. 109 of Coolidge (1959). For the current state of the problem, see Laumon (1976); for the real case, see Sect. V.15.

V.15. Real and Euclidean algebraic curves

The topology of real algebraic curves. It's the first of all questions and a problem that is typical for what we pursue in this book, i.e. requiring an ascent of the ladder, being of extreme difficulty and in the end not yet being settled. We consider a singularity-free algebraic curve C of degree n in $\mathbb{R}P^2$: what is its topology?

The complex case would be trivial: it would be a surface with $\frac{(n-1)(n-2)}{2} = g$ holes, where g is its genus. In the real case, then, we have seen that it was a circle for all the conics, and one or two circles for the cubics. Moreover, one of the circles is, in $\mathbb{R}P^2$, homotopic to a projective line, while the other is contractible. We say that a topological circle of $\mathbb{R}P^2$ is an *oval* or a *line* according as to whether it is homotopic to zero or to a projective line, respectively; there are no other possibilities. In each degree there never exists more than one line: in fact a little algebraic topology shows that two lines must intersect in at least one point (as do two true projective lines), which will thus be a double point, but these have been excluded. And, in odd degree, there always exists one line, because each real equation of odd degree possesses at least one real root: we ignore the possibility of this line and observe the ovals that remain. Two questions arise: first that of their *number*, then that of knowing how they are placed with respect to one another: they can be mutually exclusive or one might enclose another.

If at first it's just a question of the number of these ovals, we have had the perfect answer since Harnack (1876): the maximum number is given by $\frac{1}{2}(n-1)(n-2)+1$ in degree n. Moreover, it is attainable; see the elementary (but dated) exposition A'Campo (1980). We indeed find 1 for the conics, 2 for the cubics. The proof of the bound is by contradiction and amounts only to finding a curve of some degree that intersects the curve under consideration in too many points *vis-a-vis* Bezout. In the other direction, the idea works by recurrence and applies the *small parameter method*: if $C = 0$ is the equation of a curve of degree n (with already the maximum number of ovals and well chosen besides) and $D = 0$ is the equation of a line intersecting only one of the ovals in n points, we consider the curves of the equation $CD = \varepsilon$. For ε sufficiently small, we obtain a curve with at most $n - 1$ ovals. The method can be seen better for 4 and 6 in the figures below. For the first we take two conics with equations C_1 and C_2 and the curve with the equation $C_1 \cdot C_2 = \varepsilon$ with ε sufficiently small and with the appropriate sign; we then have four connected components.

For the case of degree 6, we again takes two conics C_1 and C_2 and, in addition, four lines L_i ($i = 1, 2, 3, 4$) and proceed in two stages: first with the curve K_4 with equation $C_1 \cdot C_2 + \varepsilon L_1 \cdot L_2 \cdot L_3 \cdot L_4$, then with the curve with the equation $K_4 \cdot C_1 = \eta$. We find eleven connected components:

The question of knowing how these different ovals are positioned is of a different order of difficulty, and in fact has not yet been answered. For curves of degree 6 it was already a part of the 16th Hilbert problem. Apart from some rather partial intermediate results, it is only recently that the problem has made spectacular

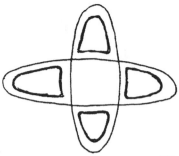

Fig. V.15.1. A' Campo (1980) © Springer and N. A' Campo

progress. We give briefly a selection of results; for complementary results see the two non-technical expository texts Risler (1993) and Itenberg and Viro (1996).

As in Sect. V.10, the most interesting curves are those called *extremals* (M-*curves* in the current literature), i.e. those that have the maximal number of ovals. As soon as three ovals are nested, the maximum number is decreased (this can't happen for $n = 4$). There remains for the extremal curves but a single integer p for which the values are unknown, specifically that of the ovals situated in the interior of an oval, the others all being exterior, as in the "symbolic" figure:

We should remark that these drawings are oversimplified; the reader will find circles intersecting the symbolic curves drawn in many more points than permitted by Bézout; in contrast, Fig. V.15.1 is realistic.

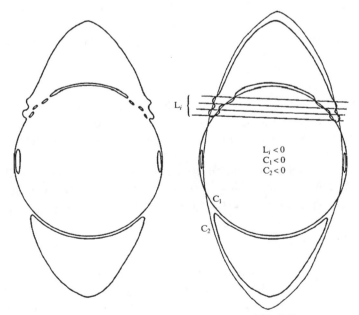

Fig. V.15.2. A' Campo (1980) © Springer and N. A' Campo

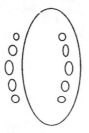 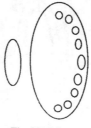 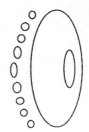

Fig. V.15.3.

We now discuss the most important case, that of even dimension $n = 2k$. The curves constructed by Harnack gave $\frac{3k(k-1)}{2}$ ovals on the exterior and thus $\frac{(k-1)(k-2)}{2}$ on the interior. On the interior it's difficult to set a value. In dimension 6 — the first really prickly case — Hilbert succeeded in getting 8 on the interior, thinking that there were but the two possibilities, 1 and 8, for p. But in 1974 Gudkov found a curve with $p = 4$ and showed furthermore that 8, 4, 1 are the only possibilities in degree 6. All these results use the method of small parameters, together with a fine analysis of Hilbert of the singularities that can be obtained upon deformation of nonsingular curves.

The true revolution is in Viro (1990a), based on his completely new method, discovered in 1983 and valid in arbitrary dimension; the essential idea is well explained in Itenberg and Viro (1996). We can start with a drawing of the triangulation type T, called *patchwork* , for which the vertices are the points with integer coordinates (i, j) of the square $\{(x, y) : |x| + |y| \leqslant m^2\}$, where m is any integer. With T we associate the polynomial $\sum_{i,j} \sigma_{i,j} x^i y^j t^{\nu(i,j)}$, where the summation is taken over the vertices of the triangulation, the $\sigma_{i,j}$ are signs, t is a real parameter and $\nu(i, j)$ an appropriate convex function on the triangulation T. It is shown that for $t > 0$ sufficiently small this yields a singularity-free curve in x, y of degree equal to m and whose connected components are directly connected with T. With appropriate T's we can construct curves with p sufficiently varied with respect to the degree $n = 2k$. Here is the patchwork yielding a curve of degree 10 for which p equals 32: this curve was the first to invalidate an old conjecture going back to Ragsdale in 1906.

To this day the realizable pairs (n, p) are not known. For example, still for the case $n = 2k$ and where $q = \frac{1}{2}(n - 1)(n - 2) - p$, do we always have for the M-curves the inequalities $p \leqslant \frac{3k(k-1)}{2} + 1$ and $q \leqslant \frac{3k(k-1)}{2} + 1$? Or, what is equivalent, is $|p - q| \leqslant k^2$?

Another problem, that of the maximum number of points of inflection, was broached without rigor by Klein: he thought this number was a third of that for the complex case, i.e. $n(n - 2)$ for the degree n. The complete proof, in (Ronga, 1998), uses — other than algebraic geometry — the theory of singularities (transversality, etc.) already encountered in the 3264 conics of Chasles in Sect. IV.9.

Fig. V.15.4. A patchwork that yields a curve of degree 10 and with $p = 32$. Itenberg,
Viro (1996) © Springer and I. Itenberg

◆

Real Euclidean curves: evolutes, involutes and caustics. For each plane Euclidean curve (not necessarily algebraic) of class C^1, we can define two notions: that of *evolute* and that of *involute*. The evolute $C^\diamond$ of the curve C is the curve that is the envelope of the normal lines to C: for a circle it reduces to a point, a completely degenerate case; for an ellipse we already find a curve of degree 6 with four cusps, etc. The four vertex theorem shows that the evolute of a simple closed curve always has at least four cusps, for our envelope always has a cusp when the curvature is maximum or minimum at the corresponding point. Pay attention − see Fig. V.15.4 − to what is obtainable with some double counting: in the case of the second figure we're dealing with a curve of constant width, whose evolute is completely traversed twice.

We have seen in Fig. V.10.14 how this is a particular case of the more general notion of wave front. Some curves are interesting to mention: the hypocycloids and epicycloids ; these are curves for which the evolutes are the "the same", which means precisely that they are obtainable from the original curve by a similitude; see 9.14.34 of [B] or p. 345 of [BG]. We define these curves (whence their name) as those obtained by letting one circle roll on another − on the exterior or the interior − tracing the path of a point fixed on the rolling circle.

The case of a circle rolling on a line is well known − we then speak of a *cycloid* − and of multiple importance in applications. The evolute is obtained by translation; it is thus essentially identical. We first show that it is a *brachistochrone* and the only one; see Sect. 1.5.4 of Dombrowski (1999), a delightful work both for its pedagogy and for the history of this theorem. That is, a ball placed somewhere along the curve and initially at rest slides to the bottom in a time that is independent of its starting point. This is perhaps difficult to believe, but it is true and explains why the pendulums of old clocks were hung by flexible metallic leaf and not about

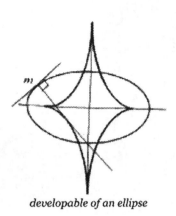

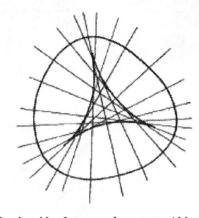

developable of an ellipse   *developable of a curve of constant width*

Fig. V.15.5.

an axis of rotation. In this way the lowest point of the pendulum would describe —
approximately, to be sure — a cycloid, and thus the duration of the swing was much
more independent of amplitude than was the case for the classical pendulum. Also,
the form of the cycloid explains why a car that is driving on a gravel road throws
gravel **forward**, and not backward, since the gravel which adheres to the tire will
describe a portion of a cycloid.

But now it is not difficult to go in the reverse direction, from $C^\diamond$ to C. It suffices
to know the formula for the first variation; see any book on Riemannian geometry,
e.g. I.1.C of Berger (2001b). In particular, the following is shown: let a, b be points
of C such that, when we pass from a to b, we don't encounter any cusp on $C^\diamond$
between the associated points $a^\diamond$, $b^\diamond$. Then the arc length of $C^\diamond$ between $a^\diamond$ and $b^\diamond$
is equal to the difference in the lengths $aa^\diamond$ and $bb^\diamond$. We can thus reconstruct C —
with some degree of caution — starting with $C^\diamond$, **unrolling** a string fixed to $C^\diamond$. The
involute of a curve is the curve obtained, starting at some point of the given curve,
by the unrolling procedure. We obtain in fact a family of curves, said to be *parallel*.
For the distance between the two curves, measured along their common normals, is
constant. If we take the involute of a curve having an odd number of cusps, we then
find curves of constant width.

♦

What more can we ask regarding the case of algebraic curves? We mention two
results from the end of the nineteenth century that seem little known to contemporary geometers. The first deals with the singularities of the evolute, which is again
clearly an algebraic curve; formulas may be found in Coolidge (1959).

The second problem is to know when the involutes are algebraic in this case,
which isn't true generally. The involute of a circle is a spiral and absolutely not algebraic, nor incidentally is an ordinary cycloid. The answer has been known since
Humbert (1888) and is not at all elementary. It is directly tied to the problem of
knowing when the arc length is an algebraic function of the endpoints, which isn't

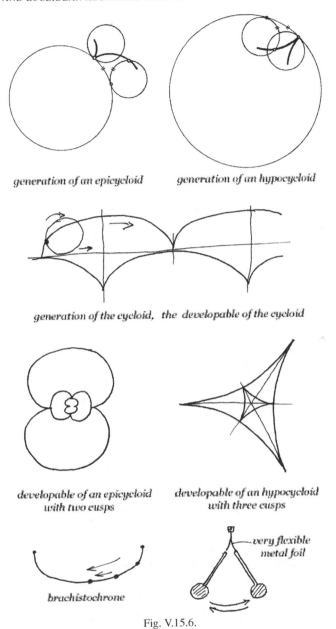

generation of an epicycloid   *generation of an hypocycloid*

generation of the cycloid, the developable of the cycloid

developable of an epicycloid   *developable of an hypocycloid*
with two cusps   *with three cusps*

very flexible metal foil

brachistochrone

Fig. V.15.6.

the case for circles or ellipses. The answer is that the desired curves are the *caustics* of algebraic curves: the caustic of C for one point source of light (or one direction if the point source is infinitely remote) is the envelope curve of the lines obtained by optical reflection from the light source considered. For example, one of the caustics of a circle is the epicycloid with two cusps. Humbert's result is not so

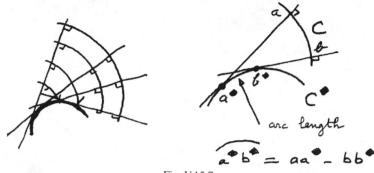

Fig. V.15.7.

surprising given what has been said about the reproductive properties of the epi- and hypocycloids. But the proof requires the whole arsenal of the so-called *Abelian* integrals and an important theorem of Abel respecting them. This problem is important in physics, in research on the totality of nonlinear electromagnetism theories; see Gibbons and Rasheed (1995).

Metric properties. Finally and as always in the Euclidean setting, it is correct to ask whether there exist **metric** definitions of algebraic curves, analogous to those of conics. This subject, not much in fashion nowadays, is well detailed in Chap. X of Coolidge (1959), but we don't know of a nice result that is really pleasant to state and leave it to readers who consult the book to make up their own minds about the various results. Here we present just one: if we intersect a planar algebraic curve whose complexification does not contain cyclic points – see Sect. II.XYZ – with a circle and the number of points obtained is equal to the degree of the curve, then the center of gravity of the points obtained depends only on the center, but not on the radius, of the circle. Already for conics, other than circles, the result isn't clear.

Let us say that we would like definitions of algebraic curves by means of distance relations, and not solely by metric properties relating certain distances or angles; we can if necessary be content with the articulated polygons below.

Here is a nice result that was pointed out to us by Chern. It uses Abelian integrals and is differential in nature: if we intersect an algebraic plane curve C by lines at the points $\{x_i\}$, and if the number of lines equals the degree of the curve, then $\sum_i \frac{k(x_i)}{\sin^3 \tau_i} = 0$, where $k(x_i)$ denotes the algebraic curvature of C and τ_i is the angle the curve makes with the line (modulo π). This result is taken from p. 84 of Segre (1957), where it serves as a very special example. It seems to us that the entire book abounds with results that are interesting but presently little in fashion.

Articulated polygons. In Sect. II.3 (see also Sect. II.XYZ and Sect. V.7) we encountered the *inverter* problem, in particular for describing a linear segment with a plane articulated system. Each point of an articulated system with a single degree of freedom describes a piece of an algebraic curve, since everything there is polynomial. The converse is true; it's the Kempe-Koenigs theorem: every algebraic curve of $\mathbb{R}^2$ can be realized by an articulated system. This realization can generally only be local, for an articulated system can get stuck after a certain time. Readers may refer to the completely elementary but partial exposition in Chap. V of Hilbert and Cohn-Vossen (1952) and the much more profound version in Lebesgue (1950).

The case of articulated quadrilaterals, even parallelograms, has unexpected depth. The profound reason lies in the mathematical structure defined as the set of quadrilaterals whose side lengths are given. One of the components can be fixed, leaving a degree of freedom. To study this structure, it is deceptive to be interested as above in curves described by points bound to the system. For example, for a trapezoidal rectangle of given dimension, the larger side being fixed, the middle of the smaller side describes a complete curve of degree 6, called *Watt's long inflection curve*, which was used for constructing a machine that imitates the human gait. But the correct description is seen by examining the relation between the points of the two circles whose centers are the extremities of the base whose distance apart is given. But we see that this is of the exact same algebraic nature of correspondence (2,2) as that of the Poncelet polygons. It is thus once more an elliptic curve that gives the sought-after structure. This was discovered a long time ago — see Darboux (1879) — but hugely ignored. The interesting dynamic geometry text (Benoist and Hulin, 2004) deals with quadrilaterals of given side lengths.

This reveals in depth the results of Emch (see Sect. IV.8), which explain the Poncelet polygons, and more, with the aid of articulated systems consisting of juxtaposed parallelograms.

◆

The theory of articulated systems, this time with several degrees of freedom and thus providing algebraic objects of higher dimensions, has recently been the object of several works that are astonishing for non-platonists. It is thus that the structure of a plane polygon with five sides (two real parameters) yields a surface, part of a complex surface discovered in the 1970s: the K3 manifolds of Calabi-Yau. Starting with six sides (three real parameters), we get so-called *mirror* manifolds, which didn't appear in mathematics until 1980 and play an essential role in mathematical physics; see Voisin (1996). All of this is contained in unpublished work of M. Kontsevich of IHES. The case of quadrilaterals was already treated from this point of view in Darboux (1879). We obtain the space of these quadrilaterals within isometry by fixing two points as explained above. In a way analogous to what was explained in Sect. IV.12, for our quadrilateral there exist two canonical involutions described in Fig. V.15.9 below:

For this point of view, see Fig. IV.8.7 and the references Emch (1900), Emch (1901) and Barth and Bauer (1996). Being involutions of an elliptic curve, their composition is a translation. This translation is always of order 6, i.e at the end of six repetitions we get the initial object, which can be shown by drawing a cube

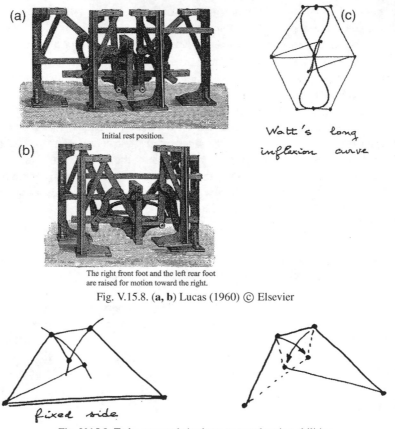

(a)

Initial rest position.

(b)

Watt's long inflexion curve

(c)

The right front foot and the left rear foot
are raised for motion toward the right.

Fig. V.15.8. (**a, b**) Lucas (1960) © Elsevier

fixed side

Fig. V.15.9. To be repeated six times to test drawing abilities

and observing the hexagons formed by the edges transversed appropriately. We already found a transformation of order 6 in Fig. II.2.10, and readers may investigate whether or not there is a link between the two cases.

For an elementary exposition of articulated systems with numerous drawings, see Rideau (1989). For articulated polygons in $\mathbb{R}^3$ made conceptual, see Kapovich and Millson (1996). The recent text (Benoist and Hulin, 2004) studies the dynamics of foldings of quadrilaterals when they are iterated indefinitely.

V.16. Finite order geometry

Here we study the converse, if it exists, of the fact that each line intersects an algebraic curve of given degree n in at most n points in the real domain, and in exactly n points (with multiplicities) in the complex domain. This type of question, which consists of studying curves and more generally submanifolds of affine or projective space which are intersected by each line in a finite given number of points (called the *order* of the object in question), is called *finite order geometry*, and was the object of numerous studies over the decade preceding 1900. The entire book Haupt and Künneth (1967) is dedicated to it; references can be found there as desired. We take from it some typical results that are not too technical, that are of two types: first, locally, it is shown that certain conditions of finite order imply that the object is more or less differentiable, that its singular points aren't too menacing; then come the global results, which use the preceding as needed. We treat the case of arbitrary dimension, so as not to disseminate relatively isolated results in this book.

At first we stay with the truly real domain. We need to realize right off, beginning with order 2, that we are at loose ends: every strictly convex curve is of order 2, valid locally or globally. We thus seem lost, but it is almost impossible to comprehend the stubbornness of mathematicians. First, we have already stated that the four vertex and the six point sextactic theorems had been proved initially by methods of finite order geometry (Sects. V.8 and V.9). There is also Möbius's theorem, which states that each noncontractible closed curve in the real projective plane possesses at least three inflection points.

A theorem of Juel from 1914 states that each connected simple plane curve of order 3 can only have the topological forms shown below:

Fig. V.16.1.

By *curve* we understand exclusively a continuous curve. In Marchaud (1965) it was proved that when a surface of order 3 (cubic from $\mathbb{R}^3$ or $\mathbb{R}P^3$) possesses lines in sufficient number (i.e. at least 8), then all these lines belong to a cubic algebraic surface; see p. 393 of Haupt and Künneth (1967) and the references mentioned there. For a generic surface (not a cone, plane or ruled surface) the number is 15 or 27. This configuration of 27 lines was encountered in Sect. I.9.

Also Marchaud (1936) seems at first very spectacular: each surface of $\mathbb{R}^3$ of order 2 is either a piece of a convex surface or a piece of a (ruled) quadric. It is, however, necessary to realize what is being called a *surface* (otherwise finite point sets, etc., will be solutions). For Marchaud a surface of order n is simply a set for which each plane section is a curve of order n.

For those interested in circles, we find this in Juel: a surface of $\mathbb{E}^3$ that is intersected by each circle in at least four points is a cyclide (surface obtainable from a torus by an inversion mentioned at the end of Sect. II.7). We also find some strictly finite order geometry in Sedykh (2000); and there is in Darboux (1880) a study of the maximum number of families of circles (more generally of conics, for it is basically an affine problem) which can belong to a surface.

Fig. V.16.2. As soon as a surface is neither convex nor a part of a (ruled) quadric, there exists an abundance of lines that intersect it in at least three distinct points

♦

But there doesn't seem to be any connecting thread when we peruse the thick book Haupt and Künneth (1967) and we strongly sense a lack of unity. How can we climb a ladder when we can't even find the ladder? It is in Thom (1969) that at last things are placed in a proper setting — the one he himself invented — of singularities and of transversality. Based on this ascent, three things from this pioneering book should be pointed out that throw an entirely new and renewed light on finite order geometry. First, the notion of finite order for each subobject of dimension N is generalized at the outset; typically it is a submanifold W of dimension N of $\mathbb{R}^M$: the condition of finite order for this object W is that each affine subspace of $\mathbb{R}^M$ of dimension M − N intersects it in at most n points. Thus Thom proved this "almost everything" type of result:

> *If* V *is a compact differential manifold of dimension* N, *then in the space of all differentiable mappings from* V *into* $\mathbb{R}^M$ *(with a* C^r *topology), there exists an everywhere dense open set* U *such that for each mapping* $g \in U$ *the image* $g(V)$ *is of finite degree.*

Thus the objects of finite order aren't rarities, but instead the general rule, but it is important to realize that the degree of g for all possible choices of g is not bounded on U.

To quote Thom's second point, it's necessary to define the notion of *local order* of a point v of a $V^N \subset \mathbb{R}^M$: it's the limit superior of the number of points of intersection of V with an (M − N)-dimensional subspace for which the distance to v tends to zero. Next we must speak of the local order of a *generic* submanifold, including for example those of the space U above. This local order grows with the codimension M − N. The first theorem implies that a compact embedding of minimum order has an order that does not exceed the generic local order. For example it is known (Thom) that the generic local order for the surfaces of $\mathbb{R}^3$ equals four. It is effectively attained for each surface; but we can, furthermore, for each number of holes, find a surface of order 4 exactly. This is due to Calabi; take a suitable tubular neighborhood of the graph drawn below:

But in return the preceding brings out that:

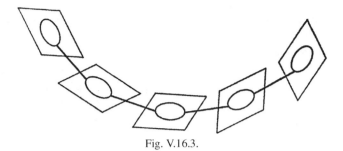

Fig. V.16.3.

The objects whose total order is less than the generic local order are likely very rigid.

That is exactly what the two results of Marchaud and Juel given above say, results that so to speak return now to the regular order.

The third result was announced exclusively by Thom, and clarified and completely proved in (Pohl, 1975); it is also Pohl who answered the most essential question that we used for motivation:

> *Let V^{2k} be a compact submanifold of real dimension $2k$ of the complex projective space $\mathbb{C}P^{k+n}$ such that each element of an everywhere dense set of complex projective subspaces of dimension n intersect it in exactly m points, where m is a fixed integer. Then V^{2k} is a submanifold, either a (complex) algebraic variety of dimension k, or an image under a projective transformation of $\mathbb{C}P^{k+n}$ of the canonic embedding $\mathbb{R}P^{2k} \subset \mathbb{C}P^{k+n}$.*

In fact this last type of subset is actually intersected in exactly one point by each complex projective subspace of dimension n. To our knowledge − Pohl's work excepted − this extraordinary breakthrough of Thom has not been followed by new discoveries. For example, Pohl's result requires that the submanifold be of class C^4. A recent text on this difficult subject is Meyer (2006).

V. XYZ

Higher dimensions. Algebraic geometry can of course be done in all dimensions. We have pursued it in degree 2 for the quadrics, but we had to remain reasonable. However, we have done a little algebraic geometry in dimension 5 while studying the 3264 conics of Chasles. We merely mention too, in the spirit of the resurgence of algebraic geometry in the **real** case, problems of topology for example: the number of connected components, but above all the topology of these components by themselves since, as soon as we go beyond surfaces, the topology of compact manifolds is an immense realm, not yet fully explored. The bulk of the progress is due to Viro; see Viro (1990b). It is incidentally from his general theory that he subsequently deduced refinements for the case of plane curves. We are not able to resist the very geometric theorem on the algebraic surfaces of order 3, called cubic here, already encountered in Sect. I.9: a cubic surface of $\mathbb{C}P^3$ which isn't ruled (doesn't consist of a one parameter family of lines) always possesses (contains) 27 lines. Their combinatorial configuration is imposed; see Schläfli's double six in Sect. I.9. Here, in the real case of cubic surfaces of $\mathbb{R}P^3$ — in contrast to the inflection points of, of which there can only be 3 in the real case as opposed to 9 in the complex case — there exist real cubic surfaces possessing 27 lines; see Fig. I.9.5. The twin books Fischer (1986a, 1986b) — from which the figure is taken — are fascinating and the photos superb.

Differentiability of functions, of mappings. The *class* C^p is that of functions having derivatives of order up to (and including) p and such that these derivatives are moreover continuous; the merely continuous case is expressed as C^0. If derivatives of all orders exist, we speak of the class C^∞. It can involve numerical functions, but more often we are in the general setting of mappings $U \to \mathbb{R}^n$ where U is an open subset of $\mathbb{R}^m$. The fundamental property of these classes is that of being closed under the **composition** of mappings. It is this property that makes what follows possible. But before all else the notion of diffeomorphism: two open subsets U, V of $\mathbb{R}^m$ will be called *diffeomorphic* if there exists a bijection $f : U \to V$ such that both f and f^{-1} are of class C^0 (or any class other than C^0 and *mutatis mutandis*). We are restricted to $m = n$, for this is imposed by the differentiability. In all these definitions that are going to come from differential objects, we omit specification of the class. As for references, many works on differential geometry are good.

Only rarely will we allude to the classes $C^{k,\alpha}$ of functions having derivatives of order through k and such that these derivatives are *Lipschitz of ratio* α. A function is Lipschitz of ratio α when it satisfies $\frac{f(x,y)}{\|x-y\|} \leqslant \alpha$ for all x, y with $x \neq y$. The notion of a Lipschitz function extends immediately to the case where the source space is an arbitrary metric space; we can also speak of Lipschitz mappings between two given metric spaces.

Of course the functions of class C^1 whose differential is of norm bounded by α are Lipschitz with ratio α. But the converse is false. We mention however, even though we won't use it, the fundamental discovery of Rademacher in the setting of functions on $\mathbb{R}^n$: a Lipschitz function is almost everywhere differentiable. This theorem has been the constant object of extensions; see for example Cheeger (1999).

There is finally the class of functions called *real analytic*, denoted C^ω. These are the functions that are infinitely differentiable and for which, moreover, the Taylor series at each point converges and coincides with the function itself over an open interval containing the point. The analytic functions are the only ones that are capable of being "predictive", in contrast to the general C^∞ functions, the classic example being the function that is zero on $]-\infty, 0]$ and which equals e^{-1/x^2} on $]0, \infty[$. From what is on the left nothing can be predicted for what is on the right!

General topology. We have already used classic notions from general topology and will use them extensively in the sequel, without specifying each time what is in-

volved: topological space, open sets, closed sets, compact set, complete spaces (typically when we deal with metric spaces), etc. Readers will be able to find references
to their liking.

Measures. The notion of *measure* is a subtle one, even in the new millennium. It
belongs on Jacob's ladder a bit above metric spaces, topology, simple connectivity,
etc. The book Oxtoby (1980) is interesting, for it also treats the notion of "almost
everywhere" in an exclusively topological setting, often called G_δ-*dense*, the only
such notion possible in those situations where there doesn't exist a measure that
is adequate for the geometry considered (especially in those cases where one can't
exist by the profound nature of things).

Manifolds. It is impossible to define a *differentiable manifold of dimension n*
quickly and well, especially since in practice it is always necessary to restrict ourselves to those that are in a sense *denumerably infinite*, as explained below. This
is likely why Elie Cartan used this notion without ever defining it. The quote Cartan (1946–1951) is famous: "the notion of manifold is rather difficult to define with
precision". We won't do much better than Cartan. Let us say first that a *topological
manifold of dimension n* is a **separable** topological space (satisfying Hausdorff's
axiom) for which each point is contained in an open set that is homeomorphic to an
open subset of $\mathbb{R}^n$. A *chart* of a topological manifold M (of dimension n) is a pair
(U, Φ) consisting of an open subset U of M and a homomorphism Φ of U onto an
open subset of $\mathbb{R}^n$; a differentiable manifold is a topological manifold that can be
covered by a set of charts (an *atlas*) such that, for each pair of charts (U, Φ), (V, Ψ)
for which $U \cap V$ is nonempty, the mapping $\Psi \circ \Phi^{-1} : \Phi(U \cap V) \to \Psi(U \cap V)$ is a
diffeomorphism.

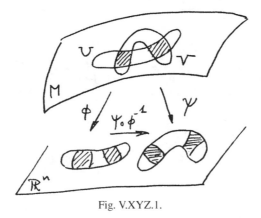

Fig. V.XYZ.1.

The *coordinates associated* with a chart Φ are the x_i $(i = 1, \ldots, n)$: $U \to \mathbb{R}$ defined by transforming by Φ the canonical coordinates $\{u_i\}$ $(i = 1, \ldots, n)$ of $\mathbb{R}^n$: $x_i = u_i \circ \Phi$.

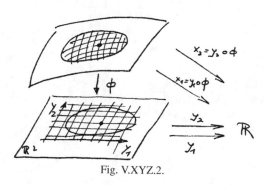

Fig. V.XYZ.2.

We consider here only differentiable manifolds that admit denumerable atlases, for which we define *tangent space* at each point and the *tangent bundle*, which is the collection of all the tangent spaces $T_m M$ for the various points m of M. This definition is consistent with the various charts thanks to the rule that gives the differential of the composition of two mappings. For geometry, the tangent vectors are the velocities of curves traced in the manifold. In a chart with coordinates $\{x_i\}$ the tangent spaces also have automatically associated coordinates; they are such that if c is a curve of M, then the coordinates of the velocity $c'(t)$ will be simply the $(x_i(c(t)))' = \frac{d(x_i(c(t)))}{dt}$.

We also have the notion of differentiable mapping $f : M \to N$ between two manifolds (*understood* to be differentiable from now on), thus also of diffeomorphism. Such a mapping $f : M \to N$ admits a differential df: this object $df(m)$ at $m \in M$ is a linear mapping $df(m): T_m M \to T_{f(m)} N$. For "pure" geometry, the nicest definition entails the effect of f on the velocities of the curves.

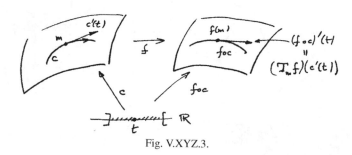

Fig. V.XYZ.3.

It is following Whitney that we know the form of this definition, but we owe to him also the essential theorem that every differentiable manifold of dimension n is diffeomorphic to a differentiable submanifold of an $\mathbb{R}^N$ of dimension N sufficiently large as a function of n. A *submanifold* of an $\mathbb{R}^N$ is defined as for the geometric curves, by equivalent conditions: being everywhere locally a graph, knowing that things are diffeomorphic to the situation of a linear subspace. We can also define the notion of submanifold of a manifold. We say that $f : M \to N$ is an *immersion* if df is everywhere injective. We say that $f : M \to N$ is an *embedding* if it is an injective immersion.

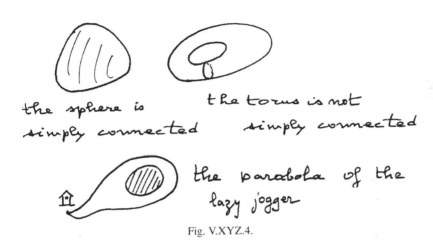

the sphere is
simply connected

the torus is not
simply connected

the parabola of the
lazy jogger

Fig. V.XYZ.4.

◆

We see that what was said in Sect. V.3 with respect to curves can now be expressed thus: there are but two manifolds of dimension 1, the line $\mathbb{R}$ and the circle S^1. A geometric curve is a submanifold of dimension 1 of $\mathbb{R}^2$; it is simple closed if it is an embedding of S^1 in $\mathbb{R}^2$; it is a kinematic curve if it is an immersion of $\mathbb{R}$ in $\mathbb{R}^2$ (it will have been noted that each open interval of $\mathbb{R}$ is diffeomorphic to $\mathbb{R}$ itself). A closed kinematic curve is an immersion of S^1 in $\mathbb{R}^2$. Here we are dealing only with plane curves, but by replacing $\mathbb{R}^2$ by $\mathbb{R}^N$ we obtain analogous objects in space (of three dimensions for $\mathbb{R}^3$, etc). For the question of the classification of manifolds of dimension greater than 1, see Sect. III.1.D of Berger (2001b).

◆

The notion of *homotopy* is very general and plays a role in numerous contexts. The simplest case is that of loops. A *loop* in a topological space T (which is not prohibited from being a differentiable manifold) is a continuous mapping $f : [0, 1] \to T$ such that $f(0) = f(1)$; the *origin* of this loop is this point $f(0) = f(1)$. Two loops with the same origin are called homotopic if we can pass continuously from one to the other; which is to say precisely that there exist a continuous mapping $F : [0, 1] \times [0, 1] \to T$ such that the restriction of F to $[0, 1] \times \{0\}$ (resp. $[0, 1] \times \{1\}$) is the first (resp. the second) loop. But we can do the same thing with two curves without a marked origin (called a free homotopy); we will encounter all this in particular in Chap. XII. But when we deal with more definite objects than just continuous mappings (topological curves, for example) we usually require that supplementary conditions be preserved along the deformation. We then speak rather of *isotopy*; we saw it in Sect. V.10 above.

A topological space is called *simply connected* if each loop (or, what amounts to the same thing, each closed curve) is contractible, i.e. is homotopic to a point (a degenerate curve); see also Sect. III.5. An essential fact is that every reasonable space, manifolds in particular, admit a *universal covering* that is simply connected (and essentially unique). In particular a covering of a simply connected space can only be trivial, that is, a bijection; we will see a typical application in Sect. VI.7. The sphere is simply connected (of dimension two), the circle (of dimension one) isn't; that was essential in Sect. V.3. For a formulation of all these notions, see in part [BG], but the complete details are always a bit long; see also works on algebraic topology.

Bibliography

[B] Berger, M. (1987, 2009).*Geometry I, II*. Berlin/Heidelberg/New York: Springer

[BG] Berger, M., & Gostiaux, B. (1987). *Differential geometry: Manifolds, curves and surfaces*. Berlin/Heidelberg/New York: Springer

A'Campo, N. (1980). Sur la première partie du seizième problème de Hilbert. In *Séminaire Bourbaki 1978–79: Vol. 770. Springer lecture notes in mathematics* (pp. 208–227). Berlin/Heidelberg/New York: Springer

A'Campo, N. (2000a). Generic immersion of curves, knots, monodromy and gordian number. Publications mathm´ atiques de líInstitut des hautes études scientifiques, 770, 208–227

A'Campo, N. (2000b). Planar trees, slalom curves and hyperbolic knots. Publications mathm´ atiques de líInstitut des hautes études scientifiques

Alias, J. (1984). *La voie ferrée*. Paris: Eyrolles

Angenent, S. (1991). On formation of singularities in the curve shortening flow. *Journal of Differential Geometry, 33*, 601–633

Angenent, S. (1992). Shrinking doughnuts. In N.G. Lloyd, L.A. Peletier, & J. Serrin (Eds.), *Nonlinear diffusion equations and their equilibrium states, 3*. Boston: Birkhäuser

Arnold, V. (1978). *Mathematical methods of classical mechanics*. Berlin/Heidelberg/New York: Springer

Arnold, V. (1994a). *Topological invariants of plane curves and caustics. University Lecture Series*. Providence, RI: American Mathematical Society

Arnold, V. (1994b). Plane curves, their invariants, perestroikas and classifications. *Advances in Soviet Mathematics, 21*, 33–91

Arnold, V. (1995). The geometry of spherical curves and the algebra of quaternions. *Russian Mathematical Surveys, 50*, 1–68

Arnold, V. (1996). Remarks on the extactic points of plane curves. In *The Gelfand mathematical seminars* (pp. 11–22). Boston: Birkhäuser

Barth, W., & Bauer, T. (1996). Poncelet theorems. *Expositiones Mathematicae, 14*, 125–144

Benoist, Y. & Hulin, D. (2004). Itération de pliages de quadrilatères. *Inventiones Mathematicae, 157*, 147–194

Bérard, P., Besson, G., & Gallot, S. (1985). Sur une inégalité isopérimétrique qui généralise celle de Paul Lévy. *Inventiones Mathematicae, 80*, 295–308

Berger, M. (1972). Enveloppes de droites. *Bulletin de l'Association des Professeurs de Mathématiques de l'Enseignement Public, 283*, 311–314

Berger, M. (1993). Encounter with a geometer: Eugenio Calabi. In P. de Bartolomeis, F. Tricerri, & E. Vesentini (Eds.), *Conference in honour of Eugenio Calabi, manifolds and geometry* (pp. 20–60). Pisa: Cambridge University Press

Berger, M. (1999). *Riemannian geometry during the second half of the twentieth century*. Providence, RI: American Mathematical Society

Berger, M. (2003). *A panoramic view of Riemannian geometry*. Berlin/Heidelberg/New York: Springer

Blaschke, W. (1949). *Kreis und Kugel*. New York: Chelsea

Bol, G. (1950). *Projektive Differentialgeometrie*. Göttingen: Vandenhoeck and Ruprecht

Boy, W. (1903). Curvatura Integra. *Mathematische Annalen, 57*, 151–184

Brieskorn, E., & Knörrer, H. (1986). *Plane algebraic curves*. Boston: Birkhäuser

Buchin, S. (1983). *Affine differential geometry*. New York: Gordon and Breach

Burago, Y., & Zalgaller, V. (1988). *Geometric inequalities*. Berlin/Heidelberg/New York: Springer

Cartan, E. (1946–1951). *Leçons sur la géométrie des espaces de Riemann* (2nd ed.). Paris: Gauthier-Villars

Chenciner, A. (2006). *Courbes algébriques*. Berlin/Heidelberg/New York: Springer

Cheeger, J. (1999). Differentiability of Lipschitz functions on a metric measure space. *GAFA, Geometric and Functional Analysis, 9*, 428–517

Chmutov, S., & Duzhin, S. (1997). Explicit formulas for Arnold's generic curve invariants. In *The Arnold-Gelfand mathematical seminars* (pp. 123–138). Boston: Birkhäuser

Coolidge, J. (1959). *A treatise on algebraic plane curves*. Dover: Oxford University Press

Darboux, G. (1879). De l'emploi des fonctions elliptiques dans la théorie du quadrilatère plan. *Bulletin des sciences mathématiques, 3*, 109–128

Darboux, G. (1880). Sur le contact des coniques et des surfaces. *Comptes Rendus, Acadèmie des sciences de Paris*, 91, 969–971

De Turck, D., Gluck, H., Pomerleano, D., & Shea Vick, R. (2007). The four vertex theorem and its converse. *Notices of the American Mathematical Society, 54*, 191–207

Dieudonné, J. (1960). *Foundations of modern analysis*. New York: Academic press

Dieudonné, J. (1985). *History of algebraic geometry*. Monterey, CA: Wadsworth

Dillen, F.J.E., Verstraelen, L.C.A. (Eds.). (2000). *Handbook of differential geometry*. Amsterdam: Elsevier

Dombrowski, P. (1999). *Wege in euklidischen Ebenen*. Berlin/Heidelberg/New York: Springer

Emch, A. (1900). Illustration of the elliptic integral of the first kind by a certain link work. *Annals of Mathematics, 1*, 81–92

Emch, A. (1901). An application of elliptic functions to Peaucellier link-work (inversor). *Annals of Mathematics, 2*, 60–63

Fabricius-Bjerre, F. (1962). On the double tangents of plane closed curves. *Mathematica Scandinavica, 11*, 113–116

Fischer, G. (1986a). *Mathematische modelle*. Braunschweig, Germany: Vieweg

Fischer, G. (1986b). *Mathematical models: Photograph volume and commentary*. Braunschweig, Germany: Vieweg

Fulton, W. (1969). *Algebraic curves*. New York: Benjamin

Gage, M. (1986b). On an area-preserving evolution equation for plane curves. *Contemporary Mathematics, 51*, 51–62

Gage, M., & Hamilton, R. (1986a). The heat equation shrinking plane curves. *Journal of Differential Geometry, 23*, 6996

Giannopoulos, A., & Milman, V. (2001). Euclidean structure in finite dimensional normed spaces. In W.B. Johnson & J. Lindenstrauss (Eds.), *Handbook of Geometry of Banach Spaces* (Vol. 1). Dordrecht: Kluwer

Gibbons, G., & Rasheed, D. (1995). *Nuclear Physics, B454*, 185

Gluck, H. (1971). The converse of the four vertex theorem. *Líenseignement mathématique, XVII*, 295–309

Gluck, H. (1972). The generalized Minkowski problem in differential geometry in the large. *Annals of Mathematics, 96*, 245–276

Greenberg, M. (1967). *Lectures on algebraic topology*. New York: Benjamin

Griffiths, P., & Harris, J. (1978). *Principles of algebraic geometry*. New York: John Wiley

Gromov, M. (1980). *Paul Levy's isoperimetric inequality.* Prepublication M/80/320, Institut des Hautes Études Scientifiques. Appears as Appendix C in Gromov (1999)

Gromov, M. (1999). *Metric structures for Riemannian and non-Riemannian spaces.* Boston: Birkhäuser

Guggenheimer, H. (1963). *Differential geometry.* New York: McGraw Hill

Guieu, L., Mourre, E. & Ovsienko, V. (1996). Theorem on six vertices of a plane curve via Sturm theory. In *Arnold-Gelfand mathematical seminars* (pp. 257–266). Boston: Birkhäuser

Harnack, A. (1876). Über Vieltheiligkeit der ebenen algebraischen Curven. *Mathematische Annalen, 10,* 189–199

Haupt, O., & Künneth, H. (1967). Endliche Ordnung. Berlin/Heidelberg/New York: Springer

Hellgouarch, Y. (2000). Rectificatif à l'article de H. Darmon. *Gazette des mathÈmaticiens (Soc. Math. France), 85,* 31–32

Hilbert, D., & Cohn-Vossen, S. (1952). *Geometry and the imagination.* New York: Chelsea

Hofer, H., & Zehnder, E. (1994). *Symplectic capacities.* Boston: Birkhäuser

Hopf, H. (1935). Über die Drehung der Tangenten und Sehnen ebener Kurven. *Compositio Mathematica, 2,* 50–62

Humbert, G. (1888). Sur les courbes algébriques planes rectifiables. *Journal de Mathématiques Pures et Appliquée, IV,* 133–151

Hurwitz, M. (1902). Sur quelques applications géométriques des séries de Fourier. *Annales Scientifiques de l'École Normale Supérieure, 19,* 357–408

Itenberg, I., & Viro, O. (1996). Patchworking algebraic curves disproves the Ragsdale conjecture. *The Mathematical Intelligencer, 18*(4), 19–28

Kapovich, M., & Millson, J. (1996). The symplectic geometry of polygons in Euclidean space. *Journal of Differential Geometry, 44,* 479–513

Kaufman, L. (1994). *Knots and physics.* Singapore: World Scientific

Kneser, A. (1912). *Bermerkungen über die Anzahl der Extreme der Krümmung auf geschlossenene Kurven und über verwandte Fragen in einer nicht-euklidischen Geometrie* (pp. 170–180). Leipzig-Berlin: H. Weber Festschrift

Knöthe, H. (1957). Contributions to the theory of convex bodies. *Michigan Mathematical Journal, 4,* 39–52

Lafontaine, J. (1996). *Introduction aux variétés différentielles.* Grenoble: Presses Universitaires de Grenoble

Laumon, G. (1976). Degré de la variété duale d'une hypersurface à singularités isolées. *Bulletin de la Société MathÈmatique de France, 104,* 51–63

Lebesgue, H. (1950). *Leçons sur les constructions géométriques* (Reprint Jacques Gabay, 1987). Paris: Gauthier-Villars

Lucas, E. (1960). *Récréations mathématiques.* Paris: Gauthier-Villars

Marchaud, A. (1936). Les surfaces du second ordre en géométrie finie. *Journal de Mathématiques Pures et Appliquée, 18,* 293–300

Marchaud, A. (1965). Sur les droites de la surface du troisième ordre en géométrie finie. *Journal de Mathématiques Pures et Appliquée, 44,* 49–69

McDuff, D. (2000). A glimpse into symplectic geometry. In V. Arnold, M. Atiyah, P. Lax, & B. Mazur (Eds.), *Mathematics: Frontiers and perspectives.* Providence, RI: American Mathematical Society

McDuff, D., & Salamon, D. (1998). *Introduction to symplectic topology.* Oxford, UK: Oxford

Mukhopadhyaya, S. (1909). New methods in the geometry of plane arcs. *Bulletin of the Calcutta Mathematical Society, 1,* 31–37

Osserman, R. (1978). The isoperimetric inequality. *Bulletin of the American Mathematical Society, 84,* 1182–1238

Osserman, R. (1979). Bonnesen-Fenchel isoperimetric inequalities. *The American Mathematical Monthly, 86,* 1–29

Osserman, R. (1985). The four-or-more vertex theorem. *The American Mathematical Monthly, 92,* 332–337

Oxtoby, J. (1980). *Measure and category*. Berlin/Heidelberg/New York: Springer

Petitot, J., & Tondut, Y. (1998). Vers une neuro-géométrie. Fibrations corticales, structures de contact et contours subjectifs modaux. *Mathématiques, Infomatiques et Sciences Humaines, EHESS, 145*, 5–101

Pohl, W. (1975). A theorem of Géométrie Finie. *Journal of Differential Geometry, 10*, 435–466

Porteous, I. (1994). *Geometric differentiation (for the intelligence of curves and surfaces)*. Cambridge, UK: Cambridge University Press

Porter, T. (1933). A history of the classical isoperimetric problem. In *Contributions to the calculus of variations*. Chicago: University of Chicago Press

Raphaël, E. (1992). Peut-on faire flotter des troncs d'arbre en apesanteur? *Bulletin de la Société Française de Physique, 92*, 21

Raphaël, E., di Meglio, J.-M., Berger, M., & Calabi, E. (1992). Convex particles at interfaces. *Journal de Physique I, 2*, 571–579

Raphaël, E., & Williams, D. (1993). Three-dimensional convex particles at interfaces. *Journal of Colloid and Interface Science, 155*, 509–511

Rideau, F. (1989). Les systèmes articulés. *Pour la Science, 136*, 94–101

Risler, J.-J. (1993). Construction d'hypersurfaces réelles (d'après Viro), Séminaire Bourbaki. *Astérisque, 216*, 69–87

Ronga, F. (1998). Klein's paper on real flexes vindicated. In B. Jakubczyk, W. Pawlucki, & J. Stasica (Eds.), *Singularities symposium – Lojasiewicz 70*. Warszawa: Banach Center Publications 44

Santalo, L. (1976). *Integral geometry and geometric probability*. Reading, MA: Addison-Wesley

Schmidt, E. (1939). Über das isoperimetrische Problem im Raum von *n* Dimensionen. *Mathematische Zeitschrift, 44*, 689–788

Schwartz, H. (1884). Beweis des Satzes, dass die Kugel kleinere Oberfläche besitzt als jeder andere Körper gleichen Volumens. *Nachrichten von der Gesellschaft der Wissenschaften zu Göttingen, ou OEuvres complètes, Berlin, 1890*, 1–13

Sedykh, V. (2000). *Discrete variants of the four-vertex theorem* (Preprint no. 9615). CERE-MADE, Dauphine: Université Paris IX

Segre, B. (1957). *Some properties of differentiable varieties and transformations*. Berlin/ Heidelberg/New York: Springer

Segre, B. (1968). Alcune proprieta differenziali in grande delle curve chiuse sghembe. *Rendiconti di Matematica, 1*, 237–297

Serre, J.-P. (1970). *Cours d'arithmétique*. Paris: Presses Universitaires de France

Silverman, J. (1986). *The arithmetic of elliptic curves*. Berlin/Heidelberg/New York: Springer

Silverman, J., & Tate, J. (1992). *Rational points on elliptic curves*. Berlin/Heidelberg/New York: Springer

Sossinski, A. (1999). *Noeuds (Genèse d'une théorie mathématique)*. Paris: Seuil

Tabachnikov, S. (2002). Dual billiards in the hyperbolic plane. *Nonlinearity, 15*, 1051–1072

Thom, R. (1962). Sur la théorie des enveloppes. *Journal de Mathématiques Pures et Appliquée, 41*, 177–192

Thom, R. (1969). Sur les variétés d'ordre fini. In D. Spencer & S. Iyannaga (Eds.), *Global analysis*. Papers in Honor of K. Kodaira. Princeton, NJ: Princeton University Press

Thorbergsson, G., & Umehara, M. (1999). A unified approach to the four vertex theorems. II. *American Mathematical Sociey Translations, 190*, 229–252

Thorbergsson, G., & Umehara, M. (2002). Sextactic points on a simple closed curve. *Nagoya Mathematical Journal, 167*, 55–94

Thorbergsson, G., & Umehara, M. (2004). A global theory of flexes of periodic functions. *Nagoya Mathematical Journal, 173*, 85–138

Turaev, V. (1994). *Quantum invariants of knots and 3-manifods*. Berlin: de Gruyter

Uribe-Vargas, R. (2004a). Four-vertex theorems, Sturm theory and Lagrangian singularities. *Mathematical Physics, Analysis and Geometry, 7*, 223–237

Uribe-Vargas, R. (2004b). On singularities, "perestroikas" and differential geometry of space curves. *L'enseignement mathématique, 50*, 69–101

Viro, O. (1990a). Real algebraic plane curves: Constructions with controlled topology. *Leningrad Mathematical Journal, 1*(5), 1059–1134

Viro, O. (1990b). Progress in the topology of real algebraic varieties over the last six years. *Russian Mathematical Surveys, 41*, 55–82

Voisin, C. (1996). *Symétrie miroir*. Paris: Société Mathematique de France

Walker, R. (1950). *Algebraic curves*. Princeton, NJ: Princeton University Press

Whitney, H. (1937). On regular closed curves in the plane. *Compositio Mathematica, 4*, 276–284

Chapter VI
Smooth surfaces

VI.1. Which objects are involved and why? Classification of compact surfaces

We will study — contemplate — the next simplest objects after curves, i.e. *surfaces*. We studied curves essentially in the plane, whereas surfaces appear in the Euclidean three-dimensional space $\mathbb{E}^3$. However, we will see soon enough the necessity of considering *abstract surfaces*; see Sect. V.XYZ. We didn't encounter this problem for curves, for the only abstract curves are the line and the circle, and we can always visualize them, with their internal geometry, as situated in the plane. This impossibility of visualizing certain surfaces has already been encountered in Sect. I.7 with regard to the projective plane and in Sect. V.14 with regard to elliptic curves. We will encounter it once more in Sect. VI.4 below with regard to hyperbolic geometry. The word *smooth* is usual for saying differentiable, having a differential, requiring the existence of a tangent plane at the very least. In another direction there are the polyhedra, that will be treated amply in Chap. VIII.

Using Sect. V.XYZ, a *surface in* $\mathbb{E}^3$ (more briefly just *surface*) is a submanifold of $\mathbb{E}^3$ of dimension 2, in contrast to an *abstract surface*, i.e. a differentiable manifold of dimension 2. A surface in $\mathbb{E}^3$ can thus also be an *embedding* in $\mathbb{E}^3$ of an abstract surface (abstract manifold of dimension 2), but we will exceptionally also consider *immersions* of surfaces in $\mathbb{E}^3$.

Here we are not going to try to present everything that is known — some of which is quite recent — but rather concentrate on what isn't known about surfaces. In particular, we will mention some rather spectacular instances of problems that are open but nonetheless simple to state, some recent. On the other hand, we won't be climbing very high on Jacob's ladder. A rather complete exposition of the basic definitions for surfaces, of synthetic nature without proofs, can be found in the last chapters of [BG], which can be supplemented for definitions and basic results by parts of Berger (2003). Readers can also consult the textbooks do Carmo (1976), Stoker (1969), Klingenberg (1973) and Montiel and Ros (1997). The rather old reference Struik (1950) remains much appreciated, Gray (1993) is written systematically using the computer algebra system **Mathematica** and we should add the recent synthetic work Burago and Zalgaller (1992). Finally Porteous (1994), in its title a bit audacious, presents an approach to curves and surfaces which is rather "orthogonal" to classic approaches, holding to the singularities and transversality of Thom, a book that is very interesting, because it uses an approach which is not very theoretical, but rather systematically geared toward practical applications and the real world.

M. Berger, *Geometry Revealed*, DOI 10.1007/978-3-540-70997-8_6,
© Springer-Verlag Berlin Heidelberg 2010

We will borrow from it numerous times in the sequel. Note too that our Chap. XII will be entirely devoted to large scale behavior, i.e. geodesic flow on a surface, an idea from Hamiltonian dynamics.

As motivations for the study of surfaces, there are first those that are, as with curves, facets of everyday life — thus de facto simple models for doing physics or for testing certain hypotheses. The most striking case is that of our planet, which is rather well represented by a flattened ellipsoid of revolution. The spherical model is very useful, but was treated amply in Chap. III. As in Sect. III.1, the essential thing is to introduce the so-called *intrinsic* (or *internal*) metric of a surface, given by the lower bound of the lengths of curves traced **on** the surface, joining two points considered. Contrary to the case of the sphere, the search for these shortest paths doesn't in general have more than a theoretical answer. Specifically, they belong among the *geodesics* but, on the one hand, it isn't possible to calculate these explicitly except in very special cases and, on the other hand, it is hard to know in practice when a geodesic *stops* being a shortest path; this is the problem of the *cut locus*. For *complete* surfaces, these shortest paths always exist between two arbitrary points but aren't necessarily unique; see the definition of cut locus in Sect. VI.3.

The problems about surfaces are twofold. There are those that concern the intrinsic metric for its own sake, then those concerning the connections between the intrinsic metric and the way in which the surface appears in space, which reveals relationships between the two *fundamental forms*.

Here is a question that may occur to readers, above all if they peruse the table of contents or do some browsing. Why do we bother with curves and surfaces when we have done geometry in arbitrary or high dimensions in preceding chapters? It is because they are essential for all of Riemannian geometry, a subject too immense for us to give more than a brief summary. Readers can refer to the synthesis Berger (1998), duplicated in Berger (1999), but above all Berger (2003). However, we should also mention that another reason concerns hypersurfaces of $\mathbb{E}^n$: as soon as we have $n > 3$, then generically the intrinsic metric determines the way in which they are embedded, an old remark that can be found in Killing (1885); thus there isn't the subtle mixture of two fundamental forms that we observe when $n = 3$. This mixture is only provisional and still under scrutiny; see for example Martinez-Maure (2000).

Here now are some *examples of surfaces*. We have devoted all of Chap. III to the sphere, so-called Euclidean spheres (sometimes called *round spheres* to distinguish them from surfaces that have only the topology of a sphere. We have also encountered the quadric surfaces (ellipsoids, etc.) at the end of Chap. IV. Very important for both theory and practice are the *surfaces of revolution*, which are invariant under the group of rotations about a fixed line (and are smooth). These are determined once we know a *meridian* (or *the meridian*, since all meridian curves are essentially the same). The *parallels* are sections by planes that are perpendicular to the axis of revolution. According to the topological classification of plane curves (see Sect. V.3)

there are thus but four (connected) topological types for surfaces of revolution: the sphere and the torus for the compact case, the plane and the cylinder for the non-compact case.

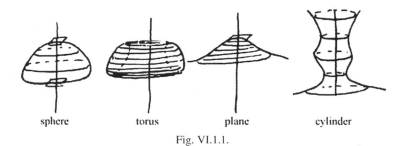

sphere   torus   plane   cylinder

Fig. VI.1.1.

We have encountered the ellipsoids and, more generally, the quadrics, in Sect. IV.10. The *ruled surfaces* are those composed of a one-parameter family of lines in space. The developable surfaces, for example the cylinders and the (pointed!) cones, are a particular case that will be seen in Sect. VI.7. Although beyond the main scope of our goals, we point out for interested readers that the tangent plane of a ruled surface, when the contact point describes one of the lines of the family (called *generators*), varies in homographic fashion, so that the cross ratio (see Sect. I.6) of four points of a generator is equal to the cross ratio of four planes tangent at these points (the set of planes passing through a given line has, by duality, a structure of the projective line). This homography evidently depends on the generator considered; moreover it is singular precisely for the developable surfaces, for the tangent plane remains constant as we move along a generator.

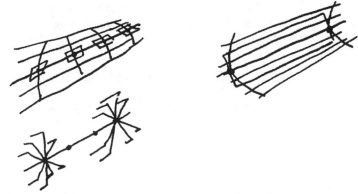

Fig. VI.1.2. For a general ruled surface, the tangent plane turns when we move along a generator, and in a homographic manner. For a developable surface the tangent plane remains constant along a generator

The hyperboloids of one sheet are ruled and can also be surfaces of revolution; see Fig. IV.10.2.

Although not an essential point for this chapter, it is impossible to talk about surfaces without mentioning their *topological classification*. We saw in Sect. V.3 that the only abstract curves are the circle and the line. An essential and natural result from algebraic topology is that the only **compact orientable** surfaces are those of the figure below; they are classified by the number of holes (called the *genus* for reasons that we have seen for algebraic curves in Sect. V.13): 0 for the sphere, 1 for the torus. The compact orientable surface with g holes is called the *torus with g holes*; it can be regarded as the connected sum of g ordinary tori.

Fig. VI.1.3. Connected sum of two tori

What has then happened with our real projective plane $\mathbb{R}P^2$? Its problem is that it is not orientable; but with it the classification of abstract compact nonorientable surfaces also complete: every compact nonorientable surface is the connected sum of g projective planes $\mathbb{R}P^2$ ($g \geqslant 1$), which defines the genus of the surface in question. It amounts to the same as saying that a compact nonorientable surface of genus g is a sphere with g holes, in which each hole is filled with a Möbius strip. We can give another, more visual, classification by introducing the Klein bottle:

Fig. VI.1.4. Construction of the Klein bottle: a nonorientable and immersed surface, with only one circle of self-intersection. Hilbert, Cohn-Vossen (1996) © Springer

In Sect. I.7 we saw sketches of the real projective plane $\mathbb{R}P^2$, but the Klein bottle seems much easier to draw.

Here is the announced classification: every compact nonorientable surface is either

— the connected sum of a torus with n holes and $\mathbb{R}P^2$, or
— the connected sum of a torus with n holes and a Klein bottle.

The connection with the preceding classification is easily made if we know that $\mathbb{R}P^2 + \mathbb{R}P^2 = $ Klein, $3\mathbb{R}P^2 = $ torus $+$ Klein (here $+$ denotes the connected sum operation).

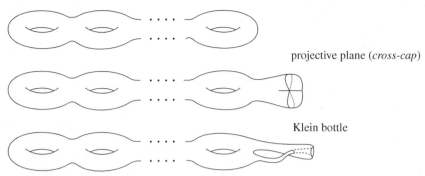

projective plane (*cross-cap*)

Klein bottle

Fig. VI.1.5. Classification of orientable and nonorientable compact surfaces

These abstract compact nonorientable surfaces, like the real projective plane, are not embeddable in $\mathbb{R}^3$. They are only immersible, an *immersion* being by definition a differentiable mapping of the given surface into the space, where the tangent mapping is of maximum rank everywhere, here of rank 2; we don't require that such a mapping be injective, which is to say that we don't prohibit self-intersections — otherwise we would have an *embedding*.

As intuitive as it may be, this classification of surfaces is never easy to prove completely, above all in the embedded case, in spite of the pedagogic efforts of many mathematicians. Readers can consult one of the references below, but should pay attention to whether Morse theory facilitates matters as Gramain (1971) claims, but the classic proofs by slicing provide more information on the fundamental group. Proofs can be found in Gauld (1982), Hirsch (1976), Moise (1977), Seifert-Threlfall (1980), Wallace (1968), Massey (1991) and Stillwell (1980).

One of the difficulties is in showing that each surface is triangulizable, and triangulizability is often assumed by the classifiers. A recent reference for triangulating a surface quickly and well is Colin de Verdière (1990). For a triangulation with S vertices, A edges and F faces, the Euler-Poincaré characteristic χ of the surface is provided by the formula $S - A + F = \chi$. In the orientable case the number g of holes is related to χ by $\chi = 2(1 - g)$. A portion of a triangulation can be seen in Fig. VI.7.7.

We encounter an amusing version of the classification of surfaces with the regular star polyhedra in Sect. VIII.5: two of these don't have the topology of the sphere, but instead that of a surface with four holes. To see this, it suffices to use the preceding formula.

VI.2. The intrinsic metric and the problem of the shortest path

We now pose for general surfaces the same problem as was posed for the sphere in Sect. III.1: what is the shortest path from one point to the other? We will see that there is no perfect answer, but only local results.

The *intrinsic* or "internal" metric d of a surface S of $\mathbb{E}^3$ is defined by

$$d_S(p,q) = d(p,q) = \inf\{\text{length}(c): c \text{ a curve traced on S joining } p \text{ to } q\};$$

i.e. the distance between two points is the lower bound of the lengths of curves connecting the two points. Observe that this distance isn't in general the one that is *induced* by the embedding $S \subset \mathbb{E}^3$, a distance which would be $d_{\text{induced}}(p,q) = \|q - p\|$, but which scarcely has any practical interest. It is however clear that $d_S \geqslant d_{\text{induced}}$, which allows us to show subsequently that d really is a metric on S: readers will have the courage to verify the three axioms. Now we watch for two essential difficulties that will occupy us: perhaps there is no curve that joins p to q or perhaps the distance isn't realized by any curve.

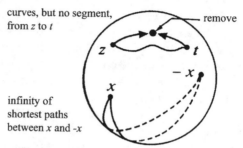

curves, but no segment, from z to t

remove

infinity of shortest paths between x and $-x$

Fig. VI.2.1. [BG] Berger, Gostiaux (1987) © Presses Universitaires de France

If a curve realizing the distance exists, it is a *shortest path* and is given this name, or that of *segment*, in analogy to the Euclidean case. The term *minimizing geodesic* is long out of use, even colloquially; see below. The first difficulty is simple: for obvious reasons it suffices (and is also necessary) that the surface be *connected*. Secondly, the problem of existence of a shortest path is of a whole other difficulty and requires ideas from general topology and study of the calculus of variations, which furnish the *geodesics* that we will see in the next section. In fact, this apparently simple question wasn't correctly posed and resolved until (Hopf and Rinow, 1931), one reason for this tardiness being that general topology only dates from the start of the twentieth century.

We give here the (perfect) answer of Hopf and Rinow: the surface S may or not be *intrinsically complete*, i.e. complete with respect to its intrinsic metric, where completeness means that every Cauchy sequence of points of S has a limit in S. The awaited answer is: *if S is intrinsically complete, then two arbitrary point of S are joined by at least one segment*.

For proofs see the references mentioned, but once this is established we must pay attention to two things. The first is that completeness isn't necessary; e.g. an open disk possesses all desired shortest paths between its points. The second is more subtle: a surface in $\mathbb{E}^3$ can be complete for its intrinsic metric without being complete for the induced metric: roll a plane into an (infinite) cylinder; the rolled-up plane is intrinsically complete, but not complete for the induced metric; see Fig. VI.2.2. Readers can construct a surface rolled-up on the interior of a torus of revolution, thus **bounded** as a subset, but unbounded for the intrinsic metric; in particular this surface isn't compact.

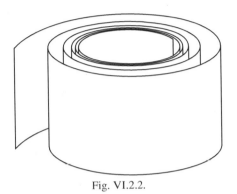

Fig. VI.2.2.

VI.3. The geodesics, the cut locus and the recalcitrant ellipsoids

In order to investigate segments of a general surface, even of revolution, we use methods of the *calculus of variations*, which makes use of the tools of differential calculus as a first step. Let us suppose the problem is solved, that is, let us specify a segment $c : [0, 1] \to S$. We make small variations in it and express that the curve c has minimal length. In particular the derivative of the length must be zero (this is only necessary and certainly not sufficient in general). It seems that the first person to have made this calculation was a Bernoulli — for the Bernoullis were a whole dynasty — specifically Johann (1667–1748), who taught it to his student Euler; see Hauchecorne and Suratteau (1996). This is perhaps the moment to say that this reference is excellent for historical data on the mathematicians whom we mention, and can be complemented by Berger (2005) with its annotated bibliography on the history of mathematics, in France in large part.

The proposition that follows appears for the first time in the writings of Euler (as in Chap. V we have parameterized the curve c by its arc length):

> *In order that the curve c of S, supposed traversed at constant speed, satisfy the condition of zero derivative for each variation, it is necessary and sufficient that the second derivative $c''(t)$ (the acceleration) is normal to the surface (i.e. orthogonal to its tangent plane).*

A *geodesic of a surface* is any curve having this property. In a chart of the surface the equation of the geodesics is a second order (vector) differential equation, but with terms quadratic in the first derivatives. Two things follow: the first is that, for a sufficiently small parameter (which may be considered as *time* or as *distance*),

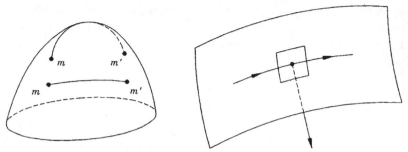

Fig. VI.3.1. Geodesics. The acceleration vector is always normal to the surface. Berger
(1994) © Università di Parma and A. Concari

there exists a unique geodesic emanating from a given point with a given velocity vector. The second is that in general we can't be sure that geodesics exist for
"large" times; we see this clearly on any (incomplete) surface with punctures or
holes.

A surface being given, we want to know graphically if there exist curves for
which the acceleration is constantly perpendicular to the surface and, if so, we want
to describe them as best possible. Intuitively, we are already convinced: for example,
if a portion of the surface is convex, we can stretch an elastic string or rubber band
between two points, which will then realize the shortest path. Physically, a ball set in
motion on the surface will have a well determined trajectory if we know the starting
point and starting velocity (in magnitude and direction).

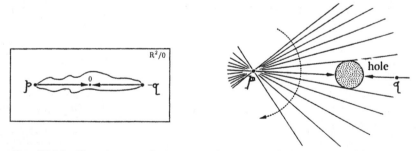

Fig. VI.3.2. Nonexistence of shortest paths. The geodesics are not defined for
"all times"

Another "practical" way of finding these curves, these paths, is to have a small
rudimentary vehicle roll (and only roll) along the surface, a "vehicle" composed
of two parallel and identical wheels fixed rigidly to an axis. The two wheels mark
two curves on the surface that enclose a curve of the type sought, the better the
approximation when the wheels are close together. On a plane, it is a way of going
straight ahead.

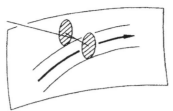

Fig. VI.3.3. A small vehicle, with two parallel wheels very close together, rolling on a surface, gives an approximation to geodesics (better when axis length is small)

Note. As the acceleration is normal to the surface, the speed (a scalar) along a geodesic is constant. From now on we **normalize** the speed by taking it constantly equal to 1.

To return to the problem of extending geodesics indefinitely, we recall that in Hopf and Rinow (1931) it was proved that:

> *There is equivalence between the property that a surface be complete (always for the intrinsic metric) and the property that all geodesics be defined for all time (from $-\infty$ to $+\infty$).*

From now on we will only consider **complete** surfaces, except where the contrary is explicitly stated. There remain at least two things to be done right away. The first is to calculate, when possible, the geodesics for certain surfaces. The second is to say when a geodesic c, restricted to an interval $[a, b]$, is a segment joining $c(a)$ to $c(b)$. It is already clear for the sphere that a geodesic is not in general a segment on all intervals for which it is defined. It is the same for all other compact surfaces and for all their geodesics.

We can say quite a bit about the geodesics for all the surfaces of revolution ; in fact, they can be calculated with the aid of integrals. For if we project a geodesic c onto $p \circ c$ on a fixed plane P perpendicular to the axis of revolution — since the normals to a surface of revolution intersect its axis — the plane curve $p \circ c$ of P (now no longer parameterized by arc length) will have a vector second derivative $(p \circ c)''$ passing constantly through a fixed point. We thus know that, in polar coordinates, this curve (ρ, θ) satisfies $\rho^2 \theta' = $ constant (where θ' denotes the derivative with respect to time). This is the "area law" for motions with central acceleration. This is not sufficient for calculating the geodesics, but already permits us to say lots of things about them. For example, when the meridian is strictly convex, the geodesics oscillate in the band enclosed between two parallels that have the same projection on P. We will encounter this in Chap. XII, but remark right away that, in such a band, there is a dichotomy: either all geodesics close up on themselves (and we can say that they are *periodic*), or else they are all everywhere dense in the band, but never close. We will come back to this question in Sect. XII.3.B.

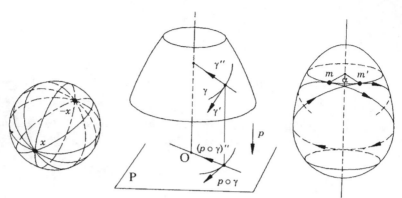

Fig. VI.3.4. *At left*, what happens on a sphere. *At right*, what happens on a surface of revolution (m' is derived from m by a rotation through an angle α). Berger (1994) © Università di Parma and A. Concari

A particular case is that of quadrics of revolution, where we will stay with ellipsoids. We have mentioned that an ellipsoid provides a good approximation of our planet. The integrals that furnish the geodesics are elliptic integrals; see Sect. V.14, and for a recent exposition see Chap. 3 of Klingenberg (1995), but pay particular attention to some of the fine details. Since elliptic functions are well known, tables were made that allowed for numerical calculations for finding shortest paths prior to the availability of modern techniques (e.g. satellites and computers), and that was fundamental in geodesy. We can consult the classic works on elliptic functions, e.g. Appell and Lacour (1922), but that does not permit us to answer the question: when does a geodesic cease to be minimizing? See below for more on this theme which we will take up again in Sect. XII.3.B.

Readers can study the different types of geodesics for a torus of revolution; this was the object of Bliss (1902–1903). It is interesting, for assessing the evolution of mathematics, to observe that what we propose as an exercise was, in 1902, an item of cutting-edge research.

♦

Since Jacobi it has also been known how to calculate geodesics for all the quadrics. Let us again stay with ellipsoids E (not of revolution, sometimes called *triaxial* ellipsoids , where the axes can have different lengths); we use hyperelliptic integrals . More than the calculation, what is interesting is the global behavior of these geodesics. To describe it we need (and this suffices) to use the family of quadrics homofocal to the ellipsoid considered: see Sect. IV.10. These homofocal quadrics slice E in two families of curves which are the *proper curves* of the second fundamental form that we will encounter in Sect. V.6, which means that their tangent is always a proper direction of this second fundamental form; these curves turn out to be (it's a theorem) the lines of curvature. In Fig. VI.3.5 we note four special points of E, called the *umbilics* of E (just two are shown). We will encounter this also in Sect. XII.3. We once again show Fig. IV.10.6 because of the quotation below.

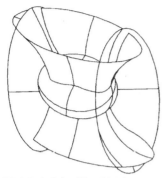
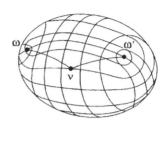

Fig. VI.3.5. *At left*: ellipsoid E intersected by hyperboloids of families homofocal to E, that of hyperboloids of one sheet and that of hyperboloids of two sheets. The traces of the hyperboloids of two sheets are the lines of maximum curvature, those of the hyperboloids of one sheet are the lines of minimum curvature. *At right*: an ellipsoid with three unequal axes has four umbilics, but we see only two of them in the figure, ω and ω'

We take from Porteous (1994, p. 260), these phrases of Monge, who conveyed his enthusiasm for the above figure and for what a room would be like that had an ellipsoidal vault whose tiles followed the lines of curvature:

> *Finally, two chandeliers suspended from the umbilics of the vault, with whose suspension the entire vault seemed to compete, served to illuminate the room during the night.*
> *We won't enter into much greater detail in this regard; for us it suffices to have indicated to artists a simple object whose decoration, although very opulent, could have nothing arbitrary, since it consisted principally in unveiling for all eyes a very gracious arrangement which is in the very nature of this object.*

Just as the geodesics of surfaces of revolution oscillate between two suitable parallels, so the geodesics here oscillate between pairs of associated lines of curvature when they are intersected by the same homofocal quadric ; the latter are of two types, according as they correspond to hyperboloids of one sheet or of two sheets . The case of geodesics that emanate from the umbilics is exceptional: they all pass through the "antipodal" umbilic , but we should pay attention to the fact that when they return to the point of departure, it is along a different tangent, and they are never periodic, with the exception of the plane section that contains the four umbilics. Thus the umbilics aren't like the poles of a surface of revolution. But, just as for the surfaces of revolution, a band of geodesics between associated lines of curvature — depending on the band — are either all periodic or else never close and are then everywhere dense in the band. A final property of the geodesics of the ellipsoid: the lines tangent to a chosen geodesic are all tangent to the same homofocal quadric; we will return to this in Sect. XII.3.C.

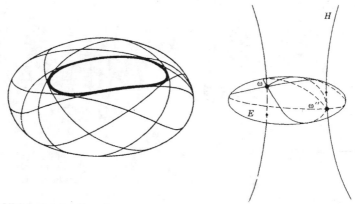

Fig. VI.3.6. *At left*: a line of curvature realized as the envelope of a family of geodesics (the rest of the envelope is symmetric and hidden). *At right*: geodesics emanating from an umbilic

◆

We can now attack the "global" problem of the shortest path, more precisely the problem of possible non uniqueness. To the best of our knowledge its study goes back to Poincaré (1905). It is not very difficult to establish what follows. Let c : $[0, \infty[\to S$ be a geodesic emanating from a point $c(0) = p$ and with initial velocity vector $c'(0) = v$: either it is minimizing up to infinity, which means that for each $T > 0$ the path $c : [0, T] \to S$ is a shortest path between $c(0) = p$ and $c(T)$, or else there exists a positive value beyond which c is no longer minimizing. This value is *denoted* cut(c) (it can also be denoted cut(v) since v and c determine one another) and is called the *cut value*.

Fig. VI.3.7. The two possibilities for a cut point: two shortest paths or/and conjugate points

What makes the subject difficult is that at cut(c) two situations can arise, possibly simultaneously: either a second geodesic d arrives at $q = c(\text{cut}(v))$, minimizing and distinct from c, or else the point q is *conjugate* to p *on* c; by definition, this means that the geodesics c_α in a neighborhood of $c = c_0$ and emanating from the chosen point p envelop a curve tangent to c at q. In other words, the curve $\alpha \mapsto c_\alpha(\mathrm{cut}(v))$ has velocity zero at $\alpha = 0$. In general the figure may show a cusp, and things may even get bad in the C^∞ case and even in the case of the sphere where all the geodesics pass through the same antipodal point.

We also need to know that if $c(t)$ is conjugate to p on c, then c isn't minimizing after t, it ceases to be minimizing **before** t or no later than at t. Whence our interest in knowing the envelope of the geodesics that emanate from p. Finally we call *cut locus* of a point p and denote by $\mathrm{cutloc}(p)$ the set of points $c(\mathrm{cut}(c))$, where c runs through all geodesics emanating from p. For standard spheres, the cut locus of each point reduces to a single antipodal point; the situation is thus very special: there is always a unique shortest path except for antipodal points, at which we may depart in any direction and always arrive at the antipode, with same distance for all starting directions.

◆

The structure of the cut locus was studied by Poincaré in the real analytic case, *denoted C^ω*; see Sect. V.XYZ for these notations. This study was completed only in Myers (1935a, 1935b) and Myers (1936). In the C^ω case the cut locus of a point is always a graph whose free extremities correspond to conjugate points; at each terminal point of this graph there occur exactly two minimizing geodesics and, at a vertex of the graph of order k, there occur exactly k.

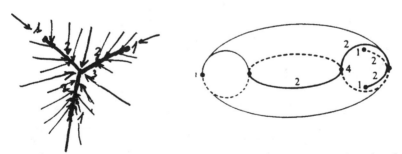

Fig. VI.3.8. *At left*: cut locus with a triple point; *At right*: the cut locus of a point of the exterior meridian of a torus of revolution; the number of minimizing geodesics reaching a point of the cut locus is indicated

In cases that are just C^∞ the cut locus can be pathological, even fractal. Its structure has recently been clarified in Hebda (1994). In contrast, it is easy to see that the cut locus is always the closure of the set of points reached by at least two distinct minimizing geodesics.

What about the cut locus for our favorite surfaces? One motivation is the following: even though the geodesics contain the segments (the minimizing curves), and even if we know their equation, we are told nothing about the cut locus values. Now it is important to know the cut loci if we want to completely understand the intrinsic metric structure of a surface, in particular for the ellipsoid of revolution of our planet. It is in fact disconcerting to have two shortest paths for traveling from one point to another: which should we choose? It is even a philosophical problem: Aristotle, Buridan and Spinoza were interested, from this point of view, in the difference between humans and animals.

It is astonishing to confirm that the only two cases, to our knowledge, where the cut locus of all points is known are very particular surfaces of revolution (e.g. the sphere and the torus; see Fig. VI.3.8) and the Zoll surfaces that we will encounter in Chap. XII, because they have the property that their geodesics are all periodic, although they are not round spheres.

For the sphere, the cut locus of each point is reduced to its antipodal point. For certain Zoll surfaces (see Sect. XII.3.B), there is a description in Besse (1978) of the structure of the cut locus, also a proof although rather dense. Except for the two cases mentioned, the cut locus is completely unknown for all other surfaces.

We can't silently pass over the antipodal property of the sphere. It's an old problem of Blaschke from 1927, when he conjectured that only the sphere has this property. This was proved only in 1963, by Leon Green. We find his proof, among others and the study of what happens in higher dimensions, in Besse (1978), which is entirely devoted to this problem and its generalizations. The proof requires at least two rungs up the ladder: the geometry of phase space, specifically the *unitary fiber*, i.e. the set formed of all the unit vectors tangent to the surface in question. The fiber of the geodesic flow of the surface considered, a manifold of dimension $3 = 2 + 1$, will be the object of Chap. XII. There we will need to apply Liouville's theorem to it; see Sect. XII.6.A.

Returning to ellipsoids: for ellipsoids of revolution, the cut locus of a pole consists just of its antipode; if there are three axes, the cut locus of an umbilic is reduced to the antipodal umbilic. And for the other points? It is in fact asserted in Braunmühl (1882) that the cut locus of ellipsoids, at each point except at umbilics, has the topology of an **interval**. This doesn't tell us where it is, but at least we would know that there aren't any triple points or worse. Unfortunately, Braunmühl's text relies on an assertion of Jacobi relating to ellipsoids: the envelope of the geodesics emanating from a non umbilical point possesses exactly four cusps (see the reference to Arnold just below). If that is true, then we have essentially intervals, for the results quoted earlier show that two of the cusps belong to the cut locus; the two others, which correspond to local maxima for the distance along the geodesic to the point of contact of the envelope, don't intervene. There thus remains an interval whose two extremities are the two cusps, that are themselves local minima. For the claim of Jacobi from 1864, already mentioned at the end of Sect. V.10, and for its modern context, see Arnold (1994, §13, p. 39).

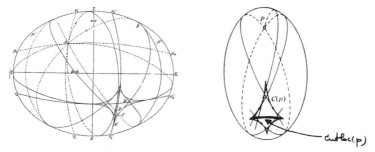

Fig. VI.3.9. An old sketch of Braunmühl (1882). A conjecture due in part to Jacobi.
Braunmühl (1882) © Springer

Fig. VI.3.10. A figure which seems evident, but remains conjectural. Braunmühl (1882) © Springer

But it isn't known how to prove Jacobi's conjecture even for ellipses of revolution, even for those that are close to being a sphere, and again even for the points of their equator. The fact of knowing the equations of the geodesics isn't sufficient in the present state of our knowledge. To realize the difficulty of the problem, which is in fact a problem of algebraic geometry, think of an ellipsoid of revolution that is very elongated vertically, and fix a point on the equator; see Fig. VI.3.11. As the curvature (see below) along the equator is close to zero, the geodesics emanating from this point and near the equator will have an envelope that turns a large number of times, which spirals about the equator. It is surely reasonable to conjecture in this case precisely that the cut locus is the part of the meridian composed of points that are between the two conjugate points which are above. More generally, it is conjectured that the cut locus is always contained in a line of curvature, and this is supported by computer studies. See Itoh and Kiyohara (2004) and Sinclair (2002).

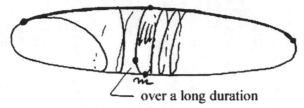

over a long duration

Fig. VI.3.11. It is difficult to know what happens for geodesics that turn very slowly

We can also realize the difficulty as follows. Take a sphere and deform it only very little. At the outset the geodesics emanating from a given point all pass through the antipode; topologically, we need to know what happens when we deform the family of lines passing through a point of a plane. We find in Arnold (1994) the following result: the families of nearby lines have an envelope with at least four cusps. But this doesn't say that there are only four; in fact it is easy to construct families of lines possessing an arbitrary number ≥ 4 of cusps.

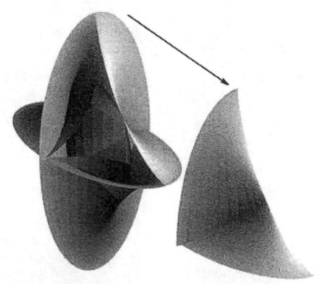

Fig. VI.3.12. Cayley's astroid, envelope of the normals to a triaxial ellipsoid of $\mathbb{R}^3$. An octant has been cut out and translated for a better comprehension of the surface. Joets, Ribotta (1999) © A. K. Peters Ltd.

Although not directly related to our problem, below is a picture of *Cayley's astroid* (the envelope of the normals to an ellipsoid), a surface studied by Cayley in the middle of the nineteenth century, taken from Joets and Ribotta (1999), where it is studied for the case of dimension four; see also Banchoff (1984). On this topic, we don't know any reference that treats the problem of intrinsic geometry of ellipsoids of dimension greater than two. For example: what are the properties that generalize what was encountered above with the umbilics in dimension 2? We will return to envelopes of normals in Sect. VI.10.

VI.4. An indispensable abstract concept: Riemannian surfaces

The following section studies the connection between abstract surfaces and surfaces of $\mathbb{R}^3$. It is scarcely possible to understand it well without the fundamental idea — introduced by Riemann in 1852 — of Riemannian surface (in fact, Riemann placed his definition in arbitrary dimension). The starting remark is that, to define the intrinsic metric of a surface $S \subset \mathbb{E}^3$, we use the Euclidean norm of $\mathbb{E}^3$ only for the tangent vectors to S. Whence the idea of Riemann: to define a geometry on an abstract manifold M of dimension two (differential surface, see Sect. V.XYZ) by placing at each point the differential $(dx_1, ..., dx_n)$ of an arbitrary Euclidean structure $\sum_{i,j} g_{i,j} dx^i dx^j$, requiring only that this structure depends continuously, or differentiably, on the point describing the surface. Such an object is called a *Riemannian surface*; if we want to specify the Riemannian metric, we denote this by (M, g). We will encounter constructions of such abstract surfaces — almost never realizable in $\mathbb{E}^3$ — for the objects called "space forms" that we will see in Sect. VI.XYZ.

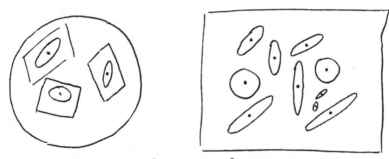

Fig. VI.4.1. An abstract ds^2 on the sphere S^2; a field of ellipses in the plane

The above sketches are given in $\mathbb{E}^3$, but this is only to remedy our difficulty in seeing "abstractly". We need to exert ourselves to think, to "see", abstractly. In the case of a surface that is "already" in $\mathbb{E}^3$, this "abstract" structure is simply — on each tangent space for this surface — the Euclidean structure induced by the embedding of the tangent plane in the space. In the abstract case, the embedded case occurs by taking for the definition of *length of a curve* $c : [a,b] \to M$ the quantity $\text{length}(c) = \int_a^b \|c'(t)\| \, dt$. In a chart $\{x, y\}$ of M (see Sect. V.XYZ) we define the Riemannian metric by $g = a \, dx^2 + 2b \, dx dy + c \, dy^2$, this expression signifying that for a tangent vector $\mathbf{v}$ of associated coordinates $\{u, v\}$ in this chart (see Sect. V.XYZ) we have $\|\mathbf{v}\|^2 = au^2 + 2buv + cv^2$. Many books use the notation $ds^2 = g$ to indicate the fact that g provides the infinitesimal arc lengths of curves (see Sect. V.6). An essential historical example of an abstract Riemannian surface is the hyperbolic plane (see Sects. II.6 and II.XYZ). It is defined on the open disk $x^2 + y^2 < 1$ by:

$$ds^2 = \frac{4}{\left(1 - (x^2 + y^2)\right)^2}(dx^2 + dy^2).$$

And it is thus that in 1854 Riemann defined hyperbolic geometry for the first time with absolute correctness and, moreover, in all dimensions, where the formula is essentially the same. The previous partial results used Beltrami's surface, which has the major disadvantage of being doubly incomplete: on the one hand we get only part of the hyperbolic plane; on the other it is necessary to truncate the surface or to cover it while turning about, which is an abstract representation. We will see right away in Sect. VI.5 that a theorem of Hilbert shows that we can never realize the whole hyperbolic plane as a surface of $\mathbb{E}^3$.

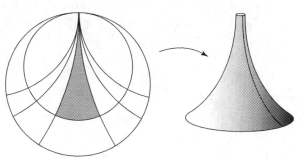

Fig. VI.4.2. The Beltrami surface, a doubly incomplete representation of the hyperbolic plane. There is a singularity at the boundary that can't be extended; see also Fig. VI.4.4, but above all the general result of Hilbert on p. 413

In this abstract setting the shortest path problem is distinctly more difficult. We proceed as above by the calculus of variations and thus define the notion of *geodesic*, but here we find only a differential equation whose solutions we can't interpret as curves with acceleration normal to the surface. It wasn't until the turn of the XXth Century that matters were first well understood, by Ricci and Levi-Civita; see for example Berger (2001) or Berger (2003). But if we want to be global we have to wait for the result of Hopf-Rinow, equally valid in the abstract setting of Riemannian surfaces. Thus finally we have geodesics: we have a proper metric space structure and the geodesics extend indefinitely if this metric is complete. Note what is not well explained in most works: the fact that the axiom $d(p,q) = 0 \Rightarrow p = q$ requires, in the definition of an abstract surface, that the topology is *separated*. We could talk about cut locus again, but we won't have need of it. An *isometry* between two Riemannian surfaces (M, g) and (N, h) is a diffeomorphism $f : M \to N$ that preserves their metrics, which is equivalent to saying that the differential $df : TM \to TN$ preserves the corresponding norms of the tangent vectors.

♦

The Gaussian curvature K. There's hardly any mathematics without invariants! We're indebted to Gauss for having found the fundamental invariant of Riemannian surfaces (in Sect. VI.6 we will return to the history and its context). We call this invariant *Gaussian curvature* — or simply *the curvature* — and *denote* it by K. We will also say occasionally *total curvature*, but that can lead to confusion with the integral formulas such as (VI.6.3) and (VI.7.1). This function $K : M \to \mathbb{R}$ hasn't a simple direct expression for abstract surfaces; if it is calculated in a chart in terms of the g_{ij}, we find a really complex expression involving the first and second derivatives of the g_{ij}.

In contrast, its geometric interpretation is simple: the curvature measures the infinitesimal rate of divergence (or convergence) of the geodesics. That will become very clear in Sect. VI.6, but we need to see it here in an abstract way. Precisely, here is an equation due to Gauss: let $c_\alpha(t)$ be a one-parameter family α of geodesics c_α defined in the neighborhood of an initial geodesic c_0 such that the curves $\alpha \mapsto c_\alpha(t)$ are orthogonal to c_α. The field of vectors tangent along the extent of c_0 is the field $\frac{\partial(c_\alpha(t))}{\partial \alpha}$. Then if we put $Y(t) = \|\frac{\partial(c_\alpha(t))}{\partial \alpha}\|$, this norm satisfies the second order differential equation $Y''(t) + K(c(t)) \cdot Y(t) = 0$.

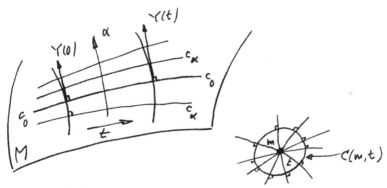

Fig. VI.4.3. Finite diagram of a one-parameter family of geodesics

A particular case consists of taking the one-parameter family α of geodesics c_α emanating from a given point $m \in M$, where α traverses the unit circle of $T_m M$. For each t, the curve $C(m; t) : \alpha \mapsto c_\alpha(t)$ can be called the *circle* of center m and of radius t. Then the equation of Gauss yields for the norm $\|\frac{d(c_\alpha(t))}{d\alpha}\|$ the partial development in t:

$$\left\| \frac{d(c_\alpha(t))}{d\alpha} \right\| = t - \frac{t^3}{3} K(m) + o(t^3).$$

First, we observe that this development does not contain a t^2 term (a Riemannian surface is Euclidean up to order two); next it does not depend on α but only on m. Thus the total length of circles of small radius equals: $\text{length}(C(m;t)) = 2\pi t - \frac{t^3}{3} K(m) + o(t^3)$. We can say that the metrics with $K < 0$ are *super-Euclidean* and those with $K > 0$ *sub-Euclidean*.

◆

Whatever the importance of the curvature, it is **not** in general a characteristic invariant: a diffeomorphism $f : M \to N$ can preserve the curvatures K_M, K_N without

being an isometry (see the definition at the beginning of the next section); it suffices
to take any two small open subsets of the surfaces M and N and take for f any
mapping that preserves the level curves of K. Few works mention this important
point; see however Spivak (1970, pp. 310, 311 of Vol. II). Still less do they treat
the problem of knowing whether it is possible to find characteristic invariants for a
Riemann surface. This problem was attacked very early by Darboux (1887, 1889,
1894, 1896) (or Darboux (1972)). We extract what follows from Chap. XIII of Car-
tan (1946–1951): in order that two surfaces for which the functions K and $\|d\,\mathrm{K}\|^2$
are independent be isometric by a mapping f, it suffices that the mapping preserve
the four functions K, $\|d\,\mathrm{K}\|^2$, $\langle\mathrm{K}, \|d\,\mathrm{K}\|^2\rangle$, $\|(\|d\,\mathrm{K}\|^2)\|^2$. If there is a relation of the
type $F(\mathrm{K}, \|d\,\mathrm{K}\|^2) = 0$, then it suffices that f preserve $\|d\,\mathrm{K}\|^2$ and the Laplacian
$\Delta\mathrm{K}$. Moreover, the surfaces of this type are diffeomorphic to surfaces of revolution.

The notation df for a function on a manifold designates its *differential*, which
means the linear form on the tangent spaces which associates $df(x) = x(f)$ with
each tangent vector x, i.e. it is the derivative of f for x. For the geometer this is the
ordinary derivative $\frac{d(f(c(t)))}{dt}(0)$, where c is any curve whose velocity at the origin is
the vector x. In coordinates, $df = \left(\frac{\partial f}{\partial x_1}, \frac{\partial f}{\partial x_2}\right)$. For a function f on a Riemann surface, by duality the scalar product allows transformation of the differential df into
a vector (in fact a field of vectors) called the *gradient/* of f and denoted $\mathrm{grad}(f)$.
The gradient is thus defined by the relation: $df(x) = \langle\mathrm{grad}(f), x\rangle$ for every vec-
tor x. We thus set $\|df\| = \|\mathrm{grad}(f)\|$. Finally, the *Laplacian* Δf of a function on
a (Riemannian) surface is the operator that generalizes the Laplacian in the plane
defined in Sect. IX.4.

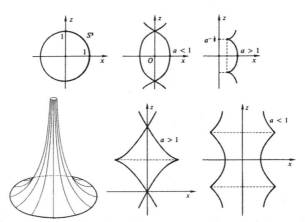

Fig. VI.4.4. The different types of meridian defining a surface of revolution of constant
curvature; only one the six types yields a true surface (i.e. without singularities). [BG]
Berger, Gostiaux (1987) © Presses Universitaires de France

If an (M, g) has **constant** curvature K, say $\mathrm{K} = k$, then its metric is –
in contrast to the variable curvature case – completely known and completely

determined by k, at least locally. This can be seen from the Gauss differential equation $Y''(t) + K(c(t)) \cdot Y(t) = 0$: if this constant k is zero, the metric is everywhere locally Euclidean, and we can thus say once again that the curvature measures the deficiency of being Euclidean. If the constant k is positive, then the metric is everywhere locally isometric to that of the sphere of radius $\frac{1}{\sqrt{k}}$. If $k < 0$ then the metric is everywhere locally isometric to a geometry of the hyperbolic plane (see Sects. II.5 and II.XYZ) appropriately normalized, the case $K = -1$ being just that of the hyperbolic geometry exhibited in Sects. II.5 and II.XYZ. Readers must then ask themselves: what about the typical global case if M is compact? It's a fundamental question of geometry called the *space form problem* which is not on our agenda; see ideas and references in Sect. VI.XYZ. It is in fact a particular case of the problem that follows. Before continuing, we mention that the determination of the surfaces of revolution of constant curvature is easy; see Sect. 3.9 of Klingenberg (1973) or Exercise 7 of Chap. 3 of do Carmo (1976).

In fact, the burning question is now this: We possess on the one hand the surfaces of $\mathbb{E}^3$ which are automatically Riemannian, and the abstract Riemannian surfaces on the other. What is the relationship between these two categories of objects?

VI.5. Problems of isometries: abstract surfaces versus surfaces of $\mathbb{E}^3$

We now consider two surfaces S and S', abstract or in $\mathbb{E}^3$: an *isometry* between S and S' is a bijective mapping $f : S \rightarrow S'$ that preserves their intrinsic metrics, that is, $d_S(p,q) = d_{S'}(f(p), f(q))$ for all p and q in S. For a long time it was not realized (or was considered obvious) that this condition implies the differentiability of the mapping f, which is at least of class C^1; see Calabi and Hartman (1970) for a very precise position on these problems, with their pitfalls. So it is good to note that the same definition of the intrinsic metric permits us to give the equivalent definition: an isometry $f : S \rightarrow S'$ is a bijective mapping, of class C^1 at least, that preserves the norms of tangent vectors. For a surface S, we let $T_m S$ *denote* the vector space (a plane) of vectors tangent to S at the point $m \in S$. Then the condition for isometry is $\| f'(v) \| = \| v \|$ for each vector v tangent to S.

To see this more clearly, we note the evidence: when we are in $\mathbb{E}^3$ each tangent plane to a surface inherits a planar Euclidean structure, i.e. a positive definite quadratic form. We need thus to see that the intrinsic metric of a surface S is constructed starting with the set of quadratic forms on the set of $T_m S$; we have seen this at the beginning of Sect. VI.4. It's the set of these quadratic forms that are called the *first fundamental form of the surface*. An isometry is a differentiable mapping that preserves the first fundamental form.

As general references on the problems of this section, consult: Gromov (1986), Vol. V of Spivak (1970), Burago and Zalgaller (1992), and various parts of Burago and Zalgaller (1992), which attaches great importance — typical of the Russian school for this area — to low order differentiability.

♦

The important question of whether the abstract surfaces are more general than those of $\mathbb{E}^3$ is not yet settled. There is the *local* problem and the *global* problem. In the local case we only want to know if, given a point m of the Riemannian surface (M, g), we can find a neighborhood U (perhaps very small) of m and an isometry between (U, g_U) (the restriction of g to U) and a surface of $\mathbb{E}^3$, i.e. a mapping $f : U \to \mathbb{E}^3$ such that $f(U)$ is a surface of $\mathbb{E}^3$ and f an isometry between (U, g_U) and $f(U)$ endowed with the induced metric. In the global case we want an isometry between all M and $f(M)$. We say, in brief, that (M, g) is *realizable in $\mathbb{E}^3$ locally (resp. globally).*

♦

The question has not been settled locally or globally. Locally we know this: if the metric g of M is real analytic (see Sect. V.XYZ) then it is locally realizable: for each point and for an appropriate neighborhood we have a local isometry (Cartan-Janet, 1926). Beyond this, without analyticity, we have a local realization at each point where K is nonzero; the solution uses standard existence theorems for partial differential equations which are obviated once K vanishes at a point. Subsequently more and more partial refinements have been found, e.g. for the case where the differential dK isn't zero. In 1971 Pogorelov found an example of a Riemannian surface (U, g) pointed at m such that no neighborhood of m, however small, can be realized in $\mathbb{E}^3$; this metric is evidently very strange, e.g. the points where the curvature is zero form a dense subset. Above all it is only of class C^2, and the problem is today totally open whether, starting with C^3, or even with C^∞, a local embedding is always possible. So we know today that, even locally, abstract Riemannian surfaces are more general, but a complete result is still awaited. In particular, the regularity of g plays a role that is not well understood, except for the class C^1; see below. The metric of class C^2 of Pogorelov has a curvature K that is Lipschitz (see Sect. V.XYZ). If we want to realize every surface, be it locally or globally, we must use an embedding in $\mathbb{E}^N$ with at least $N \geq 5$; see the references cited.

♦

Globally we certainly cannot hope to realize every compact abstract Riemannian surface (M, g) in $\mathbb{E}^3$, for the simple reason that there exist compact abstract surfaces with negative curvature and that each compact surface of $\mathbb{E}^3$ has points at which the curvature is strictly positive. To see this, take any point not on the surface and a sphere about this point that contains the surface but has smallest possible radius. It exists by compactness and is necessarily tangent to the surface at some point; at such a point of contact the curvature of the surface is greater than that of the sphere, thus positive.

But in return there exists a very satisfying — although very difficult — result that bears the names of Weyl, Nirenberg, Pogorelov, Alexandrov: let (S^2, g) be a structure of an abstract Riemannian surface on the topological sphere S^2 that is

C^∞ (resp. analytic) and such that the curvature K is everywhere strictly positive. Then it is isometrically embeddable by a C^∞ (resp. analytic) mapping into E^3. The proof cannot be done "easily"; we will in fact see in Sect. VI.9 that the surface obtained in $\mathbb{E}^3$ is unique (within a global isometry of $\mathbb{E}^3$). We only mention that Pogorelov's proof from 1969 uses the space of all Riemannian metrics on S^2 by showing that, in this space, the subset of metrics with everywhere positive curvature is connected. But we always need very subtle results on partial differential equations; see Sect. VIII.6. The case $K \geqslant 0$ is broached in Guan and Li (1994), but only an embedding of class $C^{1,1}$ is provided. The proof is achieved by studying the limit of embeddings associated earlier with metrics with strictly positive curvature that approach the given metric. See also the end of Sect. VIII.6 for a schematic key for this type of proof.

It wasn't until Gromov in 1986 and subsequent elaboration by Labourie in 1989 that the existence and uniqueness could be put in the same theoretical setting that goes to the bottom of things; see Sect. VI.7 for the setting and for references.

We mention in passing, although it is a bit off subject, a little known result of Gluck (1972) that states, for the surfaces of $\mathbb{E}^3$ having the topology of the sphere S^2, that there is no four vertex theorem (cf. Sect. V.8). Gluck proved that, for every function on the abstract sphere S^2, we can always find a sphere S^2 embedded in $\mathbb{E}^3$ for which the Gaussian curvature is this function. In particular it can very well have but a single maximum and a single minimum.

Instead of taking the curvature as a point function, we can take it as a function of the direction of the tangent plane; this is *Minkowski's problem*. The direction of the tangent plane is usually given as points of the unit sphere corresponding to unit vectors orthogonal to these tangent planes, and thus a function is specified on the sphere. This problem is today completely understood; see the brief mention in Sect. VI.7.

Which embeddings can we expect for abstract Riemannian surfaces, not necessarily compact? A typical case is that of the hyperbolic plane $\text{Hyp}^2 = (\mathbb{R}^2, g = \text{hyp})$, a geometry encountered in Sect. II.4. Can we realize this geometry in $\mathbb{E}^3$? A result of Hilbert in 1901 states that there isn't any global realization of Hyp^2 in $\mathbb{E}^3$; locally the realization is very easy with pieces of surfaces of revolution (those of Beltrami or others), but this cannot be done globally. The proof consists of following a geodesic and showing that a singularity necessarily develops in the embedding at the end of a certain distance. Hilbert's result was refined by Efimov in 1964: there is always impossibility for realizing a complete abstract surface $(\mathbb{R}^2, g)$ as soon as its curvature is bounded above by a negative constant: $K \leqslant \text{constant} < 0$. The proof is very complicated — it uses dimension two so as to make use of functions of a

complex variable; see Sect. VI.XYZ. Moreover, for low order differentiability enthusiasts — even though it is not in the spirit of this book — the result is even valid
in the C^2 setting, but then we encounter the difficulty connected with the *theorema
egregium*; see the next section. It is already difficult to show that the area is infinite;
but it is evidently of class C^3 in Gauss's equation $Y''(t) + K(c(t)) \cdot Y(t) = 0$. See
the excellent introduction in Klotz-Milnor (1972), in all of which readers will find
interest.

The case C^1. All the preceding extends to embeddings of class at least C^2, if only
for defining the curvature. But if we are content with embeddings only of class
C^1, then the so-called Nash-Kuiper theorem (see Bleecker, 1997) allows an isometric embedding in $\mathbb{E}^3$ of every abstract Riemannian surface, and with lots of
choice; the choice is in fact enormous. Once a compact Riemannian surface is embedded in $\mathbb{E}^3$ without additional conditions, in a *contraction* manner, i.e. so that the
embedding considered reduces distances, then there exist isometric embeddings of
class C^1 which are arbitrarily close in the class C^0 to the given embedding. For
example, there exist lots of surfaces of $\mathbb{E}^3$ which aren't round spheres, but for
which the intrinsic metric is nonetheless isometric to that of a round sphere (this
in strong contrast with the uniqueness provided by a theorem of Cohn-Vossen that
we will study in Sect. VI.9, where the smoothness condition requires that the embedding be C^2). To see this, take a sphere with radius just a bit smaller; this trivially
yields a contraction embedding. The proof of the Nash-Kuiper result Nash, John is
hard: it's a mixture of the theory of underdetermined partial differential equations
and geometry. Roughly speaking, the desired metrics are constructed by very in-
tense foldings and crumplings around a contraction embedding. The totally novel
theory of Nash from 1956 is in fact very well understood thanks to Gromov; see
the h-principle in his book Gromov (1986).

VI.6. Local shape of surfaces: the second fundamental form, total curvature and mean curvature, their geometric interpretation, the *theorema egregium*, the manufacture of precise balls

It is finally time to occupy ourselves with the surfaces in $\mathbb{E}^3$, no longer just
with their intrinsic metric but also with the way in which they are situated in space.
We will say the minimum needed for grasping certain important results and open
problems. This situation is described perfectly by the *second fundamental form* and
in particular by the two invariants called *total curvature* and *mean curvature*. In
brief, we look at how the surface differs from a plane, more precisely how it differs
from its tangent plane. For this it suffices to know the shape of the plane curves that
are the sections of the surface by planes containing the normal to the surface. Now
for us, after Sect. V.6, it is the curvature of these curves that we need study. The
result is quite simple — and quite old:

(VI.6.1) *When the plane turns about the normal at a given point m the algebraic
curvature of the section curve is a quadratic form.*

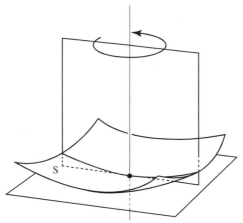

Fig. VI.6.1. Section by a normal plane

From the point of view of calculus, II involves the first two derivatives of the equation of the surface, for example as the local graph of a function $\mathbb{R}^2 \to \mathbb{R}$. The first fundamental form by itself requires only first derivatives.

In an orthonormal system having as origin a point m of the surface S, where the first two vectors are in the tangent plane at m and where the third is directed toward the normal, in the neighborhood of m the surface has an equation of the form

$$z = f(x, y) = \frac{1}{2}ax^2 + bxy + \frac{1}{2}cy^2 + \text{terms of higher order.}$$

Readers should know that the curvature of the plane curve $z = \frac{1}{2}ax^2$ is equal to a. For the same reason, the curvature of the section of S by the plane containing the normal and the vector $(\cos\theta, \sin\theta)$ equals

$$k(\theta) = a\cos^2\theta + 2b\cos\theta\sin\theta + c\sin^2\theta.$$

In other words, it is the value taken by the quadratic form $ax^2 + 2bxy + cy^2$ for the vector $(\cos\theta, \sin\theta)$.

This quadratic form can be defined in an intrinsic fashion as a quadratic form on $T_m S$, and is called the *second fundamental quadratic form*, and *denoted* II.

The *first fundamental quadratic form, denoted* I, is simply the metric on $T_m S$ $(x^2 + y^2)$ in the coordinates used above. We note that the sign of the second fundamental form depends on a choice of orientation on S (or on a preferred direction for the exterior normal).

As mentioned above, from the point of view of calculus the second fundamental form involves the first and second derivatives of the equation of the surface (for example, in the expression for S above, a, b, c are the second derivatives of f at the origin). The first fundamental form, on the other hand, requires only first derivatives and allows us to pass from the initial coordinate system to a system connected to the tangent plane and the normal.

Now that on the tangent space T_mS we dispose over the two quadratic forms II (the second fundamental form) and I (the Euclidean structure), we can diagonalize II with respect to I, i.e. find an orthonormal system of T_mS where II takes the form $k_1x^2 + k_2y^2$. The eigenvalues k_1 and k_2 are called *principal curvatures*, the eigen-directions the *principal directions* (they are of course orthogonal). It should be noted that k_1 and k_2 are the minimum and maximum values of the curvature $k(\theta)$ of the sections of S by the plane turning about the normal.

The trace and the determinant of II with respect to I are two essential invariants:

(VI.6.2) $K(ext) = k_1k_2$ is called the total curvature *despite the danger of confusing it with the integrals of this curvature;* $H = \frac{1}{2}(k_1 + k_2)$ *is called the* mean curvature.

The notation $K(ext)$ reminds us that the definition uses information exterior to S, as opposed to K, which ultimately uses distances calculated within S. Notice that $K(ext)$ does not depend on the orientation of the normal since it is a product, whereas the sign of H changes with orientation.

The points where $K(ext) > 0$ are called *elliptic*, those where $K(ext) < 0$ *hyperbolic*, those where $K(ext) = 0$ are called *parabolic* and those where $k_1 = k_2 = 0$ *planar*; moreover those where $k_1 = k_2$ are called *umbilics*. The principal directions are only well determined away from umbilics, where the principal curvatures are distinct (i.e. where II isn't a multiple of I). The umbilics are in general isolated points; apart from these points, we thus dispose over two direction fields of tangent lines, those in the direction of maximum curvature and those in the direction of minimum curvature. These direction fields are integrated into two families of curves of S, called *lines of curvature*, which we will find many times in the sequel.

It may seem surprising that the condition $k_1 = k_2$ in general defines isolated points, and not a curve on S. The explanation is that in the three dimensional space of *real* quadratic forms $ax^2 + 2bxy + cy^2$, or in that of the real symmetric matrices $\left(\begin{smallmatrix} a & b \\ b & c \end{smallmatrix}\right)$, the condition of equality of the two eigenvalues in fact corresponds to *two* independent conditions: $a = c, b = 0$. This is a remark of which Arnold is fond; see Arnold (1974).

The exemplary case where the second fundamental form can be found without any calculation is that of surfaces of revolution. In fact, each normal to such a surface intersects the axis of revolution and, by symmetry, we see that the two principal directions are those of the meridians and parallels, which are in fact the lines of curvature. We obtain immediately the principal curvatures and their reciprocals, the *radii of principal curvature*, the one being the length of the normal between the point and the axis, the other the radius of curvature of the meridian at that point.

♦

Here is a simple point, but of great practical importance, which is in the spirit of industrial fabrication (see Sects. II.2, V.7 and VII.13.C) and for which the proof is left to readers: if all points of a surface are umbilics, then the surface is necessarily a portion of a sphere, a local result; as needed, see any book on differential

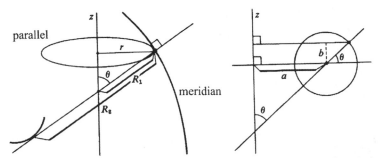

Fig. VI.6.2. Finding the radii of principal curvature is trivial for a surface of revolution

geometry, e.g. p. 147 of do Carmo (1976). It follows from this that if we rub two surface portions together (with the classic abrasive emery powder, or any other abrasive, between the two) in all directions, they will finally slide against one another when both are spherical. In fact, we can say for example that they are surfaces of revolution in more than one way, in at least two different directions for the axes of revolution, which can happen only for spheres; or again that all points are umbilics since in the rubbing process we turn in every direction about each point. Group theory enthusiasts can see here also the fact that the only groups of isometries of the Euclidean space $\mathbb{R}^3$ having at least two parameters, all finite, are those formed by rotations about a fixed point.

The reader should know that there doesn't exist at present any other way of fabricating spherical surfaces, industrially or even by high precision mechanics, metrology, etc, for example for the lenses of ordinary glasses. The case of astigmatic lenses and so-called progressive lenses are a totally different story. For astigmatic lenses, glass tori of revolution are used. As for progressive lenses, only imperfect approximations to a good optic on a very high field depth can be obtained. Progressive lens technology is improving daily and the story is not yet complete. In high precision work, the surfaces obtained must be studied by interferometry and corrected by appropriate manual abrasion; this is also done for parabolic mirrors for telescopes. The artisans who do this work have "micrometers in their hands". Nevertheless, computer aided design (CAD) and robots have recently begun to be used for managing the corrections in high precision work, e.g. for mirrors of large telescopes.

The manufacture of spherical balls is still a completely different story, simpler in a certain sense. An old method consisted of placing approximately round balls in a receptacle and intermingling them sufficiently long, using abrasives. We will see in Sect. VII.13.C why this intermingling yields true balls (approximate in practice, perfect in theory). But since 1907 a patented process of the SKF corporation has allowed quick and accurate manufacture of balls, with precision within a micron, using the above property with the following device. It begins with two plates in which grooves are cut in the form of a half tori, the difference in diameters corresponding to the diameter required for the balls. These two plates are placed against one another and turned. A suitable abrasive and some approximate balls are injected into each of the toroidal tunnels; at each turn the balls are ejected and reintroduced

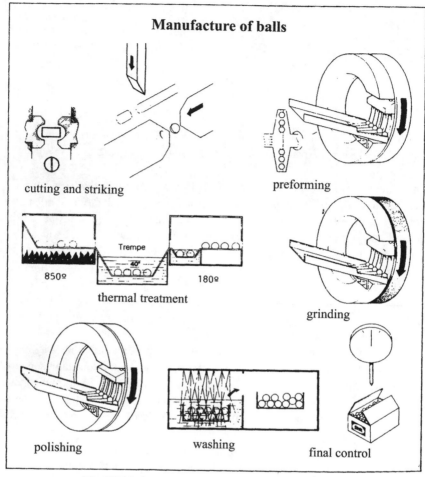

Fig. VI.6.3. Claude, Devanture (1988) © Nathan Édition

in random fashion into another "tunnel". After three machinings in a row of this type with appropriate abrasives we extremely quickly obtain balls that are almost perfect. Professionals themselves are ever astonished by the simplicity of the process and the exceptionally good result. If we want still higher precision, of the order of a micron or better, we sort the balls thus obtained according to their diameter. We will encounter this in Sect. VII.13.C. For small balls, see in Sect. VI.8 below for the mathematical formulation and more details.

We are now going to present some classical topics that are given in detail in the various references that have already been mentioned. First, the local form of a surface is **completely known** when $K(ext) \neq 0$. If $K(ext)(m) > 0$, then in a suitable neighborhood of the point m the surface is the boundary of a strictly convex subset of $\mathbb{E}^3$ (see Chap. VII for notions of convexity). If $K(ext)(m) < 0$, always in a suitable neighborhood, the situation is diffeomorphic to that of a hyperbolic paraboloid and its tangent plane. That is to say, matters are as in the figure below: the tangent plane at m crosses the surface when two (differentiable) curves cross at m, two portions of the tangent plane determined by these curves are strictly above the tangent plane and two are strictly below.

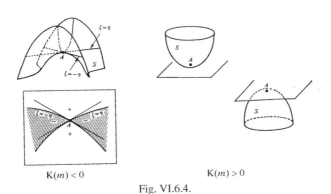

$$K(m) < 0 \qquad\qquad K(m) > 0$$

Fig. VI.6.4.

On the other hand, once $K(ext)$ is zero at m, even in the domain C^∞, we can't say more in general about the surface; the sketches clearly show that practically everything is possible, in particular crossing its tangent plane infinitely many times in each neighborhood of the point considered.

In order to be able to state things, we need to restrict ourselves to the *generic* case, i.e. to shapes that are stable under small deformations, which must occur in a precise setting. We encountered this notion of genericity in the case of plane curves in Sect. V.10; we will also encounter it for convex sets in Sect. VII.13.D, and for billiards in Sect. XI.10.C. For the genericity of surface shapes, see Porteous (1994).

◆

The *theorema egregium* of Gauss is truly one of the most remarkable theorems of mathematics:

(VI.6.3) *The Gaussian curvature* K *of the intrinsic metric of a surface of* $\mathbb{E}^3$ *coincides with the total curvature* $K(ext) = k_1 k_2$ *of its second fundamental form.*

We are going to comment on this result at some length. A fundamental documentation can be found in the historical book (Dombrowski, 1979), all the more precious in that it is not known how Gauss discovered this result, which was absolutely not "in the air", in contrast to the majority of great mathematical results that come at the end of a succession of conjectures, etc. It would have been a long time before someone else discovered it; readers will find a hypothesis on this topic below.

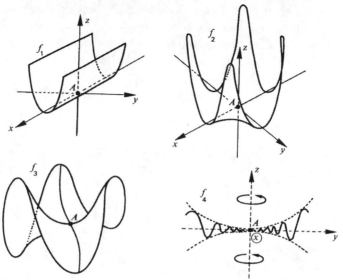

Fig. VI.6.5. The surface of revolution generated by the curve below on the right is
pathological

The preceding may make readers think that the proof of the *theorema egregium*
is never easy. No conceptual proof of it is known, making us stay at the bottom of
the ladder. The simplest proof consists of calculating ds^2 in coordinates. We thus
take locally the expression

$$(u, v) \mapsto \big(x = x(u, v), y = y(u, v), z = z(u, v)\big)$$

for our surface S. If a curve of the chart is given by $t \mapsto (u(t), v(t))$, then in S it will
give the curve $c : t \mapsto (x(t), y(t), z(t))$, and the norm of the velocity vector will be
$\|c'(t)\| = (x'^2(t) + y'^2(t) + z'^2(t))^{1/2}$. In th chart we will have $x' = x_u u' + x_v v'$,
$y' = y_u u' + y_v v'$, $z' = z_u u' + z_v v'$, where a_b denotes the partial derivative $\frac{\partial a}{\partial b}$. Thus
we can write $\|c'\|^2 = Eu'^2 + 2Fu'v' + Gv'^2$, where E, F, G are functions of (u, v).
Today the standard expression for such a ds^2 is $ds^2 = Edu^2 + 2Fdudv + Gdv^2$.
    The calculation is greatly simplified if we take for S an equation of the form

$$z = f(x, y),$$

the point m being the origin, the x, y plane being the tangent plane and the z axis
the normal at m. We thus obtain for ds^2 an expansion restricted to the neighborhood
of $m = (0, 0)$:

$$ds^2 = dx^2 + dy^2 + a(x, y)dx^2 + 2b(x, y)dxdy + c(x, y)dy^2 + o(x^2 + y^2),$$

where a, b, c are homogeneous polynomials of degree 2. From this partial expansion
we calculate the second fundamental form and we find without difficulty the value
$K(m) = \frac{1}{2}(a''_{yy} - 2b''_{xy} + c''_{xx})$. This proves the *theorema egregium*.
    This proof did not suffice for Gauss, who wanted to be so sure of his theorem
that he launched into the complete calculation of K in an arbitrary chart and for a

general expression $ds^2 = g = E du^2 + 2F du dv + G dv^2$, an impressive calculation that can be found *in extenso* in Dombrowski (1979). In Sect. II.2.A of Berger (2001) is found a more intrinsic proof in the spirit of Riemann; it consists of writing the equation of geodesics and of ds^2 in so-called *normal* coordinates, i.e. where the geodesics emanating from the origin are the lines of the chart in question.

◆

For a surface $S \subset \mathbb{E}^3$ there exists a natural mapping $f_S : S \to S^2$ with values in the unit sphere $S^2 \subset \mathbb{E}^3$. To define it, we first remark that at each point of the surface we have two opposite normal unit vectors; to choose between them, it suffices to choose a side of the space that the surface cuts locally in two, or else to have an orientation of the surface in the space. The first way is called a *co-orientation* ; it is used in an essential way for the curves of Arnold's theory seen in Sect. V.10. We suppose from now on that we have made such a choice (propagated by continuity). The mapping consists then of associating with each point m of S its chosen vector exterior normal. Practically all the books call it the *Gauss mapping* ; in fact, it was used a little before by Rodrigues. Its essential property is this: we observe the effect of f_S on the areas of S and of S^2. We see intuitively, in observing things infinitesimally along the principal directions, that f_S multiplies the lengths in those directions by k_1 and k_2 respectively, thus the areas are multiplied infinitesimally at m by $k_1 k_2 = K(m)$. It will be noted that f_S does or does not change orientation accordingly as K is positive or negative.

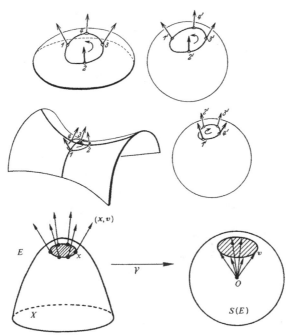

Fig. VI.6.6. the Gauss mapping. [BG] Berger, Gostiaux (1987) © Presses Universitaires de France

A consequence is: if D is a domain of S then:

(VI.6.3), $$\text{Area}(f_S(D)) = \int_D K(m)\,dm$$

where the element of integration (denoted dm) on the surface is the most natural; readers can assume the task of seeing how a measure on a surface of $\mathbb{E}^3$ is defined canonically. If S is a strictly convex surface then the mapping f_S is bijective and thus $\int_S K(m)\,dm = 4\pi$, this formula having been obtained without our having to calculate the curvature K of the surface; and so it was proved by Rodrigues — without any calculation — that for each ellipsoid we have 4π as the complete integral of the curvature.

If we extend the notion of *degree*, introduced in connection with curves for the mappings $S^1 \to S^1$, to mappings between surfaces, the general formula will be:

$$\int_S K(m)\,dm = 4\pi \cdot \text{degree}(S \to S^2).$$

For this notion of degree, see e.g. [BG].

In Hilbert and Cohn-Vossen (1952) we find a calculation of angles for the polyhedrons that can give us some intuition about the *theorema egregium*, by considering that the deficiency of the angles at a vertex is a "discrete" analogue of the "continuous" curvature. We consider a vertex of the polyhedron where n faces meet and denote the angles of these faces by β_i. These vertex angles completely determine the metric of the polyhedron. We call the curvature of the polyhedron at the point considered the quantity $2\pi - \sum_i \beta_i$; what follows shows that it is a good definition. The Gauss mapping associated with this vertex degenerates into a polygon on the sphere S^2: the vertices of this polygon are the vectors normal to the faces, sides correspond to what happens when we revolve about the corresponding edges. The figure shows that the angles of this spherical polygon are the supplementary angles $\alpha_i = \pi - \beta_i$. But the surface area of this spherical polygon equals (decompose into triangles and apply to each piece the formula III.1.2) $\sum_i \alpha_i - (n-2)\pi = 2\pi - \sum_i \beta_i$, which is our desired curvature.

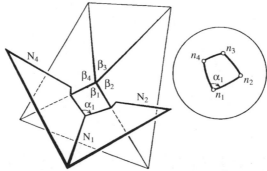

Fig. VI.6.7. The polyhedral proof suggested by Hilbert for establishing geometrically the *theorema egregium*, when $n = 4$. Hilbert, Cohn-Vossen (1996) © Springer

The proof of the *theorema egregium*, as much as the result itself, poses a problem for those who (justifiably) prefer low order differentiability: to define the total curvature of a surface S given locally as the graph of a function f, it suffices that f be of class C^2 (we derive the normal to the surface), whereas for computing the Gaussian curvature it is necessary to differentiate the ds^2 two times. Now this ds^2 uses essentially the first derivative of f, whence finally the appearance of third derivatives of f. There is a mystery here, invisible somehow to less reflective people who work in C^∞ without moral concern. This strange phenomenon seems never to be mentioned in textbooks. The dilemma was attacked by Hermann Weyl starting in 1916, to attempt to know how to define curvature with only two derivatives. Since van Kampen (1938) we have a proof of the *theorema egregium* in the setting where f is only C^2. For more subtle situations, appropriate references can be found in Klotz-Milnor (1972), to which may be added Chern, Hartman, and Wintner (1954).

A plane curve is determined within a plane isometry by its curvature as a function of arc length; see Sect. V.7. A surface in space is not determined by its Gaussian curvature, nor even by its two principal curvatures; we will see a celebrated counterexample in Sect. VI.8. On the other hand, it *is* determined − within an isometry of the space − by its two fundamental forms. The counterexample shows that it does not suffice to know the principal curvatures, the principal directions also being required; and this in fact suffices. The classic theorem isn't difficult; we break trail by showing that there are relations between the two fundamental forms. The first is surely the *theorema egregium*, but there also exist the so-called *Codazzi-Mainardi* equations, which are a bit tricky to express: we defer to the various references on surfaces, according the taste of readers and their ability to access them. These equations are useful in diverse results; we will see some further on. For a recent approach to the theorem of determination by the two fundamental forms, see Ciarlet and Larsonneur (2000).

VI.7. What is known about the total curvature (of Gauss)

Surfaces with curvature zero. In Sect. VI.4 we saw that the Riemannian surfaces of (Gaussian or total) curvature zero are locally isometric to the Euclidean plane. But this says nothing about the way that they are embedded in space. For example a cone (with the apex removed, so a not to have a singularity) and an arbitrary cylinder have zero curvature. The local and global classification of (complete) surfaces with (Gaussian or total) curvature K equal to zero is well understood, but this understanding is not so very old.

We begin with the **local**. It is a very old result that if $k_1 = k_2 = 0$ in a neighborhood of a point, then over this neighborhood the surface is a piece of a plane in space. If now $K = 0$ at a point, but $k_1 \neq 0$, then in a neighborhood of this point the lines of curvature (corresponding to the zero principal curvature) are inevitably seen as linear segments in space. But we must pay attention to the way in which the

examples of cones and cylinders can be deceptive: a ruled surface, i.e. a one parameter family of lines (or pieces of same, but we can always extend what will follow to
complete lines) is not in general a surface of zero curvature. The lines which define
a ruled surface are called its *generators*. The characteristic difference between ruled
surfaces and cones and cylinders is that the plane tangent to the ruled surface **turns**
when we move along a generator (see Fig. VI.1.2). To say that it doesn't turn is tantamount to saying that the curvature is zero, or again that the surface is the *envelope*
of a one parameter family of planes. It can be shown that this is again equivalent
to these lines being the tangents to a curve in space. But this can't be a global figure, since our surface isn't differentiable at points of the so-called *striction* curve;
see Fig. VI.7.1. In the case of a cylinder, this striction curve is entirely *removed to
infinity.*

Such surfaces are called *developable*, this because they are precisely mappable
onto a plane with conservation of the metric.

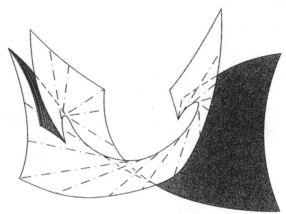

Fig. VI.7.1. A developable surface and its striction curve. Porteous (1994)
© Cambridge University Press

We can think of the surfaces with curvature zero as crumpled sheets of paper. It
is an interesting exercise to see that we can fold a piece of paper not only along a
line, but along any traced curve whatsoever. Mark off the curve and fold: this works,
but don't use too much space around the folding curve, and the curve must not be too
complicated, for otherwise the sheet will intersect itself. More on this subject can
be found in Porteous (1994) and in Fuchs and Tabachnikov (1995). Resuming our
discussion, locally the surfaces of curvature zero are pieces of developable surfaces,
but also locally we can produce pathological objects even of class C^∞: take pieces
of planes and join pieces of non planar developable surfaces to them.

But what is the case **globally** when we look for surfaces of curvature zero that
are complete, i.e. without boundary (see Sect. VI.2)? We can anticipate from the
above figure that things don't work globally. We can guess the answer: the fact
that the surface is complete says that it contains entire lines. Let us look at these

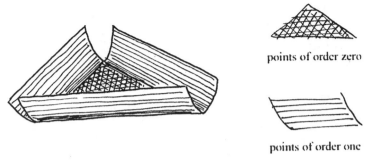

points of order zero

points of order one

Fig. VI.7.2. How to make a developable surface not containing any complete line

generators: they envelope a curve, but this curve is a singularity (see Fig. VI.7.2); the only reasonable case is that where all the points of contact of the generators with their envelope curve are at infinity, i.e. the generators are all parallel to each other. Our surface must finally be a cylinder, but a general cylinder and not necessarily a surface of revolution. As intuitive as the result may seem, it wasn't until (Hartman and Nirenberg, 1959) (see also 5.8 of do Carmo, 1976) that we had a rigorous proof. This proof however, in spite of being fine and subtle, does not introduce any new concept.

Surfaces with positive curvature. What is the general form of a complete surface S of $\mathbb{E}^3$ with curvature everywhere positive (resp. nonnegative)? We had a good answer in the case of plane curves in Sect. V.5, this answer requiring the subtle turning tangent theorem or *Umlaufsatz*. In the spatial case we find that things are simpler; the result goes back to Hadamard (1897). We consider a compact surface S, with $K > 0$ everywhere, and we want to show that it is the boundary of a convex domain of $\mathbb{E}^3$. The sole idea is to apply the normal Gauss-Rodrigues mapping (defined in the preceding section): $S \to S^2$. The condition $K > 0$ implies that this mapping is everywhere locally a diffeomorphism; we thus see easily (since S is compact) that it is a covering of S^2 (see as needed the end of Sect. V.XYZ or [BG] for the concept and what follows). But the sphere S^2 is *simply connected* (see also Sect. III.5), which implies that the covering is in fact a bijection. Analogous to the case of curves, we show subsequently that this implies that the surface is entirely on one side of its tangent plane. It is interesting to mention here a difference between curves and surfaces. For (not necessarily simple) closed curves (i.e. immersions of the circle S^1 in $\mathbb{E}^2$) the result was false: the curvature being positive throughout allowed self-intersections. This is not the case with surfaces: if the surface S is only the image of an immersion of the abstract sphere S^2 and is of positive curvature, then in fact we inevitably have an embedding (and the surface is the boundary of a convex set). The proof was given by Hadamard.

It can be anticipated that the result remains true under the weaker hypothesis $K \geqslant 0$. This says in effect that flat pieces are permitted. Such is the case, but the

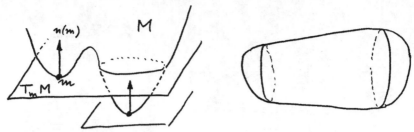

Fig. VI.7.3. Why a surface with K > 0 is necessarily convex

proof had to wait for Chern and Lashof (1958). A new tool was needed: the general formulas of Chern and Lashof, which related the integral of the curvature to topological invariants via Morse theory, specifically the Betti numbers. Details and generalization of the result of Hadamard to the noncompact case can be found in Sect. 11.3.2 of [BG], (5,6-B) of do Carmo (1976) and Chap. I of Burago and Zalgaller (1992).

In Sect. VI.9 we will review the uniqueness results given in this section for surfaces with positive curvature.

Surfaces with negative curvature. Geometrically these are both the surfaces that cross their tangent plane at each point and those for which the intrinsic metric is *hyper-Euclidean* (see Sect. I.5.D of Berger (2003) for details): the sum of the angles of each sufficiently small geodesic triangle (with sides that are geodesics) is less than π or, equivalently, each sufficiently small triangle satisfies $a^2 > b^2 + c^2 - 2bc \cos \alpha$:

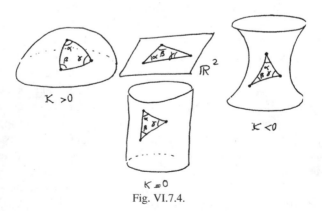

Fig. VI.7.4.

"Sufficiently small" can be replaced by the condition that the triangle is the boundary of a simply connected domain; see below. The exact default from π will be measured by the Gauss-Bonnet formula below. Readers will have noted that these are the properties of that hyperbolic geometry (Sect. II.4) for which the curvature K constantly equals -1. The surfaces with negative curvature are very important in dynamics, for then the geodesics diverge, even exponentially; we will encounter them amply in Chap. VII. We will also encounter them in considering the space forms in Sect. VI.XYZ. Finally, the minimal surfaces of the following section always have a K that is negative or zero. The basic text for their study is Hadamard (1898), much in advance of its time and where the theory of chaos was discovered. There are two important things to mention for these surfaces.

It may seem at first very intuitive that a complete surface with negative curvature cannot be contained in a bounded region of space. Hadamard was sure of it. This assertion is called *Hadamard's conjecture*. It was much studied; see for example Chap. II of Burago and Zalgaller (1992). It was refuted not long ago in Nadirashvili (1996), where the author succeeds in constructing a minimal surface without any planar point (thus with strictly negative curvature) that is contained in a bounded region of $\mathbb{E}^3$. The construction uses Weierstrass's formulas for minimal surfaces; it "suffices" to find suitable functions which are solutions. These formulas, essential for understanding minimal surfaces, will be given in the next section.

The Gauss-Bonnet formula. We consider a triangle of a surface whose sides are geodesic arcs; we suppose furthermore that the triangle determines a domain D, i.e. that it can be "filled in". The serious formulation is that this triangle is the boundary of a simply connected domain (see if needed the few lines at the end of Sect. V.XYZ). See the figure for a non simply connected counterexample, where readers will show that, the boundary curve being a (closed) periodic geodesic, we have $\alpha = \beta = \gamma = \pi$, but that $\int_D K(m)\,dm = -2\pi$. But in the simply connected case we always have the formula relating the curvature to the angles α, β, γ of the triangle:

$$(\text{VI.7.1}) \qquad \alpha + \beta + \gamma - \pi = \int_D K(m)\,dm.$$

It may be noted that this is a generalization of formula III.1.2 of spherical geometry and of formula II.4.1 of hyperbolic geometry. It should be realized that the domain can be enormous, have curves of different signs, but all that is compensated for in the end. Formula VI.7.1 is never easy to prove in an elementary way; it does not follow from the formula $\text{Area}(f_S(D)) = \int_D K(m)\,dm$ given above for the Gauss mapping. Readers may choose a proof from the classic texts. In any case, the result is concerned only with the intrinsic metric and thus with abstract Riemannian surfaces. For the (approximate) date of the discovery of this formula see Dombrowski (1979).

In fact, the above formula is due to Gauss. Bonnet proved (in Bonnet (1855) a more general formula, valid when the sides do not necessarily lie along geodesics. It is then necessary to modify the quantity $A + B + C - \pi$, by adding to it the integral of the *geodesic curvatures* of the sides. The geodesic curvature of a curve along a surface of $\mathbb{E}^3$ measures the deficiency of the curve from being a geodesic. This notion generalizes the algebraic curvature of plane curves, which measures the deficiency of a curve from being a line. The proof is no more difficult than that of Gauss's formula, but it is nonetheless difficult and non conceptual.

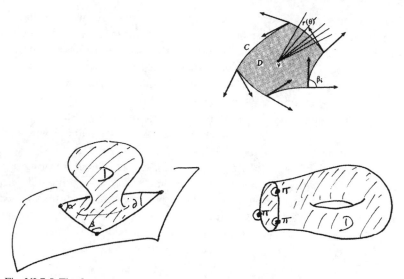

Fig. VI.7.5. The figure *lower right* shows clearly that simple connectivity is indispensable

Now, we want a conceptual proof, so why not use Stoke's formula, since it relates an integral on the interior of a domain to an integral along the boundary? This conceptual proof appeared in Chern (1944) and was the start of a revolution in differential geometry, that of "Chern classes". We use the language of differential geometry; see [BG]. The idea is to introduce the *unitary fiber* US *of the surface* S (that we will amply encounter in Sect. XII.6), i.e. the set formed of all unit vectors tangent to S. The canonical projection p : US $\rightarrow$ S consists of associating with each unit vector the point of S where it is tangent.

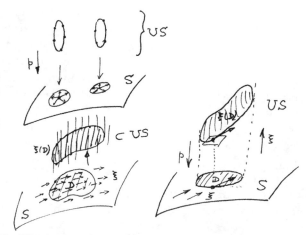

Fig. VI.7.6. The surface is lifted by a vector field into its unitary fiber

The essential thing is that the three dimensional manifold US possesses a canonical differential 1-form θ and above all that its exterior derivative $d\theta$ is the form lifted onto US by p of the 2-form of the curvature $K(m)dm$ of S (S oriented so that its canonical measure dm becomes a 2-form). Now if the domain D is simply connected, we can define a continuous field ξ of unit vectors on D; thus we can lift D into US. The Stokes formula applied to $\xi(D)$ is exactly the Bonnet formula, because the 1-form θ is nothing other than the geodesic curvature. The unitary fiber will not be confused with the direction fiber introduced in Sect. V.10 for plane curves: in the first case, the fiber is the circle formed by the vectors of length 1, in the second it is the circle formed by the linear directions, the quotient by identification of antipodal vectors of the preceding.

We note that here we have a typical example of the necessity for introducing objects "above" an object for a good understanding of what happens "below". In the present case we can say that it has to do with a ladder which has as much a physical as a conceptual nature.

The so-called Gauss-Bonnet theorem. The result concerns compact abstract Riemannian surfaces. It is intuitive that they can be triangulated, i.e. divided into a finite number of triangles with simply connected interiors (taking them sufficiently small will accomplish this). If we apply Gauss's formula to each of these triangles and take the sum, we obtain the total integral of the curvature on the surface S on the one hand, and $2\pi(s - a + f)$ on the other (because the sum of the angles at a vertex of the triangulation will be 2π in total), whence:

(VI.7.2) $$\int_S K(m)\,dm = 2\pi(s - a + f)$$

where here s (resp. a, f) denotes the number of vertices (resp. sides (edges), triangles (faces)) of the triangulation.

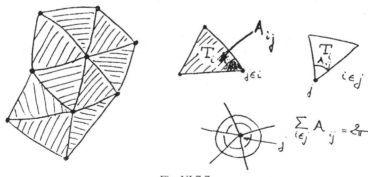

Fig. VI.7.7.

Now the quantity $s - a + f$ does not depend on the triangulation but only on the topological type of the surface; it is called the *characteristic* of the surface, often denoted by $\chi(S)$. If the surface is orientable and its genus is γ (see Sect. VI.1), then $\chi = 2(1 - \gamma)$. This formula is clearly of primary importance; here is an example: if the curvature is nonnegative (or else nonpositive) on a torus ($\chi = 0$), then it is in fact identically zero and the torus is flat (that is to say, locally Euclidean). Finally we have the formula:

$$(\text{VI.7.3}) \qquad \int_S K(m)\,dm = 2\pi\chi(S).$$

It remains to prove that the triangulations exist! We have already spoken about this in Sect. VI.1.

Why have we written "so-called Gauss-Bonnet" even though the result is in fact called "Gauss-Bonnet" in the majority of books that treat it? Certainly the proof is not difficult once the existence of a triangulation has been obtained. Historically, to the best of our knowledge, this aspect enters with its proof for the first time in Boy (1903). This is the same mathematician whose name is associated with the surface mentioned in Sect. I.7 representing the real projective $\mathbb{R}P^2$ immersed in $\mathbb{R}^3$; see Karcher and Pinkall (1997). In Berger (2003) we have opted for the name "global theorem of Gauss-Bonnet".

VI.8. What we know how to do with the mean curvature, all about soap bubbles and lead balls

We recall that the mean curvature changes sign when we reverse our choice of the unit normal. In everything that follows, we always suppose that we have chosen a normal direction once for all. The curvature K gives very roughly the local shape of the surface (convexity or crossing of the tangent plane), but more essentially the deficiency from being Euclidean: a development limited by the length of the small circles, whereas the second fundamental form is related to the way the surface is situated in space. The geometric interpretation of the mean curvature is the following.

Instead of examining the variation of the length of the small circles, we look at the variation of the area of a small domain Ω of the surface when Ω is displaced normally from the surface. We can guess the result: the lines of curvature are, in the small, pieces of circles of radii $\frac{1}{k_1}$ and $\frac{1}{k_2}$, the infinitesimal variation of the area is thus equal to $k_1 + k_2 = 2H$. More generally, if we deform the surface over a domain Ω by carrying the function $f(m)$ along the normals, we obtain the essential formula:

$$(\text{VI.8.1}) \qquad \text{derivative of the area} = 2\int_\Omega H(m)\,f(m)\,dm.$$

Minimal surfaces. The first question is: What are the surfaces of zero mean curvature? They are called *minimal surfaces*. In fact, the above formula shows that a local variation of the surface does not change the area to the first order. There is thus a

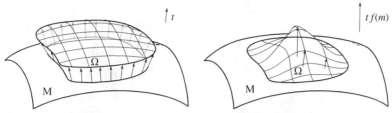

Fig. VI.8.1. Two one-parameter variations of a surface fragment, the first with constant normal displacement, the second vanishing at the boundary of a domain

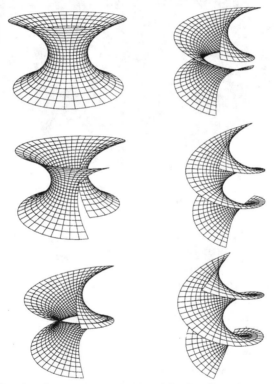

Fig. VI.8.2. In spite of appearances, all these minimal surfaces are actually isometric.
Spivak (1999) © M. Spivak

certain abuse of language: the area function is critical in the surface, which doesn't mean that this area is minimum in the class of variations considered, e.g. for those that are restricted to having a given curve as fixed boundary. We nonetheless keep the name "minimal surface". We find such surfaces (with boundary) by dipping a curve made out of metal wire into a soap solution and withdrawing it slowly; a film is formed that is a minimal surface.

We can't even give the merest idea of the very numerous and detailed results or of the open questions concerning the minimal surfaces in $\mathbb{E}^3$, it being

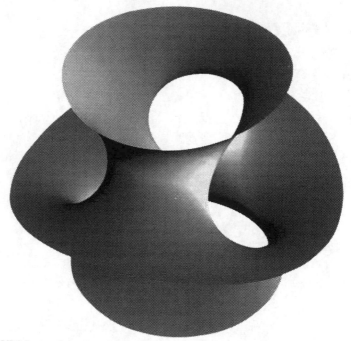

Fig. VI.8.3. A minimal surface discovered by Costa, Hofmann and Meeks. Wagon
(2009) © S. Wagon

an immense subject which has been the constant object of study, almost without
interruption subsequent to the memoir of Plateau (1873). Recent references, complete when they appeared, are the detailed books Dierkes, Hildebrandt, Küster, and
Wohlrab (1992), Nitsche (1996), the synthesis Osserman (1996) and finally Colding
and Minicozzi (1999). The richest in illustrations is by far the first of these. See
also the very recent Lazard-Holly and Meeks (2001). We can but quickly treat a few
small points.

A simple remark: a complete minimal surface of $\mathbb{E}^3$, without boundary, has negative or zero curvature and thus is never compact. In fact, we saw in Sect. VI.5 that
every compact surface necessarily has points of positive curvature. Now comes a
celebrated example, that of the one-parameter family of minimal surfaces that joins
the two most celebrated (complete) minimal surfaces, the helicoid (which is ruled)
and the catenoid (which is a surface of revolution):

This family of surfaces has the remarkable property that each of its members
is minimal and isometric to all other surfaces of the family. Of course, this for the
intrinsic metrics of these surfaces and not the restrictions of isometries of $\mathbb{E}^3$. Thus
in particular they all have the same principal curvatures k_1 and k_2 since $k_1 = -k_2$
and $K = k_1 k_2$ is invariant under the deformation (although variable on each surface). Thus we find surfaces having the same principal curvatures, but different in
$\mathbb{E}^3$. This doesn't contradict the fact that the two fundamental forms determine the

Fig. VI.8.4. Longhurst's model of an Enneper surface, in bubinga wood. Bruning,
Cantrell, Longhurst, Schwalbe, Wagon (2000) © R. Longhurst

surface within an isometry of the space (end of Sect. VI.6); what happens here is
that the principal directions "turn" during the deformation.

For explicit construction of the family of the above example, the simplest thing
is to climb the ladder with Weierstrass. Beginning in 1861, Weierstrass gave a way
for finding many minimal surfaces with the help of the theory of holomorphic functions of a complex variable. Here are his formulas set on an open subset U of the
plane $\mathbb{R}^2 = \mathbb{C}$, where $f(Z) = f(u + iv)$ is an arbitrary holomorphic function
(at least initially) that is integrated along a path going from the origin to Z. We
then define an immersion $(u, v) \mapsto (x(u, v), y(u, v), z(u, v))$ of U into $\mathbb{R}^3$ by the
formulas:

$$x(u, v) = \mathrm{Re}\left(\int_0^Z (1 - Z^2) f(Z) \, dZ\right),$$
$$y(u, v) = \mathrm{Re}\left(i \int_0^Z (1 + Z^2) f(Z) \, dZ\right),$$
$$z(u, v) = \mathrm{Re}\left(2 \int_0^Z Z f(Z) \, dZ\right).$$

It is these formulas and some refinements, steadily in progress, that have recently
made way for the construction of complete minimal surfaces of various topological
types, such as the one below:

Enneper's surface is a great classic and here is a sculpture inspired by it:

The essential point, once the function f has been chosen in the Weierstrass formulas, is knowing if the surface obtained is embedded and not just immersed, presenting self-intersections.

We recall here Nadirashvili's counterexample from the end of Sect. VI.7: there exist bounded and complete minimal surfaces in space that have no planar points.

Soap bubbles and the fabrication of small balls. If we blow a soap bubbles using soapy water and a straw, don't we obtain magnificent spheres? A physicist can show mathematically the we obtain a compact surface whose area is a local minimum among all those that envelop a domain having a given volume (notion of capillary tension), because a force is exerted that requires the area to be minimal. If additionally this were an absolute minimum, we could have settled the issue with the isoperimetric inequality; see Sect. VI.11 below. But we only know, thanks to formula (VI.8.1), that here we have $\int_D H(m) f(m) \, dm = 0$, where the functions f are to be chosen from those that model a variation conserving the volume, i.e. such that $\int_M f(m) \, dm = 0$. A classic argument of the calculus of variations implies that there are enough such f to conclude that "the mean curvature H is a constant function on the surface".

Certainly (round) spheres have constant mean curvature, but are they really the only ones? Is every soap bubble, for which we have just seen that the mean curvature is constant, a sphere? The result that they must indeed be round spheres is true, but awaited (Hopf, 1951) for a proof. And, again, Hopf assumed that the surface had the *topology* of a sphere; a theorem without this hypothesis on the genus (it must be shown that surfaces of higher genus with constant mean curvature can't exist in $\mathbb{E}^3$) awaited A.D. Alexandrov in 1955 . Alexandrov's proof is geometric; it consists of applying — to the surface and its images by symmetry — a comparison principle for surfaces that is analogous to the one already see for curves in Sect. V.7. For the sphere Hopf uses functions of a complex variable: the abstract Riemannian sphere must be given a complex structure. What is called a holomorphic differential is then constructed; see Sect. VI.XYZ for complex structures on a surface, which must be zero for reasons of the topology of the sphere, which implies by construction that all the points are umbilics, which can only happen for round spheres (an elementary local result, already expressed in Sect. VI.6). The details can be found in Hopf (1983) and do Carmo (1976).

In the industrial realm, the preceding explains how certain balls are made (small balls — for more serious balls, see Sects. VII.13.C, II.2, VI.6). There are two methods: in the first, metal is dripped into an appropriate liquid, and these drops take on a perfectly spherical form because of surface tension. In the other method, an arrangement is made for letting drops of liquid loose under conditions of weightlessness; the drops also become spheres, for the same reason. In both cases, the condition given by physics is that the surface area is a local minimum for surfaces enveloping a given constant volume, thus in particular the mean curvature is constant as we have seen above.

For all that, the preceding does not end the study of surfaces with constant mean curvature, for there remains a natural case, that of surfaces that are just immersed and not necessarily embedded. For the non embedded case, where the topological type of the sphere is excluded, the proof of Hopf with holomorphic differentials applies by the same construction. It is not the same for the topological type of the torus, where immersed tori were found for the first time in Wente (1986), contrary to what Hopf had conjectured. These matters are connected with elliptic functions; these tori are completely classified in Pinkall and Sterling (1989).

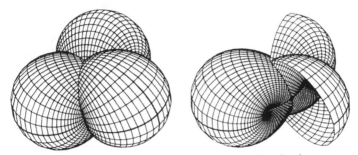

Fig. VI.8.5. The Wente Torus. *At right* a portion of the surface has been removed to gain a view of its interior

There is also the difficult problem of the double soap bubble, solved just in 2000; see Chap. 14 of Morgan (1988–2008 (3rd ed., 2000)). But that of the triple bubble problem remains open, and likewise so does the analogous planar case. For the connection between certain soap bubbles with 120 pieces and the regular polytope with 120 faces in $\mathbb{E}^4$, see Sect. VIII.11 and Stillwell (2001).

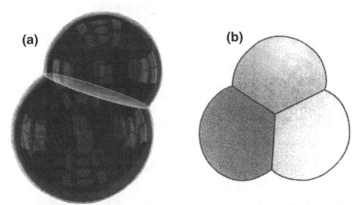

Fig. VI.8.6. The conjecture was that the double bubble that contained two given (in general different) volumes is composed of three surfaces that are pieces of spheres (that the angles at their intersections are equal to 120° follows immediately from the first variation formula. Readers may formulate a conjecture for the triple bubble). **(a)** Sullivan, Technische Universität Berlin. **(b)** Morgan (2000) © Mathematical Association of America

It's easy to construct complete surfaces of revolution with constant mean curvature, for which the meridians are provided by an ordinary differential equation; they are called *Delaunay surfaces*. Those with a geometric bent and who like unexpected relationships will appreciate knowing that they can be obtained by rotating the locus of one focus of a conic that is made to roll on a line (the catenoid with mean curvature zero corresponds to the case of the parabola). Some of these surfaces are merely immersed, but the rest are truly embedded.

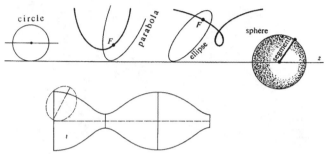

Fig. VI.8.7. Examples of meridians of a surface of revolution with constant mean curvature. Readers can study the case of ellipses in more detail

Obtaining surfaces with constant mean curvature of different topological type (noncompact in fact, but with more "pieces" than the surfaces of revolution seen above) had to await (Kapouleas, 1995). Today a classification is still needed to get these examples into a sufficiently general setting. The constructions of Kapouleas are based on a detailed analysis which allows the construction of connected sums. Recent results and a bibliography can be found in Mazzeo, Pacard, and Pollack (2000); see also the numerous computer-generated drawings in Grosse (1997), Grosse-Brauckmann and Polthier (1997) and Kilian, McIntosh, and Schmitt (2000).

VI.9. What we don't entirely know how to do for surfaces

Here we give several natural examples of open problems for surfaces in space, sometimes stated around results having positive answers.

Various isometries (the compact case). We begin with the case of results that affirm the congruence (identity within an isometry of $\mathbb{E}^3$) of isometric surfaces. Next we speak about a closely related problem, that of isometric deformation. In 1899 Liebmann showed without much difficulty that the only compact surfaces with constant positive curvature are the round spheres. In 1927 Cohn-Vossen established a much more general result: two isometric compact surfaces with everywhere positive curvature are congruent. The figures below show clearly the absolute necessity of the condition of positive curvature:

Fig. VI.9.1. How to construct two surfaces that are isometric but not congruent, i.e.
that are not obtainable from one another by a global isometry of the ambient space

These figures call for two remarks that lead to open problems. The first is that we are dealing with a counterexample of class C^∞ which is quite flexible. To the present no one has found a counterexample with two real analytic surfaces (class C^ω) which are isometric. The second is: can we find a one-parameter family (a deformation) of compact isometric surfaces between them? In at least class C^2 it is a completely open question. If such deformations were to be found, the question then arises as to whether the volume they enclose remains constant. See below for those that are merely class C^1 and Sect. VIII.6 for the case of polyhedra and the question of what happens to the volume contained by the surface.

We can find proofs of Cohn-Vossen's theorem on p. 46 of Chern (1989) and p. 103, Sect. 6.2.7 of Klingenberg (1973) (Herglotz's integral formula). In these classic arguments, by isometry, we are able to impose on one surface the second fundamental form of the other; we then compose a shrewd mixture of them that furnishes a nonnegative real number that is zero only in case the two fundamental forms coincide. By integrating this number over the whole of the first surface, a positive number is obtained. But the situation is different when we reverse the roles of the two surfaces; the attached number then changes sign, by the construction. The integral is thus finally zero, hence the coincidence of the two second fundamental forms which, along with the first form (isometry), yields congruence in the space (see the end of Sect. VI.6).

Nevertheless, the result seems completely isolated and leaves us hopelessly at the bottom of the ladder. It appears that the first person to have climbed sufficiently high to be able to see clearly was Gromov in 1986, both for the existence problem of Sect. VI.5 and for uniqueness. We find a presentation and some generalizations in Labourie (1989), but we can't in a reasonable amount of space give an outline (or even a brief insight) of Labourie's work; we simply say that the theoretical setting used is that of *symplectic geometry*, which Gromov believes will play a major role in the twenty-first century.

Isometries and local isometric deformations (bending, tennis balls and more).
If we suppress the compactness conditions while keeping (or not keeping) the completeness condition, then "there is plenty of room for nontrivial isometries and deformations". For example, there are all the cylinders, that have the topology of the plane or that of a cylinder of revolution. There are no other possibilities, since

there are only two types of curves, the line and the circle. Here is an example of iso-metric deformation, but only local (no completeness here): take a tennis ball and cut out a small piece. You can then deform it without much effort. It's a highly isometric deformation because modifying the intrinsic metric requires, as with fabric, the ex-ertion of much force tangential to the surface.

These examples lead us to think that, at least locally, every sufficiently small piece of a surface is deformable. By *deformation* we understand a one-parameter family (continuous, of course) of surfaces all isometric to each other. In fact, this de-formability is an extremely complicated and badly understood and is far from being settled at present, even when only infinitesimal deformations are studied, which are naturally a starting point. There is also the question of the differentiability classes that enter into the matter, both for the metric and for the deformation parameter. The author of a (or rather the) bible of differential geometry (Spivak, 1970) appar-ently gave up the problem in disgust while preparing the book for publication. The problem has been much studied by the Russian school. In Chap. III of Burago and Zalgaller (1992) there is a very complete synthesis that in fact has mostly to do with partial differential equations, but of a type that is generally nonclassical. Geome-try enters only little, but it is interesting to remark that mirror symmetry about a parabolic point comes into the picture (see 6.4 of this reference) and is present in an essential way in the proof of the nonexistence of embedded surfaces with constant mean curvature (see Sect. VI.8).

We will need to retain just two things: only the parabolic points (those where the total curvature is zero) cause a problem; as soon as the total curvature is nonzero, de-formation is possible. On the other hand, in the neighborhood of a parabolic point, a deformation need not exist, even infinitesimally. The other factor that we need take into account is the order to which the curvature is nullified; it's the large or-ders that cause difficulties (which was by the way the case in the counterexample of Pogorelov for an abstract Riemannian surface not isometrically embeddable in $\mathbb{E}^3$; see Sect. VI.5. To allow the complexity of the problem to be felt we give a single example, due to Efimov in 1948: the surface $(x, y) \mapsto (x, y, x^9 + \lambda x^7 y^2 + y^9)$, where the real number λ is transcendental (see Sect. IX.1 and note the meager geo-metric character of this condition!), is never deformable locally. We will now look at Bleecker's crumpling.

What occurs in the C^1 realm? Completely different things, as we have already partially seen in Sect. VI.5! Let us review the work of Bleecker (1996). There the Nash-Kuiper C^1 embedding result is extended to one-parameter families of surfaces. The author applies it to one-parameter embeddings of compact surfaces in $\mathbb{E}^3$ and thus obtains examples of surfaces of spherical type which are isometric but not as a consequence of a global isometry of $\mathbb{E}^3$. In view of the beginning of this section, it is good then that we are only in a domain that is C^1 and strictly nonconvex, even if we begin with an initial convex surface. For example, we can begin with ellipsoids. The essential interest is that the author shows, if the initial ellipsoid is sufficiently flattened, that the volume of its isometric deformations increases strictly. This result is surprising, for we will see in Sect. VIII.6 that isometric deformations of polyhedra are necessarily volume preserving. Thus the differential context is an astonishing intermediary between the C^2 setting and the purely polyhedral one. We again mention that such surfaces are truly extremely crumpled.

The Caratheodory conjecture. We recall that an umbilic is a point of the surface where the two principal curvatures are equal. A surface − a torus of revolution for example − needn't have an umbilic (see in Fig. VI.6.2 above how to find the principal curvatures of a surface of revolution). Any points of a surface of revolution that are situated on the axis of revolution are umbilics. More subtle are the umbilics of the ellipsoids encountered in Sect. VI.3. What is the situation in general? We would like to know that umbilics always exist for surfaces of topological type other than that of the torus, and still better: how many must there be? The matter isn't settled; here is what we know.

We give the initial idea and the associated concept. We consider a compact surface (with oriented normals). We suppose the umbilics are finite in number, for otherwise there is nothing to gain in this way, so that what remains after their removal is an open subset $\mathcal{O}$ of the surface and our umbilics $\{x_i\}$ are all isolated points. At each point of $\mathcal{O}$ there is a well determined principal direction that corresponds to the greater principal curvature, since these two curvatures are never equal. This allows definition on $\mathcal{O}$ of a continuous *field* ξ of *tangent directions*. For such a field we have the theory of Hopf (1926) at our disposal. It associates an *index* Ind(x) with each x of S; it is a half-integer thus defined. We recall the degree theory of mappings $S^1 \to S^1$ encountered with respect to curves in Sect. V.3 and also in Sect. V.10. We consider about x a very small circle C on the surface. At each point c of C we associate a direction $f(c)$ of $T_x S$ (this can be done in numerous ways, for example take the parallel to $\xi(c)$ in $\mathbb{E}^3$ and project it onto the space $T_x S$). The set of lines of $T_x S$ is also a circle S^1, a real projective line; see Sect. I.6. Thus f is clearly a mapping $S^1 \to S^1$. We *set* Ind(x) $= \frac{1}{2}$degree(f). By the invariance of the degree under homotopy (end of Sect. V.XYZ) this index does not depend on the small circle chosen in the neighborhood of x and is thus associated with the point x (and with the field of tangent lines considered). Readers should verify that if $x \in \mathcal{O}$, then its index is zero. The $\frac{1}{2}$ shouldn't be surprising; it is imposed because the tangent lines considered turn twice about the point when we go once around the small circle.

Hopf's result says that, whatever the continuous field of directions tangent to S and not zero at more than finitely many points $\{x_i\}$, we always have:

$$\sum_i \text{Ind}(x_i) = \chi(S)$$

where $\chi(S)$ denotes the characteristic of the surface. For a proof, see for example II.8.4 of [BG]. Thus if $\chi(S)$ is nonzero, the surface must have at least one umbilic; this is thus the case for all the types of compact surfaces of $\mathbb{E}^3$ other than that of the torus (for which we have seen that it can indeed be umbilic free). The points

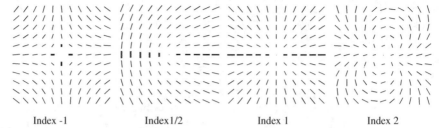

Index -1   Index1/2   Index 1   Index 2

Fig. VI.9.2. *Examples of indexes.* When the direction field can be oriented in a coherent manner so as to become a continuous vector field, its index is equal to that of the vector field – it's integral. When the direction field isn't orientable, its index is a half-integer

of the axis of a surface of revolution have index 1 and each umbilic of an ellipsoid has index $\frac{1}{2}$. See the figures of the three types of generic umbilics, as classified by Darboux, in Sect. VI.10.

The Caratheodory conjecture is that each surface with the topological type of the sphere possesses at least two umbilics. It remains open presently, but has been proved for real analytic surfaces (class C^ω). As in all attempts at proving the conjecture, the idea is to show that the index of an umbilic can never exceed the value 1. It is true in the real analytic domain, where the proof is based on the existence of complex coordinates so as to be able to use analytic function theory as well as on ideas taken from the theory of algebraic curves (the Puiseux expansions cited in Sect. V.13). The degrees of rigor of various authors (Hamburger from 1940, Bol) has been improving; see Klotz (1959), Scherbel (1993) and Gutierrez and Sotomayor (1992). But the proof still remains extremely difficult.

The Willmore conjecture. This conjecture from 1965 is the following. For each compact surface S, we introduce the *Willmore functional* $S \mapsto \mathcal{F}(S) = \int_S H^2(x)\,dx$, where H is the mean curvature of S. This functional has the remarkable property of being invariant under the conformal group of $\mathbb{E}^3$ (for example, under inversion, etc.). We calculate easily $\mathcal{F}(S)$ for tori of revolution and find that it is minimum and equals $2\pi^2$ exactly for the tori of revolution for which the conformal type is that of the equidistant Clifford torus in S^3 (see Sect. II.6 and as an exercise examine the respective radii of the torus of revolution of this type). In particular this value remains equal to $2\pi^2$ for all the conformal images of this special torus.

In Willmore (1971) it was conjectured that the inequality $\mathcal{F}(S) \geqslant 2\pi^2$ always holds for surfaces of the topological type of a torus and that equality only holds for the conformal images indicated above. This conjecture has been most intensely studied, but is not yet completely resolved. As a reference, consult Ros (1999). The problem is resolved for many types of toroidal surfaces, whether tubular neighborhoods of space curves or the tori for which the conformal type is not too far from that of the (equidistant) Clifford torus (see Sect. II.6); that is, for which the complex conformal structure is that of an abstract square flat torus. But the most astonishing thing is that we know that the functional actually attains its minimum value for a torus embedded in $\mathbb{E}^3$ that is in addition real analytic. But the value of $\mathscr{F}$ isn't known, for we don't know the torus in question. It's a surprising situation, for in the great majority of minimization problems, the greatest difficulty lies in the existence of a minimizing object; its properties are then such that we know enough to finish (see for example the isoperimetric case in Sect. V.12). The situation here is thus completely atypical.

The surfaces of constant width. In Chap. V, on curves, we discussed *curves of constant width*. These are the boundaries of convex subsets of $\mathbb{E}^2$ that are likewise said to be *of constant width*, i.e. each pair of lines that frames the curve has the same breadth. The circle isn't the only such, for we can include every curve that is the developable of a compact curve having an odd number of cusps. The *Reuleaux triangle* corresponds to the case where such a curve is degenerate at the three vertices of an equilateral triangle; see 12.10.5 of [B]. Lebesgue showed, using Fourier series, that the Reuleaux triangle is the unique curve of a given width with minimum area.

In higher dimensions, starting with $d = 3$, the convex bodies (or the hypersurfaces) of $\mathbb{R}^d$ of constant width are almost a total mystery. If in all dimensions and for a given constant width the balls are those that have a maximum volume (it's trivial!), for the minimum starting with $d = 3$ the problem is almost completely open. The spherical harmonics don't yield any conclusion presently; the state of the problem in 1991 can be found in Problem A.22 of Croft, Falconer, and Guy (1991), along with references.

VI.10. Surfaces and genericity

Let us recall (see Sects. V.10, VII.13.D, XI.6) the notion of *genericity*: the generic is what is "real" and what we actually see. An equivalent but more mathematical definition is this: the generic is what is stable under small deformations. In fact nature ignores what isn't stable, for the attached probability is zero.

The subject is not exhausted by classifying the generic objects, because throughout mathematics evolutionary phenomena are encountered, families with one or more parameters.

For example, a parameterized curve, e.g. a C^∞ mapping of S^1 into the plane, is generically an immersion (there are no local singularities), and the only global singularities are ordinary crossings. But in a *generic one-parameter family*, i.e. stable as a family with one parameter, we can encounter

  − cusps,
  − self intersections with contact.

Stability here means that we can't eliminate these singularities with a small deformation of the one-parameter family.

In this situation we speak of (unavoidable) singularities of codimension 1.

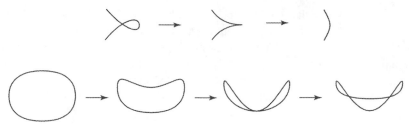

Fig. VI.10.1. Singularities of codimension 1 for plane curves (immersions of S^1): cusp
(local singularity) and self-contact (global singularity)

This idea of genericity is recent, it began with Whitney in 1955, but it was Thom
who got it completely underway in a general setting in a series of articles, then
with numerous applications in Thom (1989). Presently it is part of the foundations
of geometry and of mathematical physics. The basic reference, for a nontechnical
initial *exposé*, is Thom (1972–1977). For surfaces, genericity can't be found in the
classic works, except for Porteous (1994). For a completely mathematical version,
Demazure (2000) or Demazure (1989) can be consulted.

For problems concerning the **visualization** of surfaces, consult the basic reference Koenderink (1993) and the more recent Petitot and Tondut (2000).

In Sect. V.16 we already encountered the idea of the relation of the genericity of
surfaces to their local order, i.e. in the number of points in which they are intersected
by every line in the neighborhood of a given point. This order is 2 for elliptic points
(local convexity), 3 for hyperbolic points (called passes or saddle points), 4 for a
parabolic point (non planar), 5 for a biinflection point. In this section we will treat
three examples (two of these are intimately related) of genericity for the surfaces of
$\mathbb{E}^3$. But beforehand we should give a bit more detail about the behavior of the set of
normals to a surface.

Digression: the focal sheets of a surface. Surfaces enter in an essential way into
geometric optics as wave fronts of light rays. We need to remember this: the light
rays being (for an instant at least) the normal lines to a surface S, we need to know
what happens as time proceeds; i.e. what becomes of the surface when it is allowed
to be displaced uniformly with time along these normals (these *parallel* surfaces
are the *wave fronts*). We have encountered this idea of parallel objects for curves
in Sect. V.10. The nature of what we will later call focal sheets of a surface in
$\mathbb{R}^3$ is a fundamental problem in optics, since it is along these that light rays get
concentrated.

At a point where the two principal curvatures are nonzero and distinct, the
situation is this: the lines of curvature (here *line* means *curve*) are exactly the curves
of the surface at which the normals admit (locally, of course) an envelope curve, or
again generate a surface with total curvature zero, thus a developable surface; see
Sect. VI.7. At a generic point these envelope curves do not have singularities and
the basic generic figure is thus that shown below (two cases according to the sign of
the total curvature):

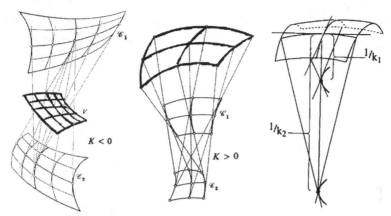

Fig. VI.10.2. The two types of disposition of the focal sheets of a surface according to
the sign of K

The two surfaces of the figure are called the *focal sheets* of S (we also say
caustics). They are common, at least locally, to all the *parallel* surfaces. The two
points where the normal touches them are the points where the light is focused; they
are situated on the normal at a distance equal to one of the two radii of principal
curvature. The fact that these are different (we are not at an umbilic) interprets the
fact that the surface is *astigmatic*, thus a very bad optical instrument. Such is the
case with the eyes of astigmatics; the defect of the surface of their eye is corrected
by appropriate lenses. What happens at umbilics or at points where the envelopes of
normals have singularities isn't describable in the single C^∞ context, but only in the
generic context, that will be defined precisely. Now we say only that singularities
are expected, fringes, etc., for the focal surfaces. The complete figure in the case of
the ellipsoid was already studied in Cayley (1873) and was shown in Fig. VI.3.12,
taken from Joets and Ribotta (1999), which studies the case of dimension four for
reasons of crystal physics.

Figure VI.10.1 is only valid in the generic case where not only are the two principal radii of curvature different, but also the radius of the associated principal curvature varies with a nonzero derivative along the lines of curvature. Beyond this, the
focal sheets have a behavior that is still poorly understood; see Porteous (1994): the
focal sheets then have cusps, etc.

Even before studying the singularities of the focal sheets, we can find where
these intersect: it is when the two radii of principal curvature are equal, by definition
at an umbilic, so we need first study the behavior of the lines of curvature in the
neighborhood of an umbilic.

The generic umbilics. This form of the lines of curvature has been studied since
Darboux (1887, 1889, 1894, 1896); in a modern book it is found in Porteous (1994).
The idea is to look at the Taylor approximation of the surface given as a graph
$(x, y) \mapsto (x, y, z = f(x, y))$. At an umbilic, the approximation begins with

$k(x^2 + y^2)$; in the generic case the third degree terms will not all be zero, thus of the form $ax^3 + 3bx^2y + 3cxy^2 + dy^3$. The problem is thus in particular to classify the cubic forms under the action of the orthogonal group for dimension 2. This first classification furnishes three types of figures:

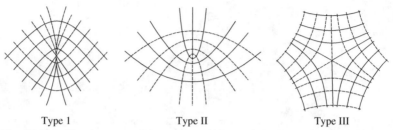

Type 1   Type II   Type III

Fig. VI.10.3. The three types of generic umbilics and the lines of curvature in the neighborhood

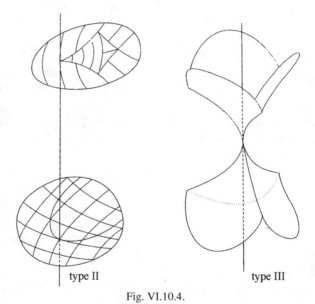

type II   type III

Fig. VI.10.4.

The index of these umbilics (see Sect. VI.9) is $\frac{1}{2}$ in the first two figures and $-\frac{1}{2}$ in the third. But the complete study is more subtle. Moreover, for the focal sheets we best use Thom's so-called *catastrophe* theory that we can't even begin an outline: see details and references in Porteous (1994), which is very concise, and in Koenderink (1993), which is much more detailed. It is noted in Porteous (p. 271) that a detailed study of umbilics had been undertaken by Gullstrand starting in 1904, with the aim of understanding the human eye, (and that this physician by training and self-taught mathematician and physicist received the Nobel prize in 1911 for this work). Moreover, the book of Porteous tries to remain elementary and doesn't enter into catastrophe theory, transversality, etc.

There is thus a clear link between the singularities of the focal sheets and those of points of the surface, i.e. the umbilics. The focal sheets can also have cusp edges. Here are some figures — for the two systems of lines of curvature and the focal sheets — that appear in color in the Porteous book, which may be consulted for more information.

Platonic readers will also appreciate in Porteous (1994) the brief and informal exposition on the connections between certain types of singularities and the polyhedra in spaces of dimensions 3 and 4, via the groups that are associated with them, called *Coxeter* groups. Such connections already appear in the theory of curves. We can't talk about this, even briefly, since we are forced, for lack of space, to treat only singularity-free plane curves.

◆

How we view a surface: apparent contours. At first we place ourselves at infinity, in a given direction δ; that is, we study two things: first, the boundary of the projection $\pi(S)$ of S, onto a plane P perpendicular to δ, by the mapping $\pi : \mathbb{E}^3 \to P$ defined by the direction δ. Then in fact the "curve" of S that is furnished by this boundary, i.e. the set of points of S where the tangent plane contains the direction δ, can be called its *rim*, whereas the projected curve is really the apparent contour. The generic case that comes up first is the one we think of naturally, whether it be an ellipsoid, a sphere, a torus (caution! — not too inclined!). Here we find the first stable singularity of the mappings $\mathbb{R}^2 \to \mathbb{R}^2$, called *folding*, which is quite natural, but see the figure below, which shows realizations in $\mathbb{R}^3$. The points of the visible plane have, outside of the apparent contour, two inverse images. On p. 221 of Porteous (1994) there is a nice little theorem, quite recent, which provides a relation between the Gaussian curvature at a point of the apparent contour of the folded-only type and the curvatures of two curves, one formed by this apparent contour and the other by the section of the surface by vertical normal plane.

Now, if we incline a torus more, the interior curve of the apparent contour will develop cusps. This situation is also stable under small deformations: it's the second stable singularity of the mappings $\mathbb{R}^2 \to \mathbb{R}^2$ and is called a *crease*. There aren't any others in the sole setting $\mathbb{R}^2 \to \mathbb{R}^2$. Here the points of the image, off the boundary curve, have one or three inverse images. But the problem of generic apparent contours doesn't stop here. The object in which we must study the stable forms is the three-parameter family of apparent contours of S obtained when we vary the viewpoint in $\mathbb{R}^3$. A fold or a crease observed from a certain point will persist if we move over a small neighborhood. Other singularities, in contrast, can only be observed from points of a certain set $\Sigma \subset \mathbb{R}^3$, which in general will be a surface, a curve or a point. Taken in isolation, they are thus unstable; but what implies the stability of the family is that a small deformation of the surface allows them to persist, all the while deforming or displacing the set Σ. The complete study of generic figures of

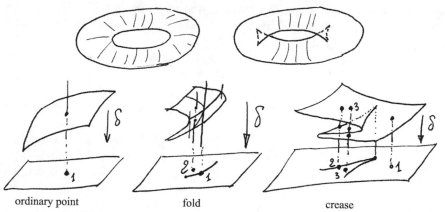

ordinary point   fold   crease

Fig. VI.10.5. The three types of singularities of the apparent contour of a surface

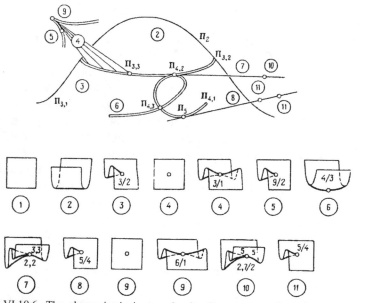

Fig. VI.10.6. The eleven intrinsic generic situations on a surface. Arnold (1983)
© London Mathematical Society and British Library

apparent contours and rims, thus comprised, only dates from 1981. We find it, placed
in a general setting, in pp. 145–152 of Arnold (1983), to which readers are referred.

We extract several figures from this study. An essential role is played by the
curve formed, on a generic surface, by the parabolic points, i.e. those where the
Gaussian curvature is zero. This curve will separate the points where the surface is

Fig. VI.10.7.

convex (K > 0) from those where it crosses its tangent plane (K < 0). Moreover, we expect to find supplementary singularities at the points (of negative curvature) for which the projection direction is also an *asymptotic direction*, i.e. a direction at a point of a surface for which the second fundamental form is zero. There clearly aren't any unless the Gaussian curvature is negative. Finally, we find **eleven** generic situations, represented in the figure:

We should emphasize that here we are concerned with the theoretical apparent contour, typically as in the case of the torus drawn above. What we really see in this situation is much less, a portion only. Just look at the drawing below. In the sketch of the torus, the apparent contour has been solidly sketched for what is *seen*, and in dashes for what exists, but is hidden from the eye. We also note that the classification presented doesn't contain the *global* generic singularities, such as the appearance of a hidden portion:

For a study oriented toward physical reality, and not just mathematics, see the very detailed Koenderink (1993).

VI.11. The isoperimetric inequality for surfaces

We will be very brief, and here is why: first, while the problem can be posed in all dimensions, and also while there aren't any particular properties connected to this problem, unlike certain problems relating the two fundamental forms (see "Higher dimensions" in Sect. VI.XYZ). The isoperimetric inequality is true for surfaces, the minimum of the ratio $Area^3/Volume^2$ being attained by spheres and only by them (see however the hairy spheres in Sect. VII.14); here we have again, as in Sect. V.12, several possible proofs. We refer to Burago and Zalgaller (1988) for a complete exposition; see also the outline in Sect. VII.14.

However, we would like to mention two other proofs. First that via Stokes' formula, given in the planar case in Sect. V.12; here we slice by planes, then the plane sections by lines, and finally the linear sections by points, all the while keeping the ratios of lengths, areas and volumes constant. For more details, if they are wanted, see Berger and Gostiaux (1987) or Berger (1990b), and again the remark on Schmidt's proof in Sect. V.12. Next the one using GMT (*geometric measure theory*): this theory shows, by being placed in a sufficiently but not too large category of objects (for example, the submanifolds aren't sufficiently general) so that there exist such objects minimizing $Area^3/Volume^2$; in fact the minimizing object is a true smooth surface (and in a higher dimension their singularities form a set of measure zero). Let S be such a minimizing surface; the differential calculus, in a manner analogous to what we stated for the plane in Sect. V.12, implies that S is of constant mean curvature H, thus it is necessarily a sphere (Hopf's theorem: see Sect. VI.8). We can also proceed, in any dimension, without this (subtle) result by filling the whole interior of the domain − of which S is the boundary − with normals, and using a simple inequality in the calculation of this volume. Here are some details, recalling what was said in Sect. V.12.

This method was introduced in Gromov (1980) in the very general setting of Riemannian manifolds; it can also be found in I.5.G of Berger (2003). It is very important to mention it, for it was Gromov who was the first to realize that GMT could be used for isoperimetric problems, the technique being that of filling. The idea is to fill the interior domain D of S by the normals. This is possible because each point d of D possesses at least one *foot* i.e. a point m of S which is closest to d among all the points of S. But then this condition of a minimum implies (at a smooth point) that the line md is normal to S at m, we thus fill all of D. In the reverse direction, when we start with an $m \in S$, it is necessary to stop when the normal ceases to minimize the distance; the length in question is called the *cut value m* (compare with the theory of the cut-locus in Sect. VI.3), denoted cut(m). We then calculate the total volume, denoted Volume(D), as an integral on the surface with the help of Fubini's theorem, by separating the variables into points of S and distance along the normals.

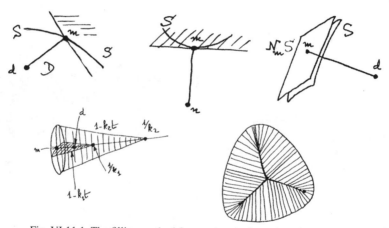

Fig. VI.11.1. The filling method for proving the isoperimetric inequality

It thus suffices to know the measure along normals, the second fundamental form works here by definition, and we have finally:

(VI.11.1)
$$\text{Volume}(D) = \int_S \int_0^{\text{cut}(S)} \left(1 - k_1(m)t\right)\left(1 - k_2(m)t\right)dt\,dm,$$

where the k_i are the principal curvatures. There is no difficulty because it is easy to see that the cuts always come before the first focal point $\frac{1}{k_1}$ (we have assumed $k_1 \leqslant k_2$): cut(m) $\leqslant \frac{1}{k_1}$. But the elementary inequality $xy \leqslant ((x+y)/2)^2$ shows that the above integrand is always bounded by $(1 - ht)^2$, where $h = \frac{k_1 + k_2}{2}$ is the mean curvature. Whence easily Volume(D) $\leqslant \frac{1}{3} \int_S \frac{dm}{h(m)}$. But we have seen that S must have constant mean curvature h (Formula VI.8.1), where finally Volume(D) $\leqslant \frac{1}{3h}$ Area(S). This isn't the inequality we were looking for. The idea for concluding is that the quotient Area^3/Volume^2 is also a minimum. We write that its derivative (first variation) is zero for the variation of the surface obtained by transporting a function f that is constant on the normals to S; we find that this derived formula (Formula VI.8.1) is zero only if we have the equality Volume(D) $= \frac{1}{3h}$ Area(S). By replacing h by the value that this equality provides, we deduce from this the isoperimetric inequality which was sought.

In the optimal isoperimetric inequalities (of all types), the case of equality is always more difficult to treat. See the case of convex sets in Chap. VII. Here the method gives round spheres instantaneously, for if there is equality in the above inequality, then there is equality at each point of the integrand and thus we have everywhere $k_1 = k_2 =$ is constant, so that all the points are umbilics and we have a round sphere (see above). More simply still, the fact that here the two principal curvatures are constant implies that all the normals pass through a fixed point, Q.E.D. We will give the "established" philosophy of the classical isoperimetric inequality in Sect. VII.13.

VI. XYZ

The fabrication of surfaces (sequel to Sect. V.7). How are planes, spheres and balls made? In fact, for planes and spheres it suffices to have **pieces** of them. The practical uses are innumerable, in optics, for example. Note however that if the lenses of ordinary eyeglasses are pieces of spheres, for astigmatics we need to use tori; as for Varilux (lenses with progressive focal length) matters are not yet well understood: it is not yet known what theoretical form will give a very good approximation to the desired result, i.e. good convergence of horizontal lines, with gradual variation of this convergence. In an absolute sense a perfect answer is impossible; readers will convince themselves of this by examining the behavior of the normals indicated in Sect. VI.10.

The fabrication of spherical lenses hasn't changed for a long time, matters are often still just as they were described in Bouasse (1917). What happens when we rub, while turning, two surfaces against one another? If we turn about a single point, we obtain only surfaces of revolution, but if we turn "most everywhere", we can't obtain anything but umbilics, and only spheres have their normals passing through a fixed point: study the subgroups of the three-parameter group of rotations of $\mathbb{E}^3$.

That is how spherical surfaces are made. That the result is sufficiently good is checked, as in Sect. V.7, by interferometry (study of *fringes*). As for the straight rulers in Sect. V.7, we obtain good planar surfaces by rubbing three pieces alternatively against one another. See more technical details in Bouasse (1917), whose reading we can't recommend too strongly, in particular the introduction for its pedagogic reflections.

Surfaces are complex curves. We have already seen in several places (twice in Sect. VI.8 and once in Sect. VI.9) that it would be interesting to attempt to introduce functions of a complex variable on a surface so as to be able to use the holomorphic

calculus. To put a complex structure on a Riemannian surface seems simple enough
at first glance. We say that a differentiable mapping from one oriented surface
M into another N ($\mathbb{E}^2 = \mathbb{C}$ is a particular case) *conformal* if the differential
$df : T_m M \to T_{f(m)} N$ is a direct (orientation-preserving) similarity. Since the holomorphic transformations on $\mathbb{C}$ coincide with the conformal transformations of $\mathbb{E}^2$,
the specification of an atlas made up of conformal charts of M allows us to endow
M with a structure of *analytic complex manifold of dimension* 1 (or, equivalently, of
a complex curve). But we can't be sure *a priori* that conformal charts of M exist,
nor that there are enough of them to compose an atlas. The coordinates associated
with a conformal chart $m \to (x(m), y(m))$ are called *isothermal coordinates*. Here
are equivalent properties:

– the lines $x =$ const. and $y =$ const. are orthogonal and we have $\|\nabla x\| = \|\nabla y\|$ (we also say that the lines $x =$ const. and $y =$ const. form a lattice of
"infinitesimal squares");
– we have $ds^2 = \lambda(x, y)(dx^2 + dy^2)$.

The existence of isothermal coordinates is a (not very difficult) classic result
(middle of the nineteenth century) in the theory of partial differential equations; see
a very quick presentation in Chern et al., 1954). For the geometer, a simple example
of the structure of an analytic complex manifold of dimension 1: for the sphere S^2
(with its canonical Riemannian structure of a round sphere), an atlas is obtained
by lifting by stereographic projection via the north pole and via the south pole, the
complex structure of $\mathbb{E}^2 = \mathbb{C}$.

Conformal representation and modules. Can we classify all the Riemannian surfaces? Yes, that has been done in two stages, by the theory of conformal representation beginning at the end of the nineteenth century and by Teichmüller in the 1930s
(theory of modules). This is one of the essential points of the theory called *Riemann
surfaces*, objects that are essential in the theory of numbers, functions of a complex variable, differential equations, etc. A Riemann surface is nothing other that an
oriented Riemannian surface that is endowed with a complex structure.

The *fundamental theorem of conformal representation* asserts that:

> On each oriented compact Riemannian surface (S, g) there exists a unique
> function $f : S \to \mathbb{R}_+^*$ such that the new metric $f.g$ is of constant curva-
> ture K.

The sign of K is imposed by the topology of S: it is positive for the sphere,
zero for the torus and negative in the other cases with more holes. For existence
(uniqueness is easy) a result from the theory of partial differential equations had
been awaited for a long time. In particular, Klein (the founder of the Erlanger Program in 1872) thought that it didn't require proof, because its physical rendering is
that of electric equilibrium on the surface. The first complete proof dates only from
the beginning of the twentieth century. We also encountered Klein in Sect. III.2,
where we used a theorem that is due to him.

A first consequence is thus that each oriented compact surface possesses metrics with constant curvature, thus locally isometric to Euclidean geometry, whether spherical or hyperbolic. This result can also be obtained by elementary geometry; see below.

The second consequence is that the classification of (oriented) Riemannian surfaces is reduced to those that are of constant curvature. These objects are often called the *space forms*. It can be shown without difficulty that these are exactly the (compact) geometries for which the metric satisfies a universal relation $F(b, c, \alpha)$ furnishing the length of the third side of a triangle $a = F(b, c, \alpha)$ as a function only of the two other sides b, c and the angle α that they form (recall the formulas of Euclidean space, of the sphere in III.1.1 and of the hyperbolic plane in Fig. II.4.2).

◆

Space forms. We can always normalize the constant of the curvature to $+1, 0, -1$. The case $+1$ is that of the round sphere; there is but one of them with curvature $+1$. We will encounter this situation in Sect. XII.5.B.

The case 0 is treated thus: we consider the universal covering of the given flat torus $\mathcal{T}$, i.e. the mapping $\mathbb{E}^2 \to \mathcal{T}$ which is everywhere a local isometry between $\mathbb{E}^2$ and $\mathcal{T}$. In the reverse direction, $\mathcal{T}$ is obtained as the quotient of $\mathbb{E}^2$ by a group of translations that we denote by Λ. This group is a lattice on $\mathbb{E}^2$. The situation is thus exactly what we studied in the digression in Sect. V.14; and there we gave the complete classification of flat tori (within isometry, and with an orientation). The result is that the space of all flat tori (within an isometry and with normalization by an appropriate homothety) is a noncompact surface, with two singular points. It is thus in particular an object of dimension 2, our flat tori depending on two parameters if they are taken normalized, on 3 parameters otherwise. Geometrically we may prefer to view the flat tori by slitting them with two cuts, obtaining thus a parallelogram, and inversely by identifying the opposite sides of a parallelogram of $\mathbb{E}^2$, two at a time. But we still don't have language at our disposal as easy as that of groups. As an exercise, readers may study, according the type of flat torus considered, the shape of the cut-locus of a point.

The negative case is quite a different story, and not completely settled. Contrary to the sphere and the flat tori the space forms with negative curvature are not homogeneous spaces, i.e. they are not endowed with a group of isometries that acts transitively, where "all the points are the same". The first method for studying a space form $\mathcal{S}$ with curvature -1 is to copy what was done for the tori. We thus start with a compact surface $\mathcal{S}$ of genus $\gamma \geq 2$ and consider its universal covering; endowed with the locally transported metric; it's thus the geometry of constant curvature -1, simply connected and complete, i.e. the hyperbolic plane Hyp^2. We obtain $\mathcal{S}$ "below" by taking the quotient of Hyp^2 by a suitable group of isometries; but the groups are very difficult to "see", even though they are fundamental and have been much studied (and it hasn't ended, Maskit, 1988). Let us say only, for the geometer, that just as the tori are obtained by identification of opposite sides of a parallelogram, here we can find in Hyp^2 polygonal domains with 4γ sides that yield $\mathcal{S}$ by appropriate pairwise identification of their sides. It is, moreover, a way of determining the difficult classification of compact surfaces (see Sect. VI.1) .

But there is a much simpler way of constructing all the space forms with negative curvature, it is the *pants* method. An important result in Hyp^2 is to show that there exist hexagons for which the six angles are right angles and for which the three sides x, y, z can be given; then we have formulas that show that the three remaining sides a, b, c are determined.

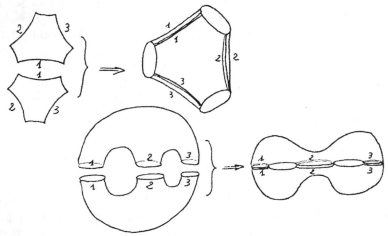

Fig. VI.XYZ.1. Construction and study of abstract surfaces with constant negative curvature by the pants method

An supplementary exercise, useless here but see Buser (1992) for a prospective proof and important applications, is the concurrence of the three *altitudes* of all the hexagons with six right angles. We return to the construction; we can then take a pair of such hexagons and glue them along their corresponding sides (see figure) to obtain *pants*, which is a surface with boundary (but always with constant curvature -1), for which the boundary is formed by three circles (which are geodesics, here hyperbolic lines) of equal lengths. Our boundary curves are finally singularity-free because the angles of these hexagons all equal $\frac{\pi}{2}$. But now we can fasten such pants together in arbitrary number to obtain all the types of surfaces of genus $\gamma \geqslant 2$, these objects being singularity-free surfaces, truly smooth (i.e. of class C^∞), because the boundary curves are hyperbolic lines and hyperbolic geometry is symmetric with respect to each of its lines.

This yields **some** space forms. But we get them all as follows: to see it, it suffices to incise some *a priori* space form by periodic geodesics that are pairwise nonintersecting; this is not too difficult. This geometric description allows a classification, for which we sketch the outline. We count the number of parameters on which the hexagons with six right angles and certain equalities involving the sides are dependent; then when we assemble along the boundary circles we can make one pair of pants turn with respect to the other. A careful study leads finally to this result: the space of all our $\mathcal{S}$ of a given genus γ is a set of dimension equal to $6\gamma - 6$. It is called the *Teichmüller space* (for genus γ) after O. Teichmüller, who was the first to study it in depth, in 1939. A detailed exposition of this description with hexagons, then pants, can be found in Buser (1992); but we can study the structure of Teich-müller spaces much further. This still remains a profound subject; see once more Buser (1992).

◆

Higher dimensions. Here is an important remark which, although classic, remains little known. The theory of surfaces we have just considered is an important subject, not only because it is classical and treats surfaces of our ordinary space, but also for the following reason. Take in $\mathbb{E}^d$ of dimension d greater than 3 any generic hy-persurface (in whatever precise sense desired). Then it is easy to see that its type of abstract isometry (as a Riemannian manifold of dimension $d - 1$) determines its second fundamental form as embedded in $\mathbb{E}^d$. This is why: at each point take a basis $\{e_i\}$ that diagonalizes the second fundamental form and let η_i be the eigenval-ues (the principal curvatures); then equations that generalize the *theorema egregium* assert that the sectional curvatures $K(e_i, e_j)$ of the abstract Riemannian metric are given by the formulas $K(e_i, e_j) = \eta_i \eta_j$ for all the $i \neq j$. But as soon as we know the three products $= ab, bc, ca$ of three *nonzero* real numbers a, b, c, we can calculate these three numbers. Thus in the case where there is practically but very little sectional curvature zero (which is certainly a generic condition), the Rieman-nian structure determines the second fundamental form (as many eigendirections as eigenvalues), as well as its derivatives. It is then classical — these are the gen-eral equations called Gauss-Codazzi — that these hypotheses completely determine the hypersurface within a global isometry of the whole space; our hypersurfaces are thus "the same". The above remark goes back to Killing (1885), it is found in Thomas (1935) and on p. 42 of Volume II of Kobayashi and Nomizu (1963–1969). Matters are completely different when $d = 3$; for we know only that $K = \eta_1 \eta_2$ and there is thus room for stimulating questions which, as we have seen, are not yet settled.

Bibliography

[B] Berger, M. (1987, 2009). *Geometry I, II*. Berlin/Heidelberg/New York: Springer
[BG] Berger, M., & Gostiaux, B. (1987). *Differential geometry: Manifolds, curves and surfaces*. Berlin/Heidelberg/New York: Springer
(1991) Biographical Dictionary of Mathematicians. New York: Charles Scribner's Sons
Appell, P., & Lacour, E. (1922). *Fonctions elliptiques*. Paris: Gauthier-Villars
Arnold, V. (1978). *Mathematical methods of classical mechanics*. Berlin/Heidelberg/New York: Springer
Arnold, V. (1983). Singularities of systems of rays. *Russian mathematica, 38*(2), 83–176
Arnold, V.I. (1994). *Topological invariants of plane curves and caustics*. Providence, RI: American Mathematical Society

Audin, M., & Lafontaine, J. (Eds.). (1994). *Pseudo-holomorphic curves in symplectic geometry.* Boston: Birkhäuser

Banchoff, T. (2004) Differential geometry and computer graphics, in *Perspectives in Mathematics*: anniversary of Oberwolfach 1984, Basel: Birkhäuser

Barany, I. (2008) . Random points and lattice points, *Bulletin of the AMS, 45*, 339–366

Berger, M. (1993). Encounter with a geometer: Eugenio Calabi. In P. de Bartolomeis, F. Tricerri, & E. Vesentini (Eds.), *Conference in honour of Eugenio Calabi, manifolds and geometry* (pp. 20–60). Pisa: Cambridge University Press

Berger, M. (1994). Géométrie et dynamique sur une surface. *Rivista di Matematica della Università di Parma, 3*, 3–65

Berger, M. (1998) Riemannian geometry during the second half of the century, *Jahresbericht der DMV, 100*, 45–208

Berger, M. (1999) Riemannian geometry during the second half of the century, *AMS University Lecture Series, 17*, Providence: American Mathematical Society

Berger, M. (2000). Encounter with a geometer I, II. *Notices of the American Mathematical Society, 47*(2), *47*(3), 183–194, 326–340

Berger, M. (2001). Peut-on définir la géométrie aujourd'hui? *Results in Mathematics, 40*, 37–87

Berger, M. (2003). *A panoramic view of Riemannian geometry.* Berlin/Heidelberg/New York: Springer

Berger, M., (2005a). *Cinq siècles de mathématiques en France.* Paris: ADPF (Association pour la diffusion de la pensée française)

Berger, M. (2005b). Dynamiser la géométrie élémentaire: Introduction à des travaux de Richard Schwartz. R.C., *Atti della Accademia Nazionale dei Lincei. Classe di, Ser. 25*, 127–153

Besse, A. (1978). *Manifolds all of whose geodesics are closed.* Berlin/Heidelberg/New York: Springer

Bleecker, D. (1996). Volume increasing isometric deformations of convex polyhedra. *Journal of Differential Geometry, 43*, 505–526

Bleecker, D. (1997). Isometric deformations of compact hypersurfaces. *Geometriae Dedicata, 64*, 193–227

Bliss, G. (1902–1903). The geodesic lines on an anchor ring. *Annals of Mathematics, 4*, 1–20

Bonnet, O. (1855). Sur quelques propriétés des lignes géodésiques. *Comptes Rendus de l'Académie des sciences, 40*, 1311–1313

Bouasse, G. (1917). Construction, description et emploi des appareils de mesure et d'observation. Paris: Delagrave

Boy, W. (1903). Curvatura Integra. *Mathematische Annalen, 57*, 151–184

Braunmühl, A.v. (1882). Geodätische Linien und ihre Enveloppen auf dreiaxigen Flächen zweiten Grades. *Mathematische Annalen, 20*, 557–586

Bruning, J., Cantrell, A., Longhurst, R., Schwalbe, D., & Wagon, S. (2000). Rhapsody in white; a victory of mathematicians. *The Mathematical Intelligencer, 23*(4), 37–40

Burago, Y., & Zalgaller, V. (1988). *Geometric inequalities.* Berlin/Heidelberg/New York: Springer

Burago, Y., & Zalgaller, V. (Eds.). (1992). *Geometry III.* Berlin/Heidelberg/New York: Springer

Buser, P. (1992). Geometry and spectra of compact Riemann surfaces. Boston: Birkhäuser

Calabi, E., & Hartman, P. (1970). On the smoothness of isometries. *Duke Mathematical Journal, 37*, 397–401

Cartan, E. (1946–1951). *Leçons sur la géométrie des espaces de Riemann* (2nd ed.). Paris: Gauthier-Villars

Cayley, A. (1873). On the centro-surface of an ellipsoid. *Transactions of the Cambridge Philosophical Society, 12*, 319–365

Chern, S.-S. (1944). A simple intrinsic proof of the Gauss-Bonnet formula for closed Riemannian manifolds. *Annals of Mathematics, 45*, 747–752

Chern, S.S. (1989). Studies in global geometry and analysis. Mathematical Association of America

Chern, S.-S., Hartman, P., & Wintner, A. (1954). On isothermic coordinates. *Commentarii Mathematici Helvetici, 28*, 301–309

Chern, S.-S., & Lashof, R. (1958). On the total curvature of immersed manifolds. *Michigan Mathematical Journal, 5*, 5–12

Ciarlet, P., & Larsonneur, F. (2000). Sur la détermination d'une surface dans $\mathbb{R}^3$ à partir de ses deux formes fondamentales. *Comptes Rendus de l'Académie des sciences, 331*, 893–897

Claude, J., & Devanture, C. (1988). *Les roulements.* Paris: SNR Roulements and Nathan Communications

Colding, T., & Minicozzi, W. (1999). *Minimal surfaces.* New York: Courant Institute of Mathematical Sciences, New York University

Colin de Verdière, Y. (1990). Triangulations presque équilatérales des surfaces. *Journal of Differential Geometry, 32*, 199–207

Corft, H., Falconer, K. & Guy, R. (1991). *Unsolved Problems in Geometry*, New York: Springer Verlag

Darboux, G. (1887, 1889, 1894, 1896). *Leçons sur la théorie générale des surfaces.* Paris: Gauthier-Villars

Darboux, G. (1972). Leçons sur la théorie générale des surfaces. New York: Chelsea

Demazure, M. (1989). *Géométrie – Catastrophes et Bifurcations.* Paris: Ellipses

Demazure, M. (2000). *Bifurcations and catastrophes.* Berlin/Heidelberg/New York: Springer

Dierkes, U., Hildebrandt, S., Küster, A., & Wohlrab, O. (1992). *Minimal surfaces I and II.* Berlin/Heidelberg/New York: Springer

Do Carmo, M. (1976). *Differential geometry of curves and surfaces.* Englewood Cliffs, NJ, Prentice-Hall

Dombrowski, P. (1979). *150 years after Gauss.* Paris: Société Mathématique de France

Fuchs, D., & Tabachnikov, S. (1995). More on paperfolding. *The American Mathematical Monthly, 106*, 27–35

Gardner, R. (1995). *Geometric tomography.* Cambridge, UK: Cambridge University Press

Gauld, D. (1982). *Differential topology.* New York: Marcel Dekker

Gluck, H. (1972). The generalized Minkowski problem in differential geometry in the large. *Annals of Mathematics, 96*, 245–276

Gramain, A. (1971). *Topologie des surfaces.* Paris: Presses Universitaires de France

Gray, A. (1993). *Modern differential geometry of curves and surfaces.* Boca Raton, FL: CRC Press

Gromov, M. (1980). *Paul Levy's isoperimetric inequality.* Prepublication M/80/320, Institut des Hautes Études Scientifiques. Reprinted as Appendix C of Gromov (1999)

Gromov, M. (1985). Pseudo-holomorphic curves in symplectic manifolds. *Inventiones Mathematicae, 82*, 307–347

Gromov, M. (1986). *Partial differential relations.* Berlin/Heidelberg/New York: Springer

Gromov, M. (1999). *Metric structures for Riemannian and non-Riemannian manifolds.* Boston: Birkhäuser

Grosse, K. (1997). Gyroids of constant mean curvature. *Experimental Mathematics, 6*, 33–50

Grosse-Brauckmann, K., & Polthier, K. (1997). Constant mean curvature surfaces with low genus. *Experimental Mathematics, 6*, 32–50

Guan, P., & Li, Y. (1994). The Weyl problem with nonnegative Gauss curvature. *Journal of Differential Geometry, 39*, 331–342

Gutierrez, C., & Sotomayor, T. (1992). *Lines of curvature and umbilical points on surfaces.* Rio de Janeiro: IMPA

Hadamard, J. (1897). Sur certaines propriétés des trajectoires en dynamique. *Journal de Mathématiques, 5*, 331–387

Hadamard, J. (1898). Les surfaces à courbure opposées et leurs lignes géodésiques. *Journal de Mathématiques Pures et Appliquée, 4*, 27–73

Hartman, P., & Nirenberg, L. (1959). On spherical images whose Jacobians do not change sign. *American Journal of Mathematics, 81*, 901–920

Hauchecorne, B., & Suratteau, D. (1996). *Des mathématiciens de A à Z.* Paris: Ellipses

Hebda, J. (1994). Metric structure of cut-loci in surfaces and Ambrose's problem. *Journal of Differential Geometry, 40*, 621–642

Hilbert, D. (1901). Flächen von konstanter Gauss'schen Krümmung. *Transactions of the American Mathematical Society, 2*, 87–99

Hilbert, D., & Cohn-Vossen, S. (1952). *Geometry and the imagination*. New York: Chelsea

Hilbert, D., & Cohn-Vossen, S. (1996). *Anschauliche geometrie*. Berlin/Heidelberg/New York: Springer

Hirsch, M. (1976). *Differential topology*. Berlin/Heidelberg/New York: Springer

Hopf, H. (1926). Vektorfelden in n-dimensionalen Mannigfaltigkeiten. *Mathematische Annalen, 96*, 225–250

Hopf, H. (1951). Über Flächen mit einer Relation zwischen den Hauptkrümmungen. *Mathematische Nachrichten, 4*, 232–249

Hopf, H. (1983). *Differential geometry in the large*. Berlin/Heidelberg/New York: Springer

Hopf, H., & Rinow, W. (1931). Über den Begriff der vollständigen differentialgeometrischen Flächen. *Commentarii Mathematici Helvetici, 3*, 209–225

Itoh, J. & Kiyohara, K. (2004). The cut loci and the conjugate loci on ellipsoids, *Manuscripta Math. 114* (2)

Joets, A., & Ribotta, R. (1999). Caustique de la surface ellipsoïdale à trois dimensions. *Experimental Mathematics, 8*, 57–62

Kapouleas, N. (1995). Constant mean curvature surfaces constructed by fusing Wente tori. *Inventiones Mathematicae, 119*, 443–518

Karcher, H., & Pinkall, U. (1997). Die Boysche Fläche in Oberwolfach. *Mitteilungen der Deutschen Math.-verein (DMV), 1997*, 45–47

Kilian, M., McIntosh, I., & Schmitt, N. (2000). New constant mean curvature surfaces. *Experimental Mathematics, 9*, 565–612

Killing, W. (1885). *Die Nicht-euklidischen Raumformen in Analytischer Behandlung*. Leipzig: Teubner

Klingenberg, W. (1973). *Eine Vorlesung über Differentialgeometrie*. Berlin/Heidelberg/New York: Springer

Klingenberg, W. (1995). *Riemannian geometry* (2nd ed.). Berlin: de Gruyter

Klotz-Milnor, T. (1972). Efimov's theorem about complete immersed surfaces of negative curvature. *Advances in Mathematics, 8*, 474–543

Klotz, T. (1959). On G. Bol's proof of Caratheodory conjecture. *Communications on Pure and Applied Mathematics, XII*, 277–311

Kobayashi, S., & Nomizu, K. (1963–1969). *Foundations of differential geometry I, II*. New York: Wiley Interscience

Koenderink, J. (1993). *Solid shape*. Cambridge, MA: MIT Press

Labourie, F. (1989). Immersions isométriques elliptiques et courbes pseudo-holomorphes. *Journal of Differential Geometry, 30*, 393–424

Lazard-Holly, H., & Meeks, W. (2001). Classification of doubly-periodic minimal surfaces. *Inventiones Mathematicae, 143*, 1–25

Martinez-Maure, Y. (2000). Contre-exemple à une caractérisation conjecturée de la sphère. *Comptes Rendus de l'Académie des sciences I, 332*(2001), 41–44

Maskit, B. (1988). *Kleinean groups*. Berlin/Heidelberg/New York: Springer

Massey, S. (1991). *A basic course in algebraic topology*. Berlin/Heidelberg/New York: Springer

Mazzeo, R., Pacard, F., & Pollack, D. (2000). Connected sums of constant mean curvature surfaces in Euclidean 3 space. *Journal für die Reine und Angewandte Mathematik, 536*, 115–165

Moise, E. (1977). *Geometric topology in dimensions 2 and 3*. Berlin/Heidelberg/New York: Springer

Montiel, S., & Ros, A. (1997). *Curvas y superficies*. Granada: Proyecto Sur de Ediciones

Morgan, F. (1988–2008). *Geometric measure theory: A Beginner's guide* (4th ed.). San Diego: Academic Press

Morgan, F. (1993). *Riemannian geometry: A Beginnner's guide*. Boston, MA: Jones and Bartlett.

Morgan, F. (November, 2000) Double bubble no more trouble. *Math Horizons, 2*, 30–31

Myers, S. (1935a). Riemannian manifolds in the large. *Duke Mathematical Journal, 1*, 39–49

Myers, S. (1935b). Connections between differential geometry and toplogy, I. *Duke Mathematical Journal, 35,* 376–391

Myers, S. (1936). Connections between differential geometry and topology, II. *Duke Mathematical Journal, 2,* 95–102

Nadirashvili, N. (1996). Hadamard's and Calabi-Yau's conjectures on negatively curved and minimal surfaces. *Inventiones Mathematicae, 126,* 457–466

Nitsche, J. (1996). *Minimal surfaces.* Berlin/Heidelberg/New York: Springer

Osserman, R. (1996). *Geometry V: Minimal surfaces.* Berlin/Heidelberg/New York: Springer

Petitot, J., & Tondut, Y. (1999). Vers une neuro-géométrie. Fibrations corticales, structures de contact et contours subjectifs modaux. *Mathématiques Informatique et Sciences humaines, 145,* 5–101, Paris: EHSS

Pinkall, U., & Sterling, I. (1989). On the classification of constant mean curvature tori. *Annals of Mathematics, 130,* 407–451

Poincaré, H. (1905). Sur les lignes géodésiques des surfaces convexes. *Transactions of the American Mathematical Society, 6,* 237–274

Porteous, I. (1994). *Geometric differentiation (for the intelligence of curves and surfaces).* Cambridge, UK: Cambridge University Press

Ros, A. (1999). The Willmore conjecture in the real projective space. *Mathematical Research Letters, 6,* 487–493

Scherbel, H. (1993) *A new proof of Hamburger's index theorem on umbilical points.* Zürich: ETH

Seifert-Threlfall (1980). *A textbook of topology.* New York: Academic Press

Sinclair, R. (2003). On the last geometric statement of Jacobi, *Experimental Mathematics, 12,* 477–486

Sinclair, R. & Tanaka, M. (2002). The set of poles of a two-sheeted hyperboloid, *Experimental Mathematics, 11,* 27–36

SNR Roulements (1988). *Les roulements.* Paris: Nathan-Communications

Spivak, M. (1970). *A comprehensive introduction to differential geometry.* Publish or Perish, Boston

Stillwell, J. (1980). *Classical topology and combinatorial group theory.* Berlin/Heidelberg/New York: Springer

Stillwell, J. (2001). The story of the 120-cell. *Notices of the American Mathematical Society, 48,* 17–25

Stoker, J.J. (1969). *Differential geometry.* New York: Wiley Interscience

Struik, D. (1950). *Lectures on classical differential geometry.* Reading, MA: Addison-Wesley

Thom, R. (1972–1977). *Stabilité structurelle et morphogénèse.* Paris: Benjamin-InterEditions

Thom, R. (1989). *Structural stability and morphogenesis: An outline of a general theory of models.* Reading, MA: Addison-Wesley

Thomas, T. (1935). Riemannian spaces and their characterizations. *Acta Mathematica, 67,* 169–211

van Kampen, E. (1938). The theorems of Gauss-Bonnet and Stokes. *American Journal of Mathematics, 60,* 129–138

Wallace, A. (1968). *Differential topology.* New York: Benjamin

Wente, H. (1986). Counterexample to a conjecture of H. Hopf. *Pacific Journal of Mathematics, 121,* 193–243

Wichiramala, W. (2004). Proof of the planar triple bubble conjecture. *Journal für die Reine und Angewandte Mathematik, 567,* 1–49

Willmore, T. (1971). Minimum curvature of Riemannian immersions. *Journal of the London Mathematical Society, Second Series, 3,* 307–310

Zamfirescu, T. (2004). On the cut locus in Aleksandrov spaces and applications to convex surfaces, *Pacific J. of Math., 217*(2), 375–386

Chapter VII
Convexity and convex sets

VII.1. History and introduction

The history of convexity is rather astonishing, even paradoxical, and we explain why. On the one hand, the notion of convexity is extremely natural, so much so that we find it, for example, in works on art and anatomy without it being defined. Below are two excerpts, one from a book on art (1985) illustrating a modern sculpture; the other, from a classic anatomy reference (Rouvière), describes the extremely subtle overlapping of menisci in the knee. We also find in the same work a description of the aortic arch, also very complex, which uses the words "concave" and "convex" several times.

Intra-articular menisci or crescent shaped fibrocartilage. "So arranged, the tibial plateaus are not adapted to the femoral condyles. Congruence is achieved by the interposition between the tibia and the femur of intra-articular menisci or crescent shaped fibrocartilage.

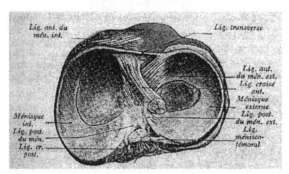

Fig. VII.1.1.

The crescent shaped fibrocartilage is differentiated into medial or lateral. Each of them is a triangular, prismatic lamella curved to form a crescent.

They are recognized as follows: a *concave upper face* in rapport with the femoral condyles; – a *lower face*, applied to the periphery of the corresponding tibial plateau; an *external face* or *periphery*) (base of the prism), convex, very dense, adhering to the articular capsule; – an *internal* (or *central*) side, concave, sharp, and whose concavity conforms to the center of the tibial plateau; – finally, two *extremities* or *horns*, whence the ligament fasciae" (Rouvière, 1973.)

M. Berger, *Geometry Revealed*, DOI 10.1007/978-3-540-70997-8_7,
© Springer-Verlag Berlin Heidelberg 2010

"Women were a favorite subject of Achipenko. This sculpture reflects the influence of cubism in the optical juxtaposition of concave and convex forms and in the use of the void to give the impression of massiveness."

Fig. VII.1.2. A. Archipenko, 1914. Woman combing her hair, bronze. Israel Museum, Jerusalem

On the other hand, the notion of convexity didn't appear "seriously" in mathematics until Minkowski at the turn of the twentieth century. A striking case is that the book Hilbert and Cohn-Vossen (1952), with its astonishing title and which is a bit our model (see its introduction), doesn't contain a single word on convexity. For a rather detailed history of the subject, see Chap. 0 of Gruber and Wills (1993). Here we give a brief summary.

We find the notion of convexity with Archimedes (250 B.C.), with Newton (1720) with respect to the classification of singularities of algebraic curves, with Poinsot (toward 1800) regarding statics (see Fig.VII.1.3) and the notion of a sustentation polygon, with Fourier where we find the seeds of linear programming. Then comes Minkowski; we will amply encounter this great mathematician in Chaps. IX and X. Apart from specific results, he had an entire program, but he died very young. Between Minkowski and the 1960s, we find little about convexity in the literature, with the notable exception of an important work on polyhedra, Steinitz and Rademacher (1934); see the next chapter. After the 1960s, convexity suddenly took on considerable importance. Here is how Grünbaum, in the introduction to Grünbaum (1993), explains this "hollow", this abyss. On the one hand, the problems that arose naturally were too difficult for their time − there were not enough powerful tools; on the other hand, the demand for such results had been very strong, from at least three directions: harmonic analysis, probability theory and linear programming. We will return to the history in the next chapter. Convexity in physics appears, among other places, in the book Israel (1979); read the introduction there by Whigtman.

The ideas and results in this domain are, however, a real jungle, we don't have the courage to traverse them completely; we will just make a quick safari, the locations chosen by two criteria: that they enter into the "Jacob's ladder" vision of our book and that they provide aesthetically satisfying examples. We need to emphasize that to a large extent we will work in arbitrary dimension and not just the visible 2 and 3; and even more: we will be interested in how various concepts behave when the dimension becomes very large. One reason for this is that function spaces, in harmonic analysis, are of infinite dimension, and yet it is anticipated that we can grasp them, attain intuition for them, by looking at spaces which are of increasingly large finite dimension. See the representative Giannopoulos and Milman (2004). Another reason is that contemporary applications of linear programming require not just hundreds, but even thousands of variables. We will encounter large dimensions, with applications in the forefront; and, in Chap. X, the very recent proof of Kepler's conjecture. We add that the idea of convexity has incredible power, its applications are legion; we will see this right away in Sect. VII.2 with applications that are simultaneously spectacular and of disarming simplicity. But we will also see some astonishing illustrations of our theme: results whose statements are elementary but whose proofs (at least to this day) very difficult, and problems whose statements are equally elementary, but which are still unresolved.

Regarding references, here first is a partial list of classic works on the subject: [B], Chaps. 11 and 12 of Eggleston (1958), Valentine (1964) and Leichtweiss (1980). But since the discipline has taken off, it is impossible to write a book that treats all aspects. We find an exception in Gruber and Wills (1993), which is of a sort that gives "reports" and is therefore without detailed treatment. It may be consulted regularly

Fig. VII.1.3.

for the nonclassical subjects about which we will speak. At the time of its appearance, this work was a perfectly up-to-date and complete reference; it remains an incomparable source of synthesis. Subsequent to the older works listed, there have appeared several monographs: Gardner (1995) and Schneider (1993). For polytopes more specifically – see Chap. VIII – we can refer to Ziegler (1995). A good part of what we will treat is very well synthesized in Giannopoulos and Milman (2001).

The structure of this chapter, the choices we have made in this immense subject, reflect our taste for this or that topic, result, problem, etc. In Sect. VII.13 ("Miscellaneous") we very rapidly survey various topics that seem difficult to pass by in silence and say again that Gruber and Wills (1993) is the ideal reference for completing our little excursion when that is needed; also Gruber (1996, 2007).

VII.2. Convex functions, examples and first applications

A function f (with real values) and defined on an (arbitrary: open, closed, infinite) interval I is called *convex* if:

$$f(\lambda x + (1 - \lambda)y) \leqslant \lambda f(x) + (1 - \lambda)f(y) \text{ for each } x, y \in I \text{ and each } \lambda \in [0, 1].$$

A function f is called *strictly convex* if we have strict inequality throughout the above (for $x \neq y$ and $\lambda \in]0, 1[$). For the geometry we read this definition on the graph of the function: for any two points of the graph, the portion of the graph between these two points lies below the segment of the line that joins them.

The opposing definitions of *concave* and *strictly concave* are left to the reader. Three trivialities:

- a convex function attains its maximum at an extremity of its interval of definition;

- a strictly convex function attains its minimum at a single point;

- the convexity of f implies $f(\sum_i \lambda_i x_i) \leqslant \sum_i \lambda_i f(x_i)$ for all finite sequences $\{x_i\} \subset I$ and $\{\lambda_i\}$ such that $\sum_i \lambda_i = 1$ and $\lambda_i \geqslant 0$ for all the i.

Two items that are a little less evident, yet simple enough with differential calculus:

- a convex function is continuous on the interior of its interval of definition (but note that at the endpoints of this interval it can attain any sufficiently large value);

- a convex function has right and left derivatives at each point on the interior of its interval of definition. Moreover, the points at which it fails to be differentiable are denumerable.

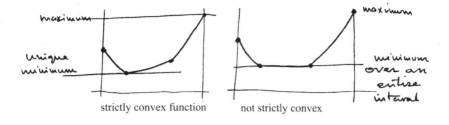

strictly convex function   not strictly convex

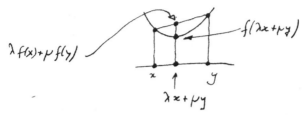

Fig. VII.2.1.

Finally, an essential criterion for convexity:

- is f is twice differentiable and if $f'' \geq 0$, then f is convex (and *mutatis mutandis* for strict convexity).

♦

Here are two classic and essential examples. The first is $f = -\log x$ on $\mathbb{R}_+^*$, whose second derivative is $\frac{1}{x^2}$ and thus clearly positive. For all λ_i as above and the $a_i > 0$ we thus have $\sum_i \lambda_i a_i \geq \prod_i a_i^{\lambda_i}$, for example $\lambda a + \lambda' b \geq a^\lambda b^{\lambda'}$ for $\lambda + \lambda' = 1$. In particular, for each integer n and taking all the λ_i equal to $\frac{1}{n}$, we find

(VII.2.1) $$\left(\frac{a_1 + \cdots + a_n}{n}\right)^n \geq a_1 \ldots a_n.$$

If this inequality (often called *Newton's inequality*) is trivial for $n = 2$, the reader will gain more respect for it starting with $n = 3$ by attempting it without convexity! We already used it in Sect. VI.11 and will encounter it again in the following section.

For the second example we take $f = x^p$ (for all real $p > 1$). A classical calculation, but a bit long and tricky (see as needed the references given) shows that the convexity of x^p implies the inequalities:

(VII.2.2) Hölder inequality:

$$\sum_i x_i y_i \leq \left(\sum_i x_i^p\right)^{1/p} \left(\sum_i y_i^q\right)^{1/q},$$

where $\dfrac{1}{p} + \dfrac{1}{q} = 1$ *(positive numbers).*

When $p = 2$ (and thus $q = 2$) we know this inequality well from Euclidean ge-
ometry: in $\mathbb{E}^n$: $\langle x, y \rangle \leqslant \|x\| \|y\|$. And here the triangle inequality is generalized by:

(VII.2.3) Minkowski's inequality:

$$\left(\sum_i |x_i + y_i|^p \right)^{1/p} \leqslant \left(\sum_i |x_i|^p \right)^{1/p} + \left(\sum_i |y_i|^p \right)^{1/p}$$

(note that the plus sign on the left may be replaced by a minus sign).

The above inequalities are nowadays used constantly in practically all of analysis.
In fact, first they extend to integrals, which replace the finite sequences, for which
we can use For example an arbitrary measure:

(VII.2.4) $\displaystyle\int fg \leqslant \left(\int f^p \right)^{1/p} \left(\int g^q \right)^{1/q}$ *(always with* $\dfrac{1}{p} + \dfrac{1}{q} = 1$).

For $p = q = 2$ we find the classical Schwarz inequality. Next, they imply that, for
functions, $\left(\int f^p \right)^{1/p}$ is a proper norm. We thus have function spaces at our disposal,
called L^p spaces, which are indispensable in numerous convergence problems, for
example for solutions of partial differential equations. The inequality (VII.2.4) extends trivially to arbitrary finite products; we will use these in an essential way for
sections of the cube in Sect. VII.11.C.

Here are three geometric examples that illustrate the power of the notion of a convex function. The first is the isoperimetric inequality for spherical triangles; see
Sect. III.1. We saw in Sect. III.5 that such an inequality is open starting with
3-dimensional spheres. Here we use the formula (see for example 18.6 of [B])

$$\tan \left(\frac{\text{Area}}{4} \right) = \sqrt{ \tan \frac{p}{2} \, \tan \frac{p-a}{2} \, \tan \frac{p-b}{2} \, \tan \frac{p-c}{2} }$$

and the fact that the function $\log \circ \tan$ is strictly concave. We obtain

$$\tan \frac{p-a}{2} \, \tan \frac{p-b}{2} \, \tan \frac{p-c}{2} \leqslant \left(\tan \frac{p}{6} \right)^2,$$

with equality if and only if $a = b = c$.

In the second example, we attempt to resolve problems of the "isoperimetric"
type for polygons with a fixed number of sides that are inscribed in a given circle
(for the considerably more difficult general problem, where there is no such circle, see Sect. VIII.3). We let α_i denote the half-angles at the center of these sides,
thus $\sum_i \alpha_i = \pi$. We look for polygons of maximal area and for those of maximal
perimeter. The area of the polygon is $\sum_i \cos \alpha_i \sin \alpha_i$ and the perimeter $2 \sum_i \sin \alpha_i$
(having normalized the radius of the circle to 1). We obtain the given result by using
the strict concavity of the function $\sin$ and that of the product $\sin \cdot \cos$ on $[0, \pi/2]$:
in the two cases the maximum is attained for regular polygons and for them only.

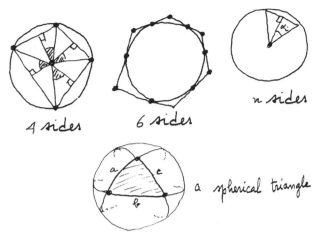

4 sides   6 sides   n sides

a spherical triangle

Fig. VII.2.2. Four sides: to maximize the perimeter or the area; six sides: to minimize
the perimeter or the area; n sides: $\sum \alpha_i = \pi$, perimeter $= 2 \sum \sin \alpha_i \leqslant 2n \sin \frac{\pi}{n}$

Thirdly, we pose the same problems for polygons circumscribed about a circle.
The area this time equals $\sum_i \left(\tan \frac{\alpha_i}{2} \right)$ and we proceed in the same fashion. Furthermore, for the perimeter it suffices to remark that, for all polygons circumscribed
about the unit circle, we have area $= \frac{1}{2}$ perimeter. As to what happens when we pass
from inscribed to circumscribed, we need to be aware — even if we are experts on
duality — that the areas of the inscribed polygon and its circumscribed dual are not
reciprocals of each other. For this essential problem of "volume and duality", see
Sect. V.8.

VII.3. Convex functions of several variables, an important example

Suppose now that we want to define convexity for functions of several variables,
i.e. for functions D $\rightarrow$ $\mathbb{R}$ where D is a subset of $\mathbb{R}^n$. We then want to write a
condition in $\lambda x + (1 - \lambda)y$, where $x, y \in$ D. This condition cannot therefore have
the sense that all $\lambda x + (1 - \lambda)y$ are again in D. Observe that these combinations
of points are an affine notion (see Sect. I.1), which leads us entirely naturally to the
definition: *a subset* D *of a real affine space* A *is called* convex *if* $\lambda x + (1 - \lambda)y \in$ D
for arbitrary $x, y \in$ D *and* $\lambda \in [0, 1]$. The notion of strict convexity for sets is
more difficult to define; see Sect. VII.10.A and classical references. In this work, we
only consider real affine spaces of finite dimension, thus the spaces $\mathbb{R}^n$. We employ
interchangeably the terms *convex set* and *convex body*. A function f defined on a
convex set D is said to be *convex* if $f(\lambda x + (1 - \lambda)y) \leqslant \lambda f(x) + (1 - \lambda)f(y)$
for arbitrary $x, y \in$ D and $\lambda \in [0, 1]$. There is an analogous definition for *strict
convexity*. Convexity is therefore an affine notion, there is no *a priori* Euclidean
structure.

As in the single variable case, a strictly convex function attains its minimum at a
unique point. A convex function is continuous at each point interior to its domain D

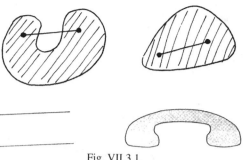

Fig. VII.3.1.

of definition. We will have no need of it, but it is important for a good understanding of the meaning of convexity: a convex function is almost everywhere differentiable in the interior of D, whereas in general there is no second derivative. With respect to convexity criteria, we have here the analogue of the second derivative criterion in the one-variable case: if the function f has on its convex domain of definition D a second derivative f'' which, in its guise as a quadratic form is positive or zero (resp. strictly positive), then the function f is convex (resp. strictly convex). In coordinate language, the condition is that the matrix of the $\frac{\partial^2 f}{\partial x_i \partial x_j}$ is symmetric with nonnegative (resp. strictly positive) eigenvalues; see also Sect. VII.13. A very complete reference for the differentiability of convex functions is Rockafellar and Wets (1998). There is also the classic Roberts and Varberg (1973), then Giles (1982), but the simplest for a first reading is the synthesis Gruber and Wills (1993, Sect. 4.2).

We should be wary that things may be worse than in the single variable case. Practically anything at all can happen at the boundary. A typical example consists in taking D to be the unit disk and a function f that is identically zero on the interior and which takes on arbitrary nonnegative values on the boundary.

♦

Here is a subtle example – essential for the geometry of convex sets – of a strictly convex function; see Sect. VII.10.C. The convex set of definition will be the set $\mathcal{Q}$ of all positive definite quadratic forms defined on $\mathbb{E}^d$, a part of $\mathbb{R}^{d(d+1)/2}$. To each $q \in \mathcal{Q}$ we can attach its *determinant* $\det(q)$. This is the determinant of the matrix representing q in the canonical basis or any orthonormal basis. It's also the product of the eigenvalues of q. We will subsequently use the geometric interpretation of this determinant:

> Within a constant (dependent only on the dimension), we have: $1/\sqrt{\det(q)} = (\det(q))^{-1/2}$ is equal to the volume of the (here always solid) ellipsoid $q^{-1}([0, 1])$.

We first verify that: if we diagonalize q in $q = \sum_i a_i x_i^2$, then $q^{-1}([0, 1])$ is an ellipsoid whose axes have lengths equal to the $1/\sqrt{a_i}$ and for which we will see in Sect. VII.6.B have volume $\beta(d) \prod_i 1/\sqrt{a_i}$, where $\beta(d)$ denotes the volume of the unit ball of $\mathbb{E}^d$. This is indeed what we wanted.

Now: *the function $q \mapsto \det(q)^{-1/2}$ is strictly convex on* $\mathcal{Q}$. To see this we diagonalize simultaneously two forms q and q'; with the obvious prime notation we have,
successively:

$$
\begin{aligned}
\left[\det(\lambda q + \lambda' q')\right]^{-1/2} &= \left[\prod_i (\lambda a_i + \lambda' a_i')\right]^{-1/2} \\
&\leq \left[\prod_i (a_i^{\lambda} a_i'^{\lambda'})\right]^{-1/2} \\
&= \left[(\det(q)^{-1/2})^{\lambda}\right]\left[(\det(q')^{-1/2})^{\lambda'}\right] \\
&\leq \lambda(\det(q)^{-1/2} + \lambda'(\det(q')^{-1/2},
\end{aligned}
$$

where we have applied the inequality $\lambda a + \lambda' b \geq a^{\lambda} b^{\lambda'}, \lambda + \lambda' = 1$ from Sect. VII.2
several times.

VII.4. Examples of convex sets

Here now are some essential examples of convex sets. For all of these we are in
an unspecified affine space.

First a cautionary example, suggested by the function above. An open disk in the
plane to which we adjoin an **arbitrary** portion of the boundary circle always remains convex. In contrast, we cannot do the same with a square (verify this for
yourself!).

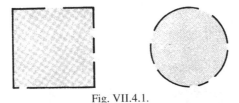

Fig. VII.4.1.

This is the moment to remark that a convex set need neither be open nor closed.
However, in the majority of cases we will work with *closed* convex sets; in fact they
will most often be compact.

Two basic examples: a *half-space* (open or closed) associated with a hyperplane, – and the intersection of two arbitrary convex sets; in fact any intersection
of convex sets from any collection, not necessarily countable in number, remains
convex.

This shows that, whatever the subset D, there exists a smallest convex set that
contains D, called the *convex envelope* of D. For more on the relation to the half-
spaces, see Sect. VII.10; we should remember for now that a closed convex set can
be realized as a denumerable intersection of closed half-spaces.

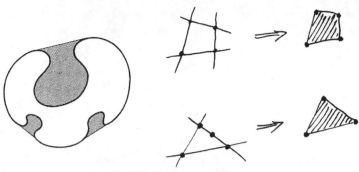

Fig. VII.4.2. Convex envelopes

♦

Two examples are important for the sequel, and in a sense they are two extremes. First, the *solid ellipsoids*: in each dimension the inequality $\sum_i \frac{x_i^2}{a_i^2} \leqslant 1$ defines a compact convex set, where all the a_i are taken positive. To prove it, we give an argument essential in itself, i.e. that from the affine viewpoint (the dimension being fixed) all ellipsoids are the same. In particular, they are equivalent to the closed ball $\sum_i x_i^2 \leqslant 1$, which is convex because of the triangle inequality. The ellipsoids play a pivotal role in the theory of convexity and in its applications. We find them notably in Sect. VII.10.C. They may also be characterized by a number of their properties, typically by having an affine symmetry for each linear direction; see Sect. VII.10.C. We can say that they are the most beautiful of convex sets. All of 11.1.3 of Gruber and Wills (1993) is dedicated to them.

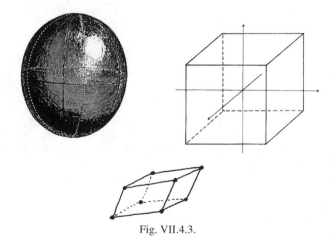

Fig. VII.4.3.

At the other extreme we find the *(solid) parallelepipeds*: we choose any basis and denote by x_i the associated coordinates; then the associated parallelepiped is the set $\{(x_1, \ldots, x_d) : 0 \leqslant x_i \leqslant 1 \text{ for all } i\} = \prod_i [0, 1] = [0, 1]^d$. To the affine eye here too all parallelepipeds are the same and identical (affine-isomorphic) to the cube $[0, 1]^d$ of the Euclidean space $\mathbb{E}^d$. We may consider them as being, in several senses, the "worst" among the convex sets; see Sect. VII.10.D.

◆

We have just suggested that the ellipsoids are in some way "the most beautiful" of the convex sets, the "best", whereas the parallelepipeds would be the least round, the "worst". We are going to show in the sequel that this is indeed the case, and in several different respects. Nevertheless, for this we will need to restrict ourselves to a fundamental class: the *center-symmetric convex sets*. These are the convex sets C for which there exists a point p of the affine space such that C is preserved by the affine symmetry with center p. Ellipsoids and cubes are center symmetric, in contrast to the worst (or the simplest and most natural) of convex sets: the *simplex*. A simplex of $\mathbb{R}^d$ is what generalizes a triangle in the plane and the tetrahedra of three-dimensional space: it is by definition the convex envelope of a set of $d + 1$ points of an affine space of dimension d which are affinely independent, which is to say that if this set is $\{x_i\}$ ($i = 0, \ldots, d$), then the vectors $x_i - x_0, i = 1, \ldots, d$, form a basis of the associated vector space. Otherwise expressed, the two inequalities $\sum_i \lambda_i x_i = p$ and $\sum_i \lambda_i = 1$ determine the λ_i uniquely.

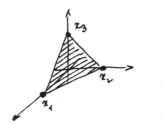 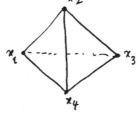

Fig. VII.4.4.

It is correct to say **the** simplex of a given dimension, for affinely they are all equivalent. The parallelepipeds and the simplexes are two particular instances of *polytopes*, which generalize to arbitrary dimensions the polygons in the plane and the polyhedra in space. The definition is trivial: a *convex polytope* is the envelope of a finite number of points of the space. We devote all of Chap. VIII to them. In fact, despite spectacular progress since 1960, there still remain plenty of open problems concerning polytopes.

There is an enormous difference between the center-symmetric convex sets and convex sets in general; we will see this in various places; the polytopes are no exception.

◆

Our final example corresponds to the L^p. We consider the set of $\mathbb{R}^d$ defined, for a real $p \geqslant 1$, by $\sum_i |x_i|^p \leqslant 1$ (denoted in the sequel by the expression Ball$(d; p)$).

For $p = 2$ we find the ordinary (closed) Euclidean balls and for $p = 1$ we find the *cocube*; we will see in Sect. VII.5 that this is the polytope dual to the cube. In the plane this amounts to a square (but with diagonal verticals), in space to an octahedron (regular in $\mathbb{E}^d$). For the other p ($p > 2$), we find sets containing the unit ball and contained therein. It isn't trivial to see geometrically that they are really convex sets (try it!), but it is an immediate consequence of the Minkowski inequality seen in Sect. VII.2. The cube corresponds to infinite p (in this case we replace $\left(\sum |x_i|^p \right)^{1/p}$ by its limit, which is $\sup |x_i|$).

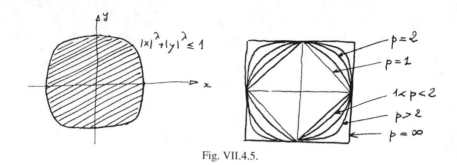

Fig. VII.4.5.

VII.5. Three essential operations on convex sets

Geometers too must be knowledgeable and have the courage to introduce invariants and concepts of a more or less algebraic nature in order to make progress. We will identify some rungs of the ladder, three beyond a "zero-th" rung, which is the sectioning of a convex set by an affine subspace (of any dimension): it always yields another convex set.

However, we will see in Sect. VII.11 that "slicing" (sectioning by hyperplanes) reveals as many difficulties as surprises, even for the simplest convex sets. Here now are three much less trivial operations.

VII.5.A. The (Steiner, Schwarz) symmetrizations

The technique of symmetrization has numerous applications; we will see it enter into the isoperimetric inequality in Sect. VII.7. With it we prove the fact (seen in Sect. I.1) that when $d + 2$ points are thrown at random at a convex set, it is when that set is an ellipsoid that the probability that they form the vertices of a simplex is greatest. The ellipsoids also have an extreme property for the behavior of the product of the volume of a convex set and that of its polar; see Sect. VII.9.

In Sect. V.11 we encountered Steiner's symmetrization for a plane curve and a linear direction. It can be defined just as well in arbitrary dimension and we present it in the affine context. We take hyperplane H and a direction δ transverse to H. To each portion P of space we associate its *(Steiner) symmetrization* $\mathrm{sym}_{H,\delta}(P)$ defined as in the figure:

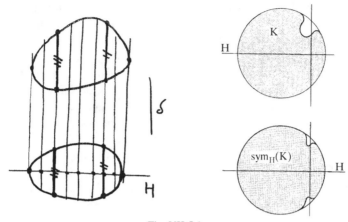

Fig. VII.5.1.

This says that, for each line D parallel to δ, the part $D \cap \mathrm{sym}_{H,\delta}(P)$ is a segment of D having its midpoint at $H \cap D$ and length equal to the length of $D \cap P$. By construction, our symmetrized set is certainly symmetric with respect to the hyperplane H for the direction δ. In the Euclidean case we always tacitly choose δ to be the direction orthogonal to H and use the notation sym_H.

It is easy to see that each Steiner symmetrization preserves convexity: if D is convex, then so is $\mathrm{sym}_{H,\delta}(P)$. A sketch and the word *trapezoid* suffice for seeing it.

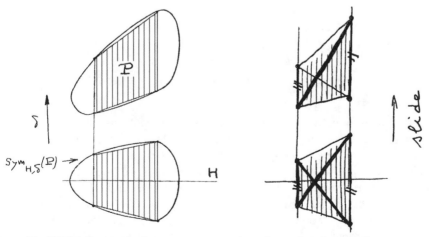

Fig. VII.5.2. Symmetrization preserves convexity and can only reduce the diameter

The formal result is that in a trapezoid, parallel to the base, the section length is an affine function. It follows that the length of the sections of the convex set P that are parallel to δ, considered as a function defined on H, is clearly a concave function. A brief remark for the purist: there is no need of a Euclidean structure in order to speak of the length of these parallel segments, it suffices to have a unit length in the δ direction.

The *Schwarz symmetrization* is less evident. We place ourselves in dimension 3 and in the Euclidean context. We take a line D of $\mathbb{E}^3$ and slice the set P considered by planes H perpendicular to D, which yields convex sets that have an area; we then construct, centered at the point D $\cap$ H, the disk which has the same area as P $\cap$ H. The union of disks so constructed is a set in $\mathbb{E}^3$, denoted $\mathrm{sym}_D(P)$ and called the *Schwarz symmetrization* about D. By construction it is a set of revolution about the axis D, but is it a convex set once P is convex? A bad surprise (that will ultimately become good) awaits us: this seems almost impossible to prove. In fact, we have to show that the function defined on D by the radius of the disks in question is concave. We thus need to see if the function given by the square root of the area of the sections P $\cap$ H is concave on the line D.

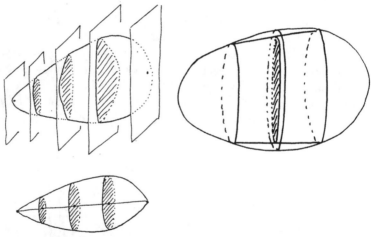

Fig. VII.5.3.

The little trapezium game seen above now becomes this: let P $\cap$ H, P $\cap$ H' be two sections corresponding to two points m, m' of D. We need to study the section P $\cap$ H'' corresponding to the point $m'' = \lambda m + \mu m'$ of D, where $\lambda + \mu = 1$, $\lambda \geq 0$, $\mu \geq 0$. The convexity of P tells us only that P $\cap$ H'' contains all the points $\lambda x + \mu y$, where x ranges over P $\cap$ H and y over the convex set P $\cap$ H'. This operation associated with λ, μ and two convex sets (here P $\cap$ H and P $\cap$ H') is due to Minkowski and will be the subject of the next subsection. Its motivation thus arises completely naturally. As for our problem of the concavity of the square root of the area, it will be treated in Sect. VII.8: it's the Brunn-Minkowski Theorem, in our opinion one of the two fundamental theorems of convexity. It remains difficult to prove, as we will amply see. The second fundamental theorem, simpler to state and deceptively evident, is that of Hahn-Banach; see Sect. VII.10.A. It implies, among innumerable other consequences, the result which states that there is equivalence between the convex envelopes of finite sets of points and the finite intersections (when they are, in addition, compact) of closed half-spaces. There would scarcely be any polyhedra without this result, nor any Chap. VIII. This correspondence is pleasantly precise thanks to polarity; see below.

Readers will be able to define without difficulty a general symmetrization sym_A in $\mathbb{E}^d$ which includes those of Steiner and of Schwarz, associated with an affine subspace of any dimension. In fact, a Euclidean structure in any affine direction K transverse to A suffices. Any of these symmetrizations preserves convexity, a consequence of the Brunn-Minkowski inequality (VII.8.1).

These symmetrizations, above all when the codimension equals 1 — so that the affine subspace is a line and the sections are those by orthogonal hyperplanes — are essential in analysis, where they are applied to the graph of a function:

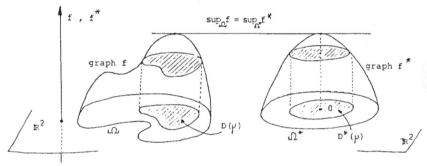

Fig. VII.5.4.

See the applications in Sect. VII.13.C: inequalities of the isoperimetric type, the "fundamental" — i.e. lowest — frequency of a drum, etc.

Observant readers will have noticed that each symmetrization of an arbitrary convex set itself possesses a symmetrization (about the subspace D). What happens if we continue *ad infinitum* to symmetrize a convex set, say in the plane, with respect to the directions of different lines? Do we obtain in the limit, supposing that is meaningful, an object that is symmetric with respect to all directions, i.e. a disk? We will see in Sect. VII.10.C that the answer is yes. Only recently has there been interest in the **rapidity** of convergence of symmetrization. It turns out in fact that very few symmetrizations very quickly produce a convex set that is close to a ball. Things get better as the dimension increases, but readers will be able to program things in the plane and see displayed the rapidity with which almost perfect disks are obtained. For more detail on the evaluation of the rapidity as a function of dimension and of a required precision ε, along with proofs, see 7.3 of Giannopoulos and Milman (2001),

Tsolomitis (1996) and Klartag and Milman (2003). This phenomenon is in particular related to that of concentration that will be encountered in Sect. VII.12.

VII.5.B. Some algebra of convex sets: Minkowski's sum

Let C and D be two convex sets in the same affine space. Their *Minkowski sum* $C + D$ is defined as the set $\{x + y : x \in C, y \in D\}$; it is trivially convex. This supposes that we have chosen an origin for our affine space so as to make it a vector space. But if we change the origin the result is the same within a translation, thus the "appearance" of the result remains the same. We see that we can define a whole algebra of linear combinations $uC + vD$, where u, v can be arbitrary real numbers, although the difference $C - D$ of two convex sets in particular is much more seldom used than the sum. The figures below give some examples of Minkowski sums.

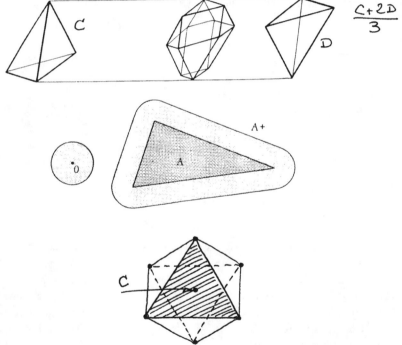

Fig. VII.5.5. $\frac{C-C}{2}$ is always center-symmetric

Note also that $\lambda C + \mu D$, with $\lambda + \mu = 1$, can be defined intrinsically. With these examples readers will begin to take into cognizance that the Brunn-Minkowski result (Sect. VII.8), which describes the behavior of the *volume* function under the Minkowski sum, is far from being evident. A basic example, that we present for simplicity in the plane, is the sum of an arbitrary convex set C and a (henceforth small) disk D(R) of radius R: then $C + D(R)$ is nothing other than the *tubular neighborhood* $\mathrm{Tub}_R(C)$ *of radius* R *of* C, that is to say the set of points in the plane a distance from C less than or equal to R (to see this, vectorize the space at the center of the disk). In fact, in this example C can be an arbitrary portion of the plane, for example a curve. We will encounter these $\mathrm{Tub}_R(C)$ in all dimensions in dealing with the isoperimetric inequality in Sects. VII.8 and VII.14. As an exercise we propose that readers try to identify the convex set $2C - C$.

The dangers of intuition. If we look at the Minkowski sum of a polygon and a disk, we see that this sum has a *regularizing* effect. Readers who know about regularization by way of convolution of functions may think that it is the same here. Well, what happens is just opposite, but this wasn't discovered until Kiselman (1987), where we find the following double assertion in the case of the plane. On the one hand the Minkowski sum of two planar convex sets, for which the boundary is C^∞, is not C^∞ in general, for example for the epigraphs of the functions $\frac{x^4}{4}$ and $\frac{x^6}{6}$; their sum is the epigraph of the function $\frac{x^6}{6} - \frac{3}{4}|x|^{20/3} + f$, where f is of class C^∞; thus the differentiability here does not exceed 6 and thus is not infinite. It is furthermore the worst possible result, since also the boundary of the sum of two epigraphs of class C^∞ is always of class $C^{20/3}$. (A function is said to be of class $C^{k+\alpha}$, with k positive integral and $\alpha \in]0, 1[$, when its k-th derivative exists and satisfies the Hölder condition of order α: $|f^{(k)}(x) - f^{(k)}(y)| \leqslant C|x - y|^\alpha$ for all x, y and a constant $C \geqslant 0$.)

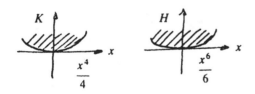

Fig. VII.5.6.

VII.5.C. A duality: polarity

In Sect. IV.2 we encountered polarity with respect to a circle in the Euclidean plane, which we may suppose to be the unit circle of $\mathbb{E}^2$. This extends in any dimension d to $\mathbb{E}^d$ without difficulty: we define a bijection between the set of all points other than the origin onto the set of hyperplanes that don't contain the origin by $x \mapsto \{y : x \cdot y = 1\}$, called the *polar* hyperplane of the point x; and x is called the *pole* of the corresponding hyperplane. We speak here of polarity with respect to a sphere, here the unit sphere. The incidence properties are trivial in view of the bilinearity of the scalar product. In algebraic geometry — see Sect. V.13 — we can thus define the dual curve of a given curve. Here we are going to apply this polarity to compact convex sets whose interior contains the origin.

For each set C of $\mathbb{E}^d$ we call the *polar* C^0 of C the part of $\mathbb{E}^d$ defined by $C^0 = \{y : x \cdot y \leqslant 1 \text{ for any } x \in C\}$. Thus:

> *If C is compact and contains the origin in its interior, then C^0 is a compact convex set containing the origin in its interior.*

It's a good duality: $(C^0)^0 = C$ for each convex set in the collection of all compact convex sets containing the origin in their interiors. Set inclusions are reversed: if $C \subset C'$, then $C^0 \supset C'^0$. For the intersection $C \cap C'$, the polar $(C \cap C')^0$ is the convex envelope of the union $C^0 \cup C'^0$.

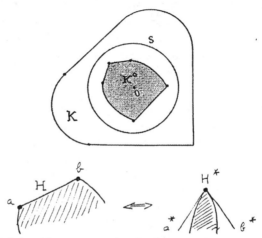

Fig. VII.5.7. Polarity. Above, we see that the flat parts of K correspond to angular points of K^0, and conversely. Below, detail of the correspondence

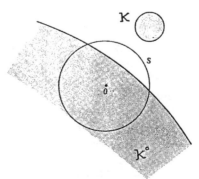

Fig. VII.5.8. Polarity. If K doesn't contain the origin in its interior, then K^0 isn't bounded

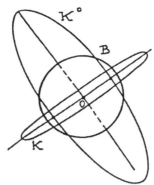

Fig. VII.5.9. $\sum_i \frac{x_i^2}{a_i} \leqslant 1$ has for polar $\sum_i a_i x_i^2 \leqslant 1$

Here are some comments on the above figures. Briefly, they borrow certain concepts from Sect. VII.6. If a set is strictly convex, then its polar is also. The essential fact is that the hyperplane polar to a point on the boundary is tangent to the boundary of the polar, and this at a point that is the pole of the hyperplane tangent to the initial convex set at the point of the boundary considered.

The situation that is completely opposite the strict case (smooth boundary, see Sect. VII.13.D) is that of (compact) polyhedra: the vertices of a polyhedron correspond to the hyperplane faces of its polar (and vice versa). In the intermediate cases (see Fig. VII.5.7) flat parts of the boundary correspond to the angular points.

Finally we give some explicit examples. The unit ball is its own polar, more generally the polar of a (solid) ellipsoid $\sum_i \frac{x_i^2}{a_i^2} \leqslant 1$ is the ellipsoid $\sum_i a_i^2 x_i^2 \leqslant 1$. This is a particular case of the fact that polarity preserves central symmetry. But it remains true that the polar of an ellipsoid not centered at the origin (but still containing it in its interior) is again an ellipsoid. It is less evident, but results from considerations (to be generalized) on duality with respect to conics; look in Sect. IV.4.

The polar of the cube $[-1, 1]^d$, centered at the origin, is the polytope that generalizes the regular octahedron in $\mathbb{E}^3$, i.e. $\sum_i |x_i| \leqslant 1$. In the planar case, this remains exceptionally a square, but from dimension 3 on it is another type of polytope – essential different, as we will see. We will call it the *cocube*; (**co** is a prefix used rather systematically for dual objects and properties); we *denote* it by Cocub^d. The correspondence between faces and vertices is easily described. As for the polar of a simplex, it's always a simplex, regularity being preserved.

This duality between faces and vertices is not difficult to establish. On the other hand, its generalization to the fact that the polar of $\sum_i |x_i|^p \leqslant 1$ (for all real $p > 1$) is $\sum_i |x_i|^q \leqslant 1$ with $\frac{1}{p} + \frac{1}{q} = 1$ requires the use of the Minkowski inequality seen in Sect. VII.2. Don't fail to observe that the case of cube and cocube correspond here to $p = 1$ and $q = \infty$. The $\sum_i |x_i|^p \leqslant 1$ are called the *unit balls* for L$_p$, because they are the unit balls for the metrics $d(x, y) = (\sum_i |x_i - y_i|^p)^{1/p}$ of

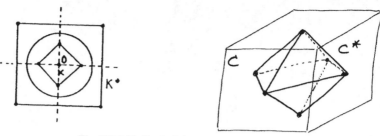

Fig. VII.5.10. Dual of the square, dual of the cube

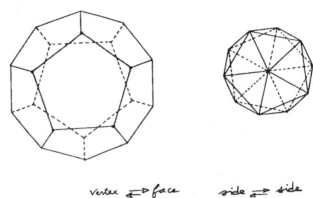

Fig. VII.5.11. Duality of the dodecahedron (20 vertices, 12 faces, 30 edges) and the
icosahedron (12 vertices, 20 faces, 30 edges)

$\mathbb{R}^d$; the classical notation for these spaces is ℓ_p or ℓ_p^d; we employ subsequently the *notation* Ball$(d; p)$ for their unit ball (so as not to overburden readers' memories; see Fig. VII.4.5).

VII.6. Volume and area of (compacts) convex sets, classical volumes: Can the volume be calculated in polynomial time?

For compact convex sets, there exists a good notion of *volume*. We have been diffident up until now about notions of volume, of measure. We will remain so, even though the problem of the notion of measure can be considered part of elementary geometry. But we need to climb a bit up the ladder in order to see things clearly. In the general case, arbitrary parts of Euclidean space (more generally of an affine space, of a sphere, of a Riemannian manifold) are not *measurable*. But all "reasonable" sets are measurable. The notion of measure languished a very long time before finally being clarified by Lebesgue in 1902 . Let us retain this: for the compact convex sets of a Euclidean space there is a good notion of measure. Moreover, it can be constructed in an elementary manner; see Sect. VII.6.C and, for more details, 12.2.5 and 12.9.3 of [B]. Furthermore, it is an affine notion as soon as we fix a unit, e.g. by choosing a vectorial basis and proclaiming that the cube constructed on it has volume 1.

On the other hand, for wild objects in $\mathbb{E}^d$, e.g. fractals, etc., various notions of measure are possible; the one most used after Lebesgue measure is that of Hausdorff (related to the notion of Hausdorff dimension). Possible references are: Federer (difficult reading), Falconer (1990), Feder (1988), David and Semmes (1997), Morgan (1988–2000). In fact, just as for convex sets, all these possible notions of measure coincide for good objects, typically surfaces in $\mathbb{E}^3$ or more generally submanifolds of arbitrary dimension in $\mathbb{E}^d$.

Technically, we use the tools and notations of integral calculus, as we have done starting with Sect. I.2. The volume of C will be $\int_C dx_1 \ldots dx_d$, provided the coordinates $\{x_i\}$ give the unit measure, and will be denoted by Vol(C). We make free use of Fubini's theorem and a change of variables, which permits us to make calculations recurrently. The typical example is that of a pyramid C of altitude h and base a convex set B in a hyperplane (but it could be any measurable subset of the hyperplane); thus

$$\text{(VII.6.1)} \qquad \text{Vol(C)} = \int_0^h \left(\frac{t}{h}\right)^{d-1} \text{Vol(B)}\, dt = \frac{h}{d}\, \text{Vol(B)}.$$

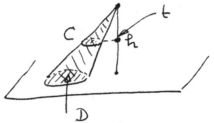

Fig. VII.6.1. $\text{Vol(C)} = \int_0^h \left(\frac{t}{h}\right)^{d-1} \text{Vol(B)}\, dt = \frac{h}{d} \text{Vol(B)}$

And, to be sure, the volume is a d-*dimensional* object if, for each scalar λ under homothety with ratio λ, we have $\text{Vol}(\lambda C) = \lambda^d \text{Vol(C)}$.

What is important, and new in spirit, is that most of the time we will be interested not in **exact** values of volumes, but in their **asymptotic behavior** when the dimension tends toward infinity. There are numerous motivations for wanting asymptotics and high dimensions. From the "pure" side, the spaces of analysis (function spaces) are of infinite dimension and we can hope to gain some insight by approximations by spaces of finite, but increasingly greater, dimension. From the "applied" side, physics and in particular statistical mechanics treats spaces whose dimensions are of the order of Avogadro's number (10^{23}). We will also encounter large dimensions in Chap. X.

VII.6.A. Volume of cubes, cocubes and simplexes

The volume of the cube $\text{Cub}^d = [0, 1]^d$ is 1 and that of $[-1, 1]^d$ is 2^d. For a general parallelepiped generated by the vectors $\{x_i\}$ of $\mathbb{E}^d$, it is useful to know that the volume is equal to $\sqrt{\det(x_i.x_j)}$, the square root of the Grammian determinant. We might think that the volume of a cube grows quickly with the dimension; this is certainly true, things are exponential, but we will see that this is fairly modest growth in this realm.

In fact the volume of Cocub^d (polar of $[-1, 1]^d$) is equal to $\frac{2^d}{d!}$ as can be seen by decomposing it into 2^d pieces associated with the various "quadrants", where each coordinate has a specific sign, and because the formula for the pyramid yields $d!$ when applied recurrently. This volume tends extremely quickly toward zero when the dimension tends toward infinity; see in fact Stirling's formula below (VII.6.B.2). We will specify this "roughly" by an appropriate notation in Sect. VII.6.B below. The term in 2^d in the formula for the volume of the cocube is thus completely devoured when d becomes very large.

For the simplex it is much worse, things are in $\frac{1}{d!}$ purely. Roughly, if the simplex is of "magnitude D", then its volume is of order $\frac{D^d}{d!}$. For the regular simplex, all of whose edges have length equal to 1, the volume equals $\frac{\sqrt{d+1}}{d!2^{d/2}}$. Readers can make the calculation; it will be good to know that very often the best way of calculating with a regular simplex of dimension d is to embed it in $\mathbb{E}^{d+1}$, using the end points of the vectors of the canonical basis as vertices (Fig. VII.4.4); for example the equilateral triangle is then seen in $\mathbb{R}^3$ as the set of points such that $x_1 + x_2 + x_3 = 1$ and $x_1 \geqslant 0$, $x_2 \geqslant 0$, $x_3 \geqslant 0$.

VII.6.B. Balls, spheres and ellipsoids

At the other extreme from the (regular) polytopes, we have the compact convex sets, the balls. We *denote* by $\beta(d)$ the volume of the unit ball of $\mathbb{E}^{d+1}$, by $\alpha(d)$ the d-dimensional area of the unit sphere (in the literature $\omega(d)$ is often seen instead of $\beta(d)$). The calculation of these constants is always subtle; the result itself will indicate one of the reasons. First note that the formula (VII.6.1) extends infinitesimally, and if the ball is the "pyramid" with base the sphere and constant altitude equal to 1, we will then have $\beta(d) = \frac{\alpha(d-1)}{d}$. It is therefore sufficient that we calculate either the volumes of balls or the "areas" of spheres.

Calculation of the volume of spheres. First a classic trick that we cannot resist mentioning even though we will end up not using it; it consists in calculating $I = \int_{\mathbb{R}^d} \exp(-x_1^2 - \cdots - x_d^2)\, dx_1 \ldots dx_d$ in two ways, first by using Fubini directly and then changing to spherical coordinates. We obtain

$$I = \left(\int_{\mathbb{R}} \exp(-x^2)\, dx \right)^d = \mathrm{Vol}(S^{d-1}). \int_0^\infty e^{-\rho^2} \rho^{d-1}\, d\rho.$$

All these integrals are known and expressed in terms of the *gamma function*, whence $\beta(d) = \pi^{d/2} / \Gamma(\frac{d}{2} + 1)$. Unfortunately the explicit expression for Γ differs according to whether the input is an integer or half an odd integer. We obtain finally:

$$\text{(VII.6.B.1)} \qquad \beta(2k) = \frac{\pi^k}{k!}, \quad \beta(2k+1) = \frac{2^{k+1}\pi^k}{(2k+1)(2k-1)\dots 3.1}.$$

A wonderful calculation, but where the geometer sees nothing. Let us instead calculate the $\alpha(d) = \text{Vol}(S^d)$ directly, as geometers and by recurrence, staying at first between the north pole and the equator. That is, we regard S^d as a sort of pyramid on S^{d-1}, but here along the sections of constant latitude. Things vary along the latitude, not linearly as in an ordinary pyramid, but as $\sin^{d-1}\theta$. As the initial directions to the north pole describe exactly the unit sphere of the tangent space, we have: $\text{Vol}(S^d) = \text{Vol}(S^{d-1}) \int_0^\pi \sin^{d-1}\theta\, d\theta$. This integral is calculated by recurrence, by integrating twice by parts and by knowing that $\text{Vol}(S^0) = $ measure of $[-1, 1] = 2$ and that $\text{Vol}(S^1) = 2\pi$. We happily get the preceding result and see again that parity plays an inevitable role.

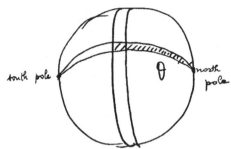

Fig. VII.6.2. The volume at distance θ from the north pole is give by $\sin^{d-1}\theta\, d\theta$, and simply by $\sin\theta d\theta$ in the case of the ordinary sphere

It seems that we owe to Herman Gluck the best procedure for calculating $\text{Vol}(S^d)$. We generalize the construction given in Sect. II.5 for S^3, obtained starting with two circles and taking all quarter circles whose endpoints trace out these two circles. Here we construct S^{d+2} starting with S^1 and S^d with such quarter circles (in algebraic topology it is called the *join* $S^1 * S^d$). We decompose the integral in dimension $d + 2$ into three coordinates: that of "x" on S^1, that of "y" on S^d and a coordinate on the quarter circles. The integral element, if α denotes the length computed on the quarter circles, is $\sin^d\alpha \cdot \cos\alpha$, for we have:

$$\text{Vol}(S^{d+2}) = \int_0^{\pi/2} \text{Vol}(S^1(\cos\alpha)) \cdot \text{Vol}(S^d(\sin\alpha))\, d\alpha$$

$$= \text{Vol}(S^1) \cdot \text{Vol}(S^d) \int_0^{\pi/2} \sin^d\theta \cdot \cos\alpha\, d\alpha = \frac{2\pi}{d+1}\text{Vol}(S^d),$$

which is the (second order) recurrence we wanted, while the integral in α was trivial to compute.

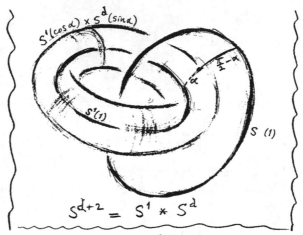

Fig. VII.6.3. Representation of the sphere S^{d+2} as the join of the sphere S^d with the
circle S^1: $S^{d+2} = S^1 * S^d$

♦

Behavior of the volume. Two things are important here, the first being the behavior
of $\beta(d)$, and thus also that of $\alpha(d)$, when d tends toward infinity. Stirling's formula
gives it immediately as an asymptotic function of $d!$, i.e.

(VII.6.B.2 Stirling's formula) $d! \sim_{d \to \infty} \sqrt{2\pi d} \left(\dfrac{d}{e}\right)^d$.

A first remark is that $\beta(d)$ tends very fast toward zero when the dimension tends
to infinity. The first values are deceptive, up until 6 they are increasing. Here is a
table for a selection of values:

d	$\beta(d)$	d	$\beta(d)$
2	3,14159	11	1,884
3	4,188	12	1,335
4	4,93	13	0,91
5	5,263	14	0,56
6	5,1677	15	
7	4,124	16	0,235
8	4,0587	17	
9	3,298	18	0,08
10	2,550	19	
		20	0,02

$\beta(100) \approx 2.10^{-40}$ $\beta(1000) = ?$

Table VII.6.1

We leave it to readers to fill the gaps with some computations that are convincing
for this "surprising?" result. The fact that there is convergence to zero lets us predict
the sketch of the graph of $\sin^d \theta$: we clearly see that the shaded area tends toward

zero when d tends toward infinity, but the sketch tells us nothing of the speed. Some analysis is necessary, here via Stirling.

As the Brunn-Minkowski inequality shows us in Sect. VII.8, the most significant quantity for the volumes of convex sets (situated in a space of dimension d) is their d-th root. Here we find that $(\beta(d))^{1/d}$ is of order $c \cdot \frac{1}{\sqrt{d}}$, where c is a constant independent of the dimension, specifically $\sqrt{2\pi e}$. To clarify this sort of thing, we use the *notation*

$$f = \bullet g$$

to signify equivalence within a constant *when the dimension tends toward infinity* (i.e. that the ratio of the two functions f and g tends toward a positive constant). We can specify an index for the constant when there is more than one: $\bullet_1$, $\bullet_2$, etc. There will be comparable notations for inequalities: $f \geqslant \bullet g$, etc.

Stirling's formula shows that $\beta^{1/d}(d) = \bullet d^{-1/2}$, whereas for the cube and the cocube we have: $\mathrm{Vol}^{1/d}(\mathrm{Cub}_d) = \bullet 1$ and $\mathrm{Vol}^{1/d}(\mathrm{Cocub}_d) = \bullet d^{-1}$. In Sect. X.5 we will need the exact constant:

(VII.6.B.3)
$$\beta^{1/d}(d) \sim_{d \to \infty} \frac{\sqrt{2\pi e}}{\sqrt{d}}.$$

The second matter, to which we will return in much detail in Sect. VII.12, is even more important. It's the *phenomenon of concentration at the equators*: the graph of $\sin^{d-1}\theta$ shows not only that the volume tends very quickly toward zero with increasing dimension, but above all that it is concentrated more and more about the equator. For a given fixed ε, the tubular neighborhood of radius ε about the equator will have more and more relative volume. This implies that what remains after the tube is removed is of smaller and smaller measure (compared to the total measure).

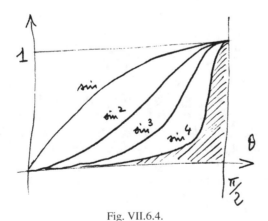

Fig. VII.6.4.

♦

Volume of ellipsoids and a natural question. A simple linear change of variables shows at once that:

$$\text{Vol}\left(\sum_i \frac{x_i^2}{a_i^2} \leqslant 1\right) = \beta(d)\prod_i a_i.$$

Here alert readers will begin to feel uneasy. In fact, since we have said that the cubes, cocubes and balls are nothing other than particular cases of the sets $\sum_i |x_i|^p \leqslant 1$ in L_p (denoted above by Ball$(d;p)$)), readers have the right to demand a general formula in p for Vol$(\sum_i |x_i|^p \leqslant 1)$. It's a classic: Vol(Ball$(d;p)$) $= \frac{2\Gamma(1+1/p)^d}{\Gamma(1+d/p)}$; see e.g. the end of the first chapter of Pisier (1989).

VII.6.C. Approximation by polytopes, areas of convex sets

An abuse of language is involved: we have seen how to define, for a compact convex subset (with nonempty interior) of $\mathbb{E}^d$ (in the Euclidean context that we need here), the generalization to all dimensions of the length of a curve ($d = 2$) and the area of a surface ($d = 3$). We have pointed out above that such a notion isn't unique for arbitrary subsets, even if we specify that we want a ($d - 1$)-dimensional measure. When the *boundary* ∂C (the difference between the closure and the interior) of the convex set C is a smooth hypersurface (submanifold of codimension 1; see Sect. V.XYZ), the area enclosed by ∂C exists and is well defined, for all possible notions of measure. It turns out that this works as well for all compact convex sets, and most importantly all this can be done in a completely elementary way. We use *two notations*, either Area(C), or else Vol$_{d-1}(\partial C)$ when C is in $\mathbb{E}^d$. We give two ways of proceeding; details can be found in [B].

The first way consists in first showing that every compact convex set can be approximated with arbitrary precision by polytopes. The approximation idea is straightforward and is left to the reader (on the other hand, we will see that the question of the number of vertices that are necessary for obtaining a good approximation is quite a different story). For a polytope P of $\mathbb{E}^d$ *its area* Area(P) is the sum of the canonical Euclidean volumes of its faces (which are polytopes in $\mathbb{E}^{d-1}$). We then define Area(C) as the limit of the areas of the polytopes that approximate it. It is of course necessary to show that this limit exists and is independent of the approximations chosen, which isn't very difficult. In the planar case, we obtain the definition of the length of a curve as the limit of the perimeter of the inscribed polygons, but starting with dimension three; however, if we are dealing not with convex sets, but with objects such as polyhedra or polytopes — see for example the Venetian lamp (Fig. VII.14.3) — difficulties emerge which are in some sense insurmountable. In fact [B] does not use approximation by polytopes in the definition (because it is difficult to show the independence of the limit), but instead that given by Cauchy's formula, which states that the area of a convex set is the mean of its projections on hyperplanes when we take all possible directions; see 12.10.2 of [B].

The second method is due to Minkowski: the starting point is the observation that, if a domain C of $\mathbb{E}^d$ has a smooth hypersurface as its boundary ∂C, then the calculation of the volume of its tubular neighborhood Tub$_\varepsilon$(C) of radius ε satisfies

$$\lim_{\varepsilon \to 0} \frac{\text{Vol}(\text{Tub}_\varepsilon(C)) - \text{Vol}(C)}{\varepsilon} = \text{Area}(\partial C)$$

.

Now for convex sets we know how to write $\text{Tub}_\varepsilon(C)$ "a la Minkowski" as $\text{Tub}_\varepsilon(C) = C + \varepsilon B$, where B denotes the unit ball of $\mathbb{E}^d$ (see Sect. VII.5.B). Minkowski thus chose *as definition*:

$$\text{Area}(\partial C) = \text{Vol}_{d-1}(\partial C) = \lim_{\varepsilon \to 0} \frac{\text{Vol}(C + \varepsilon B) - \text{Vol}(C)}{\varepsilon}.$$

This definition of the area of the boundary might be called *the painter's definition*. Just as we can calculate the number of sheep by dividing the number of their feet by four, we can calculate the area of a surface by evaluating the volume of paint that is need for covering it (the layer of paint is of almost constant small thickness).

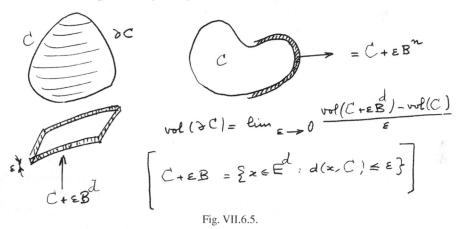

Fig. VII.6.5.

It can be shown that the two definitions given coincide, and here we have a second reason for being interested in the behavior of $\text{Vol}(A+B)$ as a function of $\text{Vol}(A)$ and $\text{Vol}(B)$ (Sect. VII.8). In the following three sections we are going to study the behavior of the volume under the three operations on convex sets introduced in the preceding section.

Interested readers may reread the preceding while noting that the area is the first term of the Taylor expansion in ε of $\text{Vol}(C + \varepsilon B)$ and then ask the significance of the subsequent terms. The answer will be given in Sect. VII.13.B.

VII.6.D. Mission impossible: calculating the volume of a convex set numerically

For an "arbitrary" convex set an explicit formula for its volume clearly doesn't come into question. In return, for practical problems, we might hope to find effective algorithms for calculating the volume of a given convex set numerically to a good degree of accuracy. Practically, convex sets come provided with two oracles: one tells *membership*, the other tells *separation*. This is to say that we have one criterion for deciding whether a point belongs to the convex set, and another for knowing whether it is in a given half-space. We will see a result that is surprising for the non expert: there does not exist a polynomial algorithm that always permits calculation of the volume of a convex set with arbitrary accuracy, even if we have both oracles at our disposal. We sketch the content of Barany and Füredi (1987), where this phenomenon was discovered for the first time.

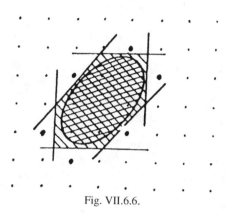

Fig. VII.6.6.

At first we concern ourselves only with the oracle of belonging. The starting idea, obligatory, is to take (many) points of C and to calculate the volume of the enveloping convex polytope of these points. It is easy to do this polynomially. The problem now is thus to know whether the volume of the inscribed polytopes provide good approximations to the volume of the convex set. If we approximate each convex set by polytopes, we will clearly have approximations, but how good are they? For approximation by polytopes, we have existence itself from 12.9 of [B]. But we can ask more, even much more, of this approximation; a complete summary of the problem is found in 1.10 of Gruber and Wills (1993). The simplest investigation concerns the number of vertices, effectively the number of points that we have "chosen" on the interior of the convex set. For this it suffices to study the *MacBeath numbers* for C, defined as:

$$\mathrm{McB}(C, n) = \sup_{P}\left\{\frac{\mathrm{Vol}(P)}{\mathrm{Vol}(C)} : P \text{ polytope inscribed in C having } n \text{ vertices}\right\}.$$

We find in MacBeath (1951) the inequality:

For every convex set C we have $\mathrm{McB}(C) \geqslant \mathrm{McB}(ball)$.

The ball is thus the **worst** among the convex sets with respect to the approximation of volume by polytopes, which isn't terribly surprising (compare with the Steiner problem in Sect. I.2). The proof is easy using the Steiner symmetrization, we proceed as in the following section: it suffices to know that a Steiner symmetrization, for any hyperplane direction H, reduces the MacBeath number: for each integer n we have $\mathrm{McB}(\mathrm{sym}_H(C), n) \leqslant \mathrm{McB}(C, n)$.

This established, it remains to find, in place of the exact value — which is completely unknown for an arbitrary integer — a good approximation for $\mathrm{McB}(ball, n)$.

And the result is a disaster: a bound that is extremely small as soon as the number of points is no longer exponential in the dimension:

$$\mathrm{McB}(\mathrm{ball}(\mathbb{E}^d), n) \leqslant \left(\frac{2ea \log d}{d}\right)^{d/2} = \left(\frac{2e \log n}{d}\right)^{d/2},$$

where we have thus set $n = d^a$.

The proof consists in reasoning that mixes combinatorics with the geometry of triangulation decompositions for polytopes. Thus we are in the exponential realm and without hope of the polynomial; this is all the more grave in that in most applications the dimensions are large.

Nonetheless, we might hope to succeed by also using the oracle of separation. This comes down to regarding the convex set as the envelope of half-spaces, thus by considering simultaneously the polytopes P inscribed in C and their polars P^0, which are polytopes circumscribed about C; see the definition of polarity in Sect. VII.5.C. Unfortunately the inequality that we will see in Sect. VII.9 relating to $\mathrm{Vol}(P) \cdot \mathrm{Vol}(P^0)$ allows the authors to show that the ratio $\frac{\mathrm{Vol}^{\mathrm{ext}}(C)}{\mathrm{Vol}^{\mathrm{int}}(C)}$ of the approximations of the volumes $\mathrm{Vol}^{\mathrm{ext}}(C)$ by exclusion and $\mathrm{Vol}^{\mathrm{int}}(C)$ by inclusion always satisfies $\frac{\mathrm{Vol}^{\mathrm{ext}}(C)}{\mathrm{Vol}^{\mathrm{int}}(C)} \geqslant \bullet \left(\frac{d}{\log d}\right)^d$. The situation thus remains catastrophic even with the two oracles at our disposal.

However, it should be pointed out to interested readers that there exist polynomial algorithms that provide a good approximation of the volume, but only in "almost all cases"; see Lovacz (1990). This result uses abstruse methods of the calculus of probabilities, in particular random walks on the interior of a convex set; see the recentBollobas (1997) and also Barany (2008), already mentioned at the end of Sect. I.2.

In Sect. VII.9 we will return amply to the behavior of the product $\mathrm{Vol}(C) \cdot \mathrm{Vol}(C^0)$ of the volume of a convex set and its polar.

VII.7. Volume, area, diameter and symmetrizations: first proof of the isoperimetric inequality and other applications

We only treat the Euclidean case. Any symmetrization whatever **preserves** the volume of convex sets (we assume — see the following section — that first of all these symmetrizations preserve the convexity property; we have seen that this isn't so obvious: see Sect. VII.5). This is trivial, because we decompose the integral as a product in the complementary directions, and by definition we then integrate functions which are equal. Thus we always have

$$\mathrm{Vol}(\mathrm{sym}_D(C)) = \mathrm{Vol}(C).$$

The essential fact, and much less easy, is that all these symmetrizations **reduce** the area. More precisely: so long as the convex set isn't symmetric with respect to the direction of symmetrization, then

$$\mathrm{Area}(\mathrm{sym}(\partial C)) < \mathrm{Area}(\partial C).$$
$$\scriptstyle D$$

From this and the preservation of volume, we see right away a scheme for inferring the isoperimetric inequality: we start with an arbitrary convex set which we symmetrize "a la Steiner", again and again about directions of appropriate lines. Finally we have to obtain a convex set that is symmetric with respect to all directions, which can't be anything other than a ball of some radius. Thus we have first:

$$\text{for any C}: \frac{\text{Area}^d(\partial C)}{\text{Vol}^{d-1}(C)} \geq \frac{\alpha^d(d)}{\beta^{d-1}(d)}.$$

For all convex sets of a given volume, the area of the boundary is greater or equal to that of a sphere whose radius is the same as the ball whose volume is that of the given convex set. Now, the strict inequality above allows us to resolve the case of equality: only the balls satisfy $\frac{\text{Area}^d(\partial C)}{\text{Vol}^{d-1}(C)} = \frac{\alpha^d(d)}{\beta^{d-1}(d)}$. For completeness, we must first prove the strict inequality which we asserted; the reader can consult [B]. The geometric idea is that symmetrization strictly reduces the area if the object considered isn't already symmetric with respect to the direction used, but the proof of this point is subtle and uses the expression of the convex set with the aid of integrals of almost everywhere differentiable functions (because of convexity; see Sect. VII.3).

But then we must make precise this symmetrization with respect to "all" directions. This was done for the first time by Blaschke in 1916; here is how.

Blaschke's *Kugelungsatz*. The term "*Kugelungsatz*" can be translated by "spherification theorem" (more precisely, *ballification*). In order to properly speak of a limit, we need a topology. On the collection of convex sets, it is known and easy; it's the *Hausdorff metric*. It is defined for arbitrary subsets of $\mathbb{E}^d$ (more generally, exactly in the same way, for subsets of arbitrary metric space). The *distance* $d_H(C, D)$ is defined for two compact subsets of $\mathbb{E}^d$ as the least ε such that both $D \subset \text{Tub}_\varepsilon(C)$ and $C \subset \text{Tub}_\varepsilon(D)$. The essential fact is that there is compactness for this metric:

Each bounded sequence of compact sets has a convergent subsequence.

We will return to the metric space of all compact sets in Sect. VII.13.D. This again has to do with a Euclidean situation; for the affine context and another metric, see Sect. VII.10.D.

For convex sets and this metric, volume and boundary area are on the one hand continuous functions. On the other hand, we can choose without difficulty a set of Steiner symmetrization directions that is everywhere dense in the set of all directions of the space. Then, starting with a given convex set, we obtain by successive symmetrizations a sequence of convex sets for which we can extract a convergent subsequence. By construction this limit will be a ball, QED. In Tsolomitis (1996) can be found information on the rapidity of convergence by symmetrization toward a sphere (a ball): it is extremely rapid (see also 7.2 of Giannopoulos and Milman (2001)).

◆

Bieberbach's isodiametric inequality says this:

Among all the convex sets of a given volume, there is exactly one of least possible diameter, i.e. the ball.

Otherwise expressed, for each convex set C we have $\mathrm{Vol}(C) \leqslant 2^{-d}\beta(d)\,\mathrm{diam}(C)$, with equality only for balls. To see this, we only need "verify" that the Steiner symmetrization strictly reduces the diameter of nonsymmetric convex sets, and apply the *Kugelungssatz*. We have already used an isodiametric inequality in Sect. II.1 — but on the sphere — for Borsuk's conjecture.

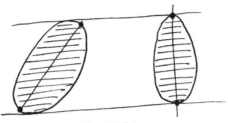

Fig. VII.7.1.

Symmetrization is used in many other places. It was used (see Sect. I.1) for establishing one of the bounds in Steiner's problem on points thrown randomly at the space. It was used for MacBeath's theorem in Sect. VII.6.D above. Finally, we can ask more about the structure of the space of all convex sets, which is a rung on the ladder; see Sect. VII.13.D for this. See also Sect. VII.13.C.

VII.8. Volume and Minkowski addition: the Brunn-Minkowski theorem and a second proof of the isoperimetric inequality

In Sect. VII.5.A we saw, regarding the Schwarz symmetrization, the necessity of knowing the behavior of the volume under Minkowski addition, i.e. of studying Vol(K + H). Let us now slice a compact convex set K by parallel hyperplanes and look for the behavior of the volume of these sections (as they are all parallel, the measure chosen has no importance; in particular, there is no need of being in a Euclidean space when we consider but one direction). It seems intuitive that convexity implies that the volume will increase from zero to a maximum value, then decrease toward zero.

For example, it seems even more intuitive, even more "certain", that for a center-symmetric convex set the maximum is obtained by the section with the hyperplane that passes through the center. All this is true, and more still; but it is far from being obvious. The above figures leave us with some doubts; for example, there isn't always continuity, typically where at an extremity we have a whole face of a polyhedron; but above all the area (volume) function of the sections isn't concave (except in the planar case — see Fig. VII.5.2), but that is what it must be if we

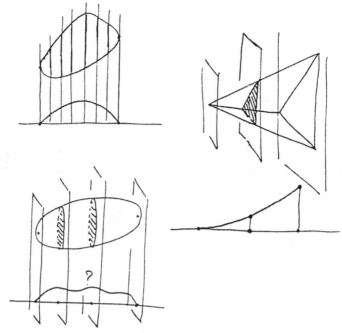

Fig. VII.8.1.

would hope for a maximum theorem for the center. The essential idea, suggested by Fig. VII.5.3 — a figure where we expect that the circular symmetrization is again a convex set — is to take, as the function of the section, the radius of the ball obtained (with the same volume as that of the corresponding section), i.e. the d-th root of the volume of the section. Then everything goes very nicely. The volume function of our sections depends on a real number, an (arbitrary) affine coordinate t for locating the mutually parallel hyperplanes $P(t)$. Thus the function $t \mapsto (\mathrm{Vol}(K \cap P(t)))^{1/d}$ is concave. This results in:

(VII.8.1) Brunn's (and Minkowski's) theorem. *For any two compact convex sets K and H in a space of dimension d the function*

$$\lambda \mapsto (\mathrm{Vol}(\lambda K + (1 - \lambda)H)^{1/d}$$

is concave.

That is to say, we have

$$\mathrm{Vol}^{1/d}(\lambda K + (1 - \lambda)H) \geq \lambda \, \mathrm{Vol}^{1/d}(K) + (1 - \lambda) \, \mathrm{Vol}^{1/d}(H).$$

We remark that, since $\mathrm{Vol}^{1/d}$ is homogeneous of degree 1, the general inequality is equivalent to the single inequality

$$\mathrm{Vol}^{1/d}(K + H) \geq \mathrm{Vol}^{1/d}(K) + \mathrm{Vol}^{1/d}(H).$$

Let us see first why this inequality implies the expected behavior for the volumes of a slicing.

When K is cut by two parallel hyperplanes P and P′, convexity implies that K contains all points of the form $\lambda x + (1 - \lambda)y$ when x ranges over $K \cap P$, y over $K \cap P'$ and λ over $[0, 1]$. In particular, for a given λ, setting $P'' = \lambda P + (1 - \lambda)P'$, the convex set $K \cap P''$ contains the entire convex set $\lambda(K \cap P) + (1 - \lambda)(K \cap P')$, which proves our assertion thanks to (VII.8.1).

Theorem (VII.8.1) had been discovered by Brunn for dimensions 2 and 3 in his thesis in 1887–1889. Its importance was recognized in 1910 by Minkowski, who used the theorem extensively for geometric number theory in all dimensions (see our Chaps. IX and X); he gave an analytic proof and in particular characterized the case of equality; i.e. we have the theorem:

(VII.8.2) Complete Brunn-Minkowski theorem:

$$\mathrm{Vol}^{1/d}(K + H) \geqslant \mathrm{Vol}^{1/d}(K) + \mathrm{Vol}^{1/d}(H)$$

where equality holds if and only if K and H are homothetic.

(a)   **(b)**

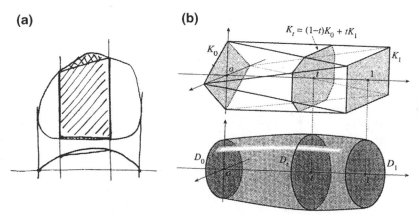

Fig. VII.8.2. **(b)** Gardner (1995) © Cambridge University Press

For more details on the history of this theorem, see Burago and Zalgaller (1988) and the notes for Sect. 6.1 on p. 314 of Schneider (1993); the latter work is entirely devoted to our theorem and various of its important extensions.

As for proofs, we know only one that is truly geometric, that can be *really seen* and will be given soon: that of Blaschke. But it is based on a limit argument and can't characterize the case of equality. The other proofs, both of (VII.8.1) and (VII.8.2), seem difficult to us and not very natural, even for the weak case of (VII.8.1) − e.g. that given in 11.8.8 of [B]. Perhaps the simplest and quickest, providing the case of equality, is that of Schneider (1993, pp. 310, 311).

In fact, it has recently been realized that the Brunn-Minkowski theorem plays a role not only in convexity, but in other domains as well, going well beyond its statement by itself. A very recent exposition that brings together these numerous ramifications is Maurey (2003), in which there are many references beyond those we give here. To begin: in Barthe (1998), the Brunn-Minkowski inequality is set in the very general and theoretical framework of *measure transport*. The result states that if we have two measures μ and ν (having densities) on $\mathbb{R}^d$, then we can always transport μ onto ν with the help of the mapping $f : \mathbb{R}^d \to \mathbb{R}^d$, which has an appropriate convexity property. This transport problem, having to do with excavating an embankment, dates from Monge in 1781 and is one of the earliest optimization problems, along with its companion, the brachistochrone of the cycloid. In Brenier, Frisch, and Hénon (2003) it is seen to arise in astrophysics, apropos the origin of the universe.

The text of Barthe proves a new inequality, in some sense the reverse of (VII.10.E.2) below, at the same time giving a new and more conceptual proof of the Brascamp-Lieb inequality, which we will find explicitly in (VII.10.E.2). Then (difficult) inequalities are proved − in a very natural way − which characterize, by their volume quotient, parallelepipeds and simplexes (much harder in the non-symmetric case). These are inequalities due to Ball, that we state in (VII.10.E.1) and (VII.10.E.3), specifically that parallelepipeds (respectively simplexes in the not necessarily symmetric case) are characterized as having the largest ratio of their volume to the volumes of the ellipsoids of smallest volume containing them. But most importantly the reverse inequality of Barthe allows proof of assertions dual to (VII.10.E.1) and (VII.10.E.3), specifically that parallelepipeds (respectively simplexes) are characterized as convex sets for which the ratio of the volume of an ellipsoid of larger volume which contains them to their own volume is smallest. We will see amply if Sect. VII.9 that it is out of the question to infer these inequalities easily from (VII.10.E.1) and (VII.10.E.3), the volume of a convex set and that of its dual having very bad relations. A large part of these results are fundamentally investigations of reverse inequalities to that of Brunn-Minkowski, clearly with supplementary conditions.

In Sect. 2.2 of Giannopoulos and Milman (2001) are found important considerations on (VII.8.2), in particular a proof (which moreover yields the case of equality) based on the Knöthe mapping and that of Brenier (1991). The Knöthe mapping is that defined, for a proof of the isoperimetric inequality, in Sect. V.12 of this work; it is presented in detail in 12.11.7.2 of [B] and in 6.6.9 of [BG]. Again in Sect. 2.2 of Giannopoulos and Milman (2001) we see how the phenomenon of spherical concentration (see Sect. VII.12.B) of (VII.8.2) enters into the picture.

Here finally is a natural proof, that of Blaschke from 1914, analogous to Steiner's proof of the isoperimetric inequality (expanded by Blaschke), which was presented in Sect. V.11. It is suggested by Fig. VII.8.2 above. We will explain matters in dimension 2, but it goes through *mutatis mutandis* in higher dimensions. The trick is to put ourselves artificially in a space of three dimensions. We place in any way whatever the two given convex sets in two parallel planes P and Q and consider the convex set U = (1/2)K + (1/2)H, which is thus located in a plane R situated midway between P and Q. Let D be a line orthogonal to these three planes. This posed, the idea of Blaschke is to symmetrize "a la Steiner" the convex set C defined as the convex envelope of K and H, an envelope which obviously contains K, H and U. We perform this symmetrization with respect to planes S containing the line D. Such a symmetrization preserves the convexity of the objects obtained, as well as the areas of the three sections by P, Q, R. If we symmetrize with the S larger and larger in number and well chosen, Blaschke's compactness theorem (given in Sect. VII.13.D) shows that we obtain a convex limit object C* which is a convex set of revolution about the line D; this final symmetrization, called Schwarz's, is a solid of revolution; it was used in the very first proof of the isoperimetric inequality. Our three sections are thus disks whose areas are equal to the respective areas of the three initial convex sets K, H, U; and these areas are equal in each case to the square of the radius of the corresponding disk (multiplied by π). But since C* is a convex set, the radii of the disks of C* are a concave function along the line D and, in particular: $\text{Area}^{1/2}(U) \geq (1/2)\,\text{Area}^{1/2}(K) + (1/2)(\text{Area}^{1/2}(H))$, which is what we needed to show.

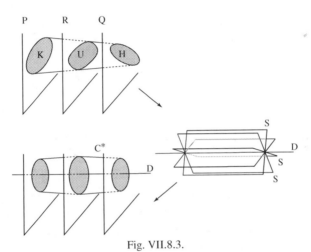

Fig. VII.8.3.

◆

The proof of the isoperimetric inequality is now immediate. With the notations of Sect. 6.C (where B is the unit ball of our Euclidean space of dimension d) we thus have:

$$\text{Area}(\partial C) = \lim_{\varepsilon \to 0} \frac{\text{Vol}(C + \varepsilon B) - \text{Vol}(C)}{\varepsilon}.$$

Then

$$\text{Vol}(C + \varepsilon B) \geq (\text{Vol}^{1/d}(C) + \varepsilon\,\text{Vol}^{1/d}(B))^d$$
$$\geq \text{Vol}(C) + \varepsilon d\,\text{Vol}^{d/d-1}(C)\,\text{Vol}^{1/d}(B),$$

whence

$$\text{Area}(\partial C) \geq d\,\text{Vol}^{d/d-1}(C)\,\text{Vol}^{1/d}(B), \quad \text{QED.}$$

Unfortunately here too we have proceeded with a limit and thus the case of equality is beyond our reach. We will return to this in the "intermezzo" of Sect. VII.14.

Finally the Brunn inequality clearly demonstrates that each Steiner symmetrization (in an arbitrary dimension) transforms a convex set into a convex set. Likewise the function associated with the volume of the sections by parallel affine subspaces (of arbitrary dimension) will be a concave function on the space of complementary dimension.

VII.9. Volume and polarity

We already cautioned at the end of Sect. VII.2 that the volume of a convex set and its polar do not have as their product the square of the volume of the unit ball; we also encountered this problem in Sect. VII.6. The behavior of the very natural product $\mathrm{Vol}(C) \cdot \mathrm{Vol}(C^0)$ is a difficult subject whose development is not yet over. Beside its geometric naturalness, results on the behavior of this product have numerous applications to functional analysis and geometric number theory (in the spirit of Chap. X). Here is the present state of the problem.

We first remark that, for reasons of homogeneity, the product $\mathrm{Vol}(C) \cdot \mathrm{Vol}(C^0)$ is an *affine invariant* attached to the convex set C and to a choice of origin. What might we anticipate on this subject? First, that there does not exist an upper bound for this product; it suffices to approach closer and closer to the boundary of C from the origin, which must be on the interior of C; thus C^0 becomes very large and its volume also, and so the product $\mathrm{Vol}(C) \cdot \mathrm{Vol}(C^0)$ isn't bounded in all generality. In the other direction it isn't clear that we can make this product as small as we wish.

We can also calculate the product for different convex sets of reference. We know that it equals $\beta^2(d)$ for ellipsoids centered at the origin. For the cube and its cocube we find $\frac{4^d}{d!}$ (see Sect. VII.6.A), which is much smaller than $\beta^2(d)$. Things are still worse for the regular simplex, we have $\frac{(d+1)^{d+1}}{(d!)^2}$. However, if we take the d-th root, which we are used to doing systematically, we see that we have, in all these three examples, $\left(\mathrm{Vol}(C) \cdot \mathrm{Vol}(C^0)\right)^{1/d} \sim d^{-1}$, which gives us some hope, at least in the center-symmetric case.

In this paragraph we restrict ourselves to the center-symmetric convex sets which are centered at the origin. By 1900 Mahler had *conjectured* that we always have $\frac{4^d}{d!} \leqslant \mathrm{Vol}(C) \cdot \mathrm{Vol}(C^0) \leqslant \beta^2(d)$, with the idea that the bounds are only attained on the one side by the ellipsoids, from the other by the parallelepipeds or their polars. We now know that we actually have $\mathrm{Vol}(C) \cdot \mathrm{Vol}(C^0) \leqslant \beta^2(d)$ for all convex sets, with equality only for ellipsoids. Blaschke began a proof, completed only in 1982 by Schneider; see below apropos the not necessarily center-symmetric case. The crux of the matter is that the Steiner symmetrization increases the product $\mathrm{Vol}(C) \cdot \mathrm{Vol}(C^0)$. For the rest, readers can consult 6.2 of Giannopoulos and Milman (2001).

It is entirely different for the lower inequality: the problem is still not resolved today. Readers will easily prove that $\mathrm{Vol}(C) \cdot \mathrm{Vol}(C^0) \geqslant \frac{4^d}{(d!)^2}$, for example with the John-Loewner ellipsoid (see Sect. VII.10.C) . But a good bound is a different story. In dimension 2, Mahler obtained the lower bound 8; the idea is to approximate by polygons and to show that the number of sides can be reduced, ending up with only four sides. In the figure, transformations are performed that keep a constant area for

C but augment strictly that of C^0. Things are then evident and it is the occasion for remarking that the square and the cosquare are one and the same object, which is not the case for cubes in higher dimensions.

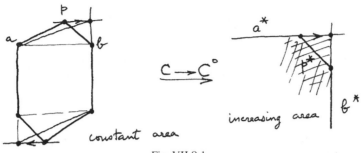

Fig. VII.9.1.

If however the Mahler conjecture is true there will be, modulo affine transformations, exactly two minima of a different nature. In fact, the conjecture is false for the case of equality; counterexamples have been found for which $\mathrm{Vol}(C) \cdot \mathrm{Vol}(C^0) = \frac{4^d}{d!}$, starting with dimension 4, simply by taking the product of a cube and a cocube of appropriate sizes. It remains to find, in desperation, a lower bound whose size is of a good order. This was not achieved until Bourgain and Milman (1987), where it was found that $\mathrm{Vol}(C) \cdot \mathrm{Vol}(C^0) \geqslant \bullet\ d^{-1}$ always holds; more precisely, there exists a universal constant $c > 0$ such that $\mathrm{Vol}(C) \cdot \mathrm{Vol}(C^0) \geqslant c^d \beta^2(d)$ for all convex sets and each dimension d. The proof is hard; it uses probabilities (choose certain subspaces at random and apply the John-Loewner ellipsoids). As for the exact optimal lower bound, the Mahler conjecture remains open; see the note 9.3 on p. 339 of Gardner (1995).

The positive solution for the problem of the upper bound may lead us to think that the quotient $\frac{\mathrm{Vol}(C) \cdot \mathrm{Vol}(C^0)}{\beta^2(d)}$ is a good evaluator of the deficiency of C for being an ellipsoid. This is not true, as can be seen by taking intersection of balls with cubes of an appropriate edge according to the dimension — precisely by studying the convex sets $C_d = B \cap [-t_d, t_d]^d$, which are the intersections of the cube of side $2t_d$ with the unit ball B, and taking the value $t_d = \sqrt{\frac{t \log d}{d}}$. We will see below that a good evaluator for the deficiency of C being an ellipsoid is very simply the smallest number λ for which there exists an ellipsoid E such that $E \subset C \subset \lambda E$. There is also another good evaluator for the John-Loewner ellipsoid (Sect. VII.10.C).

♦

In the not necessarily center-symmetric case, we have a very satisfactory answer for the upper bound. Since $\mathrm{Vol}(C) \cdot \mathrm{Vol}(C^0)$ can be arbitrarily large, it is natural to consider the point z such that, if C_z^0 designates the polar of C with respect to the unit ball with center z, then $\mathrm{Vol}(C) \cdot \mathrm{Vol}(C_z^0)$ is minimum. We then set $p(C) = \mathrm{Vol}(C) \cdot \mathrm{Vol}(C_z^0)$. In 1985 Petty obtained the perfect answer: for all the convex sets of $\mathbb{E}^d$ we have $p(C) \leqslant \beta^2(d)$ *with equality exactly for the ellipsoids*. We find in Meyer and Pajor (1990) a very beautiful proof that doesn't use Steiner symmetrization (and moreover provides generalizations of Petty's result).

It remains to discuss the lower bound in the not necessarily center-symmetric case. Milman considers that it is not impossible that we always have

$$\text{Vol}(C) \cdot \text{Vol}(C^0) \geqslant \frac{(d+1)^{d+1}}{(d!)^2}$$

with equality only for simplexes. In fact, the polar of a (regular) simplex is also a (regular) simplex, and it's thus a problem with a single minimum, which gives us more hope.

The preceding leads naturally to the question: what is the worst convex set? What is the best convex set? — with the idea that the answer is: the best are the ellipsoids, the worst the parallelepipeds. We will see at the end of the following section that this is indeed the case for certain well posed questions.

VII.10. The appearance of convex sets, their degree of badness

We now try to discover the *appearance* of convex sets, a vague term that will be made precise here. We shouldn't confound the appearance of a **solid body** with the appearance of the boundary of a convex set. Apart from the topology of the boundary in Sect. VII.10.B below, the geometry of the boundary (a surface for the convex sets of ordinary space) will be studied in Sect. VII.13.D; we will find some surprising things there, at least at first glance.

VII.10.A. How to generate a convex set

Here are a few basic and classic results that are found in standard texts and which are used in proofs. It is not in the spirit of the present work to present them in detail.

The first is that of *Caratheodory*, which says that the convex envelope $\mathcal{E}(A)$ of an arbitrary subset A can be generated by the simplexes all of whose vertices belong to A, which is to say by the barycenters with nonnegative coefficients of $d+1$ points of A and thus that *a priori* $\mathcal{E}(A)$ is the set of barycenters $\lambda_1 a_1 + \cdots + \lambda_n a_n$ with n arbitrary, $a_i \in A$, $\sum \lambda_i = 1$. A corollary that isn't completely evident otherwise is that $\mathcal{E}(A)$ is compact once A is. Another is that each point of a polytope belongs to a simplex for which the vertices are chosen among those of this polytope. For generation with the fewest points possible, see the Krein-Milman theorem on extremal points in the references cited.

The second is that of Hahn-Banach, whose proof is never obvious — we all have our favorite version — and for which we recall only the most spectacular corollary, which seems illuminating. Let C be an arbitrary convex set and a a point of its boundary ∂C. Then: *there exists at least one hyperplane H that contains a and*

Fig. VII.10.1.

such that C *is entirely contained in one of the half-spaces determined by* H. Such a hyperplane is called a *hyperplane of support of* C *at a*. It is clear that convexity is necessary. On the other hand, depending on the point of ∂C, the hyperplane of support may or may not be unique. There are two extreme cases: that of polytopes, where the vertices have enormously many hyperplanes of support, and that of convex sets whose boundary is a submanifold of codimension 1, for which each point of the boundary possesses a unique hyperplane of support, i.e. the hyperplane tangent to the submanifold.

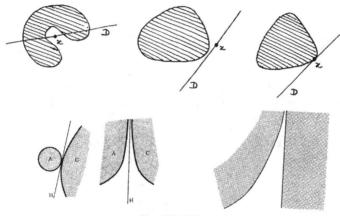

Fig. VII.10.2.

A corollary is that every convex set is equal to the intersection of the half-spaces that contain it. In the Euclidean context, another corollary is that, for each convex set C, each point p of the space has a unique *near point q* in C, i.e. a point q such that the distance between p and the points of C is minimum at q. It is interesting to know that this property characterizes convex sets, which is Motzkin's theorem; see for example Valentine (1964). We also infer the principal separation theorem (among others) for two convex sets, but we must then pay attention to varied hypotheses of compactness, openness and closedness (see Fig. VII.10.2).

VII.10.B. Topology of complex sets and their boundaries

We emphasize the solution to this problem, because it is extremely natural and yet appears in few works on convex sets, because it is difficult to prove. We have found it only in Busemann (1958), on p. 3. Only convex sets with nonempty interior are considered, because in the contrary case they are contained in an affine subspace of dimension smaller than that of the ambient space, in which they have a nonempty interior. The compact (it suffices to say bounded) case is simple: the boundary of a bounded convex set in $\mathbb{R}^d$ is homeomorphic to (has the same topology as) the sphere S^{d-1}. In the unbounded case, this boundary can still be homeomorphic to S^{d-1} but in general it will be homeomorphic to $S^{d-r-1} \times \mathbb{R}^r$. To prove this in the case where $0 < r < d - 1$, we need to show that our convex set is in fact a cylinder, i.e. the product of a line and a convex set of dimension $d - 1$. It's this part that is difficult; see 11.3.8 of [B].

VII.10.C. The John-Loewner ellipsoid and its applications

We want to know how much a convex set C differs from an ellipsoid, or else find the ellipsoid closest to C. The spirit in which this is done is much used in control theory, linear programming and optimization; see the synthesis Gruber and Wills (1993), Sects. 2.7, 2.8 (ellipsoid method of Grötschel, Lovasz, Schrijver) and 2.9.

The John-Loewner theorem − not a collaboration, but a result published independently by these two mathematicians − is of an affine nature and states:

(VII.10.C.1) *Given a bounded set C, we consider all ellipsoids centered at the origin and containing C. Then there exists one of least volume and it is unique, called the* John-Loewner ellipsoid *of C.*

When an origin hasn't been fixed, but we are working with center-symmetric convex sets, then of course we take the center of symmetry as origin. The proof of existence is a simple matter of compactness, the difficulty comes with uniqueness. But this is an immediate consequence of the end of Sect. VII.3: strict convexity of the function $q \mapsto \det(q)^{-1/2}$ for quadratic forms q.

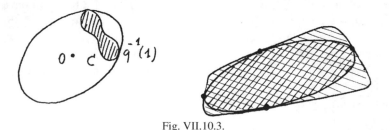

Fig. VII.10.3.

This result has numerous consequences; here are a few of them. The first answers the question posed in Sect. IV.2: are the ellipses and more generally the ellipsoids the only geometric objects that admit affine hyperplane symmetries with respect to each hyperplane direction of the space? For simplicity we look to the planar case and we let C be the convex envelope of such a geometric object. Possessing two distinct affine symmetries, it has first of all a center of symmetry. Let us consider the John-Loewner ellipse E of C associated with this center. This (solid) ellipse contains at least one point a of the boundary, since it is of minimum area. But the uniqueness of E shows that each affine symmetry σ that preserves C also preserves E, and thus C also contains the point $\sigma(a)$. Finally, the boundary of C is nothing other than the whole of E, QED.

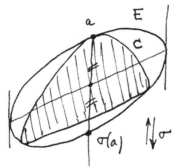

Fig. VII.10.4. Only the conics (the quadrics) admit affine symmetries for all directions

This proof is not very high on the ladder. To climb up a bit, we consider a bounded subset C in the affine space $\mathbb{R}^d$ and the group Γ of all transformations that preserve it. First, the elements of Γ necessarily preserve the center of gravity of C, supposed to be at the origin of $\mathbb{R}^d$. Because of its uniqueness, Γ preserves the John-Loewner ellipsoid attached to C and thus as a linear group preserves the quadratic form q that defines it. But this is to say that Γ is a subgroup of the orthogonal group of $(\mathbb{R}^d, q)$ and is, in particular, compact. We have thus shown − it's the second consequence − that the subgroup of the affine group that preserves a bounded set is in fact a group of isometries for a suitable Euclidean structure . Returning to the objects that admit affine symmetries for all directions − which is now to say the orthogonal symmetries for all directions − only the spheres (the balls) possess this property. Readers will find yet more applications in this spirit in 11.8 of [B].

A third consequence consists of applying polarity, always in the center-symmetric case. If polarity preserves ellipsoids and inverts their volume, we thus now know that *each convex set contains a unique (solid) ellipsoid of maximum volume*. But we should caution that the two ellipsoids, the one containing minimum volume and the other containing maximum volume, are not in general homothetic to one another. The study of their respective positions is in fact not complete; see Milman and Pajor (1989).

A final consequence:

(VII.10.C.2) *If* C $\subset$ $\mathbb{E}^d$ *is a center-symmetric convex set, then there exists an ellipsoid* E *such that we have the sandwich* E $\subset$ C $\subset$ $\sqrt{d}$E.

To see this, we look at the drawings and proceed by contradiction with appropriate ellipses, the case of larger dimensions not introducing anything more than a necessary change of constant; specifically $\sqrt{d}$ replaces $\sqrt{2}$ and the ellipses are replaced by ellipsoids of revolution.

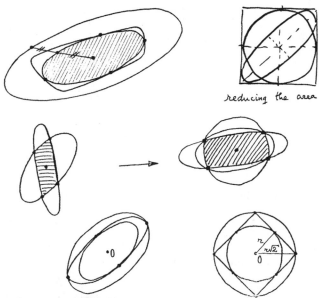

reducing the area

Fig. VII.10.5. The sandwich theorem. As the result is affine, we may suppose that the John-Loewner ellipse is the unit disk. We then proceed by the following contradiction: we suppose that C contains a point p situated a distance more than $\sqrt{2}$ from the origin; since C is center-symmetric, it also contains $-p$ and the convex envelope C' of the figure formed by $\{p, -p,$ unit disk$\}$. Then, as a simple area calculation shows, an elliptical neighborhood E' of the unit disk and inscribed in C' will have an area strictly larger than 1. In the opposite direction, the ellipse of largest area contained in a square is the inscribed disk. And it is necessary to go to $\sqrt{2}$ to contain our square entirely

Moreover, this constant $\sqrt{d}$ is optimal, as is shown by the case of the cube or the parallelepiped. We deduce from (VII.10.C.2), something essential for characterizing a bounded convex set by its boundary S. We place ourselves at a point interior to this convex set; then for each half-line $\mathbb{R}^* \cdot u$ (where u is a unit vector) passing through the origin there is a well defined point $f(u) \cdot u$ where it intersects S. The convex set is thus determined by a function $f : S^{d-1} \to \mathbb{R}^*$. The convexity of the solid implies that this function is continuous (Sects. VII.3 and VII.13). But in fact, even if it is not differentiable, we can control its variation. Recall that a function on a metric space (here the sphere, the distance denoted by $\mathrm{dist}(\cdot, \cdot)$ and it matters little whether it is the intrinsic or extrinsic distance for the sphere) is called *Lipschitz of ratio k* if we have $\frac{|f(x)-f(y)|}{\mathrm{dist}(x,y)} \leq k$ for all $x \neq y$. This condition is implied by differentiability when the modulus of the derivative is everywhere bounded by k, but it is a weaker condition generally. Here we deduce from (VII.10.C.2) and from the convexity that

(VII.10.C.3) *In dimension d the function f defining as above a center-symmetric convex set is always Lipschitz of ratio $\sqrt{d}$.*

This corollary will be used in an essential way in Sect. VII.12 for proving Dvoretsky's theorem on the existence of almost spherical sections in every convex set. For the non center-symmetric case, see the following section. In the context of Sect. VII.9 preceding, we can use the above sandwich to give a lower bound of $\mathrm{Vol}(C) \cdot \mathrm{Vol}(C^0)$, assuming central symmetry. In fact, first $\mathrm{Vol}(C) \geq \mathrm{Vol}(E)$; then polarity implies from $C \subset \sqrt{d}E$ that $C^0 \supset d^{-1/2}E^0$, whence $\mathrm{Vol}(C^0) \geq d^{-d/2}\mathrm{Vol}(E^0)$. As $\mathrm{Vol}(E) \cdot \mathrm{Vol}(E^0) = \beta^2(d)$, and we obtain the very, very bad inequality $\mathrm{Vol}(C) \cdot \mathrm{Vol}(C^0) \geq d^{-d/2}\beta^2(d)$. This allows us to better appreciate the Bourgain-Milman result from Sect. VII.9, which gives $\mathrm{Vol}(C) \cdot \mathrm{Vol}(C^0) \geqslant\bullet d^{-1}$. Thus, although the John-Loewner ellipsoid is a very good tool, it is not suitable for every problem.

Interested readers may ask whether there are other ellipsoids attached to a convex set that are obtainable by a *minimization* process. Such is in fact the case; what there is on this topic can be found in 2.3 of Giannopoulos and Milman (2001) for the case of minimum area and that of minimum width. In the same section of that text the nature of points of contact of a compact set with its John-Loewner ellipsoid is nicely clarified and we will use this in an essential way in the study of badness in Sect. VII.10.E. Throughout that text things are treated in an equivalent way: seek a Euclidean structure best suited for a given convex set (which amounts to choosing an ellipsoid).

For the non center-symmetric case, see the end of the next subsection.

VII.10.D. A first metric space formed of all center-symmetric convex sets: the compact set of Banach-Mazur

We posed the problem of measuring in one way or another the deficiency of a convex set for being an ellipsoid, the ratio $\frac{\mathrm{Vol(E)}}{\mathrm{Vol(C)}}$ for the ellipsoid of maximum volume included in C not being the only possibility. Another evaluator might be the smallest number λ for which there exists an ellipsoid E such that $E \subset C \subset \lambda E$. In pursuing this idea systematically, Banach and Mazur came to the following definition. It has to do with estimating the *difference* (the *distance*) between any two center-symmetric convex sets, but only **in the affine setting**. In fact, in the preceding the underlying idea was that all ellipsoids are the same within an affine transformation. We want to study the *affine form* of convex sets (for the Euclidean form; see Sect. VII.13.D). All this happens implicitly in the affine space $\mathbb{R}^d$ of dimension d.

We put ourselves in the space of all center-symmetric convex sets (centered for example at the origin) and we define the *Banach-Mazur distance* $d_{B-M}(C, D)$ between two convex sets by:

(VII.10.D.1) $d_{B-M}(C, D) = \inf\{\log \lambda \in \mathbb{R}$: *there exists an affine transformation* f *such that* $f(D) \subset C \subset \lambda f(D)\}$.

It can be shown without much difficulty that d_{B-M} is really a metric, not on the space of all center-symmetric convex sets, but on the quotient space $\mathcal{K}$ of these by the group of affine transformations. Only the triangle inequality requires much attention. It can be shown also that $\mathcal{K}$ is a compact topological space (of infinite dimension). For this space, see 1.9 of Gruber and Wills (1993), but updated by 7.2 of Giannopoulos and Milman (2001). The book Tomczak-Jaegermann (1989) is entirely devoted to the subject, but is now a bit dated. Note that the above definition, applied naively to the non centrally symmetric case, yields a function that is not in general a distance. Finally, for reasons of convenience, numerous works use the constant λ directly, and not its logarithm.

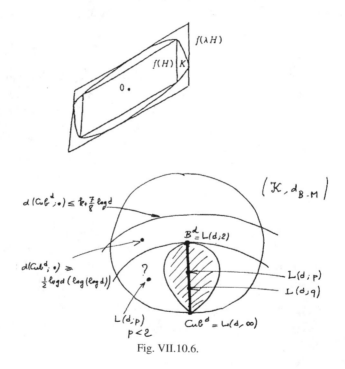

Fig. VII.10.6.

A number of questions face us regarding $(\mathcal{K}, d_{B-M})$. We first remark that the corollary VII.10.C.2 says this: if we let B_d denote the image of the ball (or of any ellipsoid) of dimension d in the quotient $\mathcal{K}$, then for each point (a center-symmetric convex set) C of $\mathcal{K}$ we have $d_{B-M}(C, B_d) \leq \frac{1}{2} \log d$ and only the cubes Cub_d (the parallelepipeds) are a distance $\frac{1}{2} \log d$ from the ball. In particular the diameter of $(\mathcal{K}, d_{B-M})$ is certainly bounded by $\log d$, but its exact diameter is not known. However, Gluskin (1981) proved the following estimate: there exists a universal constant c and for each dimension there exist pairs of convex sets in $\mathbb{R}^d$ for which the distance is equal to $\log(cd)$. The argument is probabilistic (measure-theoretic) in nature; it is known that these convex sets exist, but Gluskin has not exhibited them explicitly. We will encounter this type of result in Sect. X.5, where we will have theoretical existence without explicit construction, but this will be more serious in Sect. X.5 because it has to do with matters that are very useful in applications.

On the other hand we would like to know whether there exists an analogue of $d_{B-M}(C, B_d) \leqslant \frac{1}{2} \log d$ for the cube: what is the maximum of $d_{B-M}(C, Cub_d)$ when C traverses $\mathcal{K}$? It is a question of knowing what are the best sandwiches possible where the squeezers this time are the parallelepipeds. The answer is very much open; it is an extremely difficult problem. Essentially two things are known: first, there are examples of C due to Szarek in 1990 where $d_{B-M}(C, Cub_d) \geqslant \frac{1}{2} \log d + \log(\log d) + c$ (for a universal c). These examples of Szarek are constructed in probabilistic fashion: convex sets are chosen, in a certain sense, at random, but with the randomness well controlled. In the positive direction it is now known that for each C we have $d_{B-M}(C, Cub_d) \leqslant a + c \log d$, where c is universal. The best c isn't known; presently we have $\frac{4}{5}$. References are to be found at the end of 4.1 of Giannopoulos and Milman (2001). This type of result is very difficult to prove. Two ideas are used: the first consists, as in the proof of (VII.10.C.2), of looking at the points of contact of the John-Loewner ellipsoid, attached to the given convex set C, with that set; but here it is shown that these contact vectors form an "almost orthogonal" basis and thus the set is not so far from being cubic. With the aid of this system very many points are placed in the convex set, a set to which a combinatoric theorem is applied, the Sauer-Shelah lemma (see Chap. 15 of Pach and Agarwal, 1995), which then allows them to be placed in a not too large cube.

If the ellipsoids and parallelepipeds are the most natural convex sets as regards universality, they aren't the only ones; we will also encounter others, specifically the balls $L(d; p) : \sum_i |x_i|^p \leqslant 1$. The Euclidean balls (ellipsoids here) are the $L(d; 2)$, the cubes are the $L(d; \infty)$ and the cocubes (squares in the plane and regular octahedra in space) are the $L(d; 1)$. Where are these various $L(d; p)$ situated in $(\mathcal{K}, d_{B-M})$? Figure VII.10.6 above provides an answer, known presently only for $p \geqslant 2$. The distances $d_{B-M}(L(d; p), L(d; q))$ are known for all pairs $2 \leqslant p \leqslant q \leqslant \infty$; they can be found in Chap. 37 of the book that is entirely devoted to our compact set: Tomczak-Jaegermann (1989). We have:

$$d_{B-M}(L(d; p), L(d; q)) = \left(\frac{1}{p} - \frac{1}{q}\right) \log d.$$

And thus the points corresponding to the family of the $L(d; p)$ for $p \in [2, \infty]$ form a segment in $(\mathcal{K}, d_{B-M})$; they are *aligned* there. It may seem scandalous that the $d_{B-M}(L(d; p), L(d; q))$ are not known once p or q is less than 2, but it is a fact.

There thus remains much to be done in the geometric space $(\mathcal{K}, d_{\text{B-M}})$, even though it is the most natural possible and defines the compact center-symmetric convex sets.

For the non center-symmetric case, more difficult as we have seen and as we will see once again, consult 7.3 of Giannopoulos and Milman (2001). In any case we know that there is always sandwiching between homothetic ellipsoids, but we can't do better here than with a ratio equal to the dimension, as the regular simplex shows. Moreover, only the simplex attains this extreme bound. The inequality is already found in John (1948); the characterization of the simplexes by equality is in Palmon 1992; the proof consists, as in that of Ball for (VII.10.E.1) and also the one in Sect. VII.11.C here, of carefully analyzing the points of contact between the convex set and the John-Loewner ellipsoid; but there's no using or creating a new functional inequality.

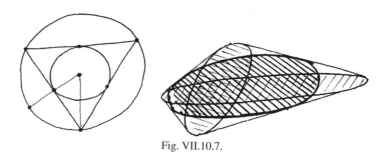

Fig. VII.10.7.

VII.10.E. The Rogalski conjecture and a mapping of isoperimetric type

We return to the idea of evaluating the deficiency of a center-symmetric convex set C for being an ellipsoid by looking at the ratio $\text{qv}(C) = \frac{\text{Vol}(C)}{\text{Vol}(E)}$, called the *volumic quotient*, where E is the ellipsoid of maximum volume contained in C. This volumic quotient is an affine invariant. In 1980 Rogalski posed the question of knowing what is the maximum possible value of the volumic quotient and the convex sets for which this hypothetical value is attained. The trivial bound (see above) $d^{d/2}$ is ridiculous. We might think, in the centrally symmetric case, that these are the worst cubes and thus conjecture that we always have a center-symmetric convex set C:

(VII.10.E.1) $\text{qv}(C) \leqslant \dfrac{2^d}{\beta(d)}$ *with equality only for parallelepipeds.*

It is true, and has been proved in Ball (1989). The proof is hard and two things are necessary, which means climbing two rungs. The first is already found in John: we need to characterize the points of contact of E with C, which is done by calculating the derivative of the volume for every compatible variation. This derivative must be zero when the volume is maximum. We suppose that E is the unit ball: perform an appropriate affine transformation, which is possible since qv is affinely invariant. The condition of maximal volume says this: there exist unit vectors u_i and reals c_i such that the $c_i u_i$ belong to all the ellipsoids considered and such for all x we have the identity (see Milman and Pajor, 1989):

$$\sum c_i \langle x, u_i \rangle^2 = \|x\|^2.$$

The points of contact are the $\pm c_i u_i$ and we observe that, in general, they are $2d$ in number, except in particular cases, such as the cube for example.

Now, there is a very subtle inequality from integral calculus, much more subtle than Hölder's and due to Brascamp-Lieb, which says this: for $\{u_i, c_i\}$ satisfying the above relation, for all x and for all integrable functions f_i, we have the inequality:

(VII.10.E.2) *Brascamp-Lieb inequality:*

$$\int_{\mathbb{R}^n} \prod_{i=1}^{m} f_i(\langle u_i, x \rangle)^{c_i} \, dx \leq \prod_{i=1}^{m} \left(\int_{\mathbb{R}} f_i \right)^{c_i}.$$

The case of equality necessarily yields cubes. Now for the not necessarily center-symmetric case, we find in Ball (1991) the awaited answer:

(VII.10.E.3) *For every convex set C we have* $\mathrm{qv}(C) \leq \dfrac{d^{d/2}(d+1)^{(d+1)/2}}{d!} \cdot \beta(d)$, *with equality only for the simplexes.*

As we have already said, things are always more difficult in the nonsymmetric case. Ball's proof is analogous to the one in the center-symmetric case, but more subtle.

In the same text Ball gives an interesting application to a sort of *inverse isoperimetric inequality*. More and more elongated ellipsoids show that the isoperimetric quotient $\dfrac{\mathrm{Area}^d(\partial C)}{\mathrm{Vol}^{d-1}(C)}$ is never bounded above. But this has to do with a result in Euclidean space, and in fact we stretch balls with affine transformations. But in a *completely affine* spirit we can look at what is the worst case possible case when we let C be **deformed** by suitable affine transformations. By simple application of the definition of the area of ∂C,

$$\mathrm{Area}(\partial C) = \mathrm{Vol}^{d-1}(\partial C) = \lim_{\varepsilon \to 0} \frac{\mathrm{Vol}(C + \varepsilon B) - \mathrm{Vol}(C)}{\varepsilon},$$

Ball derives these two optimal results directly from his inequalities for the volumic quotient:

(VII.10.E.4) *For any center-symmetric convex set C, there exists an affine image* C^* *of C such that*

$$\mathrm{Vol}(C^*) = \mathrm{Vol}(\mathrm{Cub}_d)$$

and

$$\mathrm{Area}(\partial C^*) \leq \mathrm{Area}(\partial(\mathrm{Cub}_d)).$$

For arbitrary convex sets this will be

$$\text{Vol}(C) = \text{Vol}(\textit{regular simplex}),$$

with

$$\text{Area}(\partial C^*) \leqslant \text{Area}(\partial(\textit{regular simplex})).$$

◆

In Sect. VII.8 we mentioned in advance the dual inequalities of (VII.10.E.1) and (VII.10.E.3), which are proved with a dual inequality for (VII.10.E.2), a difficult inequality which appeared recently in Barthe (1998).

VII.10.F. Badness test for a convex set using its moments of inertia: the ellipsoids of Legendre and Binet (ellipsoid of inertia), the inertial invariant and the grand conjecture on convex sets (first formulation)

We return again to the problem of "best" approximation of a convex set by an ellipsoid; for example if we think distance (metric), we can consider approximation by ellipsoids whose boundary is metrically close to the boundary of the given convex set. This type of problem regularly enjoys new contributions and is not even close to being completely understood. Here are two ellipsoids canonically attached to a convex set (in fact, to any bounded subset whatever) in the Euclidean space $\mathbb{E}^d$; they come from physics, in particular from analytical mechanics.

The *ellipsoid of inertia* is introduced naturally; it is sometimes called *Binet's*. It originates from the case of ordinary space $\mathbb{E}^3$. We seek to have the bounded set C (called a *body*) turn about a line D in space; the resistance (inertia) to doing this is expressed by a number called the *moment of inertia of* C *with respect to* D and is calculated with the integral $I^2(C; D) = \int_C \text{distance}^2(x, C) \cdot dx$. To have a convex set turn about a line and test its inertia as a function of its other invariants is a good way of testing its **badness**. Inertias are also defined with respect to points and with respect to planes. If the plane u^* contains the origin and is orthogonal to the unit vector u, then $I^2(C; u^*) = \int_C \langle x, u \rangle^2 \cdot dx$. As for the moment with respect to the origin, this is simply $\int_C \|x\|^2 \cdot dx$. When lines or planes are subjected to parallel displacement, the corresponding moment of inertia is always minimum when the line or plane passes through the *center of gravity* of the body, i.e. the point a such that $\int_C (x - a) \cdot dx = 0$. This has been known since Apollonius and is completely elementary with algebraic expression of the Euclidean structure.

Henceforth we put the origin at the center of gravity of the body considered, in arbitrary dimension d; nothing need be changed in the formulas. We then observe that $\int_C \langle x, u \rangle^2 \cdot dx$ is a quadratic form in u, whence the fact that the set

$$\text{Binet}(C) = \left\{ \frac{u}{I(C; u^*)} : u \in S^{d-1} \right\}$$

Fig. VII.10.8. The required effort is minimum when the line passes through the center of gravity of the solid

is an ellipsoid (the surface, not the full ellipsoid). The full ellipsoid will be called the *ellipsoid of inertia (or Binet's)* of the body C. Note that it is defined only when C contains the origin in its interior; otherwise $I(C; u^*)$ can be zero.

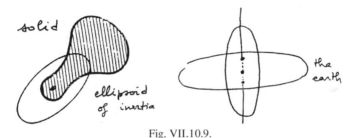

Fig. VII.10.9.

◆

The ellipsoid of inertial is anything but a good approximation for C because, if C is very flattened in a direction of the plane u^*, then the corresponding moment of inertia will be very small and in the orthogonal direction u we will thus have a point of the ellipsoid very remote from the origin. For example, the ellipsoid of our planet admits a flattened ellipsoid of revolution as a good approximation, the opposite of an elongated ellipsoid of revolution. But it remains the good object for mechanics, as we will see in Sect. VII.13.C. If we want an ellipsoid that is a good approximation for C, we need to choose something like the polar of Binet(C). In fact, the quadratic nature of inertia shows that there exists a solid ellipsoid (unique of course), called *Legendre's* and denoted Legendre(C), such that

$$\int_{\text{Legendre}(C)} \langle x, u \rangle^2 \cdot dx = \int_C \langle x, u \rangle^2 \cdot dx$$

for each vector u. And in fact this ellipsoid is homothetic with the polar $(\text{Binet}(C))^0$:

$$\text{Binet}(C) = \sqrt{d + 2} \frac{\text{Vol}(\text{Legendre}(C))}{(\text{Legendre}(C))^0}.$$

At present the optimal relations between the four ellipsoids: Binet's, Legendre's, John-Loewner (interior and exterior) remain to be discovered.

It is an amusing exercise to calculate the *moments of inertia* of our favorites: balls, cubes and cocubes (we'll stay with center-symmetric bodies). Don't forget the basic trick: by symmetry all moments with respect to coordinate planes are equal, but their sum is $\sum \int x_i^2 \cdot dx = \int \|x\|^2 \cdot dx$. We find the results:

$$\mathrm{I}^2(\mathrm{Cub}_d) = \frac{2}{3}, \mathrm{I}^2(\mathrm{Cocub}_d) = \frac{1}{(d+1)(d+2)} \cdot \frac{2^d}{d!}, \mathrm{I}^2(\mathrm{Ball}_d) = \frac{\beta(d)}{d+2}$$

(this for each direction u). The ellipsoids of inertia are balls in the three cases, which is clear since the only ellipsoids having as many symmetries as cubes (or cocubes) are balls.

We now introduce a universal inertial invariant for bodies (convex if desired). A large part of what follows in the next sections is directly inspired by the fundamental text (Milman and Pajor, 1989). The idea once again is to use the affine group for normalizing things in some way. We doubly normalize a center-symmetric bounded set C by an affine transformation in such a way that its volume is equal to 1 and its ellipsoid of inertia is a ball. Let C* be the center-symmetric convex set so obtained. Then all the moments of inertia are equal and this common value is an invariant of C that we will denote by InvInert(C), the (*inertial invariant of* C or else the *reduced inertia of* C). We find then, for our three bodies:

$$\mathrm{InvInert}(\mathrm{Cub}_d) = \frac{1}{2\sqrt{3}}, \quad \mathrm{InvInert}(\mathrm{Cocub}_d) = \frac{(d!)^{1/d}}{\sqrt{(d+1)(d+2)}}$$

$$\mathrm{InvInert}(\mathrm{Ball}_d) = \frac{1}{\sqrt{d+2}} \frac{1}{\beta(d)^{1/d}}.$$

But these results are astonishing, read with the notation "•" of Sect. VII.6.A and Stirling's formula, specifically that we have InvInert $= \bullet 1$ in the three cases; thus, asymptotically with the dimension, the invariants of inertia behave like constants. Having discovered this about our three convex types, it is thus right to ask what happens with general centrally symmetric convex sets. Might we expect, for example, a sandwich of the type $\bullet_1 1 \leqslant \mathrm{InvInert}(C) \leqslant \bullet_2 1$ for each center-symmetric convex set C? Presently we have only a single (but very good) answer for one side. This answer is, moreover, completely trivial to state and to prove: for each center-symmetric convex set C we have InvInert(C) $\geqslant$ InvInert(Ball), with equality only for ellipsoids. In particular, InvInert(C) $\geqslant\bullet$ 1. For the proof we make a sketch and write simply:

$$d \cdot \mathrm{InvInert}(\mathrm{Ball\ B}) = \int_{B \cap C} \|x\|^2 \cdot dx + \int_{C \setminus C \cap B} \|x\|^2 \cdot dx$$

$$\geqslant \int_{B \setminus B \cap C} \|x\|^2 \cdot dx + \int_{B \setminus C \cap B} \|x\|^2 \cdot dx$$

$$= d \cdot \mathrm{InvInert}(B),$$

since $\mathrm{Vol}(B) = \mathrm{Vol}(C) = 1$ implies that $\mathrm{Vol}(B \setminus C \cap B) = \mathrm{Vol}(C \setminus C \cap B)$ and that $\|x\|$ on $C \setminus C \cap B$ is always greater than $\|x\|$ on $B \setminus C \cap B$ by definition of the ball!

In contrast, it is an open question whether we always have $\mathrm{InvInert}(C) \leqslant \bullet\, 1$. Precisely, we have:

(VII.10.F.1) *(Grand conjecture on the symmetric convex sets.) We have*

$$\mathrm{InvInert}(C) \leqslant \bullet\, 1$$

for each center-symmetric convex set C.

We will return to this conjecture amply in Sect. VII.11, for it is equivalent to at least four other conjectures, all of them just as simple and natural to state (always for each center-symmetric convex set C). With the John-Loewner ellipsoid we have right away a very bad bound, $\mathrm{InvInert}(C) \leqslant \sqrt{d}$. But we know more since Bourgain (1991), specifically that

$$\mathrm{InvInert}(C) \leqslant \bullet\, d^{1/4} \log d.$$

See 7.1 of Giannopoulos and Milman (2001) for an idea of the proof. Let us say right away that the conjecture (VII.10.F.1) has been verified for numerous categories of convex sets having a very small additional property; see Milman and Pajor (1989). The non center-symmetric case is much more sophisticated, even for finding the right statements to conjecture; see Milman and Pajor (1999) and Giannopoulos and Milman (2001).

It is very pleasant to find here Sylvester's problem, encountered in Sect. I.2: let us recall that we look for the probability P(D) that, when we throw $d + 2$ points at random at a convex set D in $\mathbb{R}^d$, these $d + 2$ points are the $d + 2$ vertices of their convex envelope (that is, we don't want any one of them to be in the interior of the convex set generated by the others). This is an affine, not Euclidean, problem. We have seen in Sect. I.2 that we have an optimal upper bound (attained only by the ellipsoids). On the other hand, starting with dimension 3, the optimal lower bound isn't known. In any case we don't want it to be too small. Now precisely the following asymptotic equivalence in the dimension is proved in Milman and Pajor (1989):

$$(1 - P(D))^{1/d} =_\bullet \frac{\mathrm{InvInert}(D)}{\sqrt{d}}$$

and thus, if the grand conjecture is true, we have

$$(1 - P(D))^{1/d} =_\bullet \frac{1}{\sqrt{d}},$$

i.e. we are assured that P(D) will never be too small.

VII.11. Volumes of slices of convex sets

We have encountered slicing of convex sets several times, essentially with respect to the symmetrization operations in Sect. VII.5.A, slicing by parallel affine subspaces among them. But the problem of obtaining information about the volumes of the slices of a convex set by affine subspaces is very natural. We are going to spend a lot of time on this problem, because it harbors unsuspected difficulties and has a certain practical importance (scanners, etc.). We will first slice with lines, then with hyperplanes (the most natural among numerous problems). The following section is mostly devoted to almost spherical slices (but of small dimension compared to that of the space). Readers can confirm many things that are astonishing, either because they aren't at all intuitive or on the contrary very "obvious" but in fact extremely difficult to prove or even unsettled. Other than the references already given, for sections Gardner (1995) is a systematic source.

VII.11.A. Slicing by lines, Hammer's X-ray problem

If we think about the Steiner symmetrization, the Hammer problem is: can we reconstruct a convex set with the help of its symmetrizations? It is very clear that one alone will not suffice, but what if we symmetrize in several directions? We formulate the problem in the plane, which will extend to higher dimensions where we slice by parallel hyperplanes. In 1963 Hammer asked about how many directions in a plane are necessary for reconstructing a body by X-rays effected in these directions. Perfect reconstruction is understood to be within a translation. We are dealing with a body with constant density and thus with its form (this isn't the real problem of radiography or of a scanner, where we seek to reconstruct an interior with highly variable density). The complete answer is in Gardner and McMullen (1980) and is quite pleasing.

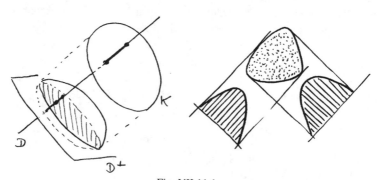

Fig. VII.11.1.

Now the basic remark is that the set S of these directions is *ill chosen* — even a large number does not suffice — on account of the following example: the two squares, or the two parallelograms, of the figure have the same symmetries for the four directions of their diagonals. More generally, if the set S is that of the diagonals of a regular polygon (or any one of its affine images), we can't reconstruct the subpolygons of the figure.

The result of Gardner-McMullen is that these figures constitute the only exception:

> *So long as the convex set S isn't an affine image of the diagonals of a regular polygon, it can be reconstructed from its own symmetrizations in the directions of* S.

The proof is very geometric; by contradiction we end up constructing an (affinely) regular polygon. Since three directions are always those of the diagonals of a hexagon which is the affine image of a (Euclidean) regular hexagon, three directions will never suffice. On the other hand, this was our introduction for the notion of the cross ratio in Sect. I.3: once S has four elements that are not in harmonic division, this suffices; see Fig. VII.11.2. We are also always successful as soon as S is infinite.

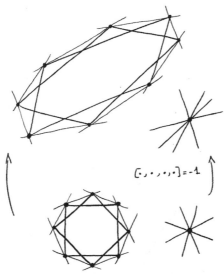

Fig. VII.11.2. Four directions of lines are in harmonic division if they are the directions of the diagonals of a regular octagon or of one of its affine images

For different viewpoints on X-rays and convex sets, see Chap. 5 of Gardner (1995).

From now on in the present section we consider only center-symmetric convex sets.

VII.11.B. Slicing with hyperplanes: general questions, the grand conjecture

Our hope for reconstruction is not without foundation. In fact, on the one hand being given a convex set C centered at the origin is equivalent to being given a function f on the unit sphere S^{d-1} with values > 0, the function which associates with u the abscissa of the point where the half-line $\mathbb{R}_+ u$ intersects the boundary of C. The convex set is thus defined, in "polar" — or rather "spherical" — coordinates by the condition $\rho \leqslant f_C(u)$, $u \in S^{d-1}$; it is symmetric if f_C is even, i.e. if $f_C(-u) = f_C(u)$ for each u. On the other hand, being given the volume of the hyperplane sections amounts as well to being given an even function on S^{d-1}, that which associates with u the $(d-1)$-dimensional volume of $C \cap u^0$, where u^0 is the hyperplane passing through the origin and orthogonal to u.

To say that C is symmetric is to say that the function $f : S^{d-1} \to \mathbb{R}^*$ is even, i.e. $f(-u) = f(u)$ for all u. In what follows, we denote by f the function thus associated with the convex set C (more precisely, f_C when necessary). Having the volume of the hyperplane sections depends also only on an (even) function on the unit sphere; specifically we associate with u the volume of $C \cap u^0$, where $u^0 = \{x : \langle x, u \rangle = 0\}$ denotes the hyperplane passing through the origin and orthogonal to the vector u.

With integrals in spherical coordinates (ρ, u) we find

$$\mathrm{Vol}(C) = \int_{u \in S^{d-1}} \int_0^{f(u)} \rho^{d-1} \, d\rho d\theta = \frac{1}{d} \int_{u \in S^{d-1}} f^d \, du$$

for the total volume of C, and for that of the section by the hyperplane $H = u^0$:

$$\mathrm{Vol}(C \cap H) = \frac{1}{d-1} \int_{v \in S^{d-1} \cap u^0} f^{d-1} \, dv.$$

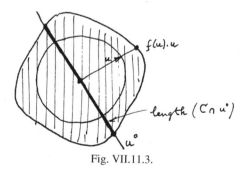

Fig. VII.11.3.

Our problem is thus to reconstruct a function f on the sphere with the aid of the integrals of its power f^{d-1} over all its equators. It is an old problem that was solved in Funk (1913) and put in a more general context in Radon (1917) (Radon incidentally wasn't aware of Funk's result). The basic work on the subject is Helgason (1980). It has to do with inverting the operator $g \mapsto g^0$ between functions on S^{d-1}, where

$$g^0 = \int_{v \in S^{d-1} \cap u^0} g \, dv.$$

This operator is called the *Radon transformation*, the work of Funk having long remained unknown, because it treated a very particular problem that we will encounter in Sect. XII.3.B, that of metrics on the sphere for which all the geodesics are periodic. We note that this function can only be good for even functions on the sphere, since for an odd function the integral over an equator is always zero.

From Sects. III.7 and III.XYZ we know that there exists a theoretical tool for treating problems on the sphere: the *spherical harmonics*. The idea for inverting $g \mapsto g^0$ is now very simple: we decompose the even function g into its spherical harmonics $g = \sum a_i \phi_i$ and since $g \mapsto g^0$ is linear it suffices to know the ϕ_i^0. But $g \mapsto g^0$ commutes with all the isometries of the sphere. The orthogonal group is sufficiently big so that this implies that the ϕ_i^0 are themselves spherical harmonics, at least within a scalar: $\phi_i^0 = c(i) \phi_i^0$. Thus the inverse of $\sum a_i \phi_i$ is $\sum a_i c(i)^{-1} \phi_i$.

However, this theoretical solution proves to be completely useless for solving any of the natural problems that we are going to pose. Apart from that, very simply: *if* Vol($C \cap H$) *is constant for all the* H, *then* C *is a ball.* Readers should attempt to show in an elementary manner (without using spherical harmonics), that an even function on S^2 for which all the integrals along the equators are zero is itself zero.

We now pose the problem that is known as Busemann-Petty conjecture (1965):

(VII.11.B.1) *Let* C *and* D *be two center-symmetric convex sets such that*

$$\text{Vol}(C \cap H) \geqslant \text{Vol}(D \cap H)$$

> *for all the hyperplanes passing through their common center. Do we have* Vol(C) $\geqslant$ Vol(D)?

The method of attack using the Radon transformation doesn't work at all, because this transformation is not generally well behaved with regard to inequalities. Thanks to our recent acquaintance with hyperplane sections of the cube, we will see later that the answer to VII.11.B.1 is **no**, starting with dimension 5. Trivially the answer is yes for the plane. It has recently been proved for dimension 3 in Gardner (1994) and simultaneously in dimensions 3 and 4 in Gardner, Koldobsky, and Schlumprecht (1999). The first proof is the conjunction of two things: on the one hand the fact that in dimension 3 all the convex sets have the particular property of being *intersection bodies*. An intersection body C is a convex set for which there exists a convex set C' such that the function $f(u)$ defined from C on the sphere S^{d-1} has, for each u, the value of the volume Vol($C' \cap u^0$). It can be shown that, in dimension 3, every convex set is an intersection body (in other words, an even function on S^2 has a special property). Now it is shown that for such functions the Radon transformation is well behaved with regard to inequalities. On the other hand the more recent proof that holds through dimension 4 is of a completely different nature. It is very conceptual and modulo a suitable formula that gives the behavior of the $(d-2)$-th derivatives of the functions of parallel sections. Now convexity never allows control over more than the second derivative, as we saw in Sect. VII.3.

The analysis into spherical harmonics doesn't do us any good, because the above formulas don't commute with the inequality relations. We need rather a relation of the sort

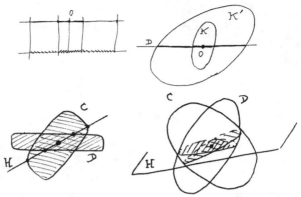

Fig. VII.11.4.

$$\mathrm{Vol}^{d-1}(C) = c(d) \int_{u \in S^{d-1}} \mathrm{Vol}^d (C \cap u^0) \cdot du$$

with a constant $c(d)$ dependent only on the dimension (and thus not on the convex set). Unfortunately, it is easy to prove that it **cannot exist** when $d \geq 3$. On the other hand, we find in Busemann (1953) the inequality

(VII.11.B.2) $$\mathrm{Vol}^{d-1}(C) \geq \frac{\beta(d)^{d-2}}{d \cdot \beta(d-1)^d} \int_{u \in S^{d-1}} \mathrm{Vol}^d (C \cap u^0) \cdot du$$

valid for every convex set (not necessarily center-symmetric), equality holding, if $d > 3$, only for ellipsoids.

It is proved by symmetrization, and Blaschke used it for the Steiner problem of Sect. I.2. From it we derive at once a partial answer to the Busemann-Petty problem:

If C is such that $\mathrm{Vol}(C \cap H) \geq \beta(d-1)$ *for each H, then* $\mathrm{Vol}(C) \geq \beta(d)$.

The Busemann-Petty conjecture is likewise true for ellipsoids.

Let us express (VII.11.B.2) asymptotically as a function of the dimension, using the asymptotic behavior of the $\beta(d)$. We find the following, where the notation "$\bullet$" is as indicated in Sect. VII.6.B:

(VII.11.B.3) $$\mathrm{Vol}^{d-1/d}(C) \geq \bullet \left(\int_{u \in S^{d-1}} \mathrm{Vol}^d (C \cap u^0) \cdot du \right)^{1/d}.$$

Whence the question: do we have an asymptotic inequality in the other sense? That is, do we have asymptotically

$$\mathrm{Vol}^{d-1/d}(C) \leq \bullet \left(\int_{u \in S^{d-1}} \mathrm{Vol}^d (C \cap u^0) \cdot du \right)^{1/d} ?$$

The problem is open, but we find in Milman and Pajor (1989) the fact that it is *equivalent to the grand conjecture* (VII.10.F1), a conjecture that is also equivalent to: $\mathrm{Vol}(C \cap H) \geq \mathrm{Vol}(D \cap H)$ implies $\mathrm{Vol}(C) \geq \bullet \mathrm{Vol}(D)$. We find only the very weak conclusion $\mathrm{Vol}(C) \geq \bullet d \cdot \mathrm{Vol}(D)$ with the help of the John-Loewner ellipsoid, which provides an answer to everything, but in all the present problems with catastrophic bounds. We begin to realize what the grand conjecture is telling us: that there do not exist **really bad convex sets.**

Recently there hasn't been much progress on the grand conjecture. However, it has just been in a certain way "displaced." Among the numerous references that treat this problem; see Milman (2000), Bourgain, Klartag, and Milman (2003), Klartag and Milman (2003) and Klartag (2005).

◆

Regarding hyperplane sections and global volume we can pose questions that are yet more naive. Moreover, they can be inspired by the formulas and inequalities above. We remain here in the context of center-symmetric convex sets, hyperplane sections being uniquely those passing through the origin (the center of symmetry). The simplest question is twofold, but in the same spirit: the volume being given, does there always exist at least one hyperplane section of not too large a volume and a hyperplane section with not too small a volume? We might well feel that if all the hyperplane sections are small, then the total volume is small (at least not too big) and *mutatis mutandis*. Naturally the John-Loewner ellipsoid gives an answer, specifically: if $\text{Vol}(C) = 1$, then there exists a hyperplane H with $\text{Vol}(C \cap H) \geqslant \bullet \, d$ and a hyperplane H with $\text{Vol}(C) \leqslant \bullet \, d^{-1}$. But this is in fact a very bad value.

On the other hand, formula (VII.11.B.3) implies trivially:

when $\text{Vol}(C) = 1$, *we have a hyperplane* H *with* $\text{Vol}(C \cap H) \leqslant \bullet$.

But in the reverse direction we see right away that the converse result for the existence of an H such that $\text{Vol}(C \cap H) \geqslant \bullet$ is equivalent to the grand conjecture.

VII.11.C. The hyperplane sections of the cube

In the realm of center-symmetric convex sets, opposite to balls (or to ellipsoids), we find the cubes. We assume the Euclidean case and look at the variation of the volume of the hyperplane sections of the "unit" cube $\text{Cub}_d = [-\frac{1}{2}, \frac{1}{2}]^d$ (the expression $[0, 1]^d$ is also alright, except that it is not centered at the origin). From (VII.8.1) we know that, for a family of parallel hyperplane sections, the maximum of the volume is obtained by the central section. We therefore consider at first only central sections of the cube. The behavior of the volume of these hyperplane sections illustrates in a representative way the purpose of our book: the simplest questions, apparently, have intuitive answers that are false, or true but the proof difficult, or use a concept seeming to have nothing to do with the problem at hand, or even remain open. *First problem*: enclose $\text{Vol}(\text{Cub}_d \cap H)$ on the left and right, i.e. find the minimum and the maximum.

In the planar case, that of the square, we see that the minimum is 1, attained solely by the sections parallel to the coordinate lines, whereas the maximum is $\sqrt{2}$, attained by the diagonals. What happens in higher dimensions?

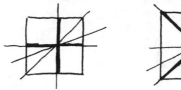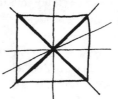

Fig. VII.11.5. Minimum of the central sections = 1 (*at left*); absolute maximum = $\sqrt{2}$
(*at right*)

◆

The problem of the *minimum* seems trivial: this volume is always greater than or equal to 1, with equality only for hyperplanes containing $d-1$ coordinate axes. This is very clear when the section is sufficiently close to a coordinate section because, if $H = u^0$ or u^0 is close to, say, the basis vector e_1, then it is elementary that $\mathrm{Vol}(H \cap \mathrm{Cub}_d) = \frac{1}{\langle u^0, e_1 \rangle}$, which is clearly greater than or equal to 1, with equality if and only if $u^0 = e_1$. But once H intersects edges other than those parallel to e_1, it is a completely different story. There does not exist a simple and geometric formula giving $\mathrm{Vol}(\mathrm{Cub}_d \cap H)$. In fact, we have:

(VII.11.C.1) *We always have* $\mathrm{Vol}(\mathrm{Cub}_d \cap H) \geq 1$, *with equality only for coordinate hyperplanes.*

The first proof was long attributed to Hensley in 1979, but in fact was to be found in Hadwiger (1972). Here is a rapid proof, but it requires concepts from probability theory. To simplify its expression, we give the proof in dimension 3, but there is fundamentally nothing to change in higher dimensions. In $\mathbb{R}^3$ we write the vector $u = (a, b, c)$ with $a^2 + b^2 + c^2 = 1$ and denote by $g(s)$ the area of the section of the cube $[-\frac{1}{2}, \frac{1}{2}]^3$ by the plane $H(s)$ with equation $ax + by + cz = s$. We want to show that we always have $M = g(0) \geq 1$.

For $t > 0$, the integral $G(t) = \int_0^t g(s)ds$ represents the volume of the portion of the cube located between the planes $H(0)$ and $H(t)$ and is interpreted as the probability that a point of the cube is located between $H(0)$ and $H(t)$. In probabilistic terms, $g(t)$ is then the density of the random variable $t = ax + by + cz$. We calculate the *variance* of this random variable (variance is for probability what inertia is for mechanics). For a symmetric random variable like t, the variance equals $V = 2\int_0^\infty t^2 \cdot g(t)\, dt$. We always have, whatever the direction given by u, $G(0) = 0$ and $G(\infty) = \frac{1}{2}$. On the other hand we have $g(t) \leq g(0)$ for each t, since $g(t) = g(-t)$ and the function g is concave, whence $G(t) \leq M.t$ for $t > 0$. Thus we have

$$V \geq \frac{1}{M^2} \int_0^\infty g(t)G^2(t)\, dt = \frac{1}{M^2} \int_0^\infty G'G^2\, dt = \frac{G^3(\infty) - G^3(0)}{3M^2} = \frac{1}{12M^2}.$$

But the random variable $t = ax + by + cz$ is the sum of the three independent random variables ax, by, cz, so the variance of their sum is equal to the sum of their variances. That of ax equals $\int_{-\frac{1}{2}}^{\frac{1}{2}} t^2\, dt = \frac{a^2}{12}$, so finally we will have

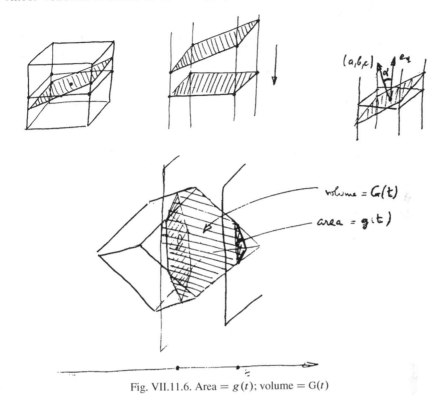

Fig. VII.11.6. Area $= g(t)$; volume $= G(t)$

$\frac{a^2+b^2+c^2}{12} = \frac{1}{12} \geqslant \frac{1}{12M^2}$, QED. Furthermore, equality is possible only if $G(t) = Mt$ for each t, which occurs only for the basis vectors.

◆

For the maximum problem, which was resolved only in Ball (1989), geometric intuition can be dangerous. In fact, the author of the present book had always thought that, for the cube in ordinary three-dimensional space, the maximum would be attained by the nice regular hexagon section and had almost proposed it as an exercise in a book. This is the section that has the most symmetries, so that Ball (1986) came as a real bombshell to your author, specifically that:

(VII.11.C.2) *We always have* $\mathrm{Vol}(\mathrm{Cub}_d \cap H) \leqslant \sqrt{2}$ *with equality for H of the type called* suspension *of the diagonal section of the square, i.e. for* $H = u^0$ *where u possesses* $d - 2$ *zero coordinates, the two others each equaling* $\frac{1}{\sqrt{2}}$.

It's a good occasion to mention here a principle that is well known to modern physicists, that of *symmetry breaking*, already considered in Sect. III.3. Nature does not alway like solutions having maximum symmetry, even if violating symmetry means admitting several solutions. In fact, Ball did not know that Ehrhart had

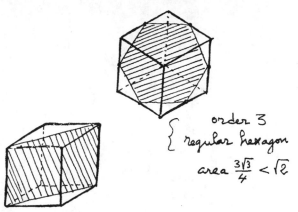

$$\left\{ \begin{array}{l} \text{order 3} \\ \text{regular hexagon} \end{array} \right.$$

$$\text{area } \frac{3\sqrt{3}}{4} < \sqrt{2}$$

Fig. VII.11.7.

already proved this in Ehrhart (1966), but only for dimension 3 and by a very tedious method and without hope of extension to higher dimensions. The problem itself had already been posed by Eggleston in 1963 in the *American Mathematical Monthly*.

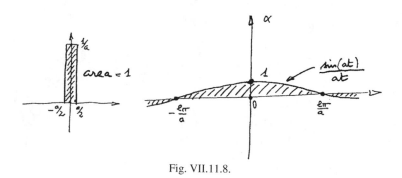

Fig. VII.11.8.

As above we write things in dimension 3, but everything carries over to arbitrary dimension *mutatis mutandis*. The proof rests first on an "explicit" integral formula for calculating, with the above notations, the $g(0) = \mathrm{Vol}(H)$ associated with $H = u^0$ and $u = (a, b, c)$. It seems that there is no direct geometric formula; any such formula would have to be different according to the various ways in which H meets the edges. We take from probability theory the notion of characteristic function of a random variable with density $f(x)$, i.e. the Fourier transform $\Phi(t) = \int_{-\infty}^{\infty} \cos(x.t) f(x)\, dx$. Taking "FourierTrans" is an automatic reflex for a physicist; see also Sect. IX.2. Here we apply this to the variable $ax + by + cz$, for which the densities are given by a function that, in the case of the first term ax, equals 0 except on $[-a/2, a/2]$ where it equals $a/2$, and similarly for by and cz. The associated characteristic functions are $\frac{2\sin\frac{at}{2}}{at}$, $\frac{2\sin\frac{bt}{2}}{bt}$, $\frac{2\sin\frac{ct}{2}}{ct}$. Now the inverse Fourier transformation yields $g(s)$ and in particular we have for $G(0) = \mathrm{Area}(\mathrm{Cub}_3 \cap u^0)$ the marvelous formula:

(VII.11.C.3) $\qquad$ $\text{Area}(\text{Cub}_3 \cap H) = \dfrac{1}{2\pi} \displaystyle\int_{-\infty}^{\infty} \dfrac{2 \sin \frac{at}{2}}{at} \dfrac{2 \sin \frac{bt}{2}}{bt} \dfrac{2 \sin \frac{ct}{2}}{ct} \, dt.$

We then apply the generalized Hölder inequality — see (VII.2.4) — and easily eliminate the factors 2 to find:

$$\int_{-\infty}^{\infty} \frac{2 \sin \frac{at}{2}}{at} \frac{2 \sin \frac{bt}{2}}{bt} \frac{2 \sin \frac{ct}{2}}{ct} \, dt$$

$$\leqslant \left(\int_{-\infty}^{\infty} \left(\frac{\sin at}{at} \right)^p dt \right)^{1/p} \left(\int_{-\infty}^{\infty} \left(\frac{\sin bt}{bt} \right)^q dt \right)^{1/q} \left(\int_{-\infty}^{\infty} \left(\frac{\sin ct}{ct} \right)^r dt \right)^{1/r},$$

with $p = \frac{1}{a^2}, q = \frac{1}{b^2}, r = \frac{1}{c^2}$ in such a way that we will have $\frac{1}{p} + \frac{1}{q} + \frac{1}{r} = 1$, as required for Hölder. To conclude, we need a lemma which is in fact very subtle (in spite of its "classical" integral aspect), for which the proof isn't conceptual but a mix of series expansions with *ad hoc* majorizations. It states:

(VII.11.C.4) *For each $p \geqslant 2$ we always have*

$$\frac{1}{\pi} \int_{-\infty}^{\infty} \left| \frac{\sin t}{t} \right|^p dt \leqslant \frac{\sqrt{2}}{\sqrt{p}},$$

with equality if and only if $p = 2$.

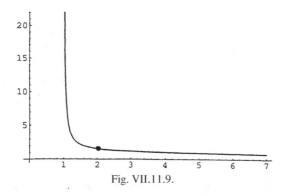

Fig. VII.11.9.

We are thus finally done, because once one of p, q, r is smaller than 2 — i.e. once one of a, b, c is greater than $1/\sqrt{2}$ — we have the trivial case seen at the beginning of the section of sections which intersect only at parallel edges and thus have an area equal to $\frac{1}{\langle u^0, e_1 \rangle} \leqslant \frac{1}{\sqrt{2}}$.

It can be shown without added difficulty that the optimal majorization of the sections of the cube by subspaces of dimension $d - k$ is furnished by $(\sqrt{2})^k$; see Ball (1989).

We have studied the volume of sections of the cube; sections of balls all have the same value in a given dimension, specifically the volume of the ball of this dimension. The case of ellipsoids is left to readers; see Fig. VII.12.1. But the Euclidean balls and cubes are particular cases of the balls Ball(d; p) (see Sect. VII.3), and the same for the cocubes. For a study of the volumes of these sections, see Meyer and Pajor (1988). A very recent and complete reference is Barthe and Nahor (2002) The proofs use inequalities that are more subtle — indeed more elusive — than that of Hölder and also the notion of logarithmic concavity. We will encounter this type of inequality below in the Sect. VII.12.C.

The result (VII.11.C.2) of Ball immediately furnishes simple, indeed trivial, counterexamples to the Busemann-Petty conjecture (VII.11.B.1), at least starting with dimension 10. For Cub_{10}, which has 1 for its total volume, all the hyperplane sections have a volume less than or equal to 1.41421.... For the sphere $Ball_{10}$, the total volume is 2.55016... and that of all the hyperplane sections is constant and equal to 3.29850.... If by homothety we normalize the ball so that its volume is equal to 1, then the volume of all the hyperplane sections will equal 1.42038... > 1.41421.... It is clear, in view of the formulas for the volume of balls, that things are still worse in dimensions higher that 10. On the other hand, finding counterexamples to (VII.11.B.1) in dimensions ranging from 5 to 9 is more difficult; this is done in Gardner (1993), but see also Gardner et al. (1999)

For applications of the volume of sections of the cube to number theory, seemingly far removed, see Bombieri and Vaaler (1983).

For the sections of the regular simplex in $\mathbb{E}^d$, see Webb (1996). First, Ball showed in 1992 that the sections of greatest volume are exactly those that contain a face; this for every sectional dimension, not just for hyperplanes. The interest is then to look for hyperplane sections of maximum volume, but which pass through the center of gravity of the simplex. Webb showed that these are the ones that contain $d-1$ vertices. This result provides an intriguing estimate in the theory of polynomial interpolation.

VII.12. Sections of low dimension: the concentration phenomenon and the Dvoretsky theorem on the existence of almost spherical sections

VII.12.A. Statement of the result

We again place ourselves in the Euclidean context. For studying the shape — the appearance — of a convex set (always center-symmetric in this section) we can observe slicing of subspaces (of suitable dimension) such that the section obtained is geometrically the most special possible. For example, although it teaches us nothing, all sections by lines passing through the origin are symmetric. For planar sections, we could look for those that are as close as possible to disks, i.e. that the plane cuts the boundary of the convex set so that the distance to the origin varies little. It is certainly hopeless to expect true circles in all generality. However, it is "almost" so for high dimensions and we find plane sections which are more and more neighbors of circles when the convex set that we are slicing is in a $\mathbb{E}^d$ with d larger and larger. Precisely the following spectacular result appeared in Dvoretsky (1961):

(VII.12.A.1) Dvoretsky's theorem. *For all d and positive ε, there exists a dimension $k(\varepsilon; d)$ such that each convex set K in $\mathbb{E}^{k(\varepsilon; d)}$ contains a section V of dimension d which is within ε of the ball Ball_d of V, which is to say that there exists λ such that*

$$\lambda \cdot \text{Ball}_d \subset V \cap K \subset \lambda(1 + \varepsilon) \cdot \text{Ball}_d.$$

We must not fail to note the corollary that, in an infinite dimensional space, every center-symmetric convex set admits ε-spherical sections for each ε (this doesn't mean we can find a spherical limit!). The original proof was extremely complicated and gave an estimate for $k(\varepsilon; d)$ of the order $\exp\!\left(c\varepsilon^{-2}d^2 \log d\right)$, where c is a constant. The result was announced in 1959, and in fact had been conjectured, we will see by what motivation, by Grothendieck no later than 1956; see Lindenstrauss and Milman (1993) and also Milman (1992). Afterwards there appeared several proofs of (VII.12.A.1) of a completely different nature; furthermore, the various authors had succeeded in notably improving the value of $k(\varepsilon; d)$, of order $\exp\!\left(c\varepsilon^{-2}d\right)$. We find the gist of the problem of the best $k(\varepsilon; d)$ in 4.1 of Giannopoulos and Milman (2001). We also find there that if after having taken a section we make a little additional effort by taking a quotient, we then obtain things that are almost spherical for extremely favorable dimensions, of the same order as the dimension we started with.

The proof that we are going to give — rather for which we are going to provide the essence of the ideas — is that of Figiel, Lindenstrauss, and Milman (1977), based on the fundamental idea of Milman (1971). It is a fact that Dvoretsky's original proof was very complicated and understood by few. That of Figiel et al. was the first really clear proof. It is extremely geometric and allows us to discover several interesting rungs of the ladder. We will see that in fact the true nature of things is still more astonishing than the statement (VII.12.A.1), specifically: there in fact exists an enormity of such *almost spherical* sections, the enormity made clear in the proof. Another proof, very rapid but less geometric, less *visible*, is that of Pisier (1989).

Here is the motivation — from analysis — of our desire for almost-spherical sections. It is not a matter of pure geometric curiosity, but a fundamental problem of functional analysis. In all the $\mathbb{E}^d$, we know classically that for the sequence $\{x_i\}$ (vectorial therefore, but in finite dimensions it suffices to know it for sequences of real numbers), there is equivalence between the three properties:

(i) $\sum_i |x_i| < \infty$;

(ii) $\sum_i x_{\pi(i)}$ converge unconditionally, i.e. for each permutation π of $\mathbb{N}$;

(iii) $\sum_i \varepsilon_i x_i$ converges for arbitrary $\varepsilon_i = \pm 1$.

(We then speak, for ordinary real numbers, of an *absolutely convergent* series.) But matters are entirely different in infinite dimension, i.e. in function spaces. There we only know that (i) implies (ii) and that (ii) and (iii) are equivalent. But (ii) does not in general imply (i), for example in Hilbert space, and thus we need a more general solution. It's for studying this problem that we need results of the type of (VII.12.A.1). Before this result, there appeared in Dvoretsky and Rogers (1950) a weaker, albeit essential, result:

(VII.12.A.2) *Let* C *be an arbitrary center-symmetric convex set in* $\mathbb{E}^d$, *with* $d = 4k^2$. *There always exists an affine transformation* f *such that both* $f(C)$ *contains the unit ball, and there exists a subspace* V *of dimension* k *such that* $f(C) \subset [-2, 2]^k$.

We can compare this "squeezing" with that furnished by the John-Loewner el-lipsoid in order to see that there has been real progress. Grothendieck (1956) ends with a conjecture stronger than (VII.12.A.2): in each dimension d we have, for a suitable sub-dimension k, a section that can be enclosed between two homothetic ellipsoids, and this with a homothety ratio less than or equal to 2. Dvoretsky's result is thus much more precise.

♦

Before proving Dvoretsky, some geometric considerations. First, despite appear-ances, it is practically just as much an affine result as Euclidean. In fact, we leave it to readers to see that each ellipsoid in dimension d admits sections which are exactly balls for dimension $[\frac{d}{2}]$ (integer part $\frac{d}{2}$) (first look at dimension 3 and find disks there).

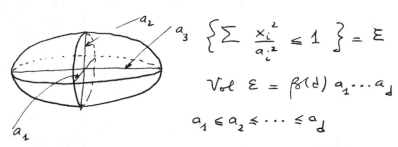

Fig. VII.12.1. Find circular sections! Find spherical sections of dimension $[d/2]$

At the other end of complexity for the convex sets, we consider the cubes. Now the cube of dimension 3 admits a plane section that is a regular hexagon, thus much closer to being a circle than is a square. It is likewise easy to construct, in the cube Cub_{2n}, a plane section which is a regular $2n$-gon, thus more and more close to a circle when n tends toward infinity. This is very encouraging, if we consider that cubes are the worst among the convex sets, but from here to a proof there is a chasm. We are going to lay all this out along the skeleton of the proof, because it is a remarkable example for ascending (and then descending) Jacob's ladder. Moreover, we will discover some explicit estimates (but not explicit sections: see the end of this section). The proof rests on two things: the concentration phenomenon of Paul Lévy and an integration over the set of all the vector subspaces of given dimension of a fixed vector space.

VII.12.B. The concentration phenomenon of Paul Lévy

This phenomenon, discovered in 1919, remained unnoticed for a long time. Paul Lévy was an author of prophetic nature, but his writing made for heavy going. We find it in rather explicit form in Lévy (1951). It says this:

(VII.12.B.1) Concentration *phenomenon. A function on the sphere* S^D *(thinking of large dimension D) varies only little from its median value on a portion of the sphere whose measure is very close to the total measure of* S^D.

The median value of a function f on the set E with measure μ is the number M such that $\mu(f^{-1}([\inf f, M])) = \mu(f^{-1}([M, \sup f])) = \frac{1}{2}\mu(E)$. Of course it is required that the function considered be controlled in some way or another. The simplest case is when it is differentiable, but this hypothesis is stronger than required. Here the function considered is that which defines a center-symmetric convex set as in (VII.10.C.3) (see also Sect. VII.11.B), for which we just saw in (VII.10.C.3) that it was always Lipschitz with ratio $\sqrt{d}$, which is good enough for our purposes.

The explanation results from the conjunction of two properties of the sphere: the first is the fact that an equator (any: they are all the "same") has tubular neighborhoods of very small radius but which nevertheless have a very large volume. The second is the isoperimetric inequality for for domains on the sphere; it is this latter property that allows us to pass from tubular neighborhoods of equators to those of an arbitrary subset (specifically the subset formed of points where the function equals its median value).

The concentration phenomenon discovered by Paul Lévy was in fact "well known" by physicists, precisely in statistical mechanics, which governs the behavior of sets of particles in a gas. It deals with spaces of very large dimensions, of the order of Avogadro's number 10^{23}, and physicists knew from Maxwell's work that physical states are observed that correspond to mean energies, i.e. only phenomena whose probability isn't too small are seen in nature. The result of Paul Lévy placed this in a rigorous form. This concentration phenomenon may evidently be interpreted as a result in the calculus of probabilities. In this direction, a recent essential reference is Talagrand (1995). The question of the concentration on the surrounding spheres, not just on equators, but over subspheres of smaller dimensions, is extremely difficult; it has been broached very recently in Gromov (2003). For an exposition of the

concentration phenomenon with connections to analysis, probability and isoperimetry, see Ledoux (2001).

However, the probabilists only employ measures, whereas geometers prefer metric spaces. But exactly for geometry this phenomenon has been cast in a geometric
setting with measure and metric simultaneously in Chap. $3\frac{1}{2}$ of Gromov (1999),
where we believe the reader will be pleased to study, among other things, the notion of *observable diameter*. In Berger (2000) there is a three page summary of that
chapter.

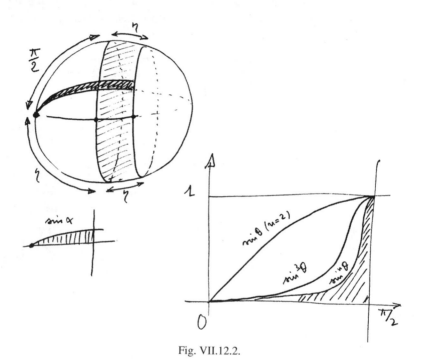

Fig. VII.12.2.

Here is how we prove precisely the phenomenon of Paul Lévy. We are on the
sphere S^{D-1}, for which we normalize the total measure so that it equals 1. We first
calculate the volume of the tubular neighborhood $\mathrm{Tub}(\mathcal{H};\alpha)$ with radius α of a solid
hemisphere $\mathcal{H}$, which equals (see Sect. VII.6.B)

$$\mathrm{Vol}(\mathrm{Tub}(\mathcal{H};\alpha)) = \frac{\int_0^{\pi/2+\alpha} \sin^{D-2} t.dt}{\beta(D-1)},$$

and by using that $\lim(1 + \frac{1}{n})^n = e$ when n tends toward infinity, we obtain the
majorization (very accurate in fact):

(VII.12.B.2) $$\mathrm{Vol}(\mathrm{Tub}(\mathcal{H};\alpha)) \geq 1 - 4\exp(-\frac{D\alpha^2}{2}).$$

(The tubular neighborhood of radius α of an arbitrary part H of a metric space is the set of points of the space located at a distance from H less than of equal to α).

Now we think of a proof of the isoperimetric inequality for the sphere in the spirit of Sect. VII.8 and the "painting" method. What we prove for the sphere is this:

(VII.12.B.3) *For each domain* $K \subset S^{D-1}$ *and each* α *we have*

$$\mathrm{Vol}(\mathrm{Tub}(K;\alpha)) \geq \mathrm{Vol}(\mathrm{Tub}(K^0;\alpha)),$$

where K^0 *denotes a spherical cap such that* $\mathrm{Vol}(K) = \mathrm{Vol}(K^0)$.

The proof is carried out by hyperplane symmetrization of Steiner type on the sphere; details can be found in Figiel et al., 1977); see also Burago and Zalgaller (1988), as well as Fig. VII.11.7 above. Now we apply (VII.12.B.2) and (VII.12.B.3) to the domain $K = f^{-1}([\inf f, M])$, where f is the function $f : S^{D-1} \to \mathbb{R}$ for which M is the median value. If additionally f is Lipschitz with ratio $\sqrt{D}$, we find directly the concentration phenomenon, which says that f is almost constant and equal to its median value M over a portion of the sphere of almost total volume:

(VII.12.B.4) $$\mu(m : |f(m) - M| \leq \alpha\sqrt{D}) \geq 1 - 4\exp(-\frac{D\alpha^2}{2}).$$

VII.12.C. The proof

We now consider an arbitrary center-symmetric convex set in E^D and define it by the function on its boundary $f_C : S^{D-1} \to \mathbb{R}$. Finding subspaces V such that $V \cap C$ is ε-spherical is the same as finding V such that the restriction of f_C to V varies at most by ε. We accomplish this by cutting V into two pieces: one piece of V will be in $\frac{\varepsilon}{2}$ by concentration inherited on V from that in S^{D-1}, the other on V itself by the Lipschitz property.

We consider the set of all vector subspaces of E^D of given dimension d (this set is called the *Grassmannian* and denoted $\mathrm{Grass}(D, d)$). We give it a measure that is invariant under the action of the isometries of E^D. Such a measure is unique if normalized. The Fubini theorem applied to a function F on the sphere S^{D-1} tells us that:

$$\int_{S^{D-1}} F.\mu = \int_{V \in \mathrm{Grass}(D,d)} \left(\int_{V \cap S^{D-1}} F.\mu' \right) \mu''$$

where μ is the (normalized) measure for S^{D-1}, μ' that (also normalized) for the sphere $S^{D-1} \cap V$ and μ'' that for the Grassmannian seen above. We apply Fubini's theorem, taking for F the characteristic function of the domain $A_\alpha = \{m : |f_C(m) - M| \leq \alpha\sqrt{D}\}$. This for the constant α such that $\alpha\sqrt{D} = \frac{\varepsilon}{2}$, which tells us that the mean value of the measures $\mathrm{Vol}(S^{d-1}(V) \cap A_\alpha)$ for the measure μ'', where V

traverses the Grassmannian Grass(D, d), is greater than $1 - 4\exp(-\frac{D\alpha^2}{2}) = 1 -$
$4\exp(-\frac{\varepsilon^2}{8})$. Thus first there exists a sufficiently large part (of the Grassmannian) of
V on which Vol(S^{d-1}(V) $\cap A_\alpha$) at least equals this mean value. We take such a V;
the function f_C varies there by $\frac{\varepsilon}{2}$ at most over a subset of sufficiently large measure.

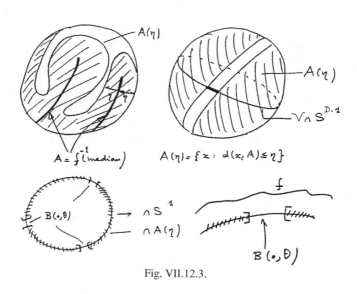

Fig. VII.12.3.

We have yet to see what happens on what remains. But f_C is Lipschitz there
with ratio $\sqrt{D}$, so that it suffices that what remains is of small diameter δ in
V. Now the volume of a spherical cap B(δ) of radius δ is easily estimated in d
and δ, because for small radii δ we have equivalence with the Euclidean case,
so let Vol(B(δ)) $\approx \sqrt{d}\,\frac{r^{d-1}}{d}$. The final condition provides us with the desired
value for the small dimension d: we want to have $\frac{r^{d-1}}{d} \geqslant 4\exp(-\frac{D\alpha^2}{2})$. We eas-
ily find the desired relation D $\approx \exp(c|\log\varepsilon|\varepsilon^{-2}d)$, where c is a universal con-
stant. The best triples (D, d, ε) are still not known, but we know that we may take
D(d, ε) = $\exp(c\varepsilon^{-2}d)$ and that this is optimal. It isn't surprising that it is the bad
cubes that provide the limit. On the other hand, it isn't known what is the best ex-
pression for ε as a function of d.

And so we dispose of "lots" of almost spherical sections (roughly more than half
of the Grassmannian). However, it is a result obtained in a theoretical way, thus
nonconstructive; but presently there does not exist an algorithm for finding such
sections when the convex set is given in one form or another, e.g. by a function on
the unit sphere. The reader shouldn't be surprised by this fact: we will encounter an
analogous situation in Sect. X.5 concerning the existence of very dense lattices. This

is not surprising because the proof by Mahler of this existence is also an argument by integration over the set of lattices.

We can also use Dvoretsky's theorem to effectively distribute points on spheres (cf. Sect. III.3); see Wagner (1993) and the end of 7.3 of Giannopoulos and Milman (2001).

VII.13. Miscellany

VII.13.A. Projections

Whether in the affine context or the Euclidean, we have projections at our disposal. Of course convexity, as well as central symmetry, is preserved. In Sect. VII.6.C we already encountered Cauchy's formula, which provided the area of a convex set as a function of the integral of the volume of its various Euclidean projections.

Very briefly, the projections and the intersections are **dual** (polar) contexts. However, in several problems projections behave better − even much better − than sections. And in any case there isn't a duality that is perfect, even less one that is explicit, apart from trivialities. If we reflect on the difficulties encountered for the hyperplane sections of the cube, we can quote with pleasure the result of McMullen (1983): we consider the unit cube in $\mathbb{E}^d$ and project it in an arbitrary direction by a subspace V and also in the orthogonal direction V^0. Then the volumes of these two projections are equal, that is to say:

(VII.13.A.1) *For each* V, *we have*

$$\text{Vol}(\text{ projection of } \text{Cub}_d \text{ on } V) = \text{Vol}(\text{ projection of } \text{Cub}_d \text{ on } V^0).$$

It's a pretty calculation from linear algebra, from determinants. The following corollary clearly shows the difference with the hyperplane sections:

(VII.13.A.2) *The maximum volume for the hyperplane projections is obtained by the directions of the principal diagonals, and thus has for its value $\sqrt{d}$; these are also those projections that have the most symmetries.*

In fact, the length of the projections onto a line is very clearly of maximum equal to $\sqrt{d}$. Beyond the references already given, for projections (and sections with a viewpoint more focused on X-rays), see Gardner (1995).

The analogue for the hyperplane projections of the Busemann-Petty conjecture (VII.11.B.1) is also classical and is called Shephard's conjecture. But, as for sections, things are true if we restrict ourselves to certain classes of convex sets. It is here that the notions of: *projection bodies, intersection bodies, zonoids (mixed projection bodies)* appear. For these categories the zonoids in particular are unavoidable; see Chap. 4.9 of Gruber and Wills (1993). A curious exception to the duality is that Shephard's conjecture is false for dimension 3 and greater, whereas we have seen that Busemann-Petty's is only false starting with dimension 4.

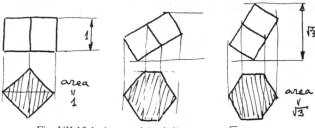

Fig. VII.13.1. Area $= 1$ (*at left*), area $= \sqrt{3}$ (*at right*)

The projections of ellipsoids, in all dimensions and subdimensions, are again ellipsoids. This is a characteristic property of ellipsoids, among a host of others: see 1.11 of Gruber and Wills (1993). Readers will form an aesthetic opinion with this reference and the purely ellipsoidal article (Petty, 1983). We have seen the characterization by the affine symmetries in Sect. VII.10.C; numerous other characterizations are very natural but with proofs that are not very conceptual. In the spirit of Chap. IV, the one we prefer and can't resist quoting is that of the so-called *false center* (see Petty): take a (not necessarily center-symmetric) convex set which possesses a point p such that each section of dimension a fixed k (between 2 and $d - 1$) admits a center of symmetry (but which of course is not necessarily the point p). Then, either p is a center of symmetry of the convex set or our convex set is an ellipsoid, whereupon for each point p the sections passing through p have centers of symmetry (in general different from p) because these sections are always ellipsoids.

VII.13.B. Steiner-Minkowski formula and mixed volume

We are now in the Euclidean setting. In Sect. VII.8 we encountered the volume of a tubular neighborhood of a convex set, $\mathrm{Vol}(\mathrm{Tub}(C, \varepsilon)) = \mathrm{Vol}(C + \varepsilon B)$, and proved the isoperimetric inequality as a consequence of the Brunn-Minkowski inequality by making ε tend toward zero. But in fact $\mathrm{Vol}(\mathrm{Tub}(C, \varepsilon))$ possesses an extremely strong property: for all positive ε, this volume is exactly a polynomial of degree d in ε:

$$(\text{VII.13.B.1}) \qquad \mathrm{Vol}(\mathrm{Tub}(C, \varepsilon)) = \sum_{d}^{0} \mathscr{L}_i(C)\varepsilon^i .$$

We know in succession that: $\mathscr{L}_0$ is the volume, $\mathscr{L}_1$ is the area and $\mathscr{L}_d$ is the volume $\beta(d)$ of the unit ball, as readers can verify. The other invariants $\mathscr{L}_2, \ldots,$ $\mathscr{L}_{d-1}$ attached to the convex sets are evidently very interesting. For an elementary (and very incomplete) exposition see 12.10 of [B]; also Gray (1990), a book entirely dedicated to tubes; see also Appendix A of Gardner (1995) as well as Gray (1993) for programming. But this polynomial is but a special case of the one that follows.

In fact the formula (and the definition) of the *mixed volumes* simultaneously extends the one of Brunn-Minkowski by on the one hand taking arbitrary convex sets and on the other by taking any finite number of them. The "simple" result is

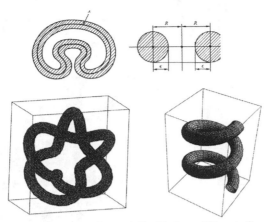

Fig. VII.13.2. Gray (1993) © Routledge/Taylor and Francis Group, LLC

that, for any finite number n of convex sets $C(i)$, the function $\mathrm{Vol}(\sum_i \lambda_i C(i))$ is a polynomial of degree d in the $\lambda_1, \ldots, \lambda_n$, for which the coefficients associated with the $(i_1, \ldots, i_n)$ are called the *mixed volumes* of the $C(i)$. There are inequalities for them that extend the one of Brunn-Minkowski, long conjectured and difficult to prove; see the book Schneider (1993), which is entirely devoted to them; see also Appendix A.3 of Gardner (1995). Their interest lies in the fact that they are useful and natural generalizations of the isoperimetric inequality. See also the indexes for Giannopoulos and Milman (2001) and Gruber and Wills (1993) for numerous references to "mixed volumes".

VII.13.C. Convex sets and mathematical physics: the floating body that loses its head, the fundamental frequency, the Poinsot motion, Newtonian gravitation, the destiny of the *rolling stones*

The setting here is Euclidean and for a general convex set. At the outset, for simplicity, we stay in dimension three most of the time, but problems can be posed (mathematically, not physically) in arbitrary dimension.

♦

The *floating body* problem finds its origins in Archimedes' principle: if we float a body C of uniform density $\delta < 1$ in water it will in general assume a position of equilibrium, which is dictated by the (weak) fact that the volume cut off in C by the plane of the water surface will be exactly $\delta \cdot \mathrm{Vol}(C)$, but above all that the center of gravity is vertically aligned with the center of pressure (or buoyancy), which is nothing other than the center of gravity G′ of the submerged portion; see 6.5.15 of [BG] for the proof.

It is clear that a sphere (a ball) does not have a privileged position of equilibrium, all directions are possible. Does this only happen for spheres? This question was

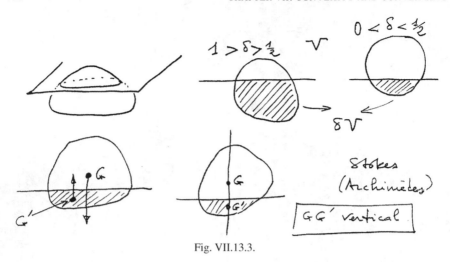

Fig. VII.13.3.

posed in 1934 by Ulam and is found again in the list of Ulam (1960). Despite the strength of this condition we will see that the answer is almost completely open; here is the status of the problem (we will see subsequently that the problem leads to some very interesting studies): the problem is found in Sect. A.6 of Croft, Falconer, and Guy (1991), p. 353 of Gruber and Wills (1993), in Schütt (1997) and above all in the most up-to-date: Gardner (1995, pp. 337, 338). For the truly physically mechanical details the didactic reference is Gilbert (1991), which cites for practice (e.g. boat construction) the basic work Appell (1952, Chap. XXI-4). But above all read Rorres (2004) which treats among other things the problem of the wandering of icebergs, with the aid of René Thom's catastrophe theory.

Precisely − in the spirit of treatments destined for naval engineers − a remark imposes itself: in the construction of boats we desire (except for certain competitive purposes, in kayaks for example) maximum stability, i.e. equilibrium positions such that, as soon as there is a displacement, the restoring force is as strong as possible. This leads to the introduction of the notion of *metacenter*. This point, located at equilibrium on the vertical containing the center of gravity and the center of pressure (buoyancy), must be as high as possible on this line, and in any case above the center of (total) gravity; see Gilbert (1991). The metacenter is subtle in its definition. We consider the surface S formed by all the centers of gravity of submerged portions. For a given position (inclination) under an infinitesimal variation, the normal to S has a point of smallest variation which is the associated metacenter. It is always situated on that normal, between the two focal points of this normal; see Sect. VI.10. Ulam's problem is a typical example of a problem that excites the curiosity of mathematicians, but does not seem, at least not at present, to have practical consequences. There are several problems of this type in our book, that is to say of characteristic uniqueness but of less speculative type: Sylvester's problem in Sect. I.2, isoperimetric problems in Sects. V.11, V.12, VII.14, VIII.7 and surface antipodals in Sects. VI.3 and XII.3.

Even for very simple bodies, such as cylinders of revolution, the determination of equilibrium positions is not so easy; see Gilbert (1991) and Fig. VII.13.5 below. The case of the regular tetrahedron is still more difficult.

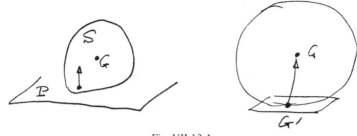

Fig. VII.13.4.

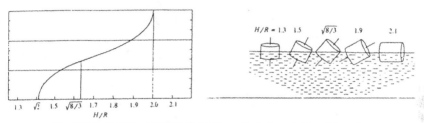

Fig. VII.13.5. Gilbert (1991) © Mathematical Association of America

Apart from the case of density $\delta = 0$, only two results are known, both in the single case of density $\delta = \frac{1}{2}$. They are obtained using the fact that the condition of flotation in all directions tells us this: consider the center of gravity G′ of a part cut off by a plane which leaves aside a volume equal to δ Vol(C). Then Ulam's condition implies that, when we take all directions, the points G′ are on a fixed sphere with center G (the total center of gravity of C), where V is the total volume of the convex set. This condition can be found physically very quickly by potential energy considerations. Then:

if $\delta = \frac{1}{2}$ *and if furthermore the convex set* C *is center-symmetric, it must be a sphere.*

This result is in Falconer (1983) (note that Gilbert 1991, although an excellent exposition, only proves a much weaker version of this result). Here is how it is proved. If $\delta = \frac{1}{2}$, then the Archimedean sections that cut the floating body C in half its volume are exactly all the sections that pass through the center. We then calculate (in polar coordinates) the depth $p(u)$ of this center of gravity for the vertical direction given by a vector u as an integral leading to the function $f : S^2 \to \mathbb{R}^*_+$ which defines the boundary ∂C of the convex set (Fig. VII.11.4):

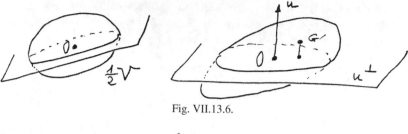

Fig. VII.13.6.

$$p(u) = \int_{x \in S^2} f^4(x)|x.u|\, dx.$$

The idea is now to use the development of f in spherical harmonics (see Sect. III. XYZ or VII.11.B); we see then that p can't be constant unless f itself is. All this is in the spirit of the Radon transformation; see Sect. XII.3.B. Recall that, on spheres, each even (i.e. antipodally invariant) function is known once we know its integral over all the great circles; in particular, if this integral is constant, then the function itself must be constant (we clearly need take only even functions, since each odd function has a zero integral on the great circles). This reasoning is valid in every dimension d; we only need replace f^4 by f^{d+1}. The case $d = 2$ can be interpreted as that of a (long) floating cylinder (always center-symmetric), long enough so that the extremities have a negligible effect.

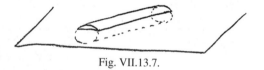

Fig. VII.13.7.

If we now suppress the condition of central symmetry, but still for the case $\delta = \frac{1}{2}$ and in dimension 2 (the case of floating cylinders – for which we wish that they should lose their heads), then we have a multitude of counterexamples. Here is how: we write the condition for the center of gravity of the submerged part always being on the vertical through the total center of gravity. If we develop the boundary curve of the convex set (supposed smooth) in a Fourier series (as in Sect. V.12 for the proof of the isoperimetric inequality "a la Hurwitz"), we find explicit conditions on the coefficients of the Fourier series with coordinates $\{x(u), y(u)\}$ of the boundary curve of C, and the expression of the general solution is:

$$x(u) = \cos u + \sum_{k=1}^{\infty} [-(a'_{2k+1} + a'_{2k-1}) \cos 2ku + (a_{2k+1} + a_{2k-1}) \sin 2ku],$$

$$y(u) = \sin u + \sum_{k=1}^{\infty} [(a_{2k+1} - a_{2k-1}) \cos 2ku + (a'_{2k+1} - a'_{2k-1}) \sin 2ku],$$

and these developments make way for a multitude of functions, which can even be made real analytic by having the coefficients decrease rapidly enough; and if we take just a finite number of terms, we obtain algebraic curves. All this is done in Auerbach (1938), where there is a geometric characterization of these curves. We give two extreme cases of sections of long floating cylinders that are in equilibrium in all directions:

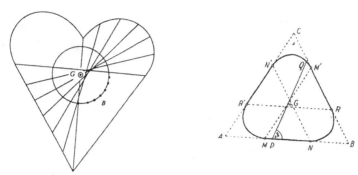

Fig. VII.13.8.

Apart from the two cases above, Ulam's question is entirely open, on the positive side as well as on the negative. We may however prefer geometry to triple integrals, at least for a first approach. It is intuitive that the position(s) of equilibrium can be obtained thus: we look at the envelope of all planes that cut off the constant volume $\delta \cdot \mathrm{Vol}(C)$. If all goes well, this again is a nice surface (convex ?) *denoted* $C(\delta)$ and called "the convex floating body", and floating equilibrium will be attained when the distance from the center of gravity of C to $C(\delta)$ measured along the vertical is minimum. The vertical passing through the center of gravity must be normal to the surface of $C(\delta)$, whence the fact that the problem is connected with the study of $C(\delta)$. This $C(\delta)$ is called a "floating body" in much of the literature on convex sets. It is here that surprises lie in wait for us.

In fact, first of all $C(\delta)$ is not in general convex; for a triangle in the plane, all the envelope curves have six cusps (even very close to the boundary, and except for $\delta = \frac{1}{2}$ where there are but three cusps):

This does not prevent us from now defining $C(\delta)$ instead as the intersection of the half spaces defined by the above sections. It is thus always a convex set by definition, still *denoted* $C(\delta)$ for simplicity, and which is called a *floating body* in the literature (even though strictly speaking this problem is different from Ulam's). The case of triangles above shows that these $C(\delta)$ are not smooth, even for small δ. On the other hand we see (in every dimension) in Meyer and Reisner (1991) that if the convex set is center-symmetric, if for $0 < \delta < \frac{1}{2}$ each $C(\delta)$ is strictly convex so that the centers of gravity of the sections are the points of contact with the hyperplanes of support, and above all if the boundary of C is of class C^1, then each point of $C(\delta)$ is of class C^2. Thus our intuition is confirmed (above all for small δ): when we cut a convex

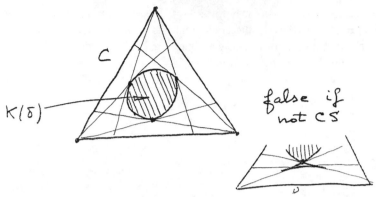

Fig. VII.13.9. Complete curve with cusps and the convex envelope alone

set in slices of volume δ Vol(C), this has the effect of making things **smoother**. This geometric smoothing may make us think of convolution in analysis (see however the end of).

The "floating bodies" $C(\delta)$ are much studied in the theory of convex sets, if only because, for small δ in the center-symmetric case, they furnish natural approximations for convex sets. See the appendix of Milman and Pajor (1989), Schütt (1997) and, in the spirit of the characterization of ellipsoids (a very practical pastime for convex set enthusiasts), see 1.11 of Gruber and Wills (1993). In Schütt and Werner (1994) is found the fact that if, for a sequence of δ_n tending toward zero, each $C(\delta_n)$ is homothetic to C, then C is necessarily an ellipsoid.

The *Poinsot motion* is that of a solid about its center of gravity, without any force or external attraction, just its own inertia; and it doesn't matter if the motion is not about its center of gravity, we then only need subtract a translation of constant velocity. This problem from mechanics is resolved explicitly with the aid of elliptic functions; see for example Appell and Lacour (1922, v. II, p. 219). Numerous works on elliptic functions treat this integration. We will retain Poinsot's very elegant geometric interpretation: the movement of the body under consideration is the same as that of its ellipsoid of inertia that is made to roll without slippage on a plane — the so-called invariable plane. A proof is found in Arnold (1976, ¶29.C). The distance from the plane to the center of the ellipsoid depends on the initial conditions. The behavior of the motion and also the stability of the equilibrium positions can then be predicted. The trajectories can be read on the ellipsoid of inertia as the intersections of this ellipsoid with the various concentric spheres, i.e. the points of the ellipsoid which are a given distance from its center. In particular, we clearly see the movement of our planet and its precession. We must not forget, in view of the definition of the ellipsoid of inertia, that if the solid is a flattened ellipsoid of revolution, then its ellipsoid of inertia is then elongated. The curve on the ellipsoid bears the pretty name *polhode*, and the curve on plane on which the ellipsoid rolls that of *herpolode*.

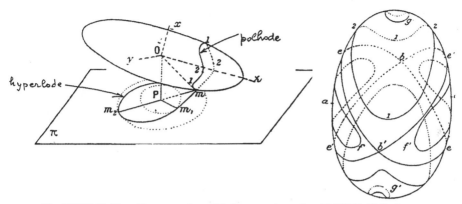

Fig. VII.13.10. Visual interpretation of Poinsot motions. Arnold (1978) © Springer

◆

Another desirable characterization of spheres is by Newtonian *gravitation*: it is classical that the Newtonian attraction of a ball B in $\mathbb{E}^3$ on an arbitrary point in space (outside the ball) is the same as that of its center of gravity charged with the total mass. In integral calculus the proof consists of the formula:

$$(VII.13.C.1) \qquad \int_B \frac{m-x}{d^3(m,x)} \cdot dm = \frac{\mathrm{Vol(B)}}{d^3(x,g)} \cdot (g-x) \qquad \text{for each } x.$$

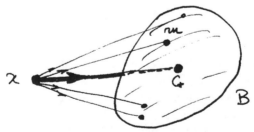

Fig. VII.13.11. For a sphere (solid and of uniform density), whatever the exterior point *x*, the total Newtonian attraction is equal to that of the center of gravity (center of the sphere) alone charged with the total mass

Is this characteristic of spheres? The answer is affirmative; it has been known only since Ahoronov, Schiffer, and Zalcman (1981). The (elementary?) proof proceeds thus. Let P be the domain satisfying (VII.13.C.1) for which we want to show that it is a ball. We first deduce from (VII.13.C.1) that $\int_{m\in B} \frac{1}{|m-x|}\,dm = \frac{c}{|x|}$ for all x exterior to P. This allows us to construct functions on the space which are harmonic on a neighborhood of P, i.e. such that their mean on the whole ball in the space is equal to their value at the center of the ball. Then the condition implies that their mean on P is equal to the value at O (the center of gravity of P). We thus construct a positive function on a ball B of maximum radius contained in P such that the integral over B is equal to the integral over P; thus P = B.

There exist numerous other problems in the same spirit, for example that of distribution of electric charge on a surface. For gravitation again, for uniform mass density on the boundary of a convex set, only the sphere has the property of not exerting any gravitational force on any point on the interior (Reichel, 2009).

The problem of the lowest (or *fundamental*) frequency, to which we append another problem, is as follows. We will explain things in the planar case, but the problem can be posed in all dimensions and the results we obtain are valid *mutatis mutandis*; see among other references: Sect. I.6 of Berger (2001) and Brascamp and Lieb (1976). We consider a convex set C (or more generally a compact domain with nonempty interior) in the plane $\mathbb{E}^2$ as a vibrating membrane (make a drum whose edge is the boundary of C). It vibrates while giving off sounds that we decompose into harmonics, the lowest sound corresponding to the so-called fundamental; its wave length is a real number $\lambda(C)$. We look for an "isoperimetric" inequality which says that, among all C of a given area, it's the circular disk (and it alone) which has the lowest fundamental frequency. This old question was resolved positively (and independently) by Faber and Krahn in 1923–1924. It suffices to show that the Schwarz symmetrization, which as we know preserves area, reduces the fundamental frequency; and this is not difficult using the Dirichlet principle which furnishes this frequency as the minimum value of a functional on the functions $f : C \to \mathbb{R}$ which are zero on the boundary of C. We apply Fig.VII.5.4 to the displacement function of the vibration of first frequency.

A more difficult question has for a long time held the experts in check. This is the *level curve* problem for the function of the displacement of the membrane that furnishes the fundamental frequency. Since the displacement is zero at the edge and varies according to a simple principle (its Laplacian is proportional to the function itself), it is reasonable to conjecture that the level curves are convex if the domain of the drum is convex. We needed to wait for Brascamp and Lieb (1976) to find out: the answer is yes, but the proof is difficult; it rests on new inequalities for functions, which are in a sense inverse to Hölder (VII.2.2). We observe that it is these new inequalities that allowed proof of the results of Ball (1989, 1991) on the volumes of sections of a cube; see Sect. VII.11.C.

In fact, the authors prove a much stronger result: a convexity space for the way in which heat is propagated in a convex domain C. From it they deduce a "gratuitous" corollary: we have convexity of the function λ for the Minkowski addition: $\lambda(aC + (1-a)C') \leqslant a \cdot \lambda(C) + (1-a) \cdot \lambda(C')$. For a recent point of view on these inequalities "a la Brascamp-Lieb", see Barthe (1998).

♦

On p. 13 of Hilbert and Cohn-Vossen (1952) we find the following assertion: the pebbles on an ocean beach are approximately ellipsoids (and the mathematical study of this phenomenon requires probability theory). Where Hilbert might have gotten this is still a mystery today. However, in Rogers (1976) we encounter the conjecture that we get ellipsoids more than spheres, but we don't know (after having read the article by Firey) why the author has made this conjecture; see below.

In Firey (1974) the study undertaken ends up with the partial differential equation given below; Firey is able to show only local existence and conjectures existence for all times t and that the limit form is spherical. The conclusion of this result is thus that if we find mostly ellipsoidal pebbles, it is because they are not anisotropic, i.e. they have privileged directions with respect to abrasion. Firey reduces the physical problem to a mathematical condition of the evolution equation type that we have already encountered in Sect. V.12, specifically deforming a surface in ordinary space along its normals and this by an amount proportional to the Gaussian curvature at any point (and renormalizing as in Sect. V.12 to a constant volume). In the introduction to Firey (1974) there are physical considerations that lead to the erosion of pebbles (or industrial balls) being governed by the above geometrical evolution. Gauss's formula (VI.6.3) enters in an essential way. Even without Firey, it remains natural enough that a surface is more strongly attacked by abrasion, erosion, at its more "acute" points. For those who like explicit equations, the equation that now replaces (V.12.1) is:

(VII.13.C.2), $$\frac{\partial x}{\partial t}(p,t) = -K(p,t)x(p,t),$$

where $S(t)$ is the family of surfaces obtained in the course of the erosion; it is described by the following family of curves which are the orthogonal trajectories $p \mapsto x(p,t)$ (where $p \in S(0)$, the initial surface) of the surfaces $S(t)$. Here $K(p,t)$ denotes the Gaussian curvature of $S(t)$ at the point $x(p,t)$.

It was not until (Andrews, 1999) that we obtained for the first time a complete proof that the limits are always round spheres (in fact a point, but spheres after appropriate renormalization). Between Firey and Andrews, (Chou, 1985) proved convergence to a point and the persistence of convexity, but not the sphericity. The proof consists in controlling the evolution equation introduced by Firey throughout time, and all that similar to what we have already encountered. We note that nowadays there is a trend toward the use of refined methods of partial differential equations in solving problems of "elementary" geometry.

However, to return to the case of the ellipsoids and the quotation from Hilbert and Cohn-Vossen (1952), the recent work of Andrews proves several things. First, if in (VII.13.C.2) we replace K by $K^{1/4}$, then there will be convergence to an ellipsoid:

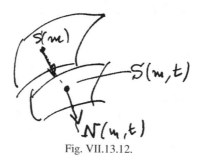

Fig. VII.13.12.

for the powers of K greater than 1/4 the curve of surfaces obtained (as a function of the time) in the space of all convex sets is tangent to the space of ellipsoids of this global space. So the assertion of Hilbert is now perhaps justified. We also find in Andrews a partial study of what happens to powers of K greater than 1. His most recent text is Andrews (2002); for example, for the bad powers of K the isoperimetric ratio does not remain bounded. The text Daskadopoulos and Lee (2004) studies the case where there are at the outset flat portions of the eroded object.

We continue to pursue the subject "geometry and manufacuring", begun in Sects. II.1, V.7 and VI.6, for the manufacture of rulers, of graduated rulers, of plane surfaces, of goniometers and here the problem of making very good spheres (balls). For the manufacture of ball bearings and cylinders, see Panaccione (2003) and review Fig. VI.6.3. In fact, in practice, for a very long time now, very good balls have been fabricated by placing a rather large number in a receptacle along with an abrasive and rotating the receptacle for a long time. Very good balls can be obtained incredibly rapidly in this way. If we want excellent balls, we repeat the process several times, decreasing the coarseness of the abrasive. However, subsequent to a patent by SKF Corporation from 1907, quality balls for rolling of all kinds have been manufactured using a technique different from the one described in IV.6; see Fig. VI.6.3. This is repeated with the use of plates of varied parameters, e.g. using various abrasives. What is inexplicable mathematically, it seems to us, by playing as is needed (its a matter of patents) with the four parameters of the pressure of the plates, the quality of the liquid, the choice of abrasive, the duration and the choices as a function of the time, we end up with a set of say a hundred balls which, upon emerging, have diameters almost all equal and equal within a micron to the diameter required by the purchaser. We leave it to readers to see precisely why rolling in circular grooves of a given radius (thus half tori of revolution) produces spheres of the same radius. In practice, for still more precision, the balls obtained are sorted by their actual size. Presently the precision obtained corresponds to disparities not exceeding 30 m of the Earth; see more details in Panaccione (2003). For the not very sophisticated manufacture of small balls, see Sect. VI.8.

Thus Firey's (partial) result, completed by Andrews, shows why the old technique, scarcely used any more, produced very good balls, a theoretical result in which metrologists and manufacturers can find amusement. As for us, it seems that two problems remain for mathematicians. The first is that (VII.13.C.2) is perhaps not the right equation for governing what happens in an industrial receptacle; a suitable coefficient may need to be added, of a sort that will give convergence (without renormalization) toward a sphere and not toward a point. It can be shown that this very quickly gives surfaces very close to spheres; all this can be found in Andrews (1999). We could perhaps relate what is happening here to the speed with which the surfaces become close to spherical: practically, we obtain an evolution equation for the deviation from sphericity. This deviation is typically evaluated by the square $(k_1 - k_2)^2 = 4(H^2 - K)$ of the difference of the principal radii of curvature (see Sect. VI.6). We can also relate this result to that of Tsolomitis (1996), discussed in Sect. 6, specifically that the Steiner symmetrizations very quickly gives objects very close to balls. The new equation will have to take into account the existence of an abrasive (see Sect. II.XYZ for the abrasion paradox).

Equation (VII.13.C.2) is what is called an *evolution equation*; such equations are playing an ever greater role in numerous areas, both in pure mathematics and in applications; see Huisken (2000).

VII.13.D. The appearance of the boundary of a convex set; the space of all convex sets

In Sect. VII.10 we amply studied the appearance of convex sets as **solid bodies.** We restrict ourselves here to bounded convex sets, let us say compact, but not center-symmetric in general. We now want to study the appearance of their boundary, which in ordinary space is a visible surface. Between the two extreme cases of balls (or ellipsoids) and cubes, what occurs? In the first case, we have a true (infinitely) differentiable surface; in the second to the contrary all the edge points are singular. What happens when we take a convex set **at random** ? Will it have great regularity or on the contrary almost always have lots of singularities? If we think of what we know about convex functions (see Sect. VII.3), we may think that the boundary of a convex set is a rather good surface from the point of view of geometry. In fact, the answer is going to be that for almost all convex sets the geometry of the boundary is an absolute disaster. For more details than in what follows, see 1.9 and 4.10 of Gruber and Wills (1993). We will give explicit references only for results that have come after the appearance of this fundamental work.

The first thing to do is to be able to define a notion like "almost everywhere" and even before that define the space of all convex sets. We consider here the set $\mathcal{K}$ = $\mathcal{K}(\mathbb{E}^d)$ of all compact convex sets in $\mathbb{E}^d$. We will not specify the dimension d in which we work when this isn't necessary, but it is always fixed. In Sect. VII.7 we defined a metric, called the *Hausdorff metric* and denoted d_H, as follows: $d_H(C, C')$ is the least ε such that we have both $C \subset \mathrm{Tub}(C'; \varepsilon)$ and $C' \subset \mathrm{Tub}(C; \varepsilon)$, that is to say that within ε we can simultaneously approximate C by C' and C' by C:

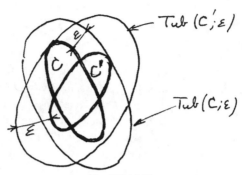

Fig. VII.13.13.

The essential fact is that $(\mathcal{K}, d_H)$ is a *complete* space and is furthermore *locally compact* if we restrict ourselves to convex sets with nonempty interior. This (local) compactness is due to Blaschke and we used it in an essential way in Sect. VII.7 for proving the isoperimetric inequality.

Note that this metric can be defined for the collection of all bounded subsets of $\mathbb{E}^d$, and indeed for the collection of all bounded sets in any metric space whatsoever. But by a simple embedding device, we can define a metric on nothing less that the collection of all compact metric spaces. This naive but revolutionary idea was introduced and used several times by Gromov in 1978. For the avatars of this metric, and others, see Berger (1998a) for a brief survey and the originals: Gromov (1999) (in particular Chap. $3\frac{1}{2}$) and Gromov (2000). Further, see 1.9.4 of Gruber and Wills (1993) for other structures on the sets $\mathcal{K}$.

Now this does not suffice for our ability to state propositions containing the condition "for almost all convex sets". The actual usage is to reserve this expression for when the context in which we are working is that of a space endowed with a measure; then "almost all" would say "within a set of measure zero". Let us therefore see whether there exist good measures on $\mathcal{K} = \mathcal{K}(\mathbb{E}^d)$, "good" meaning geometrically satisfying. The most natural condition to impose on them is their being invariant under the group of isometries of $(\mathcal{K}, d_H)$. Now, there doesn't exist any such measure; see Theorem 10 in 1.9 of Gruber and Wills (1993) or Bandt and Baraki (1986). Interested readers will ask what are the isometries of $(\mathcal{K}, d_H)$. They are all known: there aren't any, apart for obvious isometries of $\mathbb{E}^d$ itself, other than Minkowski addition with a fixed convex set; see again 1.9.4 of Gruber and Wills (1993). What should we do? Have we failed?

No, since Baire at about the beginning of the twentieth century, to whom we owe a notion that replaces that of almost everywhere (in fact he was mostly interested in a detailed study of functional singularities) in metric spaces but without a canonically attached measure. We define, in a complete metric space, the notion of a *sparse set*: a subset is called sparse if it is a countable union of nowhere dense sets. In this context we replace almost everywhere by *most*, or *typical*, or *generic*. This is used when the complementary subset is sparse. We can also even say *almost everywhere* if we are clear in its meaning, but this is not advised. Here is why: it is important to know that, if we have some space that is both metric and measure, the notions *sparse* and *of measure zero* do not in general coincide; this is already the case with the interval [0,1], for its metric and for the canonical measure. All the details, with the needed examples, can be found in Oxtoby (1980). We need to know that not a few works use the term "G_δ-dense" for the complements of sparse sets. We will use the term G_δ-dense in the sequel, among other places in Sects. XI.6 and XI.10.C. However, a way recently opened up is that of finding *valuations* on the set of all convex sets; see Alesker (2001).

♦

Here now, with this language, is what is known about the appearance, the geometry, of the boundary of generic convex sets. Some of the results have a local version that translates only the generic properties of convex functions (of several variables), which we will not discuss. Here first is the nature of the boundary as a subset of $\mathbb{E}^d$:

(VII.13.D.1) *For most convex sets, the boundary is a submanifold of class C^1 and is strictly convex (which means that its tangent hyperplane has but the one point of contact). For most convex sets the boundary is not of class C^2.*

Worse: we can easily define a notion of lower bound and of upper bound of sectional curvature, say in $\mathbb{E}^3$ for simplicity, but what follows extends to arbitrary dimension. Then:

(VII.13.D.2) *For most convex sets, at each point of the boundary: either the lower bound of the curvature is zero or else it is infinite (recall that for surfaces the sectional curvature coincides with Gaussian total curvature).*

These results are not surprising if we realize that, on the one hand, the various notions of curvature require the class C^2 (see Chap. VI) and, furthermore, that each convex set can be approximated by polytopes. Now, polytopes have a boundary where the curvature at a point is either zero or infinite. We conclude: in their appearance and in the first approximation, convex sets have generically smooth boundaries, but in the second approximation very bad ones. This will be further aggravated for the metric geometry of these boundaries.

As we have done for the sphere and for smooth surfaces, we want to study the problem of *shortest paths* (also called segments, or geodesic segments), more generally of geodesics (curves that are locally shortest paths, but extended as far as possible). In the class C^2 we have good behavior for the geodesics, as in Chap. VI, but for a generic convex set we have both better and worse:

(VII.13.D.3) *For most convex sets each point of the boundary is connected to most other points by a unique geodesic segment. But, on the other hand, still for most convex sets, for each of their points there exists a dense subset of points which are connected to it by at least three different geodesic segments.*

This last property of *triples* contrasts strongly with the case of smooth surfaces; see Sect. VI.3. cut locus enthusiasts will be disappointed here: the cut locus can be of infinite length: Zamfirescu (2004). Also, VII.13.D.3 shows that the cut locus approaches as closely as is wanted to the point, in particular the radius of injectivity at each point is always zero. Another strong contrast with Sect. XII.5.D is:

(VII.13.D.4) *On the boundary of most convex sets, there does not exist any periodic geodesic.*

The situation for polytopes is different as we will see this in Sect. XII.4, but readers can show that geodesics always exist; in return, for most there have to be double points; see Sect. XII.4.A.

VII.13.E. Immobilization of a convex set

Here is a practical problem from economics: how many points are really necessary for immobilizing a convex set? That means that we place k points on the boundary of a convex set; we fix them and see whether or not it is possible (using isometries) to move the set in such a way that the fixed points "slide" along the boundary. We easily "see" that two points in the plane, or three in space, are insufficient. Heuristically there remain $d + 1 - k$ parameters for moving the convex set. The organization of the subject is subtle; it is however studied in the section "fixing systems" of Boltyanski, Martini, and Soltan (1997): balls excepted (or more precisely the convex sets for which $d + 1$ points of the boundary have spherical neighborhoods), $d + 1$ points suffice for immobilizing the convex set.

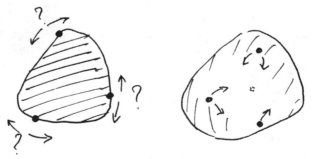

Fig. VII.13.14.

In Zamfirescu (1995) a closely related problem is studied: *how to hold (for example in a net) a convex set?* It's an old problem (see the references given by

Zamfirescu, who treats the case of a circle of support). Previous results concerned mainly *nets*, the most economical (at least in theory).

VII.14. Intermezzo: can we dispose of the isoperimetric inequality?

We have encountered this inequality numerous times and proposed numerous types of proofs: for curves in Sects. V.11, V.12, for surfaces in Sect. VI.11, for spheres S^2 and S^3 in Sect. III.5 and in the present chapter for polygons in Sect. VII.2 and for general convex sets in Sects. VII.7 and VII.8. Why such profligacy? Why treat on the one hand smooth surfaces and on the other convex sets? Why not treat more general objects, less regular in one sense or another? In particular, the convex case is terribly restrictive. For references on isoperimetric inequalities and physics, see 4.4 of Gruber and Wills (1993).

Two initial remarks. The first is that, whatever the method of proof, the final object (if it is to be smooth to exclude in particular hairy spheres) is **convex**. In the case of curves we see that taking the convex envelope improves the isoperimetric ratio, because the area increases while the perimeter decreases:

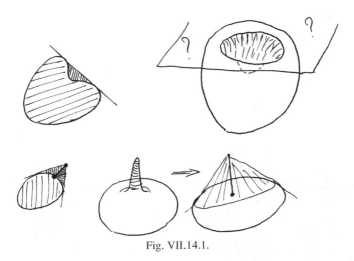

Fig. VII.14.1.

This is false for higher dimensional manifolds, beginning with surfaces; a smooth peak, but sufficiently pointed, will have a convex envelope for which simultaneously the area and the volume both increase. A remark: it is not known how to show that *a priori* the limit object for which the isoperimetric quotient is minimum, even if we know its existence, is convex. In this regard, readers can convince themselves about the lack of rigor (astonishing in retrospect) of texts such as Rouché and de Comberousse (1912); see pp. 234–236 of Vol. II.

◆

The second remark (see Osserman, 1978) consists of this paradox: it is easy, by fold-ing the boundary more and more wildly that we can increase its area much without increasing the volume it encompasses. The more we allow objects, for which we want to obtain the isoperimetric inequality, to have an irregular boundary, the more difficult is the proof.

There are several reasons for this: the first is the difficulty of giving notions of $(d - 1)$-dimensional volume (area when $d = 3$ but we continue to say "area" in each dimension) for the boundary of a not too general object (the notion of volume in itself is easy for all decent objects and we can live without knowing about the existence of nonmeasurable sets). For different possible notions of area, see 12.2 of Burago and Zalgaller (1988). We discover the other reasons for the difficulty of the problem by giving two essential examples. The first is trivial: take a ball in $\mathbb{E}^3$ and add to it some hair: whatever the definitions that we give to the volume of this ball and to the area of its boundary, the values will always be the canonical $4\pi R^2$ and $\frac{4\pi}{3}R^3$, and the isoperimetric quotient will thus be obtained by non unique objects. We will not fail to notice that our hairy ball is neither convex nor smooth.

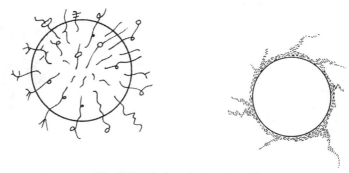

Fig. VII.14.2. Some hair, and a veil

But the second example is simultaneously more serious and more subtle to ex-hibit. We look first for a natural and geometric notion of area and discuss $\mathbb{E}^3$ for simplicity. Just as we define the length of a curve as the limit superior of the length of polygons inscribed in it, the point of departure for defining the area of a surface will be triangulating it and defining the area as the limit superior of the area of these triangulations (as we take the sizes smaller and smaller). Unfortunately the Venetian lantern (or Chinese lantern) below shows that this limit superior is infinity.

But then why not take as definition of the area the *limit inferior* of the area of the triangulations when the dimensions of all the triangles tends toward zero, more precisely triangulations that converge toward the surface? This is the definition of Lebesgue. But then we find in Besicovitch (1945) a terrifying example, specifically a cube for which we modify the boundary in such a way that the total volume remains equal to 1 but for which the Lebesgue area of the boundary surface is as small as we wish!

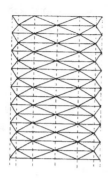
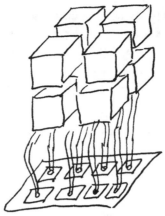

Fig. VII.14.3. The Chinese (or Venetian) lantern and Besicovitch's cube. [BG] Berger, Gostiaux (1987) © Presses Universitaires de France

Here is the beginning of the Besicovitch construction. We start with a unit cube C and in its interior we inscribe eight cubes A_i a little smaller than the eight cubes of edge $\frac{1}{2}$ which make up C, so that A_i has volume $\frac{3}{4}\cdot\frac{1}{8}$. On a face B of C we mark off eight disjoint squares B_i. Next we connect each A_i to each B_i by tubes c_i, triangulated into triangles and rectangles, and not intersecting each other. We iterate this construction, with cubes of volume $\frac{5}{8}\cdot\frac{1}{8^2}$, then $\frac{9}{16}\cdot\frac{1}{8^3}$, etc. We then show that there exists a limit object (evidently of fractal nature) that has the same topology as a cube (or a ball), of total volume $\frac{1}{2}$, whereas the area of its boundary can be made as small as we want!

The conclusion to be drawn is that there can't exist an isoperimetric inequality unless the notions of volume and area are suitably related. Such an example, when added to the other simpler ones above, explains the difficulty of obtaining an isoperimetric inequality for a class of objects large enough to include the objects that we encounter in more or less natural geometric problems. This is why the proof of Sect. VII.8 is almost dazzling, since area is defined as the limit of volumes. Also, we mustn't forget that we also need hope to be able to describe the objects that satisfy the critical value of the inequality. From among the numerous theories that exist today, we restrict ourselves to two categories: that of smooth objects (submanifolds: Sect. V.XYZ) and that of convex sets. Already for the convex sets we had to accept a certain evil. But we remark that for the smooth case, for the proof of the inequality, we had to appeal in Sect. VI.11 to profound results from geometric measure theory.

In a certain way and with commentary, we will now review everything we have encountered so far on the subject with an eye, moreover, toward possible generalizations and hopes. Practically all of what we will say can be found in Burago and

Zalgaller (1988), but Osserman (1978) can be consulted for a first approach, for this reference is full of information and motivations. We give six points of view; for the case of curves we find no less than 10 on p. 2 of Burago and Zalgaller (1988).

The method of Sect. VII.8 via the *Brunn-Minkowski* inequality is the simplest, and the most rapid. But its serious inconvenience lies in not immediately allowing the case of equality. Another major defect: it doesn't really apply except in Euclidean space.

The symmetrization method of Sect. VII.7 requires little that is sophisticated and besides permits us to easily treat the case of equality. Above all it also works in hyperbolic spaces and for spheres, for these spaces admit hyperplane symmetries in all directions. Thus the isoperimetric inequality is valid in these spaces, but of course the formula obtained cannot be of type $\text{Area} \geq c(d) \cdot \text{Vol}^{d/d-1}$, since these spaces don't admit homotheties; but the result remains essentially the same. Specifically, for all domains with smooth boundary and given volume, the minimum of the area (or the volume) of the boundary is greater of equal to the corresponding value of the area of the boundary of a ball whose volume is the same as the given volume. Furthermore, there is equality only for balls. We have used the isoperimetric inequality for spheres in an essential way in the proof of Dvoretsky's concentration phenomenon in Sect. VII.12.C.

We saw that the introduction of the *spherical harmonics* would be natural for proving the isoperimetric inequality in $\mathbb{E}^d$; unfortunately it is not known how to do this at present except for the planar case: $d = 2$.

We saw in Sects. V.12 and VI.11 that it is very natural to want to obtain the isoperimetric inequality with the aid of *Stokes' formula*. This method however is subtle once the boundary isn't strictly convex. On the other hand, it very easily provides the characterization of equality.

We find in Burago and Zalgaller (1988) some refined isoperimetric inequalities, i.e. using objects that are general enough and adequate notions of area. For example, Hadwiger's result (initially in the important work Hadwiger, 1957) furnishes, in the case of equality, balls with a *veil* in a sense similar to the definition of the one with hairs mentioned in Sect. V.12; see Fig. VII.14.2.

 We are also indebted to a whole school of *Italian geometers* for refined isoperimetric inequalities, which appear in the same book: Burago and Zalgaller (1988).

◆

We have pointed out the *descent* method by an evolution equation, on the one hand because it is very natural, on the other because it constitutes part of a whole collection of results obtained in geometry with natural evolution equations: we start with an arbitrary object and reduce its distortion with respect to an ideal object which we have in mind. For this in the realm of Riemannian geometry; see for example Berger (2001) or Berger (1999). But the great event was Perelman's proof of the Poincaré conjecture by this method; see the references toward the end of Sect. III.5.

◆

We have kept *geometric measure theory* for last, because it is nowadays the method that allows us to go the furthest in the most general context (and is moreover the most recent, dating from the beginning of the 1960s). We already used it in Sects. V.12 and VI.11 for proving the isoperimetric inequality in all the Euclidean spaces, with characterization of the case of equality. We look briefly at its power by an example on the sphere.

Notions of volume and area (i.e. the volume of the boundary, as always) are defined for very general objects, but where these two notions are necessarily related: such objects are called *integral currents*. Let us say, as a bit of a caricature, that these are objects defined by *duality*. As with distributions, we define them by their effect on functions; likewise for defining their boundary. These are the general currents and we restrict ourselves to integral currents, which are by definition precisely those for which we have simultaneously a notion of volume for the domain and area for its boundary. For domains for which the boundary is a submanifold, or for convex domains, the notions of volume and area coincide (thank goodness!) with the prior notions.

This definition works in very general spaces and in particular in the context of Riemannian manifolds. The theory provides, in the general Riemannian context, the following essential result, a particular case of a more general result for currents, but which will suffice for us here:

(VII.14.1) *In a compact Riemannian manifold, we consider all the domains which have a given volume V and for which the boundary is a nice submanifold ∂V. Then there exists a domain D, with volume equal to V, and for which the boundary ∂D has an area less than of equal to that of the boundary of all other such domains. Moreover, the boundary of D is almost everywhere (that is, on a set of full measure) a nice submanifold (in fact the more general result is that this boundary doesn't even have singularities except on a part whose codimension does not exceed 8).*

We apply this to the sphere S^d, where the idea of the proof goes back to Gromov. Having fixed the volume V, we thus let D denote a domain that realizes the minimum of the area of the boundary. Then the boundary $\partial D = S$ is almost everywhere a submanifold of the sphere (thus of dimension $d - 1$ here still). In S^d

a hypersurface (submanifold of dimension $d - 1$) will again have a second fundamental form, principal curvatures k_i and a mean curvature h. Here again the condition of minimum area of the boundary implies that this mean curvature is a constant, always denoted by h. We copy the proof of Sect. VI.11 exactly and give the proof on S^3, which extends trivially to arbitrary dimension. That is, we fill in with normals to the boundary at its nonsingular points, which suffices because each point of the interior comes from its corresponding *foot* (or near-point) on the boundary.

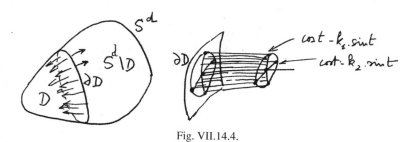

Fig. VII.14.4.

The only two things that need be changed are the following: the first is that the formula (VI.11.1) becomes

(VII.14.2)

$$\text{Volume}(D) = \int_M \int_0^{\text{cut}(M)} (\cos t - k_1(m) \cdot \sin t)(\cos t - k_2(m) \cdot \sin t) \, dt \, dm.$$

Thus we will now have:

$$V = \text{Vol}(D) \leq \text{Area}(S) \int_0^{\text{arc cot} t} (\cos t - h \sin t)^2 dt,$$

where we recall that the notation arc cot h denotes the real t defined by the equality $\frac{\cos t}{\sin t} = h$. But furthermore, we don't have to integrate explicitly (it's a classic trick), because if we calculate the volume of a spherical cap of radius arc cot h with this formula, there will be equality in the formula if all the normals intersect at the center of the cap. We will then be done (without having to supplement the derivation done in Sect. VI.11) if we know that the mean curvature (which we know is constant) of S is h_0, i.e. the mean curvature of the boundary of the spherical cap whose volume equals V; or, better still, merely if this mean curvature is less than or equal to that quantity. Here enter the incredible cleverness of Gromov: if this mean curvature h satisfies $h > h_0$, then it suffices to consider the complementary domain $S^3 \setminus D$: the mean curvature of its boundary is equal to $-h$, and all the signs change; we are finally done. The case of equality follows easily as in Sect. VI.11.

For the case of higher dimensions, we need to replace the inequality $xy \leq ((x + y)/2)^2$ by Newton's inequality: $(\frac{a_1 + \cdots + a_n}{n})^n \geq a_1 \ldots a_n$ that we proved in Sect. VII.2.

The problem of the *isoperimetric deficit* consists of saying, for a geometric object, something about the deviation from isoperimetric equality. We take for simplicity the case of the plane, where we deal with a curve of length L, which encloses a domain of area A: to evaluate the difference $L^2 - 4\pi A$ or the difference $\frac{L^2}{A} - 4\pi$ with the aid of various invariants attached to the curve in question or the domain it encloses. For the case of curves, the basic reference is Osserman (1979), complemented by the first chapter of Burago and Zalgaller (1988). For higher dimensions — it is an area of active research — see for example several sections of Gruber and Wills (1993).

We have seen in Sect. V.12 that the method of integral geometry (mentioned but briefly in Sect. V.12), as well as that of Fourier series, lets us find an expression for the deficit. But we can scarcely say that this deficit is expressed in terms of natural geometric invariants. For curves, one of the simplest deficits is that of Bonnensen-Fenchel, which asserts that $L^2 - 4\pi A \geq \pi^2(R - r)^2$, where R (resp. r) denotes the radius of the smallest circle containing (resp. contained inside) the curve. We remark that the case of equality is then treated in lightning fashion. See Osserman (1979) and Burago and Zalgaller (1988) for refinements. For higher dimensions, the answers are less satisfying at present; see 21.1 of Burago and Zalgaller (1988). But this is the moment to point out that the results on *mixed volumes*, to which we alluded in Sect. VII.13.B provide, in the case of convex sets, sandwiches for the isoperimetric inequality. The simplest case is that of double inequality (due to Minkowski):

$$S^2 \geq 3VM \quad \text{and} \quad M^2 \geq 4\pi S.$$

Here M denotes the integral of the mean curvature of the boundary (supposed smooth).

Here we have come to the transition to the following chapter. In fact, we will see there that isoperimetric inequalities for the polyhedra are a more arduous world than for smooth objects. In particular, several inequalities still retain the rank of conjecture.

Bibliography

[B] Berger, M. (1987, 2009). *Geometry I, II*. Berlin/Heidelberg/New York: Springer

[BG] Berger, M., & Gostiaux, B. (1987). *Differential geometry: Manifolds, curves and surfaces*. Berlin/Heidelberg/New York: Springer

(1985). *Treasures of the Israel museum Jerusalem*. Jerusalem and Genève: Israel Museum and Pierre-Alain Ferrazini

Ahoronov, D., Schiffer, M., & Zalcman, L. (1981). Potato Kugel. *Israel Journal of Mathematics*, 40, 331–339

Alesker, S. (2001). Description of translation invariant valuations on convex sets with solution of McMullen's conjecture. *Geometric and Functional Analysis (GAFA)*, 11, 244–272

Andrews, B. (1999). Gauss curvature flow: the fate of rolling stones. *Inventiones Mathematicae*, 138, 151–161

Andrews, B. (2002). *Positively curved surfaces in three-sphere* (Vol. 2, pp. 221–230). Proceedings of the ICM, Beijing

Appell (1952). *Traité de mécanique rationnelle*. Paris: Gauthier-Villars

Appell, P., & Lacour, E. (1922). *Fonctions elliptiques*. Paris: Gauthier-Villars

Arnold, V. (1976). *Les méthodes mathématiques de la mécanique classique*, Moscow: MIR

Arnold, V. (1978). *Mathematical methods of classical mechanics*. Berlin/Heidelberg/New York: Springer

Auerbach, H. (1938). Sur un problème de M. Ulam concernant l'équilibre des corps flottants. *Studia Mathematica, 7*, 121–142

Ball, K. (1986). Cube slicing in $\mathbb{R}^n$. *Proceedings of the American Mathematical Society, 97*, 465–473

Ball, K. (1989). Volumes of sections of cubes and related problems. In *Geometric aspects of functional analysis: Vol. 1376. Springer lecture notes in mathematics* (pp. 251–260). Berlin/Heidelberg/New York: Springer

Ball, K. (1991). Volume ratios and a reverse isoperimetric inequality. *Journal of the London Mathematical Society, Second Series, 44*, 351–359

Bandt, C., & Baraki, G. (1986). Metrically invariant measures on locally homogeneous spaces and hyperspaces. *Pacific Journal of Mathematics, 121*, 13–28

Barany, I., & Füredi, Z. (1987). Computing the volume is difficult. *Discrete & Computational Geometry, 2*, 319–326

Barany, I. (2008). Random points and lattice points, *Bulletin of the AMS, 45*, 339–366

Barthe, F. (1998). On a reverse form of Brascamp-Lieb inequality. *Inventiones Mathematicae, 134*, 335–361

Barthe, F. & Nahor, A. (2002). Hyperplane projections of the unit ball of l(n;p), *Discrete and Computational Geometry, 27*, 215–228

Berger, M. (1999). *Riemannian geometry during the second half of the twentieth century*. Providence, RI: American Mathematical Society

Berger, M. (2000). Encounter with a geometer I, II. *Notices of the American Mathematical Society, 47*(2), *47*(3), 183–194, 326–340

Berger, M. (2003). *A panoramic view of Riemannian geometry*. Berlin/Heidelberg/New York: Springer

Besicowitch, A. (1945). On the definition and value of the area of a surface. *The Quarterly Journal of Mathematics, Oxford, 27*, 141–144

Bollobas, B. (1997). Volume estimates and rapid mixing in *Flavors in Geometry*, Silvio Levy (Ed.), MSRI Publications, 31, Cambridge: Cambridge University Press

Boltyanski, V., Martini, H., & Soltan, P. (1997). *Excursions into combinatorial geometry*. Berlin/Heidelberg/New York: Springer

Bombieri, E., & Vaaler, J. (1983). On Siegel's lemma, plus addendum. *Inventiones Mathematicae, 73, 75*, 11–32, 377

Bourgain, J. (1991). On the distribution of polynomials on high dimensional convex sets. In L.A. Milman (Ed.), *Geometric aspects of functional analysis: Vol. 1469. Springer lecture notes in mathematics* (pp. 127–137). Berlin/Heidelberg/New York: Springer

Bourgain, J., Klartag B. Milman, V., (2003). A reduction of the slicing problem to finite volume ratio bodies. *Comptes Rendus de l'Académie des sciences, 1336*, 331–334

Bourgain, J., & Milman, V. (1987). New volume ratio properties for convex symmetric bodies in $\mathbb{R}^n$. *Inventiones Mathematicae, 88*, 319–340

Brascamp, H., & Lieb, E. (1976). On extensions of the Brunn-Minkowski and Prékopa-Leindler theorems, including inequalities for log concave functions, and with an application to the diffusion equation. *Journal of Functional Analysis, 22*, 366–389

Brenier, Y. (1991). Polar factorizations and monotone rearrangement of vector-valued functions. *Communications on Pure and Applied Mathematics, 44*, 375–417

Brenier, Y. Frisch, U., Hénon, M (2003). Reconstruction of the early Universe as a convex optimization problem. *Monthly Notices of the Royal Astronomical Society, 346*, 501–524

Burago, Y., & Zalgaller, V. (1988). *Geometric inequalities*. Berlin/Heidelberg/New York: Springer

Busemann, H. (1953). Volumes in terms of concurrent cross-sections. *Journal of Mathematics, 3*, 1–12

Busemann, H. (1958). *Convex surfaces*. New York: Wiley Interscience

Chou, K. (1985). Deforming a hypersurface by its Gauss-Kronecker curvature. *Communications on Pure and Applied Mathematics, 38*, 867–882

Claude J., & Davanture C. (1988). *Les roulements*. Paris: SNR Roulements and Nathan Communications

Croft, H., Falconer, K., & Guy, R. (1991). *Unsolved problems in geometry*. Berlin/Heidelberg/New York: Springer

Daskadopoulos, P., & Lee, K.-A. (2004). Worn stones with flat sides all time regularity of the interface. *Inventiones Mathematicae, 156*, 445–493

David, G., & Semmes, S. (1997). *Fractured fractals and broken dreams*. Oxford: Clarendon Press

Dvoretsky, A. (1961). *Some results on convex bodies and Banach spaces* (pp. 123–160). Proceedings of the international symposium on Linear Spaces, July 1960, Jerusalem

Dvoretsky, A., & Rogers, C. (1950). Absolute and unconditional convergence in normed linear spaces. *Proceedings of the National Academy of Sciences of the USA, 36*, 192–197

Eggleston, H. (1958). *Convexity*. Cambridge, UK: Cambridge University Press

Ehrhart, E. (1966). Sections maximales du cube. *Revue de mathématiques spéciales, 76*(10), 249–251

Falconer, K. (1983). Applications of results on spherical integration to the theory of convex sets. *The American Mathematical Monthly, 90*, 690–693

Falconer, K. (1990). *Fractal geometry*. New York: John Wiley

Feder, J. (1988). *Fractals*. New York: Plenum Press

Figiel, T., Lindenstrauss, J., & Milman, V. (1977). The dimension of almost spherical sections of convex bodies. *Acta Mathematica, 139*, 53–94

Firey, W. (1974). On the shapes of worn stones. *Mathematika, 21*, 1–11

Funk, P. (1913). Über Flächen mit lauter gescholssenen geodätischen Linien. *Mathematische Annalen, 74*, 278–300

Gardner, R. (1994). A positive answer to the Busemann-Petty problem in three dimensions. *Annals of Mathematics, 140*, 435–447

Gardner, R. (1995). *Geometric tomography*. New York: Cambridge University Press

Gardner, R., Koldobsky, A., & Schlumprecht, S. (1999). An analytical solution to the Busemann-Petty problem on sections of convex bodies. *Annals of Mathematics, 149*, 691–703

Gardner, R., & McMullen, P. (1980). On Hammer's X-ray problem. *Journal of the London Mathematical Society, Second Series, 2*, 171–175

Giannopoulos, A., & Milman, V. (2001). Euclidean structure in finite dimensional normed spaces. In W. Johnson & J. Lindenstrauss (Eds.), *Handbook of geometry of Banach spaces* (pp. 707–779). New York: Kluwer

Giannopoulos, A., & Milman, V. (2004). Asymptotic convex geometry, short overview. In S. Donaldson & M. Gromov (Eds.), *Different faces of geometry* (pp. 87–162). New York: Kluwer

Gilbert, E. (1991). How things float. *The American Mathematical Monthly, 98*, 201–216

Giles, J. (1982). *Convex analysis with application in the differentiation of convex functions*. London: Pitman

Gluskin, E. (1981). The diameter of the Minkowski compactum is approximately equal to n. *Functional Analysis and its Applications, 15*, 72–73

Gray, A. (1990). *Tubes*. Reading, MA: Addison-Wesley

Gray, A. (1993). *Modern differential geometry of curves and surfaces*. Boca Raton, FL: CRC Press

Gromov, M. (1999). *Metric structures for Riemannian and non-Riemannian spaces*. Boston: Birkhäuser

Gromov, M. (2000). Spaces and questions. *Geometric and Functional Analysis (GAFA), Special Volume*, 118–161

Gromov, M. (2003). Isoperimetry of waists and concentration maps, *GAFA , 13*, 178–215

Gromov, M., & Milman, V. (1987). Generalization of the spherical isoperimetric inequality to uniformly convex Banach spaces. *Compositio Mathematica, 62*, 263–282

Grothendieck, A. (1956). Sur certaines classes de suites dans les espaces de Banach et le théorème de Dvoretsky-Rogers. *Bol. Soc. Mat. Sao-Paulo, 8*, 81–110. Reprinted in 1998, *Resenhas, 3*, 447–477

Gruber, P. (1996). *Basic problems and recent progress in convexity theory* (pp. 22–33). In fourth international conference on Geometry, Academy Athens Thessaloniki

Gruber, P. (2007). *Convex and discrete geometry*. Berlin/Heidelberg/New York: Springer

Gruber, P., & Wills, J. (Eds.). (1993). *Handbook of convex geometry*. North-Holland: Elsevier

Grünbaum, B. (1993). *Convex polytopes*. Berlin/Heidelberg/New York: Springer

Hadwiger, H. (1957). *Vorlesungen über Inhalt, Oberfläche und Isoperimetrie.* Berlin/Heidelberg/New York: Springer

Hadwiger, H. (1972). *Gitterperiodische Punktmengen und Isoperimetrie. Monatshefte für Mathematik. 76*, 410–418

Helgason, S. (1980). *The Radon transform*. Boston: Birkhäuser

Hensley, D. (1979). Slicing the cube in $\mathbb{R}^n$ and probability. *Proceedings of the American Mathematical Society, 73*, 95–100

Hilbert, D., & Cohn-Vossen, S. (1952). *Geometry and the imagination*. New York: Chelsea

Huisken, G. (2000). Evolution equations in geometry, in *Mathematics Unlimited: 2000 and Beyond*, Engquist & Schmid (Eds.), 2, 593–604, New York: Springer Verlag

Israel, R. (1979). *Convexity in the theory of lattice gases*. Princeton, NJ: Princeton University Press

John, F. (1948). Extremum problems with inequalities as a subsidiary condition. In *Studies and essays presented to R. Courant on his 60th birthday* (pp. 187–204). New York: Wiley Interscience

Kiselman, C. (1987). Smoothness of vector sums of plane convex sets. *Mathematica Scandinavica, 60*, 239–252

Klartag, B. (2005). An isomorphic version of the slicing problem. *Journal of Functional Analysis, 218*, 372–394

Klartag, B., & Milman, V. (2003). Isomorphic Steiner symmetrization. *Inventiones Mathematicae, 153*, 463–485

Ledoux, M. (2001). The concentration of measure phenomenon, *Mathematical Surveys and Monographs, 789*, Providence: American Mathematical Society

Leichtweiss, K. (1980). *Konvexe Mengen*. Berlin: VEB Deutscher Verlag der Wissenschaften and Springer

Lévy, P. (1951). *Problèmes concrets d'Analyse fonctionnelle*. Paris: Gauthier-Villars

Lindenstrauss, J., & Milman, V. (1993). The local theory of normed spaces and its applications to convexity. In P. Gruber & J. Wills (Eds.), *Handbook of geometric convexity*. North-Holland: Elsevier, 1149–1220

Lovacz, L. (1990). *Geometric algorithms and algorithmic geometry* (Vol. 1, pp. 139–154). In Proceedings of the ICM, Kyoto

MacBeath, A. (1951). An extremal property of the hypersphere. *Proceedings of the Cambridge Philosophical Society, 47*, 245–247

Maurey, B. (2003). Inégalité de Brunn-Minkowski-Lusternik, et autres inégalités géométriques fonctionnelles. In *Séminaire Bourbaki* (Vol. 2003–2004, pp. 95–113), Astérisque 299 (2005, Société mathématique de France)

McMullen, P. (1983). Volumes of projections of unit cubes. *The Bulletin of the London Mathematical Society, 16*, 278–280

Meyer, M., & Pajor, A. (1988). Sections of the unit ball of L(n, p). *Journal of Functional Analysis, 80*, 109–123

Meyer, M., & Pajor, A. (1990). On the Blaschke-Santalo inequality. *Archiv der Mathematik, 55*, 82–93

Meyer, M., & Reisner, S. (1991). A geometric property of the boundary of symmetric convex bodies and convexity of flotation surfaces. *Geometriae Dedicata, 37*, 327–337

Milman, V. (1971). A new proof of the theorem of A. Dvoretsky on sections of convex bodies. *Functional Analysis and its Applications, 5*, 28–37

Milman, V. (1992). Dvoretzky's theorem – thirty years later. *Geometric and Functional Analysis, 2*, 455–479

Milman, V., & Pajor, A. (1989). Isotropic positions and inertia ellipsoids and zonoids of the unit ball of a normed n-dimensional space. In J. Lindenstrauss & V. Milman (Eds.), *Geometric aspects of functional analysis* (pp. 64–104). Berlin/Heidelberg/New York: Springer

Milman, V., & Pajor, A. (1999). Entropy methods on asymptotic convex geometry. *Comptes Rendus de l'Académie des sciences, 329*, 303–308

Morgan, F. (2008). *Geometric Measure Theory: A Beginner's Guide* (4th ed.). Amsterdam: Academic Press

Osserman, R. (1978). The isoperimetric inequality. *Bulletin of the American Mathematical Society, 84*, 1182–1238

Osserman, R. (1979). Bonnesen-Fenchel isoperimetric inequalities. *The American Mathematical Monthly, 86*, 1–29

Oxtoby, J. (1980). *Measure and category*. Berlin/Heidelberg/New York: Springer

Pach, J., & Agarwal, P. (1995). *Combinatorial geometry*. New York: Wiley

Palmon, O. (1992). The only convex body with extremal distance from the ball is the simplex. *Israel Journal of Mathematics, 80*, 337–349

Panaccione, G. (2003). *Sphères qui roulent...* Pour la Science, "La sphère sous toutes ses formes", octobre–décembre 2003

Petty, C. (1983). Ellipsoids. In G.A. Wills (Ed.), *Convexity and its applications* (pp. 264–276). Boston: Birkhäuser

Pisier, G. (1989). The volume of convex bodies and Banach space geometry. Cambridge, UK: Cambridge University Press

Radon, J. (1917). Über die Bestimmung von Funktionen durch ihre Integralwerte langs gewisser Mannigfaltigkeiten. *Ber. Verh. Sächs. Akad. Wiss. Leipzig Math.-Phys. Kl., 69*, 262–267

Reichel, W. (2009). Characterizations of balls by Riesz-potentials. *Annali di Matematica Pura ed Applicata, 188*, 235–245

Roberts, A., & Varberg, D. (1973). *Convex functions*. New York: Academic Press

Rockafellar, R., & Wets, R. (1998). *Variational analysis*. Berlin/Heidelberg/New York: Springer

Rogers, C. (1976). In D. Larman & C. Rogers (Eds.), Proceedings of the Durham symposium on the Relations Between Infinite-Dimensional and Finite-Dimensional Convexty. *The Bulletin of the London Mathematical Society, 8*, 1–33

Rorres, C. (2004). Completing Book II of Archimede's: On floating bodies. *The Mathematical Intelligencer, 26*, 32–42

Rouché, E., & de Comberousse, C. (1912). *Traité de géométrie (deux volumes)*. Paris: Gauthier-Villars

Rouvière, A. (1973). *Traité d'anatomie*. Pairs: Masson

Schneider, R. (1993). *Convex bodies: The Brunn-Minkowski theory*. Cambridge, UK: Cambridge University Press

Schütt, C. (1997). Floating body, illumination body, and polytopal approximation. *Comptes Rendus de l'Académie des sciences, 324*, 201–203

Schütt, C., & Werner, E. (1994). Homothetic convex floating bodies. *Geometriae Dedicata, 49*, 335–348

Steinitz, E., & Rademacher, H. (1934). *Vorlesungen über die Theorie der Polyeder*. Berlin/Heidelberg/New York: Springer

Talagrand, M. (1995). Concentration of measure ans isoperimetric inequalities in product spaces. *Publications mathm´ atiques de líInstitut des hautes études scientifiques, 81*, 73–205

Tomczak-Jaegermann, N. (1989). *Banach–Mazur distance and finite-dimensional operator ideals*. London: Pitman

Tsolomitis (1996). Quantitative Steiner/Schwarz-type symmetrizations. *Geometriae Dedicata, 60*, 187–206

Ulam, S. (1960). *A collection of mathematical problems*. New York: Wiley Interscience

Valentine, F. (1964). *Convex sets*. New York: McGraw-Hill

Wagner, G. (1993). On a new method for constructing good point sets on spheres. *Discrete & Computational Geometry, 9*, 111–129

Webb, S. (1996). Slices of the regular simplex. *Geometriae Dedicata, 61*, 19–28

Zamfirescu, T. (1995). How to hold a convex body. *Geometriae Dedicata, 54*, 313–316
Zamfirescu, T. (2004). On the cut locus in Alexandrov spaces and applications to convex surfaces. *Pacific Journal of Mathematics, 217*, 375–386
Ziegler, G. (1995). *Lectures on polytopes*. Berlin/Heidelberg/New York: Springer

Chapter VIII
Polygons, polyhedra, polytopes

VIII.1. Introduction

The *polytopes* are, by definition, the convex envelopes of finite sets of points of an affine space. When this space is of dimension 2 (a plane), we speak of *polygons*; if the dimension is 3, we speak of *polyhedra*, and from then on — or from the very beginning — of *polytopes*. We are thus dealing with objects that are simplest after triangles. Now a detailed study of polyhedra is very recent. If we exclude the fundamental book of Steinitz from 1934 and his papers from between 1906 and 1928, we find practically nothing on polyhedra before the 1960s. We can read an interesting analysis of Steinitz's book in Tucker (1935–2000), but even though the analysis is very enthusiastic, the polyhedra are qualified as Steinitz's "hobby". Why this paradox? — or rather this disinterest, for it is clear, in view of the great progress in numerous areas of mathematics, that this absence of results can only be attributed to a disaffection for the subject. Grünbaum, in his book from 1967, which was for a long time the only reference on the subject, explains this disinterest. He gives two reasons. The first is the failure of Euler in classifying polytopes (polyhedra) in dimension 3, which led to the subject being considered too difficult. The second is the normative influence of Klein during a third of the twentieth Century, all the more important in that Germany was then the great mathematical center. This influence certainly contributed to essential progress in algebra, topology, analysis, but the study of polyhedra remained a curiosity for amateurs. The exodus from this stagnation — explained in the introduction of the preceding chapter — came under the strong impetus of computation, combinatorics, algorithmics and of linear programming.

We won't lack for simple objects, for finding recent results, for finding it necessary to climb the ladder, nor for problems that are simple to state but still remain open. We have made a personal choice in the problems broached and ultimately we can only very briefly mention the necessary concepts. In short, it is the recent progress in algebra, along with that of combinatorics, that now make it possible to better understand polytopes.

We will also see that the apparent simplicity of polytopes is deceptive and creates a trap into which numerous leading mathematicians have fallen.

General references for this chapter on polytopes are Grünbaum (1993a), the more elementary (Bronsted, 1983) and the remarkable (Ziegler, 1995), which contains inspired notes at the end of each chapter. This last reference is certainly the best access to the methods and to recent concepts that have allowed for spectacular progress in the theory of polytopes. We find numerous partial references at the

M. Berger, *Geometry Revealed*, DOI 10.1007/978-3-540-70997-8_8,
© Springer-Verlag Berlin Heidelberg 2010

beginning of "lecture 0" of Ziegler and of course references are provided us along the way. To this prophetic genre we can add the millennium texts, such as Stanley (2000), Ziegler (2001) and [B], a reference that is partial and more elementary, but used often in our work here. Finally the recent Cromwell (1997) is concerned exclusively with polyhedra, i.e. with polytopes in dimension three; it is very detailed, very pedagogical and contains much remarkable historical information.

VIII.2. Basic notions

A *polytope* is by definition the convex envelope of a finite number of points of the affine space considered. In certain cases we will be placed in a Euclidean space, e.g. when we are dealing with the area of the boundary (or of the various faces or the lengths of edges), with rigidity, or with inscribing a polytope in a sphere. In dimension 1 we find closed segments, in dimension 2 polygons and in dimension 3 polyhedra. A basic result is the equivalence between this definition and the following characterization:

(VIII.2.1) *A polytope is an intersection of a finite number of closed half-spaces that is furthermore bounded (and thus compact).*

It is typically this second definition that occurs in linear programming problems, where the initial data are linear inequalities that define the half-spaces.

These two definitions may seem trivially equivalent, yet this still needs to be proved. But that this obviousness is false can be seen when we have to deal with defining the *vertices* and the *faces* of our polytope. To convince ourselves of this we only need look at the sketches:

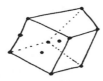 

Fig. VIII.2.1. Useless vertices (*at left*), useless half-planes (*at right*)

A first caveat is obvious from the figures: for things to be interesting, it is necessary that our polytope not be contained in any affine subspace. We could define the dimension of a polytope to be equal to the dimension of smallest affine subspace containing it, or else we could consider — **which we always do** — only polytopes whose dimension equals that of the space in which we work. This is equivalent to requiring that the interior of our polytope not be empty.

For faces as well as for vertices, we see that there may exist superfluous vertices and superfluous faces; make your own sketch in dimension 1. Points situated on the interior of faces must not be counted as true vertices; and there are also useless faces. We will not make these definitions truly precise, still less the equivalence of the definitions. The difficulty is not too great, yet should not be underestimated. The result is called the *the main theorem for polytopes* in Ziegler (1995) and is treated very carefully.

This having been achieved, we can define the *vertices* heuristically as the points that are absolutely necessary in order that their convex envelope be the given polytope. The *faces* will be the intersections of the polytope with the half-spaces that are absolutely necessary so that the intersection of these associated half-spaces be the given polytope. If we think of dimension 3, we see that it is also interesting to consider the *edges* of a polytope. More generally, in arbitrary dimension we define by recurrence the notion of k-face. In dimension d, without further qualification, the faces are by definition the $(d-1)$-*faces* and the vertices are the 0-*faces*. The $(d-2)$-*faces* are the faces of the $(d-1)$-faces, etc. So the edges are the 1-faces; in particular, in dimension 3, they are the faces of the faces. Note that, for each k, the k-faces are polytopes (of dimension k) in the affine space that supports them.

As is well noted on p. 29 of Ziegler (1995), the equivalence between the two definitions immediately implies that the intersection of two polytopes with an affine subspace, or with another polytope, is always a polytope (in the space considered); that the projection of a polytope is again a polytope; and that the Minkowski sum (defined in Sect. VII.5.B) of two polytopes is again one — properties which are not evident if we don't choose the right definition!

A notion that arises very often, because it often allows us to proceed by recurrence on the dimension, is that of the *star of a vertex*. Roughly speaking, the star of a vertex of a polytope of dimension d is the set constituted by the faces of the polyhedron that contain that vertex. More precisely, we can slice the polyhedron by hyperplanes that are sufficiently close to the vertex and of well chosen direction. Then, the star (in fact the whole polytope) is sliced by such a hyperplane in a polytope of dimension $d-1$. The combinatorial type is independent of the section; and it's this type that we can call the star of the vertex in question. In the Euclidean case we can also slice by spheres centered at the vertex studied; the star then becomes a *spherical polytope* (of dimension one less). In the particular case of regular polytopes, the star will itself be a regular polytope, made up of the vertices connected by an edge to the vertex chosen; see Sect. VIII.11. In dimension 3, the star will thus be a regular polygon.

The notion of *polarity* (or *duality*), seen for general convex sets in Sect. VII.5.C, can be restricted to polytopes without difficulty. We place ourselves in a Euclidean space $\mathbb{E}^d$ and consider polarity with respect to the unit sphere centered at an interior point P. It is then easy to see that the polar P^0 is again a polytope, called the *polar*, or *dual*, of P; and we see that polarity exchanges vertices and faces, more generally the k-faces with the $(d-k-1)$-faces. For figures, see Sect. VII.5.

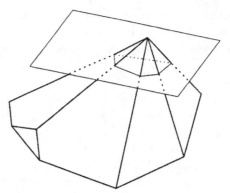

Fig. VIII.2.2. The star of a vertex

VIII.3. Polygons

In way of introduction we review some elementary problems for polygons that we will subsequently study for polyhedra and polytopes. This has to do with: *combinatorics, regularity, rigidity, isoperimetry, rationality, inscribability*. Certain problems will be treated together either because they are very simple, or else simply not to burden the text. Keep in mind that unless otherwise indicated we are working in the context of convex polygons.

The *combinatorics* of polygons are trivial: for each polygon, the number f_0 of vertices equals the number of *sides* f_1 (we may say side instead of edge), and for each integer n there exist polygons with $f_0 = f_1 = n$. The classification is thus complete.

Moreover — in the Euclidean setting $\mathbb{E}^2$ — for each integer n there exists a *regular polygon* with n sides and it is unique within a similitude, whence the language: "*the* regular polygon with n sides". We encounter these polygons in everyday life, and scarcely a year goes by without our seeing a new integer n appear on a car or some other manufactured object, not to speak of the plastic arts. We very often find the values 3, 4, 5, 6, 10, 12, 20, etc. The values 7 and 9 are found nowadays on automobile wheel covers (hub caps).

There also exist regular star polygons and we denote them by fractions $\{\frac{p}{q}\}$. See the above figures; we have also already encountered them in Sect. II.6.

But haven't we forgotten to define a regular polygon? To be sure, but the response is treacherous: it is that all the sides (their lengths more precisely) are equal and all the angles at the vertices are equal. The perfidy lies in the fact that this definition extends badly to higher dimensions; even the bible (Coxeter, 1973) doesn't give any explicit definition that isn't subtly hidden. In what follows we prefer to proceed thus: we observe that a regular polygon possesses a good group of *isometries*, i.e.

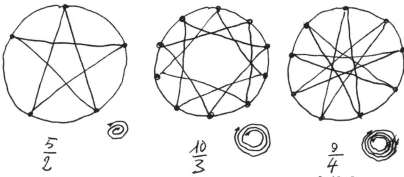

$$\frac{5}{2} \qquad\qquad \frac{10}{3} \qquad\qquad \frac{9}{4}$$

Fig. VIII.3.1. Regular star polygons, denoted respectively $\frac{5}{2}$, $\frac{10}{3}$, $\frac{9}{4}$

isometries of $\mathbb{E}^2$ which leave it globally invariant: the dihedral group of order $2n$. In fact, we can do better: the group of isometries preserving the regular polygon is *transitive* on the *flags* of the polygon. We recall that a flag of a polygon is a pair formed by a vertex and an edge issuing from that vertex, transitive meaning here that each flag can be sent into any other flag by an isometry in the group. We note finally that each regular polygon is both *inscribable* in a circle and *circumscribable* about another (see also Sects. VIII.5 and VIII.11 below).

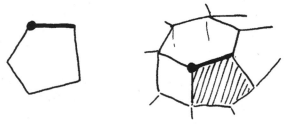

Fig. VIII.3.2. A flag in a polygon (two elements), a flag in a polyhedron (three elements)

In passing, we remark on the following: we occasionally read that it is known exactly for which integers n the regular polygon of n sides is constructible with ruler and compass. For what follows, see the very detailed Carrega (1981), Stewart (1973) or the interesting Gottlieb (1999) and the references mentioned there, completed by pp. 4, 5 of the Mathematical Intelligencer from 1999, Vol. 3. Note also that 12.4.6 of [B] contains some inaccuracies. We know how to construct regular polygons with n sides with ruler and compass when $n = ab$ when a, b are relatively prime and the construction is possible for a and for b. The construction is impossible when $n = a^2$, a an odd prime. As for the remaining cases, when a is an odd prime the construction is possible if and only if a is a Fermat number, i.e. it has the form $F(k) = 2^{2^k} + 1$. Unfortunately we don't know these Fermat prime numbers; we know that 3, 5, 17, 257, 65 537 are prime, and that F(5) is divisible by 641. For more on these numbers, see Hardy and Wright (1938, pp. 14, 15). But within knowledge of the Fermat primes, the problem is resolved. The preceding is due to Gauss, completed by Wantzel.

But mathematicians are never finished: we find in (Bainville & Genevés, 2000) the theory and examples of regular polygons constructible with the aid of conics in addition to ruler and compass. We are given the right to construct a point in the intersection of two conics already constructed, which amounts to allowing solutions of equations of third degree (with constructible coefficients), or third roots, as constructible numbers (coordinates of constructible points); see Sect. IV.6.

♦

The problem of the *rigidity* of Euclidean polygons consists of studying the possible shapes of a polygon for which the number of sides n is fixed, along with the lengths of these sides. Studying the possible shapes means trying to classify these shapes modulo the group of plane isometries. For triangles ($n = 3$) there is uniqueness, but starting with $n = 4$ it is clear that deformations are possible. It is here that it is wise to allow polygons the possibility of ceasing to be convex. We fix two vertices, which is always possible by isometry; then an articulated quadrilateral has a form that depends on exactly one parameter. More generally, we can speak of an *articulated plane system*. We encountered these articulated quadrilaterals in Sect. II.3 − inverters that can be used, at least in theory − see Sect. V.7 for the practice, both industrial and metrological − for tracing parts of (straight) lines, as in Sect. IV.8. An elementary exposition, with numerous drawings, is Rideau (1989). See the details at the end of Sect. V.15 and, for articulated polygons in space, the very conceptual references Kapovich and Millson (1996) and Foth and Lozano (2004).

For those who like open problems, we point out the following: we take an (articulated) polygon in the plane having an arbitrary number of sides. Can we deform it into a convex polygon in such a way that each side remains constantly of the same length and there are never any self-intersections throughout the deformation?

It seems it was in (Darboux, 1879) that for the first time a detailed study was undertaken of plane quadrilaterals for which the lengths of the four sides are given. Darboux shows that their totality is naturally an elliptic curve. The foldings about the diagonals are the involutions of this curve (cf. the same philosophy as for the Poncelet polygons in Sect. IV.8). What happens in space with these quadrilaterals when we perform successive foldings about the two diagonals leads to a dynamic and to very subtle questions in the theory of dynamical systems at the present highest level. These questions have gotten good answers in Benoist and Hulin (2004).

Another iterative operation for a polygon consists of joining the vertices two at a time; this again yields, on the interior of the initial polygon, a new polygon having the same number of vertices. What is the shape of the polygons of the sequence obtained by iterating this operation indefinitely? We have at the present time only partial answers − except for pentagons − and a conjecture in the spirit of completely integrable systems. The point of the problem can be found in Schwartz (1992, 2001), to be possibly completed by looking at his website at:

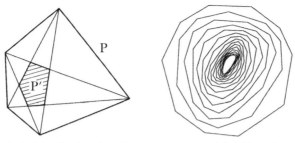

Fig. VIII.3.3. *At left*: first iteration for a pentagon, *at right*: fifteen iterations for a decagon

www.math.brown.edu/~res/. See also an introduction to the work of Richard Schwartz, which appears in portions of (Berger, 2005).

The awaited *isoperimetric inequality* for polygons is:

> *For all (convex) polygons with n sides we have $\frac{L^2}{A} \geq 4n \operatorname{tg} \frac{\pi}{n}$. Moreover this minimal value is attained only for regular polygons.*

In Sect. VII.2 we saw the special cases where the polygons considered had to be inscribed in, or circumscribed about, a circle. In the general case considered here, we need to do more; we can't be content with what was done in Sect. VII.2, nor in the opposite direction with the isoperimetric inequality for curves. But whereas we are working with the set of shapes of polygons of a fixed number of sides, and whereas this set is compact (of finite dimension) and the function $\frac{L^2}{A}$ is continuous there, we know that a minimum is attained by at least one polygon P. We can then "assemble by hand" several proofs, for example by showing first that the sides of P are equal — for otherwise we would increase the perimeter while keeping the area constant (see the figure below).

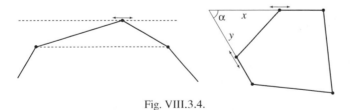

Fig. VIII.3.4.

We can also independently show that all the angles are equal by proving (left to readers) that the area of a quadrilateral with fixed base and three equal sides is maximum when it is a trapezoid. Therefore P has all its sides and angles equal, and is thus regular. Another way of completing the proof is by utilizing the inscribability in a circle of the polygon with given sides which has maximal area; see below.

A more conceptual proof is therefore required, better yet an estimate of the deficit. We find this in Fejes Tóth (1972); one of the proofs consists of filling the interior of the polygon with polygons at constant distances from the boundary:

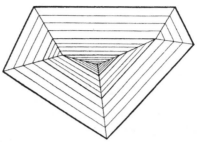

Fig. VIII.3.5. Proof of the isoperimetric inequality by the *filling* method

This filling is not in fact truly different from that of Gromov seen in Sect. VII.14. But here we proceed by remarking that, at least at the start, the **parallel** polygons keep the same number of sides and thus, by homothety, $\frac{L^2}{A}$ remains constant. In fact, subsequently we eventually lose sides, but $\frac{L^2}{A}$ still remains constant and ultimately is greater than the value for a triangle. As $4n \operatorname{tg} \frac{\pi}{n}$ decreases with n, we are done. But it can also be remarked – see Fig. VIII.3.3(b) – that this ratio is worse than for the polygon circumscribed about a circle and whose sides are parallel to those of P. Now for such circumscribed polygons the proof was given in Sect. VII.2.

We mustn't believe that we have finished with problems of isoperimetric type for polygons. The debate was revived in 1966 by Paul Lévy, see Chouika (1999). His problem is to find a formula for the maximum area of a polygon whose side lengths are given. Heron's formula $A = \sqrt{p(p-a)(p-b)(p-c)}$, where $p = \frac{L}{2}$, gives the exact value for triangles; and for quadrilaterals – but in addition inscribable in a circle – we have the formula of Brahmagupta: $A = \sqrt{(p-a)(p-b)(p-c)(p-d)}$, where again $p = \frac{L}{2}$.

Now the first thing to do is show that the polygon of maximum area is inscribable in a circle. Existence comes from compactness, and inscribability is easy: it suffices to show it for a quadrilateral. We write that the derivative of the area $ab.\sin\alpha + cd.\sin\beta$ is zero under the condition $a^2 + b^2 - 2ab\cos\alpha = c^2 + d^2 - 2cd\cos\beta$, which implies $\alpha + \beta = \pi$. We remark in passing, with Fig. II.2.13 in view, that this proves that each (articulated) polygon for which the sides are given is inscribable in a circle.

Paul Lévy's conjecture – still open – concerns the ratios between the (maximum) area A and the general expression (which is no longer exact starting with more than four sides): $\frac{L^2}{A} = (1 - \frac{2a_1}{L}) \dots (1 - \frac{2a_n}{L})$, where L denotes the perimeter and where the a_i are the side lengths. A reference for this problem is Chouika (1999).

The difficulty lies in knowing what replaces the above formulas of Héron and of Brahmagupta. Thus, for all that, dimension two and compactness have not rendered our task trivial; and this augurs poorly for the case of polyhedra, and our fears unfortunately will be seen to be fully justified in Sect. VIII.7.

The *rationality* problems for polytopes have natural motivations. We can want all the coordinates of the vertices to be rational numbers − or even integers − for reasons of practical approximation or because computers know only pixels. Thus let a polygon with n sides be given: can we always inscribe it in the lattice $\mathbb{Z}^2 \subset \mathbb{R}^2$ composed of points with integral coordinates? At first we require only that there exist a convex polygon with n sides (there is but one combinatorial type) with all its vertices in $\mathbb{Z}^2$. The response *yes* is evident, but the problem naturally gets refined: as a function of n, what is the minimum size of our polygon inscribed in $\mathbb{Z}^2$? If $F(n)$ is this minimum, i.e. if we can always inscribe a convex polygon with n sides in $[0, 1, 2, \ldots, F(n)]^2$, we can prove without too much trouble that $F(n)$ is of the order of $n^{3/2}$. In fact the final response dates from 1981: $F(n) = 2\pi(\frac{n}{12})^{3/2} + O(n \log n)$; see p. 122 of Ziegler (1995). We dedicate all of Sect. VIII.9 to what happens regarding the analogous question for polyhedra and polytopes.

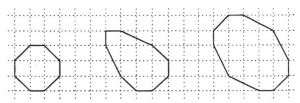

Fig. VIII.3.6.

VIII.4. Polyhedra: combinatorics

We have seen that polygons have only one combinatorial type, in fact one for each integer n, the number of sides or − what is the same − the number of vertices.

We let f_i denote the number of i-faces ($i = 0, 1, 2$) of a polyhedron P in $\mathbb{R}^3$. There are two essential things to see right away. The first is that these three numbers are not independent, because we have for each convex polyhedron:

(VIII.4.1) *Euler's formula:* $f_0 - f_1 + f_2 = 2$.

The second is that the triple $\{f_0, f_1, f_2\}$ doesn't determine the combinatoric of the polyhedron, as the figure shows:

Here the triple is $\{6, 12, 8\}$ for each of the two figures. We concern ourselves first with (VIII.4.1); historians have recently discovered that this formula was already present − although well hidden − in the work of Descartes (1680), considerably before Euler (1752); see Hauchecorne and Suratteau (1996); but above all read a

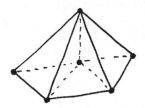 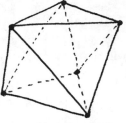

Fig. VIII.4.1. These two polyhedra don't have the same combinatoric even though they
have the same number of vertices, edges and faces: $f_0 = 6$, $f_1 = 12$, $f_2 = 8$

detailed and impassioned analysis of this history, in connection moreover with other
histories of successive incomplete proofs, in the important book (Lakatos, 1984).

A classic proof consists of placing ourselves on the interior of the polyhedron
at O and from there projecting the polyhedron onto the unit sphere with center
O (we will have introduced a Euclidean structure in the process). We obtain a
decomposition of the sphere into spherical polygons and we use the formula of
Harriot-Girard from Sect. III.1 for calculating the area of the sphere in two ways.
On the one hand it equals 4π in total; but if a face F_j has k sides it has area
$\text{Area}(F_j) = (\sum_i A_{j;i}) - (k - 2)\pi$, where the $A_{j;i}$ are the vertex angles of the
spherical polygon F_j. We take the sum of all these areas for all the faces, obtaining
the total area 4π of the sphere. But in the sum $\sum_j \text{Area}(F_j) = 4\pi$, which becomes
a double sum in i and j, we change the order of summation, yielding a summation
over the vertices. Now, at each vertex, the total sum of the angles of the faces that
contain it is equal to 2π and we find successively $4\pi = 2\pi f_0 - \sum_k \pi(k - 2)$ (number of faces having k sides) $= 2\pi f_0 - 4\pi f_1 + 2\pi f_2$, since $\sum_k k$ (number of faces
having k sides) $= 2f_1$, this because each edge belongs to two faces, QED.

We can't but reproach the strong use of Euclidean geometry in the proof of this
formula, which is purely combinatorial (or say, if need be, affine). In fact, there
doesn't exist a very simple non-Euclidean proof. The history of the proofs, as well
as of generalization to higher dimensions, is very interesting; for more details than
we will give, see pp. 141,142 of Grünbaum (1967) (but the *first* edition, in Chap. 13).
For a complete proof both of shellability and of Euler's formula, see Lecture 8 of
Ziegler (1995).

If we want an affine or even combinatorial proof we might think of this one,
which is essentially a recurrence proof, since $f_0 = f_1$ for plane polygons. We start
with a vertex and the faces that contain it; then the quantity $f_0 - f_1 + f_2$ equals 1
for the object obtained. Now we attach a new face; we can consider it evident (by
examining the new boundary) that the quantity $f_0 - f_1 + f_2$ remains constant (and
thus equal to 1). We continue to attach faces in this way until there remains only the
last face; this time the quantity $f_0 - f_1 + f_2$ increases by 1 and thus finally equals
2, QED.

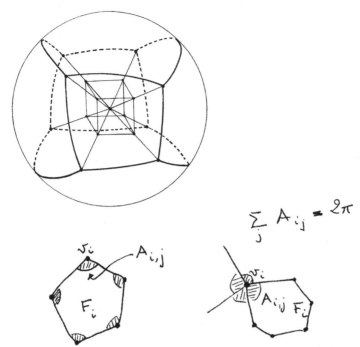

Fig. VIII.4.2. The classic proof of Euler's formula with the help of spherical geometry

This proof extends to any dimension and has been used by numerous mathematicians. It is consoling to know that, if all their proofs were insufficient − even in ordinary dimension 3 − this was discovered only in Bruggesser and Mali (1971); see the fascinating historical details in the book (Lakatos, 1984). We have used the fact that the polyhedron can be obtained by adjunction of successive faces, so that at each stage the new face is attached to the boundary of the preceding set along a polygonal line that never closes until the very last stage (simple connectivity, see the end of Sect. V.XYZ). In dimension 3, the Jordan theorem (see Sect. V.1) allows us to finish and is necessary for this; in higher dimensions we must use a result that we need to prove, whose technical statement is that each polytope is *shellable*. To realize that the difficulty exists, we can think of generating a dodecahedron by successively appending pentagons (thus making a *circuit* of the faces), but in a bad way (readers can make a drawing, taking inspiration from Fig. VIII.4.3); we can't be satisfied with adjoining new faces in merely arbitrary ways.

In fact we can dispense with shellability: the first correct proof for all dimensions that is combinatorial and elementary in nature is Hadwiger's from 1955. There is the prior correct proof due to Poincaré in 1899, but it is much more difficult and deals with a result that is valid for all compact manifolds specialized to the case of the boundary of a polytope, which is always topologically a sphere (cf. Sect. VII.10.B). Nowadays Euler's relation is proved in an essentially elementary way by recurrence. Here is the outline for polyhedra (but it extends to any dimension *mutatis mutandis*).

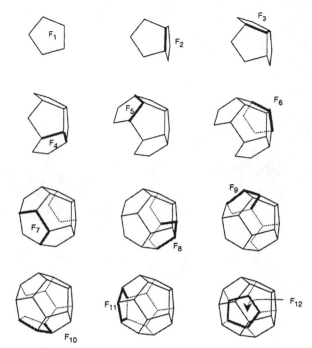

Fig. VIII.4.3. How to shell a regular dodecahedron

We cut the polyhedron by parallel successive planes, choosing their direction in such a way so as to never encounter more than one vertex. We examine the behavior of the function $f_0 - f_1 + f_2$ associated with the portion situated above the sectioning plane. At the highest vertex this function equals 1, then at the beginning (as we lower the sectioning plane) it remains 1 because (the recurrence hypothesis is trivial here) vertices and sides are of equal number in a polygon. We see then that, when we pass a vertex, $f_0 - f_1 + f_2$ doesn't change and always equals 1. When we have reached the very bottom, there is one more vertex to adjoin and thus finally $f_0 - f_1 + f_2 = 2$.

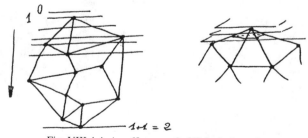

Fig. VIII.4.4. An affine proof of Euler's formula

The applications of Euler's formula are legion. We encountered it for good arrangements of points on the sphere in Sect. III.3; we will make use of it again in

Sect. VIII.5 for quickly determining all the regular polyhedra; finally it is fundamental for studying Cauchy rigidity in Sect. VIII.6. We have also used it for golf balls in Sect. III.3. There exists a formula for general surfaces; see the end of Sect. VI.1.

The possibility of shelling comes up in an essential way in the proofs of results that we merely mention in Sects. VIII.10 and VIII.12.

♦

Formula (VIII.4.1) is a necessary condition for the triple $\{f_0, f_1, f_2\}$ of a polyhedron in space. What about sufficient conditions, specifically for constructing a polyhedron having a given triple. The answer has been known since 1906 (Steinitz): it is necessary and sufficient that we have

(VIII.4.2) $\qquad 4 \leqslant f_0 \leqslant 2f_2 - 4, \quad 4 \leqslant f_2 \leqslant 2f_0 - 4, \quad f_0 - f_1 + f_2 = 2$

(in fact, either one of the first two relations will, along with the third, suffice); see below for their geometric aspect. As for proving (Sect. VIII.4.2), we need to ascend the ladder for a good understanding of the combinatorics of polyhedra. The concept used by Steinitz for his proof is that of the *graph associated with a polyhedron* (readers will be able to "see" it), specifically the one composed by the vertices and the edges. The fundamental result of Steinitz is that a graph will be associated with a polyhedron in the space $\mathbb{R}^3$ if and only if it is *planar and 3-connected*, i.e. that the graph can be drawn in a plane without intersection of edges, and that any two vertices can be joined by at least three different paths that intersect only at their extremities. We will meet this result of Steinitz again in Sect. VIII.9 for the rationality of polyhedra.

We reread the result of Steinitz thus: to each polyhedron P we attach its f-*vector* $f(P) = \{f_0(P), f_1(P), f_2(P)\}$, which we think of as being in $\mathbb{Z}^3 \subset \mathbb{R}^3$. The problem is then of knowing the nature of the set $f(\mathcal{P}^d)$ formed by the f-vectors of all polyhedra. Euler's formula says first of all that it is situated in an affine plane, the one with equation $x - y + z = 2$.

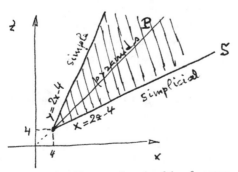

Fig. VIII.4.5. The exact domain of the f-vectors

The Steinitz conditions show that, in this plane, $f(\mathcal{P}^d)$ is precisely the intersection with $\mathbb{Z}^3$ of the convex set given by the two inequalities $x \leqslant 2z - 4$ and

$z \leqslant 2x - 4$. We have stated that the necessity is difficult; on the other hand, existence is simpler and left to readers; it can be seen well in the figure, where the letter P indicates the pyramids (for which $f_0 = f_2$ and the letter S indicates the polyhedra for which all the faces are triangles (called *simplicial* polyhedra) or else exactly three edges emanate from each vertex (called *simple* polyhedra), i.e. the polar is simplicial. Their analogues in higher dimensions will play an essential role in Sect. VIII.10. Readers will have observed that the symmetry about the bisector gives the polarity. The tetrahedron is the only case that is both simple and simplicial. The cube and the dodecahedron are simple; the octahedron and the icosahedron are simplicial.

But still today we don't know how to completely characterize the sequences $\{\Phi_3, \Phi_4, \ldots\}$, where each Φ_i denotes the number of faces of the polytope to be found that are polygons with i sides (here edges). But we know a lot of things; see for example 13.3 of Grünbaum (1967) or Grünbaum (1993a), or finally the end of 2.3 of Gruber and Wills (1993).

As pleasing as the preceding was, it is also rather primitive and avoided the fundamental definition: two polyhedra have the *same combinatorial type* if there exists a triple of bijections between their vertices, their edges and their faces, respectively, that preserve the membership relations between the vertices, edges and faces. We recall that Euler did not succeed in classifying the combinatorial types of polyhedra, even for the weak values of the number of vertices. We now know why: it is a difficult problem of combinatorics and has to do with classifying 3-connected planar graphs. Presently we know a rough estimate of the number of combinatorial types possible for a polyhedron with n vertices, specifically the order of $an^{-7/2}b^{n-1}$ as n tends toward infinity, where a and b are constants. If we know the numbers $n + 1$ of vertices and $f + 1$ of faces, then we know the asymptotic value of the number of possible combinatorial types, specifically $\frac{1}{972ij(i+j)}C_{j+3}^{2i}C_{i+3}^{2j}$; see Theorem 6.2 of Chap. 2.3 of Gruber and Wills (1993) for this result, the culminating point of numerous works by various authors. Moreover, the problem of determining the computational complexity of these types is open: is it or is it not an NP-complete problem? For an elementary exposition of algorithmic complexity, see Damphousse (2005).

VIII.5. Regular Euclidean polyhedra

The five regular polyhedra of our space of three dimensions have fascinated generations; see Cromwell (1997), a book entirely dedicated to polyhedra in $\mathbb{R}^3$, very full of simple and pleasant things, including historical information. We mention only what is apparently a rather recent development: the notion that the study of regular polyhedra would have been one of Euclid's motives for producing his *Elements*.

We repeat that the definition of regular polyhedra in $\mathbb{E}^3$ is deceptive in its simplicity. We demand of course that all the faces be identical regular polygons, but this is inadequate; readers will find their own counterexamples. It is moreover required that the same number of faces meet at each vertex. It can be shown that this is a good definition, but the proof is bizarre and it is left to readers to see its artificial nature. We proceed much faster by requiring, in addition to the first condition, that the *dihedral* angles, i.e. those formed by pairs of faces along a common edge, are all equal. Certainly this condition has the appearance of being very strong, perhaps even too strong; but then we see that the group of isometries is indeed transitive on the flags, as was required in Sect. VIII.3. Here the *flags* are triples formed by a vertex, an edge that emanates from this vertex and a face that contains this edge. In general it does not suffice that the group of isometries be transitive on the vertices; readers can produce counterexamples. Above all, the definition with transitivity on flags is the only one that generalizes immediately, without any complication, to arbitrary dimension; see Sect. VIII.11.

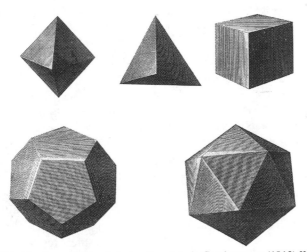

Fig. VIII.5.1. The five regular polyhedra. Rouché, de Comberousse (1912) With Kind Permission of Elsevier. Public domain

♦

We look first very quickly at why there are only a finite number of regular polyhedra, in contrast to the case of polygons. If each face has k vertices, each vertex h edges, then the Euler relation tells us:

$$\frac{1}{k} + \frac{1}{h} > \frac{1}{2},$$

since $f_1 = \frac{kf_2}{2} = \frac{hf_0}{2}$. As both k and h are greater than 2, we see that the only possible *pairs* are $\{3, 3\}$, $\{3, 4\}$, $\{4, 3\}$, $\{3, 5\}$, $\{5, 3\}$. The existence will be seen below. We note that this "classification" is purely combinatorial, the Euclidean regularity didn't enter. But this method, resting on Euler's formula, will be completely inadequate in higher dimensions. We will meet it again in Sect. VIII.11 in a truly natural way, by a Euclidean technique that works in all dimensions, including dimension 3.

◆

Before showing the existence and uniqueness of the five regular polyhedra in $\mathbb{E}^3$, a few immediate remarks, valid without change in any dimension. First, the flag transitivity implies that a regular polyhedron is always inscribed in a sphere. In fact the center of gravity of this compact set is fixed by every isometry, and thus all the vertices are an equal distance from it. The polar of a regular polyhedron is again a regular polyhedron. Thus we only need study the pairs $\{3, 3\}$, $\{4, 3\}$, $\{5, 3\}$. Finally — thanks to polarity — there will be **uniqueness** within similitude, the pair being given. We then note that a little spherical trigonometry will show that the dihedral angles are determined by the pair. With an appropriate similitude we can bring the face of one onto the face of the other. The dihedral angles being equal, we can reconstruct the polyhedron face by face. This remark will be used below for rigidity: a bijection between two polyhedra which is a face-by-face isometry and preserves all the corresponding dihedral angles can be extended to a global isometry.

◆

It remains to see the **existence** of a polyhedron corresponding to each of the five (three) pairs possible. For $\{3, 3\}$ the *regular tetrahedron*, trivial to construct, answers the question. For $\{4, 3\}$ the *cube* answers the question, and by polarity, also for $\{3, 4\}$, which is the *regular octahedron*. The one existing for $\{5, 3\}$, called the *regular dodecahedron*, is a classical pitfall. In fact, naively, we can take a collection of 12 equal regular pentagons, since k and h determine f_1. Three can be assembled in a unique way (because a spherical triangle is known if the length of its three sides is known) and we proceed step by step; everything closes up nicely as we know by looking at various public objects. Unfortunately, the mathematician wants a proof that everything closes up nicely. This is left to readers; there are two different proofs in [B], one for $\{5, 3\}$ based on stacking the faces on an appropriate cube; and one for $\{3, 5\}$, the *regular icosahedron*. We may choose whichever we prefer or find a possibly new one for ourselves. In conclusion:

> *Within similarity, there exist exactly five regular polyhedra in the Euclidean space of dimension 3.*

We find in the bible (Coxeter, 1973) all the numerical values of lengths, angles, etc. that are wanted. They are not very mysterious to calculate, however, using the formulas of spherical trigonometry (see Sect. III.1).

The figure above is intended to illustrate the correct but conceptual proof of this delicate existence; it is due — to the best of our knowledge — to Milnor. Such proofs are sometimes deprecatingly referred to as "general nonsense", which is paradoxical since they can, as here, involve conceptual ascent. If the regular icosahedron exists, then we can project it onto the sphere and see what happens. Each face becomes a spherical triangle for which each of the three angles at a vertex equals 72 °: this determines such a triangle. Then we make an abstract surface $\sum$ (manifold of dimension 2, see Sect. V.XYZ) by gluing abstractly 20 such triangles in accordance

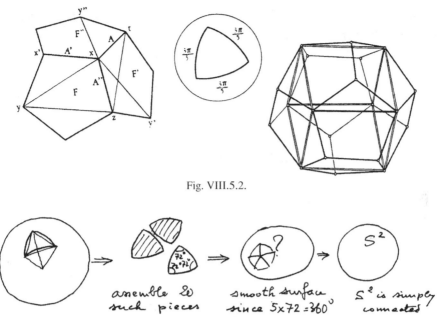

Fig. VIII.5.2.

Fig. VIII.5.3. "*Milnor general nonsense*": assemble 20 identical pieces to obtain a
smooth surface because $5 \times 72° = 360°$; S^2 is simply connected

with the combinatorics of the icosahedron. We obtain a smooth surface because, at
each vertex, the five triangles have an angle sum that is equal to $5 \times 72° = 360°$.
We define a mapping of $\sum$ onto the sphere S^2 in an obvious fashion; by construction
this is a covering (see Sect. V.XYZ). But S^2 is simply connected, thus the covering
is *a fortiori* the identity! $-$ and $\sum$ is the regular icosahedron we sought.

♦

Having defined a regular polyhedron by a transitivity property of its group of isometries, it is natural to ask about the structure of this group, which is the same both for a
given polyhedron and its polar. For $\{3, 3\}$ (the regular tetrahedron) it's the group $\mathcal{S}_4$
of all permutations on four objects, a so-called *symmetric group*. For the cube and
the regular octahedron, it's a group with 24 elements, obtained by composing the 6
coordinate permutations with their 8 *changes of sign* of the coordinates. For $\{5, 3\}$
and $\{3, 5\}$, the answer harbors a little trap. We show first that the subgroup of their
isometries which preserve orientation $-$ the *rotations* $-$ is isomorphic to the *alternating group* $\mathcal{A}_5$, i.e. to the group of even permutations of five objects. To see this,
we note that there are five orthonormal trihedrons formed by the lines joining the
center to the midpoints of the edges. It is then easy to see that our group is exactly
the group $\mathcal{A}_5$ of these five trihedra. In contrast, the total group is not isomorphic to
the symmetric group $\mathcal{S}_5$; it has indeed 120 elements, but this group is the product
$\mathbb{Z}^2 \times \mathcal{A}_5$, and this is not $\mathcal{S}_5$! Flag enthusiasts will note that the isometry group of
a regular polyhedron is *simply transitive* on the flags, i.e. that there exists a unique

element of this group sending a flag onto another and that the number of flags is indeed the order of the group.

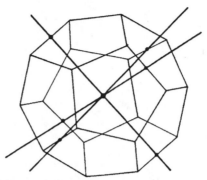

Fig. VIII.5.4. How to find orthonormal axes in a regular dodecahedron

The celebrated text (Klein, 2003) uses the geometry of $\{3, 5\}$ to study the general equation of fifth degree; see a recent exposition in Hirzebruch (1987b, pp. 656–661), and again Arnold (1990), which is a text for those who appreciate the unity of mathematics. The enthusiast for universality or unexpected encounters in mathematics should look at the table on p. 410 of Arnold (2000).

There exist four regular *star* polyhedra, called the Kepler-Poinsot solids; the definition is left to the reader (it is Poinsot who discovered all four, but it is Cauchy who proved that there don't exist others). Here they are:

The generalized symbols of Schläfli type use fractional notation, as is used for the regular polygons, some of which we have drawn in Fig. VIII.3.1. The symbols for them are $\{\frac{5}{2}, 5\}$, $\{5, \frac{5}{2}\}$, $\{\frac{5}{2}, 3\}$, $\{3, \frac{5}{2}\}$, grouped by polarity, which simply reverses the order of the symbols. Readers can complete the definition and the classification for themselves; see as needed (Coxeter, 1973). It is very interesting that two among them do not have the topology of the sphere, but that of a surface with four holes, as we see by applying Euler's formula for general compact oriented surfaces (see the end of Sect. VI.1): $f_0 - f_1 + f_2 = 2(1 - \#\text{ of holes})$. The topology here is that obtained by "unembedding" the polyhedron so as not to take into account its self-intersections. For $\{\frac{5}{2}, 5\}$ we have in fact $f_0 = f_2 = 12$ and $A = 30$, whence the number of holes $= 4$; likewise for its polar. In contrast, we indeed find $f_0 - f_1 + f_2 = 2$ for the two others, which do in fact have the topology of the sphere.

The soccer ball shows strong "regularity": the isometry group is transitive on the vertices and on the edges, but not on the faces, which are of two different types.

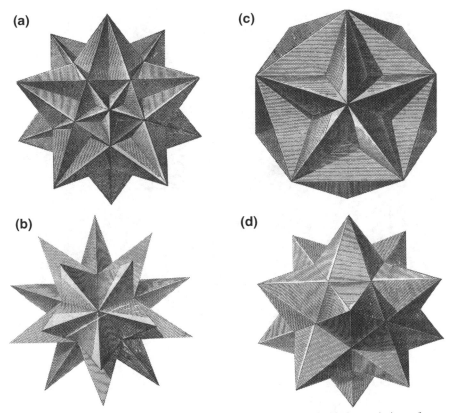

Fig. VIII.5.5. **(a,c)** Rouché, de Comberousse (1912) with kind permission of Elsevier, public domain; **(b,d)** Fischer (1986) © G. Fischer

It isn't easy to give a good definition of a *semi-regular polyhedron*; it is another trap which has snared many authors. It is also difficult not to forget them in the complete classification, especially the two *snub* polyhedra: the snub cube and the snub dodecahedron. For these two polyhedra the isometry groups contain only rotations and do not, moreover, have a center of symmetry. In 1.11 of Gruber and Wills (1993) there are numerous references for these polyhedra, some of which are called *Archimedean solids*. However, the classic books Coxeter (1973) and Fejes Tóth (1964) — incredibly — lack both clarity and details and do not even achieve the level of definitions. Apart from some articles and historical references, Gruber and Wills mention the books Cundy and Rollet (1961), Lyusternik (1963), Roman (1987), Cromwell (1997) is fascinating.

We have already encountered the snub cube in Sect. III.3, where it was characterized as furnishing the best distribution of 24 points on the sphere. We will further encounter the truncated octahedron as a Voronoi domain of the network A_3^* in Sect. X.1; see Fig. X.1.8.

Fig. VIII.5.6. Holden (1971) © Columbia University Press

VIII.6. Euclidean polyhedra: Cauchy rigidity and Alexandrov existence

Here again everything is Euclidean. We will amply see that, once the number of sides exceeds three, an articulated polygon is "flexible". In contrast, in space things are indeed different, for as early as 1813 Cauchy proved almost rigorously his *rigidity* result, as follows (as announced in fact by Lagrange, with nothing added). Here it is in modern language:

(VIII.6.1) *Let* P *and* P′ *be two polyhedra (convex as usual) and* f *a mapping between their boundaries that preserves combinatorics, i.e. sends faces onto faces while respecting incidence relations. Suppose that, restricted to each face,* f *is a Euclidean isometry. Then there exists an isometry* f* *of all space such that* f *is the restriction of* f* *to the boundary of* P.

In fact we are indeed dealing with a *congruence* theorem, the word congruence being used to designate objects that are isometric by way of a global isometry of the space. But congruence in particular implies *rigidity*, which is to say that there do not exist *deformations*, or *flexions*, of a polyhedron which are *articulated*, which says further that the faces, which are rigid pieces, remain intact but are allowed to turn about the edges that connect them. From the metric point of view, we can say that the condition on the mapping f is that it be an *isometry* between the boundaries, for their intrinsic metric, i.e. that of shortest paths traced on them; Roth (1981) is an introductory and very pedagogic text.

We can justifiably consider (VIII.6.1) as a uniqueness result. Is there a naturally associated existence theorem? Indeed so, see below.

We return to (VIII.6.1). The theorem is trivial if at each vertex only **three** faces meet, for then we have a trihedron for which the three face angles and the three dihedral angles are known: this because a spherical triangle is known once we know its three sides; see Sect. III.1. This is, for example, the case with the regular dodecahedron, cube, etc. In contrast, the fact that the regular icosahedron is rigid can only be proved via Cauchy. Recall − see Sect. VIII.4 − that polyhedra for which each vertex belongs to only three faces are called *simple*. The reason for this designation is that their polar is a polyhedron for which all the faces are triangles, which are called *simplicial*; we will see in Sect. VIII.10 why they are especially important and this in all dimensions.

We should not fail to compare (VIII.6.1) with the content of Sect. VI.9, where the same result is stated but where the polyhedral boundaries are replaced by smooth convex surfaces of class C^2. As in Sect. VI.9, Cauchy's result calls for remarks and questions. Convexity is here an absolutely necessary; see the figure below (to be compared with Fig. VI.9.1).

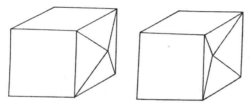

Fig. VIII.6.1. Two isometric but not congruent polyhedra, but one of them is not convex

In the context of Sect. VI.9 we saw that the obvious counterexample given by this figure leaves two problems open, for neither did it exhibit deformations nor are the surfaces seen there real analytic. Here things are doubly different, because first in Connelly (1978) there is shown a (non convex) **deformable** (only slightly, to be sure, but nonetheless truly deformable) polyhedron: the figure below presents a polyhedron due to Steffen that is easier to construct (although with cardboard it's not easy to carry out the final gluing). It seems to us that there still do not exist examples of polyhedra that are truly "very" flexible.

The second difference from the case of surfaces is that each isometric deformation of a (non convex) polyhedron preserves the volume: (Connelly, Sabitov, and Walz, 1997). The proof makes essential use of algebraic geometry. This is very natural, because having an articulated polyhedron is the same as having algebraic (polynomial) equations in the coordinates of the vertices, since distances are expressed by square roots of polynomials in the coordinates. It remains to show that the volume has the nature of an algebraic integer (i.e. a root of a polynomial with integer coefficients), and thus in particular cannot vary in a continuous fashion. A very good exposition of this can be found in Schlenker (2002).

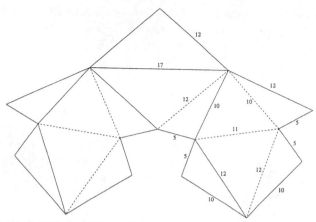

Fig. VIII.6.2. *Solid lines*: foldings of mountains; *dotted lines*: foldings of valleys. [B]
Géométrie. Nathan (1977, 1990) réimp. Cassini (2009) © Nathan Édition

◆

The proof of (VIII.6.1) does not use any new concepts, but it has some very nice geometry and is very different from the proof for surfaces. The first very careful proof
is likely that of Alexandrov (1958) (in Russian and also very recently in English
translation): (Alexandrov, 2005); but it functions in a much more general context. It
is found in some detail in [B], 12.8 and 18.7 (but less rigorously), is quickly treated
in Cromwell (1997), but Lyusternik (1963) likely remains the best reference. The
proof originally laid out by Cauchy includes some traps that we will mention in our
sketch of his proof, and is why we used the term "almost rigorously". It is necessary
to show that the dihedral angles along the edges are all equal. To do this, Cauchy
studied what happens during deformation of the *star* attached to a vertex, i.e. the set
formed by the faces that contain the vertex, for example five (equilateral) triangles
in the case of a regular icosahedron. We see and we feel that this set is deformable.
To know more about it we intersect it with a (sufficiently small) sphere, say the unit
sphere S^2: we obtain a spherical polygon whose side lengths are fixed and for which
the angles at the vertices are precisely the original dihedral angles considered. We
are thus led to study the pairs $\mathcal{P}$, $\mathcal{P}'$ of spherical polygons having their respective
sides equal in the correspondence that we are given between the two polytopes. We
attach a *sign* $+$, 0 or $-$ to a vertex of $\mathcal{P}$ according as the angle of $\mathcal{P}$ at this vertex is
greater, equal or smaller than the corresponding angle of $\mathcal{P}'$.
    The lemma used by Cauchy is:

(VIII.6.2) *If there is no sign 0, then as we move about $\mathcal{P}$ there are at least four sign
        changes.*

    Cauchy's idea was to prove this lemma by contradiction: if there are but two
sign changes, then we cut the polygon in two as in the figure, and the fact that
all the angles have increased for one demi-polygon and decreased for the other

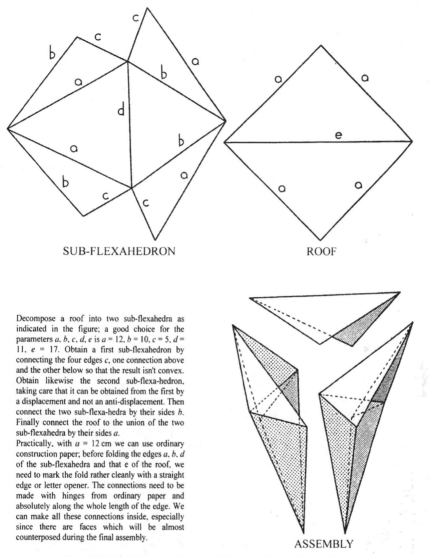

SUB-FLEXAHEDRON   ROOF

Decompose a roof into two sub-flexahedra as indicated in the figure; a good choice for the parameters a, b, c, d, e is $a = 12, b = 10, c = 5, d = 11, e = 17$. Obtain a first sub-flexahedron by connecting the four edges c, one connection above and the other below so that the result isn't convex. Obtain likewise the second sub-flexa-hedron, taking care that it can be obtained from the first by a displacement and not an anti-displacement. Then connect the two sub-flexa-hedra by their sides b. Finally connect the roof to the union of the two sub-flexahedra by their sides a.

Practically, with $a = 12$ cm we can use ordinary construction paper; before folding the edges a, b, d of the sub-flexahedra and that e of the roof, we need to mark the fold rather cleanly with a straight edge or letter opener. The connections need to be made with hinges from ordinary paper and absolutely along the whole length of the edge. We can make all these connections inside, especially since there are faces which will be almost counterposed during the final assembly.

ASSEMBLY

Fig. VIII.6.3. Following Klaus Steffen, drawing by Benoit Berger

demi-polygon is impossible. Cauchy "showed" by deformation that if all the angles of an open spherical polygon increase, then the length of the chord that closes it increases also. His proof was inadequate and was filled in by numerous authors, starting with Hadamard and Lebesgue ; see the references given, but above all Stoker (1968) for the history. We have already seen in Sect. III.4, for the kissing number, that spherical geometry is often full of pitfalls. Cauchy's error was in assuming that, in increasing an angle while keeping the sides constant, a spherical polygon will always remain convex.

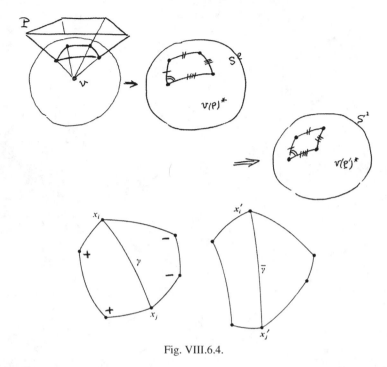

Fig. VIII.6.4.

We now complete the proof in the case where all the dihedral angles are unequal by showing that this is impossible. We evaluate in two ways the number v of pairs of edges of the polytope having a vertex in common and for which there is a change of sign for the comparison of the dihedral angles between the two polyhedra. On the one hand the Lemma VIII.6.2 says exactly that $v \geq 4s$, where s is the number of vertices. But on the other hand, in turning about the faces, we have trivially $v \leq 2\Phi_3 + 4\Phi_4 + 4\Phi_5 + 6\Phi_6 + 6\Phi_7 + \cdots$, where Φ_i denotes the number of faces having i vertices. We have thus:

$$4a - 4f = 2(3\Phi_3 + 4\Phi_4 + 5\Phi_5 + \cdots) - 4(\Phi_3 + \Phi_4 + \Phi_5 + \cdots)$$
$$= 2\Phi_3 + 4\Phi_4 + 6\Phi_5 + \cdots \geq v \geq 4s,$$

which contradicts Euler's formula $s - a + f = 2$.

Thus certain dihedral angles are necessarily equal. We give ourselves up to a rather subtle game: we remove from the boundary of the polyhedron P the edges having dihedral angles equal to the corresponding ones of P'. A graph remains, but we can apply once more the preceding method to this graph and find a contradiction once again. Finally all the dihedral angles must be equal; see the details in the references mentioned.

We do not treat rigidity for polytopes in higher dimension, because in fact it follows in all dimensions by recurrence from the case for dimension 3, as can be seen by regarding their stars. This is to say that, for a polytope in dimension 4, there is already rigidity about a single vertex, because its star is a polyhedron in S^3 and it is immediate that (VIII.6.1) extends to polyhedra in S^3. In dimension 3, there is only rigidity for stars having only three elements.

♦

Cauchy rigidity appears as a uniqueness result. But what is the corresponding existence theorem, and how do we state it and resolve it, if possible? The double merit accrues to Alexandrov; see Alexandrov (1955, 1958, 2005), Pogorelov (1973). If we think of the result cited in Sect. VI.5 apropos surfaces, we can thus formulate a natural existence problem of polyhedra realizing a given geometry. The surface geometry of a polyhedron doesn't pose any problem on the interior of the faces; it is locally Euclidean. But in fact the edges don't pose any problem either — they are easily traversed because we can "flatten" two adjacent faces; and thus finally, on the whole surface (i.e. boundary) of the polyhedron **less the vertices** $\{s_i\}$, we have a geometry that is everywhere locally Euclidean. As for the geometry around the vertices, it is locally that of a cone of revolution for which the total angle at the vertex s_i is equal to the quantity $curv(s_i) = \sum_k f_{k;i}$, where the $f_{k;i}$ designate the angles at the vertices s_i of the polygonal faces of the polyhedron that contains s_i. Finally, the geometry on the polyhedral surface is a metric geometry that is everywhere locally Euclidean on the sphere S^2, except at a finite number of points where the geometry is locally that of a cone of revolution with total angle at the vertex s_i is equal to the quantity $courb(s_i) = \sum_k f_{k;i}$, where the $f_{k;i}$ denote the angles at the vertices s_i of the polygonal faces of the polyhedron that contain s_i. Finally, the geometry on the polyhedral surface is metric and everywhere locally Euclidean on the sphere S^2, except at a finite number of points where the geometry is (locally) that of a cone of revolution with total angle at the vertex belonging to $]0, 2\pi[$, a condition that completely defines the geometries considered in the statement (VIII.6.3). The following fundamental result is due to Alexandrov:

(VIII.6.3) *Whatever geometry of the type just considered is defined on the sphere S^2, there exists a unique polyhedron in $\mathbb{R}^3$ that realizes it.*

We call this result fundamental because it yields an important number of generalizations in different contexts, in particular in hyperbolic geometry. We cite the very recent (Schlenker, 1998) and the basic reference (Pogorelov, 1973). In the union of these two works can be found all the necessary intermediary references. We finish with three observations.

The first is that, at the end of the proof, we constructed a polygonization of the sphere and we have a graph at our disposal whose vertices are the given s_i, the edges being now replaced by segments. The graph wasn't given in the hypothesis, where only the distances between the s_i mattered.

The second is that Alexandrov used (VIII.6.3), by a passage to the limit, for proving a weak version of the final theorem cited in Sect. VI.5 for surfaces, a theorem that is one of the major achievements of the geometry of the twentieth Century.

The third point consists of giving the basic scheme for a whole series of proofs of results of the same type. It consists of considering the mapping $f : \mathcal{E} \to \mathcal{F}$ from the space $\mathcal{E}$ of all geometries formed by polyhedral surfaces into the space $\mathcal{F}$ of all geometries on the sphere of the type described above. These spaces are of finite dimension, after we fix the number of vertices. We then show that f is a *proper* mapping (a very important concept already encountered in Sect. V.2), the technical term for stipulating that the inverse image of a compact set is compact. This result is consequent, heuristically, from the fact that a vertex cannot recede to infinity, the fixed geometry prevents it. Next the Cauchy theorem, seen as a non-deformability assertion, implies that this mapping is a covering. Classic topology shows that f is a surjective covering, and finally a bijection from (VIII.6.3). This scheme appears also in the proof of VIII.8.1. For use of polyhedra in physical materials where rigidity is wanted, see the very surprising Kanel-Belov, Dystein, Estrin, Pasternak, and Ivanov-Pogodaev (2004).

VIII.7. Isoperimetry for Euclidean polyhedra

Despite of the appearance of simplicity for polyhedra, the isoperimetric questions are far from being resolved for them. Section III.3 indicated that. In fact, there are several types of problems, depending on how the combinatorics are imposed, either the number of vertices, or faces or edges. See the Florian report in 1.6 of Gruber and Wills (1993). A first question of Steiner in 1842: *is it the case that, in their combinatorial class, the regular polyhedra are those having the best isoperimetric ratio?* It has to do with the ratio $\frac{A^3}{V^2}$ between the area of the boundary and the enclosed volume. The question is natural in the sense that the regular polyhedra play the role of the regular polygons.

Steiner's question has not been completely resolved, although many have attacked it. Contrary to the case of polygons, we can't proceed by compactness, because a limit will have to have a different combinatoric. Another difficulty (see the case of polygons) is the fact that not all the combinatorial types are inscribable (or circumscribable) in a sphere; see the following section. Still another difficulty (see Sect. VII.9) is that polarity, which can make it possible to treat just half the cases, does not respect volumes. And above all we lack a conceptual tool, even imagining that any exist. It is of course possible that one day we will be able to reduce isoperimetric problems to a computer program; but here is where we are today, to the best knowledge of the author.

Steiner's conjecture is true for the combinatorial types of the regular polyhedra, with the exception of the icosahedron, for which the problem still remains open. For the tetrahedron, it is classic (the proof is left to readers); for the octahedron Steiner proved it as early as 1842, by symmetrization.

♦

For the cube and the octahedron, it's a consequence of a general result of Fejes Tóth (1948):

(VIII.7.1) *For each polyhedron with f faces, we always have*

$$\frac{A^3}{V^2} \geq 54(f - 2)\,\mathrm{tg}(\omega_f)(4\sin^2\omega_f - 1),$$

where $\omega_f = \frac{\pi f}{6(f-2)}$, *equality holding only for the regular tetrahedron, cube and dodecahedron.*

The proof of (VIII.7.1) is difficult and was preceded by several incomplete proofs. But it is also interesting to know that for $f = 8$ and $f = 20$ we can do **better**, for $\frac{A^3}{V^2}$, than the value for the regular octahedron and icosahedron (compare with Sect. III.3): (Goldberg, 1934).

Still with a fixed number of faces, we know since Minkowski (1897) that there always exists at least one optimal polyhedron; Lindelöf (1869) showed that such an optimal polyhedron is always circumscribed about a sphere and that, moreover, the faces touch this sphere at their centers of gravity. This provides a good transition to the following section.

If we pass to the polar (dual) situation, specifically with a fixed number v of vertices, there is only the:

(VIII.7.2) *Conjecture: for all polyhedra with v vertices we have*

$$\frac{A^3}{V^2} \geq \frac{27\sqrt{3}}{2}(v - 2)(3\,\mathrm{tg}^2\,\omega_v - 1)$$

where ω is defined as above, equality holding only for regular tetrahedra, octahedra and icosahedra.

We also know that the cube and dodecahedron are not the best for $v = 8$ and $v = 20$.

Finally, in the case where the number a of edges is fixed, we know since (Steinitz, 1927) that there exists at least one polyhedron realizing the minimum and since Fejes Tóth (1948) that it is always simplicial.

There will be no section for the isoperimetry of polytopes when ($d \geq 4$), not just because the preceding involves immense difficulties (we will later see that other properties radically distinguish the case $d \geq 4$ from $d = 2$ and $d = 3$), but also because we don't know of any reference on the subject apart from Boroczky and Boroczky (1996).

VIII.8. Inscribability properties of Euclidean polyhedra; how to encage a sphere (an egg) and the connection with packings of circles

The problem of knowing if every polyhedral combinatoric can be realized by a polyhedron inscribed in a sphere was posed by Steiner in 1832. The fact that it was resolved (in the negative) only in 1927, by Steinitz, well illustrates the disdain in which the study of polyhedra was held (and perhaps the fear that they inspired). It is all the more surprising to find a counterexample that is trivial, especially by polarity. We are therefore looking for a type of polyhedron that can never be circumscribed in a sphere; the most spectacular (and also the simplest) is the truncated dodecahedron; see Fig. VIII.8.1.

The remark of Steinitz is that, to each edge of a polyhedron circumscribed about a sphere, there is attached a real number, specifically the angle under which we see this edge from the point of contact with the tangent sphere of these two faces: in fact this angle is the same for each of these two faces, which can be seen for example by symmetry about the plane containing the edge in question and the center. Now the sum of these numbers for all the edges of a face evidently equals 2π. The truncation figure shows that each triangle of the polyhedron will contribute $\frac{2\pi}{3}$ in each pentagonal face. When we take the sum of these contributions of our invariants for the 12 pentagonal faces, we find a contribution of $12 \cdot 5 \cdot \frac{2\pi}{3} > 12 \cdot 2\pi$, QED.

It is easy to see that an abundance of counterexamples exist. Steinitz's example was a cube truncated just once. And by *polarity* we obtain polyhedra (the combinatoric alone is given) that are never inscribable in a sphere. But that does not answer the question of knowing which are the inscribable or circumscribable types. We have just seen that a necessary condition is that we be able to associate with each edge a positive real number so that the sum of these numbers for each face equals 2π. But what is a sufficient condition? We had to wait for Hodgson, Rivin, and Smith (1992), which is in fact the presentation of a result of Riven from 1992, to have the complete answer; see also Hodgson and Rivin (1993).

(VIII.8.1) *A necessary and sufficient condition for the circumscribability of a combinatorial type of polygon is the existence, for each edge e, of a number $w(e)$ which satisfies the three conditions: 1) $0 < w(e) < \pi$ for each e, 2) $\sum_{e \in F} w(e) = 2\pi$ for each face F, 3) for each simple circuit of edges $\{e_i\}$ that does not bound a face of the polyhedron, $\sum_i w(e_i) > 2\pi$.*

The conditions are clearly necessary, but the proof of (VIII.8.1) is exceedingly hard. It consists of lots of geometry, of which a part is the technique used for the proof of (VIII.6.3); existence is proved by an argument in the space of metrics of a certain type on the sphere S^2. However, here we need several innovations for finding the right context, which is double: on the one hand it is that of hyperbolic geometry in three dimensions, realized here as the ball bounded by the sphere S^2 in $\mathbb{R}^3$: see Sect. II.4, but especially also Sect. II.XYZ. We will consider, thanks to the linear model (also know as Klein's) of Hyp^3, a polyhedron P **inscribed** in the sphere

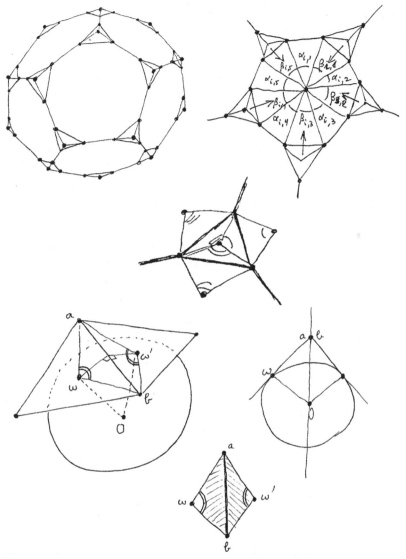

Fig. VIII.8.1. On the interior of a face with 10 sides, truncation forces an angle sum
less than 2π

S^2 as an *ideal* polyhedron of Hyp^3, i.e. for which all the vertices are at infinity. We
search for the combinatorics that are possible for such polyhedra. Rivin resolved this
problem by studying those of the polyhedra P* polar (dual) to the P. Those are the
ones we are looking for; they are, in $\mathbb{R}^3$, polyhedra circumscribed about the sphere
S^2; they are if we wish, on the **exterior** of S^2. We must thus work in the geometry
from the exterior of S^2, which is the counterpart of the hyperbolic geometry Hyp^3

from its interior. This geometry exists, it is well known in general relativity as the *De Sitter sphere* S_1^2. It is obtained in the same way as hyperbolic geometry — see Fig. II.XYZ.18, except that this time we take the infinitesimal geometry on each of its tangent planes induced by the quadratic form $x^2 + y^2 + t^2 - z^2$ on the hyperboloid (in dimension four) of one nappe with the equation $x^2 + y^2 + t^2 - z^2 = -1$. The quadratic form isn't positive; it is of type $(2, 1)$, as in relativity in two spatial dimensions. This is a geometry which isn't metric in the strict sense. It is in the geometry in which Rivin worked, by extending the whole technique simultaneously of generalized rigidity "a la Cauchy" and by considering the space of all the possible metrics (counting parameters); and then we need to finish by using the *properness* of a suitable mapping.

For those who know about algorithmic complexity, the conditions of (VIII.7.1) are verifiable in polynomial time. The result (VIII.7.1) implies first that most (in a precise sense, left to readers to determine) combinatorial types permit neither inscribability nor circumscribability. The second thing concerns the preceding section: for most combinatorial types the minimum isoperimetric ratio $\frac{A^3}{V^2}$ will never be attained, by Minkowski-Lindelöf; see Sect. VIII.7 above. The limit of this function on the set of polyhedra for which the combinatorial type is given is attained, but for a type which has in general degenerated in the limit (and thus isn't the same).

But between inscribing and circumscribing there is a middle course: ensure that all the edges of the polyhedron are tangent to a single sphere. This possibility occasionally gets presented as an "open" problem — see e.g. B18 in Croft, Falconer, and Guy (1991) — but in fact in Sect. II.8 we find this in the context of circle packings with imposed combinatoric, which will be seen to be equivalent. We will say that the polyhedron P is *midscribed in a sphere* S if all the edges of P are tangent to S. If we conceive of — or indeed produce — the graph of the polyhedron simply by way of edges made of metal wire, we see that the sphere is *encaged* in the polyhedron.

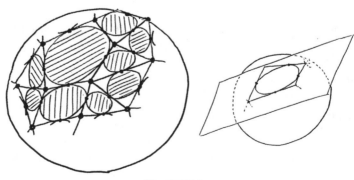

Fig. VIII.8.2.

But what could this possibly have to do with circle packings — disks in fact — in the plane, say on the interior of a given circle and with a given combinatoric, for example hexagonal on the interior and prescribed on the boundary circle?

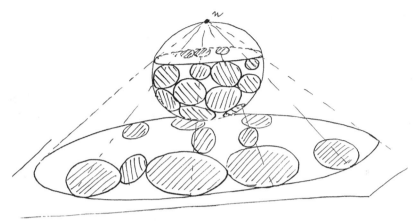

Fig. VIII.8.3.

But it's quite simply the same thing, thanks to two remarks. The first is that the faces of the midscribed polyhedron cut out circles on the surface of the sphere (small circles, not in general equators), which are tangent to one another exactly when they correspond to two faces having an edge in common. The combinatorics of this circle packing on the sphere is thus equivalent to the combinatorics of the polyhedron. The second remark is that we can pass to the plane by a stereographic projection, by choosing the center of this projection at the center of one of the circles that is cut out. This posed, we have the result:

(VIII.8.2) (Encagement, by multiple authors.) *each type of (convex) polyhedral combinatoric is realizable by a polyhedron midscribed in a sphere. Moreover this realization is unique within a conformal transformation of the sphere, i.e. within an element of the Möbius group of S^2. In particular, the realization is unique if we impose the center of gravity.*

It seems this result was featured in its entirety for the first time in Thurston (1978), who considered it as a corollary, or rather a particular case, of a result of Andreev (1970) that concerned the existence of certain polyhedra in the hyperbolic space of dimension 3. The connection with hyperbolic geometry is worth explaining; see Sects. II.5 and II.XYZ for hyperbolic geometry. The corollary is that obtained when we deal with ideal polyhedra, i.e. with vertices at infinity. The result can be found in a somewhat different context in pp. 117–119 of Ziegler (1995).

In fact Thurston made use of conformal representation for his approximations, amply described in Sect. II.8; and we saw there also that during a conference a specialist remarked to the speaker that (VIII.9.1) follows directly from a result of Koebe from 1936. The proof of Koebe belongs to a totally different domain, i.e. conformal representation , and there is no possibility of giving the ideas here. As for Andreev's proof, it is purely geometric. On the one hand, it is one of the proofs of (VIII.6.3) and (VIII.8.1). On the other hand, in Sect. IX.10 we will see an essential point: in hyperbolic geometry, in a triangle, the angles are subjected to the single condition that their sum be strictly less that π; apart from that, we can take any triplet of angles (including three zeros for the ideal polyhedra) and always find a corresponding triangle; and the same thing is valid for the dihedral angles of polyhedra in Hyp^3 and extends to the ideal polyhedra, whence an enormous flexibility in the types of polyhedra that are possible.

Because of its numerous applications (see e.g. Sect. II.8) and its aesthetic beauty for geometers, the fact that every combinatoric is possible for circle packings – in the plane as on the sphere – has subsequently been the object of several proofs that we will now discuss in some detail.

The proof of Schramm (1992) goes furthest, is very conceptual and of rare elegance, and provides the features that interest us, is that. The progression of thought is simple: in an arbitrary polyhedron P with given combinatoric, we can trivially inscribe a convex set C which is tangent to each of its edges: choose small pieces of convex surfaces tangent on the interior to these edges (at arbitrary points) and join them by a convex surface, roughly an egg C, which is thus encaged in P.

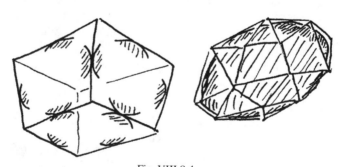

Fig. VIII.8.4.

Now we deform C (continuously of course) into a sphere S, in any way whatever, obtaining a one parameter family $C(t)$ of convex sets such that $C(0) = C$ and $C(1) = S$, and we see whether we find a continuous family $P(t)$ of polyhedra with the imposed combinatoric such that each $P(t)$ is midscribed in $C(t)$. Now there exists a very general conceptual tool, developed by Thom, for knowing whether this is possible; it's the concept of *transversality*. Schramm shows that the conditions necessary for applying this tool are fulfilled in our problem; we have only to count correctly the number of parameters that come into play. The proof so far resembles the one of Cauchy for (VIII.6.1). In fact he actual composition of the proof yields the following general result:

(VIII.8.3) *(Encagement of an arbitrary convex set): given an arbitrary convex set whose boundary is of class* C^∞ *and has everywhere positive Gaussian curvature, each polyhedral combinatorial type in* $\mathbb{R}^3$ *can be realized by a polyhedron all of whose sides are tangent to this convex set. Better: the set of all these realizations is a* C^∞ *manifold of dimension 6. Better still: if we impose the center of gravity, there is uniqueness within an isometry of* $\mathbb{R}^3$.

♦

Other proofs: Colin de Verdière (1991) and Brägger (1992), both based on a variational principle. A functional is constructed on the triangulations for which the sides are fixed initially in an arbitrary way, but not the angles. But then these angles in general do not have any reason to be such that their sum at each vertex is equal to 2π. The functional chosen is such that these sums are indeed equal to 2π when the derivative of the functional is zero. Now it is shown that the functional constructed is strictly convex and thus attains its minimum at some point − which turns out to be unique − and thus this minimum describes the desired configuration.

♦

In Lecture 4 of Ziegler (1995) it can be seen how the Koebe-Andreev-Thurston theorem provides the quickest proof of (VIII.9.1) below. Finally, in 5.2 of Lovász (2000), the theorem is put in very good perspective as a bridge between the discrete and the continuous.

♦

It will henceforth be pointless to dedicate an entire section to inscribability properties for polytopes in higher dimensions. In fact, we easily see by a suitable recurrence that for all pairs (m, d) − other than the pair $(1, 3)$ − for which $0 \leq m \leq d$ and $d > 2$, there exist polytope combinatorics in dimension d that can never be realized by polytopes whose m-faces are all tangent to the same sphere; see Schulte (1987). So our case $(1, 3)$ above is the sole exception to inscription never being possible, and we have seen the extent to which it is rich in resonances. The idea for seeing all these impossibilities is to notice that, if a polytope admits a non inscribable star, then the polytope itself will also not be inscribable.

VIII.9. Polyhedra: rationality

We have seen in Sect. VIII.3 that every polygonal combinatoric can be realized by polygons whose vertices belong to the lattice $\mathbb{Z}^2$. Can we do the same with polyhedra in $\mathbb{Z}^3$? Otherwise stated: can every polyhedral combinatoric be rendered *integer*? The answer is **yes**:

(VIII.9.1) *(Steinitz): every polyhedral combinatoric is realizable by a polyhedron whose vertices belong to $\mathbb{Z}^3$.*

This result of Steinitz is a particular case of his theory of realization of graphs by polyhedra and no proof to this day is truly simple and quick. This "yes" is theoretically satisfying because it implies that we can enter each polyhedral combinatoric into a computer exactly; but we will see later that the precise answer is more subtle. It amounts in fact to assigning integer coordinates (modulo the resolution of the computer) to each of the vertices, but moreover in a way so that the combinatoric is respected and so that we don't leave the screen! In the whole Steinitz theory it is essential that each abstract polyhedral combinatoric is realizable in $\mathbb{R}^3$ if and only if the associated graph is 3-connected; see Sect. VIII.4.

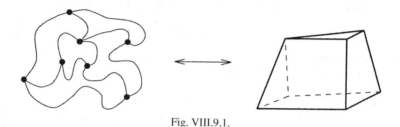

Fig. VIII.9.1.

It's the moment to remark that, just as Euclidean rigidity was trivial for co-simplicial polyhedra, the problem is trivial here too for simplicial polyhedra. In fact we can always perturb each vertex of such a polyhedron a little without changing its combinatoric. We are then done, since each real number is approximable by rational numbers as closely as we wish. We multiply by their common denominator and obtain thus points in $\mathbb{Z}^3$. In contrast, as soon as four vertices have to be in the same face, there are linear relations that must be respected. This is the difficulty of the problem: we have to find sufficient mobility while simultaneously respecting all the conditions.

Alert readers will have noticed that we do not yet know enough to be able to enter every polyhedron in a computer because here, just as for polygons in Sect. VIII.3, we need to know an estimate of the function $G(n)$ which describes the smallest integer $G(n)$ such that each polyhedron with n vertices can be inscribed in $[0, 1, 2, ..., G(n)]^3$. In fact, the computer can only deal with calculations of limited size that are expressed in whole numbers or by rationals in $[0, 1]$. In contrast to polygons, where the answer was in terms of $n^{3/2}$, things are here catastrophic and badly understood. What is known can be found in Ziegler (1995, p. 123), i.e. that $G(n) \leq \exp(\log n \cdot 169 \cdot n^3)$; but there exists no result in the reverse sense (this would be minorization); we hope some day, however, to know a bound (optimal in the two senses) that is quadratic in n. See however, at the end of Sect. VIII.12, the very recent result of Richter-Gebert. In any case, in the present state of the problem, it seems out of reach that to numerically enter a polyhedron having many vertices into a computer, at least in the primitive sense of the term; this is very different from entering it by the expedient of a system of relations between the coordinates of the vertices (just as the golden ratio $\frac{\sqrt{5}-1}{2}$ can be entered, not exactly numerically, but formally as s root of the algebraic equation $x^2 - x - 1 = 0$).

Steinitz deduced many things from his result on graphs and their explicit construction. In fact we can say that he resolved all the natural questions that present themselves concerning the combinatorics of polyhedra in space. We begin with this: the combinatoric being given, what is its *space of realization*? This term has to do with describing the family of all polyhedra in $\mathbb{R}^3$ that realize this combinatoric. Steinitz, thanks to the power and flexibility of his description of embeddings with given combinatoric, knew how to show that this space is connected, and better: it is *contractible*. It thus has neither holes nor nontrivial algebraic topology. Readers who want a precise definition of this space can associate, with each polyhedron in $\mathbb{R}^3$ having n vertices, a point of the space $\mathbb{R}^{3 \times n} = \mathbb{R}^{3n}$ by associating with each vertex its three coordinates. The connectivity of the space is often called *isotopy*, which means that any two realizations can be joined continuously by a path of polyhedra whose combinatoric doesn't vary.

Recently the experts have realized that the midscribability result (VIII.8.2) (see also (VIII.8.3)) gives the quickest known proof of (VIII.9.1); see why in Ziegler (1995, p. 116). Moreover, we obtain a bonus: each merely combinatoric automorphism can in fact be realized by an isometry of $\mathbb{R}^3$, e.g. a symmetry. From dimension 4 on this is false; see Sect. VIII.12.

VIII.10. Polytopes ($d \geqslant 4$): combinatorics I

For this section, which contains recent results and others that are extremely recent — true *tours de force* both conceptually and practically — the general references are Grünbaum (1967), Bronsted (1983), Ziegler (1995), 2.3 of Gruber and Wills (1993), but most of all the fundamental text (Richter-Gebert, 1996), which is the culmination of 40 years of incessant research (even omitting the Steinitz intermezzo). The success is due to the fact that sets of polytopes, as well as classes of combinatorially equivalent polytopes, can be given algebraic structures for which available algebraic results have yielded the conclusions that we will see. The book (Ziegler, 1995) abundantly provides very good historical notes as they are needed and to which readers can refer. In 2.3.3 of Gruber and Wills (1993) there is a sketch of the algebraic structures that appear in the combinatorics of polytopes; but also add the references of Sect. VIII.13. The relatively recent notion of *matroid* is unavoidable. See also the recent expositions Stanley (2000), Ziegler (2001) and 3.1 of Lovasz (2000).

The number Comb(v, d) of possible combinatorics of polytopes in $\mathbb{R}^d$, when the number n of vertices is fixed, is incredibly large; presently it is known how to estimate it on both sides rather well; see 2.3 of Gruber and Wills (1993). We will let Comb$_s(n, d)$ denote the number of them that are in addition simplicial. Then:

(VIII.10.1)

$$\left(\frac{n-d}{d}\right)^{nd/4} \leqslant \text{Comb}_s(n, d) \leqslant \text{Comb}(n, d)$$

$$\leqslant \exp\left(\log(\frac{n}{d}) \cdot nd^2\left(1 + \text{O}\left(\frac{1}{\log(n/d)} + \frac{\log\log(n/d)}{d \cdot \log(n/d)}\right)\right)\right)$$

and
$$\text{Comb}(n, d) \leqslant \exp(\log 2 \cdot n^3 + \text{O}(n^2)).$$

Apart from algorithmic questions and pure combinatorics, these results are not very exciting; they show only the enormous complexity of the matter and we happily leave all that; but nonetheless the proofs entail no small number of geometric considerations.

We now study first the number f_k of k-faces of a polytope as a function of the dimension of the space and the number of vertices. We possess an optimal result, i.e. an optimal upper bound. To state it we need to introduce the *cyclic polytopes* $C_d(n)$, which are in a way **the** generic polytopes in $\mathbb{R}^d$. We consider the generic algebraic curve in $\mathbb{R}^d$ given by the "generic" parametric representation $t \mapsto x(t) = (t, t^2, \ldots, t^d)$. Then $C_d(n)$ is by definition the convex envelope of the n points $x(t_1), x(t_2), \ldots, x(t_n)$, where the t_i ($i = 1, 2, \ldots, n$) is any strictly increasing, but otherwise arbitrary, sequence.

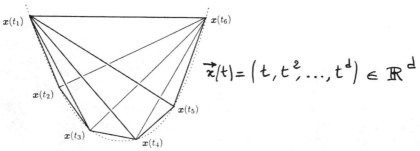

Fig. VIII.10.1. The cyclic polytope $C_d(6)$

The *upper bound conjecture*, unproved until McMullen (1971), says that the $C_d(n)$ are precisely the generic polytopes in dimension d, in the sense that they are the ones which have the maximum number of faces for a given number of vertices, and this in each dimension:

(VIII.10.2) *(The upper bound theorem): if* P *is a polytope in dimension d with n vertices, then for each integer k the number $f_k(P)$ of its k-faces does not exceed the number of k-faces of the cyclic polytope $C_d(n)$.*

The calculation of the $f_k(C_d(n))$ is left to readers. The cyclic polytopes are *simplicial* (see (VIII.10.3) below), which means that all their faces (their $(d-1)$-faces of course) are simplexes. For the *upper bound conjecture*, an elementary and very visual remark is that if we slightly displace the vertices of a non simplicial face, the number of faces can't increase when the number of vertices remains fixed. In proving this conjecture we can thus limit ourselves to working with simplicial polytopes which are sufficiently generic.

A detailed proof of (VIII.10.2) can be found, for example, in §18 of Bronsted (1983). In that book it is not necessary to ascend very high on the ladder, it is only required to introduce polytopes of adequate type (called the *neighborly* polytopes). But most of all it is great technical ability (in the best sense of the term) that is needed, and much perseverance. Finally, the case of equality is characterized by these neighborly polytopes. The method of proof of (VIII.10.2), as it appears in (Bronsted, 1983), is provided by (VIII.10.4) below. In contrast, McMullen's original proof used two concepts: the notion of *Gale diagram* and that of *Schlegel diagram*, in addition to the concept of shellability mentioned in Sect. VIII.4. The Gale and Schlegel diagrams are indispensable for the study of polytopes, both *individual* and "*ensemble*", as well as for algorithmic questions. A very good exposition is that of Ziegler (1995).

There also exists a *lower bound theorem*, which dates only from Barnette (1973); see for example Bronsted (1983).

Contrary to what happened for (VIII.4.2), starting with $d = 4$, exact knowledge of all the possible d-tuples $\{n = f_0, f_1, f_2, ..., f_{d-1}\}$, i.e. the f-vectors $f(P) \in \mathbb{Z}^d \subset \mathbb{R}^d$ of polytopes of dimension $d \geqslant 4$, is a completely open subject. Worse: it is not even known whether anything reasonable can be said about the subset of $\mathbb{N}^d$ considered!

Here now, very briefly and in rough form, is what we know today about the combinatorics of our objects. First, Euler's formula (VIII.4.1) is extended to any dimension by:

(VIII.10.2) *For each polytope of dimension d we always have*

$$\sum_{i=0,1,...,d-1} (-1)^i f_i = 1 + (-1)^{d-1},$$

where f_i denotes the number of i-faces (the 0-faces are the vertices).

For $d = 2$ this indeed restates the fact that the number of sides equals the number of vertices for every polygon. By 1852 Schläfli had announced this formula, which he proved by recurrence, but we have seen in Sect. VIII.4 above that it implicitly assumes constructibility of polytopes, which isn't at all evident in dimension 3. The first proof of constructibility dates only from Bruggesser and Mali (1971), whence the purely polytopal recurrence proof seen in Sect. VIII.4, not using Poincaré duality for compact manifolds. In fact, Euler's formula can be proved by recurrence, without using constructibility, by taking a hyperplane direction not containing any edge of the polytope considered and by regarding what happens at the passage of each vertex as the hyperplane descends from the highest vertex to the final vertex below, as we saw in Sect. VIII.4 for dimension 3.

For perhaps too curious minds it is important to know that we can show that there does not exist any linear relation beyond Euler's that is satisfied by all polytopes of a given dimension. It can be shown that such a relation is identically zero by substituting in the values of f_i for pyramids and for double pyramids; in dimension 3 these are the triples $(k + 1, 2k, k + 1)$ and $(k + 2, 3k, 2k)$, etc. For those who are fond of language that is a bit more learned, we can express the nonexistence more precisely as follows. We *define* $f(P) = \{f_0(P), f_1(P), \ldots, f_{d-1}(P)\}$ in $\mathbb{R}^d$, called the *f-vector* of P. We then consider in $\mathbb{R}^d$ the subset of $f(\mathcal{P}^d)$ of all the $f(P)$ when P ranges over all the polytopes of dimension d. The result is then that the smallest affine subspace of $\mathbb{R}^d$ that contains $f(\mathcal{P}^d)$ is the hyperplane for which an equation is given by Euler's formula.

◆

So if we want "general" relations, in the present state of the problem, they can thus only be for a restricted class of polytopes. One class indeed suggests itself, that of *simplicial polytopes*:

(VIII.10.3) *A polytope of dimension d is called* simplicial *if all its faces (faces of dimension $d - 1$) are simplexes (thus of dimension $d - 1$).*

Each polygon is simplicial and the cyclic polytopes $C_d(n)$ are all simplicial. The naturalness of this notion comes predominantly from the fact that these are the polytopes that we encounter with probability 1 if the vertices are chosen at random in $\mathbb{R}^d$. In fact, when we perturb the vertices of an arbitrary polytope a little in a general way, we destroy the faces that aren't simplexes; precisely, the convex envelope of the new set of points in $\mathbb{R}^d$ will be simplicial.

And thus a simplicial polytope doesn't change combinatorial type when we perturb it in any fashion (but sufficiently slightly); and also a simplicial combinatoric can always be realized in $\mathbb{Z}^d$. We will return amply to the rationality of polytopes in Sect. VIII.12. Simpliciality mustn't be confused with *simplicity*; a polytope in $\mathbb{R}^d$ is called *simple* if each vertex belongs to exactly d faces. It is the quite simply same as requiring that the polar itself be simplicial.

For the simplicial polytopes just defined, the question of possible values of their set $(f_0, f_1, \ldots, f_{d-1})$, i.e. of knowing the exact nature of the set $f(\mathcal{P}_s^d) \subset \mathbb{Z}^d \subset \mathbb{R}^d$, where $\mathcal{P}_s^d$ denotes the set of simplicial polytopes of dimension d, is today completely resolved, after numerous efforts, by the two results (VIII.10.4) and (VIII.10.5) below. We have then first:

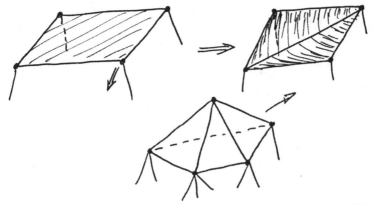

Fig. VIII.10.2. Make a polytope simplicial and then the vertices become as mobile as we want

(VIII.10.4) *(Dehn-Sommerville relations): for each simplicial polytope* P *of dimension* d *we always have the relations*

$$\sum_{j=i,\dots,d-1} (-1)^j C_{i+1}^{j+1} f_j(P) = (-1)^{d-1} f_i(P)$$

for all $i = -1, 0, 1, \dots, d-2$ *(with the convention* $f_{-1} = 1$); *moreover there do not exist other relations valid for each simplicial polytope other than the ones generated by these.*

These relations had already been found by Dehn by 1905 for the dimensions 3, 4, 5; they were known in all dimensions by Sommerville by 1927 but were then forgotten until they were rediscovered by Klee in 1963.

For $i = -1$ we recover Euler's formula; for $i = d - 2$ we find $d \cdot f_{d-1}(P) = 2 \cdot f_{d-2}(P)$, which expresses simpliciality. The other relations are not too difficult to prove; we can proceed by recurrence, using Euler's formula. The incidence relations between the faces of various dimensions should be written carefully; see Bronsted (1983). To express seriously that there are no relations beyond those of Dehn-Sommerville for the simple polytopes, we need to prove that these form a system of linear equations for the affine subspace of $\mathbb{R}^d$ generated by the set $f(\mathcal{P}_s^d)$ formed by all the f-vectors of the simple polytopes of given dimension d. But this does not say that it is the dimension of this affine subspace; it can be seen in Bronsted (1983) for example (or another of the given references) that this dimension is equal to $[\frac{d}{2}]$, the integer part of $\frac{d}{2}$. In particular, roughly half of the equations of (VIII.10.4) are superfluous, as is found in the reference mentioned.

◆

We now deal with knowing what $f(\mathcal{P}_s^d)$ is exactly in the affine subspace of dimension $[\frac{d}{2}]$ obtained above. The matter is of a whole other order of difficulty in comparison with (VIII.10.4). To state the principal double result, we define first the h-vector $(h_0, h_1, \dots, h_d)$ of a polytope as the set of integers determined, starting with the f-vector $(f_0, f_1, \ldots, f_d)$, by the equations: $h_i = \sum_{j=0,\ldots,i} C_{d-i}^{d-j} f_{j-1}$. Then:

(VIII.10.5) *(McMullen conditions): A vector $(h_0, h_1, \ldots, h_d)$ is the h-vector of a simplicial polytope if and only we have simultaneously:*
  (i) $h_i = h_{d-i}$ for each $0 \leqslant i \leqslant d$;
  (ii) $h_0 = 1$ and $h_i \leqslant h_{i+1}$ for each $0 \leqslant i \leqslant \frac{1}{2}d - 1$;
  (iii) $h_{i+1} - h_i \leqslant (h_i - h_{i-1})^{\langle i \rangle}$ for each $1 \leqslant i \leqslant \frac{1}{2}d - 1$,
  where for the integers h and i we define a new integer $h^{\langle i \rangle}$ in an algebraic manner with the help of the binomial coefficients; see p. 494 of Gruber and Wills (1993).

These complicated conditions were discovered by McMullen in 1971 and at the same time he conjectured that they are necessary and sufficient. We had to wait for Billera-Lee in 1981 for the sufficiency and Stanley in 1980 for the necessity. The sufficiency is done by exhibiting some very clever and very technically refined conditions.

The necessity is of considerable difficulty. The ideas and the materials needed for the first proof are analyzed in Ziegler (1995, p. 278). This proof requires a single high and difficult ascent of Jacob's ladder. It uses nothing less than the construction of toroidal manifolds (complex tori) associated with a polytope and the necessity then resulted, applied to these manifolds, from the *hard Lefschetz theorem* of algebraic geometry; see Oda (1988). We next obtain still more "technical" proofs, because there was a gap in the algebraic geometric machinery used by Stanley. Recently McMullen obtained a proof based on algebraic structures that can be formed directly with polytopes, with the help of various natural geometric operations, such as addition and product; see the references in Ziegler. It seem that it was necessary to wait still some time before having completely, and in a simple enough way, algebraized the set of polytopes. See the following Sects. VIII.12 and VIII.13, as well as Stanley (2000).

VIII.11. Regular polytopes ($d \geqslant 4$)

A polytope is called *regular* if its isometry group is transitive on (all) its flags (see Sect. VIII.5). We owe to Schläfli from 1850 the complete classification of regular polytopes in all dimensions. We give the result right away:

(VIII.11.1) *In dimension 4 there exist six of them, for which the Schläfli symbols (that we are going to define) are $\{3, 3, 3\}$, $\{3, 3, 4\}$, $\{3, 3, 5\}$, $\{3, 4, 3\}$, and for their polars $\{4, 3, 3\}$, $\{5, 3, 3\}$. Starting with dimension 5, there are never more than three: the regular simplex (with symbol $\{3, \ldots, 3\}$), the cube (with symbol $\{4, 3, \ldots, 3\}$) and the cocube (polar of the cube, with symbol $\{3, \ldots, 3, 4\}$).*

For more details on what follows, consult Chap. 12 of [B]. We may well be astonished by this result: the greater the dimension, the fewer the regular polytopes (an infinity in dimension 2). In fact, it is an illustration of a general (but heuristic, if not vague) principle, that we owe to Thom: "the richest structures become less and less numerous as the dimension increases, whereas the poorest structures are more and more numerous". We illustrate this principle by some examples, not all of which are explained in this book; readers can ignore those with which they are not familiar and add some of their own.

Among the *rich* structures, we find:

- simple Lie groups;

- division algebras;

- quadratic fields;

- regular polytopes;

- the symmetric group $\mathcal{S}_n$ on n elements, which is simple for $n \geqslant 5$;

- the orthogonal group $O(n)$, which is simple for $n \geqslant 5$;

- conformal geometry (Liouville's theorem, end of Sect. II.5);

- manifolds of dimensions 2, 3, 4.

Among the *poor* structures, we find:

- topological vector spaces, which are not all isomorphic in infinite dimension;

- the generic singularities of differentiable mappings;

- differentiable structures on a given manifold and their Riemannian structures, starting with dimension 5; see Berger (2003);

- finite fields.

This genre of philosophy may be compared with that of Arnold (2000).

♦

Here is Schläfli's proof, which is very simple and works by recurrence with respect to the **possible** symbols. As for existence in dimension 4, for $\{3, 4, 3\}$ and $\{5, 3, 3\}$, it is done by hand as in dimension 3. Moreover, dimension 3 shows very well the phenomenon which is at the root of the rarity of regular polytopes. Starting with dimension 5 there is thus no longer a problem of existence.

Before that, some trivial generalities. If P is regular, being compact it has a center of gravity O that is thus invariant under the isometry group G: we speak of the *center* of P. As G is transitive on the vertices, the whole polytope is inscribed in a sphere S with center O. We consider a vertex x of such a polytope and its star $St(x)$, here more precisely the set of vertices that are joined to x by an edge. As G is transitive on the directed edges at x and preserves the center O, we have that $St(x)$ is contained in a single hyperplane $H(x)$ perpendicular to the line xO. Better: the transitivity of G on all the flags shows in fact that $St(x)$ is a regular polytope of $H(x)$ of dimension $d - 1$. The center O' of $St(x)$ is the intersection of the hyperplane in which it is situated with the line xO.

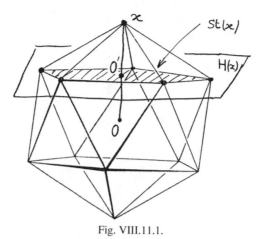

Fig. VIII.11.1.

♦

We first look at all of this in dimension 3 to exploit the visual. We let y and z be two vertices of $St(x)$, where a face (a 2-face here) Q of P intersects $St(x)$.

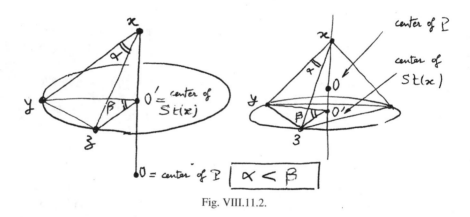

Fig. VIII.11.2.

This face is a regular polygon with k sides; its angle α at the vertex x thus equals $\pi - \frac{2\pi}{k} = \pi\frac{k-2}{k}$. Let h be the number of vertices of $St(x)$, which is thus a regular polygon with h sides, and the angle β with center $0'$ of the triangle $yO'z$ thus equals $\frac{2\pi}{h}$. The pair $\{k, h\}$ is called the *Schläfli symbol* of P. The **essential**, but nonetheless elementary, remark is that orthogonal projection onto H can only strictly increase the angle α, and thus $\alpha < \beta$. But each face and each star possess at least three sides; thus β can only take on the values $\beta = 120°, 90°, 72°, 60°$, whereas α can only take on the values $\alpha = 60°, 90°, 108°, 120°$. So we are sure that the only possible values for the pairs $\{k, h\}$ are: $\{3, 3\}, \{3, 4\}, \{3, 5\}, \{4, 3\}, \{5, 3\}$. We indeed find our five regular polyhedra of Sect. VIII.5, for which we recall the need of proving existence. But we can say that they have been constructed starting with the configuration given by a vertex x and its star $St(x)$ by installing, starting from x, regular k-gons which touch on $St(x)$.

♦

We next examine dimension 4. Now $St(x)$ is a regular polyhedron in the hyperplane H. If the 2-faces of P are all regular k-gons, then the angle at the vertex α, which here plays the role of the preceding β, is the angle at the center of $St(x)$, i.e. the angle β at which from O' we see the edges(1-faces) of $St(x)$. The symbol of P is by definition $\{k, p, q\}$ if $\{p, q\}$ is the symbol of $St(x)$ as above. We thus need to know the five values of β corresponding to the five types of regular polyhedra. We find (exact calculations are left to readers with calculators)

$$\beta(\{3, 3\}) \approx 108°5 > 108°, \quad \beta(\{3, 4\}) = 90°,$$
$$90° > \beta(\{3, 5\}) > 60°, \quad 90° > \beta(\{4, 3\}) > 60°, \quad \beta(\{5, 3\}) < 60°.$$

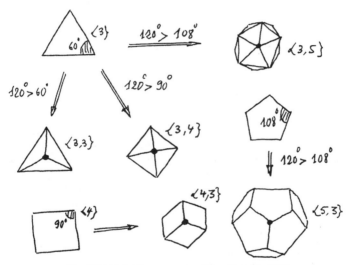

Fig. VIII.11.3. The progeny of the plane in space

Thus *a priori* the only possible symbols are $\{5, 3, 3\}$, $\{4, 3, 3\}$, $\{3, 3, 3\}$, $\{3, 3, 4\}$, $\{3, 3, 5\}$, $\{3, 4, 3\}$. Only $\{5, 3\}$, the regular dodecahedron, finds itself without progeny, because its angle at the center is too small. The regular simplex is $\{3, 3, 3\}$, the cube is $\{4, 3, 3\}$ and the cocube $\{3, 3, 4\}$. We will have noticed that polarity consists of reversing the order in the symbol. As in dimension 3, and with scarcely more trouble, we show that the $\{3, 3, 5\}$ exists, thus also its polar $\{5, 3, 3\}$. The existence of $\{3, 4, 3\}$ is trivial, but it is interesting to remark that it is the only regular polytope which has a center of symmetry and is autopolar. We will see it appear quite naturally with the optimal network D_4 in Sect. X.4. These three rare polytopes give rise to very pretty figures under plane projection:

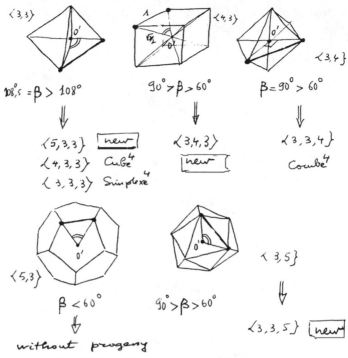

Fig. VIII.11.4. The progeny of space in dimension 4

In the figure on the black background we indeed see that the faces of $\{5, 3, 3\}$ are regular dodecahedra. There are thus 120 of them, and this is the subject of the very interesting article Stillwell (2001), which describes a whole array of situations in mathematics where this polytope is encountered, in particular in soap bubbles.

In arbitrary dimension we define as above the Schläfli symbol by recurrence: $\{k, \{., , , , , .\}\}$. Readers will have guessed that the Schläfli symbol of the polar of a regular polytope is simply the same symbol reversed from one end to the other. The regular simplex is thus $\{3, 3, \ldots, 3\}$, the cube $\{4, 3, \ldots, 3\}$ and the cocube $\{3, \ldots, 3, 4\}$. Why aren't there others starting from $d = 5$? Because the angles at the center of the St(x), thus regular polytopes, are too small. For the regular simplex, we have an angle always greater than $90°$ but which exceeds $108°$ as soon as $d = 4$. For the cocube the angle at the center always equals $90°$. For the cube, beginning with dimension 4, the angle at the center is smaller than $60°$. So the only descents permitted are those for the simplexes $\{3, \{3, \ldots, 3\}\}$ and $\{4, \{3, \ldots, 3\}\}$, and for the cocube $\{3, \{3, \ldots, 3, 4\}\}$. The cube itself admits no progeny. And this completes the announced classification.

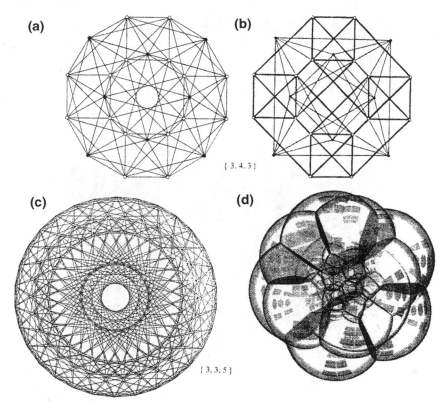

(a)

(b)

{ 3 . 4 . 3 }

(c)

(d)

{ 3, 3, 5 }

Fig. VIII.11.5. (**a**, **b**, **c**) Coxeter (1973) © Dover Publications Inc.; (**d**) © John M. Sullivan, Technische Universität Berlin

♦

In Sect. VIII.5 we encountered the notion of regular *star*; we found four of them (in dimension 3). For polygons we leave the classification to readers. For polytopes in higher dimensions, the situation is analogous to that for true regular polyhedra: there exist ten regular star polytopes in dimension four, but none more starting with dimension 5. Here are the symbols (within reversal, which yields the polars): $\{\frac{5}{2}, 5, 3\}$, $\{5, \frac{5}{2}, 5\}$, $\{\frac{5}{2}, 3, 5\}$, $\{\frac{5}{2}, 5, \frac{5}{2}\}$, $\{5, \frac{5}{2}, 3\}$, $\{\frac{5}{2}, 3, 3\}$. They are found in Coxeter (1973); the figures below are from Coxeter (1974).

There exist no more, starting with dimension 5; see Coxeter (1973), Fejes Tóth (1964). There also exists a weaker notion of *semi-regular polytopes*. For dimension 3, we find what are called the Archimedean solids, of which we spoke in Sect. VIII.5. For all these, apart from the references already mentioned, see 1.11 of Gruber and Wills (1993). For a fresh vision of these exceptional polytopes, see Iwasaki, Kenma and Matsumoto (2002).

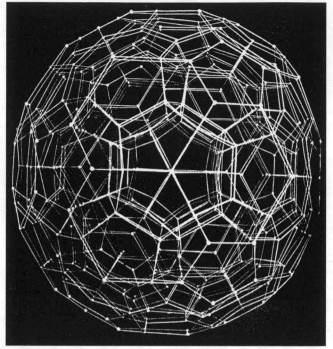

Fig. VIII.11.6. The soap bubble with 120 cells: $\{5, 3, 3\}$. Coxeter (1973) © Dover
Publications Inc.

VIII.12. Polytopes ($d \geqslant 4$): rationality, combinatorics II

The problem of knowing the substance of Steinitz's result for polytopes in dimensions $d \geqslant 4$ is today completely resolved in Richter-Gebert (1996). We will exhibit — in brief, and with plenty of technical simplifications in the assertions — the
final result of Richter-Gebert; but this is just to say that it comes as the completion of
the efforts of numerous predecessors, as is well explained in (Richter-Gebert, 1996).
We omit all the previous history. We are going to state the fundamental result of
that text, clarify what it means, then talk about some of its consequences; finally, we
will say very briefly what is involved in the proof. In essence, the theorem asserts,
starting with dimension four, that all the results of Steinitz for dimension 3 not only
still persist, but — worse — that the spaces of possible realizations of the polytopes
can be the worst possible objects. The precise definition of the *realization space* of
a polytope P of dimension d and $n > d$ vertices is the following: we enumerate the
vertices $\{x_1, x_2, \ldots, x_n\}$ in such a way that the sequence $\{x_1\}$, $\{x_1, x_2\}$, $\{x_1, x_2, x_3\}$,
$\ldots, \{x_1, x_2, \ldots, x_{d+1}\}$ forms a flag. The realization space of P is by definition the
subset
$$\mathcal{R}(P) \subset \mathbb{R}^{d \times n}$$

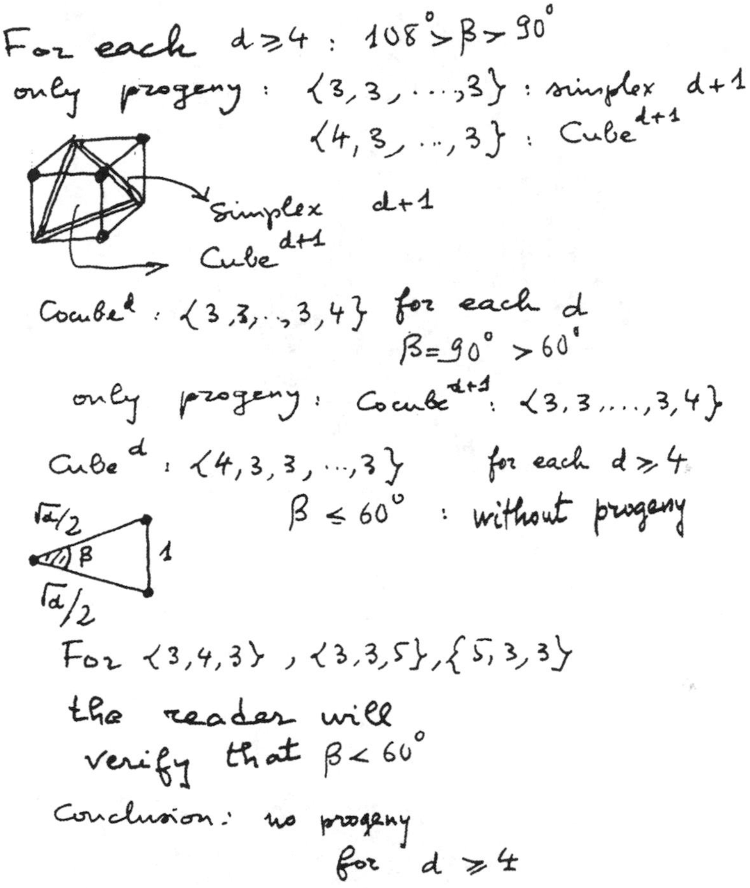

For each $d \geqslant 4$: $108° > \beta > 90°$

only progeny : $\langle 3, 3, \ldots, 3 \rangle$: simplex $d+1$

$\langle 4, 3, \ldots, 3 \rangle$: Cube^{d+1}

Simplex $d+1$

Cube^{d+1}

Cocube^d : $\langle 3, 3, \ldots, 3, 4 \rangle$ for each d

$\beta = 90° > 60°$

only progeny : Cocube^{d+1} : $\langle 3, 3, \ldots, 3, 4 \rangle$

Cube^d : $\langle 4, 3, 3, \ldots, 3 \rangle$ for each $d \geqslant 4$

$\beta \leqslant 60°$: without progeny

$\sqrt{a}/2$

β 1

$\sqrt{a}/2$

For $\langle 3, 4, 3 \rangle$, $\langle 3, 3, 5 \rangle$, $\langle 5, 3, 3 \rangle$

the reader will

verify that $\beta < 60°$

Conclusion: no progeny

for $d \geqslant 4$

Fig. VIII.11.7. Study of progeny

formed by all the matrices $Y \in \mathbb{R}^{d \times n}$ where, if $Y = \{y_1, \ldots, y_n\}$; we have first $y_k = x_k$ for $1 \leqslant k \leqslant d + 1$, then the essential condition: under the correspondence $x_i \mapsto y_i$ the polytope P is combinatorially equivalent to the convex envelope of $\{y_1, \ldots, y_n\}$.

(VIII.12.1) *(Richter-Gebert, 1996): for each primary semialgebraic set $\mathcal{V}$, there exists a polytope of dimension 4 for which the realization space is stably equivalent to $\mathcal{V}$.*

A *semialgebraic set* is a portion of $\mathbb{R}^N$ which can be described as the set of of points satisfying a finite number of equalities and inequalities (strict or not) given by real polynomials in N variables. These sets are much studied, especially nowadays, where (real) algebraic geometry is a subject in full development (for example, it provides the context for robotics, etc.); see the basic reference Bochnak, Coste, and Coste-Roy (1987). We need to include, besides the polynomial equalities, some inequalities, in order that there be stability under projection, union and intersection.

$\{\frac{5}{2}, 3, 5\}$, $\{\frac{3}{2}, 5, \frac{5}{2}\}$
$\{3, \frac{5}{2}, 5\}$, $\{3, 3, \frac{5}{2}\}$

Fig. VIII.11.8. A plane projection common to the four $\{\frac{5}{2}, 3, 5\}$, $\{\frac{5}{2}, 5, \frac{5}{2}\}$, $\{3, \frac{5}{2}, 5\}$, $\{3, 3, \frac{5}{2}\}$. Coxeter (1974) © Cambridge University Press

Fig. VIII.11.9. A plane projection of $\{\frac{5}{2}, 3, 3\}$. Coxeter (1974) © Cambridge University Press

A semialgebraic set is called *primary* if the set of its definition polynomials does not contain strict inequalities. The notion of *stable equivalence* is more subtle to define, and varies among authors. Here it says that we allow stable (in a sense to be made clear) projections and rational equivalences (i.e. reparametrizations by rational functions). See Exercise 4.22 of Ziegler (1995) for some examples of pathological semialgebraic sets.

This clarified, we easily imagine that the strict semialgebraic sets can be not only not connected, but can in fact have an algebraic topology that is practically as complicated as we wish.

The starting remark is that, a polytope being given, its realization space is **by defini-tion** a semialgebraic set. The whole proof of (VIII.12.1) "simply" consists of finding constructions of polytopes that are "basic pieces" of $\mathbb{R}^4$ and allow − by appropriate combinations − realization of 4-polytopes which subsequently prove (VIII.12.1) by operations of the types multiplication and addition. Now (VIII.12.1) allows us to re-spond to (negatively, we must say) all the questions that Steinitz resolved positively for polyhedra in $\mathbb{R}^3$. Here is a part of what can be found in Richter-Gebert (1996), which is truly very well written.

First there exist combinatorics of 4-polytopes with 34 vertices which are not realiz-able in $\mathbb{Z}^4$ (non rational polytopes). Historically the first counterexample to a result of Steinitz for polyhedra was that of Perles − encountered in Sect. I.4 − who gave a non rational polyhedron in dimension 8 and with only 12 vertices; see Richter-Gebert (1996) for the historic succession of non realizability. The number 34 is astonishingly small for a polytope of dimension 4 (see your favorite polytopes). But we can do more. A whole series of geometric configurations are used that **force** var-ious situations. For example, Pappus's theorem (see Sect. I.4) forced a collinearity; the figure below, thanks to Pascal's theorem (see Sect. IV.2), forces points to be on a conic, etc.

Let's get back to our sheep. Secondly, the example of Perles in Sect. I.4 shows that if we consider the subfield $\mathbb{Q}(\sqrt{5})$ of $\mathbb{R}$ obtained by adjoining $\sqrt{5}$ to the field $\mathbb{Q}$ of the rationals, then we can realize this polytope in S^8. From (VIII.12.1) we deduce that to be able to realize all combinations of polytopes it doesn't suffice to adjoin this or that algebraic number (i.e. a root of a polynomial with integer coefficients) to $\mathbb{Q}$, but we need to adjoin **all** algebraic numbers.

Thirdly, we can ask for the 4-polytopes having n vertices that are realizable in $\mathbb{Q}^4$, thus in $\mathbb{Z}^4$, which is the smallest integer $f(n)$ such that all these polytopes are realizable in the lattice $\{1, 2, \ldots, f(n)\}^4$. The answer is that $f(n)$ must be at least doubly exponential in n.

Fourthly a question for those who know something about algorithmic complex-ity: what complexity is required by Richter-Gebert's practical realization of a poly-tope starting with a given semialgebraic set? The answer is that it has to do with an

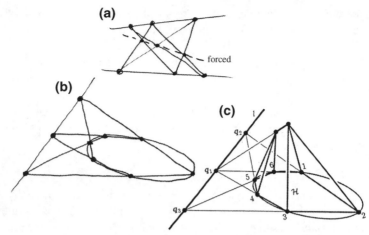

Fig. VIII.12.1. Pappus's theorem forces collinearity, Pascal's theorem forces co-conicity. (c) Richter-Gebert, Ziegler (1995) © American Mathematical Society

NP-hard problem. Moreover, this problem is equivalent, in polynomial time, to that of the "existential theory of the reals"; see again Richter-Gebert and Ziegler (1995).

In all that precedes, some uncertainty remains for the meaning of the word *realization*. In fact, if we assume the spirit of the result (VIII.6.3) on the existence of locally Euclidean metrics on the sphere S^2, we can pose for ourselves the problem(s) of realization, not for a polytope already existing in $\mathbb{R}^n$, but for a decomposition of the sphere S^{n-1} into simplexes (triangulation). Steinitz's characterization of polyhedra in $\mathbb{R}^3$ by their graph shows immediately that each triangularization of S^2 can be realized by a polyhedron. But starting with dimension 4, this is no longer the same. First, it is the moment for saying that we know of no characterization of graphs of polytopes starting with dimension 4. Next, we know that "most" triangulated spheres S^{d-1} for $(n \geqslant 3)$ are never "polytopable", i.e. realizable as a polytope in $\mathbb{R}^d$. Presently we have this result of Kalai (1988): for a given dimension n, the number of combinatorial types of triangulations of S^{d-1} with n vertices is sandwiched between $\exp(b \cdot n^{\lfloor (d-1)/2 \rfloor})$ and $\exp(c \cdot \log n \cdot n^{\lfloor d/2 \rfloor})$ (where b and c are positive constants), whereas for polytopes we have the encagement given in (VIII.10.1). In fact, Richter-Gebert and Ziegler (1995) studies more precisely the problem of this realizability, in explicit fashion.

With perfect technique, other things too are shown:

- *each combinatorial $(d-1)$-sphere having only $d+3$ vertices is polytopable;*

- *each d-polytope having $d+3$ vertices can be realized with integer vertices;*

- *the realization space of each d-polytope having $d+3$ vertices is contractible.*

The seminal idea is that we can code the combinatorics of such polytopes by arrangements of points on a line; see the references in Richter-Gebert (1996).

◆

Always thanks to his technique, Richter-Gebert even refines the case of polyhedra in $\mathbb{R}^3$. So, for example, he proves that each polyhedron in $\mathbb{R}^3$ having n vertices can be inscribed in an integer cube of side smaller than $\exp(\log 2 \cdot n^2)$. Moreover, as soon as it contains at least one triangle in its combinatoric, this integer can be lowered to 43^n.

VIII.13. Brief allusions to subjects not really touched on

First the tools. The *Schlegel* and *Gale* diagrams are classic. The idea is to reduce the dimensions to "see more clearly". The Schlegel diagram is the older and simpler, used everywhere in naive fashion, without explanation. It consists of correctly projecting the whole polytope onto one of its faces; it is therefore a diagram in a space whose dimension equals that of the polytope minus one. This will be a plane diagram for polyhedra, a "polyhedron" in the space for the polytopes of dimension four. We have already used it implicitly; see 5.2 of Ziegler (1995). Here are some other figures:

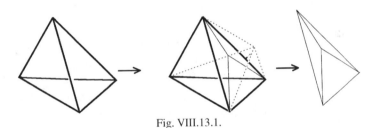

Fig. VIII.13.1.

The Gale diagram is much more difficult to define; see Lecture 6 of Ziegler (1995). It codes the vectors defining the vertices as well as the affine relations between them (e.g. if they are in the same plane, it's the notion of *circuits*). We obtain a matroid, something too complicated to be described here even briefly; the book of Ziegler is the best introduction. If the polytope is in $\mathbb{R}^d$ and has n vertices, we define a *Gale diagram* for it which is an object consisting of n points (each alloted a sign) in an affine space of dimension $n - d - 2$. Just one figure for two octahedra, in one four vertices in the same plane, but not in the other, which can be seen in the corresponding Gale diagrams.

The purpose of such diagrams is for simplifying the complete combinatorial representation of the polytope, i.e. the set of all its faces and the way they are included, one in the other. Even for a plane pentagon the drawing is complicated:

◆

The preceding, at the level of matroids, is useful but is not a sufficiently powerful algebraization. Recently some impressive ascents of Jacob's ladder have taken

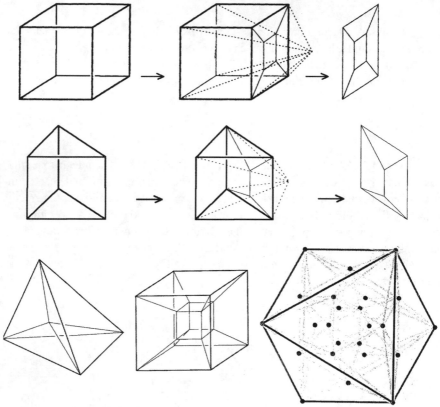

Fig. VIII.13.2. Three Schlegel diagrams of polyhedra of dimension four: for the cube,
the simplex and the $\{3, 4, 3\}$

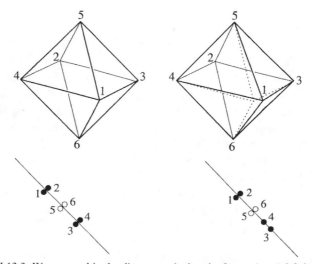

Fig. VIII.13.3. We can read in the diagrams whether the four points 1,2,3,4 are in the
same plane or not

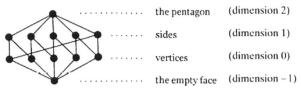

.... the pentagon   (dimension 2)

.... sides   (dimension 1)

.... vertices   (dimension 0)

.... the empty face   (dimension – 1)

Fig. VIII.13.4. The lattice of faces of a pentagon

place. They are the doing of McMullen (1993), and Morelli (1993a). We add Morelli (1993b) for the subject of calculation of volumes by slicing, a subject that is important for its naturalness but in fact also for its connections with the theory of numbers and some other very sophisticated matters. We present two algebraizations. The first is already old and consists of associating with **one** simplicial polytope an algebraic object, of the type studied in algebraic geometry, i.e. the study of manifolds – with possible singularities – defined as the set of points annulling one or several polynomials. We will thus have at our disposition all the tools of this essential discipline, which we have encountered only for the conics, the quadrics and the plane curves. We recall that the simplicial polytopes are those for which all the faces are simplexes. This restriction is not too severe, in the sense that, if we consider the polytopes as convex envelopes of finite sets of points, in the case of a **generic** point set, the convex envelope will be a simplicial polytope.

Now we give a simplicial polytope P having n vertices $v_1, v_2, \ldots, v_n$. Just as with an algebraic variety we associate the quotient of the ring of all polynomials by the polynomials which define it as the set of their common zeros, here we introduce the ring R_P which is the quotient of $K[X_1, \ldots, X_n]$ by the ideal generated by all the monomials $X_{i_1}, \ldots, X_{i_s}$ such that $\{v_{i_1}, \ldots, v_{i_s}\}$ is an s-face of P (we do not specify the field K). We obtain thus a *graded ring* A (that is, a decomposition as a finite direct sum of A_r such that the products $A_r \cdot A_s$ satisfy $A_r \cdot A_s \subset A_{r+s}$ for all the integers indexing the direct sum for the addition. It turns out that the degrees correspond to the dimensions of the faces, and the dimensions of the components of this gradation thus furnish exactly the f-vector of the polytope. We call the structure thus obtained the *Stanley ring* of the polytope. The conditions of Sect. VIII.10 are then obtained by considerations of pure algebra and of algebraic geometry, for example the toroidal varieties; for more see 2.3 of Gruber and Wills (1993).

The preceding consists of associating canonically an algebraic object (a ring) with each simplicial polynomial and, using tools of algebra and algebraic geometry for studying these objects in order to find relations for the f-vectors. Recently we engaged in a conceptual ascent by a rung up the ladder by creation of an algebraic object that represents, not just one, but the set of all polytopes. Here we are working with simple polytopes, those for which each vertex belongs to exactly d faces. We remark that they are, when defined as (compact) intersections of hyperplanes, the generic polytopes.

We summarize briefly the essence of the basic text (McMullen, 1993). We fix the dimension d of the space $\mathbb{R}^d$ where we consider the simple polytopes, beginning with the abstract Abelian group Π for which the generators are elements

Fig. VIII.13.5.

[P] associated with all the simple polytopes (in $\mathbb{R}^d$), the neutral element being associated with the empty polytope. This object is very big, we need to take the quotient by all the relations of the form $[P \cup Q] + [P \cap Q] = [P] + [Q]$ each time that $P \cap Q$ is again a simple polytope. We add the invariance under translation: $[P] = [P + v]$ for all the vectors v of $\mathbb{R}^d$. We next make a vector space of Π by setting $\lambda[P] = [\lambda P]$; then above all an algebra of of Π. We recall that an algebra is a vector space endowed moreover with a third operation — a product — with of course axioms of compatibility between the three operations now at hand. The multiplication is defined by $[P] \cdot [Q] = [P+Q]$, where "+" is the Minkowski addition (see Sect. VII.5.B). Finally, we need to associate with each polytope P, not [P], but the subalgebra $\Pi(P)$ of Π generated by all the $[P + Q]$, where Q is arbitrary.

It is shown in (McMullen, 1993) that Π has a canonical structure as a *graded algebra* direct sum of vectorial subspaces Ξ_r with the conditions (as above *chez* Stanley) $\Xi_r \cdot \Xi_s \subset \Xi_{r+s}$. This gradation is interpreted, but subtly, with not only the polytopes themselves, but also the dimensions of the faces as well as the various notions of volumes and of mixed volumes (see Sect. VII.13.B). With the aid of results from pure algebra, this very general context permits us to recover the results of Sect. VIII.10 — and also Brunn-Minkowski (see Sect. VII.8) — as well as several new things. An essential point is that each $\Pi(P)$ also inherits a gradation in $\Xi_r(P)$ and that then the number of faces r of P is nothing other than the dimension of the vector space of $\Xi_r(P)$.

Reading McMullen (1993) is difficult, reading Morelli (1993a) still more so, but the latter contains an impressive panoply of tools coming from various domains and raises the theory of polyhedra to a new level on the ladder.

Let us return to earth by ending this chapter with an open problem of disarming simplicity; see Stanley (2000):

> Is it true that the f-vector $(f_0, f_1, \ldots, f_{d-1})$ of a center-symmetric polytope always satisfies the inequality $1 + f_0 + \cdots + f_{d-1} \geqslant 3^d$?

Bibliography

[B] Berger, M. (1987, 2009) *Geometry I,II*. Berlin/Heidelberg/New York: Springer
[BG] Berger, M., & Gostiaux, B. (1987) *Differential Geometry: Manifolds, Curves and Surfaces*. Berlin/Heidelberg/New York: Springer
Alexandrov, A. (2005). *Intrinsic geometry of convex surfaces*. Chapman & Hall: CRC Press

Alexandrov, A. (2005). *Convex polyhedra*. Berlin/Heidelberg/New York: Springer

Andreev, E. (1970). On convex polyhedra in Lobachevskii spaces. *Mathematics of the USSR, Sbornik, 12*, 413–440

Arnold, V. (1990). *Huyghens and barrow, Newton and Hooke*. Boston: Birkhäuser

Arnold, V. (2000). Polymathematics: Is mathematics a single science or a set of arts? In V. Arnold, M. Atiyah, P. Lax, & B. Mazur (Eds.), *Mathematics: Frontiers and perspectives*. Providence: American Mathematical Society, 403–416

Bainville, E., & Genevés, B. (2000). Constructions using conics. *Mathematical Intelligencer, 22*(3), 60–72

Barnette, D. (1973). A proof of the lower bound conjecture for convex polytopes. *Pacific Journal of Mathematics, 46*, 349–354

Benoist, Y., & Hulin, D. (2004). Itération de pliages de quadrilatères. *Inventiones Mathematicae, 157*, 147–194

Berger, M. (2003). *A panoramic view of riemannian geometry*. Berlin/Heidelberg/New York: Springer

Berger, M. (2005a). Dynamiser la géométrie élémentaire: Introduction à des travaux de Richard Schwartz. *Atti della Accademia Nazionale dei Lincei. Classe di Scienze Fisiche, Matematiche e Naturali. Rendiconti Lincei. Serie IX. Matematica e Applicazioni. Accad. Naz. Lincei, Rome, 25*, 127–153

Bochnak, J., Coste, M., & Coste-Roy, M.-F. (1998). *Real algebraic geometry*. Berlin/Heidelberg/New York: Springer

Boroczky, K., & Boroczky, K. (1996). Isoperimetric problems for polytopes with a given number of vertices. *Mathematika, 43*, 237–254

Brägger, W. (1992). Kreispackungen und Triangulierungen. *L'enseignement mathématique, 38*, 201–217

Bronsted, A. (1983). *An introduction to convex polytopes*. Berlin/Heidelberg/New York: Springer

Bruggesser, H., & Mali, P. (1971). Shellable decomposition of cells and spheres. *Mathematica Scandinavica, 29*, 197–205

Carrega, J.-C. (1981). *Théorie des corps:la règle et le compas*. Paris: Hermann

Chouika, R. (1999). Problems on polygons and Bonnesen-type inequalities. *Indagationes Mathematicae, 10*, 495–506

Colin de Verdière, Y. (1991). Un principe variationnel pour les empilements de cercles. *Inventiones Mathematicae, 104*, 655–669

Connelly, R. (1978). A counterexample to the rigidity conjecture for polyhedra. *Publications mathm´ atiques de l'Institut des hautes études scientifiques, 47*, 333–338

Connelly, R., Sabitov, I., & Walz, A. (1997). The bellows conjecture. *Beiträge zur Algebra und Geometrie, 38*, 1–10

Coxeter, H. (1973). *Regular polytopes*. New York: Dover

Coxeter, H. (1974). *Regular complex polytopes*. Cambridge: Cambridge University Press

Croft, H., Falconer, K., & Guy, R. (1991). *Unsolved problems in geometry*. Berlin/Heidelberg/New York: Springer

Cromwell, P. (1997). *Polyhedra*. Cambridge: Cambridge University Press

Cundy, H., & Rollet, A. (1961). *Mathematical models*. London: Oxford University Press

Damphousse, P. (2005). *Petite introduction à l'algorithmique*. Paris: Ellipses

Darboux, G. (1879). De l'emploi des fonctions elliptiques dans la théorie du quadrilatère plan. *Bulletin des sciences mathématiques, 3*, 109–128

Fejes Tóth, L. (1964). *Regular figures*. London: Pergamon

Fejes Tóth, L. (1972). *Lagerungen in der Ebene, auf der Kugel und im Raum* (2nd ed.). Berlin/Heidelberg/New York: Springer

Fejes Tóth, L. (1948). The isepiphan problem for n-hedra. *American Journal of Mathematics, 70*, 174–180

Fischer, G. (1986). *Mathematische modelle, mathematical models*. Germany: Vieweg

Foth, P., & Lozano, G. (2004). The geometry of polygons in R5 and quaternions. *Geometriae Dedicata, 105*, 209–229

Goldberg, M. (1934). The isoperimetric problem for polyhedra. *The Tohoku Mathematical Journal, 40*, 226–236

Gottlieb, C. (1999). The simple and straightforward construction of the regular 257-gon. *Mathematical Intelligencer, 21*(1), 31–41

Gruber, P. (2007), Convex and Discrete Geometry. Berlin/Heidelberg/New York: Springer

Gruber, P., & Wills, J. (Eds.). (1993). *Handbook of convex geometry*. Amsterdam: North-Holland

Grünbaum, B. (1967). *Convex polytopes* (1st ed.). London: Wiley Interscience

Grünbaum, B. (1993). *Convex polytopes* (2nd ed.). Berlin/Heidelberg/New York: Springer

Hardy, G., & Wright, E. (1938). *An introduction to the theory of numbers*. Oxford: Clarendon Press

Hauchecorne, B., & Suratteau, D. (1996). *Des mathématiciens de A à Z*. Paris: Ellipses

Hirzebruch, F. (1987b). *Collected papers*. Berlin/Heidelberg/New York: Springer

Hodgson, C., & Rivin, I. (1993). A characterization of compact convex polyhedra in hyperbolic 3-space. *Inventiones Mathematicae, 111*, 77–111

Hodgson, C., Rivin, I., & Smith, W. (1992). A characterization of convex hyperbolic polyhedra and of convex polyhedra inscribed in a sphere. *Bulletin of the American Mathematical Society, 27*, 246–251

Holden, A. (1971). *Shapes, space and symmetry*. Columbia: Columbia University Press

Iwasaki, K., Kenma A., & Matsumoto, K. (2002). Polynomial invariants and harmonic functions related to exceptional regular polytopes. *Experimental Mathematics, 11*, 313–319

Kalai, G. (1988). Many triangulated spheres. *Discrete & Computational Geometry, 3*, 1–14

Kanel-Belov, A., Dystein A., Estrin, Y., Pasternak, E., & Ivanov-Pogodaev, I. A. (2008). Interlocking of convex polyhedra: Towards a geometric theory of fragmented solids. *Publications mathématiques de IHES*

Kapovich, M., & Millson, J. (1996). The symplectic geometry of polygons in Euclidean space. *Journal of Differential Geometry, 44*, 479–513

Klein, F. (2003). *Lectures on the icosahedron*. New York: Dover

Lakatos, I. (2008). *Proofs and refutations*. Cambridge: Cambridge University Press

Lindelöf, L. (1869). Propriétés générales des polyèdres qui, sous une étendue superficielle donnée, renferment le plus grand volume. *Bulletin of Academy of Sciences St. Petersburg, 14*, 257–269

Lyusternik, L. (1963). *Convex figures and polyhedra*. New York: Dover

McMullen, P. (1971). The number of faces of simplicial polytopes. *Israel Journal of Mathematics, 9*, 559–570

McMullen, P. (1993). On simple polytopes. *Inventiones Mathematicae, 113*, 419–444

Minkowski, H. (1897). Allgemeine Lehrsätze über die konvexen Polyeder. In Gesmmaelte Abh. Leipzig, 1911: Teubner

Morelli, R. (1993a). A theory of polyhedra. *Advances in Mathematics, 97*, 1–73

Morelli, R. (1993b). Translation scissors congruence. *Advances in Mathematics, 100*, 1–27

Oda, T. (1988). *Convex bodies and algebraic geometry: An introduction to the theory of toric varieties*. Berlin/Heidelberg/New York: Springer

Pogorelov, A. (1973). *Extrinsic geometry of convex surfaces*. Providence: American Mathematical Society

Richter-Gebert, J. (1996). *Realization spaces of polytopes*. Berlin/Heidelberg/New York: Springer

Richter-Gebert, J., & Ziegler, G. (1995). Realization spaces of 4-polytopes are universal. *Bulletin of the American Mathematical Society, 32*, 403–412

Rideau, F. (1989). Les systèmes articulés. *Pour la Science, 136*, 94–101

Roman, T. (1987). *Reguläre und halbreguläre Polyeder* (2nd ed.). Germany: Harry Deutsch, Frankfurt am Main

Roth, B. (1981). Rigid and flexible frameworks. *The American Mathematical Monthly, 88*, 6–21

Rouché, E., & de Comberousse, C. (1912). *Traité de géométrie* (deux volumes). Paris: Gauthier-Villars

Schlenker, J.-M. (1998). Métriques sur les polyèdres. *Journal of Differential Geometry, 48*, 323–405

Schlenker, J.-M. (2002). La conjecture des soufflets (d'après I. Sabitov). *Séminaire Bourbaki* nı 912, novembre 2002

Schramm, O. (1992). How to cage an egg. *Inventiones Mathematicae, 107*, 543–560

Schulte, E. (1987). Analogues of Steinitz's theorem about non-inscribable polytopes. In *Intuitive Geometry, Colloquia Soc. Janos Bolyai* (pp. 503–516). Amsterdam: North-Holland

Schwartz, R. (1992). The pentagram map. *Experimental Mathematics, 1*, 71–81

Schwartz, R. (2001). The pentagram map is recurrent. *Experimental Mathematics, 10*, 519–528

Stanley, R. (2000). Positivity problems and conjectures in algebraic combinatorics. In V. Arnold, M. Atiyah, P. Lax, & B. Mazur (Eds.), *Mathematics: Frontiers and perspectives, American Mathematical Society*, 295–320

Steinitz, E. (1927). Über isoperimetrische Probleme bei konvexen Polyedern I, II. *Journal fr die Reine und Angewandte Mathematik, 158*, 129–153

Stewart, I. (1973). *Galois theory*. London: Chapman and Hall

Stillwell, J. (2001). The story of the 120-cell. *Notices of the American Mathematical Society, 48*(1), 17–25

Stoker, J.J. (1968). Geometrical problems concerning polyhedra in the large. *Communications on Pure and Applied Mathematics, XXI*, 119–168

Thurston, W. (1978). *Geometry and topology of 3-Manifolds* (course notes), Princeton: Princeton University

Tucker, A. (1935–2000). Book review of Steinitz's book on polyhedra. *Bulletin of the American Mathematical Society, 41–37*, 458–471, 92–93

Ziegler, G. (1995). Lectures on polytopes. Berlin/Heidelberg/New York: Springer

Ziegler, G. (2001). Questions about polytopes. In S. Ensquist (Ed.), *Mathematics unlimited – 2001 and beyond* (pp. 1095–1211). Berlin/Heidelberg/New York: Springer

Chapter IX

Lattices, packings and tilings
in the plane

Preliminary note. For the arrangement of Chaps. IX and X, we have adopted the following
policy concerning the interplay between dimension 2 and dimensions 3 and higher. In the present
chapter — in spite of its title — we will treat higher dimensions for those subjects that will not be
treated in detail in the next chapter, either because we have chosen not to discuss them further or
because we want to treat matters extremely fast. Finally, we will encounter practically no open
problems in the case of the plane, contrary to the spirit of our book; this will be entirely different
in Chap. X.

IX.1. Lattices, a line in the standard lattice $\mathbb{Z}^2$ and the theory of continued
fractions, an immensity of applications

A *lattice* in the real affine plane A is a subset Λ that can be written as the set of
all $a + m \cdot u + n \cdot v$, where a is a point of A and — having vectorized A — where u, v
are lineally independent vectors and where m, n range over the set $\mathbb{Z}$ of all integers.
In the affine realm, there is essentially but a single lattice, i.e. $\mathbb{Z}^2 \subset \mathbb{R}^2$, which is to
say the set of pairs (m, n) of integers. If we have in addition a Euclidean structure,
then there will be different lattices; this will be the subject of Sect. IX.4, although
we will encounter these "other" lattices starting with Sect. IX.3 below.

The lattice $\mathbb{Z}^2$ and more generally the lattices of the Euclidean plane appear
abundantly in mathematics: in number theory, which isn't surprising; in analysis
when there is anything to do with doubly periodic functions in two variables (Fourier
series in two variables), but also in the theory of functions of a complex variable
(elliptic functions, which we encountered through Poncelet's theorem in Sect. IV.8);
in "pure" geometry when we study compact locally Euclidean geometries (a particular case of the *space forms*, see Sect. VI.XYZ), which arise naturally in billiards in Sect. XI.2.A; in physics, even though crystal structures are situated in three
dimensional space, when these are cut by planes or we study their diffraction patterns by X-rays or other processes, we obtain planar lattice images (Brillouin zones,
see for example 3.7 of Gruber and Wills (1993)). It is for readers to find still other
resonances for lattices.

♦

Decimal calculation is a marvelous invention; in particular, the operations of addition and multiplication are done "trivially". But it is not the best adapted for certain
problems involving rational numbers, since most of these have an infinite decimal
expansion ($\frac{1}{3} = 0,33333...$, etc.). Above all, it is not obvious from the decimal

M. Berger, *Geometry Revealed*, DOI 10.1007/978-3-540-70997-8_9,
© Springer-Verlag Berlin Heidelberg 2010

expansion how an irrational number can be approximated by rationals other than the crude $\frac{k}{10^n}$. Now such approximation is fundamental in number theory; the Greeks, who approximated π by $\frac{22}{7}$, then by $\frac{333}{106}$ and $\frac{355}{113}$, understood this well. It involves the *theory of continued fractions* (a term that will be explained right away), that has been well understood since the middle of the twentieth Century. If we are to be reproached for attaching so much importance to this part of number theory, numerous motivations arising from problems in geometry can be found below. The elementary theory of continued fractions, as simple as it may be, is always a bit subtle and hard to visualize.

This is why, even though algebra is always necessary, it is agreeable here to present the beginnings of this theory in a geometrical fashion. Here we follow Chap. 7 of Stark (1970), but see also, for a summary, the Klein polygons of Erdös, Gruber and Hammer, (1989, p. 101). The basic idea for approximating a real by rationals is surely that the approximation can be better when the denominator is larger. The other idea is to consider the reals α of $\mathbb{R}^2$ as slopes of half lines emanating from the origin in the quadrant $\mathbb{R}_+ \times \mathbb{R}_+$. We feel confident that the farther we go in the direction of α, since there are always points at a distance less than 1 from the given half line D of slope $\alpha > 0$, the more we can find rational slopes as close as we want to α. The whole interest of continued fractions is to end up doing this both as systematically and as economically as possible; i.e. we would like that the $\frac{p}{q}$ chosen at the stage considered minimizes $|\alpha - \frac{p}{q}|$ among all the $|\alpha - \frac{p'}{q'}|$ for which the denominator q' is $\leq q$.

It turns out that we can do both things simultaneously. In all that follows, we assume that α is irrational, i.e. the half line D does not contain any point of $\mathbb{Z}^2$ other than the origin. We now proceed by recurrence. It is nice to use vectorial language and to consider the vectors $V_n = (p_n; q_n)$, where the p_n/q_n denote the desired approximations for α. The first vector $V_0 = (1; [\alpha])$ is of course that given by the integer part $[\alpha]$ of α. As a point it is situated below D, but as close as possible among the vectors of the form $(1, q)$. The next vector will be $V_1 = (0, 1) + a_1 V_0$, where a_1 is the *last* integer such that V_1 is above D. Then V_2 will be $V_0 + a_2 V_1$, which is the last of this form below D, etc. We show (however we do it, it is never obvious) that the sequence $\{V_n = V_{n-2} + a_n V_{n-1} = (p_n, q_n)\}$ thus defined by recurrence is indeed automatically the best approximation possible in the sense of minimizing with bounded denominator.

We have thus an alternating sandwiching:

$$\frac{p_0}{q_0} < \frac{p_2}{q_2} < \frac{p_4}{q_4} < \cdots \alpha \cdots < \frac{p_3}{q_3} < \frac{p_1}{q_1}$$

with $\lim_{n\to\infty} \frac{p_n}{q_n} = \alpha$. By construction

$$\left|\alpha - \frac{p}{q}\right| \geq \left|\alpha - \frac{p_n}{q_n}\right| \quad \text{if } q \leq q_n,$$

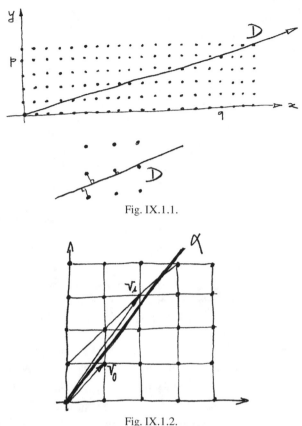

Fig. IX.1.1.

Fig. IX.1.2.

but then the sandwiching is also good in an unexpected way, as we always have

$$\left| \alpha - \frac{p_n}{q_n} \right| \le \frac{1}{q_n^2}.$$

See Sect. IX.3 for a lightning fast and optimal proof!

This provides an explanation of the term "continued fraction", as the preceding shows (proceed with gcd's) that we can simply define the pairs $(p_n ; q_n)$ with the aid of the sequence of integers $\{a_n\}$ that are defined by recurrence by an **integer part** condition on the sequences $\{\alpha_n, a_n\}$, where we have already seen that $a_0 = [\alpha]$, and further:

$$\alpha_1 = \frac{1}{\alpha - a_0}, \ a_1 = [\alpha_1], \ \alpha_2 = \frac{1}{\alpha_1 - a_1}, \ a_2 = [\alpha_2], \ldots$$

Then α is perfectly defined by the infinite sequence $\{a_0, a_1, a_2, \ldots\}$ and the convergent sequence of successive fractions

$$a_0, \quad a_0 + \frac{1}{a_1}, \quad a_0 + \cfrac{1}{a_1 + \cfrac{1}{a_2}}, \ \text{etc.,}$$

whence the expression *continued fraction*, wanting in fact to say "infinite fraction".
We write by notational convention:

$$\alpha = \langle a_0, a_1, a_2, \ldots \rangle.$$

For the details of the proofs, see Stark (1970).

We now owe our readers some examples. Here are two extremes: the first is
$\pi = \langle 3, 7, 15, 1, 292, \ldots \rangle$ for which the associated continued fractions are thus:

$$3, \quad 3 + \frac{1}{7} = \frac{22}{7}, \quad 3 + \cfrac{1}{7 + \cfrac{1}{15}} = \frac{333}{106}, \quad \frac{355}{113}, \quad \frac{103\,993}{33\,102}, \quad \ldots$$

Then $\sqrt{3} = \langle 1, 1, 2, 1, 2, 1, 2, \ldots \rangle$ for the fractions

$$1, \quad 2, \quad \frac{5}{3}, \quad \frac{7}{4}, \quad \frac{19}{11}, \quad \frac{26}{15}, \quad \frac{71}{41}, \quad \frac{97}{56}, \quad \frac{265}{153}, \quad \ldots;$$

the continued fraction that represents $\sqrt{3}$ is truly periodic (to infinity).

These examples call for remarks, for questions. The obvious fact that never
ceases to surprise the author is that π, which is supposed to be a very bad irrational
since it is even *transcendental* — it is never a root of a polynomial with integer coefficients — is very well approximated by the rationals, and very quickly! However,
it can't be *so* very well approximated, since we know from Mahler in 1953 that
$|\pi - \frac{p}{q}| \geq \frac{1}{q^{42}}$ for each q. In contrast, it is an open problem to know the best (small-
est) power k such that $|\pi - \frac{p}{q}| \geq \frac{1}{q^k}$ for each integer q. Presently it is known through
$k = 7$; that is, π is surely a transcendental number, but not by so much! Also known
is the best constant c for approximation in the form $|\pi - \frac{p}{q}| \leq \frac{c}{q^k}$. For references
about the number π, the bible is Berggren, Borwein and Bortwein (2000), but it
is necessary to be immersed in it and it is perhaps better to first read the informal
exposition (van der Poorten, 1979).

It has been known since Lagrange in about 1766 how to completely characterize
numbers with periodic continued fractions; these are exactly the so-called *quadratic*
numbers, i.e. which are roots of an equation of second degree with integer coefficients (see e.g. Stark).

This does not yet tell us which is the worst number by this theory, which is
then the one with continued fraction $\langle 1, 1, 1, \ldots \rangle$, and this is nothing other than the
golden ratio, the most beautiful irrational. We encountered it with Perles in Sect. I.4
and in the study of non rational polytopes in Sect. VIII.12; we will encounter it
again in Sect. XI.2.A.

Here, as promised, are quite different places where continued fractions appear. They
are found first in the Poncelet polygons, seen in Sect. IV.8. Then will soon mention,
without detail, their appearance in the shortest path geometry in the space (called a
modular domain) which classifies all the lattices in the Euclidean plane: Sect. IX.4.

But they also appear in more general domains, for example the tori: (Series, 1982), (Series, 1985), very pedagogic texts. We also find continued fractions in Sect. I.6 and 4.1 of Oda, (1988). They are used in the very difficult proof of the conditions (Sect. VIII.10.5). Finally, but this isn't exhaustive, in the classification of Penrose tilings of Sect. IX.9; see p. 563 of Grünbaum and Shephard (1987).

We will see them appear in the billiards of Sect. XI.2.A, almost by definition, first in the square billiard. They appear also in concave (also called hyperbolic) billiards, but in a much more subtle way; see p. 252 of Sinai (1990). In fact, we will dedicate Chap. XI entirely to billiards, because beyond their intrinsic and natural geometric interest, they are models for mathematical physics: for statistical mechanics, the study of dynamical systems and yet other areas.

Finally, we recall − see Sect. II.XYZ − the vernier caliper, which furnishes a tenth of a millimeter, and the micrometer, which furnishes a hundredth of a millimeter. We are dealing exactly with the first two terms of a continued fraction for approximating a thickness to be measured; see Fig. II.XYZ.4.

♦

Apart from this book, continued fractions can be found first in number theory; besides Stark, see Hardy and Wright (1938), which still remains the old testament of the subject. See also the appendix to Lang, 1983) and Berggren et al. (2000). For their use in the study of surfaces in algebraic geometry, see Hirzebruch and Zagier (1987); and for the study of toroidal varieties; see Oda (1988).

In dynamics, in the study of diffeomorphisms of the circle (whose study constitutes one of the simplest dynamical models), see Douady (1995); in rational mechanics, see p, 97 of Gallavotti (1983); in statistical mechanics, see Lanford III and Ruedin (1996); and for the stability of the solar system, see (Marmi, 2000). The partly historical article (Flajolet, Vallée and Vardi, 2000) is fascinating.

♦

There exists a theory of continued fractions in several variables, one of the goals being simultaneous approximation of several irrationals. We refer readers to works on geometric number theory: Gruber and Lekerkerker (1987), (Erdös et al. (1989), Cassels (1972). We mention only two things: first that the natural problem of knowing which pair (or pairs) replaces the golden ratio − for being approached in the worst possible way − remains open; then the existence of simultaneous approximations$|\alpha_i - \frac{n_i}{N}| \leqslant \frac{\bullet}{N}$ of $\{\alpha_1, \ldots, \alpha_k\}$ with increasingly large integers N is an immediate consequence of Minkowski's theorem; see Sect. IX.3.

IX.2. Three ways of counting the points $\mathbb{Z}^2$ in various domains: pick and Ehrhart formulas, circle problem

The *Pick formula*, which dates from 1899, see Grünbaum and Shephard (1993), furnishes the number of points of the standard lattice $\mathbb{Z}^2$ contained by a (not necessarily convex) plane polygon P with vertices in $\mathbb{Z}^2$ and its interior:

$$\#\{P \cap \mathbb{Z}^2\} = \text{Area}(P) + \frac{1}{2}\#\{\partial P \cap \mathbb{Z}^2\} + 1$$

where ∂P denotes the boundary of P.

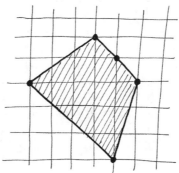

Fig. IX.2.1. Counting trees by touring the domain: $\underbrace{\frac{1}{2}6 + \frac{1}{2}4 + \frac{1}{2}3 + \frac{1}{2}12}_{\text{Area}(P)}$

$$+ \quad \underbrace{\frac{1}{2}5}_{\#\{\partial P \cap \mathbb{Z}^2\}} \quad +1 = 16$$

There does not exist any clear proof of this most simple result. The basic reference presently is Grünbaum and Shephard (1993). An explanation in depth can be found there; but an heuristic proof, based on the way that heat diffuses in the plane (and called a *Gedankenexperiment* by its author), is interesting to read: Blatter (1997). We see in the first reference that the Pick formula is used by foresters in Canada for counting the number of trees in a given domain; the area is known from the land registry, we only need make a tour while counting the trees on the periphery, there being no need of counting those on the interior!

In the following chapter we will not treat at all the various generalizations of Pick's formula, which seem to be neither very conceptual nor spectacular as they appear in 3.2.4 of Gruber and Wills (1993). In contrast, we find in Savo (1998) a conceptual context for study: as with Blatter, the *heat content* of a general domain of a general Riemannian manifold. We find moreover in 3.2.5 of Gruber and Wills (1993) some motivations for studying polytopes whose vertices belong to a lattice; they arise in a natural way in problems of combinatoric optimization.

The *Ehrhart formula* (ca. 1964) studies the behavior in λ of the number of points of $\mathbb{Z}^2$ in polygons that are homothetic in the ratio λ to an initial polygon P (always with vertices in $\mathbb{Z}^2$). It says that the behavior is quadratic in λ. Let P be a convex polygon whose vertices are in $\mathbb{Z}^2$. Then:

$$\#\{\lambda P \cap \mathbb{Z}^2\} = a(P)\lambda^2 + b(P)\lambda + c(P),$$

where: $a(P) = \text{Area}(P)$, $b(P) = \frac{1}{2}(\text{perimeter of P})$, $c(P) = 1$. The proof is left to readers; it can be done without any conceptual apparatus. It is used in number theory and in fact it was motivated by research there.

But we mention it especially for reasons that concern its generalization to $\mathbb{Z}^d$, for arbitrary dimension d, first because of the proof that, for each polytope with integer vertices, we have a polynomial formula in λ:

$$\#\{\lambda P \cap \mathbb{Z}^d\} = \sum_{i=0,1,...,d} a_i(P)\lambda^i, \quad \text{polynomial of degree } d \text{ in the integer } \lambda,$$

which is not at all simple. The basic idea is to decompose the polytope into simplexes and to use a certain additivity. Moreover this formula was not discovered and proved until (Ehrhart, 1964), which may be consulted for all this. The second reason is that there does not exist today a good interpretation of the coefficients $a_i(P)$. We certainly have $a_d(P) = \text{Volume}(P)$, $a_{d-1}(P) = \frac{1}{2}\text{Area}(\partial P)$, $a_0(P) = 1$, but the other terms are not well known. They are connected with mixed volumes, encountered in Sect. VII.13.B, but seem to still have an additional hidden depth.

Returning to polygons in the plane, can we say something if their vertices are arbitrary in the plane, no longer necessarily with integer coordinates? We can then study their *discordance*, that is to say the function

(IX.2.1) $$\text{Disc}(P) = \#\{P \cap \mathbb{Z}^2\} - \text{Area}(P).$$

In 2.3 of Huxley (1996) we find a detailed study of this discordance, and it is not surprising to see continued fractions appear. The transition with what follows is entirely natural; we look at the discordances of convex sets more general than polygons, the first case being of course that of the (circular) disk.

The *circle problem*, or "Gauss circle problem", is famous among number theorists, as well as among analysts. Here it has to do with counting how many points of $\mathbb{Z}^2$ are in the circle of radius R centered at the origin:

In fact, no one would even dream of an exact formula for $F(R) = \#\{(m,n):$ $m^2 + n^2 \leq R^2\}$. What is investigated is a good (close) asymptotic evaluation of $F(R)$ when R tends toward infinity, which is to say the behavior of the discordance

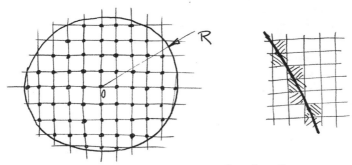

Fig. IX.2.2. $F(R) = \#\{(m,n) : m^2 + n^2 \leq R^2\}$

$\mathrm{Disc}(\mathrm{B}(0, \mathrm{R})) = \#\{\mathrm{B}(0, \mathrm{R}) \cap \mathbb{Z}^2\} - \pi \mathrm{R}^2$. Of course F(R) isn't a smooth function; it's a step function with integer values.

We get a good, though not very good, estimate by the following elementary geometric argument. Among the squares with side 1 that interest us, there are two types: those entirely interior to the circle $x^2 + y^2 = \mathrm{R}^2$ and those which straddle the circle. It is clear that the number of the latter is of order equal to the perimeter of the circle, namely $2\pi \mathrm{R}$, whereas those interior are in number of order the area of the circle, namely $\pi \mathrm{R}^2$. Since $2\pi \mathrm{R}$ becomes negligible compared to $\pi \mathrm{R}^2$ when R goes to infinity, we then have the coveted estimate:

$$\mathrm{F(R)} \sim_{\mathrm{R}\to\infty} \pi \mathrm{R}^2,$$

where the symbol "$\sim$" means that the ratio F(R) to $\pi \mathrm{R}^2$ tends to 1 as R tends toward infinity.

In view of the Ehrhart formula we might think that we can't do better, since the perimeter of a circle is of the order of R, but we are ever greedy. In the applications to number theory, there is a need for the second term of the asymptotic development of F(R). It has to do with controlling $\mathrm{F(R)} - \pi \mathrm{R}^2$. The *circle problem* is to prove (in case it's true) that we have

$$\mathrm{F(R)} - \pi \mathrm{R}^2 =_{\mathrm{R}\to\infty} \mathrm{O}(\mathrm{R}^{\frac{1}{2}+\varepsilon}) \quad \text{for each } \varepsilon > 0,$$

where the classic notation "large O" ("O") means that the quotient with $\mathrm{R}^{\frac{1}{2}+\varepsilon}$ is bounded independent of R when R tends toward infinity. It is one of the very great open problems of mathematics. For information on the circle problem, the most recent and up-to-date reference is Huxley (1996), but it is difficult to get into; for things relatively less detailed and less recent consult II § 4 of Gruber and Lekerkerker (1987). However, we know since 1915 that $\mathrm{O}(\mathrm{R}^{\frac{1}{2}})$ is false, but subsequently there have been successively better valid exponents. The power $\mathrm{R}^{1/2}$ for this second term appears unlikely at the outset: if we proceed by ignoring R in comparison with R^2, we would then think of requiring a second term in O(R). The Ehrhart formula shows that for the homotheties of the square we effectively have such a term in R. What happens for the circle — the sketch can give an idea of it — is that when we turn about the circle of radius R certain compensations are produced. Such is indeed the case, since nowadays we know that it was then realized very quickly, in 1920, that $\frac{2}{3}$ is possible. It is now $\mathrm{O}(\mathrm{R}^{46/73+\varepsilon})$. The numbers show clearly that the argument is not yet completely conceptual, at least not in the successive improvements.

We need to climb the ladder to prove such a formula; the classic and basic tool is analysis, specifically the Fourier transform already encountered for the sections of the cube in Sect. VII.11.C; see there the "physicist's reflex": when in doubt or if all else fails, take the Fourier transform!

We have counted the points of $\mathbb{Z}^2$ in polygons, more precisely the behavior of the number of points in their homotheties as a function of the homothety ratio, then we looked at the same problem for the circle. But between the polygons and the circle there are all the convex sets. Moreover, the proofs for the circle all extend more or less to the smooth convex sets. Not only a natural problem, but in the case of the circle it is strongly motivated by number theory. The more general convex domain case is motivated by the needs of numerical analysis in problems associated with these convex sets, the simplest of which is to look for a good approximation to the integral $\int_{\mathbb{R}^2} f(x)\,dx$ of a numerical function with compact support. In fact, all the results established for the circle carry over to the convex sets with smooth boundary of class C^∞. Furthermore, we would like to obtain results with the exponents varying between $\frac{1}{2} + \varepsilon$ and 1 (the case of exponent 1 is that of the polygons, and so we cannot do better). All these things are treated in great detail in Huxley (1996), where the complete panoply of diverse tools used in attacking this problem can be found. They range from geometry to analysis (Fourier series, exponential sums, oscillatory integrals), passing through number theory to the geometry of curves.

Some results very briefly, taken from Gruber and Lekerkerker (1987). They deal with convex sets C for which the boundary is a curve of class C^∞. If, at points where the curvature vanishes it has an order at most equal to the integer k, then for each homothety factor λ (tending toward infinity) we find:

$$\#\{\lambda C \cap \mathbb{Z}^2\} = \lambda^2 \operatorname{Area}(C) + O(\lambda^{2/3}) \quad \text{if } k = 0 \text{ or } 1$$

$$\#\{\lambda C \cap \mathbb{Z}^2\} = \lambda^2 \operatorname{Area}(C) + O(\lambda^{k+1/k+2}) \quad \text{for the other } k.$$

We should not fail to observe a certain optimality of the large k, by considering the convex set with equation $x^n + y^n \leqslant 1$: when n tends toward infinity, on the one hand the convex set C tends toward the square $\{|x| \leqslant 1, |y| \leqslant 1\}$; on the other hand the curvature at the points of intersection with the axes vanishes with an increasing order (since equal to $n - 2$); see as needed Fig. VII.4.5.

For the proofs, the simplest method, that gives

$$|\#\{(m,n) : m^2 + n^2 \leqslant R^2\} - \pi R^2| \leqslant 164000\, R^{2/3}$$

and that is valid more generally for every convex C^∞ curve, is in Chaix (1972). It uses the approximation of the smooth curve by polygons for which the sides have rational slope. Working with polygons is one of the cornerstones of Huxley (1996). For the exponent $\frac{2}{3}$ and improvements, the proofs use among other things the Poisson formula (IX.4.3) below. If we apply it to the characteristic function of the disk of radius R and study the evaluation of the second term — which is not very difficult — we find $\frac{2}{3}$ as stated.

The problem is truly natural; what is interesting for the number theorist is not so much the estimation of the number of pairs of integers (m, n) such that $m^2 + n^2 \leqslant R^2$, but more the number of those such that $m^2 + n^2 = R^2 = N$, i.e. the number of ways in which we can decompose a given integer N as the sum of two squares (recall the sums of 24 squares in Sect. III.3).

Another very natural variant in number theory consists of keeping only those points of $(p,q) \in \mathrm{B}(0,\mathrm{R}) \cap \mathbb{Z}^2$ for which p and q are relatively prime, usually written $(p,q) = 1$ or $\gcd(p,q) = 1$. They are sometimes called *visible points*:

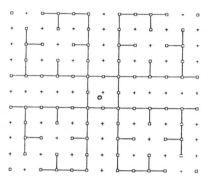

Fig. IX.2.3. Graph of the visible points of $\mathbb{Z}^2$. Erdös, Gruber, Hammer (1989) © John Wiley & Sons, Inc.

We deal with visibility starting at the origin, a problem we encounter in square billiards in Sect. XI.2. For these visible points, also called *primitives*, it has recently been proved (see Huxley and Nowak (1996)) — modulo the Riemann hypothesis — that we indeed have $O(\mathrm{R}^{\frac{1}{2}+\varepsilon})$ (always for each $\varepsilon > 0$). We recall that the Riemann hypothesis is presently the most important open conjecture in mathematics; it predicts that all the nontrivial zeros of the $\zeta(z)$ function of the complex variable z, given by the sum of the series $\sum_{1 \leqslant n} \frac{1}{n^z}$ and continued analytically to the entire plane, are on the line formed of complex numbers z for which the real part is equal to $\frac{1}{2}$.

We also need to make room for the results that consider what happens when we perturb the center of the ball $\mathrm{B}(\mathrm{O}, \mathrm{R})$, initially O, to a point p that belongs to the square $[0,1]^2$. We obtain in this way a function of R and p; we then obtain the desired $\mathrm{R}^{1/2}$ and better: see Bleher and Bourgain (1996).

Apart from number theory, we find motivation and application of the above results in mathematical physics in the theory of *percolation* — much studied presently — in Vardi (1999).

We mention briefly general dimensions greater than 2, because we won't mention them in the next chapter. They have been much less studied than dimension 2, except for some generalizations of majorizations of moderate difficulty in the planar case, where we encounter some ultrafine techniques. Several statements in this regard can be found in Sect. II.4 of Gruber and Lekerkerker (1987).

Let d be the dimension; we seek to estimate asymptotically, as a function of R, the number of points F(R) of the unit lattice contained in the ball with center the origin and radius R. The principal part is very easy; it equals $\beta(d)\mathrm{R}^d$, where $\beta(d)$ denotes the volume of the unit ball that we calculated in Sect. VII.6.B. The essential point is that the $(d-1)$-dimensional volume of the sphere of radius R is of the order of R^{d-1}, and we have $F(R) - \beta(d)R^d = O(R^{d-1})$. Can we do better? In the case of the plane $d = 2$, it is $R^{1/2}$ that we sought to approach, because contrary to the case of polygons (Ehrhart formula) where the original exponent was 1, the circle offers compensations leading toward $1/2$. In arbitrary dimension, we can also hope for strong compensations with respect to R^{d-1}. Here the work turns about $R^{(d-1)/2}$. The texts Walfisz (1957) and Fricker (1982), already rather old, are entirely devoted to this subject, which itself is still rather young. For what happens in arbitrary dimension with the estimation of *visible* points, a folkloric result − see 13.3 of Erdös et al. (1989) − is that their number is asymptotically the number of points without the condition but divided by the value for d, namely the value $\zeta(d)$ of the Riemann ζ function.

IX.3. Points of $\mathbb{Z}^2$ and of other lattices in certain convex sets: Minkowski's theorem and geometric number theory

Minkowski's theorem is very intuitive. If we take a line in $\mathbb{R}^2$ of irrational slope that passes through the origin, it will never intersect $\mathbb{Z}^2$ except at the origin. But if we thicken it ever so little into a band of width ε, then we will definitely find points of $\mathbb{Z}^2$ in this band (if only with the help of the theory of the continued fractions of Sect. IX.1). But we would like to determine the length of a partial version of such a band that is required to be certain of intersecting $\mathbb{Z}^2$. Minkowski's theorem gives a simple and complete response to the problem for every convex set C that is symmetric with respect to the origin:

(IX.3.1) (Minkowski, 1898): $C \cap (\mathbb{Z}^2 \setminus 0)$ *is nonempty provided* Area(C) ≥ 4.

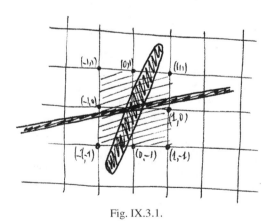

Fig. IX.3.1.

Those who are fond of epistemological remarks will appreciate the fact that Minkowski's theorem is a link between the discrete and the continuous. This does not keep the proof from being completely elementary; we will have noted that the area of the initial square of $\mathbb{Z}^2$ equals 4 and so this value 4 is evidently necessary.

However, this proof requires a bit of care; the works that treat it give slightly different versions. The most conceptual one consists of making formal (but abstract) the following reasoning (producing the figure is left to readers): the horizontals and verticals which bear the points of $\mathbb{Z}^2$ cut the plane into squares of side 1 and the convex set C, assumed compact, into a finite number of pieces that we will take as including their boundary. We relocate each of these pieces in the square $[-1, 1]^2$ by a translation by a vector pair $(2u, 2v)$, with u, v integers. If we suppose that C does not intersect $\mathbb{Z}^2$ outside of the origin, the hypothesis of symmetry, along with that of convexity, shows that the translated pieces are mutually disjoint. In fact, if this were not the case, this would mean that there would be two distinct points $x, y \in C$ such that $x - y \in 2\mathbb{Z}^2$. We would then have $-y \in C$ and $\frac{x-y}{2} \in C$ at the same time that $\frac{x-y}{2} \in 2\mathbb{Z}^2 \setminus 0$, which is a contradiction. Since translation conserves area, this shows that Area(C) $\leqslant 4$. We have in fact Area(C) < 4, because the distance between two pieces in $[-1, 1]^2$ is positive by compactness, which leaves room for a disk outside of the union of the pieces.

Instead of using slicing, the contemporary mathematician prefers applying the quotient mapping of $\mathbb{R}^2$ into the flat torus $\mathbb{R}^2/2\mathbb{Z}^2$ and considering the image of the convex set C under this mapping. The symmetry hypothesis, together with that of convexity, shows that this mapping is injective; as the quotient mapping preserves area, we are done. For flat tori see also Sects. V.14 and XII.5.A.

For applications of the study of lattices to numerical analysis (calculation of multiple integrals), see Chap. 15 of Erdös et al. (1989). This book offers a very interesting panorama of the many places in mathematics where lattices appear.

♦

Minkowski's theorem is archetypical of Minkowski's results in the discipline he completely founded around 1900, called *geometric number theory*. It is an immense area, and we will catch a glimpse why. To get an idea of the extent of this steadily flourishing theory, we can look at the very pedagogical Cassels (1972) and the very complete Gruber and Lekerkerker (1987).

Theorem (IX.3.1) furnishes the natural transition to arbitrary lattices in the Euclidean plane from those that are just in $\mathbb{Z}^2$. In fact, (IX.3.1) is an *affine* result; an affine transformation of the plane transforms a lattice into a lattice and transforms a symmetric convex set into a symmetric convex set; as for areas, they are multiplied by a fixed constant. We now place ourselves in the Euclidean plane and consider lattices Λ. The basic, albeit elementary, remark for what follows is:

(IX.3.2) *Each linear automorphism f of $\mathbb{R}^2$ such that $f(\Lambda) = \Lambda$ is given in a basis of Λ by a matrix $\begin{pmatrix} a & b \\ c & d \end{pmatrix}$ with integer coefficients whose determinant is equal to ± 1. We let $\mathrm{SL}(2; \mathbb{Z})$ denote the group of matrices with integer coefficients and determinant 1.*

In fact, the matrix considered must be invertible and have integer coefficients, and so must its inverse; thus its determinant and its reciprocal are integers; and only

1 and -1 are invertible in the set on integers $\mathbb{Z}$. Such a lattice therefore possesses an *area*; the algebraist will see it as the absolute value of the determinant: $\text{Area}(\Lambda) = |\det(u, v)|$, for any two vectors u, v that form a basis; and according to (IX.3.2) this value does not depend on the basis for Λ chosen, so that it is an invariant of the lattice. The geometer sees $\text{Area}(\Lambda)$ as the area of a parallelogram generated by two basis vectors.

We thus obtain as a consequence of (IX.3.1) that:

(IX.3.3) (Minkowski) *Each center-symmetric convex set* C *whose area satisfies* $\text{Area}(C) \geqslant 4\,\text{Area}(\Lambda)$ *contains at least one point of* $\Lambda \setminus 0$.

We illustrate immediately — and with an archetypical case — the force of this result of simplest appearance by proving the existence of continued fractions such that $|\alpha - \frac{p_n}{q_n}| \leqslant \frac{1}{q_n^2}$. In fact, let α be our number to be approximated by rationals. To each $\varepsilon > 0$ we attach the lattice $\Lambda = \{(\frac{\alpha n - m}{\varepsilon}\,;\,\varepsilon n)\colon m, n \in \mathbb{Z}\}$. It is of area equal to unity, and we consider the convex set $C = \{(x, y) \in \mathbb{R}^2\colon |x| \leqslant 1 \text{ and } |y| \leqslant 1\}$, of area equal to 4. There exists a point (m, n) of $\Lambda \setminus 0$ in C such that $|\alpha n - m| \leqslant \varepsilon$ and $|\varepsilon n| \leqslant 1$, whence $|\alpha - \frac{m}{n}| \leqslant \frac{1}{n^2}$, QED.

The theorem (IX.3.3) is used most often, as in the beginning by Minkowski himself, for studying quadratic forms $ax^2 + 2bxy + cy^2$ with integer coefficients and the values that these can take on as a function of $ac - b^2$. We can also treat certain solutions of Diophantine equations (equations in integers), whence the term *geometric number theory*; see the given references. For example, the celebrated Lagrange theorem that each integer is the sum of four squares can be proved very simply (see the proof in Cassels).

The result (IX.3.3) generalizes in an enormity of ways; see the given references. Here are just some brief indications of natural directions. First, we can seek to find more points in $C \cap \Lambda$ if the area of Λ is still greater. In the spirit that we will see subsequently in the investigation of *dense* lattices, we can proceed in the reverse direction: take a convex set C that is center-symmetric about 0 and look for so-called *critical* lattices for C, that is to say those which are of the smallest area possible among those for which all the points, except the origin, are outside of C or on its boundary. This is an optimization problem. We will dedicate an important part of the following chapter to it, in arbitrary dimension. In the planar case here, we mention the spectacular case of the critical lattice for a domain that is center-symmetric and not convex, but *starred*, given in $\mathbb{R}^2$ by the equation $|xy| \leqslant 1$. The corresponding critical lattice exists, which is a general fact about so-called Mahler compactness; see Sect. X.6. It is the *golden lattice* generated by the two vectors $(1, 1)$ and $(\frac{1}{2} - \frac{\sqrt{5}}{2}, \frac{1}{2} + \frac{\sqrt{5}}{2})$.

From the analogue of (IX.3.3) for starred sets we deduce the result:

(IX.3.4) *For each real number* α *there exist rational approximations satisfying*
$$|\alpha - \frac{m}{n}| \leqslant \frac{1}{\sqrt{5}n^2}.$$

We cannot do better, as is shown precisely by choosing α to be the golden ratio. For the algebraic approach to (IX.3.4) by number theory, see Hardy and Wright (1938) (attention: this is the *third* edition).

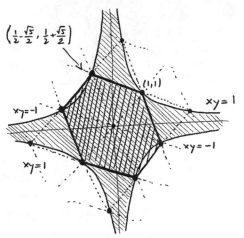

Fig. IX.3.2. The golden lattice

♦

The theorem (IX.3.3) can be extended, without any modification of the proof, to:

(IX.3.5) (Minkowski) *In each dimension d and for each lattice Λ of $\mathbb{R}^d$, if a center-symmetric C is such that* Volume(C) $\geq 2^d$ Vol(Λ), *then C contains at least one point of $\Lambda \backslash 0$.*

The *volume* of a lattice still exists, because only 1 and -1 among the integers are invertible. Geometric number theory, with the help of (IX.3.5), allows us to study other notions related to lattices that we will see in Chap. X: quadratic forms — and even polynomials — in an arbitrary number of integer variables; and readers can cobble together for themselves the existence of simultaneous approximations to a finite set $\{\alpha_i\}$ of reals by rationals of the form $\frac{m_i}{N}$ such that $|\alpha_i - \frac{m_i}{N}| \leq \frac{\circ}{N}$, where the value of the "circle" is universal (generalized continued fractions). It is still unknown today which pairs, etc., of real numbers generalize the golden ratio, i.e. the "worst" of all generalized continued fractions.

IX.4. Lattices in the Euclidean plane: classification, density, Fourier analysis on lattices, spectra and duality

The preceding section has prepared us for verifying that, even if all plane lattices are the same in the affine sense and isomorphic to the canonical $\mathbb{Z}^2$, it is not at all the same if $\mathbb{R}^2$ is considered as a Euclidean space. We have seen an example of this with the golden lattice above. But this is not the most beautiful of the lattices, as we will see. Moreover, crystallography is Euclidean: crystals do not all have the same chemical nature, and even though they are three-dimensional objects, their analysis by diffraction leads to plane lattices. We first need to classify the plane lattices to within Euclidean isomorphism, i.e. within an isometry of $\mathbb{R}^2$. For a technical reason that will appear in what follows, it is more effective to classify by the isometries that preserve orientation. In fact, we go further: we consider two lattices as geometrically the same if they only differ by a homothety; we thus ultimately take the quotient of the set of plane lattices by the group of direct similitudes (see Sect. II.XYZ).

But we have already seen this classification carried out in the digression Sect. V.13, for it amounts to the same thing as classifying cubic curves in the complex plane. We re-sketch the *modular domain* which furnishes this classification. In order to get the true space of all our lattices we need to identify the two vertical half-lines and identify by symmetry the two halves of the arc of the lower circle.

Here in detail is how we make this classification. Each lattice Λ possesses a basis $(a; b)$, where a is a vector of smallest norm, and we can moreover assume that the basis $(a; b)$ is "direct", i.e. that $det(a, b) = 1$. In fact, let $(e_1; e_2)$ be a basis of Λ. If $a = p_1 e_1 + p_2 e_2$ of smallest norm, the integers p_1 and p_2 are relatively prime, and thus there exist integers q_1 and q_2 such that $p_1 q_2 - p_2 q_1 = 1$; we then take $b = q_1 e_1 + q_2 e_2$. If the transition matrix from $(e_1; e_2)$ to (a,b) has determinant 1, we indeed have a basis for the lattice.

We identify the Euclidean plane with $\mathbb{C}$. We work within similitude; we suppose $a = 1$ and $\operatorname{Im} b > 0$, and it remains to classify the lattices possessing a basis of the form $(1; b)$ with $\operatorname{Im} b > 0$ and where all the vectors are of norm ≥ 1.

If these two conditions are satisfied, we necessarily have $|b - n| \geq 1$ for each integer n, that is to say that b belongs to the domain Δ represented in the figure.

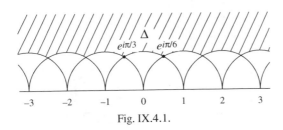

Fig. IX.4.1.

Conversely, for b in Δ, the conditions are indeed satisfied, for we have $|pb+q| \geq 1$ if $p = \pm 1$ and $\operatorname{Im}(pb + q)| = |p| \operatorname{Im} b \geq |p| \frac{\sqrt{3}}{2} \geq \sqrt{3}$ and thus $|pb + q| > 1$ if $p \geq 2$.

We observe now that if $(a; b)$ is a basis for Λ, then $(a; b - a)$, $(a; b - 2a)$, ... and $(a; b + a)$, ... are also bases for Λ. The classification can thus be done by the points of the modular domain D:

$$\operatorname{Im} z > 0, \quad |z| \geq 1, \quad |\operatorname{Re} z| \leq \frac{1}{2},$$

provided that we identify corresponding points of similar lattices. It involves only points of the boundary of D, i.e.:

– the two vertical half-lines $\operatorname{Re} z = \pm\frac{1}{2}$, $\operatorname{Im} z \geq \frac{\sqrt{3}}{2}$. The identification is done by the mapping $z \mapsto z+1$ and translates the fact that if $(a; b)$ is a basis of a lattice Λ, then $(a; b + a)$ is also one.

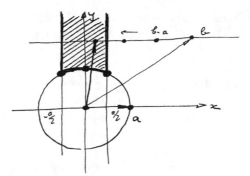

Fig. IX.4.2. The modular domain

– the arcs of the circle joining $i = \sqrt{-1}$ to $-j^2 = \frac{1}{2} + i\frac{\sqrt{3}}{2}$ and i to $j = -\frac{1}{2} + i\frac{\sqrt{3}}{2}$. The identification is done by the mapping $z \mapsto -1/z$ and corresponds to replacement of the basis $(a; b)$ by the basis $(b; -a)$ we have the case $|a| = |b| = 1$).

The space obtained by making this identification in the modular domain is called the *modular space*, and is denoted by *SpMod*.

In the modular domain, the points on the vertical axis represent the rectangular lattices and those on the lower boundary the rhombic lattices, the *square* lattice corresponding to the point $(0, 1) = i$ and the *regular hexagonal* lattice corresponding to one of the two lowest points (but one point only in *SpMod*). This gives the torus, which is the richest in symmetries and which we call the *regular hexagonal lattice*.

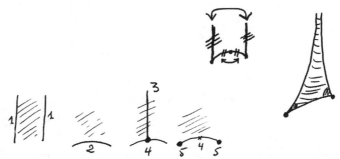

Fig. IX.4.3. *At left*, the special lattices: rhombic lattices (2), rectangular lattices (3), square lattice (4), regular hexagonal lattice (5). *At right*, identification of the modular domain and modular space

As geometers we mustn't remain where we are; we absolutely need to *see* the modular space, the space of all lattices; even more, we need to define a good geometry. As has been said, to define a lattice within (direct) similitude, we only need fix the first basis vector as the point $(1, 0)$ of $\mathbb{R}^2$, or the point 1 of $\mathbb{C}$. Then the lattices are at first classified by a point z of the plane, i.e. the second basis vector. If

$\begin{pmatrix} a & b \\ c & d \end{pmatrix}$ is in SL$(2;\mathbb{Z})$, then $(cz + d; az + b)$ is another basis of the lattice generated by $(1; z)$, and the lattice $(1; (az + b)/(cz + d))$ is similar to the lattice $(1; z)$. We thus define an action of SL$(2;\mathbb{Z})$ on the upper half-plane $\mathcal{H} = \{y > 0\}$. From the fact, seen in (IX.3.2), that each change of basis is in SL$(2;\mathbb{Z})$, it follows that two lattices are similar if the points z that are associated with them correspond under this action of SL$(2;\mathbb{Z})$. Our space *SpMod* is thus also the quotient of the upper half-plane $\mathcal{H} = \{y > 0\}$ by the group SL$(2;\mathbb{Z})$. This group does not at all leave the Euclidean structure invariant; on the contrary it turns out, by a miracle that is very pleasant for us, that it leaves invariant the hyperbolic geometry of the upper half-plane (Poincaré half-plane; see Sect. II.4), the total group of this geometry being SL$(2;\mathbb{R})$, acting by homographies. Since SL$(2;\mathbb{Z})$ acts by isometries on $\mathcal{H}$, the space *SpMod* = SL$(2;\mathbb{Z})\backslash\mathcal{H}$ is endowed with a quotient metric. For this completely natural metric, *SpMod* has the form of the figure above: a *point at infinity*, and for the rest a good (abstract) surface, apart from two singular points, having a conical singularity (see Sect. VIII.6 for the case of spherical geometry) with angle equal to π for the point corresponding to the *square* lattice, and the other with angle equal to $\frac{2\pi}{3}$ for the point corresponding to the hexagonal lattice. According to Sect. II.4 the figure shows that the total area of *SpMod* is finite and equal to $\pi - 2\frac{\pi}{3} = \frac{\pi}{3}$.

We now advance in the study of the geometry of a single lattice. For a long time now the geometric object associated with a lattice has been its *Voronoi cell* Vor(Λ), also called the *Dirichlet domain* (not to be confused with a *fundamental domain* of $\mathbb{R}^2$ for the action of Λ. The Dirichlet domain is a fundamental domain, but there are indeed others, such as the fundamental parallelogram of the lattice). It is by definition *the set of points closer to the origin than every point of the lattice other than the origin*. For the geometer, it is obtained by drawing some perpendicular bisectors. Although it is not completely obvious, it can be shown without too much trouble that only six perpendicular bisectors are needed in general, only four in the rectangular case.

Practically, we prefer to consider the compact set which is its closure, corresponding to inclusive rather than strict inequalities; we denote it in the same fashion. The set of Voronoi domains associated with all the points of a lattice form a *tiling* of the plane; we will return to tilings very soon. The one here is doubly periodic, since by construction it is invariant under the translations of the lattice.

◆

For those interested in Riemannian geometry, we can interpret the Voronoi cell and its polygonal boundary in the language of *flat tori*, briefly mentioned in Sect. V.14. This is to say that we consider the geometric quotient space **Tor**$(\Lambda) = \mathbb{R}^2/\Lambda$, which has the topology of a torus, but the metric obtained by passage to the quotient of the Euclidean metric of $\mathbb{R}^2$, which is by no means that of a torus of revolution. We see that the quotient is always a topological torus by identifying the opposite sides, pairwise parallel, of a generating parallelogram. But to see the quotient metric well,

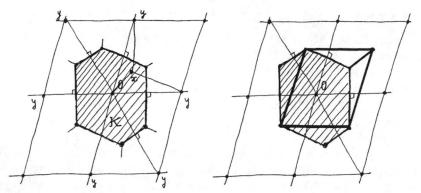

Fig. IX.4.4. *At left*, a generic Voronoi cell $K = \{x \in \mathbb{R}^2 : d(x, y) \geqslant d(x, 0) \, for$ each $y \in \Lambda\}$. *At right*, comparison of the Voronoi cell and a fundamental parallelogram of the lattice

we need to proceed as follows. We let p denote the projection $p : \mathbb{R}^2 \to \mathbf{Tor}(\Lambda)$. Geometrically we can also "see" $\mathbf{Tor}(\Lambda)$ as the domain $\mathrm{Vor}(\Lambda)$, where we identify the points of the boundary polygon $\partial(\mathrm{Vor}(\Lambda))$ for which the difference is an element of Λ. In the hexagonal case — which is the generic case — we have three identifications to make as in the figure below:

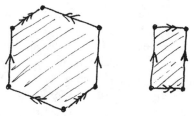

Fig. IX.4.5. Definition of $\mathbf{Tor}(\Lambda)$ by identification of the sides of the Voronoi cell

The distance between two points x, y of $\mathbf{Tor}(\Lambda)$ will be the length of the shortest path between their inverse images $p^{-1}(x)$, $p^{-1}(y)$ in $\mathbb{R}^2$, i.e. the smallest distance between two arbitrary points of these inverse images (which are translates of Λ). The translations, pulled back onto the flat torus, are isometries. To study the metric and the shortest paths, we may suppose that $x = p(0)$ and that $y = p(u)$, where $u \in \mathrm{Vor}(\Lambda)$. Then these points will be joined by a unique shortest path if the second is in the interior of $\mathrm{Vor}(\Lambda)$. They will be joined by two different shortest paths if y is on the polygonal boundary $\partial(\mathrm{Vor}(\Lambda))$, and by three shortest paths if y is a vertex of this polygon. This in the hexagonal case; in the rectangular case, there will be two or four shortest paths.

In the language introduced in Sect. VI.3 for the recalcitrant ellipsoids, we see in contrast here that the cut locus problem for the flat tori is completely resolved. In the hexagonal case the cut locus, as seen in the flat torus, is made up of three segments, only two in the rectangular case.

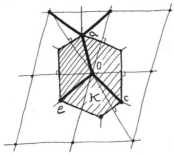

Fig. IX.4.6. Two ways of seeing the shortest paths from $p(0)$ to $p(a)$: shortest paths from 0 to the translates of a by Λ, or shortest paths from a to translates of 0

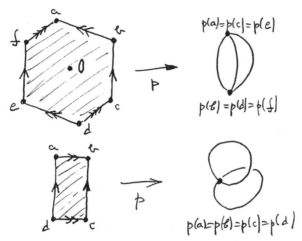

Fig. IX.4.7. The cut locus of $p(0)$ is the image under p of the boundary polygons of **Tor**(Λ). Figure left to readers: on a generic but embedded flat torus — in your imagination — sketch the cut locus and the six connections to two special points; same problem for a rectangular torus

♦

Can we find regular polygons for which the vertices are in a lattice? We find a sequence of squares in $\mathbb{Z}^2$, and of equilateral triangles and regular hexagons in the regular hexagonal lattice. It is a fact that none other is possible, neither pentagons nor any with more than six sides. The algebraist will proceed thus: a polygon with k sides in a lattice Λ implies the existence of an element of the group SL$(2; \mathbb{Z})$, which is of order k, since iterating k times has to give the identity: the rotation matrix with angle $\frac{2\pi}{k}$ is not *a priori* in this group; but in fact it is, because the center of the polygon isn't in Λ, but in $k^{-1}\Lambda$, and it suffices to place ourselves — if necessary by translating the origin — in the sublattice of $k^{-1}\Lambda$ generated by the points of the polygon. This rotation matrix, say $\begin{pmatrix} a & b \\ c & d \end{pmatrix}$, satisfies $ad - bc = 1$, whose eigenvalues are of the form $\lambda, \bar{\lambda}$, with $|\lambda|^2 = 1$ and $\lambda + \bar{\lambda} = a + d \in \mathbb{Z}$. But $|\lambda|^2 = 1$

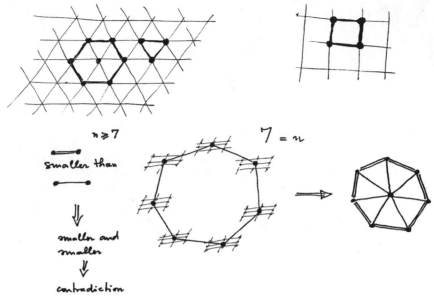

Fig. IX.4.8. Regular polygons in a lattice. At the bottom, Gromov's proof of impossibility for $k \geq 7$

implies $-2 \leq \lambda + \bar{\lambda} \leq 2$, whence the only possible values are $0, \pm 1, \pm 2$, which correspond to $k = 1, 3, 4, 6$.

Geometers will surely appreciate the following proof, which seems to be due to Gromov. Figure IX.4.8 shows that, from a regular polyhedron with seven sides inscribed in a lattice Λ, we derive a regular polygon having the same number of sides but of strictly smaller size. Since the size can't strictly decrease infinitely many times, we end up with a contradiction. The case $k = 5$ remains; it is left to the readers as a very entertaining exercise. We can consider this proof as being joyous, simple and pure, worthy of figuring in Aigner and Ziegler (1998). Our interest in Gromov's proof is that it applies to much more general geometrical situations, where an algebraic method is unavailable. It is profound in nature and belongs to what are called *descent methods*, the first ever encountered being due to Fermat in his proof that there exist nontrivial triples of integers (a, b, c) for which $a^3 + b^3 = c^3$ (see any book on number theory). Descent methods are found in numerous recent problems in group theory and geometry; see for example Wang (1972), Margulis (1977).

A geometric invariant associated with a lattice in the Euclidean plane (it will be the same thing in space, but this case will be discussed amply in the following chapter) is its *density*. First of all, we know that a lattice possesses an *area*, Area(Λ), that of a parallelogram constructed on some basis. Next, the simplest parameter is the lower bound of the norms of the vectors of $\Lambda \setminus 0$, called the *minimal norm* of the lattice and denoted NormMin(Λ). Geometrically, we can see that the minimal norm is equal to twice the radius of a maximal open disk that fits into the lattice without intersecting it.

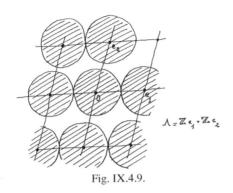

Fig. IX.4.9.

In the figure we see that the lattice *density* is a natural notion, equal by definition to

$$\text{(IX.4.1)} \qquad\qquad \mathbf{Dens}(\Lambda) = \frac{\pi}{4} \frac{(\text{NormMin}(\Lambda))^2}{\text{Area}(\Lambda)},$$

because this quantity is equal to the ratio of the area filled by the disks to the total area of the plane. To see this, it suffices to look at this quotient (which, strictly speaking, doesn't make sense as just described) in a basic parallelogram of the lattice: the total area contained in the disks is πr^2, with $r = \frac{1}{2}$ NormMin(Λ). A formal definition of the density will be given in Sect. IX.5.

A main problem, of considerable practical importance in any dimension (the definition of the density extends trivially) and which will occupy us for a good part of the following chapter, is to determine the lattice of maximum density, assuming it exists. In the present planar case, the modular domain gives an immediate answer.

(IX.4.2) *The maximal density is that of regular hexagonal lattices and thus equals*
$$\frac{\pi}{2\sqrt{3}} = 0.9069...$$

In fact, the density is invariant under similitude. By construction, the minimal norm in the figure is equal to 1; the area is that of the parallelogram constructed on the vectors $(1, 0)$, (x, y), and thus equals y. The density is thus maximal at the two lowest points of the modular domain (Fig. IX.4.2), both of which correspond to the regular hexagonal lattice. This does not prove that the best way of packing disks of the same radius in the plane is in regular hexagonal fashion, since there might exist better packings not associated with the lattice. See Sect. IX.5 for the answer.

One of the motivations invoked for studying the canonical lattice $\mathbb{Z}^2$ was the theorem of Fourier series for doubly periodic functions, which states that each sufficiently regular doubly periodic function f can be developed in a Fourier series

$$f = \sum_{m,n} a_{m,n} e^{2\pi i m x} e^{2\pi i n y} \quad \text{where} \quad a_{m,n} = \int f(x,y) e^{2\pi i m x} e^{2\pi i n y} \, dx \cdot dy.$$

It is understood that the sum is taken over the $(m,n) \in \mathbb{Z}^2$ and the integral over $\mathbb{R}^2/\mathbb{Z}^2$ (i.e. over a motif, or mesh, of the lattice). Less classic is the *Poisson formula*, which says that (still with the sums over $\mathbb{Z}^2$):

$$(\text{IX.4.3}) \qquad \frac{1}{t} \sum_{m,n} \left(\exp\left(-\pi \frac{m^2 + n^2}{t} \right) \right) = \sum_{m,n} \exp(-\pi(m^2 + n^2)t)$$

(t real and positive). The Poisson formula quoted here is a particular case of the general formula for the Fourier transform of a rapidly decreasing function f; see, for example, V., §6 Serre (1970):

$$\sum_{m,n} f(m,n) = \sum_{m,n} \widehat{f}(m,n).$$

We obtain (IX.4.3) by applying the general formula to $f(x) = \exp(-\pi \|x\|^2/t)$, for which the Fourier transform is $\widehat{f}(\xi) = \exp(-\pi t \|\xi\|^2)$. (We should however note that, according to the references, the Fourier transform my be given variable coefficients.) We now would like a Fourier analysis that is valid not just on $\mathbb{Z}^2$, but on any lattice Λ of $\mathbb{R}^2$. We need to find the Λ-periodic functions that here play the role of the $e^{2\pi i m x}$ and $e^{2\pi i n y}$ and that respect the geometry of Λ. Since all lattices are affinely equivalent, we need thus to find exponentials of the form $\exp(k(x,y))$, where k is a linear function on $\mathbb{R}^2$. We write $k(x,y)$ in the form of a scalar product $v \cdot \xi$. The condition of the invariance under Λ of $e^{2\pi i v \cdot \xi}$ is thus written $v \cdot \xi \in \mathbb{Z}$ for any $v \in \Lambda$. We thus introduce the set Λ^* defined by

$$(\text{IX.4.4}) \qquad \Lambda^* = \{ \xi \in \mathbb{R}^2 : v \cdot \xi \in \mathbb{Z} \text{ for any } v \in \Lambda \}.$$

We see right away that this is a "new" lattice in $\mathbb{R}^2$, called the *dual lattice* of Λ. This is a proper duality: $(\Lambda^*)^* = \Lambda$ for each Λ; we also have $\text{Area}(\Lambda) \cdot \text{Area}(\Lambda^*) = 1$. Note that $\mathbb{Z}^2$ is self-dual.

The Poisson formula (IX.4.3) remains valid, but here in fact shows better its profound geometric nature since it becomes:

$$(\text{IX.4.5}) \qquad \frac{1}{t} \text{Vol}(\Lambda) \sum_{v \in \Lambda} \left(\exp\left(-\pi \frac{\|v\|^2}{t} \right) \right) = \sum_{\xi \in \Lambda^*} \exp(-\pi \|\xi\|^2 t).$$

In particular, by acting on all values of t, we see that this formula determines the set $\{ \|\xi\| : \xi \in \Lambda^* \}$, considered as a subset of $\mathbb{R}$, if we know the set $\{ \|v\| : v \in \Lambda \} \subset \mathbb{R}$.

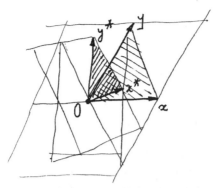

Fig. IX.4.10. A lattice and its dual are similar

The fact that this determination is possible is theoretically evident; moreover, in the present planar case, the dual lattice Λ^* is similar to Λ, as follows from a simili-tude with angle $\frac{\pi}{2}$ and ratio $\frac{1}{\text{Area}(\Lambda)}$. For example, the square lattice $\mathbb{Z}^2$ is self-dual, whereas the dual of the regular hexagonal lattice is a regular hexagonal lattice which is only similar to the original: we need to rotate and change scale.

But all this is very special to dimension 2; in general, starting with dimension 3, the dual lattice is no longer at all similar to the initial lattice; on the other hand, it is still true — because (IX.4.5) extends without change to all dimensions — that one of the sets of norms determines the other.

Why attach such importance to the set $\{\|\xi\| : \xi \in \Lambda^*\}$? The set $\{\|v\| : v \in \Lambda\}$ is itself of arithmetic nature, since the set of its squares is the set of $m^2 + n^2$, with m, n integers. For the dual, very briefly, a little mathematical physics: an important part of the physics associated with a lattice Λ — whether in crystallography, in vibrations or in heat propagation — comes from the study of partial differential equations in which the essential term is the Laplacian Δ, i.e. the operator on functions f defined by $\Delta f = -(\frac{\partial^2 f}{\partial x^2} + \frac{\partial^2 f}{\partial y^2})$. The main thing is that the functions $v \mapsto \exp(2\pi i v \cdot \xi)$ ($\xi \in \Lambda^*$) are the eigenfunctions of Δ for the eigenvalues $4\pi^2 \|\xi\|^2$, and that there aren't any others. This means that the **frequencies** (whose reciprocals are the wave lengths) of the periodic system defined by Λ — no matter whether we are dealing with a crystal or any other physical system — are the $2\pi\|\xi\|$. The set frequencies is called the *spectrum* of Λ.

With the eigenfunctions of Δ, i.e. with the functions $v \mapsto \exp(2\pi i v \cdot \xi)$, $\xi \in \Lambda^*$, we have a complete Fourier analysis associated with Λ. A classic question of math-ematical physics is "*can we hear the shape of a drum?*"; see Berger (1999) or Berger (2003). In the particular case of flat tori, we want to know whether with sole knowledge of the spectrum we can reconstruct — within an isometry of $\mathbb{R}^2$ — the complete geometry, i.e. the lattice Λ. In the general Riemannian case, where what plays the role of the $\|v\|$, $v \in \Lambda$, is the set of lengths of the periodic geodesics, the matter is not yet well understood. In the present case the response is *yes*, and here is why. We first observe that the notation $\{\|\xi\| : \xi \in \Lambda^*\}$ is imprecise. We not only need to know the real numbers in the spectrum but also their multiplicities, i.e. the number of elements of Λ^* whose norm is equal to the number in question. For example, in the lattice $\mathbb{Z}^2$, the number 1 is of multiplicity 4, the number $\sqrt{2}$ is of multiplicity 8, the number 5 is of multiplicity 12, etc. From the spectrum thus specified we extract the set of lengths $\|v\|$, $v \in \Lambda$, with its multiplicities, that is often called the *length spectrum* . By homothety we reduce to the case where $u = (1,0)$ and where v is a point of the modular domain. The vector $u = (1,0)$ corresponds in fact to the smallest value of the length spectrum (whatever its multiplicity). If v is the second basis vector in *SpMod*, we see geometrically that the second value of the length spectrum, once removed from those multiples of 1, is $\|v\|$, and similarly the third is $\|v - u\|$ (this could be $\|v + u\|$, but as we can anyway only determine Λ within isometry − and not within direct isometry − we can assume that the angle between u and v is at most a right angle). The triangle $(0, u, v)$ is thus known completely. It is always a bit hard to correctly prove what is visually obvious for the third value, using inequalities involving scalar products, etc., although there is no conceptual ascent here that would simplify the task. We could escape with the subterfuge of observing that the spectrum determines the asymptotic behavior in R of the number of points of Λ in a disk of radius R, which is the density of the lattice. But thanks to (IX.4.1) it is exactly equal to: $\mathbf{Dens}(\Lambda) = \frac{1}{4} \frac{\pi (\text{NormMin}(\Lambda))^2}{\text{Area}(\Lambda)}$. As NormMin^{-1} is the smallest nonzero eigenvalue in the spectrum, the area is thus known. But this area, looking at *SpMod*, is the coordinate y of $v = (x, y)$. As $\|v\|$ is known, we are finished. However, we had to go to infinity for it; it was very costly!

This seems very natural: reformulating the preceding, we look at how many points of the lattice there are on each circle about the origin, as a function of the radius. The answer is zero for almost all radii, and otherwise, when the radius is a value of the spectrum, the number is equal to the corresponding multiplicity. We thus obtain a step function with integer values. It seems inconceivable that two different lattices could have the same number of points on each circle with center the origin. In Sect. XII.9 we will see that the analogous problem has been resolved in dimension 3, but above all, starting with dimension 4, there exist counterexamples: see Sect. X.8.

IX.5. Packing circles (disks) of the same radius, finite or infinite in number, in the plane (notion of density). Other criteria

We have just seen that the best way of packing disks of the same radius in the plane is when their centers form a regular hexagonal lattice. But this was for packings with the aid of a lattice. Moreover this result seems stunningly true for arbitrary packings. But is it really, in general, when we pack disks "as best possible", for example in a physical manner? The answers are not so immediate, even for the problem of forming a definition of the *density*; we will show why.

The photographs below, Fig. IX.5.2 and especially Fig. IX.5.3, show that an optimal packing is not regular hexagonal in certain very simple cases. What do we

Fig. IX.5.1.

do then? We first set the radius of the disks equal to unity. We then need to make a dichotomy: put into place a theory that shows, for the large domains, that it is indeed the regular hexagonal packing that gives optimal density; and for domains of small or medium size, to manage "by hand" and at least give other criteria, which is what we do in Sect. X.6. We will see that a certain contradiction remains and how to eliminate it thanks to some completely new ideas.

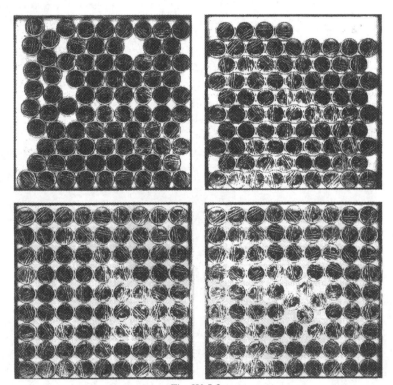

Fig. IX.5.2.

The primitive notion of density is relative to a domain D containing the disks (of radius 1): it is the ratio of the sum of the area of the disks to the area of the domain, or $\frac{\pi n}{\text{Area}(D)}$ if we have succeeded in packing n disks in D. The best density, for a given domain, is thus when the number is the largest possible. The drawings indeed show that the problem of maximum density, of optimal packing, doesn't make much sense without restrictions on D. For example we can always take a regular hexagonal packing and the convex set that it generates. Things are then optimal.

It is here that the modern readers will say: take a computer; you will easily find an optimization program for your problem that works in any domain. The trouble is that the optimization problem we have to treat is exponential and impracticable with present computers as soon as we have more than thirty parameters (at the moment of this writing, but the number increases constantly in proportion to the power of computers), say here about fifteen disks if the position of the disk depends on two parameters. We can only find optimal solutions for very small domains. Note however that the conditions of the program are very simple; we need only write that the mutual distances between the n points (the centers of the disks) are greater than or equal to 2. The functional to be minimized is the area of the domain considered less the area of the disks. A most serious criticism is that the computer only provides us ultimately with an experimental result; a proof is of a whole other order.

A very complete exposition can be found in Melissen (1997). We don't require more about this problem at present because it doesn't, at least for the moment, bring us higher on the ladder. The difficulty can be measured by the fact that we had to wait for Fodor (1999) to know the optimal disposition of 19 disks, i.e. the minimum radius of circles that contain them. It was easy to see that the regular hexagonal configuration could be improved upon. From an aesthetic point of view, the varied configurations that are found in Melissen (1997) are not thrilling; above all they do not show any symmetry, any general law. In contrast, if we give criteria other than density, there is now a very good theory that will be explained in Sect. X.6.

For those who like practical applications, manufacturers of metal cables (which in practice have several strands, 7, 19, etc.) have known for an eternity that, for cables with a large section, either we begin with subcables of 19 strands or else by "compressing" a large number of individual strands. Of recent crucial importance are the packings formed by *fiber optics*; see the references in Chap. 1 of Conway and Sloane (1999).

So we need to define a notion of density for a large packing; in fact, we consider a packing of the entire plane. What is its density? The examples above suggest a definition:

> *The density is the limit when the radius R tends to infinity, if it exists, of the density of the restriction of the packing to a disk of radius R.*

It isn't necessary to be more precise. The disks are taken with fixed center.

The above definition is the one found in Fejes Tóth (1972). It is necessary to show its independence with respect to the choice of center, which is not difficult but certainly requires going to infinity. The more recent and accessible Rogers (1964) prefers taking squares. No work shows that the two definitions are equivalent, the one with circles and the other with squares, and in Rogers' book it is difficult to find that the orientation of the squares is unimportant. Practically, to be very precise we would need to take limits superior and inferior, as is shown by examples of packings with big holes "throughout". The bible (Conway and Sloane, 1999) treats mostly packings associated with lattices and does not give a formal definition.

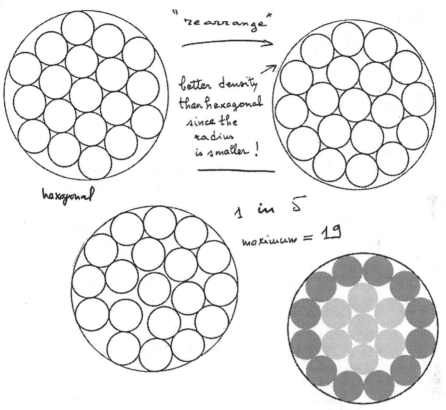

Fig. IX.5.3. Rearrange, whence a better density than hexagonal. After three tries, here is the final optimal configuration for 19 circles (after Ford)

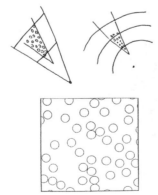

Fig. IX.5.4. How to define the density: not by exhaustion in any case

This posed, we have:

(IX.5.1) *The maximal density for a plane packing by circles of the same radius is that of a regular hexagonal lattice* $\frac{\pi}{2\sqrt{3}} = 0.9069...$

We should note well that there is no uniqueness: take a regular hexagonal lattice and remove from it several disks or many (but not too many). In the limit these holes will "vanish", they won't matter any more for sufficiently large radii. On the other hand, existence is assured. Back in 1892 the Dutch naturalist Axel Thue professed to prove (IX.5.1) rigorously and took the matter up again in 1910 with a new method of attack, but the first complete proof is due to the Hungarian Fejes Tóth; had it not been in time of war, his achievement would have been loudly celebrated. We will analyze two proofs, that of Fejes Tóth and that of Rogers, for a reason that will appear in the following chapter when we attempt to see what happens in higher dimensions.

For the first proof: with a packing of disks we associate a decomposition of the plane into *Voronoi cells* (a notion already encountered in Sect. IX.4). If x is the center of a disk of our packing, the cell Vor(x) is by definition made up of points which are closer (or at equal distance) to x than to any other center. The set of these cells forms a tiling of the plane. A material realization, in theory, is that of the "post office". We consider each disk as having a post office at its center. We slice up the plane into parts as follows. At each post office we associate all letter boxes which are closer to it than to any other post office. A postal employee naturally refuses, and rightly so, to make a long trajectory when it would be shorter for a colleague. The plane is then divided into cells, called Dirichlet or Voronoi.

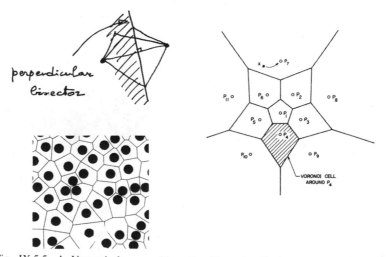

Fig. IX.5.5. A Voronoi decomposition; the Voronoi cell about P_4; the Voronoi decomposition of the disks of Fig. IX.5.3

The proof of Fejes Tóth consists of two essential stages. In the first, he proves that the area of a cell is always greater than or equal to that which corresponds to the cell of the regular hexagonal packing, i.e. a regular hexagon circumscribed in a circle of radius 1, namely $2\sqrt{3}$. Equality occurs only for this hexagon, and then the disk in question is tangent to six neighbor disks. We thus need to show that a disk surrounded by neighboring disks (of the same radius) furnishes a cell of area greater or equal to that in the regular hexagon case. This isn't obvious. If there are only six neighbors, this results from the isoperimetric inequality proved in Sect. VIII.3, but even if the contrary seems absurd, it is still not so easy to show that if our disk has seven neighbors, we have a greater area (for eight or more it's easy). But the proof does not bring with it anything conceptual.

The second stage is completely analogous to what was done for the circle problem in Sect. IX.2: we look at a disk of radius R, and observe how it intersects the decomposition into Voronoi cells of the packing considered. We then define the local density as the ratio of its area to that of its associated domain.

the best

too much
space lost

Fig. IX.5.6. Too much space lost; the best configuration

Fig. IX.5.7.

There are the cells that are completely contained in the disk, and those that intersect the boundary circle. But the contribution of the latter is of the order of the perimeter of the circle, namely $2\pi R$, and is negligible compared to πR^2 as R tends toward infinity; as for the internal contributions, we know they don't contribute more than $\frac{\pi}{2\sqrt{3}}$ (we have already used this reasoning in Sect. IX.2), QED.

The second proof, due to Rogers, uses another decomposition of the plane associated with a packing, the so-called *Delaunay* (sometimes written Delone) decomposition, named for a Russian mathematician who was interested in crystallography. It is simply *one* triangulation of the plane naturally associated with the set of centers of the packing disks. Precisely, consider the graph obtained by joining two centers if their Voronoi cells have a common side. But for $\mathbb{Z}^2$, for example, the graph defines squares. We then have the choice: either accept polygons other than triangles in the triangulation, or find a triangulation by slightly perturbing the points of E, in such a way as to put them in general position and obtain a triangulation, and then finally putting the points of E back in their places. In this latter case, as their are many ways of putting the points of E in general position, we lose uniqueness. Rogers then shows that the density of each triangle is always $\leqslant \frac{\pi}{2\sqrt{3}}$, with equality if and only if the triangle is equilateral. As with Fejes Tóth, we need to pay a bit of attention in order to proving that, in a triangle such as in the figure, the sum of the areas of the three angular sectors formed by the disks of radius 1 and centers at the vertices never exceeds the value for the equilateral triangle. The second stage is as in Fejes Tóth.

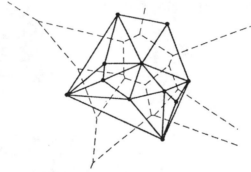

Fig. IX.5.8. For the same points and in the same figure, the Voronoi decomposition (*dashed lines*) and that of Delaunay (*solid lines*) are shown

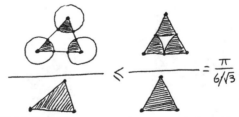

Fig. IX.5.9. Rogers' argument, using a triangulation of Delaunay

The advantage of Rogers' method is that it extends to all dimensions (but starting with dimension 3 it gives only a non optimal upper bound for the density; see Sects. IX.2 and IX.6 of the following chapter. The triangulation is unique in the generic case, genericity that can be obtained in any situation by an arbitrarily small perturbation. If we wanted, we could then pass to the limit in this technique, but that isn't necessary here.

♦

We have sought optimal packings in disks and in squares. But rather recently, in fact since Fejes Tóth in 1974, we can also define an optimal density for *n* disks by requiring that the volume of the convex envelope of these *n* disks be simply the minimum volume. But we can also require, whatever the perimeter of this convex set, that it be minimum. We will return amply to the analogous problem in higher dimensions in Sect. X.6, where several oddities and even some catastrophes await us, even though it is in dimension 3 that the classical practical problems are posed.

For this section and the following, an excellent reference is 3.3 of Gruber and Wills (1993) for packings in the plane and, for finite packings, 3.4 of the same work. Like the third edition of Conway and Sloane (1999), these two chapters are very up to date and convey very numerous references. The number of articles which are syntheses and the number of books cited gives some indication of the extent of the problems, problems that we can but touch upon in our two chapters, encouraging interested readers to consult the references.

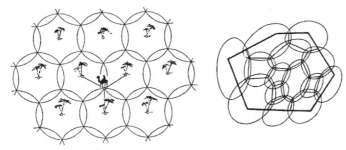

Fig. IX.5.10. A problem of the desert and an oasis. Fejes Tóth (1972) © Springer and G. F. Tóth

♦

In this section, as in others in this chapter and in all of the following chapter, we treat only packing problems. Now there exists a natural **dual** notion, that of **covering**. It is also very important practically. In 1939 Kershner proved that the *finest* covering, the most efficacious by disks of the same radius, is regular hexagonal. A proof is given in the not easy-to-read (Fejes Tóth, 1972).

Covering problems with few disks are, like packing problems, very difficult, but don't seem to lead to new concepts; we cite them, in the spirit of our introduction, rather as foils for drawing our attention to the main problems.

IX.6. Packing of squares, (flat) storage boxes, the grid (or beehive) problem

Optimal packing of disks of the same radius is a very particular, again most natural, case (cables, fiber optics, etc.) of the problem of packing of domains by congruent objects (isometric, obtained by isometry from a **model**). We might also

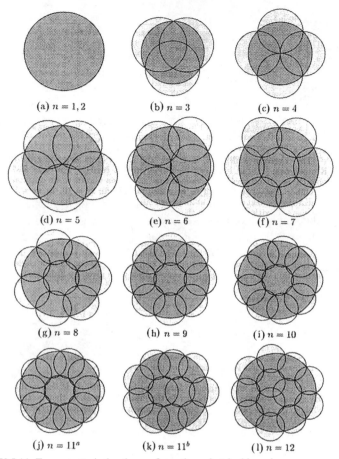

(a) $n = 1, 2$ (b) $n = 3$ (c) $n = 4$

(d) $n = 5$ (e) $n = 6$ (f) $n = 7$

(g) $n = 8$ (h) $n = 9$ (i) $n = 10$

(j) $n = 11^a$ (k) $n = 11^b$ (l) $n = 12$

Fig. IX.5.11. For up to 9 circles the configurations sketched have been proven optimal,
starting from 10 they are only conjectured

want to pack varied objects. It's an optimization problem, called the *bin problem*, because we think above all of practical packings of objects in three dimensions (which
can if required be of different types) into given boxes (readers: pack your bags and
go read (Stewart, 1979)!). The practical importance is colossal, but the algorithmic
complexity is known for being *NP-hard*, as is the case of disks in particular.

The case of squares of the same edge length, say unity, is just as spectacular
for small domains. In fact, it is just as unsolvable (or semi-solvable) by a computer
for a number exceeding fifteen or twenty as it is for disks. We are dealing here
with choosing the number n of squares and looking for the smallest square that
contains the n squares. The figures can be seen below. Presently the problem is
open beginning with $n = 10$ (at this moment of writing, but this type of question
progresses regularly). We can consult, for this and for what follows as well, Prob.
D.4 of Croft, Falconer and Guy (1991), 3.4 of Gruber and Wills (1993), and the
popular exposition Stewart (1979).

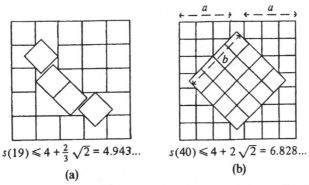

$s(19) \leqslant 4 + \frac{2}{3}\sqrt{2} = 4.943...$ $\qquad$ $s(40) \leqslant 4 + 2\sqrt{2} = 6.828...$

(a) $\qquad\qquad\qquad$ (b)

Fig. IX.6.1. Two examples, conjectured to be optimal. Croft, Falconer, Guy (1991) ©
Springer and K. Falconer

But here too there exist studies of the behavior for very large n which lead to very intricate results in number theory, problems that have been attacked by great mathematicians; see the section mentioned of Croft et al. (1991).

♦

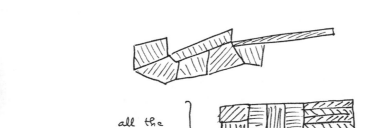

Fig. IX.6.2.

Another problem is one suggested by bees. We consider partitions of the plane into domains of unit area and think of this plane partition as a *grid*. As always, we are prepared to go to infinity if necessary in order to have a general and well-posed problem; we might seek to optimize the length of wire in the grid, i.e. the sum of the lengths of the curves separating the domains two at a time. The problem has long since been solved by fencing manufacturers and by the bees, namely the regular hexagonal lattice:

In contrast, mathematicians have taken much longer, but very recently Hales (2000) announced a positive answer to the conjecture that the regular hexagonal grid alone is optimal. The proof is conceptually relatively simple, but it uses in a subtle way a new isoperimetric inequality for which the regular hexagon is the unique minimum. The details of the proof are treated in the remarkable and pedagogic (Morgan, 2000).

Fig. IX.6.3. Photo by I. Kitrosser, Réalités, 1950

The analogous problem in dimension 3 has been studied even more, but is of colossal difficulty: it deals with partitioning (ordinary) space into cells of equal volume in such a way that the total area of their separating surfaces is minimum. At present, it is not even known which shape for the 3-dimensional cells might yield a minimum. An ideal introductory reference to the subject is Hales (2000).

IX.7. Tiling the plane with a group (crystallography). Valences, earthquakes

We will be extremely brief; the subject of tilings connected with a group is treated in great detail in Chap. I of [B] and subsequently there has not been any new fact that is mathematical and very conceptual. However, two new directions — both very young — will be pointed out at the end of this section. This subject *tiling groups* is also treated in enormously many more or less elementary books on geometry; we mention (Nikulin & Shafarevich, 1987), who indeed speak of flat tori, but not enough about the hyperbolic case; see Sect. IX.10. A *tiling* of the Euclidean plane is not a partition in the strict sense, because of common boundaries. It *is* a partition in the sense that the plane is completely covered by a family of compact sets (called a *tiles*), all with nonempty interior, for which the interiors are mutually disjoint. We are interested mostly — although not always — in *profinite* tilings, i.e. what is obtained by using **copies** of only a finite number of given tiles (*tile types*, sometimes called *prototiles*). By a copy of a tile we of course mean an $\mathbb{R}^2$ domain that is isometric to this tile.

In this section we consider the case of tilings where not only is there but a single tile type, but above all the set of isometries needed for obtaining all the tiles of the tiling form a **group** (thus a subgroup of the group $\mathrm{Isom}(\mathbb{R}^2)$ of all isometries of $\mathbb{R}^2$) which will be the *tiling group*. For relations with crystallography in the spirit of the physicist, see 3.7 of Gruber and Wills (1993). The essential result is that the possible groups are completely known and classified; there are but 5 if we only want orientation preserving isometries (no mirror line), and 12 more that require reversal of orientation. From the point of view of decorative tiles, in practice the reversal

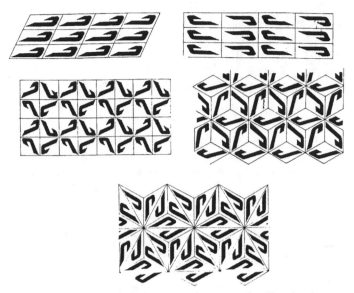

Fig. IX.7.1. The 5 orientation-preserving groups for tiling the plane

requires a second tile type. The shape of the tiles themselves can be extremely varied; only the operations to which they are subjected are classified. For the history, see the first two sections of 3.5 of Gruber and Wills (1993), which are entirely dedicated to tilings of all types.

Three points are essential; once they are acquired, the classification is easy. The first is that the tiling group must necessarily contain two linearly independent translations (so a lattice is established for the situation). The second is that there are only finitely many transformations to add to these translations for generating the whole group. These first two points require a little topology, compactness arguments and some discretion. The third point is that the possible orders for rotations are limited to 2, 3, 4, 6. More precisely, for an assumed rotation order we look at the triangle abc formed by the centers a and b of two rotations of order $\frac{2\pi}{\alpha}$ and $\frac{2\pi}{\beta}$ and the center c of a rotation composed from the first two. Then the triangle abc has angles at its vertices which are $\frac{2\pi}{\alpha}$, $\frac{2\pi}{\beta}$ and $\frac{2\pi}{\gamma}$, where α, β, γ are integers. But as the sum of these angles equals 2π, we find the condition $\frac{1}{\alpha} + \frac{1}{\beta} + \frac{1}{\gamma} = 1$, which leaves only two possibilities (and we find without difficulty all the tilings that realize them):

♦

We have to clarify, indeed rectify, historical information from 1.7.1 of [B], and we are not the only ones who have needed the clarifications that follow. The problem is knowing if the 17 possible types of tiling groups (or mural decorations) exist at the l'Alhambra of Grenada. As references, we can consult p. 56 of Grünbaum and Shephard (1987), to be completed by the more recent Grünbaum (1990). From between 1984 and 1990 there exists a whole dossier of ten articles and several letters; we thank Grünbaum for providing it.

The avatars of this history are a perfect illustration that mathematicians, the author of this book foremost, sometimes write about things which are not taken from primary sources. A widely held belief that all 17 types are effectively present at the l'Alhambra was contested for the first time in Müller (1944). A long history ensued. In 1984–1985 Grünbaum and Shephard found that there were only 13 groups effectively present at that time, followed by an argument with a mathematician from Granada, who maintained that all 17 are indeed present; then a whole literature followed. The conclusion of Grünbaum to this day is that, although we can be certain of 13 types, to be in accord with the others we need to refine the definitions, because of difficulties due to the colorings. In any case there would be at most 16 at l'Alhambra; the 17th exists at Toledo.

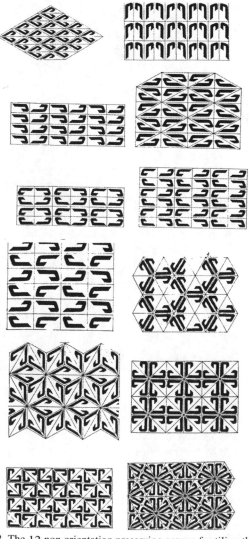

Fig. IX.7.2. The 12 non orientation preserving groups for tiling the plane

♦

We are now interested in the possible shapes of the tile type (always with a single tile type and a transitive group of isometries). For this question and numerous others that we don't broach because the classification of tilings is immense, see the old testament (Grünbaum and Shephard, 1987). The new testament is Senechal (1995) and the subject of the next section; see also p. 467 and 3.5 of Gruber and Wills (1993).

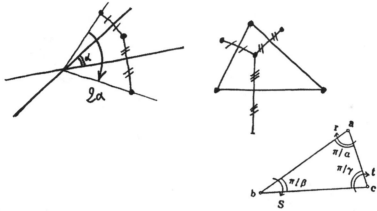

Fig. IX.7.3. The composition of two symmetries with respect to two lines is a rotation (of double angle), the composition of three symmetries with respect to the sides of a triangle is the identity. [B] Géométrie. Nathan (1977, 1990) réimp. Cassini (2009) © Nathan Édition

For simplicity, we suppose that the tile type P is a convex polygon with r sides. We are interested first in the general possibilities of the *valences* of P; we suppose that the tiling has *constant valences*, which is a regularity of a **combinatorial** nature. The *valence* of a vertex (a point where at least three tiles meet) is the number of tiles to which this point belongs. We suppose that in the tiling considered, each vertex of a tile which is a copy of the tile type has a valence equal to that of the corresponding vertex of the tile type. We moreover suppose that, at each vertex, all the angles of the tiles that contain it are equal, which is called a *regular* tiling in Grünbaum and Shephard (1987). We have thus a sequence $\{v_1, \ldots, v_r\}$ of r integers, all ≥ 3. The essential fact is that there is a very limited number of sequences possible. Because in effect the sum of the angles of a tiling equals $(r - 2)\pi$ as in each polygon with r sides; but this sum also equals $\sum_i \frac{2\pi}{v_i}$ by hypothesis, whence the relation:

(IX.7.1)
$$\sum_i \frac{1}{v_i} = \frac{r}{2} - 1.$$

As the v_i are all > 2, we see immediately that $r \leq 6$. We next establish by hand a list of the possible sequences, i.e. satisfying (IX.7.1), and we determine in a rather hard way those which are geometrically realizable and which are not; see the details in Chaps. 3 and 4 of Grünbaum and Shephard (1987), where the final list is that of the Fig. IX.7.4.

This classification remains valid under the solely combinatorial hypotheses (that is, forgetting about the isometries); we evidently need however to assume that the tiles are of bounded diameter, and carry out some arguments that go to infinity, as was done for the circle problem in Sect. IX.2 and the density studies in Sect. X.6. Otherwise we could do almost anything, for example tile with heptagons, but for which we clearly see that they are not of bounded diameter. Readers can show that we can even tile – but only therefore in a *combinatorial* manner, with a polygon having any number of sides. See Fig. IX.7.5.

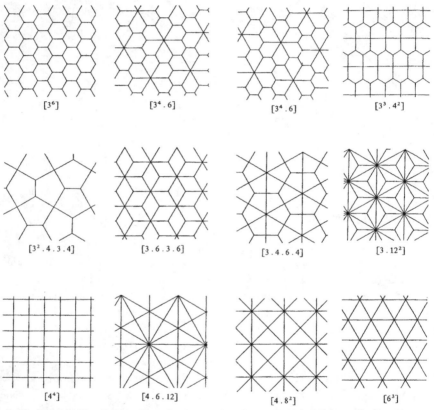

$[3^6]$ $[3^4 . 6]$ $[3^4 . 6]$ $[3^3 . 4^2]$

$[3^2 . 4 . 3 . 4]$ $[3 . 6 . 3 . 6]$ $[3 . 4 . 6 . 4]$ $[3 . 12^2]$

$[4^4]$ $[4 . 6 . 12]$ $[4 . 8^2]$ $[6^3]$

Fig. IX.7.4. The eleven types of valences, but there are twelve figures, because the valence $[3^4, 6]$ has two mirror forms, what chemists call enantomorphs. Grünbaum, Shephard (1987) © B. Grünbaum

If we now return to polygons that can tile in the ordinary sense (by isometries), we then need to limit ourselves to 6 sides at most. What are the possible shapes? The figures show that each triangle tiles; it suffices to carry out successive symmetries with respect to the midpoints of sides. The same thing is again valid for quadrilaterals, even if they are not convex. For 5 and 6, things are completely different. The tiling hexagons have been completely classified; see p. 481 of Grünbaum and

Shephard (1987). Numerous tilings are known for pentagons, but to this day they have not been classified, see p. 492 of the same reference. It seems to us that this genre of problem is one of curiosities, i.e. it is contrary to the philosophy of the present book, which is: citing open problems, but only when their solution would seem to require an ascent of the ladder and not just some work by hand, as difficult as that might be. A recent example for pentagons is the covering of Stein (1985),

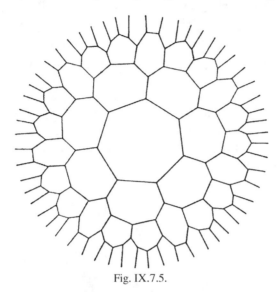

Fig. IX.7.5.

for which a computer program was used. It is interesting to note the extent to which the fact that it isn't known how to find all the pentagons that tile "proves" with a vengeance that the problem is NP-hard.

A whole series of criteria for the classification of periodic tilings can be found in Grünbaum and Shephard, (1987), but there are no new concepts; but things will go quite the other way in the next section.

Here are two viewpoints on tilings that are recent and completely new. They have nothing to do with the two basic books mentioned above. The first deals with earthquakes, with **theoretic seismology**. Here is what is proved in Kenyon (1992a): we consider a tiling of a portion of the plane, not necessarily the whole plane. We consider perturbations, if they exist, of a certain number of tiles of the tiling, by isometries; that is, we transform the tiling by isometries. The *discontinuity set* of the perturbation is the set of points which are effectively displaced. We have an *earthquake* when the discontinuity set consists of curves from one place or another right up to the boundary of the tiled domain. Then we have the *rigidity theorem* for tilings:

(IX.7.2) Each sufficiently small perturbation is an earthquake.

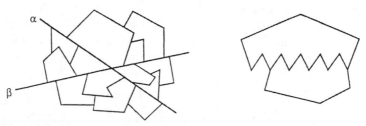

Fig. IX.7.6. Each sufficiently small perturbation is an earthquake

The figures show the necessity of the smallness of the perturbations considered. The discontinuity curves are either lines or circles, since these are the only plane curves of constant curvature (see Sect. V.7). It is also crucial to note that we are dealing here with a plane phenomenon; in dimension 3 we could have a suitable set of well chosen tiles slide continuously; see Fig. IX.7.7.

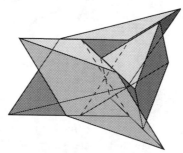

Fig. IX.7.7. Readers can construct such a sliding (non rigid) assemblage out of cardboard. Kenyon (1992a) © Springer and R. Kenyon

The proof of (IX.7.2) is a succession of nice geometric ideas; it isn't possible to summarize them rapidly, but the new ideas are sufficiently general to allow treatment of the perturbations by local diffeomorphisms, thus much more general than isometries. Above all, the study (Kenyon, 1992a) also opens new perspectives in higher dimensions.

In Kenyon (1999b) tilings with dominoes are used to study random walks, or Brownian motions. More precisely, after n steps, a random walk will be located on average a distance $n^{1/2}$ from its starting point. In numerous cases a constraint is placed on geometric nature of the random walk, as for example not allowing the path to cross itself. There are several natural ways for creating self-avoiding random paths.

One of them consists of starting with a true random walk and eliminating the loops in chronological order. These are walks with loops removed, which have been studied in Kenyon (1999b). In dimension 2 it is shown that the mean diameter of a walk with loops removed with n steps is order $n^{5/4}$. Here we need to ascend the ladder several rungs, using the relation between the paths and tilings of the plane by dominoes, asymptotic expansion of the determinant of the Laplacian, etc. The notion of conformal invariance also appears in this work.

A direct way of constructing self-avoiding random paths combines an idea from complex dynamics (the Loewner equation for encoding bidimensional curves) with probability theory. So-called SLE (Schramm-Loewner evolution) processes are created, introduced in Schramm (2000). These curves appear as the limit of numerous models from physics, e.g. percolation and the Ising model. One of these curves, of fractal dimension 4/3, is conjectured to be the limit of "uniform" self-avoiding random walks obtained by fixing n and uniformly choosing a self-avoiding path of length n from among all the possibilities. This same curve also appears as the limit of other natural self-avoiding random paths: e.g. the outer boundary between a long plane random walk and the outer boundary in a large percolation field; see Lawler, Schramm and Werner (2003), which gives a proof of Mandelbrot's conjecture concerning the exterior boundary of a plane Brownian curve.

Dynamics plays an essential role in all this work. The fact that dynamical notions are appearing more often in geometry is unavoidable and we are devoting the two final chapters of this book to two especially simple and archetypical cases: plane billiards and geodesic flow on a surface.

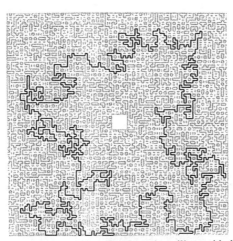

Fig. IX.7.8. We see a cycle in the union of two random tilings with dominoes in a ring. Kenyon (2000) © Annals of Probability, Institue of Mathematical Statistics

IX.8. Tilings in higher dimensions

Following the introductory note here, we will be "done" with crystallographic groups in arbitrary dimension and with tilings, which illustrates in exemplary fashion how we don't understand much in dimension 3, and again that higher dimensions are still richer in unexpected phenomena. A very up-to-date synthesis at the time of its appearance is given in 3.5 of Gruber and Wills (1993). First of all, there is nothing that need be changed in the definition of the crystallographic groups. The basic theorem is Bieberbach's , which says that the group considered must contain linearly independent translations — in number equal to the dimension, not more nor less — and thus yields a lattice. Then, roughly speaking, we adjoin a finite group (the technical term is *extension*). The quotients of $\mathbb{R}^d$ by a lattice are *flat* tori — for which the geometry is thus locally Euclidean — in dimension d. We mention just two things. On the one hand the classification of the crystallographic groups increases spectacularly with the dimension: there are 230 types in dimension 3, for which the basic book is Burckhardt (1966), then 4895 in dimension 4; see Brown, Bülow, Neubüser, Wondratchek and Zassenhaus (1978). The dimension 3 case is fundamental in physics. The case of dimension 4 does not seem to hold special interest from our point of view. For geometric crystallography, see the synthesis in 3.7 of Gruber and Wills (1993). Of very special interest among the crystallographic groups are those *generated by reflections*, because they are encountered in many other places in mathematics; see Sect. IX.10 below and above all Bourbaki (1968).

The geometer wants to investigate all the possibilities for generalizing Euclidean geometry, the simplest of which is the locally Euclidean case. We look for the geometries that are defined on spaces without singularity, i.e. differentiable manifolds. For simplicity, we restrict ourselves to the compact case. The simple idea is that the universal covering, the largest that covers the desired object, is the Euclidean space $\mathbb{R}^d$. The covering, or in the reverse direction the quotient, is obtained by finding a (discrete) group on $\mathbb{R}^d$ which acts without fixed point so as not to have a singularity but a true smooth manifold. This group is thus a crystallographic group, but it doesn't matter which. But Bieberbach's theorem says before anything else that our manifold is a finite quotient of a flat torus $\mathbb{R}^d/\Lambda$, where Λ is a lattice. We have first to classify the flat tori, i.e. what generalizes the modular space is the quotient $\mathrm{SL}(d;\mathbb{R})/\mathrm{SL}(d;\mathbb{Z})$, where $\mathrm{SL}(d;\mathbb{R})$ denotes the linear group of determinant equal to 1; then to look for the possible finite quotients (i.e. the quotients of the flat tori by finite groups that are differentiable manifolds). This classification can be carried out theoretically with an algorithm; see Wolf (1972). In dimension 2, apart from tori, we find only the flat Klein bottles, which are classified as rectangles and which are quotients of rectangular tori by an involution without fixed point.

For the general problem of geometries that generalize the classic Euclidean, spherical and hyperbolic geometries, there is the *space form problem*, which is not resolved today and involves numerous exciting and fundamental concepts; see for example the synthesis in Chap. II of Berger (1999) or Berger (2003). See Sect. IX.10 below for the hyperbolic case.

However, in numerous situations we will need to take a quotient that has singularities; the notion of *orbifold* has appeared rather recently and can be found for example in Bridson and Haefliger (1998).

We now broach the problem of tilings that are not necessarily crystallographic. We are only going to consider those that are *face-to-face*, for even then things are rather complicated. However, to minimize the difficulties for our intuition in large dimensions, we mention a result right away: if we want to tile $\mathbb{R}^d$ with equal cubes, we can see at once that in dimensions 2 and 3 there are at least two squares (cubes) that are face-to-face.

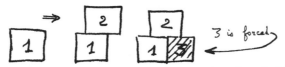

Fig. IX.8.1. The third square is "forced". Readers will investigate in $\mathbb{R}^3$ how two cubes are "forced" to be face-to-face

It can be seen that this persists through dimension 6, but an astonishing fact is that there exist tilings with cubes that never have an entire face in common, this for every dimension greater than or equal to 10. An example is found in dimension 10 (one for dimension 10 suffices, because by tiling by "displaced" layers the example extends to all dimensions) in Lagarias and Shor (1992). The unit cubes are all centered in the lattice $\frac{1}{2}\mathbb{Z}^{10}$ and the tiling admits the periods of the lattice $2\mathbb{Z}^{10}$. The idea of the construction consists of extending to the case of cubes an analogous combinatorial result, based on the construction of graphs with lots of *cliques*: a clique of a graph is a complete subgraph, i.e. one for which the elements are linked two-by-two. The notion of *code*, see Sect. X.7, is also used. In this way they destroyed a conjecture of Keller that dated from 1930.

We return to dimension 3 and look for convex polyhedra that can tile with the aid of isometric copies and face-to-face. In the plane, the number of sides was strictly limited to 6. Don't forget to observe that all this assumes convexity, otherwise the number of sides is unlimited:

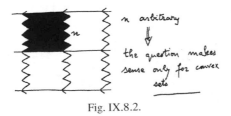

Fig. IX.8.2.

Does there exist a bound for the number of faces of polyhedra that tile $\mathbb{R}^3$? To this day only two partial results are known: we owe to Engel in 1981 the construction of a polyhedron with 38 faces that tiles face-to-face; see Fig. IX.8.2. It is also known that the maximum number cannot exceed 360, but this bound is probably absurd. As a reference, see 3.5 of Gruber and Wills (1993) and Grünbaum (1980).

The Engel polyhedron, just as its precedents which were less good, is obtained in a trivial way (at least as far as the starting principle goes): it is a Voronoi domain for a set of points invariant under a suitable crystallographic group on $\mathbb{R}^3$. But, if in the plane, such a domain is in general a hexagon, if not a rectangle; dimension 3 is a killer. These domains can have a surprising number of faces. The only computer programs for finding them are exponential and unusable presently. The Engel polyhedron has extremely small faces that are difficult to see with the eye; it is a very flat

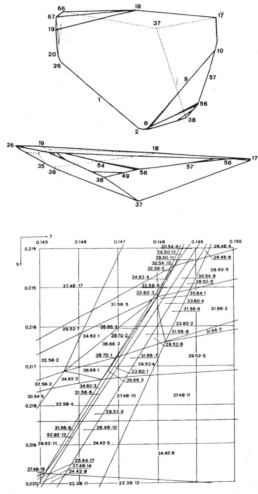

Fig. IX.8.3. Two views of the polyhedra with 38 faces and 70 vertices discovered by Engel in 1980. The graphic in $\{x, y\}$ represents the different polyhedra calculated by Engel, where x and y are two parameters of the lattice that gives rise to them and where the notation 29.52-5 indicates that there are 29 faces, 52 vertices and 5 possible types. We can appreciate the fineness, the near coincidence of certain features.
Grünbaum (1980) © American Mathematical Society

diamond and it needs to be turned a lot for tiling and, in fact, only deals with a finite number of plane intermediaries between the origin and the points of the invariant subset associated with the group considered. It is not clear to us whether the present problem is susceptible to a conceptual ascent, but it clearly shows the difficulty of spatial visualization. The calculations were completed on a computer, and the figure representing Engel's work shows the thinness of certain faces.

We now forget isometries and seek to tile ordinary space with polyhedra all having the same combinatoric (see the case of the plane just above). The problem of knowing if we can do with just any type is open; we find a positive answer in Grünbaum, Mani-Levitska, and Shephard (1984), but only for the case of simplicial polyhedra, i.e. all the faces are triangles (see Sect. VIII.10).

It seems to us justified to stay with dimension 3, which is still far from being mastered. Interested readers should see the synthesis in 3.5 of Gruber and Wills (1993).

IX.9. Algorithmics and plane tilings: aperiodic tilings and decidability, classification of Penrose tilings

For this whole section, the basic reference is Senechal (1995). However, we can read with profit the preface of Carbone, Gromov and Prusinkiewicz (2000), a book that typically illustrates some of the new orientations that parts of mathematics is taking, all in a very geometric spirit.

In 1960 Wang studied the problem of the *decidability* of a tiling of the entire plane with a finite set of tile types, i.e. the possibility of deciding in a finite time — for example with a very powerful computer — whether it is possible to tile the plane with copies (of which an unlimited number are available) of the tiles from this set. He furthermore conjectured that, from the point of view of logic, this decidability is equivalent to the nonexistence of at least one set of tiles, a set called *aperiodic*, which is to say on the one hand having the property of being able to tile the plane, but on the other hand only aperiodically. Precisely, a tiling is called *periodic* if it admits at least one nontrivial translation that preserves it. It is well to remark that here we are not requiring, as before, invariance under two linearly independent translations.

In 1967 Robert Berger responded doubly and positively to Wang's problem. He showed on the one hand this logical equivalence, on the other hand he exhibited a set of 10 000 tiles having this property of a necessarily non- periodic tiling. From a practical point of view these tiles were squares with numbered sides, and paving with them meant assembling them so that the numbers of two sides placed together are the same (cf. the game of dominoes). The paving problem is then undecidable and this by very simple geometric examples. Then in 1971 Robinson reduced the number of tiles to 6, but above all in 1974 Penrose achieved reduction of the set to the two tiles below.

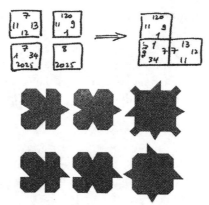

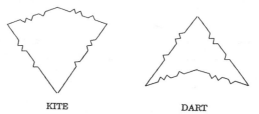

Fig. IX.9.1. Assembling squares with numbered sides; the six Robinson tiles

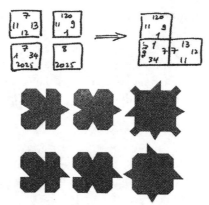

KITE   DART

Fig. IX.9.2. The two Penrose tiles (labeled by objects that fly)

The two tiles of Penrose type are nicely presented with conditions for assembly, for otherwise we could make a rhombus, which can then tile periodically. The coloring or coupling conditions can be replaced by indentations along the boundary, which is done in the figure. For a large tiling and for belief in its existence; see Figs. IX.9.7 and IX.9.9.

For later purposes, it is preferable to work with rhombi (two here) rather than triangles. But not just any rhombi! If the length of all the sides is equal to unity, the angles are those of the figure, where $\theta = \frac{\pi}{5}$. And here again is the golden ratio: the long diagonal of the large rhombus equals $\tau = \frac{\sqrt{5}+1}{2}$, whereas the short diagonal of the small rhombus equals τ^{-1}. These two rhombi don't come out of a hat; the figures show how they are obtained, starting with a Penrose tiling, by decomposition-recomposition of/into triangular pieces (the important idea is kept for the sequel).

Without coupling conditions, the two rhombi can clearly tile in a periodic fashion. These conditions can be expressed in four different ways: by serrations, by arrows, by small black and white circles at vertices and finally by segments: in the last case, the condition is that the segment couple to form a larger segment, without fracture. This would show — but we won't elaborate — that the Penrose tilings possess infinite, complete, lines called *Ammann bars*.

(a)   **(b)**

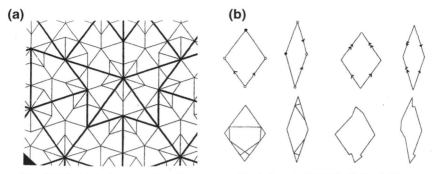

Fig. IX.9.3. **(a)** A decomposition (or recomposition). Senechal (1995) © Cambridge University Press. **(b)** The different ways of forcing non-periodicity

♦

We will devote some time to Penrose tilings, with the elements of proof for their existence and their classification. These tilings in fact add a radically new element to the traditional tiling of Euclidean geometry. Apart from the algorithmic motivation that we have just seen, we find a mixture of elementary geometry and more conceptual notions, where the setting is of a *dynamic* nature (iterations). There is also a partial notion of autosimilarity, but of a nature opposite to that of fractal objects. The best reference is Senechal (1995), all the others that we have consulted are too brief for the basic results, which are intricate. These tilings have numerous properties apart from those we mention; see the references.

We see first a completely elementary existence with our two rhombi. We look at the star from a vertex, the conditions of recomposition permit *a priori* 9 possibilities; it is shown that exactly 7 of these are prolongable (Fig. IX.9.4). It is thus shown that we can continue and so tile the entire plane. Aperiodicity results simply from the type restraints of vertices (there just isn't more to color).

This existence and aperiodicity does not render account of the equally surprising double property of the Penrose tilings. The first property is the following: we can appropriately regroup the tiles (after temporarily breaking them) to form a new Penrose tiling but for which the tiles are derived from the tile types by a similarity with ratio τ^2. These fantastic properties were discovered by Robinson in 1975, along with the complete classification. Observe well that this regroupment operation yields a tiling that is never globally homothetic to the original (this results from something we will see soon).

The second property is that, in the opposite direction, we can also make the tilings finer. These operations are repeatable: we see iteration − dynamics − appear on the horizon! This will also be an essential fact for the classification of tilings. To discover these two properties, called *composition* and *decomposition* in Grünbaum and Shephard (1987), *substitution* and *up and down* in Senechal (1995), we cut up the rhombi into four types of triangles (where here it is τ itself that appears):

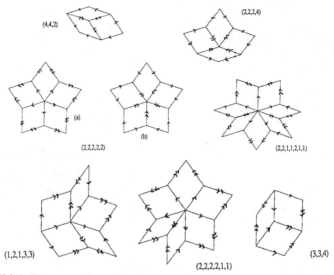

Fig. IX.9.4. The seven types of prolongable stars. Senechal (1995) © Cambridge
University Press

Fig. IX.9.5. How to expand a tiling. Senechal (1995) © Cambridge University Press

The operation *up and down* represented shows right off how to cut up a triangle into smaller triangles. This proves existence at once because we obtain a
triangle which will be able to fill out the whole plane, because this triangle can
be reproduced by symmetry along its sides. Note henceforth an essential fact in
the history of aperiodic tilings to which we will return later, i.e. the existence
of tilings having symmetries of order 5, which were prohibited above for crystals; see Sect. IX.7. The consideration of triangles above also shows why we
can always group with a ratio τ^2. This grouping provides the best explanation
of aperiodicity, because in a periodic tiling it hard to see how a grouping of
tiles could have a dimension that is an irrational multiple of the size of the basic
tiles.

The preceding allows us now to **classify** the Penrose tilings, of course within an
isometry of the Euclidean plane. That was done by Robinson in 1975, and the complete result, logically satisfying, is geometrically astonishing. Starting with a given
tiling P we have seen that we can group the tiles in such a way that they again
form tiles of two types only, i.e. the two tiles derived from the two initial tiles by a

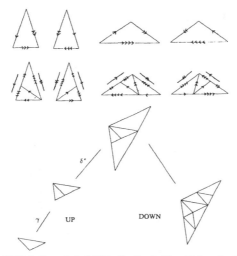

Fig. IX.9.6. Senechal (1995) © Cambridge University Press

homothety with ratio the square τ^2 or the golden ratio τ. Whence, after a homothety with ratio τ^{-2}, a new tiling P^1. From this results first the necessary non-periodicity, since we will obtain greater and greater tiles in the P^n obtained by iterating this procedure. We should be aware that the new tiling is never globally homothetic to the initial tiling; this is one of the essential facts.

For the classification, we do this: we take an arbitrary point in the interior of a tile of the initial tiling, and we look at its type (small or large rhombus) in the successive P^n; we thus obtain an infinite sequence of 0's and 1's. The sequence thus obtained obeys a single condition: a 1 is always necessarily followed by a 0. Let us say it is a *sequence $\mathscr{P}$*.

We then show two things. First every such sequence $\mathscr{P}$ with this property corresponds to a tiling (existence). For the uniqueness, we show that it depends on the choice of the initial point, but the essential remark is that, since the size increases each time by τ^2, two different points will always end up being in the same large tile, thus the associated sequences $\mathscr{P}$ differ at most by a finite number of terms. Finally:

(IX.9.1) *The set of isometry classes of the Penrose tilings (within an isometry of the plane) is isomorphic to the set of sequences $\mathscr{P}$ modulo the equivalence relation of having at most a finite number of terms that are different.*

The set of sequences $\mathscr{P}$, realized as dyadic developments in [0, 1], is of Cantor type, but the quotient is a non separated space, really pathological. Geometrically we can well see why, because after what precedes the existence result also shows a fact that is astonishing for a naive geometer (but perhaps less so for a logician): i.e. that each Penrose tiling contains − in **any** ball with any center and sufficiently large (but uniform) radius − any bounded portion of any other Penrose tiling. It is a way of realizing geometrically the subtlety of the logical problem of the non denumerable (*continuous*) axiom of choice, because what precedes shows that it

Fig. IX.9.7. For the classification we look at which type of tile, in the successive expansions, contains a point fixed initially. Senechal (1995) © Cambridge University Press

is impossible to reasonably exhibit a representative tiling of each type. Here, for interested readers, is a tiling for which the sequence is one of the two most beautiful and natural, i.e. the sequence $\{0, 0, 0, 0, 0, \ldots\}$, called the *cartwheel* by Penrose; see p. 569 of Grünbaum and Shephard (1987) for this characterization:

Readers who are even more interested will want to see a tiling for which the sequence is the most complicated: $\{0, 1, 0, 1, 0, 1, 0, 1, \ldots\}$. But then their surprise (and ours too) is that, at the present moment and to the best of our knowledge, no such tiling has ever been constructed. But now let us return to logic. The preceding seems to indicate that any geometric study (attaching any canonical metric, measure, etc.) of this set is impossible. We find, however, measures more or less adapted to $\mathcal{P}$; see the references in Senechal (1995). It can be seen in Connes (1994) how non-commutative geometry allows endowing this set with an adapted mathematical structure, i.e. an operator algebra.

Fig. IX.9.8. The cartwheel. Gardner (1977) © Scientific American

The Penrose tilings have still other surprising properties, for example (for an appropriate measure) in the space of all tilings a tiling is almost completely determined by two of its vertices; see 6.6 of Senechal (1995).

Among the many works created by the Penrose tilings, first note that they are endowed with a dynamical structure that we announced, i.e. that generated by the discrete operation which sends P to P^1 (here of zero *entropy*). By construction, this operation on the associated sequences of 0's and 1's is nothing other that the *shift* — a displacement to the right by one unit of the entire sequence. It is known that every tiling of this sort — that is to say of aperiodic type and with the *up and down* operations — cannot admit homotheties by algebraic numbers, see for example (Kenyon, 1992 b) (an algebraic number is a real root of a polynomial with integer coefficients), and once again number theory has shown itself!

There remain open questions, but are they good or bad? Readers will form their own opinions after seeing what follows. First, does there exist **only** one tile leading only to non periodic tilings? See Girault-Beauquier and Nivat (1991) for partial results, supplemented by Kenyon (1992a) and the recent Penrose (1997). In contrast, it is known that such a tile exists in dimension 3; it's a sort of rhomboid for which the faces create angles of irrational multiples of π; thus a tiling with copies of this polyhedron (whose existence needs to be shown) forces an infinite set of directions, which proscribes all periodicity.

♦

The fact that certain Penrose tilings, discovered in 1974, admit a symmetry of order 5 (which is proscribed for the "true" crystallographic groups) has been a chief

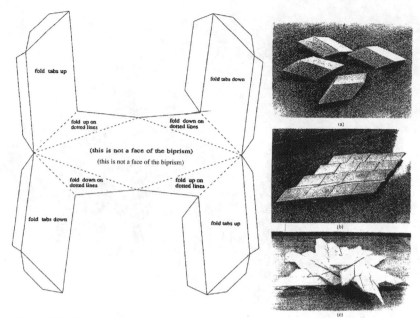

Fig. IX.9.9. How readers can make their aperiodic biprism with the model shown.
Senechal (1995) © Cambridge University Press

surprise. More surprising still was the discovery in 1984 by physicists of physical objects for which the diffraction properties showed effectively a symmetry of order 5. This discovery, joined by purely mathematical studies of Penrose tilings, revived those studies in spectacular fashion and explains the richness of the results mentioned above. With regard to dimension 3, where this sort of crystal is found, the symmetries admitted are those of the icosahedron. The basic reference remains (Senechal, 1995). The two tilings below with symmetry of order five are characterized by the sequence of 0's and 1's in Grünbaum and Shephard (1987, p. 569).

These symmetries proscribed by crystallography, of order 5 for example, which exist without the presence of a lattice in dimension 2 or 3 – regular polygons or polyhedra, and in the Penrose tilings – much intrigued geometers until they found an explanation by increasing the dimension. A systematic treatment is found in Senechal (1995). The model works by generalizing to higher dimensions things that are visible in dimension 2, where we consider the lattice $\mathbb{Z}^2$ and projection on a line D of irrational slope, its a little game to which the theory of continued fractions (see Sect. IX.1) has accustomed us. Here we choose a *box* in the direction orthogonal to that of D and project onto the line D only the lattice points in the band defined by the line and the direction of D.

To obtain the Penrose tilings we consider the lattice $\mathbb{Z}^5$ in $\mathbb{R}^5$ and choose a subspace of dimension 2 of suitable irrational slopes and a suitable box. The symmetry of order 5 sought will be that which arises, by projection, from the permutation of the 5 basis vectors for $\mathbb{R}^5$. An essential fact is that, when we look for the in-

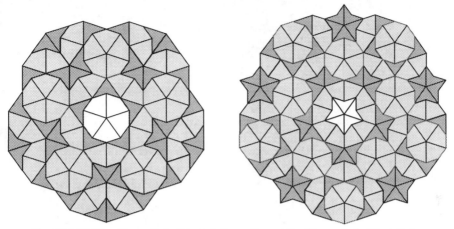

Fig. IX.9.10. The tiling called ≪the infinite sun≫ and the tiling called ≪the infinite star≫. Gardner (1977) © Scientific American

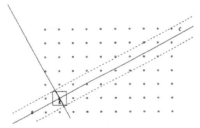

Fig. IX.9.11. A window in a cylinder. Gardner (1977) © von Scientific American

variant subspaces (here planes) under this rotation of order 5, they have irrational directions. See Chap. 3 of Arnold (1990) for connections between quasicrystals, the regular icosahedron, wave fronts and obtaining quasicrystals by projection of a lattice in dimension 5, the coefficients of Fourier series in this space. Attention: this book is exciting, but difficult beyond a superficial reading.

Penrose tilings, and other aperiodicities, have yet other properties. Thus they have a strong rigidity; it can be shown for example that, even if holes are permitted, these holes can only have a finite number of possible shapes, which are moreover completely classified.

For a recent exposition in a spirit directed toward applications, see the article by Kari, pp. 83–95, in Carbone et al. (2000).

Whether the Penrose tilings and much that is analogous belonging to the *aperiodic zoo* (Chap. 7 of Senechal (1995)) are aperiodic because it is possible to make certain recompositions, or for other simpler reasons as for the six tiles of Robinson, it remains the case that all the tiles take on but a **finite** number of directions. We are indebted to Conway for first establishing the existence of aperiodic tilings where the

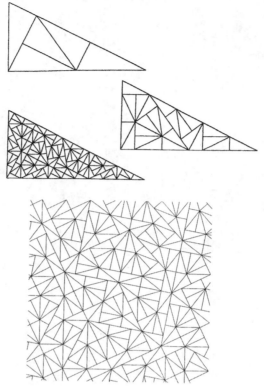

Fig. IX.9.12. How to construct the pinwheel tiling.

tile take on an infinite number of directions. The *Gears* (*pinwheel*) of Conway-Radin is the tiling obtained by iterating the construction of the figure below:

The tiling thus obtained, by iteration until the entire plane is filled by successive decompositions and homotheties, is certainly non periodic since the tiles take on a denumerably infinite number of orientations. But the initial triangle alone clearly tiles also in a periodic fashion. To overcome this, we need to take several isometric triangles and give recomposition rules that force the non-periodicity. This is possible, but incredibly complicated: Radin (1994). A conceptual interpretation remains to be found.

In Radin (1991, 1993), Radin and Sadun (1998), we find a whole program that is just as interesting for logicians as for geometers and physicists. We recall that sometimes we can't comprehend at all how crystals can form which have a regular structure on a large scale in relation to the mechanical forces in crystals that are completely local (of atomic order). The work of Radin, starting with tilings, puts them in a conceptual framework where the notions of dynamics arise in an essential way. In the language of the theory of dynamical systems, for example, the Penrose tilings are not *uniquely ergodic* (see Sect. XI.XYZ), there is but one invariant measure under the operation which defines the dynamical structure (it has been seen to be a shift, but partial). In contrast, the gears tiling above is uniquely ergodic, which is due to the fact that we have a set of rotations generated by a rotation through an irrational multiple of π. See also (Ormas, Radin, and Sadun, 2002).

But for the fundamental invariant of dynamical systems, i.e. the *entropy* defined in Sect. XI.XYZ.B, the entropy of all known aperiodic tilings is zero. As asked by Ruelle: *"Does there exist a turbulent crystal?"*

IX.10. Hyperbolic tilings and Riemann surfaces

There is no reason not to consider tilings in geometries other than Euclidean. The case of the sphere is special, because it is compact and thus each tiling is finite, so that tilings on the sphere lead naturally to the problem amply treated in Sect. III.3, i.e. that of distributing points on S^2 in a harmonious way; see however Chap. 7 of Ratcliffe (1994). For more general things on hyperbolic packings and in arbitrary dimension, see 3.2 of Gruber and Wills (1993) as well as Chap. II of (Vinberg, 1993); we treat here but a small special case.

The hyperbolic plane is of considerable interest. Even though there have been studies of aperiodic tilings in Hyp^2 (see for example Sect. II.4), we won't speak about them. We proceed exactly as in Sect. IX.7, looking for possible rotation orders that correspond to our tiling group. We have always to do with a triangle abc and three integers p, q, r for their orders; that is to say that the angles are the numbers $\frac{\pi}{p}, \frac{\pi}{q}, \frac{\pi}{r}$. But here the condition on the angles becomes

$$(IX.10.1) \qquad \frac{\pi}{p} + \frac{\pi}{q} + \frac{\pi}{r} < 1$$

because the sum of the angles of a triangle in Hyp^2 is always smaller that π; see Sect. II.4. Thus we are here in a situation contrary to that of Sect. IX.7; all values for the triples (p, q, r) are possible with the exclusion precisely of those of Euclidean crystallography. Well, for an arbitrary triple (p, q, r) satisfying (IX.9.1), there always exists a triangle in Hyp^2 for which that angles are $\frac{\pi}{p}, \frac{\pi}{q}, \frac{\pi}{r}$, this because it suffices to go far enough toward infinity and, using the continuity to obtain all intermediate values, since the *ideal triangles* have all their angles equal to zero. Finally, and mainly, with each triangle of this type we can tile the hyperbolic plane, simply be effecting hyperbolic symmetries with respect to the sides and iterating the procedure. We need to show that this really works, even if it is intuitive; numerous old and more or less classic texts are in fact rather imprecise. The good proof is conceptual and we don't know a single good reference. But it is the same as explained in Sect. VIII.5 for the regular icosahedron; it's a nice climb up the ladder, even if by a single rung. We assemble copies, infinite in number, of the triangle type by joining them along corresponding sides and about vertices in numbers respectively equal to $2p, 2q, 2r$. We obtain in this way an abstract space associated with this triangle, which is finally a topological plane endowed with a metric which is locally isometric throughout to the hyperbolic plane with its metric, because along the sides there is no problem because we have taken symmetries about the lines which bear the sides; not at the vertices either, since the total sum of the angles is always equal to 2π. Here again we can thus finally construct a mapping of this topological and everywhere locally hyperbolic plane onto the hyperbolic plane itself, and it is necessarily bijective because the plane is simply connected (see V.XYZ). Figure IX.10.1 shows the tilings associated with the triples $(2, 4, 6)$, $(4, 4, 4)$, $(5, 5, 5)$,; (∞, ∞, ∞).

For a systematic and very complete treatment, see Chap. 7 of the remarkable book (Ratcliffe, 1994). It is important to remark that the groups generated by symmetries (reflections) arise in numerous parts of mathematics and are, for example, at the heart of Lie groups, if only at first for classifying them; see for example the classic Helgason (1978) and the very general Bourbaki (1968). For a synthesis and more − for example that we can tile with polygons having an arbitrary number k of sides − see Vinberg (1993).

We attempt now a parallel with the last part of Sect. IX.7. We seek locally hyperbolic geometries and begin with dimension 2. In the same, completely general, fashion these are the compact quotients under a group without fixed point of the Poincaré half-plane H. These are Riemann surfaces; see Sect. III.XYZ. The trouble here is that here there is no longer the analogy, as with the flat tori, of Bieberbach's theorem. For example, in the group that gives rise to a tiling by reflection about the sides of a triangle (p, q, r) above, it isn't at all evident how to extract a group operating without fixed point (when the group Γ has fixed points, H/Γ is an object called *orbifold* and possesses singular points). This is why in Sect. III.XYZ we gave the construction of Riemann surfaces of arbitrary genus starting with hexagons in Hyp^2. We won't say more here; the world of Riemann surfaces belongs to the realms of number theory, analytic function theory, of algebraic geometry, and still others.

Finding space forms, that is to say geometries, that are locally hyperbolic in dimensions 3 and larger, is difficult; see Berger (1999) or Berger (2003) for a synthesis, and also Ratcliffe (1994). In fact, such space forms were not found in dimension 3 until 1931. The idea was indeed natural: it suffices to find a polyhedron in Hyp^3 which tiles by reflection (extracting next a group acting without fixed point is difficult, but possible). The idea was to take a dodecahedron whose dihedral angles are all equal to $\frac{2\pi}{n}$ for n an integer. Let us stop here, having only given a feeling for the difficulty of the subject: we see in Vinberg (1984) − see also the easier and recent (Vinberg, 1993) − that for the dimensions ≥ 30 there **can't** exist a tiling generated by reflections about the faces of any polytope in Hyp^d. To this day it isn't known by how much 3 must be increased and 30 can be lowered. Vinberg extends what is done in the Euclidean case with the theory of the so-called *Coxeter groups*; see Bourbaki (1968)). He considers the tiling polytope for which the angles between the faces are quotients of π by integers. The basic fact is that, when we compose

Fig. IX.10.1. Tilings with triangles of types $(2,4,6)$, $(4,4,4)$, $(5,5,5)$, (∞,∞,∞).
Rigby (1998) © Springer and J. Rigby

the reflections about two hyperplanes that contain the same subface, we obtain a rotation of finite order by construction. But it is necessary that things fit well when we compose two such rotations. This gives rise to a diagram. In the Euclidean case, for regular polytopes or Lie groups, there are very few possible types of diagrams and we are completely finished with the classification (the angles may never be too small). In the hyperbolic case, we can have larger orders (of very small angles). It is not known how to find all possible diagrams, but by using Dehn-Sommerville seen in (VIII.10.4) for the polytope considered we manage to show that the number of possible types of diagrams is finite and this then limits the dimension.

◆

Packing problems can also be studied in Hyp^2, indeed in Hyp^d; consult 3.3 of Gruber and Wills (1993).

Bibliography

[B] Berger, M. (1987, 2009) *Geometry I,II*. Berlin/Heidelberg/New York: Springer

[BG] Berger, M. & Gostiaux, B. (1987) *Differential Geometry: Manifolds, Curves and Surfaces*. Berlin/Heidelberg/New York: Springer

Aigner, M., & Ziegler, G. (1998). *Proofs from the book*. Berlin/Heidelberg/New York: Springer

Arnold, V. (1990). *Huyghens and barrow, newton and hooke*. Basel: Birkhäuser

Berger, M. (1999). *Riemannian geometry during the second half of the twentieth century*.Providence: American Mathematical Society

Berger, M. (2003). *A panoramic introduction to Riemannian geometry*. Berlin/Heidelberg/New York: Springer

Berggren, L., Borwein, J., & Bortwein, P. (2000). *Pi: A source book*. Berlin/Heidelberg/New York: Springer

Blatter, C. (1997). Another proof of Pick's area theorem. *Mathematics Magazine, 70*, 200

Bleher, P. & Bourgain, J. (1996). Distribution of the error term for the number of lattice points inside a shifted ball. In Berndt et al. (Eds.). *Analytic number theory* (Vol. 1). *Progress in Mathematics, 138*, Basel: Birkhäuser

Bourbaki, N. (1968). Groupes et algèbres de Lie, chapitre IV. *Éléments de mathématiques, 34*. Hermann

Bridson, M., & Haefliger, A. (1998). *Metric spaces of non-positive curvature*. Berlin/Heidelberg/New York: Springer

Brown, H., Bülow, J., Neubüser, R., Wondratchek, H., & Zassenhaus, H. (1978). *Crystallographic groups of four-dimensional space*. New York: Wiley

Burckhardt, J. (1966). *Die Bewegungsgruppen der Kristallographie*. Basel: Birkhäuser

Carbone, A., Gromov, M. & Prusinkiewicz, P. (Ed.). (2000). *Pattern formation in biology, vision and dynamics*. Alghero: World Scientific

Cassels, J. (1972). *An introduction to the geometry of numbers*. Berlin/Heidelberg/New York: Springer

Chaix, H. (1972). Démonstration élémentaire d'un théorème de van der Corput. *Comptes Rendus, AcadÈmie des sciences de Paris, 275*, 883–885

Connes, A. (1994). *Noncommutative geometry*. San Diego: Academic Press

Conway, J., & Sloane, N. (1999). *Sphere packings, lattices and groups* (3rd ed.). Berlin/Heidelberg/New York: Springer

Croft, H., Falconer, K., & Guy, R. (1991). *Unsolved problems in geometry*. Berlin/Heidelberg/New York: Springer

Douady, A. (1995). Présentation de J-C. Yoccoz. (TheWork of the Fields Medalists). In *Proceedings of the International Congress of Mathematicians, 1994*. Birkhäuser, 11–16

Ehrhart, E. (1964). Sur un problème de géométrie diophantienne linéaire. *Journal fr die Reine und Angewandte Mathematik, 226*, 1–29 and *227*, 25–49

Ehrhart, E. (1977). *Polynômes arithmétiques et méthode des polyèdres en combinatoire*. Basel: Birkhäuser

Erdös, P., Gruber, P., & Hammer, J. (1989). *Lattice points*. New York: Longman Scientific and Technical, John Wiley

Fejes Tóth, L. (1972). *Lagerungen in der Ebene, auf der Kugel und im Raum* (2nd ed.). Berlin/Heidelberg/New York: Springer

Flajolet, P., Vallée, B., & Vardi, I. (2000). *Continued fractions from Euclid to the present day*. Preprint IHES

Fodor, F. (1999). The densest packing of 19 congruent circles on a circle. *Geometriae Dedicata, 74*, 139–145

Fricker, F. (1982). *Einführung in die Gitterpunktlehre*. Basel: Birkhäuser

Gallavotti, G. (1983). *The elements of mechanics*. Berlin/Heidelberg/New York: Springer

Gardner, M. (1977). Extraordinary non-periodic tiling that enriches the theoery of tiles, Mathematical Games. *Scientific American, 1977*, 110–121

Girault-Beauquier, D., & Nivat, M. (1990). Tiling the plane with one tile (polyominoes). In *Proceedings of the Sixth Annual Symposium on Computational Geometry (Berkeley, 6–8 June 1990)*. New York: Association for Computing Machinery, 128–138

Gruber, P., & Lekerkerker, C. (1987). *Geometry of numbers*. Amsterdam: North-Holland

Gruber, P., & Wills, J. (Ed.). (1993). Handbook of convex geometry. Amsterdam: North-Holland

Grünbaum, B. (1980). Tilings with congruent tiles. *Bulletin of the American Mathematical Society (New Series), 3*, 951–973

Grünbaum, B. (1990). Periodic ornamentation of the fabric plane: Lessons from Peruvian fabrics. *Symmetry, 1*, 45–58

Grünbaum, B., Mani-Levitska, P. & Shephard, G. (1984). Tiling three-dimensional space with polyhedral tiles of a given isomorphism type. *Journal of the London Mathematical Society, Second Series, 29*, 181–191

Grünbaum, B., & Shephard, G. (1987). *Tilings and patterns*. New York: Interscience

Grünbaum, B., & Shephard, G. (1993). Pick's theorem. *The American Mathematical Monthly, 100*, 150–161

Hales, T. (2000). Cannonballs and honeycombs. *Notices of the American Mathematical Society, 47*(4), 440–449

Hardy, G., & Wright, E. (1938). *An introduction to the theory of numbers*. Oxford: Clarendon Press

Helgason, S. (1978). Differential geometry, lie groups and symmetric spaces. San Diego: Academic Press

Hirzebruch, F., & Zagier, D. (1987a). Classification of Hilbert modular surfaces. In *Collected works of Hirzebruch*. Berlin/Heidelberg/New York: Springer

Huxley, M. (1996). *Area, lattice points, and exponential sums*. Oxford: Clarendon Press

Huxley, M., & Nowak, W. (1996). Primitive lattice points in convex planar domains. *Acta Arithmetica, 76*, 271–283

Kenyon, R. (1992a). Rigidity of planar tilings. *Inventiones Mathematicae, 107*, 637–651

Kenyon, R. (1992b). Self-replicating tilings. *Contemporary Mathematics, 135*, 239–263

Kenyon, R. (2000). Conformal invariance of domino tiling. *The Annals of Probability, 28*, 759–795

Lagarias, J., & Shor, P. (1992). Keller's cube-tiling conjecture is false in high dimension. *Bulletin of the American Mathematical Society (New Series), 27*, 279–283

Lanford, O., III., & Ruedin, L. (1996). Statistical mechanical methods and continued fractions. *Helvetica Physica Acta, 69*, 908–948

Lang, S. (1983). *Fundamentals of diophantine geometry*. Berlin/Heidelberg/New York: Springer

Lawler, G.F., Schramm, O., & Werner, W. (2003), Conformal restriction properties. The chordal case. *Journal of the American Mathematical Society, 16*, 917–955

Margulis, G. (1977). Discrete groups of motions of manifolds of non-positive curvature. *American Mathematical Society Translations, 39*, 33–45

Marmi, S. (2000). *Chaotic behavior in the solar system* (following J. Laskar), *Séminaire Bourbaki* 1998–1999. In *Astérisque, 266*, 113–136

Melissen, H. (1997). Packing and covering with circles. The Netherlands: Proefschrift Universiteit Utrecht

Morgan, F. (2000). *Geometric measure theory* (4th ed.). San Diego: Academic Press

Müller, E. (1944) *Gruppentheorische und Strukturanalytische Unterssuchungen der Maurischen Ornamente aus der Alhambra in Granada*. Thèse Univ. Zürich. Rüschlikon: Baublatt

Nikulin, V., & Shafarevich, I. (1987). *Geometries and groups*. Berlin/Heidelberg/New York: Springer

Oda, T. (1988). *Convex bodies and Algebraic geometry: An introduction to the theory of toric varieties*. Berlin/Heidelberg/New York: Springer

Ormes, N., Radin, C., & Sadun, L. (2002). A homeommorphism invariant for substitution tiling spaces, *Geometriae Dedicata, 90*, 153–182

Penrose, R. (1997). Remarks on tiling. In R. Moody (Ed.), *The mathematics of the Long-Range Aperiodic Order* (pp. 467–497) Dordrecht: Kluwer

Radin, C. (1991). Global order from local sources. *Bulletin of the American Mathematical Society*, 25, 335–364

Radin, C. (1993). Symmetry of tilings of the plane. *Bulletin of the American Mathematical Society*, 29, 231–217

Radin, C. (1994). The pinwheel tiling of the plane. *Annals of Mathematics, 139*, 661–702

Radin, C., & Sadun, L. (1998). An algebraic invariant for substitution tiling systems. *Geometriae Dedicata, 73*, 21–37

Ratcliffe, J. (1994). *Foundations of hyperbolic manifolds*. Berlin/Heidelberg/New York: Springer

Rigby, J. (1998). Precise colourings of regular triangular tilings. *The Mathematical Intelligencer, 20*, 4–11

Rogers, C. (1964). *Packings and coverings*. Cambridge: Cambridge University Press

Savo, A. (1998). Uniform estimates and the whole asymptotic series of the heat content on manifolds. *Geometriae Dedicata, 73*, 181–214

Senechal, M. (1995). Quasicrystals and geometry. Cambridge: Cambridge University Press

Series, C. (1982). Non-Euclidean geometry, continued fractions and ergodic theory. *The Mathematical Intelligencer, 4*, 24–32

Series, C. (1985). The geometry of Markoff numbers. *The Mathematical Intelligencer, 7*, 20–29

Serre, J.-P. (1970). *Cours d'arithmétique*. Paris: Presses Universitaires de France

Sinai, Y. (1990). Hyperbolic billiards. In *Proceedings of the International Congress of Mathematicians (Kyoto, 1990)*, Vol. 1. Tokyo: Mathematical Society of Japan & Springer, 1991, 249–260

Stark, H. (1970). *An introduction to Number Theory*. Cambridge: MIT Press

Stein, R. (1985). Cover page. *The math. magazine, 58*, Nr. 5 (November)

Stewart, I. (1979). On ne peut pas bien ranger une valise seulement en s'asseyant dessus. *Pour la Science, 26* (décembre 1979)

van der Poorten, A. (1979). A proof that Euler missed... Apéry's proof of the irrationality of zeta(3), an informal report. *Mathematical Intelligencer, 1*, 195–204

Vardi, I. (1999). Deterministic percolation. *Communications in Mathematical Physics, 207*, 43–66

Vinberg, E. B. (1984). The absence of crystallographic groups of reflections in Lobachevskii spaces of large dimension. *Transactions of Moscow Mathematical Society, 1985*, 75–112

Vinberg, E. B. (1993a). *Geometry II*. Berlin/Heidelberg/New York: Springer

Walfisz, A. (1957). Gitterpunkte in mehrdimensionalen Kugeln. Panst. Wyd. Naukowe, Warschau

Wang, H.-C. (1972). Topics on totally discontinuous groups. In W. Boothby & G. Weiss (Eds.), *Symmetric spaces* (pp. 459–487). New York: Marcel Dekker

Wolf, J. (1972). *Spaces of constant curvature* (2nd ed.). Berkeley: J. Wolf

Chapter X
Lattices and packings in higher dimensions

X.1. Lattices and packings associated with dimension 3

A lattice in $\mathbb{R}^3$ is a Λ that can be written as the set of integer combinations of three linearly independent vectors $\{a, b, c\}$, say $\Lambda = \mathbb{Z} \cdot a + \mathbb{Z} \cdot b + \mathbb{Z} \cdot c$. As in Sect. IX.4, two Euclidean invariants are immediately associated with a lattice; they are practically dictated when we seek to pack balls of like radius in the densest possible way, thus the most economical for practical life; see more in Sect. X.4. If we want to place the centers of these balls of given radius R at the vertices of a Λ, then it is necessary that $\|\lambda - \mu\| \geq 2R$ for all the $\lambda, \mu \in \Lambda$ and $\lambda \neq \mu$, whence the notion of *minimal norm* of a lattice:

$$\mathrm{NorMin}(\Lambda) = \inf\{\|\lambda - \mu\| : \lambda, \mu \in \Lambda \text{ and } \lambda \neq \mu\}.$$

The *volume* of a lattice is the absolute value $|\det(a, b, c)|$ of the determinant of any basis $\{a, b, c\}$ of Λ. Then the density of the lattice will be, as in (IX.4.1):

$$\text{(X.1.2)} \qquad \mathbf{Dens}(\Lambda) = \frac{\pi}{6} \frac{(\mathrm{NorMin}(\Lambda))^3}{\mathrm{Vol}(\Lambda)},$$

in other words, the quotient of the volume of the ball of radius R (i.e. $4/3\pi R^3$), for $R = \frac{1}{2}\mathrm{NorMin}(\Lambda)$ (the maximum value of R), to the volume of the basic parallelepiped of the lattice.

Of course, our quantity being invariant under change of scale, we suppose henceforth that $R = \frac{1}{2}$, i.e. that $\mathrm{NorMin} = 1$. What lattice yields the largest value for the density? The result was conjectured by Seeber in 1831 and proved by Gauss the same year. Gauss wanted to answer this question, because it had come up in number theory in studying quadratic forms with integer values, well before the Minkowski's general theories; see Sect. IX.3. This is a problem of elementary geometry concerning the tetrahedron in $\mathbb{R}^3$ formed by a basis $(0, a, b, c)$: which is the one of smallest volume under the condition that all 6 edges have length greater or equal to unity. Readers will have guessed the answer: it is the regular tetrahedron and we leave the corresponding proof (which isn't so easy) to them; see also a little further below or Sect. VIII.7.

What does our optimal lattice look like? A small difficulty is that there are two distinct ways of seeing it, of realizing it. Even though the equivalence is very easy

M. Berger, *Geometry Revealed*, DOI 10.1007/978-3-540-70997-8_10,
© Springer-Verlag Berlin Heidelberg 2010

(and a good exercise in spatial visualization), its realization is important for the following section and can be seen in the to figures below:

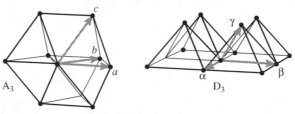

Fig. X.1.1. A_3 and D_3

We can view the two associated ball packing, seemingly quite different, in Fig. X.1.2.

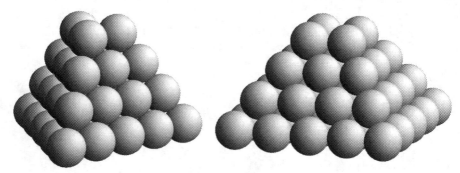

Fig. X.1.2.

We don't give coordinates of a basis for the lattice A_3, leaving this for readers to do, but in the fastest way, by placing themselves in dimension 4 and observing that the set Λ is the intersection of $\mathbb{Z}^4 \subset \mathbb{R}^4$ with the hyperplane with equation $x_1 + x_2 + x_3 + x_4 = 0$. The second figure suggests the basis $\{(1,0,0), (0,1,0), (\frac{1}{2}, \frac{1}{2}, \frac{1}{\sqrt{2}})\}$, but this representation isn't pleasant because it isn't very symmetric; after a change of scale by the factor $\sqrt{2}$ we replace it by $\{(1,1,0), (1,-1,0), (1,0,1)\}$. But above all we remark that the lattice generated can be defined as $\{(x,y,z) \in \mathbb{Z}^3 : x+y+z$ even $\}$.

We emphasize yet again that these two definitions lead to identical lattices as regards their Euclidean structure (see Fig. X.1.3) and so we obtain one and only one optimal packing of balls of the same radius, of the same **lattice type**.

Owing to crystallographers, this lattice − and of course any similar lattice − is called the *cubic face centered lattice*. It's the lattice that appears in the structure of carbon, but some atoms need to be added to the basic cubes. For pure mathematicians, it's *denoted* A_3 or D_3. We will see why in Sect. X.4. Thus:

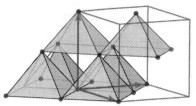
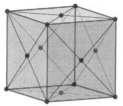

Fig. X.1.3. In the orthogonal basis $\{(1, 1, 0), (1, -1, 0), (0, 0, \sqrt{2})\}$, for which the vectors have length $\sqrt{2}$, the elements of the lattice D_3 appear as $\{(x, y, z) \in \mathbb{Z}^3 : x + y + z$ is even$\}$ and can thus be seen as the sites (atoms) of a cubic face-centered lattice. Warning: the crystallographic definitions are not the same as ours. The "unit cell" in general contains several atoms, whereas one of our basic parallelepipeds Λ (or its translates) contains but one; the cubic unit cell here contains 4 atoms

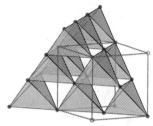

Fig. X.1.4. How to see D_3 in A_3 (and vice versa). We of course see equilateral triangles in the packing on the *right* in Fig. X.1.2. But we need to go deeper: the void *left* by the tetrahedral packing in gray contains *two* pyramidal packings with square base analogous to that of Fig. X.1.1. These pyramids can be joined two by two to form regular octahedrons

(X.1.3) (Gauss). *The densest lattice packing by balls of the same radius in* $\mathbb{R}^3$ *is that obtained with the cubic face centered lattice (the lattice* A_3*) and this density is equal to* $\pi/3\sqrt{2} = 0.74048...$.

The value of the density (calculating the volume of a parallelepiped) is left to readers. The optimality of this lattice can be considered heuristically in two ways, which are not true proofs, but provide important ideas for the sequel. The first is that packing done starting with the square lattice is optimal because this is how everyone

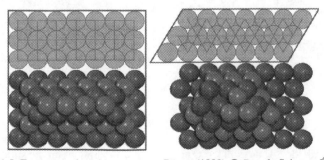

Fig. X.1.5. Two ways of stacking oranges. Pöppe (1999) © Pour la Science, Éditions Belin

does it in practice, whether on fruit stands or else — in a time now long past — stacks of cannonballs.

Mathematicians, trained systematically with the notion of recurrence, seek to construct an optimal packing in dimension 3 starting with an optimal packing in the plane, i.e. regular hexagonal . Balls are stacked on a plane in this way, then a second layer is placed on the first in analogous fashion. And things go well; the balls in the second layer fall a bit to just touch those in the first. In arbitrary dimension, this recurrence operation of optimal placement of successive layers is called a *lamination*; see Sect. X.5.

Gauss's proof is exhibited in Hales (2000): first two balls at least have to touch, but since we are dealing with a lattice, this leads to a line of tangent balls. A second line has to touch the first, thus we have right away an hexagonal packing in the plane. In three dimensions, the layer above the preceding has also to be regular hexagonal and is dropped onto the first as low as possible.

♦

With the idea of preparing for what is to come, i.e. for investigating densest packings, but without necessarily associating them with lattices, we must now look — as in the planar case of the preceding chapter — at the Voronoi decomposition and the Delaunay triangulation associated with the lattice A_3.

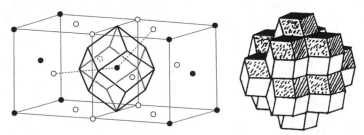

Fig. X.1.6. The Voronoi domain at a vertex of A_3, with the twelve neighboring vertices of the given vertex that enters into the definition (all located at the same distance). The packing of the Voronoi domains. Fejes Tóth (1972) © Springer and G.F. Tóth

The Voronoi domain (**the** because all are translates of this one from the origin) is that of the above figure and not at all a regular polyhedron, which explains among other things the drama of the next section. As for the Delaunay triangulation, as in the plane it is clearly not unique on account of the squares that are interposed. We could take it as invariant under the lattice; it is then made up of regular tetrahedrons **and** regular octahedrons ; this again is one of the reasons for the drama to come.

If we want to have a true triangulation available, then we have to break symmetry by a small variation that provides a way of decomposing the polyhedra that occur into tetrahedra. The result isn't unique, but is dependent on the small variation chosen.

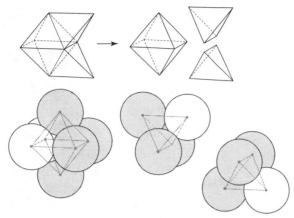

Fig. X.1.7. The packing A_3: a regrouping of three Delaunay cells (one octahedral et two tetrahedral) tiles the space. In crystallography they are called octahedral and tetrahedral sites, while thinking of potential inclusions that might accept these sites

♦

Here the space of all the lattices, modulo a similitude, is the double quotient space $SO(3;\mathbb{R}) \setminus SL(3;\mathbb{R}) / SL(3;\mathbb{Z})$, that we should see as a portion of $\mathbb{R}^5$ with identifications on the boundary. The quotient on the left by $SO(3;\mathbb{R})$ indicates that we are working within isometry in the Euclidean space $\mathbb{R}^3$, the quotient on the right represents the change of basis operations for the lattice. In fact, fixing one vector with norm equal to 1, choosing the second vector in the plane thus making two parameters, and finally three parameters for the choice of the last basis vector. If we work only with isometries, thus taking scales into account, we find six parameters. The description of this space is a very difficult problem, the difficulty being with the identifications on the boundary of the domain in $\mathbb{R}^5$; see the references in Note 2 in the Preface to the Third Edition of Conway and Sloane (1999). The study is geometric; it consists of an attempt to best describe which polyhedron constitutes the Voronoi cell.

♦

As in Sect. IX.4 we define without any change the dual lattice Λ^* of the lattice Λ by the *entirety* (or *integrity*) of scalar products. Here $\Lambda = \mathbb{Z}^3$ is yet again its own dual, but the dual A_3^* of A_3 isn't at all the same as A_3. In crystallography it is called the *cubic body centered lattice*, and is very familiar to chemists. Here is its Voronoi domain:

This time it's a semiregular polyhedron, the truncated octahedron. The lattice A_3^* provides us an excellent opportunity to talk about problems of optimal covering by balls; cf. Sect. X.5 for the plane. Since Bambah in 1954 the best covering of $\mathbb{R}^3$ by balls of the same radius that is known is where the centers are in the lattice A_3. We know moreover that it is the best possible among all lattice packings. in contrast

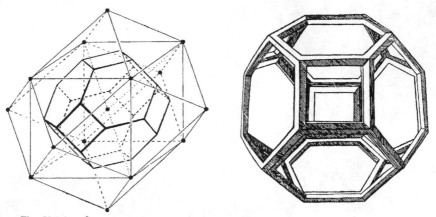

Fig. X.1.8. A_3^*, the cubic body centered lattice and its Voronoi domain, which is the truncated regular octahedron; drawing by Leonardo da Vinci. Fejes Tóth (1972)
© Springer and G.F. Tóth

the problem remains wide open for arbitrary packings. For coverings, the thing that plays the role of density is the *mesh*, which readers will easily define independent of whether it is in a lattice, e.g. it has to do with covering the plane most efficiently with equal disks or — as here — space with equal balls. All texts on packing discuss this problem, not pursued further here for lack of space.

Covering problems, just as much as those of packing, are of considerable practical utility. In dimensions 2 and 3 they are visible (planting of trees, irrigation, coverage by radar, etc.); in higher dimensions there are telecommunication problems; see the bible (Conway and Sloane, 1999) for references and a whole array of applications, some of which we will discuss below.

We have of course, thanks to Λ^*, a Fourier analysis like that for periodic functions in three variables. The Poisson formula extends unaltered:

$$(\text{X.1.3}) \qquad t^{-3/2} \operatorname{Vol}(\Lambda) \sum_{v \in \Lambda} \exp\left(-\pi \|v\|^2 / t\right) = \sum_{\xi \in \Lambda^*} \exp(-\pi t \|\xi\|^2)$$

(t real > 0) and shows here too that knowledge of the set of norms of the elements of Λ (taking their multiplicities into account) is equivalent to that of norms of elements of Λ^* (with multiplicity accounted for), that is as always called the *spectrum* of the flat torus $\mathbb{R}^3 / \Lambda$ of dimension 3. An important problem is that of knowing whether or not there exist two non congruent lattices (non isomorphic) that admit the same spectrum, which for the geometer is to say each has as many points as another, for each R, in any circle about the origin of radius R. It has just been resolved negatively; see Sect. X.7.

X.2. Optimal packing of balls in dimension 3, Kepler's conjecture at last resolved

We broach the problem of optimal packing, more precisely of the optimal density of a packings of balls of the same radius in ordinary space of three dimensions. Before discussing this exact problem, we give a few broad remarks on more general packings.

Efficient packing of objects in a given receptacle is a familiar but difficult problem. Some of us are more adept than others at packing a suitcase or the trunk of a car. It is a subject teeming with naive (and false) ideas: dig a new hole to hold the dirt (in the words of cartoonist and mathematics teacher Georges Colomb, alias Christophe, 1856–1945).

These arrangement problems are also mathematically very difficult and the solutions are most often very primitive or unknown, because the number of parameters has risen too quickly for present day power of computers . This is a reason for considering the simplest case, that of equal spheres, because only the position of the center (three parameters) matters; a sphere has no orientation − all directions are the same. In spite of this, we can't much exceed ten spheres. Our problem entails arranging balls of the same radius in a cubic box or more generally a parallelepiped. The practical importance is considerable. The manufacturer of balls wants to minimize packing and shipping costs. For packings of spheres, and for heaps of sand − the dunes − and their evolution; see Guyon (2001).

As in the planar case, the problem only really makes sense for a large number of balls. The cases of small domains are practically just as horrible as for disks in the plane; in space we expect worse. However, we mention in passing Besicovitch's problem, because of its importance for analysis; see Mattila (1995):

> *How many balls of the same radius can be placed inside a larger ball of five times the radius?*

Today 67 is the best number known (in dimension 3, because the problem is general). The figure below gives the known distribution, it is from Sullivan (1994), where all the necessary explanations and the connection with analysis can be found; in particular this question arises in the convergence of algorithms for finding minimal surfaces; see Sect. VI.8.

We now take up the problem of the best −densest − packings of balls in three dimensional space, called *Kepler's conjecture*. There is no unique solution, as we will see. It has an exciting history; Hales (1999) is a summary with all the old historical references. Kepler, in a little book entitled *A new years present: the hexagonal snow crystals*, wrote:

(X.2.1) *The face centered lattice packing is the tightest possible, in the sense that no more balls can be packed into the given receptacle.*

We won't give the precise definition of the density; we amply indicated how to proceed in Sect. IX.5 and repeat here − to spare reader from having to go back −

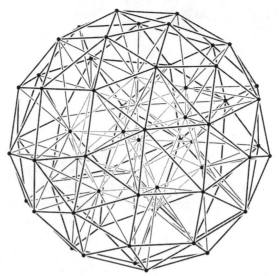

Fig. X.2.1. The centers of 67 spheres inside a ball of radius five times greater. For seeing the structure, all segments are drawn which are of length less than 1.4 and connect two points have been drawn. The thick lines connect the centers of the 52 spheres of the exterior layer, while the medium lines connect these points to the 14 of the interior layer, and finally the thin lines connect these 14 points among themselves as well as with the central point. Sullivan (1994) © Journal of Geometric Analysis

what was said about the planar case. A correct statement of this definition can be found in Rogers (1964), of course with notions of limit: lim sup and lim inf need to be considered from the start, even if they ultimately vanish in favor of ordinary limits. Rogers exhausts the space with cubes; he doesn't show that his definition is equivalent to that of Fejes-Tóth (1972), who exhausts the space with balls (and, as in all the writings of that author, is rather imprecise and difficult to read). It isn't at all obvious that it is possible to choose either cubes or balls, nor that the orientation of the cubes matters little. We indeed do need to work with infinity. Readers who would like to show the equivalence of these definitions can take inspiration from the Apollonian gasket encountered in Sect. II.9.

This conjecture (X.2.1) harbors, and has engendered, several dramas. The first is this: in a strict sense the conjecture is false, and we explain why. In fact, Harriot (the same person as in Sect. III.1) commented that "even Kepler was mistaken". We can obtain packings of the same density in the following way. We start with an plane regular hexagonal packing seen in Sect. X.1. In the above figure, a second layer has been placed. Now what happens when we place a third layer? There are two choices possible (which wasn't the case when we packed square layers); in fact there are two choices at each new layer, call them 0 and 1. We obtain the cubic face centered packing if and only if we always choose 0 or always 1. But whatever the manner of choosing each time, the density remains always $\pi/2\sqrt{3}$; this isn't so difficult to see. There are thus as many packings as the power of the continuum (the "number" of infinite sequences of 0's and 1's).

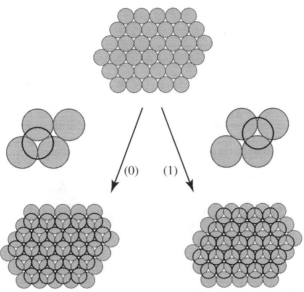

Fig. X.2.2. Balls can only be placed over one set of holes or the other. There are but two ways of proceeding. In a plane hexagonal packing, there are twice as many holes as balls

We point out that a large number of metals crystallize in the lattice A_3 (or D_3), or cubic face centered lattice, also called cubic dense: Al, Ni, Cu, Ag, Au, while others crystallize in a hexagonal dense packing: Mn, Ca, Sr, Ti. But there are some that crystallize in still other forms that aren't always of maximal density.

Here, in the case of three layers, are figures that show the comparison of the two associated Voronoi domains such that the twelve balls surround a central ball (because things are difficult to visualize, we strongly advise readers to try these arrangements with ping-pong balls and neoprene glue):

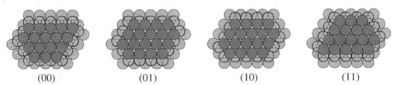

Fig. X.2.3. *Possible packings with three hexagonal layers. Center:* hexagonal dense packing (assuming periodic extension: every second layer is directly above the other); on the sides is the cubic face centered packing

Even though Kepler was occasionally mistaken, his conjecture was:

(X.2.2) (Kepler conjecture.) *The maximum density of a packing of balls of the same radius in* $\mathbb{R}^3$ *is* $\pi/2\sqrt{3}$.

This conjecture was recently proved by Hales and Ferguson (end of 1999), a proof begun in Hales (1997), followed by a series of articles by Hales and finished by him with the contribution of Ferguson:

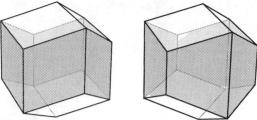

Fig. X.2.4. *Voronoi domains At left*, cubic face centered packing, also called A_3, D_3 or 000... *At right*, hexagonal dense packing, also denoted 010101...

Fig. X.2.5. *The twelve neighbor balls of a ball shown from various viewpoints. At left*, cubic face centered packing. *At right*, hexagonal dense packing

(X.2.3) *The Kepler conjecture is true.*

It isn't that numerous mathematicians weren't connected with it, but the proof encountered intrinsic difficulties, about which we will speak, and then required — once the right programs had been devised and edited — calculational resources on an extremely powerful computer, in fact at the very limit of present capabilities. We ultimately just give quickly a few brief insights about Hales' proof, emphasizing what keeps the proof from being simple, even with new concepts, as powerful as they are. We will see that things will perhaps be proved in the future in a conceptual and simple manner in dimensions 8 and 24, as they were in dimension 2. Historically, we mention only that Hilbert put this conjecture on his famous list of problems in 1900, and in Rogers (1964) we find the comment: *many mathematicians* **believe** *and all physicists* **know** *that (X.2.2) is true.* And Milnor added in (Milnor, 1994): *this is scandalous situation since the (presumably) correct answer has been known since the time of Gauss.* For fascinating historical considerations and an analysis of the proof see — besides Oesterlé (1999) — the text Hales (2000), which reads like a novel.

First of all, one difficulty is heuristic, in view of what has transpired: we have to prove a result where a minimum is attained — even locally — in different ways. Problems in the calculus of variations with several minima are notoriously difficult, e.g. because we can't say that we follow in canonical fashion a trajectory increasing the density, to converge finally toward **the** top (or bottom). Two other dramas await us if we attempt to apply the two methods of proof in the planar case, seen in Sect. IX.5.

The first route, that of Fejes-Tóth, uses the decomposition of the space into Voronoi cells attached to the centers of the packing disks, and he proves that the local density attached to a disk, i.e. that of the disk in its Voronoi domain , is maximum for the regular hexagonal case, attained moreover when the disk has six neighbors that touch it. Here we also decompose a la Voronoi and seek the minimum of the volume of such a cell. We first need to express the volume algebraically and then to calculate:

(X.2.4) *The minimum volume of the domain* $\mathrm{Vor}(x_1, \ldots, x_i, \ldots) = \{x : \|x\| \leq \|x - x_i\| \text{ for each } i\}$, *where we consider all sets of points* $\{x_i\}$ *of* $\mathbb{R}^3$ *such that* $\|x_i - x_j\| \geq 2 \ (i \neq j)$ *and* $\|x_i\| \geq 2$.

We note a first difficulty: the number of x_i isn't fixed. This, however, is apparently a naively simple optimization problem, but unfortunately the number of parameters exceeds what can be handled with present day computers. We can also hope to proceed geometrically without such means. The problem (X.2.3) admits a natural subsidiary problem: the smallest volume has a good chance of being attained when all the neighboring balls are tangent to the central ball. But we have resolved this problem in Sect. III.4, the problem of the thirteenth sphere: there are in this case at most twelve balls. We recall also that in the plane Fejes-Tóth needed to show two things: that with six disks the minimum is the regular hexagon and with seven or more the Voronoi area was greater (than the regular hexagonal cell), and that this requires a more subtle proof than is apparent, in spite of visual evidence. Here we will put our finger, and our eye, on the fact that there is a radical leap in difficulty from the planar problems to the spatial ones. In fact, here is what has happened. Fejes-Tóth made the:

(X.2.5) (Dodecahedral conjecture.) *The minimal volume required in (X.2.4) is that of the regular dodecahedron circumscribed about the unit ball, i.e.* $10\sqrt{130 - 58\sqrt{5}}$, *and is attained solely by a regular dodecahedron.*

Let us suppose that (X.2.5) is true. The drama is that this has no immediate rapport with the Kepler conjecture . In fact, in the plane case of Sect. IX.5 we end the proof because the regular hexagons indeed tile the plane. Well, the regular dodecahedron **does not tile** space, this simply because the dihedral angle (between two faces) is about 116 degrees (calculation left to readers), a value less that $360/3 = 120$ degrees!

We however end here with (X.2.5), which is interesting in itself. It was proved by Fejes-Tóth in the case where there are exactly 12 x_i, an inequality of the isoperimetric type encountered in Sect. VIII.7. Unfortunately, even if it seems outrageous at first glance, the assumption that there would be 13 or more spheres would not immediately imply that the volume is greater than that of the regular dodecahedron. The difficulty of this optimization problem is partly related to what we encountered in Sect. IX.8: there are polyhedra that tile and which have a large number of faces. Thus, in our case, the Voronoi domain could have something like forty faces, thus an optimization in dimension 150, completely out of the reach of present-day computational resources. An essential remark: in all the calculations that have been

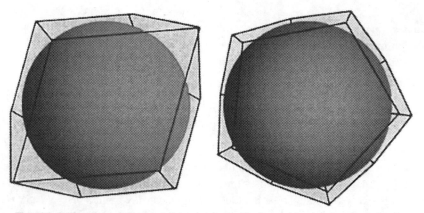

Fig. X.2.6. The Voronoi domain of a vertex in the Kepler packing is a *rhombododecahe dron*. A Voronoi domain in the form of a regular dodecahedron would have a lesser volume. But the regular dodecahedron does not tile the space. Pöppe (1999) © Pour la Science, Éditions Belin

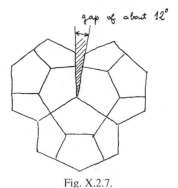

Fig. X.2.7.

attempted or achieved by computer, it was not a question of obtaining a logical proof or proving a value to be exact, but of eliminating configurations as giving a bad value and thus reducing the possibilities. In fact:

(X.2.6) (Hales and McLaughlin). *The dodecahedral conjecture is true.*

But the proof is posterior to that for (X.2.3) and uses numerous points of that proof in an essential way. We finish this intermezzo with an open problem, for those who like them; we could also again call it the *thirteenth sphere problem*. Since it is known how to prove (again this is not obvious) that there are never thirteen spheres (balls) of unit radius mutually exterior to each other and tangent to the unit ball, we are certain that the mean $\frac{1}{13}\sum_i \|x_i\|$ of the distances to the origin of the thirteen centers x_i is greater than 2 for each appropriate configuration of thirteen points. But no one to the present has succeeded in finding an explicit lower bound for the centers of the spheres of such a configuration; see Hales (1994) and compare with Fig. IX.5.5.

Fig. X.2.8. The thirteenth sphere problem: determining the minimum of $\frac{1}{13}\sum_i \|x_i\|$ under the constraints $\|x_i\| \geqslant 2$ and $\|x_i - x_j\| \geqslant 2$ $(i \neq j)$. Hales (1994) © Springer and T. Hales

♦

Since the Fejes-Tóth method didn't work, let's try the second approach, which worked for us in the plane, that of Rogers, based on the Delaunay triangulations and the notion of *local density*. But in the case of the cubic face centered lattice and in other cases, this graph defines but the Delaunay polyhedra, that need to be decomposed into tetrahedra). For this we use the pattern furnished by a small generic variation, but several patterns are possible.

The local density of the packing in a domain D is the ratio volume of the intersections of D with the packing balls to the total volume of D. Rogers (1954 or 1958) then shows:

(X.2.7) (See also (X.6.1)) (Rogers 1954). *The local density of a packing of $\mathbb{R}^d$ in a simplex is always less than or equal to the value $\sigma(d)$ (Rogers constant) obtained when the simplex is regular, of side 2 and its vertices are the centers of the packing balls. Equality occurs only in the regular case.*

This result is intuitive enough: readers should make a sketch in the plane with a triangle. In fact, the proof requires careful decompositions, based on calculations of norms and scalar products.

So we have here another known optimal local density, where unfortunately again the regular tetrahedron **does not tile** space, because its dihedral angles equal about deg 7032′, and $5 \times$ deg 7032′ = deg 360 − deg 720′. Aristotle made the famous error of declaring that the regular tetrahedron tiles space! An important point: note that $\sigma(3) = \sqrt{18}\left(\arccos\left(\frac{1}{3}\right) - \frac{\pi}{3}\right) = 0.77797...$ is just less than $\pi/2\sqrt{3}$. For the other $\sigma(d)$, see Sect. X.6 below. A simple reason is that the basic parallelogram of the hexagonal lattice is composed of two equilateral triangles, whereas from dimension three on the basic parallelepiped of the cubic centered lattice decomposes into 6 pieces surely, but of which only two are regular tetrahedra of density $\sigma(3)$, whereas the four others have a higher density, even if only because they are not equilateral.

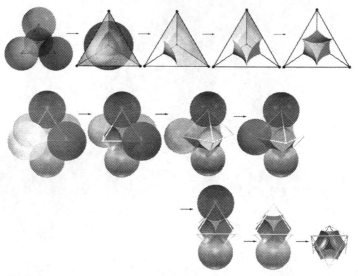

Fig. X.2.9. Representation of the local density of the packing A_3 in a Delaunay tetrahedron: in succession each of the parts borne by the spheres centered at the vertices is transferred to the tetrahedron. *Below*: the same thing for a Delaunay octahedron. Pöppe (1999) © Pour la Science, Éditions Belin

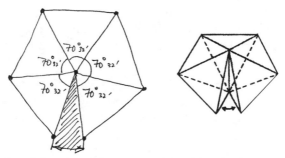

Fig. X.2.10. Gap of about deg 720′; contrary to Aristotle's claim, five regular tetrahedra do not fit together well

We point out that, despite the simplicity of its definition, the Rogers constant is not explicitly known apart from dimensions 2 and 3; readers can calculate $\sigma(2)$, and perhaps also $\sigma(3)$, finding $\sigma(3) = \sqrt{18}\left(\arccos\left(\frac{1}{3}\right) - \frac{\pi}{3}\right)$. But after that there is only an asymptotic estimate; see Sect. X.5 below and p. 569 of Gruber and Lekerkerker (1987). We can surmise that the reasons for this difficulty are of much the same sort as for the open problems on simple volumes of tetrahedra that we saw in Sect. III.5.

Before beginning with some glimpses into the proof of (X.2.3), a point of history. The long text of Hsiang (1993) claimed to prove (X.2.3); the critical analysis carried out in Hales (1994) shows that the text is to a large extent replete with "holes" – even if it presents some new elements – and is far from being decisive.

Next, the existence of an infinite packing, but in a single spatial direction, is given in Boerdijk (1952). Its density is greater than $\pi/\sqrt{18}$. This packing does not seem to have had many successors, by which we mean trying to stack spheres in successive cylinders about this first cylinder. Although those who thought the conjecture is false have typically leaned on his work, it is rather the case that he — among others — showed that the conjecture is very difficult, not false.

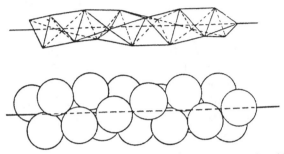

Fig. X.2.11. The infinite packing of Boerdijk, constructed with the aid of regular tetrahedra wound into spirals

From the **experimental** side, attempts have been made to pack balls by pouring them gently into a container while shaking. The result is given below. Your author can't help but mention for the benefit of contemporary readers that when he tested his "thoughts" before writing, a listener treated them with sophistry, refusing to believe that, when the balls are carefully poured into a box, bad packings were possible. The experimental results are very instructive. In 1959 Bernal compressed spheres of plasticine rolled in powdered chalk and found an average of 13.3 preponderantly pentagonal faces. Notice that, in contrast, the optimal packings always have 12 faces, which are quadrilaterals. G.D. Scott on the other hand poured thousands of balls. He found that if there is no agitation during the pouring, a density of about 0.60 is obtained, whereas with agitation he got 0.6366, still far from the theoretically optimal $\pi/\sqrt{18} = 0.74$....

We will complete these asides with this, even though it may be classic. It has to do with explaining two things: first, the trick of the fakir, who plunges and replunges a knife into a jar of rice; this ends up with the knife pressed enough to allow picking up the jar by the knife alone. The second fact is that, when we walk on wet sand at the seashore, under certain conditions the sand dries around each footprint. In the two cases the explanation is the same: the grains of rice or sand are packed fairly densely, even though they are not Kepler packings; they are packed in some good way as in the experiments we just mentioned. Both the fakir's knife and the footprint disturb this good packing, with the effect of increasing the volume of the surrounding air. In the case of the fakir, this increases the pressure on the knife more and more; in the case of the beach sand, this creates more space under the foot, and the water rushes in, drying the surrounding sand.

♦

Finally, let us go to the proof of (X.2.3). What then do we do if the local density —
whether it be that of Voronoi or of a triangulation tetrahedron — doesn't suffice? We
need to go beyond the central sphere and its twelve (or possibly thirteen) neighboring spheres and consider also the local densities associated with the neighbors of
these twelve; in other words, we need to show that the average of the twelve local
densities is greater of equal to $\pi/2\sqrt{3}$. The idea is that some compensations have
to be produced, that we surely discern in a Delaunay triangulation of the cubic face
centered lattice: we find on the one hand regular tetrahedra (with sides equal to 2),
but then four pieces of regular octahedron that are composed of tetrahedrons having
5 sides of length 2 but for which the sixth has length $2\sqrt{2}$. It's what Hales called
a *very bad score*. He associates a *score* with each star of the triangulation considered and succeeds in eliminating myriad configurations that produce a bad score.
For a brief but interesting exposition, see Oesterlé (1999). We can say in caricature
that Hales' method is a mixture of Voronoi and Delaunay decompositions. Another

Fig. X.2.12. Nature is wise

idea is to introduce tetrahedrons of various types, starting with those that are *quasi-regular*: for these the maximum ratio between side lengths never exceeds 2.51. But other types also appear. With the prerequisite of a very good knowledge of spherical trigonometry (we recall from Sects. III.4 and VIII.5 that spherical geometry is very treacherous) programs have been implemented for treating more than 100 000 linear equations in 100 to 200 variables and with 1 000 to 2 000 constraints. It is thus right at the limit of the capabilities of present day computers.

♦

The Kepler conjecture has an affine version: pack ellipsoids as best possible, each obtainable from the others by translations (it isn't required that they form a lattice). In contrast, things are different if we want to pack ellipsoids in the Euclidean space $\mathbb{R}^3$ that are always congruent (derived from one another by isometries), but where we permit them to **change** direction. Then the answer is completely different; we now know how to produce packings of density better than $\pi/3\sqrt{2} = 0.74048....$ The last that we know of appears in Wills (1991), who in fact has for density 0.7549... . We work with ellipsoids that are very flattened; at the moment the Wills packing is done with ellipsoids for which the ratio of the extreme axes is 41.78.... The initial idea is (see Wills for references to intermediate studies) arrange the ellipsoids in such a way as to bore out kinds of *tunnels* where we pack supplementary ellipsoids. We can do still better, but then the ellipsoids become extremely flattened. The tunnels are made with the help of the so-called *hexagonal dense* packing. This is no mystery for us; it is constructed with the help of hexagonal layers as indicated at the beginning of the section, but in a way that the sequence of 0's and 1's is periodic, e.g. 0110011001100... . That means that the centers of the spheres in every other layer are on the same vertical lines. These are the lines provided by the sought-for tunnels.

On the other hand, in the plane each tiling by ellipses of the same area, whatever their direction, is optimal when they are all congruent and arranged in a lattice: this is due to Fejes-Tóth; see p. 88 of Fejes-Tóth (1972).

♦

As for the dual problem of the finest covering of the space by disks, for which we saw in Sect. IX.5 that it is solved by the regular hexagonal lattice (even if a lattice wasn't initially required), it is completely open in each dimension greater than 2 (see also Sect. III.3). If we know that it is by a lattice, then it is with the lattice A_3^*. For this subject, and in any dimension, see Chap. 2 of Conway and Sloane (1999) and 3.3, 3.4 of Gruber and Wills (1993).

X.3. A bit of risky epistemology: the four color problem and the Kepler conjecture

The four color conjecture is strikingly absent from our book. It's statement is so simple: *Can a map, however complex, always be colored with four colors?*

This problem was clearly posed by Cayley in 1879: we decompose the plane into bounded regions — we can also talk about graphs — and we want to color so that any two adjacent regions always have different colors. In spite of the simplicity of its statement, its answer, affirmative, was found only in 1976 by Appel and Haken, in spite of a great number of important, but never completely conclusive, prior contributions. The proof is a mixture of old ideas that allow limiting the test of its truth to a finite number of particular maps. But here, as it has to do with precise colorings, the computer truly performs part of the proof. A good nontechnical exposition on the architecture of the proof is to be found in Fournier (1977). In brief, in no case is any new concept produced in solving this problem. Besides, it is stated in Chap. 25 of Aigner and Ziegler (1998), that there is: *no proof in The Book in sight*. To explain this quotation, we recall that Paul Erdös liked to talk about *The Book*, in which God keeps the perfect proofs. The book of Aigner and Ziegler — of which Erdös would surely have been a coauthor if he hadn't died in 1997 — assembles those proofs deemed worthy of inclusion; but this is merely an imperfect "first approximation" to *The Book*'s essence, which readers can discover for themselves.

The four color problem comes up in an anecdote concerning Minkowski — creator of concepts that can be seen in several places in our book — that we can't fail to mention and can be found in Reid (1970). We point out that this is one of the best books we know of about a mathematician, the balance between the historical, personal, anecdotal and the strictly mathematical is extraordinary. We might add the remarkable book on Fourier (Dhombres and Robert, 1998). To fully appreciate the piquancy of this story, we need to know that Minkowski was extremely modest. During a lecture on topology he mentioned the four color problem and said: "*Here is a theorem that has never been proved because only mathematicians of third rank would work on it*". And in a rare demonstration of arrogance announced: "*I believe that I can prove it.*" He began the proof on the blackboard at once, but at the end of the lecture he still wasn't finished. The project continued in the remaining lectures of the week, then continued for several more weeks. Finally one soggy morning Minkowski entered the classroom, followed by an enormous clap of thunder and, with a profoundly serious expression on his gentle round face, said: "*heaven is angry because of my pretension. My proof of the four color theorem is also incomplete.*"; and he took up again his topology lecture that he had left several weeks before. Needless to say, Minkowski ultimately never found the new concept — if it has ever existed — that would have allowed a proof.

We think that this story remarkably illustrates the fact that numerous (but not all, of course) problems of "elementary" geometry are still without solution because of their intrinsic difficulty, and not just because they are without interest or have passed out of fashion. We have attempted to indicate throughout the present book, with all the risks entailed, whether a problem is a simple curiosity or instead harbors deep concepts. For more of this, consult Berger (2001a).

♦

So, at least to this day, the positive solution of the four color theorem has not contributed any new concept and has not renewed the mathematical landscape; the proof is computerized and formal. In view of the preceding section, it is the much the same — whatever the virtues of its principal author — with the proof of Kepler's conjecture . Nowadays we are finding more and more proofs of a completely formal nature, among others for the fundamental theorem of algebra and the Jordan curve theorem.

This opinion is shared by numerous mathematicians, but not by all. But for the many of us, neither of those two solutions has really contributed to substantial progress. However, for others the four color problem has been an important motivator in the development of graph theory; for this subject, see Thomas (1998).

Again a bit of epistemology. Presently the Kepler conjecture is regarded as "proven" by the mathematical community, even if some aren't completely satisfied. In fact, it is the same with the classification of finite simple groups, with the Smith conjecture on the fixed points of group operations on the sphere S^3, and with Perelman's proof of the Poincaré conjecture , object of a Clay prize. The Clay Mathematics Institute has paid honoraria for the verification of this proof (see the end of Sect. III.5). It seems that there are increasingly more theorems for which no mathematician can grasp the proof in its entirety. Even journalists have even picked up on this circumstance; see for example an article in *Nature*, volume 42/43, July 2003: "*Does the proof stack up?*", and above all an article in *The Economist* "*Mathematics: Proof and Beauty*", April 2, 2005, pp. 69,70.

X.4. Lattices in arbitrary dimension: examples

Important remark. We are entering into the domain of packing of balls in arbitrary dimension, at the outset with or without a lattice; but having a lattice (and if possible quite dense) is easier for packing, for stacking. The general problem is of great importance, but has not received much attention until rather recently. This problem lies at the intersection of number theory, algebraic geometry, analysis, geometry (for its beauty in itself), combinatorics, and theoretical physics and possesses (or is connected with) a host of practical applications, about which we will speak a bit. Computers have permitted significant progress, but we will see that computers themselves use results from the lattice packings.

We recall too that lattices come up in a natural way in number theory (quadratic forms with integer values), in algebra (theory of finite groups), in algebraic geometry in the theory of Lie groups and Coxeter groups (generated by reflections), in analysis (e.g., numerical integration of functions of several variables), in algebraic geometry (complex tori, Jacobians), in geometry (space forms, flat tori); see in particular the error detecting codes in Sect. X.7.

The literature is thus immense, still in full effervescence, and so we have had to make drastic choices. The bible nowadays is Conway and Sloane (1999), which has the quality (essential, in pedagogy) of being redundant; because everything is proved there, sometimes in various ways. We also find introductions and motivations in the prefaces and beginnings of chapters. But we can also find interest in reading the syntheses in 3.3 and 3.4 of Gruber and Wills (1993) (very up to date as of its appearance). The fact that 3.3 entails a bibliography with more than thirty treatises and syntheses on the subject makes it already rather long. We will therefore be parsimonious with references, referring mainly to these works. Once again, the essential quest (the holy grail) is that of finding the densest packings possible. We will always consider only packings of balls of the same radius. The references Erdös, Gruber, and Hammer (1989), Gruber and Lekerkerker (1987), 3.2 of Gruber and Wills (1993) are also important.

Even though the geometry here is very difficult and anti-intuitive, we should keep in mind that, since in most of the applications we deal with high dimensions, asymptotic estimations of invariants (such as the density) are the essential goal rather some precise calculation in some dimension. Moreover, it is clear from reading the literature that mathematicians who have worked in this theory have very often appealed to geometric visualization.

It is natural to begin with packings of lattice type. It is the subject of this section, where we describe certain lattices that are very important in numerous branches of mathematics, and also of the following section, where we will study the densities of various lattices.

We thus consider *lattices* Λ, *a priori* arbitrary, in $\mathbb{R}^d$, i.e. the set of points with integer coordinates with respect to any basis $\{a_1, \ldots, a_d\}$ (i.e. d linearly independent vectors) in $\mathbb{R}^d$, considered as a Euclidean space. We can also write $\Lambda = \mathbb{Z} \cdot a_1 + \cdots + \mathbb{Z} \cdot a_d$. We have two premier fundamental geometric invariants, the volume (determinant) and the minimal norm. The *volume of Λ, denoted* $\det(\Lambda)$, is the absolute value of the determinant $\det(a_1, \ldots, a_d)$. The changes of basis of a lattice are given exactly by the elements of the group $\mathrm{SL}(d; \mathbb{Z})$, for which the elements are written as matrices with integer coordinates and for which the determinant is equal to 1. This is simply the volume of the parallelepiped constructed on the basis $\{a_1, \ldots, a_d\}$. We need to be on guard that what the bible (Conway and Sloane, 1999) calls the determinant is **the square** of our determinant. The reason is that the volume of the parallelepiped constructed on $\{a_1, \ldots, a_d\}$ has a square that is the associated Grammian determinant , which is by definition the determinant of the symmetric matrix whose elements are the scalar products $a_i \cdot a_j$. It is advantageous for calculating in a truly Euclidean manner, without having to take the absolute value.

The second basic invariant is the *minimal norm* NorMin(Λ), which is by definition the smallest distance between two distinct arbitrary points of the lattice (verify that this is always a positive number). Take care that we cannot interpret NormMin for a single basis, as the representation of $\mathbb{Z}^2$ by the basis $\{(2,3),(1,2)\}$ clearly shows. Note also that for Conway and Sloane (1999) the minimal norm is **the square** of our minimal norm. With regard to algorithmic complexity, no polynomial program is known for calculating the minimal norm of a lattice given by a basis.

Having in mind that we are systematically looking for dense lattices in the following section, we immediately define here the corresponding notions:

(X.4.1) *The* density *of a lattice* Λ *of dimension* d *is the quotient*

$$\Delta(\Lambda) = \frac{\beta(d)(\mathrm{NorMin}(\Lambda))^d}{2^d \det(\Lambda)}$$

(where $\beta(d)$ denotes — we recall — the volume of the unit ball in the space of dimension d).

This definition reflects the fact that the balls centered at the points of Λ and which have radius $\frac{1}{2}\mathrm{NorMin}(\Lambda)$ are the largest that can be packed in the lattice. On the other hand, $\Delta(\Lambda)$ is certainly the ratio of the volume of a ball of radius $\frac{1}{2}\mathrm{NorMin}(\Lambda)$ to the volume of a parallelepiped generated by a basis. We can also say that it is the volume of a Voronoi cell associated with the lattice considered. We often make use of a density called the *central density*, the same notion as defined in Conway and Sloane (1999).

(X.4.2) *The* central density *is:*

$$\mathrm{DensCentr}(\Lambda) = \frac{\Delta(\Lambda)}{\beta(d)} = \frac{(\mathrm{NorMin}(\Lambda))^d}{2^d \det(\Lambda)}.$$

The interest in this notion lies in the fact — recall Sect. VII.6.B — that the number $\beta(d)$ becomes extremely small as d becomes large, and that, even with very dense lattices, the density is a ridiculously small number. It is thus interesting to "see" things in a different way. This often is not sufficient for seeing clearly in large dimensions; see Sect. X.5 where, as we often found in Chap. VIII, there is interest in taking nothing less than the d-th root of the quantities considered.

We note that we have lost the 2^d in taking the d-th root, but this is not very serious because asymptotically $\sqrt[d]{2}$ tends toward 1 as d tends toward infinity.

The standard lattice (cubic unless otherwise stated) is thus $\mathbb{Z}^d$, with volume 1 and for minimal norm $\frac{1}{2}$. It is clear that it is very bad, from dimension 2 on, but in fact we will see (when we will have enough lattices at our disposal) that things get worse very quickly.

The lattices A_d and D_d. Motivated by the dimension 2 and 3 cases, we introduce in arbitrary dimension d the lattices *denoted* A_d, defined as those whose basis is the regular simplex, which is practical to describe, as we saw early in Sect. X.1 (see also the end of Sect. VII.6.A) by adjoining a dimension :

$$A_d = \left\{ \sum_i x_i = 0 \mid x_i \in \mathbb{Z} \text{ for each } i \right\} \subset \mathbb{Z}^{d+1}$$

(we can also give the basis, but always in $\mathbb{Z}^{d+1}$: $\{(1, -1, 0, \ldots, 0), (0, 1, -1, \ldots, 0), \ldots, (0, \ldots, 1, -1)\}$). But starting with dimension 4, it is no longer our A_d that is the densest.

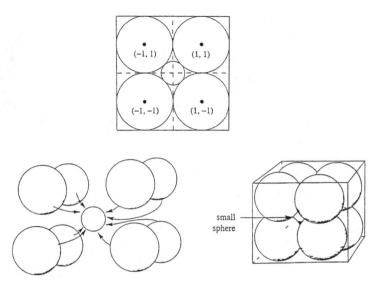

Fig. X.4.1. A "small" sphere on the interior can become larger and larger

We return to the lattice $\mathbb{Z}^d$. Interested readers will notice that, starting with dimension 5, we not only have to inscribe a ball of radius $\frac{1}{2}$ in it, but furthermore to have it slide within. Here is why, with a first idea (which will very quickly prove insufficient) for finding dense lattices: fill the holes as much as possible. In the square lattice, we see that the largest holes are fillable by balls of radius $\frac{\sqrt{d}-1}{2}$; thus beginning with dimension 4 we can add throughout balls of radius equal to $\frac{1}{2}$. In dimension 4, this gives a lattice *denoted* D_4 generated by $\mathbb{Z}^4$ and the point $(\frac{1}{2}, \frac{1}{2}, \frac{1}{2}, \frac{1}{2})$; that is, we have doubled the lattice $\mathbb{Z}^4$ by adjoining the translated $\mathbb{Z}^4 + (\frac{1}{2}, \frac{1}{2}, \frac{1}{2}, \frac{1}{2})$. We can rewrite it, after a change of basis, as generated by $(1, 1, 0, 0)$, $(0, 1, 1, 0)$, $(0, 0, 1, 1)$, $(1, 0, 0, 1)$, (the change of scale is in $\sqrt{2}$). More generally, we define:

$$D_d = \left\{ (x_1, \ldots, x_d) \in \mathbb{Z}^d \text{ with } \sum_i x_i \text{ even} \right\}.$$

This notation comes from the theory of Lie groups, where the lattices A_d and D_d are those associated (via the technical language of the *root systems* of Lie algebras) with the complex unitary and real orthogonal groups, here in even dimension.

We can very easily calculate for the two families of lattices (leaving it to readers to acquire some intuition): volume, minimal norm, central density, *kissing number* of the ball (centered at the origin and of radius half the minimal norm). We thus find in this order:

$$\text{for } A_d : \sqrt{d+1}, \ 2, \ \frac{1}{2^{d/2}\sqrt{d+1}}, \ d(d+1)$$

$$\text{for } D_d : 2, \ \frac{1}{\sqrt{2}}, \ \frac{1}{2^{(d+2)/2}}, \ 2d(d-1).$$

We recall (see Sect. III.7) that the kissing number in Euclidean geometry of dimension d is the maximum number of balls of radius 1 that can touch the unit ball, this of course without interpenetration. The above values thus show that, starting with $d = 4$, the density of D_4 is better than that of A_4. What corroborates the fact that the kissing number of D_4 is 24, against only 20 for A_4 (so therefor the regular simplex isn't a universal panacea). Two important remarks for the lattice D_4: it yields the best that is known today for the kissing number; see Sects. III.4 and III.7. It is very amusing that among all lattices, it's the only one, along with the regular plane hexagonal lattice, for which the Voronoi cell is a regular polytope, i.e. the $\{3, 4, 3\}$ encountered in Sect. VIII.11.

The lattices E_6, E_7, E_8. Those who know lie groups also know these lattices, corresponding to root systems of five exceptional Lie groups in low dimensions. For us, it's E_8 which comes up geometrically right away. In fact, we will try to repeat the device we used for constructing D_4 starting with $\mathbb{Z}^4$. We begin again with D_4; we see that, starting with dimension 8, we can again slide into D_8 the copy $D_8 + (\frac{1}{2}, \ldots, \frac{1}{2})$. The whole forms a lattice E_8, and defined thus:

$$E_8 = \left\{ (x_1, \ldots, x_8) \in \mathbb{Z}^8 \cup \left(\mathbb{Z}^8 + \left(\frac{1}{2}, \ldots, \frac{1}{2}\right)\right) \ \middle| \ \sum_i x_i \text{ even} \right\}.$$

The minimal norm is $\sqrt{2}$, the volume equals 1, whence the densities: ordinary $\pi^4/384 = 0.25367\ldots$ and central $\frac{1}{16}$, which are excellent; compare them with those of D_8. This is corroborated by a very large kissing number. We calculate it: the vectors of minimal norm are thus of four types: on the one hand those of type $(\pm 1, \pm 1, 0, \ldots, 0)$, on the other hand those of type $(\pm\frac{1}{2}, \ldots, \pm\frac{1}{2})$, with three possibilities: 8 times the same sign, 6 times the same sign and 4 times the same sign. Whence a total of $4C_8^2 + 2 + 2C_8^2 + C_8^4 = 240$. He have seen in Sect. III.7 that the kissing number 240 is optimal (and even better, it determines the location of the 240 balls within an isometry of the space). The lattice E_8 was considered well before it was used by Elie Cartan in Lie groups in 1894, in fact already in 1867 in number theory, in connection with a quadratic form with integer values having very special properties, which we will see in Sect. X.8 along with the nice properties. We will see in Sect. X.7 that E_8 is identical to its dual E_8^* (which proves trivially that its volume is equal to 1), but of course a lot more besides.

We find the lattices E_6 and E_7 more easily starting with E_8 than directly, using the following device: E_7 is simply the orthogonal complement in E_8 of any vector of minimal norm of E_8. As for E_6, this is the orthogonal in $\mathbb{R}^8$ of any subspace whatever containing a lattice of E_8 which is of type A_2. To learn more, see Chap. 4 of Conway and Sloane (1999), which we are following in any case.

The Leech lattice Λ_{24}.

In contrast, here we are dealing with a very recent object, discovered by Leech only in 1965. Moreover, the circumstances of its discovery are interesting: Leech was looking for very dense lattices by lamination, specifically here in dimension 25, by putting one layer above the lattice of dimension 24 as densely as was known at the time. Now, he perceived (this by the way isn't completely clear) that the second layer **fell, slid** (which is still better than only being able to "place inside") toward the interior of the first (see what we did with $\mathbb{Z}^4$ to obtain D_4 and with D to obtain E_8), whence in fact a new lattice in dimension 24, and evidently denser than the one then known. This construction of Leech is systematized in Chap. 6 of Conway and Sloane (1999); they basically use the very geometric study of the "deep holes" of the lattice. This is to say that they consider the packing of balls of radius $\frac{1}{2}$ NorMin(Λ) constructed with the lattice Λ considered and investigate the points of the space where it possible to place in the complement of the stacking in $\mathbb{R}^d$ a ball of maximum radius centered at such a point. This lattice is still more extraordinary than E_8: like E_8, it is self dual; thus of volume equal to 1; the minimal norm equals 2, thus the balls of the associated packing are of radius 1; whence the value of the densities: the (ordinary) density is $\pi^2/12! = 0.001930...$ and the central density equals 1. If only for this reason, the Leech lattice is something extraordinary. Its kissing number is 196560, as we will soon see (see also Sect. III.7). In the lattice zoo, the Leech lattice deserves for its rarity and elegance the title of **panda**; as much as in Riemannian geometry this title falls to the projective plane of the Cayley octaves; for the latter see for example 0.G of Berger (1999) or/and III.E of Berger (2003).

We don't know any really rapid and "clear" definition of Λ_{24}; in 4.11 of Conway and Sloane (1999) there is an impressive number of them. We will choose a few; readers will appreciate why the definitions always remain difficult. The briefest consists of saying that Λ_{24} is **the** *laminated lattice of dimension* 24. See the following section for the definition of the laminated lattice of given dimension, but in brief it is simply the lattice constructed step by step from the lattice layer of the preceding dimension and having the best density. This provides existence and uniqueness, but only in a completely theoretical way.

The second construction − explicit − requires knowledge of the *Golay* error correcting code . This code Gol_{24} is a (very well chosen) set of 2^{12} sequences of 24 elements which are all 0 or 1, which is the same as saying that $\text{Gol}_{24} \subset \{0, 1\}^{24}$; we will study it briefly in Sect. X.7. This known, the Leech lattice is generated (in redundant fashion) by all vectors of the form $\frac{1}{\sqrt{8}}(\mp 3, \pm 1^{23})$, where: 1^{23} denotes 1 written 23 times, where ∓ 3 can appear anywhere among the 24 possibilities, and finally the essential condition: the choice of the 24 plus and minus signs must be made in accordance with the Golay code. We obtain a basis:

$$\frac{1}{\sqrt{8}}$$

8 0 0 0	0 0 0 0	0 0 0 0	0 0 0 0	0 0 0 0	0 0 0 0
4 4 0 0	0 0 0 0	0 0 0 0	0 0 0 0	0 0 0 0	0 0 0 0
4 0 4 0	0 0 0 0	0 0 0 0	0 0 0 0	0 0 0 0	0 0 0 0
4 0 0 4	0 0 0 0	0 0 0 0	0 0 0 0	0 0 0 0	0 0 0 0
4 0 0 0	4 0 0 0	0 0 0 0	0 0 0 0	0 0 0 0	0 0 0 0
4 0 0 0	0 4 0 0	0 0 0 0	0 0 0 0	0 0 0 0	0 0 0 0
4 0 0 0	0 0 4 0	0 0 0 0	0 0 0 0	0 0 0 0	0 0 0 0
2 2 2 2	2 2 2 2	0 0 0 0	0 0 0 0	0 0 0 0	0 0 0 0
4 0 0 0	0 0 0 0	4 0 0 0	0 0 0 0	0 0 0 0	0 0 0 0
4 0 0 0	0 0 0 0	0 4 0 0	0 0 0 0	0 0 0 0	0 0 0 0
4 0 0 0	0 0 0 0	0 0 4 0	0 0 0 0	0 0 0 0	0 0 0 0
2 2 2 2	0 0 0 0	2 2 2 2	0 0 0 0	0 0 0 0	0 0 0 0
4 0 0 0	0 0 0 0	0 0 0 0	4 0 0 0	0 0 0 0	0 0 0 0
2 2 0 0	2 2 0 0	2 2 0 0	2 2 0 0	0 0 0 0	0 0 0 0
2 0 2 0	2 0 2 0	2 0 2 0	2 0 2 0	0 0 0 0	0 0 0 0
2 0 0 2	2 0 0 2	2 0 0 2	2 0 0 2	0 0 0 0	0 0 0 0
4 0 0 0	0 0 0 0	0 0 0 0	0 0 0 0	4 0 0 0	0 0 0 0
2 0 2 0	2 0 0 2	2 2 0 0	0 0 0 0	2 2 0 0	0 0 0 0
2 0 0 2	2 2 0 0	2 0 2 0	0 0 0 0	2 0 2 0	0 0 0 0
2 2 0 0	2 0 2 0	2 0 0 2	0 0 0 0	2 0 0 2	0 0 0 0
0 2 2 2	2 0 0 0	2 0 0 0	2 0 0 0	2 0 0 0	2 0 0 0
0 0 0 0	0 0 0 0	2 2 0 0	2 2 0 0	2 2 0 0	2 2 0 0
0 0 0 0	0 0 0 0	2 0 2 0	2 0 2 0	2 0 2 0	2 0 2 0
-3 1 1 1	1 1 1 1	1 1 1 1	1 1 1 1	1 1 1 1	1 1 1 1

Fig. X.4.2. Conway, Sloane (1999) © Springer

We see quickly which are the vectors with minimal norm 2; they have to be of one of the three forms: two 4's and the rest 0's, eight 2's and the rest 0's, one 3 and the rest 1's. We find indeed each time $2.4^2 = 4.2^2 = 9 + 23 = 32$ (to be divided by 8). To calculate their number, we need to know the values of the various types of elements of the Golay code (Sect. X.7); we find at first $2^{12} \cdot 24 = 98304$ vectors with one 3 and twenty-three 1's, because — apart from the place of the 3 — all the elements of Gol_{24} agree. Next, for the vectors with two 4's, Golay imposes no condition, so that we find $4 \cdot C_{24}^2 = 1104$ of them. For the four 2's, the number of elements of Gol_{24} of "weight" equal to eight is 759, whence $2^7 \cdot 759 = 97152$ vectors, making a grand total of 196560.

Here now is an ultrarapid definition that will please those keen on (Lorentzian) hyperbolic geometry ; it is taken from Chap. 26 of Conway (1995). In the Lorentz space $E^{24,1}$ of dimension $25 = 24 + 1$, thus by definition endowed with the quadratic form $x_1 y_1 + \cdots + x_{24} y_{24} - x_{25} y_{25}$, we consider $\mathbb{Z}^{25}$. Then the Leech lattice is isomorphic to the lattice in $\mathbb{Z}^{25}$ **decomposed** by the vector space of dimension 24 which is the orthogonal complement, for the Lorentz quadratic form, of the vector in $\mathbb{R}^{25}$ for which the coordinates are $(3, 5, 7, \ldots, 45, 47, 51; 145)$ (it is amusing to calculate its norm; we end up with -1). A recent presentation is de la Harpe and Venkov (2001).

We will see in Sect. X.8 that the Leech lattice is important in number theory because it furnishes a quadratic form with even integer values and has volume 1 (is unimodular). It is perhaps still more important in group theory, because it is what permitted the construction in 1981 of the largest *sporadic* simple finite group, "sporadic" meaning that it doesn't correspond to the classic infinite series, just as there are exceptional simple Lie groups.

This group is called the **Monster** because of its order and its sophistication; it has in fact a 54 digit number of elements defined by the product:

$$2^{46} \cdot 3^{20} \cdot 5^{9} \cdot 7^{6} \cdot 11^{2} \cdot 13^{3} \cdot 17 \cdot 19 \cdot 23 \cdot 29 \cdot 31 \cdot 41 \cdot 47 \cdot 59 \cdot 71.$$

Attention: the Monster is not the group of automorphisms of the Leech lattice. The matter is considerably more complicated and we end up in a space of dimension 196884; see the numerous places in (Conway and Sloane, 1999) the Monster group is mentioned.

X.5. Lattices in arbitrary dimension: density, laminations

We recall that the essential goal pursued here – which implies a host of applications in effective communications (whether for speed or the use of moderate-sized frequency channels) – is to find the **densest possible** lattices (that we can describe and with which we can calculate in a real way). And this mostly in high dimensions, although low and intermediate dimensions are also interesting and useful. This seems indeed to indicate the difficulty of the subject. In fact, the collection of results – both theoretical and constructive – on the density of these lattices is immense and is complicated by the diversity of the nature of the results. The absolute bible remains (Conway and Sloane, 1999). Here we will give only an outline for orienting readers in this jungle, in such a way as to give an introduction to this bible. We have also already pointed out the references (Gruber and Lekerkerker, 1987) – a bit dated for certain matters – and 3.3 of (Gruber and Wills, 1993). The very brief reference (Oesterlé, 1990) is quite illuminating. For readers who want to look at a single source, the very recent (Zong, 1999) is a marvel of conciseness, economy of resources and finally for giving lots of results.

We thus seek, in a given dimension d, dense lattices, very dense lattices, etc. We introduce the natural constant:

(X.5.1) *We denote by* $c(d)$ *the lower bound of the central densities* DensCentr(Λ) *when* Λ *ranges over the set of lattices of dimension equal to* d.

We will quickly see that this constant is always positive (this will be trivial). As we have seen several times in Sect. VII.6.9 for getting a sense of things there is interest, especially in high dimensions, in taking the d-th root of the quantities considered, which leads to the following definition (going back to 1850), but which doesn't introduce any new entity.

(X.5.2) *The* Hermite constant γ_d *is, by definition,* $\gamma_d = 4(c(d))^{2/d}$.

From the geometric point of view, modulo taking the d-th root of the volume of the unit ball, the Hermite constant represents the square of the minimal norm of a lattice of unit volume. The bible prefers to work most often with a constant of comparable order of magnitude, which is $\frac{1}{d} \log_2 \Delta(d)$, where $\Delta(d)$ denotes the minimal density of lattices of dimension d. Those keen on arithmetic will be happy to know that the Hermite constant γ_d is always the d-th root of a rational number. For this, as well as a whole introductory and pedagogic exposition on some of what we are discussing, see (Oesterlé, 1990).

The basic question would be:

(X.5.3) *What are the values of γ_d for different dimensions d, and above all what are the lattices that realize them, if they exist?*

It is likely that we will never know the complete answer, but we will see in fact that lots of things are known that give us a good bit of satisfaction. We have thus the question of the existence of optimal lattices, also called *critical* (we call attention to the fact that the terminology varies with authors: *critical, extreme* lattices). The bible says simply **densest**. Then we have the problem of calculating the Hermite constant. Beyond dimension 8 the answers are unknown, but we are well-trained to try at the very least to sandwich it between bounds as best we can. There will thus be two things to do: first to find upper bounds, and to know what is useless to hope for. Then we find lower bounds, which are more difficult, because it means exhibiting lattices of high density.

We are working here on high rungs of the ladder, indeed on a platform connected to the high rungs of several ladders. Because of this importance number theorists, when knowledge of the critical lattices is lacking, are interested in optimal lattices in a weaker sense. They have introduced the notions of *eutactic lattice* and *perfect lattice*. For example, the perfect lattices allow good calculation of the topology of the group $\mathrm{GL}(d;\mathbb{Z})$, its cohomology, its K-theory; see for example (Lee and Szczarba, 1978). This point of view is systematically absent from (Conway and Sloane, 1999), but is systematically covered in the very complete book (Martinet, 1996) and also figures in (Gruber and Lekerkerker, 1987). These two definitions are about the structure of the set of vectors of minimal norm of the lattice; these vectors have to be "sufficiently numerous". An important result of Voronoi from around the beginning of the twentieth Century asserts that a lattice is critical if and only if it is both perfect and eutactic.

It is above all remarkable that lattice density gives rise to two extraordinary phenomena. The first is that we know there exist plenty of dense lattices, even very dense. But this existence is theoretical. It does not yield lattices explicitly that have these good densities and that could then be used in practical applications. The second phenomenon is that we obtain in almost trivial fashion bounds that provide, asymptotically in the dimension, an excellent sandwiching of the Hermite constant. This "trivial" sandwiching has been, by research still in progress, successively improved by means of difficult constructions. Finally the explicit construction of very dense lattices, above all in intermediate dimensions − of order about one hundred − and in very high dimensions, has seen work flourish that uses concepts coming from algebraic geometry (elliptic curves), the worst thing being that we obtain very good lattices in very high dimensions simply by choosing them **at random**, which would really seem scandalous if we hadn't already encountered this sort of result in Sects. III.3, VII.6.D and VII.10.D.

Dimensions less than or equal to 8.

Here we have the perfect answer:

(X.5.4) *For all dimensions from 1 to 8 we know the Hermite constant and the densest lattices, which are unique and are, respectively:* $\mathbb{Z}$, A_2, A_3, D_4, D_5, E_6, E_7, E_8.

The complete proof is long and difficult, and it is presently non conceptual, which explains why the books we have cited avoid it. There are references in Sect. 1.5 of Chap. 1 of the bible; in any case the first correct proof did not turn up until 1957. It is important to point out that there is no dimension higher that 8 for which we know γ_d and, *a fortiori*, a densest lattice (there might be more than one). From 9 on this is a subject full of traps, into which several good mathematicians have fallen.

Two trivial bounds. Here are two simple arguments that provide sandwiching for the Hermite constant. The sketch below shows that for every lattice Λ of dimension d we always have: $\mathrm{Vol}(\Lambda) \geq \left(\frac{\mathrm{NorMin}(\Lambda)}{2} \right)^d \beta(d)$. Readers who aren't content with the drawing should consider the Voronoi domain of the origin or a parallelepiped on a basis and proceed to repatriations by translation. But the great conceptual luxury is to use the projection of the space onto the flat quotient torus of the space modulo the lattice considered and then proceed by a measure (volume) argument; flat tori were encountered in Sect. IX.4.

We now take the d-th roots and range over all lattices of dimension d and find, by applying Stirling's formula to the calculation of $\beta(d)$ as in Sect. VII.6.B:

(X.5.5) *For each d we have*

$$\gamma_d \lesssim \frac{d}{\pi e},$$

DIMENSION	1	2	3	4	5	6	7	8	12	16	24
DENSEST PACKING	$\mathbb{Z}$	A_2	A_3	D_4	D_5	E_6	E_7	E_8	K_{12}	Λ_{16}	Λ_{24}
HIGHEST KISSING NUMBER	$\mathbb{Z}$	A_2	A_3	D_4	D_5	E_6	E_7	E_8	P_{12a}	Λ_{16}	Λ_{24}
	2	6	12	24	40	72	126	240	840	4320	196560
THINNEST COVERING	$\mathbb{Z}$	A_2	A_3^*	A_4^*	A_5^*	A_6^*	A_7^*	A_8^*	A_{12}^*	A_{16}^*	Λ_{24}
BEST QUANTIZER	$\mathbb{Z}$	A_2	A_3^*	D_4	D_5^*	E_6^*	E_7^*	E_8	K_{12}	Λ_{16}	Λ_{24}

Fig. X.5.1. Tabulation of packing densities, kissing numbers, and the covering densities. In a box: optimal. To the *left* of the double line: optimal for the lattices.
Conway, Sloane (1999) © Springer

where the symbol $\lesssim$ signifies an asymptotic inequality, that can be written more explicitly

$$\lim_{d \to \infty} \sup \frac{\gamma_d}{d} \leq \frac{1}{\pi e}.$$

On the right is the first bound we are looking for; to obtain the left bound we will make a slight detour and work provisionally with packings not necessarily of lattice type. We begin with a densest possible lattice (it exists — see below — but this is not the essential point) and saturate it with balls if there are holes where we can add a ball of radius $\frac{1}{2}$ NorMin. We then apply the universal metric trick; see e.g. Sect. IX.11 of (Berger, 2003) for its application to Riemannian geometry. By construction, if we double the radius of all the balls of this "stuffed" packing, we necessarily obtain a **covering** of the whole space. In fact, by contradiction: the triangle inequality shows that if a point is not covered, then we may add a ball to the covering by centering it at this point, which would then not have been stuffed completely.

From the preceding we deduce the minorization of density (for arbitrary packings): *density* $\geq 2^{-d}$. We further deduce (always with Stirling's formula) for a Hermite constant γ_d^* defined *for general packings* (defined in the same way as for lattices) that we have:

(X.5.6) $$\gamma_d^* \gtrsim \frac{d}{2\pi e}.$$

It's essential and perhaps upsetting to remark that to this day no one has been able to explicitly construct packings corresponding to the above trivial lower bound: 2^{-d}. See below for the best that has been done.

Two general theorems.

The first result is that of Minkowski-Hlawka, but with regard to modern topology it is no longer really difficult: we consider the space $SL(d;\mathbb{R})/SL(d;\mathbb{Z})$ of all lattices of given dimension d, over which we consider the density. It is continuous, but above all **proper** (see this notion in Sect. V.2), thus we can apply compactness. It attains a minimum for at least one true lattice, for we have just seen that it is bounded below by a strictly positive number. So:

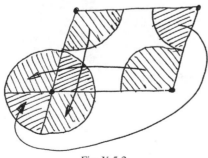

Fig. X.5.2.

In each dimension, there exists at least one lattice of greater density.

The second result — Minkowski's — is more subtle, more conceptual:

(X.5.7) *For each dimension* d *there exists at least one lattice* Λ *of density* $\Delta(\Lambda) \geqslant$ $\frac{\zeta(d)}{2^{d-1}}$, *where* $\zeta(d) = \sum_k k^{-d}$ *is the Riemann zeta function.*

By taking the d-th root and by Stirling's formula, we can now substitute $\gamma_d^* \gtrsim \frac{d}{2\pi e}$ into (X.6.5):

(X.5.8) (Corollary): $\gamma_d \gtrsim \dfrac{d}{2\pi e}$ (*i. e.* $\lim\limits_{d\to\infty} \inf \dfrac{\gamma_d}{d} \geqslant \dfrac{1}{2\pi e}$).

Unfortunately it's a theoretical estimate and — elegant as it may be — non constructive; it is based on an admirable method of averaging densities over the set of all lattices in given dimension. But there are now elementary proofs; that of (Ball, 1992) consists of integrating what happens for explicit and well chosen sets of vectors, in the process improving Minkowski's result to $\frac{(d-1)\zeta(d)}{2^{d-1}}$. Unfortunately, at the level of d-roots, we see little difference.

It seems that no one has the least idea what happens to γ_d/d when d goes to infinity: is there a limit?

♦

In contrast, between the bounds $\{\frac{1}{2\pi e}, \frac{1}{\pi e}\}$ there have been successive tightenings. Here is the state of the problem, which needs to be brought up to date regularly. We moreover need to distinguish, for the lower bounds, whether they are purely theoretical or whether lattices that improve them have been explicitly constructed. Chapter 9 of the bible is entirely devoted to this topic.

On the upper bound side, we have already seen the Rogers bound (denoted $\sigma(d)$ in the literature and defined here in (X.2.7)) that we will encounter in the next section: asymptotically it yields only $\gamma_d \lesssim \frac{d}{\pi e}$. This is, for the Hermite constant, exactly the analogue of the trivial bound (X.5.5) above for the true density. Then we have that of (Kabatianski and Levenshtein, 1978), valid for any packing and *a fortiori* for lattices. The method uses majorizations of the packing densities on spheres (cf. Sect. III.3) and spherical harmonics; see Chap. 9 of the bible. It says that:

(X.5.9) $\gamma_d \lesssim 0,872\dfrac{d}{\pi e}.$

Effective construction of dense lattices, laminations and more As it is important to have explicit constructions, those by all "mixed methods" have been the subject of extensive research. The basic reference is always (Conway and Sloane, 1993). Notice that the scandal is worse than just the existence if the proof depends on a mean value; we know then that "many" — at least "half" — of all lattices are extremely dense.

Lamination is a method that works well in low dimensions; we have already explained it above. We take the densest known lattice in dimension d and, in the space $\mathbb{R}^{d+1}$, we take a parallel copy that we let fall onto the first layer "as low as possible". It is in this way that the plane hexagonal lattice is the laminate of the lattice of dimension 1 (there is but one). There may be several choices, but when they don't all give the same density we choose the best. We then show (see Chap. 6 of the bible) the several things that follow.

First, the lattices obtained by laminating are explicitly known through dimension 48. For dimensions through 8 we obtain densest lattices; for dimension 24 we obtain the Leech lattice. On the other hand, in dimensions 10, 11, 12 denser lattices are known than those obtained by lamination. This fact alone shows, it seems to us, the extreme difficulty — or rather the sophistication — of the panorama of dense lattices.

♦

In intermediate dimensions (up to about several thousand) the discovery of methods based on number theory mixed with the algebraic geometry of elliptic curves (think Fermat theorem and ask for the relation with our plane curves of Sect. V.14) was one of the remarkable mathematical events of the 1990's. For these studies of Elkies and Shioda, see in the preface to the third edition of the bible the progress in density obtained over the classical methods of the first edition of (1988) of (Conway and Sloane, 1999). We find there explicit values and comparison with old tables; we simply need to point out that these methods are effective in the intermediate dimensions from 51 to 4096. For still higher dimensions, the best densities known have been obtained with lattices chosen at random. This is the place to point to (Mumford, 2000), who phophecises that this new century will contain a large amount of research using stochastic methods, where probability theory enters in an essential way. ♦

For large dimensions, we have excellent asymptotic estimates with lattices that are more explicit in a certain sense, but which require computer calculations that are practically impossible to carry out. What we get is the "explicit" existence of lattices of density

$$2^{-d+o(d)} \text{ for } d \to \infty$$

which is optimal in a certain sense; see the bound 2^{-d} above (recall that the notation $o(d)$ characterizes numbers for which the quotient with d tends toward zero as d tends toward infinity).

Still in very large dimensions, we find in (Rosenbloom and Tsafsman, 1990) — for an infinity of dimensions — lattices which show that for these dimensions we certainly have $\gamma_d > 0.3816 \frac{d}{\pi e}$. In the figure below we show in a graph what is known about the Hermite constant, but by using the central density $c(d)$, for which the chosen scale is obtained by dividing by $\frac{d}{\pi e}$, and also a graph taken from the bible, to which we refer readers for detailed explanations about notations, etc. Again in the bible, readers will find on pp. 15–17 details of known results in the tabulations (as of the date of appearance of the third edition). Clear additional progress is shown in (Torquato and Stillinger, 2006).

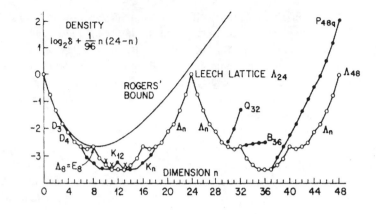

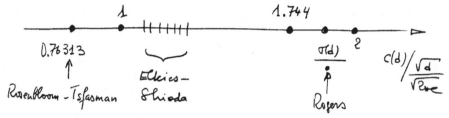

Fig. X.5.3. Conway, Sloane (1999) © Springer

We leave to readers the task of scrutinizing some values corresponding to various lattices mentioned above — low dimensions in the bible — and those of Elkies and Shioda in (Oesterlé, 1990).

We can also look for lattice packings by (center-symmetric) convex sets that are more general than balls, for example the L_p balls; see Sect. VII.6.B. But the problem is practically unapproachable, seeing the difficulty we already have had with spheres. There exists a general theorem of optimality and existence of critical lattices for the convex sets considered, which is proved in the same way as (X.5.6). It can be viewed as a converse of Minkowski's theorem (IX.3.5). These packings are useful in the theory of numbers; a synthesis is found in Chap. 3 of Gruber and Lekerkerker (1987). In a sense it can be said that the Euclidean balls are the worst, since at the extreme the cubes, or any Voronoi cell, tiles with density 1!

X.6. Packings in arbitrary dimension: various options for optimality

Infinite packing (but not necessarily over lattices). We still pack with balls of equal radius 1 (it's already difficult enough in this way; see Wills' example at the end of Sect. X.2) over **all** the space. In this case we have notions of density as explained above, and we need to go to infinity for them to make sense. The first thing to do is find a bound, in a given dimension, for the maximal packings. The first general bound, very geometric and elementary, is due to Rogers from 1954. Let $\text{Simplex}(d)$ be the regular simplex with sides all of length 2 and consider, at each of its vertices d, the portion $\text{Sect}(d)$ which is the intersection the unit ball centered at this vertex with $\text{Simplex}(d)$. Then (see Figs. IX.5.7 and X.6.2 below):

(X.6.1) *For each packing of balls of the same radius in dimension d, the density is always bounded by the constant* denoted $\sigma(d)$, *where*

$$\sigma_d = \frac{(d+1)\,\text{Sect}(d)}{\text{Vol}(\text{Simplex}(d))}$$

The proof consists of starting with a Voronoi decomposition (see Sect. X.2) of the set of centers of the packing balls considered. Then we adeptly decompose this triangulation into simplexes by a simple but skillful use of norms and scalar products, under the condition that all distances be greater than or equal to 2; see pp. 74–85 of (Rogers, 1964). It is not known how to effectively calculate Rogers' bound (recall that simplexes starting with dimension 3 pose calculational difficulties; see Sects. I.2 and III.5). There is an asymptotic estimate (see (Gruber and Lekerkerker, 1987), p. 389): $\sigma(d) \approx_{d \to \infty} 2^{-(d/2)}$.

In dimension 2 this bound is optimal, since there the regular simplex (equilateral triangle) in fact **tiles** the plane. On the other hand, we saw in Sect. X.2 that this is no longer the case in higher dimensions. So Rogers' bound can never be attained. We have seen that, for high dimensions, it is without interest. On the other hand the bound of (X.5.9), $\gamma_d \lesssim 0{,}872\frac{d}{\pi e}$, is a strong result.

If we look at the graphs in (Conway and Sloane, 1999) and of course the book itself for the constructions of the packings cited, we see that there exist packings that are **not** of lattice type (in dimensions 10, 11, 13), that are denser than the known densest lattice packings (recall that the densest lattices are not known starting with dimension 9). This is supported by the (at least at first sight) astonishing conjecture:

(X.6.2) (Rogers' Conjecture) *Starting with a certain dimension, there exist packings not of lattice type which are denser than any lattice packing.*

We end (see the following section for the definitions) by pointing out that there exist very strong connections between the density of lattices (and more general packings), the kissing number (see Sect. III.3), the spherical codes and the error correcting codes. These connections are intuitive enough, but to see how they function in detail reader will need to read a good part of (Conway and Sloane, 1999).

For generalizations of Kepler's conjecture of Sect. X.2 in higher dimensions, we remark on what is proved in Sect. III.3 about the kissing number in dimensions 8 and 24, i.e. that the configuration of spheres in maximum number touching a given sphere is fixed within an isometry, specifically that of the lattice E_8 and Λ_{24} of Leech. If then we can prove that a Voronoi cell of a packing in dimensions 8 or 24 always has a minimum volume equal to that obtained for the lattices E_8 and Λ_{24} of Leech, then the Kepler conjecture will be true in these dimensions. Moreover, that would provide the critical (densest) lattices in dimension 24 (for dimension 8, we know that it is E_8).

Finite packings I: sausage phenomena.

We have stated in Sects. IX.5 and IX.6, in part wrongly, that the packing problem for disks (still of equal radius here) scarcely makes sense if we don't cover the whole plane, more generally the space. Otherwise there are, at least for small numbers of them, configurations that have neither interest nor underlying concept, or are unclassifiable. There was a perceived necessity to go to infinity. Recently finite packings have been rehabilitated, often while considering large numbers of objects packed, and even asymptotic behavior, but involving things that are truly different from infinite packings. An informal exposition is (Wills, 1998), by which we have been guided in an essential way; and 3.4 of (Gruber and Wills, 1993) is a synthesis. For the proofs, apart from the references in Wills, some of them can be found very well explained in two books already cited (Zong, 1996) and (Zong, 1999). They deal with optimal densities, but with criteria that are more or less natural; in fact, the basic idea is to introduce a parameter which, when varied, gives account of the various problems posed. We quote Wills:

However, in contrast with the classic study of infinite packings, for finite packings there does not exist a single theory, but various approaches are used according to the problems considered.

Here we only consider the case of finite packets of balls of radius 1, in the Euclidean space $\mathbb{R}^d$, and of course without any interpenetration. With such a set of balls we associate two invariants. The first, $\dim(C)$, is the dimension of the affine sub-space of $\mathbb{R}^d$ generated by the set of centers $\{c_i\}$ of our balls. The generic case is where $\dim(C) = d$, we say then that C is a *heap*. The extreme opposite case is where $\dim(C) = 1$, which means that all the centers lie on the same affine line; we say then that C is a *sausage* (strictly speaking, if we are interested in maximum density, a true sausage is connected). Between the two cases, i.e. when $2 \leqslant \dim(C) < d$, we say that C is a *pizza*; these designations are optimal (unambiguous) in dimension 3.

The essential invariant is the *natural density* $\delta_{nat}(C)$, which is the obvious "density" of C relative to the interior of the convex set generated by C. If we *denote by* Conv(C) the convex envelope of C, i.e. the smallest convex set containing the set C (see Sect. VII.4), then:

$$(IX.6.3) \quad \delta_{nat}(C) = \frac{\text{Vol}(C)}{\text{Vol}(\text{Conv}(C))} = \frac{m\beta(d)}{\text{Vol}(\text{Conv}(C))} \ \text{if C contains m balls.}$$

Our (personal? natural?) intuition is that the sausages have a very bad natural density. This density is calculated trivially: for the strict sausage $\text{Sauc}_{d;m}$ of $\mathbb{R}^d$ with m elements we have (see the drawing):

$$(X.6.4) \qquad \delta_{nat}(\text{Sauc}_{d;m}) = \frac{m\beta(d)}{2(m-1)\beta(d-1) + \beta(d)}.$$

Fig. X.6.1. A sausage of five balls. Wills (1998) © Springer and Wills

The basic remark, even though very crude, amounts to comparing (X.6.1) with (X.6.4) when m is very large. For then we obtain, for the corresponding sausages, densities of order $\frac{\beta(d)}{2\beta(d-1)}$ and we verify that Rogers' bound is much smaller (especially for d large). So then, "sausages are the best"! It remains to see this in terms of the relative values of d and of m. Fejes-Tóth in 1972 was the first to launch this problem about sausages. Even though the problem is not completely settled, it is − in essence − completely resolved by the conjunction of the two following results.

(X.6.5) (Sausage conjecture.) *For each $d \geqslant 5$ and each m and each packet of balls C on a $\delta_{nat}(\mathrm{C}) \geqslant \delta_{nat}(\mathrm{Sauc}_{d\,;m})$, with equality only when C is a sausage (obviously strict).*

(X.6.6) (Phenomenon of Fejes-Tóth, Betke, Henk, Wills.) *The sausage conjecture is true for all $d \geqslant 13\,387$.*

The proof is an adept minorization of Voronoi domains of an arbitrary packet, see the details in Chap. 2 of Zong (1996).

The best possible dimension (5?) is still awaited; in any case we can't do better than 5 in consequence of the sausage catastrophe due to Wills in 1983, which says that the optimal configuration in dimensions 3 and 4 jumps dramatically, as a function of the number of balls considered, from sausages to heaps.

(X.6.7) *Let $d = 3, 4$. There exists an integer m_d such that for each packet C of m balls of $\mathbb{R}^d$, if $\delta_{nat}(\mathrm{C})$ is optimal, then $\dim(\mathrm{C}) = 1$ for each $m < m_d$ and $\dim(\mathrm{C}) = d$ for $m = m_d$.*

We do not know the optimal m_d; we know presently only that $m_3 \leqslant 56$ and $m_4 \leqslant 377\,000$.

Finite packings II: wrapping paper and diameter.

Can we recover from the shock of the sausage? Yes, with a criterion, in itself natural enough, for optimal packing. We deal still with finite packets C of balls of radius unity. If we think of economical wrapping, it isn't the volume of Conv(C) that matters, but rather the surface (area in dimension 3, $(d-1)$-dimensional volume for the general $\mathbb{R}^d$) of its boundary $\partial(\text{Conv}(C))$ (wrapping paper). The solution to this problem dates only from 1994–1995:

(X.6.8) (Böröczky, Zong.) *In any dimension, the* Conv(C) *for which the surface of the boundary* Vol(∂(Conv(C))) *is minimum becomes more spherical in shape as the number of balls in C increases.*

The proof makes adept use of Blaschke compactness (see Sect. VII.7), the isoperimetric inequality (see Sect. VII.14) and the Brunn-Minkowski inequality (VII.8.1).

But now a profound malaise grips us: in view of the preceding results, optimality is contradictory between the volume of the convex envelope and the area of the boundary. At one extreme we have sausages and at the other, heaps. What do we do?

Finite packings III.

We are indebted to Wills for having found in about 1995 the theory, the concept, that encompasses these apparently contradictory results. The idea is practical: wrapping is never done with infinitely thin paper, it must stand up to the job. Wills' idea is to introduce a parameter, specifically the *thickness* ε of the wrapping paper (in reality, for technical reasons, we work with $\rho = 1 + \varepsilon$). Then the *ρ-natural density* $\delta_{nat;\rho}$ of a packet of balls C (always of the same radius equal to unity) is defined starting with the set C^* of the centers of the balls that comprise C:

$$(\text{X.6.9}) \ \delta_{nat;\rho}(C) = \delta_{nat}(C) = \frac{\text{Vol}(C)}{\text{Vol}(\text{Conv}(C^*) + \rho B)} = \frac{m\beta(d)}{\text{Vol}(\text{Conv}(C^*) + \rho B)}$$

if C *is comprised of* m *balls and where* Conv(C)$+\rho$B *denotes the Minkowski convex sum of* Conv(C) *and the unit ball* B.

That is to say that Conv(C^*) + ρB is the tubular neighborhood of Conv(C^*) of radius δ; i.e. when $\rho \geqslant 1$, exactly the (very tight) wrapping of C with paper (or entirely different material) of thickness equal to $\rho - 1$.

The essential result in the spirit of Wills is the existence of two *critical* parameters, $\rho_{\text{sausage}}(d;m)$ and $\rho_{\text{heap}}(d;m)$ depending on the dimension d of the packing space and the number m of balls, such that:

(X.6.10) (Betke, Henk, and Wills, 1994). *If* $0 < \rho < \rho_{sausage}(d;m)$, *then the best packings are sausages; if* $\rho_{sausage}(d;m) < \rho < \rho_{heap}(d;m)$, *the best packings are pizzas; if* $\rho_{heap}(d;m) < \rho$, *the best packings are heaps.*

Presently the "details" are far from being complete: precise values, etc. The connection with (X.6.6) is direct; that with (X.6.8) is present if we recall that in the

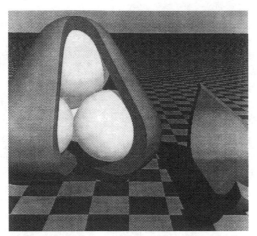

Fig. X.6.2. Tetrahedral packing by four balls. Wills (1998) © Springer and Wills

formula of VII.13.B.1 the second term represents precisely the area of the boundary. The proofs use the Brunn-Minkowski inequality (VII.8.1) in an essential way.

The work of Wills and his school provides the beginning of an answer to basic problems of physics on the shape and the formation (growth) of crystals. It deals with obtaining results on the shape of the boundary of the optimal packings (with a parameter ρ) for large values of m, typically in dimension 3. The papers mentioned in (Wills, 1998) seem to indicate that shapes close to the ones called *Wulff shapes* are obtained. These shapes are those of Dirichlet domains of the dual of the lattice associated with the crystal considered. It seems likely that the case of pizzas could explain certain physical formations of non crystallographic type, such as snow crystals and dendrites. It also seems that for certain solids it is necessary to study packings of more general objects than balls, typically ellipsoids, to take account of the *anisotropism* of the situation.

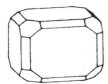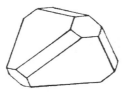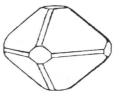

Fig. X.6.3. The Wulff shapes, from *right* to *left*: table salt, sphalerite, diamond. Wills (1998) © Springer and Wills

X.7. Error correcting codes

The error correcting codes are one of the tools, of purely mathematical provenance, that are essential in practical **communications**. The technical term is *signal* *processing.* Two other purely mathematical tools are also very important: the fast Fourier transform and wavelets. Historically suggested by Shannon in about 1940, the first relatively subtle codes constructed were by Hamming in 1950, but they were not put into really systematic form until 1969 and in a spectacular setting: the transmission of images of the space probes Mariner 6 and 7. Earlier however, at the time of the first computers working in binary, a *parity* control was used , i.e. a *rule of* 2. When your computer has frozen, it is often because a parity test has failed. The error correcting codes are essentially algebraic tools, but they are in part linked to lattices and to spherical geometry, all in rather large dimensions. As such we think that they have a place in this book, simultaneously as an example of fundamental mathematics in applications and as an example of geometric thought and visualization in problems of purely algebraic aspect.

The importance of the subject, always in full development, in ever growing perfection under the pressure of the demands of practical users, makes for an immense literature. For a nonexhaustive choice of works dedicated to the subject, see (Conway and Sloane, 1999) throughout its chapters. For an introduction, in part historical, the small book (Thompson, 1983) is likely best. For a very recent introduction, see (Elkies, 2000).

But the reference (Conway and Sloane, 1999), which we have already used in numerous places, can be important for readers for the synthesis of facts that it brings. We present here only a motivational introduction to the subject, together with some examples of error correcting codes and of heuristic considerations for deepening the geometric methods for seeing and talking about these codes.

The absolute necessity of using error correcting codes in the transmission of information is due to the enormous increase, with the appearance of faster and faster computers, of the quantity of information to be transmitted. In the "old", "classical" transmissions, Morse code, telephone, radio, the typical case is that of Morse: a message is transmitted that is composed of letters (or symbols) by associating with each symbol a sequence of *five* dots or dashes: · or −. One − even several − errors in the transmission isn't very serious. Such an error could have various causes, but was infrequent and easily reparable (e.g., failure of a radio tube, a reason that the radio tubes in undersea cables were made to very high standards).

Let us place ourselves at the other end of the historical progression, with a computer functioning at a speed of 500 megahertz, which is common nowadays. Even if our computer is so robust that there is but one error in 10^9 operations, there is the chance of making several errors in less than a second, which is obviously not acceptable. It is therefore necessary to adjoin to these operations a verification system, which is done at the price of lengthening the message; but if this allongation is reasonable, we are rescued. In the case of space probes, the major problem wasn't just the amount of information per second, but also the weakness of the signals received, thus necessarily very tainted by reception or reading errors. It is fascinating to learn that one of the space probes, created for analyzing Mars, its mission accomplished, could be reprogrammed to go further to view Jupiter, reprogramming being necessary because the code used initially wasn't effective enough. It was a young NASA staff member who proposed reprogramming the probe by implanting a new, more effective, code, discovered meanwhile. What was done permitted reading of much weaker signals than those initially viable. A favorite example of popularizers of error correcting codes is the fact that you can scratch your best music CD and nonetheless enjoy perfect listening.

Here is the underlying principle. Each message can be rewritten as a finite sequence (sometimes called a *word*) of 0 and 1 bits, for example the ASCII (American Standard Code for Information Interchange) code, universally adopted for keyboards, uses an octet for each keystroke, i.e. a sequence of eight bits, lowercase *a* is written 01010001.

But let us begin with the simplest message: to send a bit, whether *yes* or *no*, or 0 or 1. Instead of one bit, let us send three of them; precisely, for 0 we send $(0, 0, 0)$ and for 1 we send $(1, 1, 1)$. If the transmission makes at most one error we are saved: the results will always be different. So, at the price of extending a single bit to a set of three, we can correct an error. This lengthens the total message considerably, by tripling its length. But we will see that if we work with messages which are decomposed into sequences of longer bits, we can correct more errors, and this by at most doubling the total length of the message. How is this possible?

We now look at things as a geometer (but one who has to live in a space of arbitrary dimension). We observe that the eight possible triples of 0's and 1's are just simply the vertices of the unit cube $[0, 1]^3$ of $\mathbb{R}^3$:

We see here that the three neighbors of each of the two opposite vertices $(0, 0, 0)$ and $(1, 1, 1)$ are distinct from each other. The geometer loves the notion of distance: on the set $\{0, 1\}^3$ of vertices of the cube we define a distance, the so called *Hamming distance*, which by definition is equal to the number of coordinates that are different. Thus $(0, 0, 0)$ and $(1, 1, 1)$ are a distance three from each another.

We jump to arbitrary dimension: we work with messages formed of N bits, being thus the 2^N vertices $\{0, 1\}^N$ of the unit cube $[0, 1]^N$ of $\mathbb{R}^N$. In this (discrete) space, the Hamming distance between two points is again the number of coordinates that are different. The **trivial** remark is:

(X.7.1) *In order that a subset of $\mathcal{S}$ de $\{0, 1\}^N$ with k errors be correctable, it suffices that the mutual Hamming distances between distinct points of $\mathcal{S}$ be all greater or equal to $2k + 1$.*

In fact, to say that we don't make more than *k* errors about a point of $\mathcal{S}$ is to say that we stay in the ball (ball for the Hamming metric) with center this point and of radius *k*. Two such balls do not intersect if their two centers are a distance apart greater to or equal to $2k + 1$.

An *error correcting code* will be for us a subset $\mathcal{S}$ of $\{0, 1\}^N$. But in practice we want $\mathcal{S}$ to be associated with a complete subset of $\{0, 1\}^n$ with 2^n elements. So this

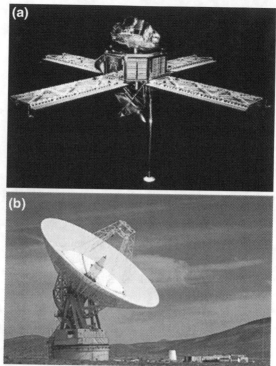

Fig. X.7.1. In 1969 the space probes Mariner 6 and 7 sent back more than 200 photographs of Mars, each divided into 658 240 squares, each affected by a brilliance coefficient 1 to 2^8. Thus each photograph required about five million bits of information. These were codes that used an error detection code and were transmitted to earth with a frequency of 16 200 bits per second, then were decoded and recomposed into photographs. (a) 69H427– Kennedy Space Center, FL – Mariner (b) 66HC215 – NASA Tracking – Antenna at goldstone© NASA

will in effect be a suitable injection $\{0, 1\}^n \to \{0, 1\}^N$, where $\mathcal{S}$ is now the image under this arrow (to be found) of $\{0, 1\}^n$ into $\{0, 1\}^N$. The notation, classic, used in (Conway and Sloane, 1999), is:

(X.7.2) $[N, n, k]$ *to denote a* $\mathcal{S} \subset \{0, 1\}^N$, *where* $\mathcal{S}$ *has* 2^n *elements and where the mutual distance of* $\mathcal{S}$ *in* $\{0, 1\}^N$ *are all greater than or equal to* k.

Such a code can thus correct $[k/2]$ errors, where $[\cdot]$ denotes the integer part. Above we only saw the caricatured example $[3, 1, 3]$. The first efficient code – from 1950 – was the $[7, 4, 3]$ code of Hamming shown here:

The idea for constructing it is to make three parity tests on three well-chosen sets of three of the four possible places of $\{0, 1\}^4$, specifically the places $\{1, 2, 3\}$, $\{1, 3, 4\}$, $\{1, 2, 4\}$. At first glance this code does not seem more interesting than our initial $[3, 1, 3]$. Indeed, it only corrects a single error with certainty, but the "volume" for transmitting information of length 4 bits is accomplished at the price of

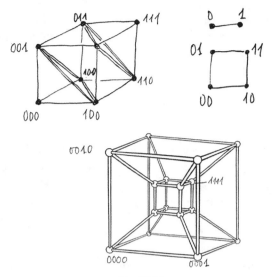

Fig. X.7.2.

		Position					Decimal
1	2	3	4	5	6	7	Value
0	0	0	0	0	0	0	0
1	1	0	1	0	0	1	1
0	1	0	1	0	1	0	2
1	0	0	0	0	1	1	3
1	0	0	1	1	0	0	4
0	1	0	0	1	0	1	5
1	1	0	0	1	1	0	6
0	0	0	1	1	1	1	7
1	1	1	0	0	0	0	8
0	0	1	1	0	0	1	9
1	0	1	1	0	1	0	10
0	1	1	0	0	1	1	11
0	1	1	1	1	0	0	12
1	0	1	0	1	0	1	13
0	0	1	0	1	1	0	14
1	1	1	1	1	1	1	15

Fig. X.7.3. The Hamming code $[7, 4, 3]$

transmitting 7 bits; the length isn't even doubled except for $[3, 1, 3]$, where it is tripled.

Thus the interesting thing we need to see in the triple of integers $[N, n, k]$ is that N not be too large with respect to n or k. The Hamming code example can be surprising, e.g. the length 7 is much longer than 4. But we can do much better with the *Golay code*:

(X.7.3) *There exists a code, called the Golay code, of type* $[23, 12, 7]$.

We will indicate below how it is constructed, because the method is in fact rather general. Its discovery goes back to 1954 and it is one of the first really sophisticated error correcting codes. It thus corrects three errors; if we think in terms of probabilities of the order 10^{-9} for one error, as at the beginning of this section, for three errors — even at high frequency — we obtain an excellent result. Because the length of the message (or more precisely the pieces into which the message is decomposed, but this amounts to the same thing) isn't even doubled. There are now lots of questions suggested by this code and by error correcting codes generally. We will make a very brief tour of the horizon (and get some geometric glimpses) of this immense subject; see as always (Conway and Sloane, 1999) for most of the answers, looking toward the prefaces for orientation. We mainly study the triples $[N, n, k]$.

First problem: n and k being given (think large), find codes such that N is as small as possible, and *let* $A(n, k)$ be this optimal value. We are here in exactly the same situation as for the dense lattices of Sect. X.5. On the one hand, we know bounds for $A(n, k)$, even if we don't know the exact values. These bounds are the subject of all of Chap. 9 of Conway and Sloane (1999). But here too the difficulty is to effectively construct codes for which the cardinality N is close to the value $A(n; k)$. In fact, a great part of the progress arises from the construction of very dense lattices, see Sect. X.7. For the codes, themselves, see as always (Conway and Sloane, 1999).

There is an obvious heuristic connection between three things: the density of lattices, the optimality of codes $[N, n, k]$, and the placement of points on spheres (in the spirit of Sect. III.3, but for spheres in arbitrary dimensions, here rather extensive). Readers should try some drawings to help visualize these connections, but we should add that they are neither automatic nor absolute. To master them, it is best first to read (Thompson, 1983), then to imbibe progressively from (Conway and Sloane, 1999).

The idea is the analogy between the unit sphere $S^{N-1} \subset \mathbb{R}^N$ and the unit cube $[0, 1]^N \subset \mathbb{R}^N$. We consider for example the Hamming code $[7, 4, 3]$: the 16 vertices $\mathscr{S}$, part of the cube $[0, 1]^7$, can be considered as the centers of balls of radius 1 for the Hamming distance. These balls have eight elements, one for their center and seven for the points situated a distance 1 from the center. We can lay out the cube on a plane, whence the (barely realistic) drawing:

Fig. X.7.4. Planar symbolic representation of the perfect Hamming code $[7, 4, 3]$

We can effectively pass from the boundary of the cube $[0, 1]^7$ to the sphere S^6 by a central projection starting from the center of the cube $(1/2, \ldots, 1/2)$. We note that this projection does not change distances much, since furthermore the Euclidean distance is equal to the square root of the Hamming distance. We can thus say that we can distribute well − in an ideal fashion − 16 points on S^6. Note the marvelous fact that these seven balls fill out completely the vertices $\{0, 1\}^7$ since $16 \cdot 8 = 2^7$. Such a code is called *perfect* ; but the perfect codes have been classified and there are but few of them, all in low dimensions; the Golay code $[23, 12, 7]$ is also perfect. In this analogy, the number of vertices of the cube in a Hamming ball corresponds to the volume of the spherical cap (balls for distance on the sphere) of corresponding radius. The connection between the density of the lattice and good distribution on the sphere can be measured with the kissing number: for the very, very bad lattice $\mathbb{Z}^8$ it equals only 16 against 240 for the lattice E_8. For the Leech lattice it is $196\,560$ against 48 for $\mathbb{Z}^{24}$. The fact that we can find lattices of high density, as well as very efficient codes in high dimensions results (at least theoretically and somewhat in caricature) from the fact that on a sphere (of large dimension), the volume of the spherical caps of given radius is very small compared to the total volume of the sphere; cf. the concentration phenomenon encountered in Sect. VII.12.B.

Algebraic construction of codes. We climb the ladder for constructing codes; we abandon the geometric cube $[0, 1]^N$ and replace it with the vector space of dimension N over the two element field $\mathbb{Z}_2$ (denoted $\mathbb{F}_2$ in the theory of finite fields, see Sect. I.XYZ). We consider the so-called *linear* codes, i.e. that $\mathcal{S}$ is required to be a vector subspace of dimension n of $(\mathbb{Z}_2)^N$. We can, however, define $\mathcal{S}$ quite simply as an injective linear mapping and thus it can be defined by a $N \times n$ matrix made up of zeros and ones. It is called the *generating* matrix . Here are the generating matrices for the Hamming code $[7, 4, 3]$ and the extended Golay code $[24, 12, 7]$ (the latter code is derived from the Golay code $[23, 12, 7]$ simply by adjoining a 0 or a 1 to each element in such a way that the total sum is always zero); it is more used in practice because of the number 24:

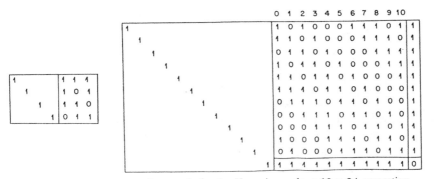

Fig. X.7.5. A 4×7 generating matrix for the Hamming code, a 12×24 generating matrix for the Golay code

The code has thus been constructed. Coding is obtained by associating with the initial n-line vector U the N-line vector V which is the product UG, where G is the generating matrix; the coding is the operation $U \mapsto V = GU$.

Next, the *cyclic* codes are relatively easy to construct, i.e. for each word $(c_0, c_1, \ldots, c_{N-1})$ the word $(c_{N-1}, c_0, c_1, \ldots, c_{N-2})$ must again be in the code. Now, the fundamental concept consists of associating with the word $(c_0, c_1, \ldots, c_{N-1})$ in $(\mathbb{Z}_2)^N$ the polynomial $c_0 + c_1 x + \cdots + c_{N-1} x^{N-1}$, which lives in the space $\mathbb{Z}_2[X]$ of all polynomials in one indeterminate X over $\mathbb{Z}_2$. To say that the code is linear and cyclic is to say that the set of polynomials associated with the code considered forms an **ideal** of $\mathbb{Z}_2[X]$, formed by elements all of which are divisible moreover by $x^N - 1$. Each such ideal is generated by a single element, called its *minimal polynomial*, denoted $g(x)$ (it divides $x^N - 1$). We say that $g(x)$ is the *generating polynomial* of the code. The initial words thus have a length equal to $n = N - \text{degree}(g(x))$. For the Hamming code [7, 4, 3] it is $g(x) = 1 + x + x^3$. We note that explicit presentations of the Hamming codes (and others as well) varying in form, cyclicity and generating polynomial are not immediately apparent.

We now deal with finding a polynomial of small degree (for Golay we want 12) to divide $x^N - 1$ (for Golay this will be $x^{23} - 1$). The method is very general and we exhibit it for the Golay code; see (Conway and Sloane, 1999) for an ultra-brief presentation and the references given there. We thus need to work in the algebraic closure of $\mathbb{Z}_2[X]$; there we can then write $x^N - 1 = \prod_{i=0}^{i=N-1}(x - \alpha^i)$, where α is a primitive N-th root of unity. The generating polynomial of the Golay code is $g(x) = \prod_i (x - \alpha^i)$, where we take only the nonzero powers i which are squares modulo 23. There are thus $\frac{23+1}{2} = 12$ of them.

This technique is very general; we can work over any finite field $\mathbb{F}_q$ by replacing the condition of being a square (modulo N) by other conditions from number theory in such a way so as to obtain good corrections in k. The number k is calculated with the α^i used, which are called the *zeros* of the code.

In all codes we need to know how many 0's and 1's are contained in the words of the code. We define in an obvious way the *weight distribution*. For example, that of the Hamming code [7, 4, 3] is $0^1\ 3^7\ 4^7\ 7^1$, which means that there is a word (with entries zero) of length zero, seven elements of weight three (i.e. each composed of three ones and thus four zeros), etc. For the extended Golay code [24, 12, 7] the distribution is $0^1\ 8^{759}\ 12^{2576}\ 16^{759}\ 24^1$.

The preceding codes, of an essentially algebraic nature, have been intensely studied and entire works are dedicated to them. One of their attractions is the connection between their practical utilization and number theory; and with algebraic geometry as well, e.g. elliptic curves; see Sect. V.14.

However. the linearity condition, *a fortiori* that of cyclicity, considerably limits the possibilities; see Sect. 2.12 of Chap. 3 of Conway and Sloane (1999).

But in order to use these codes in reasonable (real) time, we have to know how to **decode** a message, written in the above language as the product $V \mapsto UG$. But an algorithm for "general matrix inversion" is much too costly. This is why, in the construction of effective error correcting codes, we need to concern ourselves simultaneously with efficient rapid decoding of messages.

X.8. Duality, theta functions, spectra and isospectrality in lattices

Each lattice Λ of $\mathbb{R}^d$ has a dual Λ^*, formed by the points of $\mathbb{R}^d$ whose scalar product with every vector of Λ is an integer. This dual is easily seen to be again a lattice and we always have $(\Lambda^*)^* = \Lambda$; and the product of the volumes of the lattice and its dual equals 1.

In number theory, quadratic forms (with integer values) play an essential role. The relation with lattices is made in this way: with each lattice Λ we associate canonically the quadratic form $Q(\Lambda)$ defined thus: we take any basis $\{e_i\}$; then the value $Q(\Lambda)(v)$ of a vector $v = \sum_i x_i e_i$ is provided, starting with the scalar products of the elements of the chosen basis, by definition equal to $Q(\Lambda)(v) = \sum_{i,j} x_i x_i$ $(e_i \cdot e_j)$. For example, for the lattice A_2 we find the quadratic form $x^2 + xy + y^2$. The interesting lattices in number theory are those that take on integer values over all elements of the lattice considered; we call them *integral lattices*. An equivalent condition is that the dual lattice contain the original lattice: $\Lambda^* \supset \Lambda$. The majority of geometric lattices are of this type and in any case have been normed so that they will be so when possible. But of course the game is to take the smallest normalization possible. The ideal case is that of *unimodular* lattices, i.e. those with volume equal to 1, which is equivalent to $\Lambda = \Lambda^*$. But we sense that it is difficult for a lattice to be simultaneously integral and unimodular. There are certainly the $\mathbb{Z}^d$, but neither the A_d nor the D_d. On the other hand, we find E_8 and Λ_{24}; better still, these two lattices are *even*, i.e. the coordinates of its points are even integers. For number theory and algebraic geometry in particular, finding all the unimodular integral lattices is an important problem; see references in (Conway and Sloane, 1999), but a manifold useful reference is (Serre, 1970).

The importance of E_8 and Λ_{24} is clear once we realize that each even unimodular integral lattice is necessarily of a dimension that is a multiple of 8, and that in dimension 8 there is none other than E_8, whereas in dimension 16 there are but two, specifically $E_8 \oplus E_8$ and E_{16}. In general, the *notation* E_{8k} denotes the lattice constructed starting from D_{8k}, in the same way as E_8 was constructed starting from D_8, i.e. by adding a copy translated by the vector with coordinates all $\frac{1}{2}$. This lattice is *denoted* E_{8k} by Serre but D_{8k}^+ by (Conway and Sloane, 1999) for recalling its construction. As for the notation $A \oplus B$, it simply denotes the *direct product* lattice in $\mathbb{R}^{d+e}$ of the lattices A in $\mathbb{R}^d$ and B in $\mathbb{R}^e$. The even unimodular integral lattices have been classified theoretically. In dimension 24 it is known that there are 24 (Leech included), but their number grows enormously with the dimension; there are 80 million in dimension 32. We will soon see the concept that allows, among other things, their study.

As in the plane, the various norms of Λ^* provide the frequencies, the behavior of the Fourier series associated with the study of Λ-periodic objects in $\mathbb{R}^d$. Moreover, the functions $e^{2\pi i \xi \cdot x}$ $(\xi \in \Lambda^*)$ form a basis for the Λ-periodic functions on $\mathbb{R}^d$ in the various senses used in the theory of Fourier series. The Laplace operator Δ is diagonalized by this basis: $\Delta e^{2\pi i \xi \cdot x} = 4\pi^2 \|\xi\|^2 e^{2\pi i \xi \cdot x}$. The solutions of the wave equation $\frac{\partial u}{\partial t^2} = \Delta_x u$ is written

$$u(x,t) = \sum_{\xi \in \Lambda^*} a_\xi e^{2\pi i \xi \cdot x} e^{-2\pi i \|\xi\| t}.$$

A physicist would say that these are superpositions of eigenvibrations of frequency $\|\xi\|$.

The Poisson formula (IX.4.5) is written:

(X.8.1) $$\frac{1}{(4\pi t)^{d/2}} \text{Vol}(\Lambda) \sum_{v \in \Lambda} \left(\exp\left(-\frac{\|v\|^2}{4t} \right) \right) = \sum_{\xi \in \Lambda^*} \exp(-4\pi^2 \|\xi\|^2 t)$$

for each t.

We recall that for lattices in dimension 2 a knowledge of the norms of various elements of the lattice (norms counted with their multiplicities) determines a lattice within isometry: see Sect. IX.4. We will see that in a higher dimension matters are quite different.

We want to study the set of reals r such that there exists $v \in \Lambda$ with $\|v\| = r$, but while adjoining the multiplicity, yielding pairs $(r, \text{multiplicity}(r))$. The multiplicity is an essential ingredient for integral quadratic forms, beginning with the simplest among them: $\sum_i x_i^2$. Knowing in how many ways a number can be decomposed as a sum of squares is crucial; we have encountered this problem in Sects. III.3 and IX.2.A. The treatment of these questions is a magnificent example of an ascent high up the ladder, by introducing concepts of algebra and analysis seemingly outside of geometry; but they were introduced into number theory in the middle of the twentieth century. We are going to give the principal ideas and the description of the rungs, as well as what allows passage from the one to the others.

The first concept is that of the *theta function*. The knowledge of the pairs $(m\|\lambda\|, r(m) = $ multiplicity of $m)$ is evidently equal to that of the function:

$$z \mapsto \Theta_\Lambda(z) = \sum_{v \in \Lambda} e^{\pi i z \|v\|^2} = \sum_{v \in \Lambda} q^{\|v\|^2}$$

where $q = e^{\pi i z}$, as is customary in number theory.

We thus have what interests us. For each lattice Λ:

(X.8.2) $\Theta_\Lambda(z) = \sum_0^\infty r(m) q^m$ *where* $r(m)$ *is the multiplicity of the* $m = \|\lambda\|$, $\lambda \in \Lambda$.

Now we will see that the theta functions can be calculated easily for many lattices, for some basically because the theta functions lend themselves well to various operations on lattices, for others because they are they are *modular forms*.

◆

We begin by introducing the Jacobi theta functions, defined by:

$$\theta_2(z) = \sum_{-\infty}^{\infty} q^{\left(m+\frac{1}{2}\right)^2}, \; \theta_3(z) = \sum_{-\infty}^{\infty} q^{m^2}, \; \theta_4(z) = \theta_3(z+1) = \sum_{-\infty}^{\infty} (-1)^m q^{m^2}.$$

We note that $\theta_3 = \Theta_{\mathbb{Z}}$ (the lattice of dimension 1), then above all for the direct products $A \oplus B$ we have $\Theta_{A \oplus B} = \Theta_A . \Theta_B$, whence we have immediately

$$\Theta_{\mathbb{Z}^d} = \theta_3^d,$$

which yields (in the expansion of θ_3^d) the number of possible representations of an integer as a sum of d squares.

The theta function for D_d is calculated in an elementary way. We think of θ_3 as being obtained by associating q^{m^2} with the integer m, and θ_4 as obtained by associating $\pm q^{m^2}$ with m, $+$ when m is even and $-$ when m is odd. Then the function θ_4^d is obtained by adjoining a term $\pm q^{m_1^2 + \cdots + m_d^2}$ for each integer $(m_1, \ldots, m_d) \in \mathbb{Z}^d$, the sign corresponding to the parity of $m_1 + \cdots + m_d$. Since D_d is the lattice formed by points of $\mathbb{Z}^d$ for which the sum of the coordinates is even, we obtain the theta function of D_d by replacing the -1 entries by zeros. Otherwise expressed,

$$\Theta_{D_d} = \frac{1}{2}(\theta_3^d + \theta_4^d).$$

We have, for example,

$$\Theta_{A_3 = D_3} = 1 + 12q^2 + 6q^4 + 24q^6 + \cdots .$$

It can likewise be shown that the translate $D_d + (\frac{1}{2}, \ldots, \frac{1}{2})$ (which isn't a lattice) has $\frac{1}{2}\theta_2$ as its theta function, and we deduce that

$$\Theta_{D_d^+} = \frac{1}{2}(\theta_2^d + \theta_3^d + \theta_4^d).$$

Since, as readers can easily verify, D_4^+ is isometric to $\mathbb{Z}^4$, we extract the relation

$$\theta_3^4 = \theta_2^4 + \theta_4^4.$$

◆

In contrast, to calculate the theta function for the Leech lattice, it's better to appeal to the notion of *modular form*. We have already encountered the modular group $SL(2, \mathbb{Z})$ en Sect. II.4, then in Sect. V.14 in connection with elliptic curves. For what follows we need to combine Chap. VII of Serre (1970) — for modular forms — with Conway and Sloane (1999) for the explicit list of theta functions for various lattices.

The theory of modular forms brings a definition and an extremely strong result.

(X.8.3) *By a* modular form *of weight 2k we mean each function f that is holomorphic (also at infinity) in the half-plane* $H = \{z : \text{Im}(z) > 0\}$ *and which satisfies, for each z of* H, *the relation*

(∗)
$$f(z) = (cz + d)^{-2k} f\left(\frac{az + b}{cz + d}\right)$$

for each element of $SL(2; \mathbb{R})$. *The space of modular forms of a given weight is completely known, because the algebra of these modular forms is generated by the two forms* G_2 *and* G_3, *the monomial* $G_2^{\alpha} G_3^{\beta}$ *being of weight 2k, with* $k = 2\alpha + 3\beta$.

The essential fact is that the modular group is generated by the two transformations $z \mapsto z + 1$ and $z \mapsto -\frac{1}{z}$. The condition (∗) above is thus equivalent to the two conditions $f(z) = f(z + 1)$ and $f(-\frac{1}{z}) = z^{2k} f(z)$.

Every function f that is holomorphic in H and of period 1 can be written

$$f(z) = \sum_{n=-\infty}^{\infty} a_n e^{2\pi i n z} = \sum_{n=-\infty}^{\infty} a_n q^{2n}.$$

Under these conditions, we say that f is *holomorphic at infinity* in H if the coefficients a_n are zero for $n < 0$, the definition being justified by the fact that the change of variables $q = e^{\pi i z}$ provides a correspondence between functions that are holomorphic in H and of period 1 and even functions that are holomorphic for $0 < |q| < 1$, the point $q = 0$ corresponding to $z = i\infty$.

A modular form of weight $2k$ is thus a function of the form $f(z) = \sum_{n=0}^{\infty} a_n q^{2n}$, confirming the relation $f(-\frac{1}{z}) = z^{2k} f(z)$.

The theta functions for even unimodular lattices are modular forms, with weight equal to half the dimension of the lattice: they have the desired form and the condition $f(-\frac{1}{z}) = z^{2k} f(z)$ follows directly from the Poisson formula (X.8.1), which becomes here

$$\Theta_{\Lambda^*}(z) = \text{Vol}(\Lambda)\left(\frac{i}{z}\right)^{d/2} \Theta_{\Lambda}\left(-\frac{1}{z}\right)$$

(we have $\Lambda = \Lambda^*$, $\text{Vol}(\Lambda) = 1$, and d is a multiple of 8), in particular for the cases Θ_{E_8} and $\Theta_{\Lambda_{24}}$. In contrast, the theta function for the lattices A_d and D_d are not in general modular forms (they are however under a more general notion of modular form).

We already know the theta function for E_8:

$$\Theta_{E_8 = D_8^+} = \frac{1}{2}(\theta_2^8 + \theta_3^8 + \theta_4^8) = 1 + 240q^2 + 2160q^4 + \cdots,$$

but we can obtain another expression from the fact that it's a modular form of weight 4. From theorem X.8.3, we in fact infer that the vector space of modular forms of weight 4 is of dimension 1. Since $G_2 = \pi^2/45$ for $z = i\infty$, and $\Theta_{E_8} = 1$ for $q = 0$, we have

$$\Theta_{E_8} = \frac{\pi^2}{45} G_2,$$

a more useful expression in fact for calculating G_2 than for calculating Θ_{E_8}.

The theta function for the Leech lattice is a modular form of weight 12 and (X.8.3) shows that the vector space of modular forms of weight 12 has dimension 2, generated by G_2^3 and G_3^2. But in the representation $\Theta_{\Lambda_{24}} = a G_2^3 + b G_3^2$, the constants a and b are irrational, of little use for determining what interests us, the integer coefficients of the series in q. We thus use the basis of the space of modular forms of weight 12 obtained by adjoining to $\Theta_{E_8}^3$ the modular form *notée* Δ_{24}, which is called the **Ramanujan function** and related to decompositions of an integer as a sum of 24 squares (see also Sects. III.3 and IX.2), specifically

$$\Delta_{24} = \left\{ \frac{1}{2} \theta_2 \theta_3 \theta_4 \right\}^8 = q^2 \prod_1^\infty (1 - q^{2m})^{24}.$$

In this basis, the theta function of the Leech lattice is expressed (this is almost immediate):

$$\Theta_{\Lambda_{24}} = \Theta_{E_8}^3 - 720 \Delta_{24}.$$

We extract from this the expression for $\Theta_{\Lambda_{24}}$ as a function of θ_2, θ_3, θ_4, as well as the expansion

$$\Theta_{\Lambda_{24}} = 1 + 956560 q^4 + 16773120 q^6 + \cdots.$$

In a general fashion the coefficients of the theta series allow us to determine the invariants of the lattice, such as the kissing number k, the minimal norm v and the density: we have $\Theta_\Lambda = \sum_{m=0}^\infty r(m) q^{m^2} = 1 + k q^{v^2} + \cdots$ and **Dens**$(\Lambda) = \lim_{r \to \infty} \left(\frac{v}{r} \right)^d \sum_{m \leq r} r(m)$.

In the case of the Leech lattice, it is shown that

$$r_{24}(2m) = \frac{65520}{691} (\sigma_{11}(m) - \tau(m))$$

where $\sigma_{11}(m)$ is the sum of the 11-th powers of the divisors of m, and $\tau(m)$ is the coefficient of q^{2m} in the series development of the function Δ. It follows from the Ramanujan conjecture, proved by Deligne, that $\tau(m) = O(m^{11/2+\varepsilon})$ for each $\varepsilon > 0$. Since we moreover know that $r_{24}(2m)$ is of the order of m^{11}, we obtain

$$r_{24}(2m) \sim_{m \to \infty} \frac{65520}{691} \sigma_{11}(m)$$

with an excellent approximation.

We find the theta functions of all the important lattices in the bible, where they are accompanied by information about explicit values (up to $m = 100$) and the arithmetic properties of their coefficients

We deduce from the above theta function formulas the result known in another form by Witt in about 1937:

$$\Theta_{E_{16}=D_{16}^+} = \frac{1}{2}(\theta_2^{16} + \theta_3^{16} + \theta_4^{16}) = \frac{1}{4}(\theta_2^8 + \theta_3^8 + \theta_4^8)^2 = (\Theta_{E_8})^2 = \Theta_{E_8 \oplus E_8}.$$

It suffices to expand the square, using the relation $\theta_3^4 = \theta_2^4 + \theta_4^4$.

The two lattices E_{16} and $E_8 \oplus E_8$ thus have the same theta function, and thus equally many points on each sphere that is centered at the origin; readers can verify that these two lattices are not isometric in $\mathbb{R}^{16}$. For the Riemannian geometer this implies that the two flat tori $\mathbb{R}^{16}/E_{16}$ and $\mathbb{R}^{16}/E_8 \oplus E_8$ have the same spectrum — the same vibrational frequencies — even though their geometry is not the same. Also, their eigenfunctions are different. We can say that these two lattices, or the two tori, are *isospectral*. We owe to (Milnor, 1964) this Riemannian counterexample.

The isospectrality problems began in the 1960's and have inspired two generations of geometers. The question is whether a compact Riemannian manifold is characterized, i.e. determined within isomorphism, by the spectrum of its Laplacian. If we stipulate in addition the equality of the corresponding eigenfunctions of the Laplacian the answer is easy and affirmative. It was Milnor who first found two isospectral compact Riemannian manifolds that are not isometric, specifically the two flat tori associated as above with the two lattices discovered by Witt. The subject has remained without great progress, perhaps due to its difficulty. The subject was re-launched in a spectacular way in the article (Kac, 1966) by a mathematical physicist.. The title "can you hear the shape of a drum?" treated the case (simultaneously particular and general) of the spectrum musical vibrations of a planar domain, i.e. of functions which vanish on the boundary. These are in fact the harmonics produced by a drum of the same shape, where the listener is challenged to divine this shape by what is heard. This, to our knowledge, is where we are for the general torus problem; for Riemannian manifolds, consult Chap. IV of the synthesis (Berger, 1999) or else (Berger, 2003); for flat tori — alias lattices — see the preface to the third edition of the bible.

A general result of M. Kneser, based on consideration of theta functions and of algebra, states that in a given dimension the sets of non isometric isospectral lattices are always finite, the idea being that the modular forms of given weight form a vector space of finite dimension. In dimension 2, we have shown above that the spectrum determines the lattice. It is still true in dimension 3, after (Schiemann, 1997). The work consists first of looking at the minimal vectors (of successive minimal norms), then in working with those given in the modular domain of $SL(\mathbb{R}; 3)/SL(\mathbb{Z}; 3)$, which is rather well known and can be presented in the form of two convex cones. There is no need to go to infinity; and it is even shown that it suffices to know a finite number of values of the spectrum.

Again by calculating their theta function, we find in the preface to the third edition of (Conway and Sloane, 1999) an incredibly simple family of pairs of isospectral lattices , a family that depends on four vector parameters x, y, z, w and consists of the lattices in $\mathbb{R}^4$ for which the bases are, respectively:

$$\{3w - x - y - z, w + 3x + y - z, w - x + 3y + z, w + x - y + 3z\},$$
$$\{-3w - -x - y - z, w - 3x + y - z, w - x - 3y + z, w + x - y - 3z\},$$

where it is simply required that the four vectors x, y, z, w be of different lengths and mutually orthogonal. It would be hard to ask for anything simpler.

Bibliography

[B] Berger, M. (1987, 2009) *Geometry I,II*. Berlin/Heidelberg/New York: Springer

[BG] Berger, M., & Gostiaux, B. (1987) *Differential Geometry: Manifolds, Curves and Surfaces*. Berlin/Heidelberg/New York: Springer

Aigner, M., & Ziegler, G. (1998). Proofs from THE BOOK. Springer

Ball, K. (1992). A lower bound for the optimal density of lattice packings. *International Mathematics Research Notices, 10*, 217–221

Berger, M. (1999). *Riemannian geometry during the second half of the twentieth century*. Providence: American Mathematical Society

Berger, M. (2001a). Peut-on définir la géométrie aujourd'hui? In B. Engquist, W. Schmid (Eds.), *Mathematics unlimited- 2001 and beyond*. Berlin/Heidelberg/New York: Springer

Berger, M. (2003). *A panoramic introduction to Riemannian geometry*. Berlin/Heidelberg/New York: Springer

Betke, U., Henk, M., & Wills, J. (1994). Finite and infinite packings. *Journal fr die Reine und Angewandte Mathematik, 453*, 165–191

Boerdijk, A. (1952). Some remarks concerning close-packing of equal spheres. *Philips Research Reports, 7*, 303–313

Christophe *Le sapeur Camembert*. Albin Michel

Conway, J. (1995). Sphere packings, lattices, codes and greed. In *Proceedings of the International Congress of Mathematicians (Zürich, 1994)*, Vol. 1. Birkhäuser, 45–55

Conway, J. H., & Sloane, N. J. A. (1999). Sphere packings, Lattices and Groups (3rd ed.). Berlin/Heidelberg/New York: Springer

Conway, J. H., & Sloane, N. J. A. (1993). Sphere packings, lattices and groups (2nd ed.). Berlin/Heidelberg/New York: Springer

De la Harpe, P., & Venkov, B. (2001). Groupes engendrés par des réflexions, designs sphériques et réseau de Leech. *Comptes Rendus, AcadÈmie des sciences de Paris, 333*, 745–750

Dhombres, J., & Robert, J.-B. (1998). *Fourier, créateur de la physique mathématique*. Paris: Belin

Elkies, N. (2000). Lattices, linear codes, and invariants, Part I. *Notices of the American Mathematical Society, 47*(10), 1238–1945

Erdös, P., Gruber, P., & Hammer, J. (1989). *Lattice points*. New York: Longman Scientific and Technical, John Wiley

Fejes Tóth, L. (1972). *Lagerungen in der Ebene, auf der Kugel und im Raum* (2nd ed.). Berlin/Heidelberg/New York: Springer

Fournier, J.-C. (1977). *Le théorème du coloriage des cartes* (ex-conjecture des quatre couleurs), Séminaire Bourbaki, 1977–78 *Lecture Notes in Mathematics, 710*, Springer, 41–64

Gruber, P., & Lekerkerker, C. (1987). *Geometry of numbers*. Amsterdam: North-Holland

Gruber, P., & Wills, J. (Eds.). (1993). Handbook of convex geometry. Amsterdam: North-Holland

Hales, T. (1994). The status of the Kepler conjecture. *The Mathematical Intelligencer, 16*, 47–58

Hales, T. (1997). Sphere packings. *Discrete & Computational Geometry, 17*, 1–51

Hales, T. (1999). *An overview of the Kepler conjecture*, http://arxiv.org/abs/math/ 9811071

Hales, T. (2000). Cannonballs and honeycombs. *Notices of the American Mathematical Society, 47*(4), 440–449

Hsiang, W.-Y. (1993). On the sphere packing problem and the proof of Kepler's conjecture. *International Journal of Mathematics, 4*, 739–831

Kabatianski, G., & Levenshtein, V. (1978). Bounds for packings on a sphere and in a space. *Problems of Information Transmission, 14*, 1–17

Kac, M. (1996). Can one hear the shape of a drum? *The American Mathematical Monthly, 73*(4), part II, 1–23

Lee, R., & Szczarba, R. (1978). On the torsion in K4(Z) and K5(Z). *Duke Mathematical Journal, 45*, 101–129

Martinet, J. (1996). *Les réseaux parfaits des espaces euclidiens*. Paris: Masson

Mattila, P. (1995). *Geometry of sets and measures in euclidean spaces*. Cambridge: Cambridge University Press

Milnor, J. (1964). Eigenvalues of the Laplace operator on certain manifolds. *Proceedings of the National Academy of Sciences of the USA, 51*, 542

Milnor, J. (1994). Hilbert's problem 18: on crystallographic groups, fundamental domains, and on sphere packing. In J. Milnor Collected papers, Publish or Perish, Houston, 173–187

Morgan, F. (2005). Kepler's conjecture and Hales proof – a book review, *Notices of the American Mathematical Society, 52*(1), 44–47

Mumford, D. (2000). The dawning age of stochasticity. In Arnold, Atiyah, Lax, Mazur (Eds.), *Mathematics: frontiers and perspectives* (pp. 199–218). Providence: American Mathematical society

Oesterlé, J. (1990). Empilements de sphères, *Séminaire Bourbaki* 1989–1990. In *Astérisque 189–190*, 375–398

Oesterlé, J. (1999). Densité maximale des empilements de sphères en dimension 3 (d'après Thomas C. Hales et Samuel P. Ferguson), *Séminaire Bourbaki* 1989–1990. In *Astérisque 266*, 405–413

Pöppe, C. (1999). La conjecture de Kepler démontrée. *Pour la Science, 259*, mai 1999, 100–104

Reid, C. (1970). *Hilbert*. Beriln/Heidelberg/New York: Springer

Rigby, J. (1998). Precise colourings of regular triangular tilings. *The Mathematical Intelligencer, 20*, 4–11

Rogers, C. (1964). *Packings and coverings*. Cambridge: Cambridge University Press

Rosenbloom, M., & Tsafsman, M. (1990). Multiplicative lattices in global fields. *Inventiones Mathematicae, 101*, 687–696

Schiemann, A. (1997). Ternary positive definite quadratic forms are determined by their theta series, *Mathematische Annalen, 308*, 507–517

Serre, J.-P. (1970). *Cours d'arithmétique*. Paris: Presses Universitaires de France

Sullivan, J. (1994). Sphere packings give an explicit bound for the Besikovitch covering theorem. *Journal for Geometric Analysis, 4*, 219–231

Thomas, R. (1998). An update on the four-color theorem. *Notices of the American Mathematical Society, 45*(7), 848–859

Thompson, T. (1983). *From error-correcting codes through sphere packings to simple groups*. Washington: Mathematical Association of America

Torquato, S. & Stillinger, F., (2006). New conjectural bounds on the optimal density of sphere packings. *Experimental Mathematics, 15*(3), 307–332

Wills, J. (1991). An ellipsoid packing in E3 of unexpected high density. *Mathematika, 38*, 318–320

Wills, J. (1998). Spheres and sausages, crystals and catastrophes – and a joint packing theory, *The Mathematical Intelligencer, 20*(1), 16–21

Zong, C. (1996). *Strange phenomena in convex and discrete geometry*. Berlin/Heidelberg/New York: Springer

Zong, C. (1999). *Sphere packings*. Berlin/Heidelberg/New York: Springer

Chapter XI
Geometry and dynamics I: billiards

XI.1. Introduction and motivation: description of the motion of two particles of equal mass on the interior of an interval

We consider the simplest possible but nontrivial problem of particle mechanics: on an interval, say [0, 1], we consider two particles of the same mass that oscillate. They move about at constant velocity unless there is a collision, either at an endpoint or in encountering each other. The conditions in case of collision are these: when the particle on the left encounters the wall on the left at 0, it rebounds with the same speed and of course in the opposite direction; likewise when the particle on the right rebounds from the wall on the right at the point 1. When they encounter each other (with opposed directions) the condition is what is called an *elastic collision*, physically one where, after the deformation caused by the collision, the two solids reassume their shapes and retain their combined kinetic energy and momentum. It can be shown that the particles emerge in the opposite directions, but exchange their velocities. The well known and spectacular case is where one is fixed; then the other remains fixed at the point of contact while the first leaves with the same velocity as the particle that hit it. If they encounter each other while going the same direction the result is still the same: the particles exchange their velocities; the particle that was moving faster loses speed to the benefit of the other. The fundamental problem is to describe this simultaneous movement of the two particles over time, but especially when things continue indefinitely or at least for a long period of time. This consideration of infinity is natural in physics when the particles are excited with extremely large velocities and where extremely many collisions occur in a short time interval. In particular, will the motion ultimately be periodic or, to the contrary, will the particles ultimately occupy practically all possible positions?

As good geometers we look for an interpretation of this simultaneous movement, which is easy and miraculous. We give the particles the coordinates x and y respectively and determine their position in the plane. The imposed conditions show first that the descriptive point must be in the triangle of Fig. XI.1.2 above. The particles are constantly changing velocity from the three types of impacts; the trajectory is along a line. When this line encounters a vertical or horizontal side of the square, the rebound is along a new line, as in billiards. One of the two velocities is replaced by its opposite: see the figures below. Two particles encounter each other precisely when the trajectory line encounters the hypotenuse of the triangle, and the condition of exchange of velocities (including algebraic change of sign) means once again a

M. Berger, *Geometry Revealed*, DOI 10.1007/978-3-540-70997-8_11,
© Springer-Verlag Berlin Heidelberg 2010

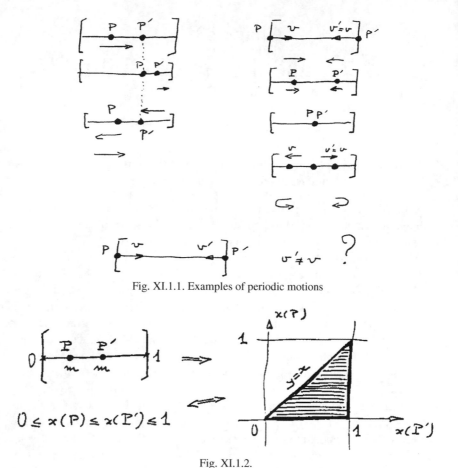

Fig. XI.1.1. Examples of periodic motions

Fig. XI.1.2.

rebound as in billiards. Such a trajectory was the essential idea of the logo for the International Congress of Mathematicians in Zürich in 1994.

Finally, in this geometric interpretation the motion of the pair of particles is read as a trajectory in billiards (with perfectly elastic cushions) in an isosceles right triangle. Readers who draw well will be able to draw such a trajectory, with ten or so rebounds; they will then perceive that at most eight possible oriented directions appear in the general case. We explain this, once again as geometers, by thinking of using the notion of symmetry.

We unfold the triangle about its hypotenuse and obtain a square. The unfolding of the trajectory at a diagonal amounts to ignoring the diagonal and continuing the trajectory as a straight line. Finally, in the square, the completely unfolded trajectory will be the same as the trajectory in an ordinary square billiard table. And so our problem of mechanics is reduced to a problem of geometry: studying the trajectories for a square billiard table, which we will do in the next section. It involves a natural and pleasant geometric problem. Motivated by this example, in the present chapter

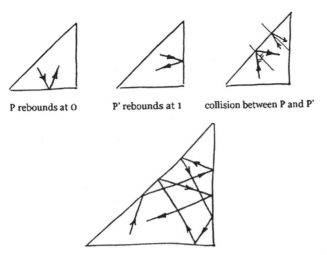

P rebounds at 0   P' rebounds at 1   collision between P and P'

Fig. XI.1.3. Three types of elastic impacts: rebounds at the two ends and collisions
between the particles

Fig. XI.1.4.

we are going to study billiard trajectories in an arbitrary bounded domain in the
plane. This problem has recently seen spectacular progress, having remained long in
obscurity. As with polytopes (see Sect. VIII.1) we can advance at least two reasons
for this redress from obscurity. The first is the strong demand of physicists, since

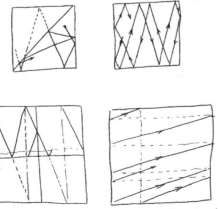

Fig. XI.1.5. Eight different directions; unfolded three times we have but a single
direction

the general billiards represent simple geometric models for the study of statistical mechanics. The second is that the means, the tools, for this study were until recently only partly available.

We begin with polygons, then pass to curved domains and finish with a brief glimpse of billiards in space. Along the way we will see how well billiard problems illustrate the essential theme of this book. Very simple problems remain unresolved; others have been resolved recently but at the cost of a strong ascent of Jacob's ladder and the use of sophisticated mathematics, in particular the subtle theory of Riemann surfaces.

Billiards are a particular case of *dynamical systems*, i.e. a mathematical situation where time or systematic iteration appears. We will thus use, more or less systematically, the language and concepts of the theory of dynamical systems. We present these briefly in Sect. XI.XYZ; they will moreover be equally essential in the following chapter, which treats the dynamical system corresponding to geodesic flow on a surface. Readers may prefer to wait until the end of the chapter for this Sect. XI.XYZ; we prefer not to parachute right away — without motivations — all the technical language of dynamical systems and ergodic theory. There exist several books entirely devoted to billiards: Katok and Strelcyn (1986), Tabachnikov (1995) and Galperin and Zemlyakov (1990). An elementary exposition is given in Berger (1991). A recent synthesis for the case of polygons is Gutkin (1996) and a partial synthesis is given in the article by Bunimovich in Sinaï (1987). There are also things about billiards in Sinaï (1976) and in Katok and Hasselblatt (1995). Our text can be considered as an introduction to the book of Tabachnikov, for we treat only a small portion of the problems that billiards pose.

Very recently "mushroom billiards" have been studied; see Porter and Lansel (2006). This is a model for studying dynamical systems that are neither chaotic nor integrable, which unfortunately is the typical case. We encounter these two extreme cases in the present chapter. For mushrooms here is a typical figure, where we indeed see a mixture of the two extremes that will be encountered in the sequel.

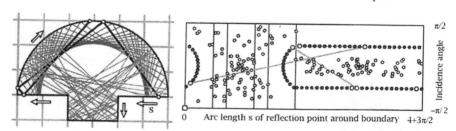

Fig. XI.1.6.

◆

Readers will appreciate the quotation from Ulam (1976) which concerns von Neumann (Johnny) and the physicist Eugene Wigner, told to Ulam by Johnny: *"Eugene and Johnny wanted to play billiards. They went to a café where billiards were played and asked an expert waiter there if he would give them lessons. The waiter says, "Are you interested in your studies? Are you interested in girls? If you really want to learn billiards, you will have to give up both". Johnny and Wigner conferred briefly and they decided they would give up one or the other but not both. They did not learn to play billiards".*

We can also be sensitive to the artistic aspect of some billiard trajectories – below for example, where we deal with the a boundary formed by three circular arcs based on an equilateral triangle:

Fig. XI.1.7.

XI.2. Playing billiards in a square

The drawings of Fig. XI.1.5 show that now, in a square, a given trajectory presents at most four possible (oriented) directions. But we can repeat the unfolding so as to have finally only a single direction, but we then have to be able to use the whole plane; see drawing (c) of Fig. XI.1.5. We can also assume that the trajectory starts from the origin. It then becomes an entire half-line and then we repatriate all the pieces that it traces out in the squares traversed so as to obtain, strictly speaking, the original trajectory. This is in a way analogous to the passage from, say, the right side to the left side in video games.

We mustn't underestimate this exercise, for in physics it is that of an "harmonic" system with two periods ω_1, ω_2. If α denotes their ratio $\frac{\omega_1}{\omega_2}$, then we have the classic result of periodic motion for rational α, and a quasi-periodic motion in the contrary case. We know that Fourier analysis allows us to detect these two types perfectly.

XI.2.A. The dichotomy and continued fractions

We can carry Sect. XI.1 forward. Before even mentioning continued fractions, we discover at the outset the fundamental dichotomy for billiards in a square:

> *Either the slope of the line (that is, the ratio of the initial speeds in the problem of two particles of equal mass) is rational and the trajectory is periodic, or this slope is irrational and the trajectory never closes.*

But this cannot satisfy us, for two problems are natural and essential in all of dynamics. The first is the study of the *counting function* FC(L), whose value is the number of periodic trajectories of length less than or equal to L. Physical motivation. These lengths represent the energy level, and there is a need to know them or at least to have a good asymptotic estimate for them when L is large. The second problem is that of knowing, in case a trajectory isn't periodic, how it is situated in the domain (here a square) as it is extended indefinitely. We will see that there are perfect answers for these problems, i.e. a very good estimate of the counting function for periodic trajectories, and for the others the fact that not only are they everywhere dense in the square but, even better, are *uniformly* so, which means that the time they spend in any given portion of the square is proportional to the area of that domain. In fact, even better still, the continued fraction development of the slope will indicate how, and at what rate, this everywhere density darkens the square.

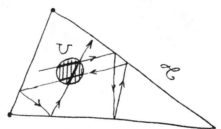

Fig. XI.2.1. There is uniform distribution if for each domain U we have
$$\lim\nolimits_{T \to \infty} \frac{\text{time spent in U}}{T} = \frac{\text{area of U}}{\text{total area}}$$

We see first why the trajectories with irrational slope are uniformly distributed. The trajectory being composed of parallel segments, it suffices to study what happens at each vertical side. Figure XI.1.5 shows exactly the set to be studied; if the slope is *denoted* α, it is the set of residues $na - [na]$ modulo 1 of all the integer multiples $n\alpha$, where $n = 1, 2, 3, \ldots$ and where as usual the brackets denote the *integer part* of a real number.

We will prove the classic result, due to Hermann Weyl:

(XI.1.2) *When α is irrational, the set of residues modulo 1 (i.e. the $n\alpha - [n\alpha]$) of the $n\alpha$, where $n = 1, 2, 3, \ldots$, is not only everywhere dense in the interval* [0, 1], *but is in fact equidistributed, i.e. in each interval* [a, b] *the mean*

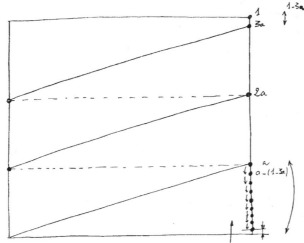

Fig. XI.2.2. $[\frac{1}{a}] = 3 = n_1$, $a = \frac{1}{n_1+a_1}$, write: $a_1 = \frac{1-3a}{a}$, fill $[0,a]$ as much as possible with pieces of length $1 - 3a$, i.e. take $[\frac{a}{1-3a}] = [\frac{1}{a_1}] = n_2 = 8$, whence $a = \frac{1}{3+\frac{1}{8+a_2}}$, etc., where $a < \frac{1}{3}$, then $a > \frac{8}{25}$, etc

value of these residues tends, as n goes to infinity, toward the length $b - a$ of this interval. Precisely:

$$\lim_{n\to\infty} \frac{\#\{k \mid k\alpha - [k\alpha] \in [a,b] \;:\; k = 0,1,2,\ldots,n\}}{n} = b - a.$$

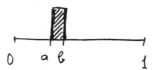

$$0 \qquad a \quad b \qquad\qquad 1$$

Fig. XI.2.3. The characteristic function of an interval $[a,b]$

The proof below was kindly communicated to us by our friend Hans Samelson of Stanford University. For each complex-valued and 2π-periodic function, each positive integer n and each irrational multiple α of π, we set:

$$A_n(f) = \frac{1}{n}\sum_{j=1}^{n} f(j\alpha).$$

Observe that:

1. $A_n(1) = 1$.

2. $\lim_{n\to\infty} A_n(e^{ikt}) = 0$ for $k \neq 0$.

3. $A_n(f)$ is linear in f.

All this shows that, for each trigonometric polynomial P we have:

$$(*) \qquad\qquad \lim_{n\to\infty} A_n(P) = \frac{1}{2\pi} \int P(t)\, dt.$$

But since:

4. $|A_n(f)| \leqslant \|f\|_\infty$

the sequence of the $\{A_n(f)\}$ is equicontinuous in f, and thus the formula $(*)$ remains valid not just for trigonometric polynomials, but for each continuous function f.

Finally, since $A_n(f) \geqslant 0$ for each function $f \geqslant 0$, if we approximate the characteristic function F of an arbitrary interval (or in fact any Riemann integrable function) by two continuous functions g_1, g_2 on the left and on the right:

$$g_1 \leqslant F(t) \leqslant g_2 \quad \text{with} \quad \int g_2\, dt < \int g_1\, dt + \epsilon$$

it follows that, for large n:

$$\left| A_n(F) - \frac{1}{2\pi} \int F(t)\, dt \right| < \epsilon.$$

Remark. It is only in item **2** that we use the sequence denoted by $j\alpha$, thus we can replace it by any sequence ξ_j for which **2** remains valid. This is the more general theorem of Weyl.

If now we want to see things in more detail, what is it that we might actually want? For discovering these wishes, and what we can expect to observe — for example the speed and nature of the darkening of the square — we advise readers to write a short computer program which displays the sequence $k\alpha - [k\alpha], k = 0, 1, 2, \ldots$, on the segment $[0, 1]$. It is just a bit harder to program the trajectory of the billiard in a square to obtain the figures given here. But it will be noticed that a knowledge of the points on any side provides the global picture in the square, since then we only need draw the lines of slope $\pm\alpha$ passing through these points. Once the program is written, it can be used for different values of α. Of course first the rationals, then the golden ratio, then π, etc.

We take first an example in caricature, i.e. the continued fraction

$$\cfrac{1}{5 + \cfrac{1}{10000 + \cfrac{1}{10000000 + \cdots}}}$$

With this number, we can well imagine what will happen. We will see first the appearance of the five points dividing $[0, 1]$ into five, then throughout there will be progressive darkening, but very slowly while approaching 10000 divisions of $[0, 1]$, except that things will be well dark before then if you cannot zoom, etc.

Fig. XI.2.4. The column on the *left* is where we choose for α the golden ratio $\frac{\sqrt{5}-1}{2}$, the one on the *right* is where $\alpha = \pi$. We get comparisons of the darkening, in spite of only 1000 rebounds for the left, against 5000 rebounds for the right

With the number π we see seven points appear, then about a hundred. This isn't at all apparent from the decimal expansion. On the other hand, the successive approximations of π by rationals are: $3, \frac{22}{7}, \frac{333}{106}, \frac{335}{113}, \frac{109333}{33102}$; see Sect. IX.1. After the seven points, we see darkening when we approach a hundred points, but since 106 and 113 are very close together, things are not going to be so very clear. Finally, with α equal to the golden ratio, the darkening is ultrarapid.

It's the moment, if our readers have not already done so, to recall the theory of continued fractions studied in Sect. IX.1. We see then that this theory answers our questions completely, since it gives the best — the fastest — approximations by rationals. In particular, the worst number to approximate — the golden ratio — is the one that gives the fastest (lightning fast) darkening. Because in some way, when we take residues modulo 1 of its multiples, they **fall** each time as close as possible **toward the middle** of the remaining intervals: the sequence is $1, \frac{1}{2}, \frac{2}{3}, \frac{3}{5}, \frac{5}{8}$, etc.

We might suspect a relation between the continued fractions associated with α and the moduli of the complex numbers $e^{2\pi i n \alpha} - 1$. These moduli play an essential role in the study of the stability of the solar system; see for example Marmi (2000). Readers will not be astonished that bad or good approximation of real numbers by rationals reemerges in numerous areas. One of the recent results is the "KAM theorem" (Kolmogorov-Arnold-Moser) which, under conditions of "non resonance" between certain frequencies (conditions of the same nature as bad approximation by rationals) provides us with a proof of the existence of caustics in smooth

convex billiards (Sect. XI.9). Generalizing this theorem, Arnold was able to show
stability for a planetary system for which the planets have extremely small masses
in comparison to the sun (much smaller than in our solar system).

Finally, we may feel like rolling the interval [0, 1] up into a circle, which is the
geometric way of taking residues modulo 1. This is practically the same as asking
what happens when we iterate a rotation angle α, except that here (for α measured in
radians) what matters is knowing whether $\frac{\alpha}{\pi}$ is irrational or not. Readers may have
guessed that this little game on a circle is precisely that of billiards on the interior
of a disk; see Section XI.9.

♦

For those who like abstract concepts, the identification of the opposite edges of
a square amounts to introducing the **abstract torus** $\mathbb{R}^2/\mathbb{Z}^2$ and to looking at the
trajectory of the projection of a line of $\mathbb{R}^2$ in this torus. This in fact appears in the
following chapter; the trajectories in the flat torus in question are none other than its
geodesics; see Sects. VI.3 and XII.2.

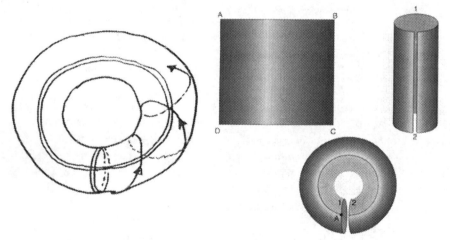

Fig. XI.2.5. Flow of a constant vector field on an (abstract) flat torus (the figure is
wrong, but an aid to the imagination). Berger (1991) © Pour la Science, Éditions Belin

♦

We can ask further questions about billiards, even in the simple case of the square.
For example, what about the natural coding of trajectories? It involves associating
with each trajectory the sequence of 0's and 1's determined by whether the trajectory encounters a horizontal side or a vertical side. We note what has already been
pointed out: contemporary geometry participates in dynamics, and also in combinatorics, algorithms, complexity theory, etc. This study was done by Morse and
Hedlund (1940) and is one of the first appearances of *symbolic dynamics*, which
had been initiated by Hadamard in 1898. Of course, it is connected with continued
fractions. We see first that if the trajectory is periodic, then the associated sequence

is itself periodic. In the nonperiodic case (irrational slope), we introduce the notion of the *complexity* of a sequence (always of 0's and 1's, but the idea is more generally valid for any infinite sequence of symbols). The *complexity function* Comp(N) of an infinite sequence is the number of distinct N-element sequences that it contains. For a periodic sequence, it is bounded. The result quoted says that the complexity for the trajectories of square billiards is the weakest possible: Comp(N) = N + 1. This means that our sequences are always the closest possible to periodic sequences; see Tabachnikov (1995).

XI.2.B. Counting periodic trajectories

The first observation is that the trajectories come in *bands*.

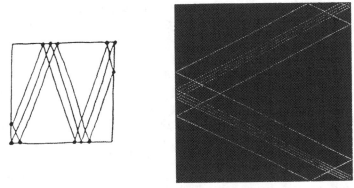

Fig. XI.2.6.

If we displace a periodic trajectory by translation we can get a trajectory that goes from one vertex to another, for which we can say that it is the boundary of the band. This is the time to remark that we haven't said what happens when a trajectory arrives at a vertex. In one sense we can only give a parachuted solution, or at least what we can guess by continuity. We displace the trajectory slightly: whether from one side or the other it starts off again parallel.

Fig. XI.2.7. What happens when a trajectory arrives precisely at a vertex of the square

We thus agree to say that when a ball arrives at a corner, it leaves along the same trajectory by which it arrived but in the reverse direction!

In what follows we know that there is a certain redundancy with parts of Chap. IX, but we think readers will appreciate not having to refer back to that chapter repeatedly. The majority of real numbers being irrational, our trajectories are everywhere dense with probability one. This involves the probability for the choice of a starting direction. But we have already said that the periodic trajectories – those that remain – are of fundamental interest. Before saying how to describe them and how to count them, we explain their interest, apart from what is obvious geometrically. The sketches of bands above makes us think of waves that are propagated on the interior of a rectangle. The periodic trajectories are what correspond to the resonance vibrations of the rectangle: its frequencies, its fundamental sounds. We can also speak of energy levels. This comparison is both profound and wrong, but it is the basis for a detailed and realistic study. In this sense a purely rectangular concert hall is a disaster because a whole series of frequencies seem amplified and in any case are never well dampened.

We can also consider a square as a vibrating membrane for which the edges are fixed. This is a seemingly different problem. From now on, as above, we consider for simplicity a square of side 1, but this doesn't change the profound nature of the problem. We know from Sect. IX.4 that the possible frequencies are given (and this is very easy to see) by functions which at the point (x, y) are products $\sin(\pi px) \cdot \sin(\pi qy)$, where p and q are positive integers. An arbitrary vibration is a Fourier series based on the preceding functions. The corresponding proper frequencies (eigenfrequencies) are $\frac{1}{2}\sqrt{(p^2 + q^2)}$. Let us now see why the periodic trajectories are associated with pairs (p, q) of integers. An initial idea is to think of the notation of the slope α as a rational fraction $\alpha = \frac{p}{q}$. But it is still more useful to use the technique of unfolding. The study above showed that the bands (packets) of periodic trajectories are in correspondence with the trajectories that go from one vertex to another. As we may always assume that one of the vertices is at the origin $(0, 0)$, the figure below shows that these trajectories are in exact correspondence with points in the upper right quadrant of the unit lattice in the plane.

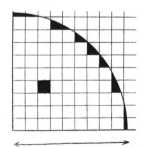

Fig. XI.2.8. How to count the bands of periodic trajectories

But we need to add (think of equal fractions $\frac{p}{q} = \frac{2p}{2q}$) that the integers (p, q) must be relatively prime. The point $(2, 2)$ in fact gives the same trajectory as $(1, 1)$! What is important for physics isn't so much the exact description of the trajectories but rather the evaluation of the energy levels of the system, i.e. the number of trajectories with energy less than a given value, which is for us here the *counting function* of the length FC(L):

$$FC(L) = \{\text{number of periodic trajectories of length} \leq L\}.$$

The solution for the system of the unit square is thus (Pythagorean theorem!):

$$FC(L) = \{(p,q) \mid p,q \text{ relatively prime integers such that } p^2 + q^2 \leq \left(\tfrac{L}{2}\right)^2\},$$

with $p \geq 0$ and q of arbitrary sign (because a trajectory has two possible orientations and there is division of the length by two when it arrives at vertices). So this is in fact an arithmetic problem and is connected with the problem of knowing in how many ways an integer can be written as the sum of two squares: $25 = 5^2 + 0^2 = 3^2 + 4^2$, etc. The order of magnitude has long been known. It should be remarked first that it is very easy to give a first approximation for the number of (p,q) such that $p^2 + q^2 \leq \frac{L^2}{4}$; because geometrically these are the points of the circle with center $(0,0)$ and radius $L/2$ that are located in the upper half-plane. We then think of its area: it equals $\frac{\pi L^2}{2}$. On the other hand the (p,q) are (almost) equal in number to the squares of side one contained in this quarter circle. Since their area always equals 1, we see that, "roughly"

$$N(L) = \text{ number of } (p,q) \text{ such that } p^2 + q^2 \leq \frac{\pi L^2}{8}$$

$$\text{with } p \neq 0, \ q \text{ of arbitrary sign} \approx \frac{\pi}{8}L^2.$$

There is the error made in the areas darkened above and the fact that we shouldn't count points of the form $(0,q)$. But it is elementary to see that both these errors are of the order of L when L is very large, because the sum of these darkened areas will be close to the length, equal to $\frac{\pi}{2}$, of the quarter circle. In mathematical language, we can now show rigorously that

$$N(L) \sim \frac{\pi}{8}L^2 \quad \text{when L tends toward infinity.}$$

Can we go further? For more than a century mathematicians have attempted to resolve the *circle problem*, about which we spoke in detail in Sect. IX.2.

We now need to return to the relatively prime (p,q). For the principal order, this is easy. We only need divide by what is overcounted: That is,

$$(p,q), \ (2p,2q), \ (3p,3q), \ \text{etc.}$$

only count as one. A bit of finesse indeed shows

$$FC(L) \sim \frac{N(L)}{1 + \frac{1}{4} + \frac{1}{9} + \cdots + \frac{1}{n^2} + \cdots} \quad \text{(when L tends toward infinity).}$$

Now it has been known since Euler that the infinite sum in the denominator equals $\frac{\pi^2}{6}$. Thus $FC(L) \sim \frac{3}{4\pi}L^2$ when L tends toward infinity.

◆

We now interpret the preceding study for the motions of our two particles, which was our original motivation. We thus see that, for a given initial position and velocity, the motion will almost always be transitive, which means that the positions of the two particles will be everywhere dense in the set of possible positions. In contrast, the motion will be periodic when the ratio of the initial velocities v'/v is a rational number $\frac{p}{q}$; and then the period will equal $p/v' = q/v$ if p, q are relatively prime.

XI.2.C. Introduction of the language of dynamical systems

According to the above remark, with the language of the abstract torus $\mathbb{R}^2/\mathbb{Z}^2$, the periodic trajectories of the billiard correspond to *periodic geodesics* of this abstract **flat** torus, and we will return to this amply in the following chapter. But there is something more important: if we want a really deep understanding in order to progress to the results below and put our affairs in a general context, we need to use the language of dynamical systems. We explain this in detail in order to introduce readers to the literature on polygonal billiards, which makes for difficult reading in spite of its naive geometric origins. It is difficult, it seems, because two types of dynamical systems are considered and that these tend to get confounded.

We want to use the language of dynamical systems – which is introduced in Sect. XI.XYZ at the end of this chapter – with the square billiard game, or even in the isosceles right triangle we started with. We start from a point in a certain oriented direction. Then we obtain a well-defined trajectory, even if we proceed to a vertex. We see in fact that the slope of the direction is what determines what happens, whatever the starting point. We thus fix an oriented direction θ, and we create a one-parameter dynamical system $f(t) : X \to X$ where X is the triangle (or the square) and where, for each $x \in X$, $f(t)(x)$ is the point of the trajectory with the initial point x, starting in the direction θ, at the end of time (of length) t. The function $f(t)$ preserve areas, i.e. the canonical measure of the triangle, and thus we indeed have what is called a *dynamical system*. The preservation of measure is trivial on the interior of the triangle, because only translations are involved. But it is also true when an edge is encountered, because we are dealing with symmetries about lines.

The set of $\{f(t)\}$ constitutes the *flow* of this system (in fact we use the word *flow* more generally for one-parameter dynamical systems; see Sect. XI.XYZ and Chap. XII). The nature of the flow has been studied above as a function of α. One conclusion was that for α irrational – in particular for almost all α – all trajectories of the flow are everywhere dense in Chap. X. This is what will be called a *minimal* dynamical system (flow) in Sect. XI.XYZ.

But we have seen something better, a uniform darkening throughout; this is a particular case of the fact that, for each irrational θ, the flow is *uniquelyergodic* – in particular ergodic – which means a lot more than equal darkening. Our proof with Fourier series implies this minimal ergodicity at once.

♦

Above we associated our billiard with a family of dynamical systems, indexed by the slope (oriented direction) θ. It is tempting to want to consider but a single dynamical system, and in fact this will be crucial for the sequel and the case of general billiards to consider more than one. Physicists have done this for a long time. It suffices to introduce a larger space, that of *phases*, which is quite simply the set $X \times S^1$ of pairs (x, θ), where x ranges over the billiard table — which can be an arbitrary domain in $\mathbb{R}^2$ — and θ the unit circle of oriented directions, which is the same thing as the set of unit vectors. At this moment there is but a single flow $\{f(t)\}$ on $\mathrm{Phases}(X) = X \times S^1$, that which associates with the pair (x, θ) the pair $f(t)(x, \theta)$ which is made up of the point and the velocity, at the end of time t, of the trajectory of the billiard that starts at time zero at the point x with the slope θ.

The basic hypothesis of statistical physics is that the dynamical systems considered are ergodic. This is certainly not the case for the billiard in the square. But we remark that the square is not a **generic** object. We pursue the essential goal of showing that the flow on phase space for the billiard game in a generic domain in the plane is ergodic. We are going to see that this fundamental question — fundamental because it is the simplest model for mathematizing statistical mechanics and lots of other more or less complex dynamical systems — has recently had some excellent partial answers, but that important questions still remain open.

XI.3. Particles with different masses: rational and irrational polygons

We return to our two particles in an interval: what if their masses are now different? We embark on this study, which will lead us more generally to that of polygonal billiards, and we will resume the results obtained in Sect. XI.6. We will see that, in spite of spectacular recent progress, the problem is far from being completely resolved.

Let m, m' be the respective masses of our particles, and v, v' their velocities. The elastic collision condition translates into conservation of linear momentum and kinetic energy, i.e. of $mv + m'v'$ and of $\frac{1}{2}mv^2 + \frac{1}{2}m'v'^2$. We won't write the result; it suffices to know that the calculation shows that the motion of the pair of particles coincides with that of billiards in a right triangle with sides of respective lengths $\sqrt{m}$ and $\sqrt{m'}$. The calculation concerns reflection through the hypotenuse; the reflections through the sides are trivial, as in the case of equal masses.

Thus, if we want to study the trajectories of a billiard in any right triangle, we know how to describe the motion of a pair of particles in an interval. A symmetry with respect to the hypotenuse (partial unfolding), when $m = m'$, led to the study of the case of the square. But what can we now do with a right triangle for which an angle (say the smaller) equals an arbitrary α between 0 and $\pi/2$? We first carry out a symmetry about the hypotenuse. There seems to be some hope if α is of the form π/k, k an integer, because by successive symmetries about this vertex we finally obtain a regular polygon with k sides. Unfortunately, with the sole exception of the case where $\alpha = \frac{\pi}{6}$ (or $\frac{\pi}{3}$ or course), good things stop there: in fact our regular polygon with k sides never tiles the plane with successive symmetries about the sides. We are thus lost, except in the case where $\alpha = \frac{\pi}{3}$ or when the masses are such that $m' = 3m$. We leave the study of the counting function to readers.

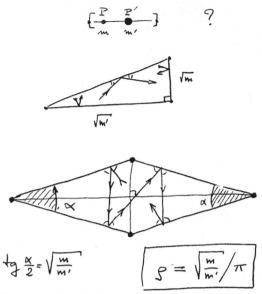

$$tg \frac{\alpha}{2} = \sqrt{\frac{m}{m'}} \qquad \boxed{\, \rho = \sqrt{\frac{m}{m'}} \Big/ \pi \,}$$

Fig. XI.3.1. Expressing that the linear momentum $mv + m'v'$ of the motion and the kinetic energy $\frac{1}{2}mv^2 + \frac{1}{2}m'v'^2$ are conserved; the billiard game in a right triangle of sides $\sqrt{m}$ and $\sqrt{m'}$

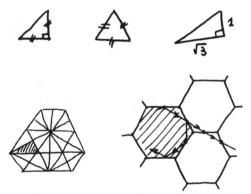

Fig. XI.3.2. Why the method doesn't work for a regular hexagon

We finish with a little (big?) booby trap, which readers will manage to explain. If the above method allows complete resolution of billiards in the square, isosceles triangles and halves of equilateral triangles, it does not allow us to know what happens for billiards in a regular hexagon. In fact, about a vertex, we find three hexagons, but the composition of three symmetries reverses orientation and thus when returning to the hexagon of departure we are in some way on the wrong side and there is no longer coherence. We will however see in the following section that billiards in the regular hexagon is completely understood, but only at the price of an ascent of Jacob's ladder, which will be the most costly and highest ascent of this book.

◆

However, what can be done in the general case? What follows is the result of spectacular progress, obtained with a time separation of ten years, first in 1975, then in 1985, at least in the case of polygons that we will call *rational*. The first basic idea can be guessed by looking at the initial elements of a trajectory in a right triangle with angle $\frac{\pi}{k}$, k an integer. Readers who can sketch well will realize that the trajectories have directions which at the end of a certain time are parallel to the initial elements. Now the directions of these trajectories are obtained, starting from the first one, by symmetries about the directions of the sides of the triangle. But then, if there are but a finite number of possible directions, we can reduce (as we have done for the right isosceles triangle) our two-parameter dynamical system in the plane to a discrete dynamical system in a single parameter: we take the set of the three sides of the triangle, but repeat a number of times equal to the possible directions.

In the Euclidean plane we thus consider a polygon P. In conformance with the definition adopted in Chap. VIII, we are dealing with convex sets. For more general "polygons", see the references given. We consider the billiard trajectories on the interior of P.

(XI.3.1) *A polygon in the Euclidean plane is called* rational *if all its vertex angles are rational multiples of* π.

This in place we have the following elementary result:

(XI.3.2) *The billiard trajectories of a polygon P all have a finite number of directions if and only if the polygon is rational.*

We in fact consider the set of directions of the sides and the group G generated by the set of symmetries associated with these directions. Once two sides form an angle irrational to π, the group G contains an infinite number of elements. We can thus show, with a little work, that there always exist trajectories taking an infinite number of directions. If on the contrary all the angles of the pairs of lines are rational, then we quickly see that G is a finite group, moreover exactly the group of isometries of a regular polygon.

For example we indeed find typically eight directions for the isosceles right triangle, ten for the triangle with angles $(\frac{\pi}{5}, \frac{2\pi}{5}, \frac{2\pi}{5})$ and six for the regular hexagon, which explains why — with only three copies — we cannot achieve reduction to the case of a single direction.

◆

Referring to Sect. XI.2.C, we see that the present context is that of a family of dynamical systems indexed by one direction θ. The total case of the phase space will be considered primarily for non rational polygons, since the preceding says among other things that, in the phase space for the rational case, we are at the opposite end from ergodicity (see Sect. XI.XYZ): there is a systematic decomposition associated with the parameter θ.

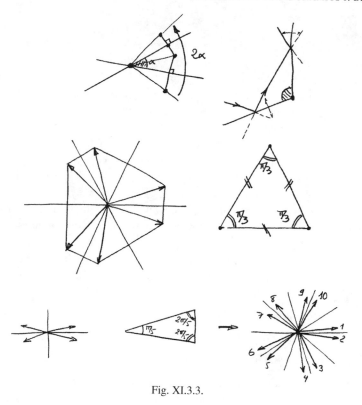

Fig. XI.3.3.

A rather recent synthesis on the subject of polygonal billiards is Gutkin (1996), to which we add the book Tabachnikov (2005), which treats billiards in general.

XI.4. Results in the case of rational polygons: first rung

In this and the following sections we consider a convex rational polygon P. We give here the first series of results obtained for them in the years around 1975. The idea consists of extending what has been done for the square. It involves an effort toward abstraction typical for mathematics, the attempt to associate a conceptually simple set with the construction of the successive pieces of a trajectory, and with the notion of trajectory a mapping − a transformation − of this set into itself. The idea for discovering the construction to be made can be found in shifting a trajectory a little parallel to itself.

The arrows change, but the **width of the band** doesn't change. It is *invariant*. We already mentioned several times that the notion of *invariance* is one of the sacred ideas in mathematics.

Second idea: if we forget the segments which make up the trajectory, we see that we can nonetheless reconstruct them with the aid of the points where these segments meet the sides and from the direction of the rebound. This shows that for one trajectory with given initial arrow, everything can be transported into a set made

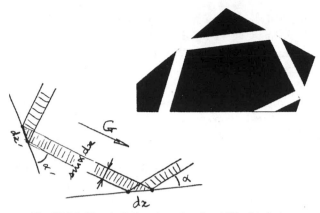

Fig. XI.4.1. How to find an *invariant* in a billiard trajectory

up of the pairs (p, f) formed by a point p on a side C and an arrow f (taken from a finite number of possibilities); and it then suffices to know how to describe the operation associated with a reflection, i.e. the passage from (p, f) to (p', f') (see the figure). But when p varies over the side C, p' does not vary over the same distance on the side C′ associated with a (p, f). Now we can remedy this by thinking of the width of the band as invariant. We thus normalize the pairs (p, f) where p belongs to the side by changing the arrow for evaluating p on C, the change of arrow being given by the *sine of the angle* of f with C (to find the width of the band). Thus the mapping between the pairs (p, f), (p', f') to be studied preserves a metric (here it has to do with lengths in a well defined sense): we say that it is *isometric*.

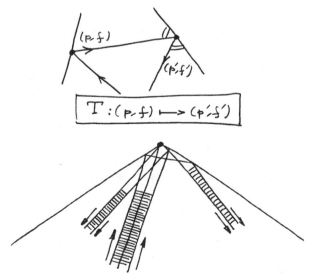

Fig. XI.4.2. What happens at a vertex

There is difficulty that must be overcome in order to arrive, for our efforts, at our goal: we cannot be content with pairs formed from points from all sides and arrows of all possible directions, because in fact we have to remove elements (p, f) when the associated segment terminates at a vertex (the band is cut in two, except in the case of the rectangle). It is time to study what happens when a trajectory arrives at a vertex of the polygon. In the case of a right angle, we have seen for the rectangle that it can only emerge in the opposite sense (time reversal). But for any other angle it seems there is no coherent definition. We can only move the trajectory a little to the left, then a little to the right, to see that there are two different responses. If we consider one band, we then see that it is divided into two bands, but we are saved just because the sum of the widths of these two bands is again the width of the initial band. However, the case where the angle at the vertex equals $\frac{\pi}{n}$, n an integer, allows definition of a unique extension. This can be seen by hand and can also be read on the abstract surface we are about to define.

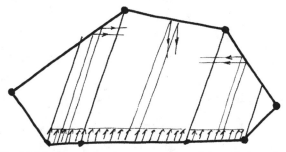

Fig. XI.4.3. Why we must suppress some points on the boundary

So the arrow in the above figure compels the removal of two points from C and the cutting of C into three pieces C′, C″, C‴. But this must be done for all the possible directions of the trajectory. We see that at the end of this construction we will have a **finite** collection of intervals (each interval is associated with a piece of a side and with one direction) and that the operation of going from the one side to the other is an exchange of intervals: we simply *arrange them differently*. The case of just two intervals is sketched below. This is the case of the square (multiplied by four); it doesn't illustrate a general rational polygon for which many more intervals would be required. Nonetheless it is exemplary, for when we iterate this exchange of two intervals we see that it is exactly the study of the fractional part of $n\alpha$, n an integer and α irrational.

In 1975–1978 the mathematicians Zemlyakov and Katok, then Boldrighini, Keane and Marchetti carried out the above construction using a known property of *interval exchanges*, which are among the simplest dynamical systems considered in ergodic theory (as models for physics). The case of two intervals $(\alpha, 1 - \alpha)$ is nothing other than that of the behavior of the $n\alpha$ modulo 1. This property generalizes what happens to the fractional parts of $n\alpha$. The conclusion is that, modulo a denumerable set, either successive points obtained give a periodic sequence, or they "darken" everywhere.

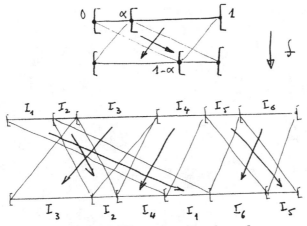

Fig. XI.4.4. What are interval exchanges?

Thus mathematicians have proved a good part of what we wanted to know subsequent to the rectangular case.

(XI.4.1) *With the possible exception of a denumerable number of trajectories, billiard trajectories in a rational polygon are either periodic or darken everything.*

Moreover, the probability is 1 that a trajectory darkens everything, for the periodic trajectories are very special: they are denumerable in number. See Fig. XI.5.6 for designing billiards where darkening does not occur throughout.

Note that we have thus **classified** the trajectories. Each trajectory has three possible behaviors: either it darkens everything, or it is periodic, or it darkens just a proper subset of the polygon. The latter two cases occur only with probability zero. In summary, we have essentially a dichotomy, modulo denumerability: periodicity or everything being darkened. In the language of ergodic theory we have what is called *topological transitivity*; see Sect. XI.XYZ. Regarding the final case, partial darkening, see the lines following the theorem XI.5.1 below.

But there remains the question of the *equality* of darkening, which was valid for the rectangle. The sketches provided above allow us to anticipate that the question is difficult; we even have the certainty, in figures such as the one below, that there is definitely not equal darkening, i.e. not a good distribution (which can't happen in rational polygons, as we will see). Before broaching this, readers can find the definition of equal darkening in Sect. XI.XYZ (see also Figure XI.2.1). For more on darkening (density) in rational polygons, see the next section.

◆

Historically it is rather recently that the above results have been obtained in a completely geometric way. We develop a trajectory as in the figure and look at a band about a line. For reasons of area, by looking at a band of maximum width we end up seeing that, with probability 1, everything is darkened in the initial polygon, or

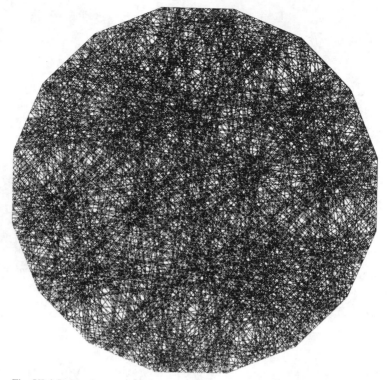

Fig. XI.4.5. A trajectory in the regular polygon with sixteen sides, 900 rebounds

else we have periodicity. All this can be found, very well explained and in a most geometric way, in Tabachnikov (1995) or Tabachnikov (2005).

Readers who want to deepen their understanding might review Sect. XI.2.C to be able to access the results below and to put matters in a more general context.

XI.5. Results in the rational case: several rungs higher on the ladder

Interval exchanges do not permit us to answer several questions about rational polygonal billiards that have been left hanging: the exact density of the trajectories, the counting function of periodic trajectories. The essential reason is that the interval exchanges introduced depend on the initial direction; we therefor lack a conceptual object that takes account of the whole polygon at once. But such an object was already indicated in the case of the square at the end of Sect. XI.2.A, where the billiard trajectories in the square were "developed" as geodesics in $\mathbb{R}^2/\mathbb{Z}^2$, i.e. projections of lines of $\mathbb{R}^2$ in the flat torus $\mathbb{R}^2/\mathbb{Z}^2$, a definitely beautiful object, compact and smooth. Can we do something analogous with an arbitrary convex rational polygon P? We are in fact going to do two things: the first is to construct an abstract object where we can study what happens, i.e. an abstract surface. And the second step begins with a remark already made: as we have described, the study of billiards in a square, more generally in a polygon, is not really a dynamical system in itself in the sense of the definition in Sect. XI.XYZ, i.e. a (compact) topological space endowed with a measure and a transformation that is a measure-preserving homeomorphism, or again a one-parameter group of such homeomorphisms. We have only introduced such a notion when the initial direction was fixed, and thus the dynamical system considered depended on this initial direction. We are going to have to find a dynamical system associated with all the billiard motions in a rational polygon. This will be simple, we have only to introduce the **phase space** of the billiard, as at the end of Sect. XI.2.C. This amounts simply to adjoining directions to points, i.e. to considering the set of pairs formed by the points of the billiard and all the oriented directions (or unit vectors) at these points; for the physicist these correspond to **velocities**.

The climb up the ladder that will be accomplished to attack (with great success) billiard flow in rational polygons (and irrational ones as a consequence), is the most expensive, the most abstract and seemingly most complicated topic in the entire book. Readers will form their own opinions. We can also ask if all the theory that will follow below, which consists of using "all" the hyperfine theory of Riemann surfaces, is ultimately really necessary for resolving a problem consisting of the study of the geometry of a space that is everywhere locally Euclidean except at a finite number of points, each of which is equipped with a conical singularity with an angle that is an integral multiple of 2π. Is it possible to find a setting, perhaps very abstract but very geometric, where we could carry out this study in an easier way? We won't neglect emphasizing the complete contrast with Sect. VIII.6, which involves geometries that are again locally Euclidean, but where the singularities correspond to angles less than 2π.

For these recent and very profound results, an introductory reference is (Arnoux, 1988); detailed references are Kerckhoff, Masur, and Smillie (1986), Masur (1990), Veech (1992), Boshernitzan, Galperin, Krüger and Troubetzkoy (1998), as well as the synthesis Masur and Tabachnikov (2002).

XI.5.A. The nature of nonperiodic trajectories

We first construct our abstract object, a surface in fact. For the square we first unfolded it into four squares by symmetries with respect to the sides, then subsequently identified the opposite sides in this block of four squares. In fact we took four squares, obtained by symmetries, and identified sides numbered as in Figure XI.5.1.

We need to correctly define the assemblage, the gluing (or sewing) to be made with copies of our polygon in the general case of a rational polygon. In Sect. XI.3 we denoted by G the group generated by the symmetries with respect to the sides of P: we index the sides, denoting them C_i, and let S_i be the (vectorial) symmetry that C_i defines. We define the abstract object sought, denoted $\mathcal{S}(P)$, by the equivalence relation on the product $P \times G$ that identifies (x, g) with $(x, S_i g)$ for $x \in C_i$. Practically, we take a number of copies of P equal to the order of G and assemble them algebraically as indicated. Readers will do the work of describing what happens in the regular pentagonal and hexagonal cases, it being remarked that the order of G is only half the order of the group of all isometries that preserve an arbitrary regular polygon P when the number of sides is even, because of parallel sides. When P has very small angles (always rational) the order of G can be enormous.

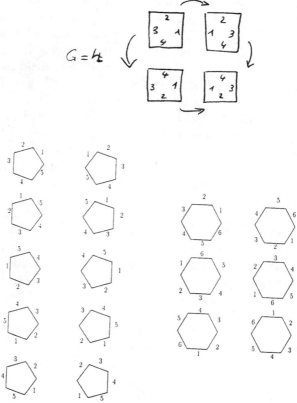

Fig. XI.5.1. The copies that need be assembled for: a square, a regular hexagon, and a regular pentagon

But a horrible thing happens (except for the square): at each vertex of the polygon considered, there are as many polygons to glue as the order of G, namely a total angle of $10 \cdot \frac{3\pi}{5} = 6\pi$ for the pentagon and $6 \cdot \frac{2\pi}{3} = 4\pi$ for the hexagon. Only the square and the isosceles right triangle yield a **flat** object, because the sum is indeed 2π. When we carry out the (absolutely) necessary gluing, we obtain an object that is doubly abstract, first (like the flat torus) because it is very likely (open question!) that it cannot be drawn with a good geometry, like a surface in $\mathbb{R}^3$, but moreover because its geometry, although abstract, is not locally Euclidean throughout. It *is* locally Euclidean, but with the exception of a finite number of points, equal in number to the vertices of the polygon considered. At these points the geometry is that of a Euclidean cone, but the vertex angle equals $2\pi k$, where k is an integer, in general greater than 1. We note that this construction goes through with any convex polygon, which need not be at all regular; the identifications are those given by the symmetries with respect to the sides. Of course, the integers k depend in general on each vertex of the rational polygon considered; the case of the regular polygons used in the examples is completely special.

For the record, but in a completely different spirit, we have already encountered such conical singularities, but where the angle was a real number between 0 and 2π, in connection with polyhedra and their rigidity, as well as with surfaces of positive curvature in Sect. VIII.6.

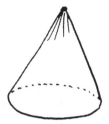

Fig. XI.5.2. Conical singularities. Angle $< 2\pi$ and angle $> 2\pi$

Readers can try to assemble polygons made of cardboard with an angle total such as 4π; in certain cases this is possible, but doesn't go through except at a single point: it is very likely not possible to make a complete compact surface, and it seems (even with just two points) that this is an open question. Here is a figure from Berger (1991), but where the surface visualized is not locally Euclidean; it is just a possible aid to the imagination. Here it involves the case of an isosceles triangle P with angles $\frac{\pi}{5}, \frac{2\pi}{5}, \frac{2\pi}{5}$, for which we need to glue ten copies, and comes down to making the indicated identifications for the regular decagon of the figure. For the angles at the vertices A, B, C we find that we have angles that are equal to 2π for A (thus no problem), but $5 \cdot \frac{4\pi}{5} = 4\pi$ for B and C, which are thus singular points.

We obtain an abstract surface $\mathcal{S}(P)$ associated with the triangle P. As curious geometers, we know that since, by construction, $\mathcal{S}(P)$ is a compact orientable surface, according to Sect. VI.1 it has a *genus*, the number of holes. We may want to calculate it, even if strangely enough this genus doesn't enter into the various formulas that we will encounter. There are 3 vertices, 10 faces and $3.10/2 = 15$ edges (we divide by two because we glue two at a time), giving an Euler characteristic equal to $3 - 15 + 10 = 2(1 - \gamma)$, the number of holes γ thus being equal to two. For the hexagon, we find four holes; for the regular pentagon, six holes, etc. The number of holes in the above figure for the case of the pentagon is thus correct. It should be mentioned that this general construction appeared in Gutkin (1984).

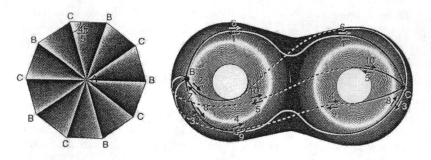

$$\text{at } A : 10 \times \frac{\pi}{5} = 2\pi \quad \text{O.K.}$$

$$\text{at } B : 10 \times \frac{2\pi}{5} = 4\pi \quad \text{a problem}$$

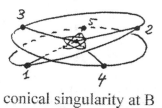

conical singularity at B

Fig. XI.5.3. The difficult problem of *seeing* the abstract surface being studied. The trajectories on the torus with two holes joining the singular points and representing different periodic trajectories of the triangle. Berger (1991) © Pour la Science, Éditions Belin

♦

But what is new here is the study of the Riemann surface associated with this object, which gave Kerckhoff, Masur and Smillie the necessary elements for making progress in about 1985 in the problem of rational polygonal billiards. It is also important to remark (an amazing phenomenon, occasionally documented in the history of mathematics) that Kerckhoff, Masur and Smillie published their work on billiards because a colleague who had listened to their lectures on their results on flows on surfaces said to them: "but what you have just shown must be useful for billiards!". And in fact, these three mathematicians are specialists on Riemann surfaces, more particularly experts on quadratic differentials on Riemann surfaces, in connection with Teichmüller theory (see below). What is then the connection with our objects?

Before answering this question, we need to construct our dynamical system, in fact (see Sect. XI.2.C) a family of dynamical systems indexed by a direction θ (say a point of the circle S^1). In fact, we are going to construct a whole series of increasingly abstract objects. We fix the rational polygon P and its associated abstract surface $\mathscr{S}(P)$, which has a finite number of vertices (singular points). We first associate with $\mathscr{S}(P)$ a *Riemann surface* $\mathscr{R}(P)$.

We first see that our construction of $\mathscr{S}(P)$ can give us a (smooth) Riemann surface $\mathscr{R}(P)$; see Sect. VI.4 as needed for this category of surface. Apart from the singular vertices, the complex structure is clear, coming from the Euclidean structure. And at the singular points, if the conical angle at such a point is $2k\pi$ we need to smooth this point by dividing everything around it by k: this operation is well known in Riemann surface theory and is called *ramification*; and it goes well because the complex mapping $z \mapsto z^k$ about the origin fits the quotient well. The situation — seen in billiard trajectories — is that of Fig. XI.5.2 for the case where k is equal to two:

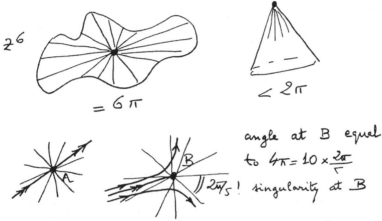

Fig. XI.5.4. Singularity at B: angle at B equal to $4\pi = 10 \times \frac{2\pi}{5}$

The divergence of the trajectories when arriving at a vertex translates the geometric fact discovered in the above figures (see also Fig. XI.4.2) on the splitting of a band into two. What interests us more here isn't the complete Riemannian structure, but only the rotation by $\frac{\pi}{2}$ associated with each of the (planar) tangent spaces, and the fact that this rotation is in fact multiplication by $i = \sqrt{-1}$ of a **complex** (holomorphic) structure: we have available locally the calculus analogous to that of holomorphic functions on $\mathbb{C}$ (for this mindset see Sects. VI.8 and VI.XYZ). On the smooth surface $\mathscr{R}$ what is known as a *conformal structure* $\mathcal{T}(P)$ is one given by such rotations on each tangent space. The space of all conformal structures (appropriately quotiented by diffeomorphisms) is called a *Teichmüller space* Teich(R) of $\mathscr{R}$ and has been much studied — incessantly in fact — since the 1930s. We briefly mentioned them in Sect. VI.XYZ. It is surely an abstract manifold, of dimension $6\gamma - 6$ if γ is the genus of $\mathscr{R}$. We are now going to associate with pairs $(\mathscr{R}(P), \theta)$ — which correspond to the flows with the directions θ in the polygon — flows in nothing less than this modular ultra-abstract space Teich($\mathscr{R}$).

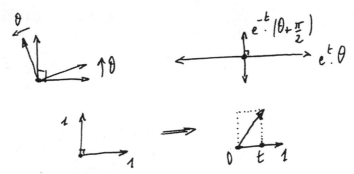

Fig. XI.5.5. The three continuous operations that act on the modular domain

In fact, if in addition we give a direction θ, this provides at each nonsingular point of $\mathcal{R}(P)$ an orthogonal marking of the everywhere locally Euclidean structure of $\mathcal{S}(P)$, but we retain only a complex structure coming with — in particular — a pair of (orthogonal) directions, i.e. at x a pair $(d(x), e(x))$. Although we don't really need the term here, such an object is called a *quadratic differential* on our Riemann surface. It is a holomorphic quadratic differential form that can be written — except perhaps at finitely many points — in a suitable chart, as $f(z)dz^2$, with $f(z) \neq 0$; and at the "singular" points, where f is zero, then it must definitely be written in the form $z^k dz^2$, where k is a positive integer.

The super-abstruse thing now is to create a family in a single parameter t of the conformal structures defined by the pairs $(e^t.d(x), e^{-t}.e(x))$: we extend the d and restrict the e. When t varies we thus obtain a curve in Teich$(\mathcal{R})$, for which the tangent vector at $t = 0$ can be considered a unit tangent vector to Teich$(\mathcal{R})$. Thus the direction θ determines a *curve* $c(\theta)$ in Teich$(\mathcal{R})$, which is in fact a geodesic of the classical dynamic metric structure that can be put on Teich$(\mathcal{R})$. To study such curves (as a trajectory of a dynamical system in Teich$(\mathcal{R})$) associated with quadratic differentials , we need to introduce a further stage, specifically we need to have the modular group PSL$(2; \mathbb{R})$ act, typically by the three one-parameter actions that generate this group:

$$(u(x), u(x)) \mapsto (e^t.u(x), e^{-t}.u(x)),$$
$$(u(x), v(x)) \mapsto (\cos\theta.u(x) - \sin t\theta.v(x), \sin\theta.u(x) + \cos\theta.v(x)),$$
$$(u(x), v(x)) \mapsto (u(x) + t.v(x), v(x)).$$

This is called a *Teichmüller flow*, and the associated trajectories have been intensely studied. It will be noted that the second operation is nothing other than Euclidean rotation through an angle θ and corresponds to various directions θ. Now we know enough about Teich$(\mathcal{R})$ and its curves $c(\theta)$ to be able to prove that, after being brought down to the polygon itself:

(XI.5.1) (Kerckhoff et al. (1986.) *For almost every direction θ the associated flow in* P *is* uniquely *ergodic (see Sect. VI.XYZ).*

♦

This is the principal result (and technique) on rational polygonal billiards. However, the study cannot stop here. In fact, it can be shown (see the precise references in our bibliography) that there is in general at best only **trichotomy**. In the case of the square, we have a dichotomy, the trajectories being either periodic or everywhere dense (and of uniform density). Here we know that there exist (rational) polygons for which there are at least three types of trajectories: the periodic, the everywhere dense with equal density, but also a set of Cantor type (and even rather large, because of positive Hausdorff measure) for which the trajectories are either everywhere dense but of nonuniform density or else worse: everywhere dense but only in a strict subdomain of the polygon. Before (Galperin, 1983) the existence of such trajectories was for a long time considered impossible. The figure below shows how to create such a trajectory very simply in a rational parallelogram, because it is shown on the one hand that the trajectory is dense in the hexagon and on the other hand that it doesn't emerge from it. Thus it misses the two triangles at the ends of the parallelogram:

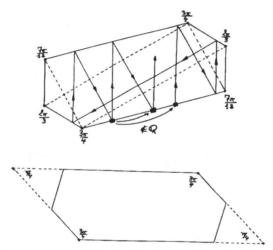

Fig. XI.5.6. How to construct a billiard table where certain trajectories avoid a whole portion of the table

However, in Veech (1992) it is proved that for all regular polygons (as well as for all isosceles triangles with base angles $\frac{\pi}{k}$) there is the good dichotomy; cf. (XI.4.1). Figure XI.4.5 doesn't allow a prediction. An exact characterization of the rational polygons with dichotomy remains an open problem.

XI.5.B. Counting periodic trajectories

The periodic trajectories in the rational polygon P correspond, on the associated abstract surface $\mathcal{S}(P)$, to the trajectories (ill-defined in a sense, see Sect. XI.2.B; think of the bands) which join two vertices (singularities). In the language of quadratic differentials, these are the points where they cancel. Masur's mastery of quadratic differentials allowed him to successfully study the counting function FC(L) for the periodic trajectories:

(XI.5.2) (Masur, 1990.) *The counting function for the periodic trajectories of each rational polygon is always sandwiched quadratically, i.e. there exist constants a and b such that $aL^2 < FC(L) < bL^2$ for all L.*

It isn't known presently whether in general there exists a limit for the quotient $\frac{FC(L)}{L^2}$, which is known for the case of regular polygons with n vertices (and of course for the isosceles triangles that give rise to them) and those mentioned at the end of this section. Again, in the text (Veech, 1992) it was not only proved that such a limit exists, but above all it was calculated by a *tour de force* of number theory. The basic idea is that in this case the abstract Riemann surface introduced above is the one corresponding to the algebraic curve $y^2 + x^n = 1$ (see Sect. V.13), one of the quadratic differentials that comes up being $\frac{dx^2}{y^2}$.

The proofs of (XI.5.2) are not easy. Geometrically they simply express the number of segments of $\mathcal{S}(P)$ of a given length that connect two vertices. If we think of Riemann surfaces, it is rather surprising that the result is quadratic, given that the metrics for Riemann surfaces are locally hyperbolic, and thus that in the universal covering (the hyperbolic plane) a ball of given radius is going to contain a number of vertices that grows exponentially as a function of the radius, if we suppose that these vertices get accounted for like copies of the initial polygon. But in fact the actual geometric structure on $\mathcal{S}(P)$ is locally Euclidean, and the result translates the fact (see the case of billiards in a square, a circle, etc. and how we proceeded in Sect. XI.2) that the area of a disk is quadratic in the radius and thus will contain a number of copies of the initial polygon that grows quadratically as a function of the radius.

As for evaluating the difficulty and subtlety of Veech's result, for interested in number theory we give the pleasure of seeing the limit: $\lim_{L \to \infty} \frac{FC(L)}{L^2} = \frac{C(n)}{p_n}$, where p_n denotes the area of the regular polygon with n sides considered and $C(n)$ a constant to be determined. This constant depends on another constant $c(n)$ which is a *multiplicative* function, i.e. $c(mn) = c(m) \cdot c(n)$ for each relatively prime pair m, n. It is then sufficient to know the c's for numbers of the form $n = 2^\nu$ and those of the form $n = p^\nu$ (with p a prime > 2). We find then that

$$c(2^\nu) = 7n^3 - 6n^2, \quad \text{and for } p > 2, \quad c(p^n) = n^4\left(1 + \frac{1 - n^{-2}}{p(p+1)}\right),$$

then:

$$C(n) = \frac{c(n) - n^3}{48\varepsilon(n)(n-2)\pi} \quad \text{if } n > 4,$$

with $\varepsilon(n) = 1$ or 2 according as n is odd or even. Readers can prove that $C(3) = \frac{1}{2\pi}$ and $C(4) = \frac{3}{4\pi}$.

Recently an exact formula for the asymptotic behavior of FC(L) was found for a large category of rational polygons; see Eskin and Masur (2001). Finally, an elaboration of periodic trajectories in rational billiards, specifically their density in phase space, can be found in Boshernitzan et al. (1998).

XI.6. Results in the case of irrational polygons

The present situation for non rational polygonal billiards is rather astonishing; see however the important note at the end of the present section. In fact, we have first a spectacular result for generic billiards; for the notion of genericity see below and, as needed, Sects. V.10, VII.13.D and also XI.10.C:

(XI.6.1) *For the dynamical system that it defines in its phase space, a generic polygon is ergodic.*

Recall that genericity must be defined with the help of general topology, because there doesn't exist a good measure on the space of all polygons; cf. Sect. V.10, VII.13.D and XI.10,C. But there exists a complete metric, which suffices.

We say that a property of polygons is generic (or that a generic polygon satisfies it) if the set of polygons that satisfies it is G_δ-dense. It suffices that this be true for the polygons with n vertices, seen as points of $\mathbb{R}^{2n}$). Thus "almost all" polygons are ergodic in phase space. In particular, for such a generic billiard and for almost all initial directions, the trajectories are everywhere dense in phase space, i.e. each point and each direction at this point can be approximated arbitrarily closely if we wait long enough, i.e. if we follow the trajectory far enough (after a sufficient number of reflections at the sides). Moreover, this everywhere density is uniform.

The proof of (XI.6.1) is not too onerous once the very difficult (XI.5.1) is known: it suffices to make somewhat deft use of the fact that integral multiples of an angle irrational to π are dense in the set of angles. We then approximate a polygon with irrational angles by rational polygons: we see that we thus obtain both spatial density − by (XI.5.1) − and that in phase space by the preceding remark.

What is astonishing is that at present it is not known how to exhibit explicitly even **a single** polygon that can be proved to be phase-ergodic; the preceding result remains purely theoretical.

◆

The second great mystery for non rational polygonal billiards, of which a good portion has been elucidated very recently, is that of their periodic trajectories. Not only is it still a question of evaluating (asymptotically) their counting function, but such a function presently makes no sense. In fact, it is not known how to show (even though everyone believes it) that in each polygon there exists at least **one** periodic trajectory (perhaps infinitely many?). See the important note at the end of this section. We first see why things are so difficult. In fact, beginning with a triangle, everyone knows a periodic trajectory: it's the one obtained by joining the feet of the three altitudes.

Unfortunately things don't work out if the triangle is obtuse. The difficulty of the problem can even escape − to go to extremes − the professional geometer. There are many among them that we have encountered who, at the end of a lecture, have proposed the following solution for constructing a periodic trajectory in each polygon: the unfolding technique while effecting an odd number of symmetries with respect to the sides.

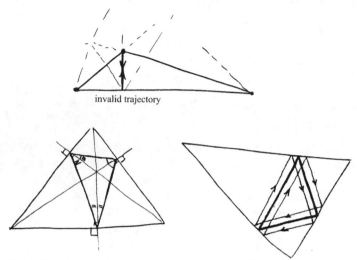

Fig. XI.6.1. The classic trajectory in an acute triangle and the invalid trajectory in an obtuse triangle

Well, the effect in the plane of an odd number of symmetries with respect to lines is a transformation called translation-reflection (it changes the orientation). It always has an invariant line D (the place in the middle of the pairs formed by a point and its image). If we trace this line D, it will necessarily be the unfolding of a periodic trajectory with an odd number of reflections (rebounds). The only trouble — nullifying — is that this line D in general doesn't intersect the sides *where it should*. This unfolding technique can be found in detail in 9.4 of [B]. Another simple remark: if we have a periodic trajectory with an odd number of rebounds, it clearly generates a band of periodic trajectories, but with twice the number of rebounds. We must pay attention to counting; see Fig. XI.6.3!

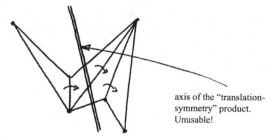

Fig. XI.6.2. Three symmetries about the sides, the axis of the translation-reflection generated, ultimately unusable!

◆

It was necessary to wait for Cipran, Hanson and Kolan (1995) to finally be sure of periodic trajectories in right triangles. The matter was immediately extended in Gutkin and Troubetzkoy (1996) to numerous (but not all, still today) types of polygons. In most of these cases these trajectories are not very pretty because they are obtained by starting orthogonally from a side, to which it is shown it is necessary to return orthogonally at a later time. It is a rather simple argument concerning the band width — whose invariance we know — and the area. These are in fact the trajectories for which we can say that they get **folded** onto themselves.

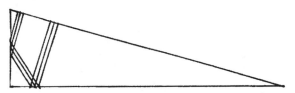

Fig. XI.6.3. How to find "two-footed" periodic trajectories in an arbitrary triangle

In particular it isn't known even how to imagine of a counting function. However, we know that if there is one, then when there is an infinity of periodic trajectories, they aren't very numerous. In fact we find in Katok (1987) the fact that the counting function is always *sub-exponential*, i.e. that

$$\lim_{L \to \infty} \frac{\log(FC(L))}{L} = 0.$$

Katok's proof is very geometric; we once again look at the parallel bands in an unfolding; the inescapable fact that the trajectories that arrive at a vertex "diverge" there complicates the affair substantially. Moreover, to conclude, it is necessary to use a theorem from ergodic theory (see (XI.XYZ.3)), i.e. the variational principle.

We have never heard any "reasonable" conjecture for an FC(L) for non rational polygons. We might think that polynomial growth is likely, but the degree of the polynomial remains a question: is it always two as in the rational case, or is the degree dependent on the exact nature of the polygon?

We also need to mention that all polygonal billiards always have null topological entropy (see Sect. XI.XYZ); this is a rather old result of Sinai. This nullity translates the geometric fact that polygonal billiards **don't disperse** the trajectories: there is no dispersion on the interior of the polygon because we are in Euclidean geometry, nor are there reflections on the sides, because these are lines and thus we only have to do with Euclidean symmetries. We will soon see something quite different in billiards where the sides are curves.

For computer enthusiasts, if we want to further understand trajectories of irrational polygonal billiards and their dynamic, computer graphics are very interesting to carry out. Programs can be written that sketch trajectories; our figures have been done thanks to programs provided the author by John Hubbard and Louis Michel. But the figures darken very quickly and seem incomprehensible in the majority of cases. The figures below show, on the other hand, that sometimes we need to wait very long to fill the triangle.

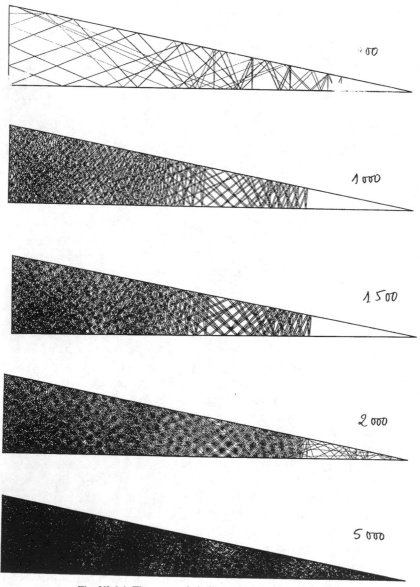

Fig. XI.6.4. The numerals indicate the number of rebounds

Well, just as we reduced the square case in Sect. XI.2 and the rational case in Sect. XI.4 to a dynamical system in a single real parameter, for irrational polygonal billiards we find our graphics have two parameters. The trajectory is **coded** by what happens at the sides: we only need know the point and the direction at that point, which provide a cloud of points in the plane. Experiments seem to show exciting things, but we know of no more advanced study. Here are some plane figures with the "whole" trajectory and the associated coding. The gain in understanding is glaringly obvious, as shown in pairs of figures below where, typically, the discrete drawing allows undecipherable phenomena to appear for the trajectory proper. We don't know of any study of this genre of figures.

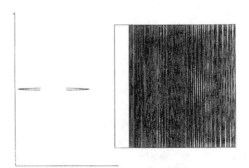

Fig. XI.6.5. Almost square billiards, with the complete figure and the coding for seeing it more clearly!

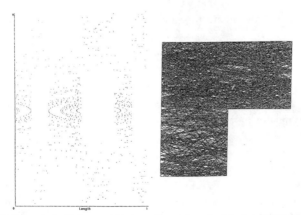

Fig. XI.6.6. Some trajectories, and their coding, where to this day it is unknown what is going on mathematically!

Important note. Very recently a major advance was made by Richard Schwartz, who succeeded in computerizing the geometry of unfoldings in a manner sufficiently detailed for realizing that the structure of triangular billiards (nonrational or course) is, for an object apparently so simple, extremely subtle. This approach, very computerized, permits him to obtain spectacular results. But the proofs use a lot of analysis, e.g. Fourier series and are perhaps close to being renormalization theories. For this work, which is in full development, see Schwartz (2006a), b, c.

XI.7. Return to the case of two masses: summary

The case of two masses $\{m, m'\}$ on the interior of an interval with elastic collisions is the problem from which we have taken motivation. The preceding furnishes the following results. The essential fact (which comes from Sect. XI.6.1) is that for a generic ratio $\frac{m'}{m}$ of the masses, if it is taken in a subset of the set of real numbers which is G_δ-dense (see Sect. VII.13.D), the system is ergodic: for almost every initial situation (origins of the particles and ratio of their velocities) the set of these positions and ratios of velocities will be everywhere dense in the (three parameter) set of all possibilities. It is the situation considered obvious in statistical mechanics. On the other hand, no one has any idea about what these $\frac{m'}{m}$ are, nobody can exhibit a single one! It is known that there exist periodic solutions, i.e. stable states; but they are (to this day) of the replicated type where one of the two masses remains fixed for a certain time. Still worse: for these periodic stable states, even if they exist in infinite number, it is not known how to asymptotically estimate their energy levels. At the very most it is known that they don't grow exponentially.

Of course it is known, thanks to (XI.5.1), that when the angle α defined by $\tan \alpha = \frac{m'}{m}$ is rational to π, then there are stable states and the energy levels grow quadratically. But it seems to us that these mass ratios and angles rational to π don't show any physical peculiarity that would interest a physicist; but the results of Section 5 at least remain very interesting for the geometer.

XI.8. Concave billiards, hyperbolic billiards

In order to study the motions of statistical mechanics, Sinai (inspired by Anosov) and other mathematicians have introduced billiard models for describing motions with impacts, typically molecules in a gaseous state. The reality is that of a three dimensional cube and of the motion of particles (with impacts) in large number (of the order of Avogadro's number). The true motion space is thus a domain of $\mathbb{R}^N$, where N is of order 10^{23}. The image to visualize is that of a cube with tube-like regions removed to reflect the fact that the particles have a positive diameter and that there is no interpenetration. Such a model is thus not mathematically describable in a simple way. This is what has led physicists to first test the simplest case that is analogous to the theoretical situation. The idea of Sinaii is that these complicated spaces, with a flat boundary but with obstacles to be removed on the interior, are in fact very similar to objects with negative curvature, and this from the case of billiards in a square with a barrier formed on the inside by a small disk. The idea is somehow to remove the disk from the square and to transform the square into a domain D of the plane for which the boundary is strictly **concave**.

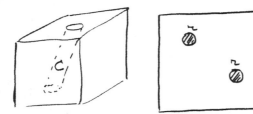

Fig. XI.8.1. Two particles of radius r in a square are represented in a cube of dimension 4 with the condition $\{(x, y, z, t) : (z - x)^2 + (t - y)^2 \geq 4r^2\}$, which amounts to removing a sort of tube in the cube

We see that the boundary ∂D will have to have some singular points. The hypothesis is that it is made up of a number of plane curves with strictly negative curvature, joined with different tangents. We thus consider the billiard trajectories in D. The dynamical system considered is that of the phase space of D, thus in three dimensions (a point and a unit vector at this point). The invariant measure is always — as in Sect. XI.2.C — the product of that of the plane with the unit circle. It is elementary that it is still preserved by reflection at the boundary. We see from the figure that concavity introduces a nonzero dispersion; the intuitive idea is then that such a billiard is going to be ergodic because successive reflections will disperse the trajectories (with their directions) more and more. All the points and all the directions are going to be approximated by a trajectory (except for periodic motions, if there are any). Note also that pieces are permitted on the interior, but then they must be **convex**, as seen from their interior!

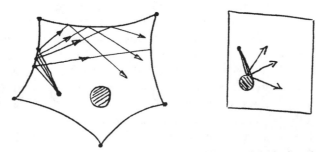

Fig. XI.8.2. How concavity forces dispersion, and how an obstacle does the same

A rigorous proof seems simple enough: ergodicity — see Sect. XI.XYZ — is assured as soon as we know that there is no subset of positive measure that is invariant under the flow; but the dispersion clearly prohibits such a set. The form remains delicate because once again, as with polygonal billiards, the difficulty comes with trajectories arriving at singular points (at vertices) of our boundary. Finally, we know practically everything about these billiards; see Sect. XI.XYZ for the language:

(XI.8.1) *Concave billiards are ergodic. They have a positive Liouville entropy h, an infinity of periodic trajectories for which the counting function is exponential and given precisely by the entropy:* $\lim_{L\to\infty} \frac{\log(FC(L))}{L} = h$.

This result goes back to Sinai in 1970; a complete proof is found in theory in Katok and Strelcyn (1986), but all the experts — out of politeness — say it is unreadable, which indicates that a really complete proof awaits the authors. On the other hand, we find an excellent presentation of the ideas in Chap. 5 of Tabachnikov (1995). Those who like the unity of mathematics will appreciate finding here the continued fractions of Sect. IX.1: the successive reflections at parts of the boundary yield dispersions that are evaluated in terms of continued fractions for which the terms are connected with the curvature of the boundary curves. The language belongs also to the discipline called *geometric optics*; it arises naturally here since the billiard trajectories are also those of light in a domain; here there are but mirrors (pieces of the boundary) and no lenses.

♦

And so concave billiards are perfectly understood and thus models of choice for physicists. But even in two dimensions their study is inadequate for having sufficiently numerous models for studying certain problems of mathematical physics, in particular those linked with **chaos** in two-parameter dynamical systems. There is also the very subtle question of connections between the counting function of the periodic trajectories and the spectrum of the domain for the Laplacian; see e.g. p. 73 of Berger (1999) or VI.7 of Berger (2003). Physicists are very interested in these connections; this has to do with the discipline called *semi-classical mechanics*, which studies what happens when the Planck constant tends to zero: do we, in the limit, get classical mechanics from quantum mechanics? The billiard trajectories are those of classical mechanics; the vibrations and their frequencies are those of quantum mechanics.

Well, concave billiards lend themselves badly to explicit calculations. The idea for remedying this is to translate the existence of negative curvature (motivated above by the nature of impact in statistical mechanics), not by the negativity of the curvature of the boundary curves, but on the interior or the domain itself. Now, we possess just the tool for this: it's hyperbolic geometry; see Sect. II.4. The simplest object imaginable is a triangle in the hyperbolic plane, the boundary made up of three lines in the hyperbolic geometry. The necessary **dispersion** is here assured by the behavior of lines in the hyperbolic plane rather than by reflections at the boundary: lines issuing from a point diverge exponentially.

If in particular we take triangles that tile the hyperbolic plane by reflections about their sides we can make explicit calculations of practically all the objects attached to such a triangle. We have mentioned, without any detail, the word *chaos*, certainly

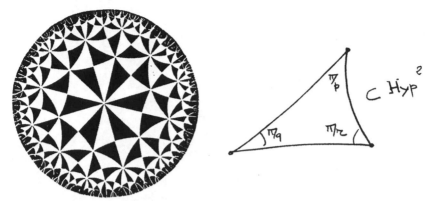

Fig. XI.8.3. Tiling of the hyperbolic plane

much in fashion. It is the moment to point out that we now know, after having practically forgotten during the course of almost a century, that the first study of chaos goes back to Hadamard (1898): motivated by his study of geodesic flow on surfaces with negative curvature — see Sect. XII.6.A) — he studied the very concrete case of billiards in a triangle of the hyperbolic plane. For billiards in different geometries, see Tabachnikov (2002). For the polygons in the hyperbolic plane, see Dogru and Tabachnikov (2002).

The semi-classical point of view in billiards wasn't studied in Tabachnikov (1995); we refer readers to the stimulating text Sarnak (1995) and — for more references — Chap. 4 of Berger (1999) and Chaps. 6 and 7 of Berger (2003).

We now leave polygons and concave boundaries to study billiards of a nature which are in a way opposite, specifically those with convex boundary. We will see that numerous surprises await us, that there are plenty of traps, and that they open the way for many interesting results, for the most part very recent. We follow an order that seems natural to us.

XI.9. Circles and ellipses

Circular billiards is indeed a banal object of study, but nonetheless we visualize it rapidly. Each trajectory is well defined by an angle α, and its nature depends on whether the ratio $\frac{\alpha}{\pi}$ is rational or irrational. In the rational case we obtain a regular polygon (a star polygon in general), otherwise a trajectory that is everywhere dense, not in the entire disk, but only in a ring. Note that, as in the square for each slope, here each angle provides a well determined type of trajectory: all are derived from one among them by rotations about the origin. To learn more about the darkening of rings, we need to know the development of $\frac{\alpha}{\pi}$ as a continued fraction, but things are not as simple as in the square, as the figures show.

We do not know any reference for such a study, but it presents no major difficulty for readers who may be interested in pursuing it.

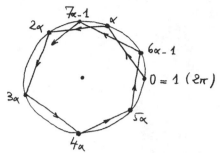

Fig. XI.9.1. The circular billiard: theory

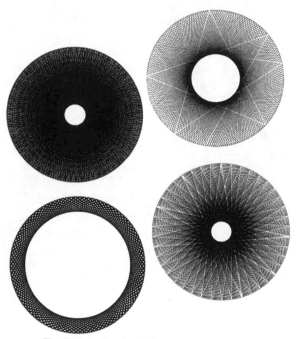

Fig. XI.9.2. Circular billiards: some examples

♦

The billiard problem on the interior of an ellipse is completely resolved thanks to what we know from Sect. IV.2 about the family of homofocal conics — see Eq. IV.2(∗) — and about Poncelet's theorem in Sect. IV.8. At the outset we have the following phenomenon: the tangent to an ellipse at one of its points is the exterior bisectrix of lines joining the point to the two foci, and more generally the two tangents issuing from a point of the ellipse making equal angles with the lines joining this point to the two foci.

And so a trajectory that starts on the interior of the ellipse will remain tangent to the ellipse homofocal to the given ellipse and tangent to this trajectory at the start:

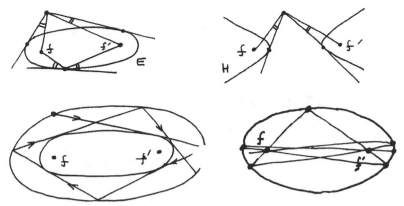

Fig. XI.9.3. Elliptical billiards

attention, this only if the trajectory is exterior to the interval joining the two foci. Otherwise it is a homofocal hyperbola to which the trajectory remains constantly tangent. Note the case of a trajectory that passes through a focus: it will oscillate while alternately passing through these foci. Readers will show that the trajectory will more and more approach the major axis of the ellipse no matter where it starts and also that it crashes when we go back in time, and can be compared with what happens for the umbilics of ellipsoids in Sect. XII.6.C. In the sequel we eliminate this probability zero case. In the language of optical geometry, the ellipse to which the trajectory remains tangent is called a *caustic* because the light rays accumulate in a certain sense (and can even burn!).

This does not suffice for knowing what will come of these trajectories: do they close on themselves or are they everywhere dense in the ring (or the double domain) contained between the two ellipses (or the ellipse and the hyperbola)? As in the cases of the circle and the square, we also have here a perfect dichotomy, but to prove it we need to appeal to Poncelet's theorem of Sect. IV.8: if a billiard trajectory (between two ellipses or an ellipse and a hyperbola) closes (is periodic), then it is the same for all trajectories which are tangent to such a conic (ellipse or hyperbola). and simultaneously that if for one of these conics (which are very particular cases of the caustics of Sect. 10.A) the trajectory doesn't close, then it never closes. Thus in conclusion: we have periodic trajectories or trajectories that are everywhere dense (but not in an equal manner) in the rings (of two possible types, topologically one disk or two). Thus, as in the case of the circle, the trajectories group themselves into one-parameter families of trajectories of the same dynamic nature.

At least three natural questions arise right away to the mathematizing mind. The first is knowing how to **calculate** when the trajectory is periodic (and then even to calculate its period) or when it never closes. To do this we need to parametrize the trajectories. But for this we have Eq. IV.2.(∗): the parameter λ is exactly what we need, because we have seen that thanks to Poncelet's theorem the type of trajectory

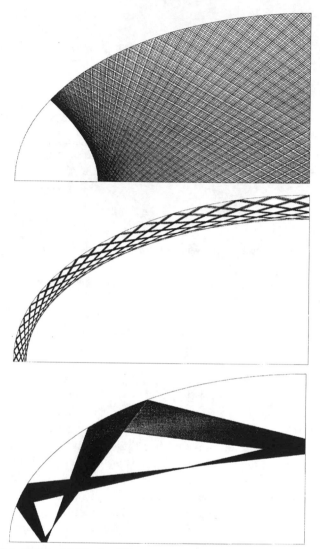

Fig. XI.9.4. Elliptical billiards, different possible cases. To avoid having too big
a figure, trajectories have been folded by reflection through the axes of symmetry

is the same for any given λ. But this problem of periodic closure (more precisely, of determining the period) was precisely treated in Sect. VI.8 in the very general setting of two conics with arbitrary equations. There is thus a calculational answer and thus there exists a truly simple formula, although we confess that we know of no reference where the calculation is actually carried out as a function of the only parameters $\{a, b; \lambda\}$; for billiards and Poncelet's theorem see 4.3 of Tabachnikov (1995).

The second question is that of the study of the darkening in the "rings". Already for the circle the question doesn't seem to have been studied in detail; here too we will need to appeal to the theory of continued fractions, but applied in the context of parameterizations by elliptic functions.

A third question is of the type that excites mathematicians and scarcely interests physicists at all. In the case of the circle, it is filled by the family of caustics composed of concentric circles. In the case of the ellipse, it is filled by the homofocal conics (for hyperbolas, by pieces of hyperbolas only), but in each case everything is filled by a one-parameter family of caustics. Here we thus call *caustic* a closed curve, situated on the interior of the billiards domain considered, and such that each tangent to this curve remains tangent to it under whatever reflections to which it is subjected. We will soon see that certain convex billiards admit an infinity of caustics, but they never form families of one real parameter. The question is thus of knowing if the ellipses are the only plane curves to admit an infinity of caustics depending continuously on one real parameter. The problem is still not resolved; it has been the subject of several works, which presently seem insufficient for drawing a conclusion without very strong additional hypotheses; see for example 2.4 of Tabachnikov (1995).

XI.10. General convex billiards

XI.10.A. Very smooth and strictly convex billiards: caustics

We study the simplest but not too special billiards, specifically those where the domain is bounded by a strictly convex curve. We even want a good amount of smoothness, that is to say that the boundary curve is strongly differentiable; we call C^k its class; see Sect. V.XYZ. In 1973 Lazutkin obtained the spectacular result that for strictly convex billiards of class at least 558 (C^{558}: believe it or not!) there is always a family of caustics. They are in a neighborhood close to the boundary, but sufficiently numerous in the following sense: they are parameterized by a set of real numbers of Cantor type, but of positive measure. The dynamic situation can thus be succinctly described.

We stand at a point on the boundary and look at what happens for directions sufficiently close to tangential. For the other directions, no one knows what happens in general; perhaps a situation of this nature isn't susceptible to exact analysis. In this set of directions, there are those that yield caustics; see the figure. The directions without caustics furnish **chaotic** trajectories that oscillate between caustics, but don't have envelopes (shaded region in the figure).

In 1989 R. Douady showed that the result of Lazutkin remains valid for a class of order only $6 + \varepsilon$; see Douady (1982) (unfortunately this text is difficult to find). The idea of Lazutkin is to apply the theory of KAM tori to the case of billiards, but for which reflection at the boundary complicates mathematical modeling. We find these tori in Sect. XII.5.B; see the book Lazutkin (1993). The best differentiability order isn't known; is it 4? We will soon see that in any case three derivatives by themselves don't suffice. We only add that, for the above caustics, the trajectories are never periodic. Now, these billiards have periodic trajectories, infinite in number; see 10.D. They are not well understood in the general case.

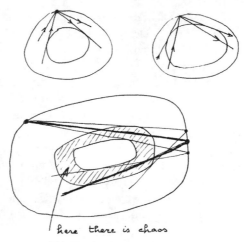

here there is chaos

Fig. XI.10.1. Lazutkin's caustics

We can give an idea of the proof, because we have already encountered the necessary tools: Fourier series, for studying the closed curves, and continued fractions. The difficult part is to show that we can globalize an easier result that concerns the linearization (close to the boundary) of the equation we need to solve to obtain a caustic. We thus work with functions on the circle, represented by Fourier series of period 2π on the line, denoting the Fourier parameter by x. We can parameterize the curve in such a way that the reflection of the trajectories is given by a function f. We will have a caustic (at the least an infinitesimally close version) corresponding on the circle to a rotation through an angle equal to α if we can find a periodic function g that is a solution of the equation:

$$f(x) = g(x) - g(e^{i\alpha}.x)$$

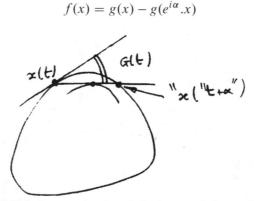

Fig. XI.10.2. An approximation of what happens during a reflection

If the Fourier series expansions are $f = \sum a_n e^{inx}$ and $g = \sum b_n e^{inx}$, the a_n being known, the unknown b_n are given by the formula $b_n = \frac{a_n}{1-e^{in\alpha}}$. We see that the Fourier series with coefficients b_n converges since we know that a_n, thus also

$\frac{a_n}{1-e^{in\alpha}}$, converges to zero fast enough. This risks not being the case if certain among the $n\alpha$ come too close to being integer multiples of 2π.

Although not our primary interest, for billiards we have already encountered frequencies and spectra; see e.g. Sects. IX.4 and 2.B above. In the present case, by 1973 Lazutkin was capable of finding connections between the caustics and certain parts of the billiard spectrum considered as a vibrating membrane. The problem is treated in detail in the second part of Lazutkin (1993).

XI.10.B. Three strange phenomena

We are now going to see that if the differentiability is reduced and/or if strictness is removed for convexity, then things happen that we wouldn't have foreseen. The first phenomenon can be found in Halpern (1977). It arises in strictly convex billiards for which the boundary is thrice differentiable (but the third derivative is not continuous everywhere) and for which the billiard (i.e. light) trajectory **isn't defined** at certain points. This is to say that we can find a trajectory that reflects at points that get closer and closer and converge to a definite point of the curve. This trajectory will reach the point in a finite time but after an infinite number of reflections. This shows in some way that everything that was said until now about billiards was imprecise. So long as we only dealt with polygons, everything was fine, except at vertices. But even for strictly convex billiards we may have to exclude certain points — or rather certain trajectories — which are no longer well defined at the end of a certain time.

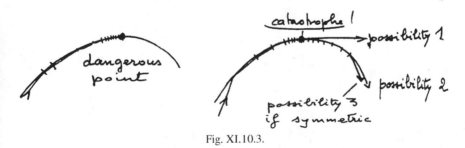

Fig. XI.10.3.

Such an example is not difficult to construct: we start with a circle, then from points $\{x_n\}$ on the circle indexed by positive integers and converging toward a point x of the circle. We trace the corresponding segments $\{x_n, x_{n+1}\}$, then the exterior bisectors of the angles that they form at these vertices. It can easily be shown, with the help of smoothing functions, that there exists a three times differentiable curve passing through the x_n and having tangents imposed at these points. It can be shown furthermore that these don't end up in class C^3 and that the third derivative is not even bounded.

Now the trajectory isn't defined afterwards, except at x. Certain humorists like to say that it flies off the billiard table at this moment; it is more serious to say that a natural extension would be to remain on the curve. Notice also that if the curve admits a Euclidean axis of symmetry about its normal at x, then the trajectory continued symmetrically is perfect. But in the general case we are forced to recognize that the trajectory is no longer really defined. In all that follows, when the billiards are not strictly convex or not at least C^3, we will thus exclude such trajectories when they exist. This is why it will be important to know, in Sect. XI.10.C to come, what happens for generic billiards.

The second surprising fact is of a different order, but it illustrates the falseness of the naive idea according to which convexity implies that, at the very least in the neighborhood of the boundary, the trajectories **turn** and this always in the same sense. However, the result of Sect. XI.10.A on caustics could have let us to suppose so, since the trajectories starting on the exterior of a caustic remain there. But in Mather (1982) this is proved: once we have convex billiards which may be of class C^2 but possess at least one point where the curvature of the boundary is zero, then for each $\varepsilon > 0$ there exist trajectories which make at least once an angle less than ε with the boundary and at least once (for the same trajectory!) an angle greater than $-\varepsilon$. Of course this implies that at some moment the trajectory will **reverse** course in some way; and this does not inhibit the billiard from being strictly convex in the sense of convexity alone. In fact Mather proved much more, i.e. that for any integer N (think very large) and the real ε (think very small) there exist trajectories that remain a distance less that ε from the boundary of the convex set, both for N bounces in a sense of traversing the boundary and subsequently, after this "turning", have again N bounces in the opposite sense: they skip (or glance) along the boundary, first in the one direction, then in the other.

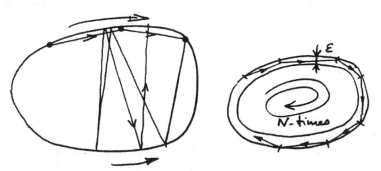

Fig. XI.10.4. On the right all bounces are *glancing* along the boundary

The proof uses the description given by Birkhoff in the 1930s for studying billiards (among other dynamical systems); see Sect. XII.5.D, where we encounter the Poincaré-Birkhoff annulus in connection with periodic geodesics). We give the ideas of the proof because it is extremely geometric, and very conceptual. The ideas are "anti-Lazutkin", "anti-caustic": the proof consists of showing two things: there can never exist a curve sufficiently close to the boundary to which a trajectory remains constantly tangent; this is a local result. The global result is by contradiction: if there is no ε-reversing, for all ε, then the existence of at least one caustic is implied.

The portion of elementary geometry of curves consists in this: we consider a caustic in a neighborhood of a point of our convex set C and let x_0 be a point where the curvature is zero, x_{-1} its antecedent and x_1 its successor. In the neighborhood of x_0 let y_{-1}, y_0, y_1 be three successive points of the boundary curve of C. By definition, along the boundary ∂C, when y_{-1} is displaced toward x_0, then y_0 and y_1 are displaced in the same sense. Unfortunately, since the curvature is zero and since at y_0 there is a reflection (which amounts to the fact that the sum of the distances to y_{-1} and y_1 is minimum and that the curvature is zero at x_0), it is implied to the contrary that y_0 regresses when y_{-1} progresses: it's the contradiction sought. The complete proof contains a calculation of the second derivatives of the function giving the distance of the two points from ∂C.

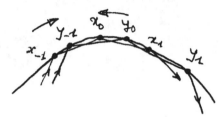

Fig. XI.10.5. The idea of the proof of Mather's result

We now need to show that if for each sufficiently small ε no trajectory reverses itself, then there exists a caustic. Everything that is needed can be found in work of Birkhoff from the 1920s. The starting point is description of the dynamical system of our billiards in the convex set C by an object with only two parameters. This is possible because a trajectory is well defined by a point of the boundary and an angle of departure (if it isn't zero) We thus introduce the **ring** $A = \partial C \times]-1, +1[$ made up of pairs (x, u), where x is a point of ∂C and $u = \cos \theta$, where θ denotes the angle of the trajectory with the boundary. This ring in indispensable in dynamical systems and will be encountered again in Sect. XII.5.D. The billiard trajectory thus furnishes, by its reflections, a mapping $f : A \to A$ that is bijective. Nothing can be done with it without an **invariant**. But precisely for this reason it is trivial to see that the measure (area) $ds \cdot du$ (where ds is the measure of arc on the boundary curve ∂C) on the ring is preserved by f. A theorem of Birkhoff, improved by Mather, shows this: if for each ε no trajectory is reversing, then we obtain in the ring A a subset $\mathcal{E}$ not containing an open set about $\partial C \times \{-1\}$ (which is one of the two elements of the boundary completion of A for the "null" angles). The fact that we have a ring and that measure is preserved then implies that the boundary of $\mathcal{E}$ contains a curve, clearly preserved by f. Such a curve is exactly a family of lines $\{x, \gamma(x)\}$ and the envelope is the caustic sought.

In fact Mather found this result while trying to understand what happens with the Bunimovich **stadium**. What's this about? It's our third strange phenomenon, strange at the very least when it appeared in 1979. We saw above that concave billiards are ergodic, the "natural" reason being that the concavity implies successive dispersions which force ergodicity: "everything turns more and more and this happens everywhere". The common belief was that convexity to the contrary actually prohibited ergodicity — until Bunimovich announced in 1979 that the football stadium of suitable dimensions yields ergodic billiards, in phase space of course:

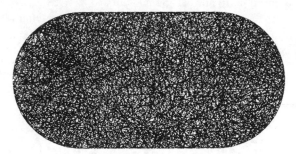

Fig. XI.10.6. The Bunimovich stadium

Moreover, computer graphics show that things move very fast, although this is not a proof. Indeed, Lazutkin's proof of the existence of caustics showed that billiards on a domain with a very smooth boundary of positive curvature can never be ergodic. It is this point that Bunimovich's proof explains to the contrary. His example was very special and his proof, to say the least, obscure. Subsequently several authors simultaneously explained Bunimovich's proof and gave more systematic, and therefore numerous, examples of convex billiards that are ergodic. For readers who like open problems, we note that to our knowledge all the convex sets found until now having ergodicity are never either strictly convex or smooth. Here now is the very geometric idea that appeared in Wojtkowski (1986) for proving ergodicity in certain cases. We work in the context of geometric optics, a discipline that will permit us to show that when the boundary curve has a radius of curvature r that satisfies (as a function of the arc length s, see Sect. V.6) the condition $\frac{d^2 r}{ds^2} \leq 0$, then **two** successive reflections bring **dispersion**. This condition on the curvature is difficult to visualize if we don't know what is going to follow. It astonished us quite a bit — we had never seen the like before. For the ellipse, see the figure below. We also see there an ergodic stadium of a different type.

Here now is why (looking above all at the figures) two reflections provide divergence: we look at how parallel rays coming from infinity get focused at a point by a curved mirror. The answer is that focusing always takes place on the circle tangent to the curve of radius **half** the radius of curvature (i.e., half the radius of the osculating circle, for which the center corresponds to focusing of the normals): see the

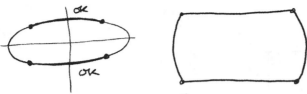

Fig. XI.10.7. The parts of the ellipse where $\frac{d^2r}{ds^2} \leqslant 0$ (*darkened parts*); another

ergodic stadium where the curves are piecewise $\frac{d^2r}{ds^2} \leqslant 0$

hashed portions. From this it can be deduced that there will be dispersion after two reflections if pairwise these circles of radius half the radius of curvature intersect **on the exterior** of the line joining the two reflection points. But it is easy to see that this happens exactly when we have $\frac{d^2r}{ds^2} \leqslant 0$.

To construct ergodic convex billiards we compose boundary curves by piecing together portions of the preceding type. More details and more examples can be found in Tabachnikov (1995); see also Ghomi (2004).

XI.10.C. Generic billiards

We have just encountered three phenomena, three particular types of billiards. But this doesn't tell us anything at all about what happens with "general" billiards. We want then at all costs to know about *generic* convex billiards. We first need to give a meaning to the word. We know that convex sets in the Euclidean plane form a set $\mathcal{C}$ that has a canonical metric, i.e. the one given by the Hausdorff metric; see Sect. VII.7. When we say that some property holds for *generic* billiards, we mean that this property is satisfied by the convex sets belonging to a G_δ-dense (see Sect. VII.13.D) subset of $\mathcal{C}$. We find in Gruber (1990) and 4.10 of Gruber and Wills (1993) the following, obtained by applying results obtained above, whether directly for smooth billiards or by appropriate polygonal approximation.

(XI.10.1) *For convex generic billiards:*

i)   *there does not exist a trajectory of Halpern type which terminates at a point after infinitely many reflections in a finite time;*

ii) *no caustics exist;*

iii) *most (this involves a suitable G_δ-dense set) trajectories are dense in the phase space;*

iv) *most trajectories are of the type that reverse in the Mather fashion, and this for each ε.*

It seems to us, however, that there is still much more study needed in order to know which convex billiards are ergodic in phase space; and more generally for studying generic behavior in greater detail.

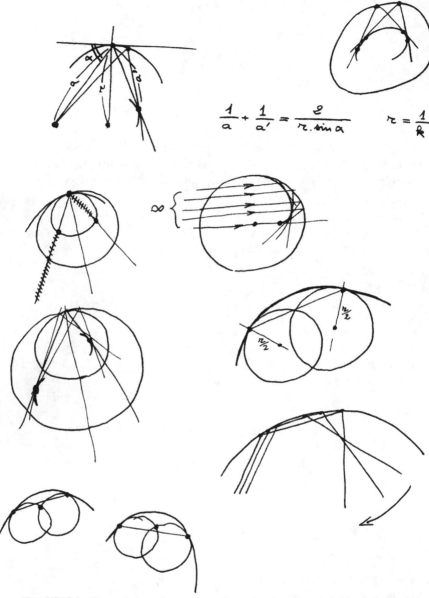

Fig. XI.10.8. These figures from geometric optics are intended for readers not familiar with the discipline for understanding the condition on the curvature discussed above. The parallel lines correspond to a point at infinity and provide a "focal image". For points at a finite distance, the corresponding images are indicated by envelopes of light rays. The essential relation is the condition of geometric optics for the distances a, a', the angle α and the radius of curvature $r = \frac{1}{k}$: $\frac{1}{a} + \frac{1}{a'} = \frac{2}{r \cdot \sin \alpha}$; there is indeed divergence if $\frac{d^2 r}{ds^2} \leqslant 0$.

XI.10.D. Periodic trajectories

It is very easy to see that smooth billiards have plenty of periodic trajectories, in contrast to polygonal billiards (which aren't smooth!). On the other hand, counting them is much more subtle. It was Birkhoff who, around 1920, proposed the following method: we seek, among the convex polygons with q sides, with q fixed, that have their q vertices $m_1, \ldots, m_q$ on the boundary, those that have the largest perimeter. Pay good attention that each vertex m_i must be *between* m_{i-1} and m_{i+1}.

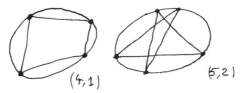

Fig. XI.10.9. A trajectory of type $(4; 1)$; a trajectory of type $(5; 2)$

It's easy to see that there actually exists a polygon for which the perimeter is the largest possible. Then it is necessarily a billiard trajectory. We can see this as follows: let m_1 and m_2 be fixed; we use the fact that $m_1 m_2 + m_2 m_3$ is maximum. Then the tangent at m_2 to the convex set must make equal angles with the lines $m_2 m_1$ and $m_2 m_3$ (mirror principle). In taking larger and larger q's, we see that there are an infinite number of periodic trajectories and the limit of their lengths equals the length L of the boundary curve. We thus can't define a counting function FC(L) in a naive sense.

But more generally, for each integer p in fact, there exists an infinity of periodic trajectories whose lengths have limit pL. It suffices to consider star polygons with q sides that turn about the convex set p times (p and q must be relatively prime). However, the periodic trajectories remain still more interesting in smooth convex sets. In Sect. 2.B above we sketched a motivation: the set of lengths of these trajectories has a connection (very subtle and still poorly understood) with the vibrations of a membrane with the shape of the convex set, a membrane vibrating for instance with fixed boundary. An asymptotic estimate is known for the counting function of the frequencies: if $N(\lambda)$ is the number of frequencies less than or equal to λ then $N(\lambda) \approx \lambda^2 \cdot$ (area of the convex set) as λ tends toward infinity.

A brief history of the discovery of this estimate furthermore shows that prophecy is no easier in mathematics than anywhere else. Recall first that we previously proved this estimate for the square and pointed out that physicists are interested in it. It is thus that in 1910 the great physicist Lorentz gave a lecture at the University of Göttingen, after which, motivated by physical considerations, he posed the problem of knowing if there is a good estimate

$$N(\lambda) \approx \lambda^2 \cdot \text{(area of the vibrating membrane)}$$

(with λ tending toward infinity). Incidentally, the funds that supported the lecture also provided prize money for the solution of Fermat's theorem! At the end of the lecture, it is reported that Hilbert remarked that the problem was very nice, but so difficult that it wouldn't be solved in his lifetime. But Hermann Weyl solved it two years later, at age 27!

But the function itself clearly jumps, since it is integer-valued. The result then says that this step function is approximated by the curve $N \mapsto \mathrm{area}(-1/2)\sqrt{N}$.

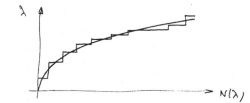

Fig. XI.10.10. The step function and its smooth approximation

It is important (as in the circle problem encountered in Sect. 2.B and in Sect. IX.2) to do better. Today we know only that the periodic trajectories can be used for giving oscillatory compensatory terms for the jumps. But much more remains to be done. We have already seen it previously, in the case of the square. Counting the pairs (p, q) of integers such that $\pi(p^2 + q^2) \leqslant \lambda^2$ is very difficult, even asymptotically. Thus we would like all the more to be able to count this number $G(\lambda)$ exactly! It is fascinating to realize that an apparently more delicate problem, that of counting the number of primes less than or equal to a given N is much better known. We even have, in a sense, explicit formulas at our disposal. For this problem, see Zagier (1977). In any case, we need to find an object to replace the counting function which, as we have seen above, has no meaning.

Mathematically, the solution is simple. Appeal is made to the *theory of distributions*. It is known that the Dirac distribution at the point a of the real line is a generalized function called the Dirac distribution: it equals zero everywhere except at a, where it is infinite. It is *denoted* $\delta(a)$. In a sense, it is the limit of functions which are concentrated more and more at a and whose graph encompasses an area always equal to 1.

We can evidently add such generalized functions and compute with them. With the set of lengths $L_1, L_2, \dots, L_n, \dots$ of the periodic trajectories we associate the infinite sum distribution

$$\delta(L_1) + \delta(L_2) + \cdots + \delta(L_n) + \cdots .$$

We can then study this distribution and there are good results on connections with different elements of the convex set; see the references below. Recall the phenomenon which seems unique to ellipses: we saw that periodic trajectories in polygons always come in *bands* (of the same length of course). In contrast, the periodic trajectories of smooth billiards (constructed above by Birkhoff's method) are in general isolated (except for elliptical billiards).

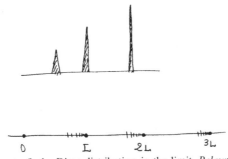

Fig. XI.10.11. How to find a Dirac distribution in the limit. *Below:* "vague" shape of the spectra of the λ

There are finally some very strong connections between three things: the lengths of periodic trajectories, the eigenvalues of the Laplacian (vibrational frequencies) and caustics. These connections exist in the more general context of Riemannian manifolds and we have seen that these are important for theoretical physics. Other than Lazutkin (1993), for the Riemannian manifold context, see Berger (1999) or the more detailed version Berger (2003).

In reality numerous dynamical systems are neither integrable nor chaotic; "mushroom billiards" have been introduced to model these intermediate cases, where along the ellipse the situation is integrable but becomes chaotic when the stem is entered; see the very recent (Porter and Lansel, 2006).

XI.10.E. Billiards and duality

The notion that is dual to billiards is that of exterior billiards. A convex set C having been given, a ray emitted from a point x_0 exterior to C is a half line "tangent" to C. The reflected point x_1 is the symmetry of x_0 with respect to the point of contact. This notion was proposed by Bernhard Neumann in the 1950s and was presented by Moser as a toy model for planetary motion; see Moser (1978).

Around 1950, B.H. Neuman asked whether all orbits of an external billiard are bounded, not ever able to escape to infinity.

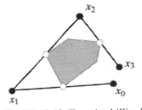

Fig. XI.10.12. Exterior billiard

In response, Moser outlined a proof of this: if the boundary of the convex set under consideration is a curve of class at least C^6, this billiard possesses a family

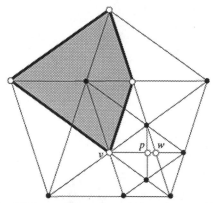

Fig. XI.10.13. Richard Schwartz's counterexample: a Penrose *kite*. The orbit of the point *p* isn't bounded

of invariant curves, i.e. curves which somehow correspond to the dual of the notion of caustic in Sect. XI.10. The complete proof is in Douady (1982). Moreover, this family extends to infinity, whence the stability that was sought. The technique of the proof remains that of the tori of KAM.

After long years, Richard Schwartz has recently confirmed the conjecture that all its orbits are stable and can't escape to infinity; Schwartz's website for these "*outerbilliards*" and the direct connection with Penrose tiling: precisely the orbit that escapes to infinity is that defined by the point *p* of the figure. Finding here the aperiodic tilings of Sect. IX.9 is a nice example of those surprising similarities (in some cases we can say fusion, coalescence) in the mathematical universe.

XI.11. Billiards in higher dimensions

Billiards in Euclidean spaces of dimension three or more are essentially in an infantile state. In brief: practically nothing is known in general. Here first is what is known, some very sparse things, although the table of contents of Tabachnikov (1995) is eloquent.

The only well-understood case is that of hyperbolic billiards, whether for polyhedra in hyperbolic spaces or for domains with concave boundary in Euclidean spaces. In the Euclidean case this means that the boundary is made up of pieces of surfaces (hypersurfaces in dimensions greater than three) with negative second fundamental form. These pieces have thus to be joined by curves (surfaces of codimension two), and also by vertices, all finite in number. The precise description is rather vague. Moreover, the complete proof that these billiards are ergodic and replete with other ideal properties (for example, the growth of the number of periodic trajectories is exponential) remains as delicate as the planar case because of the boundary singularities; see Katok and Strelcyn (1986) and Sinai (1990). Readers interested in cosmology can read in Damour, De Buyl, Henneaux, and Schomblond (2002) about "Einstein billiards" in general relativity.

◆

The polyhedral case (polytopes more generally) is a complete mystery. Exceptional is the "trivial" case of polyhedra that tile space by reflection at the faces, let us say parallelepipeds; in fact, the unfolding technique is always applicable. But such polyhedra are rarities; but this isn't to say that they aren't interesting! The parallelepipeds in three dimensions are studied like the rectangle in two, by reference to a different unit. This means that we study, for a vector (a, b, c), the vectors formed by its fractional parts $(\{na\}, \{nb\}, \{nc\})$ when the integer n increases. In mechanics, this study comes up immediately in systems with three periods. We have seen, however (for example in Sect. IX.1) that this multidimensional theory of continued fractions isn't complete. For example, an analogue of the golden ratio is unknown. There are surprising results in the rectangular case on the complexity of the sequence of faces encountered ; see Baryshnikov (1995) and Hubert (1995).

Readers who find the expression "complete mystery" excessive should realize that today no one has the slightest idea of what happens in a regular tetrahedron, the fundamentally simplest polyhedron that there is. Is there ergodicity or not? − or transitivity for most trajectories: in space or in phase space? Nor should we neglect to notice that the dihedral angles of the regular tetrahedron are irrational to π. It might be objected that it should be very easy to run computer simulations. But this turns out not to be the case, not at least in the present state of the problem. One of the reasons is that the trajectories must be coded most by four parameters at least, i.e. a point of the boundary (two parameters) and the direction at this point (two parameters). Even by taking two or three dimensional projections, the study of dynamical systems with four parameters still seems beyond reach.

However, several mathematicians have found periodic orbits in regular tetrahedra and in other polyhedra, but it is open whether all polyhedra have them, and we saw this ignorance already in the case of plane triangles. Readers can consult Stenman (1975) and Gardner (1984). But we are still far from a study with any generality. A forerunner is Stenman (1975).

As for periodic trajectories, the only reference we know is Katok (1987); it is proved there that, as in the planar case mentioned in Sect. XI.6, the counting function of the periodic trajectories of the convex polytope billiards (even assuming that such trajectories exist!) is always sub-exponential.

◆

It remains finally to mention a special result on caustics. We have seen that ellipses admit a continuous infinity of caustics, that fill their entire domain; then that there exists an important set of caustics for billiards with very smooth boundary with strictly positive curvature; and finally that points of zero curvature preclude caustics locally. In Sect. IV.10 we saw that ellipsoids have the same property as ellipses: their interior is completely filled by caustic surfaces, which are homofocal quadrics. Every tangent line to a homofocal quadric F associated with a fixed ellipsoid E, which reflects from E when encountered, yielding a new line which is again tangent to F.

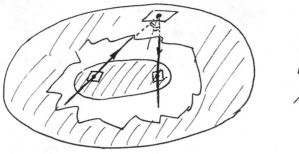

Fig. XI.11.1.

The question has remained open for some time whether Lazutkin's result generalizes to three dimensions or more, specifically of knowing whether sufficiently numerous caustics exist for domains that have smooth boundary and are strictly convex. The answer is negative and this already locally. Here is the philosophy. In the plane, we can easily locally fabricate caustics at points of positive curvature of the boundary (take a one-parameter family of lines and the family generated by the reflections of these). The reason is that tangents to a curve form a one-parameter family, and each one-parameter family has an *envelope*, and thus is the family of tangents to a curve. On the other hand the tangents to a surface in three-dimensional space form a three parameter family (two for the point of contact and one for the direction). Now in general, for a family of lines in space, even of only two parameters, there does not exist a surface to which they are all tangent. Thus the preceding local construction can't go through in general. In Sect. IV.10 we encountered homofocal quadrics, which is in fact the only case possible, the sole exception. For three parameters things become so special that it can be shown by elementary differential calculus that it can only happen — for the family of lines tangent to surface fragment F (smooth and arbitrarily small) and reflected as billiards (or light) from a surface fragment E — if both F and E are fragments of (global) quadrics E^* and F^* which are obviously members of the same family of homofocal quadrics. This local result can be found in Berger (1995) along with a global result, but without smoothness condition, in Gruber (1994).

XI.XYZ Concepts and language of dynamical systems

XI.XYZ.A. Ergodicity and mixing

A very good introduction to ergodic theory can found in Walters (1982), but it isn't complete. More complete are Arnold and Avez (1967), Mané (1987) (denser, but more complete for our Chap. XII). See also Katok and Hasselblatt (1995) (very complete and very recent, but difficult to read), Anosov and Arnold (1988), Sinaï (1976). Investing in the study of dynamical systems is a good thing for the geometer; the word *chaos* invades the geometric language in diverse situations.

By definition a (discrete) *dynamical system* is a triplet (X, μ, f) formed of a topological space X, a measure μ on X and a mapping $f : X \to X$ that preserves the measure μ, i.e. $f * \mu = \mu$ in classical notation, precisely $\mu(f^{-1}(x)) = \mu(x)$ for each measurable set x of X. In practice not only must X have finite measure, but we implicitly assume the measure normalized so that the total measure is unity (we are thus using the language of probability). In fact X is most often compact. We don't suppose that f is continuous or bijective (typical in the case of billiards, because of the corners at the boundary). However, we can essentially and morally think of f as bijective and continuous, that we are thus dealing with a homeomorphism. What is really of interest in ergodic theory is the behavior of the *iterates* f^k of f when the integer k becomes very large, let us say tends toward infinity.

But we also have encountered, for example in the case of billiards, *flows*, which means that (X, μ) is no longer equipped just with a unique mapping f, but with a one-parameter group $\{f(t)\}$ of homomorphisms $f(t)$ that preserve μ, the notion of a one-parameter group being simply the condition that when t traverses $\mathbb{R}$ we always have $f(0) = $ the identity on X, and that $f(s + t) = f(s) \circ f(t)$ (and thus $f(-t) = f^{-1}(t)$ as a consequence). Here we are interested in what happens as t tends toward infinity. Numerous works treat this subject, dealing first with the discrete case and then passing to the case of flows, neglecting sometimes to say that any flow yields a discrete case by setting $f = f(1)$. This might seem to be a specialization of the hypothesis, but since we are in fact interested in the iterates f^k with k tending toward infinity, there is no essential difference between the two situations. Incidentally, f^t is often written in place of $f(t)$. In what follows we will use both notations.

For example, in the case of billiards, we have also encountered the discrete case (by looking at the boundary, which replaces working with $f(1)$) as well as the case of flows. The following chapter will be entirely devoted to the geodesic flow $\{G^t\}$ on surfaces. For a surface S, whether in $\mathbb{R}^3$ or abstract (i.e. Riemannian), we consider its *unitary fiber* U(S), i.e. the set of all tangent vectors of unit length. The triple that by definition forms the *geodesic flow on* S will be composed of $X = $ U(S), of the canonical measure μ (called Liouville on U(S)) that will be defined in Chap. XII, and where $G^t(v)$ will be the vector tangent at time t to the geodesic of S which starts at time 0 from v as point of origin with speed one.

The simplest nontrivial dynamical system is the *Bernoulli shift* . This is the transformation $(x_n)_{n\in\mathbb{Z}} \mapsto (y_n)_{n\in\mathbb{Z}}$ with $y_n = x_{n+1}$, on the infinite product $\prod_{-\infty < n < \infty}$ (X, μ) of a measure space (X, μ). It is thus defined in trivial fashion and easy to study, thus applied in every situation where there is coding. In the case of $X = \{0, 1\}$ it is a game of heads or tails. Moreover, Bernoulli shifts are classified exactly by their metric entropy (see Sect. XI.XYZ.B below). Classifying evidently means identifying ergodic systems under the notion of *isomorphism*, which readers may have the pleasure of describing in detail.

◆

For the physicist two notions are interesting because natural, specifically that of *time average* and that of *spatial average*. We consider for $(X, \mu, f(t))$ arbitrary numerical functions $F : X \to \mathbb{R}$ (for the physicist these are the observables, because they are the sole entities that give access, information, with the aid of experimental measurement).

The time average *is* $F^*(x) = \lim_{t \to \infty} \frac{1}{t} \int_0^t F(f^s(x)) \, ds$, $x \in X$, *the* spatial average *being of course* $\int_X F(x) d\mu(x)$.

For their existence and more, we have a very general result of Birkhoff, which today probabilists consider as a special case of the strong law of large numbers:

(XI.XYZ.1) *For each measurable function* F *on* X *the average value* $F^*(x)$ *exists almost everywhere (i.e. except perhaps on a subset of measure zero) and moreover we have* $\int_X F^*(x).\mu = \int_X F(x).\mu$.

For discrete systems (X, μ, f) the average value is defined as

$$\frac{1}{n} \sum_{0 \leq i < n} F(f^n(x))$$

and we thus set

$$F^*(x) = \lim_{N \to \infty} \frac{1}{N} \sum_{k=1,\dots,N} F(f^k(x)).$$

It is fundamental for the physicist, at least for sufficiently prolongated measures, that the average value yield the spatial value, which is the desired result of the calculation, whereas the results of observation are time averages at a point. This is precisely what is assured with the notion of ergodic system:

(XI.XYZ.2) *A dynamical system is called* ergodic *if the time average* $F^*(x)$ *is equal to the spatial average for almost all* x. *An equivalent definition is that the only measurable subsets of* X *which are invariant under* f *are of measure zero or of total measure. A dynamical system is called* uniquely ergodic *if it possesses only one finite invariant measure.*

If we look at the *orbits*, i.e. the subset of X that are the $\{f^k(x); k \in \mathbb{Z}\}$ (or the $\{f^t(x) : t \in \mathbb{R}\}$), which are the objects that interest us for billiards and which for surfaces are the geodesics (seen "below" on the surface itself, for the "primitive" geometer), we see that ergodicity implies that almost all the nonperiodic trajectories are everywhere dense, and are moreover distributed in equal fashion. That is to say demagogically that we have *uniform and total chaos*. The proof consists simply in applying the definition (XI.XYZ.1) to the characteristic function $F = \chi_U$ of an arbitrary subset U of X, the equality of the spatial and temporal means says exactly that the time passed by the trajectory in U is proportional to the measure $\mu(U)$, QED.

There exists a still stronger notion than ergodicity, which is that of a *mixing flow*. We say that system (X, μ, f) is *mixing* if $\lim_{n \to \infty} \mu(f^n(A) \cap B) = \mu(A).\mu(B)$ for each pair of subsets A and B of X. It is a notion that says much: if X is a bowl of water and we add gin to it and mix with a spoon, after a reasonable time the gin will be regularly distributed throughout. The Bernoulli shifts are mixing, and convex billiards are likewise, because concave billiards are in fact isomorphic to Bernoulli shifts.

XI.XYZ.B. The various notions of entropy

The notion of **entropy** is difficult; it was discovered only in 1958 by Kolmogorov. Your author thinks that when a notion has appeared only recently, it is because a breakthrough was needed for the relevant concept to be discovered. For example, the notions of an abstract group, of a vector space and of a Euclidean axiomatic structure (defined as a real vector space endowed with a positive definite quadratic form. Heuristically the notion of entropy is trivial: it is the measure of the exponential factor which, in a dynamical system, measures the dispersion of the trajectories, or again the mixing rates, as time increases. On the other hand a formal definition is a much less simple matter. Moreover, there are several notions of entropy; and worse, the literature has endeavored to complicate the task of the reader. So-called *metric* entropy requires a measure and is in fact a notion belonging to the realm of measure, whereas so-called *topological* entropy requires a metric (although on a compact space, it is of little consequence which metric is specified because all metrics are essentially equivalent). Recently a third notion, that of *volumic* entropy, was introduced, but it only makes sense for spaces X that admit a "large" universal covering and are typically not at all compact. The references were given above. Here briefly are the definitions, where the limits are to be taken over the integers in the discrete case and over the reals in the case of flows. The definitions that we are going to give say a lot to a geometer; they are due to Bowen for topological entropy and to Brin-Katok for that of measure. They are taken from Mané (1987). In the particular case of Riemann surfaces we will see that there exists a trivial definition of entropy that is still more geometric: see Sect. XII.6.D.

In fact it is the *volumic entropy* which is by far the simplest to define. Here X is a Riemannian manifold (think of a surface) and X^* its universal covering, which we assume is endowed with inverse image Riemannian metric (see the very end of Sect. V.XYZ). We look at the volumes of the metric balls $B(p, R)$ in X^* and observe whether the quantity

$$\frac{1}{R} \log(\text{Vol}(B(p, R)))$$

has a limit when R goes to infinity, where p is a point of X^*. One proves that if X is of negative curvature, then this limit exists and is independent of the point p chosen.

We then denote this unique limit by $h_{vol}(X, g)$. Note that the geodesic flow is well hidden in this notion. It will be observed that, for defining the volumic entropy, there is no need of a Riemannian manifold: any metric and any measure suffice, as long as they are compatible, as for example in the mm-spaces of Gromov; see Chap. $3\frac{1}{2}$ of Gromov (1999) or an introductory exposition in Berger (2000a).

To define the *topological entropy* — the simplest to define after the volumic entropy of (X, f) — we place on the compact set X any metric d (assumed to yield a topology equivalent to that of X). The iterates f^n allow definition for each integer n an *associated metric* d_n defined by $d_n(x, y) = \sup_{0 \leq k \leq n} d(f^k(x), f^k(y))$. Now let $N_n(\varepsilon)$ be the minimum number of balls of radius ε for the metric d_n that are necessary for covering X and define the *topological entropy* h_{top} by:

$$h_{top}(X; f) = \lim_{\varepsilon \to 0} \lim_{n \to \infty} \frac{1}{n} \log(N_n(\varepsilon)).$$

We verify the independence with respect to the metric chosen.

Finally the most difficult notion is that of *measure-theoretic entropy* (called metric by folly), the first to appear historically. Moreover, it is practically impossible to calculate explicitly in a given situation. The definition historically uses the function $x \mapsto x. \log x$ which only becomes natural after more than a little reflection on the probability for the maximum likelihood for a game of drawing at random from among given objects and dividing into packets (it's a "memory matching game"). Here is the most recent definition, now classic. We again use an auxiliary metric, for which we can show (without very great difficulty) the ultimate independence. Here the measure is needed for defining a number $N_n(\varepsilon, \delta)$, which is the minimum number of balls of radius ε, for the distance d_n defined as above, necessary for being able to cover, instead of X in its entirety, a subset of X of measure $1 - \delta$. We then put:

$$h_\mu = h_{meas} = h_{met}(X; f) = \lim_{\delta \to 0} \lim_{\varepsilon \to 0} \lim_{n \to \infty} \frac{1}{n} \log(N_n(\varepsilon, \delta)).$$

In the case of the geodesic flow G on a surface, below we write it $h_{Liouville}$.

Of these two definitions, we see why the positivity of the measure-theoretic entropy is stronger than that of topological entropy. If the integral is positive, it is necessary that there exist a set of positive measure of points x that give an exponential divergence for f^n as n tends toward infinity.

The essential connection between the set-theoretic and topological entropies is called the *variational principle*; it is valid for every dynamical system and says that:

(XI.XYZ.3) *When we consider all invariant measures on the same* (X, f), *we always have* $h_{top} \geq h_{meas}$, *in fact more:* $h_{top} = \sup\{h_{meas}\}$ *when we consider all possible measures which are invariant under* f.

In the next chapter we will see other inequalities, determinations of connections with the positivity of an entropy, as well as the connection with the periodic trajectories in the case of geodesic flow on surfaces.

Bibliography

[B] Berger, M. (1987, 2009) *Geometry I,II.* Berlin/Heidelberg/New York: Springer

[BG] Berger, M., & Gostiaux, B. (1987) *Differential Geometry: Manifolds, Curves and Surfaces.* Berlin/Heidelberg/New York: Springer

Anosov, D., & Arnold, V. (Eds.). (1988). *Dynamical systems I.* Berlin/Heidelberg/New York: Springer

Arnold, V., & Avez, A. (1967). *Problèmes ergodiques de la mécanique classique.* Paris: Gauthier-Villars

Arnoux, P. (1988). Ergodicité générique des billards polygonaux, Séminaire Bourbaki 1987–1988. In *Astérique 161–162,* Société mathématique de France, 203–222

Baryshnikov, Y. (1995). Complexity of trajectories in rectangular billiards. *Communications in Mathematical Physics, 174,* 43–56

Berger, M. (1991). *Les billiards mathématiques. Pour la Science 163,* mai 1991, 76–85

Berger, M. (1995). Seules les quadriques admettent des caustiques. *Bulletin de la Société MathÈmatique de France, 123,* 107–116

Berger, M. (1999). *Riemannian geometry during the second half of the twentieth century.* Providence: American Mathematical Society

Berger, M. (2000a). Encounter with a geometer I, II. *Notices of the American Mathematical Society, 47*(2), 47(3), 183–194, 326–340

Berger, M. (2003). *A panoramic introduction to Riemannian geometry.* Berlin/Heidelberg/New York: Springer

Boshernitzan, M., Galperin, G., Krüger, T., & Troubetzkoy, S. (1998). Periodic billiard orbits are dense in rational polygons. *Transactions of the American Mathematical Society, 350,* 3523–3535

Cipran, B., Hanson, R., & Kolan, A. (1995). Periodic trajectories in right triangle billiards. *Physical Review, E52*(3), 2066–2071

Damour, T., De Buyl, S., Henneaux, M., & Schomblond, C. (2002). Einstein billiards and overextensions of finite-dimensional simple Lie algebras. *Journal of High Energy Physics, 8,* 2002

Dogru, F., & Tabachnikov, S. (2002). On polygonal dual billiards in the hyperbolic plane. *Regular and Chaotic Dynamics, 8,* 67–82

Douady, R. (1982). Applications du théorème des tores invariants. *Thèse Paris VII 1982*

Eskin, A., & Masur, H. (2001). Asymptotic formulas on flat surfaces. *Ergodic Theory and Dynamical Systems, 21,* 443–478

Galperin, G. (1983). Non-periodic and not everywhere dense billiard trajectories in convex polygons and polyhedrons. *Commentarii Mathematici Helvetici, 91,* 187–211

Galperin, G., & Zemlyakov, A. (1990). *Mathematical billiards (in Russian)*

Gardner, M. (1984). Bounding balls in polygons and polyhedrons. In Gardner, M. (Ed.), *The sixth book of mathematical games from Scientific American* (pp. 29–38, 211–214). Chicago: University of Chicago Press

Ghomi, M. (2004). Shortest periodic billiard trajectories in convex bodies. *Geometric and Functional Analysis, 14,* 295–302

Gromov, M. (1999a). *Metric structures for Riemannian and Non-Riemannian spaces,* J. Lafontaine & P. Pansu (Eds.), Basel: Birkhäuser

Gruber, P. (1990). Convex billiards. *Geometriae Dedicata, 33,* 205–226

Gruber, P. (1994). Only ellipsoids have caustics. *Mathematische Annalen, 303,* 185–194

Gruber, P., & Wills, J. (Eds.). (1993). *Handbook of convex geometry.* Amsterdam: North-Holland

Gutkin, E. (1984). Billiards on almost integrable polyhedral surfaces. *Ergodic Theory and Dynamical Systems, 4,* 569–584

Gutkin, E. (1996). Billiards in polygons: survey of recent results. *Journal of Statistical Physics, 83,* 7–26

Gutkin, E. (2003). Extremal triangles for a triple of concentric circles. *Preprint IHES, M03-46*

Gutkin, E., & Tabachnikov, S. (2002). Billiards in Finsler and Minkovski geometries. *Journal of Geometry and Physics, 40,* 277–301

Gutkin, E., & Troubetzkoy, S. (1996). Directional flows and strong recurrence for polygonal billiards. *International Conference on Dynamical Systems (Montevideo, 1995)*, Pitman Res. Notes Math. Ser., 362, Longman: Harlow, 21–45

Hadamard, J. (1898). Sur le billard non-euclidien. *Proc. verb. soc. sci. phys. et nat. Bordeaux, 5 mai*

Halpern, B. (1977). Strange billiards tables. *Transactions of the American Mathematical Society,* 232, 297–305

Hubert, P. (1995). Complexité des suites définies par des billards rationnels. *Bulletin de la Société MathÈmatique de France, 123*, 257–270

Katok, A. (1987). The growth rate of singular and periodic orbits for a polygonal billiard. *Communications in Mathematical Physics, 111*, 151–160

Katok, A., & Hasselblatt, B. (1995). *Introduction to the modern theory of dynamical systems*. Cambridge: Cambridge University Press

Katok, A., & Strelcyn, J.-M. (1986). *Invariant manifolds, entropy and billiards, smooth maps with singularities*. Berlin/Heidelberg/New York: Springer

Kerckhoff, S., Masur, H., & Smillie, J. (1986). Ergodicity of billiards flows and quadratic differentials. *Annals of Mathematics, 124*, 293–311

Lazutkin, V. (1993). *KAM theory and semiclassical approximations to eigenfunctions.* Berlin/Heidelberg/New York: Springer

Mané, R. (1987). *Ergodic theory and differentiable dynamics.* Berlin/Heidelberg/New York: Springer

Marmi, S. (2000). Chaotic behavior in the solar system, *Séminaire Bourbaki* (1998–1999). *Astérisque, 266*, 113–136

Masur, H. (1990). The growth rate of trajectories of a quadratic differential. *Ergodic Theory and Dynamical Systems, 10*, 151–176

Masur, H., & Tabachnikov, S. (2002). Rational billiards and flat structures. In H. A. Katok (Ed.), *Handbook of dynamical systems* (pp. 1015–1089) Amsterdam: Elsevier

Mather, J. (1982). Glancing billiards. *Ergodic Theory and Dynamical Systems, 2*, 397–403

Morse, M., & Hedlund, G. (1940). Symbolic dynamics 2, Sturmian trajectories. *American Journal of Mathematics, 62*, 1–42

Moser, J. (1978). Is the solar system stable? *The Mathematical Intelligencer, 1*, 65–71

Porter, M., & Lansel, S. (2006). Mushroom Billiards, *Notices of the American Mathematical Society, 53*, 334–337

Sarnak, P. (1995). Arithmetic Quantum Chaos. In *The schur lectures* (Tel-Aviv 1992), Israel Mathematics Conference Proceedings, 8, Bar-Ilan University, Ramat-Gan, Isral, 183–236

Schwartz, R. (2006a). *Near isosceles billiards: scaling limits and fourier series.* http://www.math.brown.edu/∼ res/Billiards/index.html

Schwartz, R. (2006b). Obtuse triangle billiards II: 100 degrees worth of periodic trajectories. *Experimental Mathematics, 18*, 137–171

Schwartz, R. (2006c). Obtuse triangle billiards I: near the.2; 3; 6/ triangle. *Experimental Mathematics, 15*, 161–182

Sinaï, Y. (1976). *Introduction to ergodic theory, Math. notes 18*. Princeton: Princeton University Press

Sinaï, Y. (Ed.). (1987). *Dynamicals systems II, encyclopaedia of mathematical sciences* (Vol. 2). Berlin/Heidelberg/New York: Springer

Sinaï, Y. (1990). Hyperbolic billiards. In *Proceedings of the International Congress of Mathematicians, Kyoto 1990*, Springer, 249–260

Stenman, F. (1975). Periodic orbits in a tetrahedral mirror. *Commentationes Physico-Mathematicae, 45*, 103–110

Tabachnikov, S. (1995). *Billiards*. Paris: Société mathématique de France

Tabachnikov, S. (2002). Dual billiards in the hyperbolic plane. *Nonlinearity, 15*, 1051–1072

Tabachnikov, S. (2005). Geometry and Billiards, *AMS Student Mathematical Studies* (Vol. 30). Providence: American Mathematical Society

Ulam, S. (1976). *Adventures of a mathematician*. New York: Charles Scribner's Sons

Veech, W. (1992). The billiard in a regular polygon. *Geometric and Functional Analysis, 2,* 341–379

Walters, P. (1982). *An introduction to ergodic theory.* Berlin/Heidelberg/New York: Springer

Wojtkowski, M. (1986). Principles for the design of billiards with nonvanishing Lyapunov exponents. *Communications in Mathematical Physics, 105,* 391–414

Zagier, D. (1977). The first 50 millions prime numbers. *The Mathematical Intelligencer, 0,* 7–19

Chapter XII

Geometry and dynamics II: geodesic flow on a surface

XII.1. Introduction

We will be interested in the geometry **on** a surface (not **of** a surface, see in Sect. VI.2 for an explanation of this important distinction) and simultaneously in mechanics on it (Arnold, 1978). There are at least three motivations for this. First motivation: our planet is to a first approximation a surface, rather well described as an ellipsoid of revolution (see below). It is thus important to comprehend the nature of the geometry of such a surface; a typical question: what is the shortest path from one point to another? This is the aspect of **living** on a surface. Now physicists — who are interested in much more complicated mechanical systems — need to study simple systems because these provide good tests for general hypotheses. For example, here is what Poincaré wrote in 1905:

"*In my* New methods for Celestial Mechanics *I studied the details of solutions of the three body problem and in particular periodic and asymptotic solutions. It suffices to refer to what I wrote on this subject to comprehend the extreme complexity of the problem; beside the principal difficulty, that which lies at the very bottom of things, there is a horde of secondary difficulties that come to complicate the task of the researcher. There would thus be interest in first studying a problem where this principal difficulty is encountered, but where there would be relief from all the secondary difficulties. This problem has already been found, it is that of the geodesic curves on a surface; this is again a problem in dynamics, so that the principal difficulty persists; but it is the simplest of all dynamical problems; first there are but two degrees of freedom, and then if we take a surface without singular points, we have nothing comparable to the difficulty encountered in problems of dynamics at the points where the velocity is zero; in the problem of geodesic curves, in fact, the velocity is constant and can thus be regarded as part of the hypotheses of the problem.*

Hadamard understood this well, and it is this that directed him to study the geodesic curves of surfaces of mixed curvature; he gave a complete solution of the problem in a memoir of highest interest. But it isn't geodesic curves of surfaces with mixed curvature to which the trajectories of the three body problem are comparable, it is on the contrary the geodesics of convex surfaces."

Finally, the third motivation is: describe the movement of particles on a surface when the time becomes very large; this is the aspect of **prediction**.

M. Berger, *Geometry Revealed*, DOI 10.1007/978-3-540-70997-8_12,
© Springer-Verlag Berlin Heidelberg 2010

In the present text we are essentially interested in the mechanical − dynamical − aspect, rather than "shortest paths," but this last harbors once again subtle problems (already encountered in Chap. VI), for which for example Chavel (1993) can be consulted.

Finally, we restrict ourselves to the case of surfaces, i.e. to the case of Riemannian manifolds of dimension two. For those of higher dimensions, natural and large open problems are still more numerous, even irritating. There does not exist, to our knowledge, a synthetic treatment of the subject, not a book or even a paper. It seems to us simplest to give some recent references from which readers can, if they wish, go further back as needed: (Rademacher, 1994), (Besson, Courtois and Gallot, 1995).

To end this introduction, we say why the case of surfaces is the first to be considered seriously. In fact, the case of a curve is simplest, but not really very interesting. Consider any closed curve in ordinary space. Then the shortest path from one point to another consists of traversing the curve between these two points, and this without going backwards. For a specialist in mechanics, matters are more precise. A point mass moving along the curve, attached in any way you want, e.g. you may think of the curve as a very narrow tube on the interior of which a tiny ball moves − but which is not subjected to any exterior force, in particular neither to a **gravitational force**, nor to friction − has a trajectory necessarily of constant speed. This is proved thus: the only force to which the point is subjected is the force of reaction of the small tube, a force that is perpendicular to the tube when there is no friction. The fundamental law of mechanics then says that the acceleration is perpendicular to the curve. Well, the simplest differential calculus shows that the part of the acceleration that isn't perpendicular to the trajectory is the derivative of the

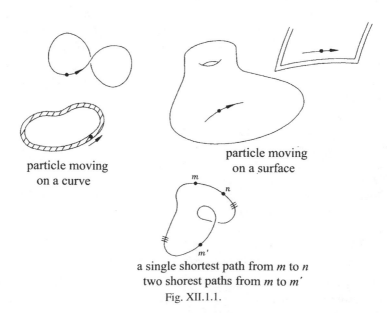

particle moving
on a curve

particle moving
on a surface

a single shortest path from m to n
two shorest paths from m to m'
Fig. XII.1.1.

speed, which is therefore zero. This says exactly that the speed is constant (see as needed Chap. V). Thus in our small tube the movement of the mass is very simple: it moves in the tube (along the curve) with constant speed. Prediction is trivial; for example the mass will return to a given point at times $t, t + k, t + 2k, t + 3k, \ldots,$ where k depends only on the length of the curve and the initial speed. Notice the two differences with the shortest path problem: here we obtain the shortest path, but under the condition that we don't continue too long at the constant speed. More precisely: two points of the curve are joined by a well determined shortest path if and only if their distance is strictly less than the half-length of the curve. Otherwise these are two *antipodes*: there are exactly two choices for going from one to the other.

XII.2. Geodesic flow on a surface: problems

Here now are the things that happen on a given surface in space, that can be a very nice round sphere or just as well an ellipsoid of revolution or much more generally a well dented "potato". A case considered separately by some authors is that of a convex surface, i.e. located on just one side of any of its tangent planes, but we will see that things are not much simpler for these.

We saw in Sect. VI.2 that it was Jacob Bernoulli — Euler's teacher in Basel — who at the beginning of the Seventeenth Century taught Euler how to find the shortest path from one point to another, in particular that a curve joining two points p and q of our surface S can only be a shortest path if, when we traverse it at constant speed, its acceleration is always perpendicular to the surface. It is simply going **straight ahead**; or again, **fleeing** most quickly if you are being pursued. We can also say: going straight ahead is tantamount to not turning, neither to the right, nor to the left. It involves a **straight** curve on the surface, which has no proper curvature "on" the surface (of course our curve possesses a curvature, but in space, not on the surface). Or again: we remain erect, meaning perpendicular to the surface, throughout the motion. Because, if you are pursued and if you turn, your pursuer will gain on you if clever enough to go straight ahead.

And thus the shortest paths are furnished by the study of mechanics on the surface. We might think of a point suitably attached to the surface (in any way we want), of a very tiny ball moving without any friction between two sheets parallel to the surface and distance apart equal to the diameter of our ball. Be well aware in all that follows that the ball has a mass but **there is no gravitation** and that there is **no exterior force** being exerted on the ball other than the reaction of the surface, a reaction that is perpendicular to it because there is no friction.

We saw in Sect. VI.3 the existence and uniqueness of geodesics, for which we gave a pragmatic interpretation; and the result of Hopf-Rinow which guarantees an extension for an arbitrarily long time.

In this entire chapter we consider only smooth surfaces without boundary and that are bounded, i.e. contained in a finite region of the space. These are

called CLOSED SURFACES in classic terminology (used also in French and German), COMPACT SURFACES in the jargon of topology.

It is thus easy to see that geometry on the surface is good (at least in principle and if we don't want to know too much about it), which is to say locally: two arbitrary points are always connected by a shortest path. But we see clearly that a shortest path isn't always unique for closed surfaces. This problem was treated in detail in Sect. VI.3, where we saw that practically all the natural problems still remain open.

To return to geodesics, i.e. to mechanics on the surface: for a sufficiently short time a geodesic is a shortest path, but subsequently "it is nothing more than a trajectory". This is exactly what will interest us: what do these trajectories **become** when we follow them indefinitely?

From now on we will only be interested in mechanics on our surfaces, i.e. in describing what the geodesics are, how they are allocated, what they become over long time periods.

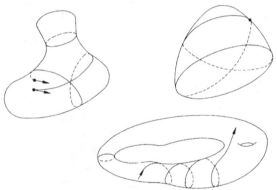

Fig. XII.2.1. Periodic geodesics and geodesics that get lost.... Berger (1994). ©
Università di Parma and A. Concari

We can say right away that there is a principal and absolutely general **dichotomy** (that arises from the uniqueness of a geodesic from its starting point with given initial vector velocity). First possibility: at the end of a certain time k, we return to the same point and with the same vector velocity; then we begin again along the same curve, then at the end of time $2k$ return to the point of departure, etc. Such a geodesic is called **periodic**. For example, for the round sphere, these are all of them: they are the great circles. For a surface of revolution, all the meridians will be periodic geodesics. Our hiker can in fact stop at the end of time k, there is nothing more to learn concerning the future. Or else the second possibility: the geodesic **gets lost** in some way on the surface, our hiker walks indefinitely into the future. But new questions arise immediately: does the geodesic remain in a confined part of the surface? How is it apportioned in this part: is it evenly distributed? These questions are fundamental in physics; they are questions of "**ergodic theory**"; see XI.XYZ.

To return to periodic geodesics, the questions that come immediately to mind are: do any/many exist? Can we count those with a given length? Is the set of them evenly distributed on the surface? Readers have no doubt noticed analogies with the billiard problem studied in the preceding chapter. We recall that the important question for physics is not whether the geodesics darken the whole surface (this is ergodicity in the space of positions), but whether they darken the space of the set of tangent directions at different points of the surface (this is a space of dimension 3, two for the surface and one for the angle of the direction, called *phase space*). This is ergodicity in **phase**.

The term **darkening** can be interpreted on a computer screen or in physical space, but in mathematical jargon as an **everywhere dense** set in a given topological space.

Before saying what is presently known about these problems — for which there has been much recent progress in several areas, but with a number of problems with an elementary statement still remaining open — it helps to dispose over a reservoir of examples of varying difficulty for both not believing just anything and for appreciating the difficulty of the problem.

Here is something very important in conclusion: inevitably we cannot remain entirely with surfaces that figured in the preceding descriptions. As is normal for Jacob's ladder, we will thus be compelled — so as to understand well what is happening — to introduce *abstract* surfaces. In modern language, these are the differentiable manifolds of dimension equal to two; see V.XYZ. In way of simplification, we consider only those that are compact and orientable.

XII.3. Some examples for sensing the difficulty of the problem

XII.3.A. The spheres

A first surprising result is the phenomenon called "Zoll surfaces". The fact that the geodesics of a *round* sphere are all periodic is not characteristic of the spheres in general. It's a story that began with Darboux in the second half of the nineteenth Century. Darboux attempted, with only partial success, to find surfaces of revolution for which all geodesics are periodic. It wasn't until 1903 that Zoll succeeded in constructing them and simultaneously demonstrated the difficulty of the problem. There are even some nonconvex sets among them, which isn't very intuitive. We will return to these surfaces right away.

It is natural to lend our attention to surfaces of revolution, for they are practically the only ones where calculation is easy.

XII.3.B. The surfaces of revolution: the Zoll surfaces

We are first going to describe the behavior of the geodesics on general surfaces of revolution, which will be useful in the sequel for more than just Darboux's problem. It is something that Clairaut discovered in the first half of the eighteenth Century. The idea is to project the surface onto a plane perpendicular to the axis of revolution. What becomes of the geodesics? Since their acceleration is normal to the surface and the normal to a surface of revolution always intersects its axis, our projected curve will be a plane curve for which the acceleration passes constantly through the center of this plane. These are called curves "with central acceleration" in mechanics. The most celebrated model is that of the movement of a single planet about the sun: all the trajectories are conics having the center as focus. Another case familiar to physicists is that of the harmonic oscillator, where a point is attracted by a force proportional to its distance to a center: here again all the trajectories are ellipses. This example by the way represents the case of plane projections of the great circles of the sphere.

For the convenience of readers, we reproduce Fig. VI.3.4:

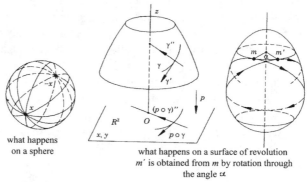

what happens
on a sphere

what happens on a surface of revolution
m' is obtained from m by rotation through
the angle α

Fig. XII.3.1. Berger (1994) © Università di Parma and A. Concari

Apart from some obvious exceptions, the projections of the geodesics for the sphere onto a plane perpendicular to the axis will be curves that oscillate in an annulus formed by two circles, one interior, the other exterior. The interior circle contains the projection of the points where the geodesic considered has a horizontal tangent, arising from two parallels (circles of a certain latitude). The exterior circle is the projection of the equator. On the surface our geodesic will oscillate between those two parallels that project onto the interior circle, and they will cross the equator between their passages to those two parallels.

This permits us to effectively describe the behavior of the geodesics on general surfaces of revolution, say convex — again for simplification — but this is not absolutely essential. Since there is revolution we see that a given geodesic, starting from the parallel above (implicitly understood to be tangent to it, i.e. with horizontal initial velocity) returns to that parallel at the end of a certain time k, but at a point which in general has turned through an angle α with respect to the initial point. As in rectangular billiards, there is a perfect **dichotomy**. Two cases and two only are possible. In the first, α is a rational multiple of π: $\alpha = \frac{p}{q}\pi$ with p and q integers. Then at the end of time qk we will have returned to the original initial conditions, and thus the geodesic will be periodic. If on the other hand α is an irrational multiple of π, then the geodesic will continue unceasingly its oscillation in the zone of the surface contained between the two parallels. As in the case of rectangular billiards and it is easy to see that it darkens (is everywhere dense in) the entire zone, and this in a regular, well distributed, way. We say that there is ergodicity (in space, but as it turns out not in phase and only in this zone between the two parallels).

The preceding study furthermore provides a way — by letting the two parallels vary — of constructing an infinity of periodic geodesics. This answers, in the "general" case, one of our questions. There can be exceptions, one of which is described below.

Returning to Darboux's problem, the geodesics will all be periodic if the meridians of the surface have the special property that, from whichever upper parallel we start, we return to the same point after exactly one departure, traveling "up-down-up", i.e. such that $\alpha = 0$, where we have made a complete rotation about the axis or revolution. Darboux succeeded in finding pieces of such meridians, but he never knew how to join them. In spite of the efforts of several geometers at the end of the nineteenth Century, it wasn't until 1903 that Zoll, a student of Hilbert, succeeded in obtaining a complete closed surface. Among these are real analytic surfaces and, better still, nonconvex surfaces, which is contrary to our first intuition; see the figures below and more generally the whole chapter of Besse (1978) that is devoted to this subject.

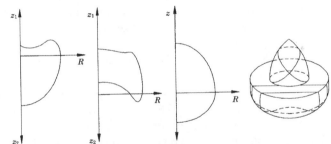

Fig. XII.3.2. Various meridians on Zoll surfaces. Observe that they are not necessarily convex. Berger (1994) © Università di Parma and A. Concari

Darboux's examples are easy to find by using Clairaut's equation. In Besse (1978) there is a complete description of all the Zoll surfaces of revolution, due to René Michel: Corollary 4.16 on p. 104. There are as many of them as there are numerical functions taking $[-1, 1]$ onto itself. It involves well expedited integral calculus.

In fact, at the present moment not all the surfaces that have this property are known: within the specialty it is called a **classification** problem, a type problem of which mathematicians are very fond. Of course two surfaces are not considered different if they obtained from each other by a translation or reflection of the space. But since Guillemin in 1976 we know many other completely periodic surfaces which are not surfaces of revolution and may even not have *any* symmetry; and he classified all those that are not too bumpy, i.e. those that are close enough to being a round sphere. Once again, the basic reference is Besse (1978) and the most recent to our knowledge is Gromoll and Grove (1981). Guillemin proved that for each *odd* numerical function on S^2 — i.e. taking on opposite values at antipodal points — there exists a continuous family of conformal deformations, produced from the canonical metric of S^2 by a function, that are Zoll. Funk had only established infinitesimal existence. Progress has recently been made in understanding Zoll surfaces in Le Brun and Mason (2002). Historically, it is interesting to note that the work of Guillemin proved a conjecture announced in Funk (1913), i.e. local verification of an infinitesimal result of Funk, a result that was nothing other than the cornerstone of the Radon transform (1917), where Funk's work was ignored. The Radon transform says in effect that we can recover a function (on the sphere, in all space, etc.) if we know the value of its integrals on, for instance, all hyperplanes. This transformation is the basis for the construction of scanners, or again the reconstruction of the density of the terrestrial core on the basis of seismographic measurements.

XII.3.C. Weinstein's counterexample

We now construct some **bad surfaces** following Weinstein. It will involve invalidating a conjecture that said that not only are there lots of periodic geodesics on a surface, but better still that, all together, they darken all the tangent directions for the surface: ergodicity in phase. Recall that here this means to darken in the strictest sense: each direction at any point on the surface can be approximated arbitrarily closely by a suitable velocity vector of a periodic geodesic, even if we must wait a long time for this to happen! Weinstein in 1970 was the first to notice that it is not difficult to construct surfaces that violate the conjecture; the matter is in fact quite simple.

If we look for a surface of revolution that is of constant curvature — normalized say to 1 — and thus in fact having everywhere locally the same geometry as the unit sphere (see Sect. VI.6), we obtain a differential equation for the equation of the meridian. It has solutions that depend on one parameter and furnish the surfaces we sought earlier. The corresponding meridians mentioned below have already been given in Fig. VI.7.5. They are spindles or bobbins; in both cases we can remove singular points. We now take such a bobbin and smooth it at the two ends, and so obtain a smooth surface of revolution composed of three parts: the two smoothings (note that these can be as small as we wish) and the central part. What is the behavior of a geodesic that departs horizontally from a parallel situated in the central part?

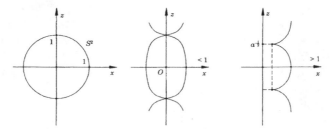

Fig. XII.3.3. Berger (1994) © Università di Parma and A. Concari

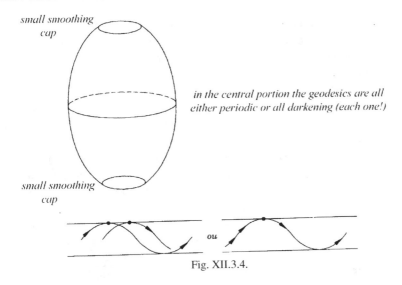

small smoothing cap

in the central portion the geodesics are all either periodic or all darkening (each one!)

small smoothing cap

ou

Fig. XII.3.4.

Here we have the whole beauty of the Weinstein construction: our central portion has the local geometry of a sphere of radius 1 throughout, but its equator has a different length, precisely equal to $2\pi D$, where D denotes the distance from this equator to the axis of revolution. If we recall the Clairaut property — see Sect. VI.3 — this then implies that our geodesic will return horizontally and along its parallel of departure, having turned through $2\pi D$, (and not 2π, as on the sphere of radius 1). And, once again, two cases present themselves. The first is that where there exist two (relatively prime) integers p and q such that $D = p/q$. Then **all** the geodesics considered, not only remain in the central portion, but are all periodic (and of common length equal to $2q\pi$). If on the contrary — and it is badness that we have already seen — D is an irrational number, then none of the geodesics leaving from the central portion with horizontal direction are periodic. As each geodesic leaving from a point of the central portion with a velocity sufficiently close to the horizontal possesses a point where the velocity is horizontal (it is true in the case of S^2, thus here too), we have a whole open set of pairs (point, direction) above the central portion that can't belong to a periodic geodesic. Better yet, this central portion can be given an area as close as we wish to the total area of the surface, since the two smoothings above can be very, very small.

XII.3.D. Ellipsoids with three axes

And if we return to our good planet, i.e. to an investigation of the shortest paths on an ellipsoid of revolution, a study which requires discovery of the geodesics? Since we are dealing with a surface of revolution, the **global** study is complete, but if we want **exact** calculations, we find that elliptic functions appear. We owe the complete study to Gauss; it provides exact formulas for constructing geographical charts. Recall that the ones most used "on the planet" are not those with Lambert projection — dear to the French — but the "universal tranverse" Mercator projection (UTM charts) that Gauss already used! Inquisitive readers will want to know about the uniqueness problem for shortest paths: they can refer to Sect. VI.3 and will find there that it remains open.

Here a small miracle occurs in the course of a curious history. In the middle of the nineteenth Century, Weierstrass asked the question about finding geodesics on an ellipsoid not of revolution, i.e. with three unequal axes. His motivation was the shape of Earth. Measurements showed that the hypothesis of an ellipsoid of revolution didn't coincide with experiment. Some thus put forth the hypothesis of an ellipsoid with three unequal axes. In fact it is now known that the right hypothesis is indeed that of an ellipsoid of revolution, but with some additional flattening at the poles. We restate here, in the context that interests us, what was already said in Sect. VI.3 about geodesics of ellipsoids.

It was Jacobi in 1839 who first succeeded in finding these geodesics. A little like with surfaces of revolution they divide into **bands** (for the specialist in mechanics, this comes in each of the two cases from the existence of a "first integral", i.e. a function defined on phase space, constant along the trajectories of geodesic flow). In the case of surfaces of revolution, these bands are those determined by pairs of parallels situated the same distance from the axis of revolution. Here these are pairs of curves of intersection on our ellipsoid by homofocal quadrics. The latter play a very important role in several problems in analysis, in dynamics and geometry that are interrelated, because among other reasons they give a coordinate system in which the fundamental differential operator, that of Laplace-Beltrami , separates. The figures for this have already been given several times; see Figs. IV.10.6, VI.3.5, and VI.3.6.

We review this construction, emphasizing here the behavior of geodesics. We start with any ellipsoid $\frac{x^2}{a^2} + \frac{y^2}{b^2} + \frac{z^2}{c^2} = 1$ and envelop it, so to speak, with the family of quadrics in a single parameter λ defined by the equation $\frac{x^2}{a^2+\lambda} + \frac{y^2}{b^2+\lambda} + \frac{z^2}{c^2+\lambda} = 1$. We assume here that $a > b > c$. On the respective intervals $[-c^2; \infty]$, $[-b^2; -c^2]$, $[-a^2; -b^2]$ for λ the quadric is: an ellipsoid, a hyperboloid of one sheet, a hyperboloid of two sheets. Through each point of space there pass in general three of these quadrics, one of each type, and their three tangent planes are mutually perpendicular. Their curves of intersection are, on each of them individually, so-called "lines of curvature", which play an important role in the theory of surfaces (but for surfaces in space these are almost never geodesics). Finally, the geodesics of our ellipsoid have the property that the lines in space that are tangent to them remain constantly tangent also to one of these homofocal quadrics. There are four very important points on the ellipsoid, its **umbilics**. These are the four points of intersection of the ellipsoid with the "degenerate" homofocal quadric consisting of the hyperbola $\frac{x^2}{a^2-b^2} - \frac{z^2}{b^2-c^2} = 1$ situated in the plane $y = 0$ of z and x. Every geodesic starting at an umbilic pass some time later through the "antipodal" umbilic. But — attention! — contrary to the case of the sphere, when we iterate this operation, we definitely return to the same umbilic, but not in general in the same direction! For readers who are curious about the elementary geometry of these surfaces, the preceding lines of curvature can be obtained as loci of points for which the sum of the distances on the ellipsoid to two umbilics is constant. We have encountered this family of quadrics twice before: in Fig. IV.10.6, then in Sect. VI.3.

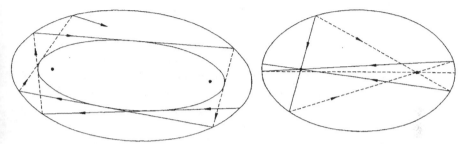

Fig. XII.3.5. How to visualize an ellipsoid by flattening it into a double ellipse. But pay attention that here this doesn't entail projection onto the plane except for the limiting case. Berger (1994) © Università di Parma and A. Concari

Note. Following Birkhoff, some like to visualize geodesics on an ellipsoid with three axes as billiard trajectories in an ellipse. These envelop homofocal ellipses or hyperbolas, the particular case of umbilics corresponding to the trajectories passing through the two foci. We leave to readers the joy of discovering what happens for the latter. Pay attention that, in reality, we are dealing with a **double-faced** full (boundary and interior) ellipse that changes face with each reflection. This corresponds, on the ellipsoid, to passing from the upper half ellipsoid to the lower half; see Fig. IV.2.9 and Sect. XI.9 on billiards.

The last example we will give is due to Morse in 1934 and shows in exemplary fashion the difficulty of the study − the investigation − of periodic geodesics (and thus of the behavior of geodesics in all generality). It involves nonetheless very simple surfaces, ellipsoids. In what was said above about geodesics of ellipsoids, we omitted the trivial case, that of the three "basic" geodesics, i.e. those which are the intersections of the ellipsoid with the three coordinate planes. These are geodesics because their accelerations are always normal to the surface, if not for reasons of symmetry.

Morse showed this:

(XII.3.D.1) *For any number L (but think of a very large number) there exist ellipsoids for which the three axes are unequal but as close to being equal as desired, and which are such that their geodesics are: the three basic geodesics (and of course their iterated overlappings) and those whose length is at least equal to L, i.e. extremely long.*

In other words, except for the three basic geodesics it is necessary to "circum-navigate" such an ellipsoid an enormous number of times before returning to the point of departure in the original direction.

In using the description given above of geodesics of an ellipsoid as oscillating between two symmetric lines of curvature, we can surmise rather easily Morse's result. For a complete proof in a recent work, see Lemma 3.4.7 of Klingenberg (1982).

XII.3.E. The flat tori

All the surfaces that we have considered until now were visible, defined in the ordinary space of three dimensions. We can surely include tori, as in the figure below, the "best" being — let's just accept it for the moment — the tori of revolution, i.e. generated those by the rotation about an axis of a circle situated in a plane that includes this axis.

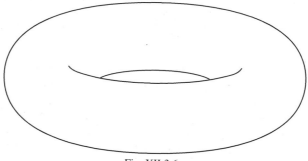

Fig. XII.3.6.

We leave it to readers to study for themselves the complete behavior of geodesics on a torus of revolution and to classify them into different types — the only reference we know on this subject is Bliss (1902–1903) — and perhaps even to study their cut-locus. They will want, of course, primarily to make use of Subsection B above.

But it is the tori, still geometric but abstract — see Sect. VI.4 — on which the behavior of geodesics is almost as simple as on the sphere; these are the flat tori, where "flat" means having throughout the same local geometry as Euclidean space, i.e. without curvature. They are constructed thus: we know that, topologically, we obtain a torus by identifying the opposite sides of a square. Think of a video game where the rule is that, if we go off at one side, we reappear on the opposite side, at the "same place". In this identification, the angles at the vertices equal $\pi/2$; we don't introduce any singularity and thus the locally Euclidean structure of the square is extended to this toroidal surface. Of course, we can never realize it as a torus in space, which must have points of strictly positive curvature , as can be seen by taking a point fixed at a good distance from the torus and then examining the point of contact with the smallest sphere that contains the torus and is centered at this point.

More abstract still, and simpler still: we consider the quotient set of the Euclidean plane by the lattice of integer coordinate translations (in a given basis, not necessarily orthogonal). The translations preserve the Euclidean structure, which gets passed on to the quotient. But pay attention that the flat torus quotient has a geometry that depends on the lattice chosen; this geometry is not in general isometric to that of the "square" torus. The description of the geodesics is however the same in all cases; the difference appears when we are interested in the actual lengths.

We have already described the behavior of geodesics in square billiards in Sect. XI.2. That of flat toroidal geodesics is similar. We have: periodic geodesics, which are grouped in continuous bands that fill the (abstract!) flat torus; or the everywhere dense geodesics, each of which is equally distributed in the torus. On the other hand, there isn't any ergodicity in phase space. We also know the asymptotic behavior of the *counting function* of the periodic geodesics; see XI.XYZ and XI.2.

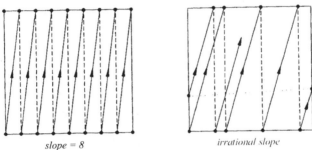

slope = 8   *irrational slope*

Fig. XII.3.7. Berger (1994) © Università di Parma and A. Concari

Otherwise expressed, there is a perfect dichotomy for the behavior of the geodesics. We saw above that this is not the case generally; but is it perhaps still the case with some other surfaces?

In the sequel we only consider, for simplicity, abstract surfaces that are compact and orientable.

The fundamental theorem on classification of (orientable — see Sect. VI.1) surfaces then says that the only topological types possible are the sphere, the torus and the surfaces of yet higher genus (with several holes).

XII.4. Existence of a periodic trajectory

XII.4.A. The torus and surfaces of higher genus

If our surface involves at least one hole, it suffices to encompass it with a curve as in Fig. XII.4.1. Then we take, from among all curves of the same topological nature, i.e. obtainable from each other by continuous deformation, the shortest. It can't be reduced to a point unless there isn't really a hole. Showing that the minimum of the lengths in this family is really attained is easy, particularly if we use

Birkhoff's method of geodesic polygons (described below) in order to reduce it to a finite dimensional situation. Physically we can think of a rubber band that is initially sufficiently small. We will treat the counting function problem — even though it isn't yet settled — for arbitrary abstract tori in Sect. 5.A.1 below.

XII.4.B. The sphere, Birkhoff's result

But if our surface has no hole, does this mean that it is a surface having the same form as the sphere, dented surely and perhaps warped, but "with the same topology"? And how can we still find a periodic geodesic?

On a surface without a hole, but having roughly the shape of that below (i.e. having a "**waist**"), we can find a periodic geodesic by taking the shortest closed curve "around the waist". As a practical matter we can find it with a sufficiently small rubber band. The proof of existence is then the same as in the case of surfaces with holes.

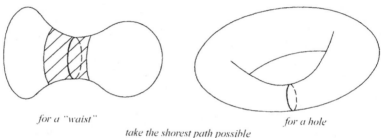

for a "waist"   *for a hole*
take the shorest path possible
Fig. XII.4.1. Berger (1994) © Università di Parma and A. Concari

But if our surface doesn't have a "waist" — typically if it is **convex** and if we know nothing more about it — the rubber band method won't work: it will contract more and more, say to a point, then "fall off the surface". Here then is the fundamental problem facing the mathematician: find at least one periodic geodesic on a surface of spherical and typically convex type. The above examples seem to show that existence is rather intuitive.

But why, apart from a very natural esthetic motivation, attach such importance to periodic geodesics? First, because they are, for the physicist, the stationary states of motion; they in some way give an energy level. Who doesn't like a bit of stability? But here we can also quote Hadamard in (1898), speaking about Poincaré and of negative curvature:

"*Secondly, the importance that this geometer accords periodic solutions in his Treatise on Celestial Mechanics is manifested equally in the present problem. Here again they have shown themselves to be "the only breach through which we might attempt to penetrate into a place until now held to be inaccessible""*.

In a more precise way, for us they have played the role of a sort of coordinate system with respect to which we can refer all other geodesics."

Poincaré broached this problem in his article from 1905, mentioned in Sect. V.3. But whatever his wish he was only able to find some methods of attack — no less than three — without being able to "get them going" rigorously.

Observe this little paradox: the round spheres have the simplest behavior for their geodesics. For each surface topologically different from the sphere, the behavior is always more or less complicated; in contrast, we have seen that for the sphere itself the existence of periodic geodesics is trivial.

To return to the existence of at least one periodic geodesic for a surface whose topology is that of the sphere, it was Birkhoff who in 1917 was the first to establish this existence rigorously. And here is his method, for which we explain the principal points in detail, for they are the basis for the majority of later developments in research on periodic geodesics in the very general context of compact Riemannian manifolds in arbitrary dimension. In all that follows the surface considered has the topological type of the sphere, i.e. it is "without holes".

First idea: continuously cover the surface with a continuous family of curves beginning with a curve reduced to a point m and ending with a curve reduced to a point m'. All this so that all the surface is completely covered. In each such covering there will be a curve for which the length is greatest. Now the fundamental idea is that, when we consider all such coverings, there is at least one for which the curve of maximum length is of minimum possible length in the set of all possible coverings. In some way this is the most economical covering, the most effective, the best possible. This technique, when it succeeds, is called the **minimax** principle. For it is then rather obvious that the longest curve of such a covering will be a periodic geodesic.

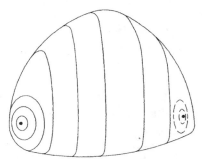

Fig. XII.4.2. Covering the surface completely. Berger (1994) © Università di Parma and A. Concari

If this isn't already obvious, it will be when we now show how to obtain such an optimal covering. The complete theory is very subtle, in fact the set of all coverings of the surface is **immense** — of infinite dimension, as are all function spaces — and finding a minimum is much more delicate than in spaces of finite dimension. Refined analytical techniques are always required. We will spare readers the subtleties and give Birkhoff's second idea, which consists of constructing, starting with any covering, a second covering in which all the curves have shorter lengths, with the sole exception of curves that are periodic geodesics (if any exist). If we assume the existence of a "minimum" covering, then this necessarily contains a periodic geodesic, for otherwise by applying Birkhoff's transformation we can produce a covering for which all the curves will have a strictly shorter length, so that our original covering wouldn't be a "minimax" covering.

The procedure Birkhoff discovered is very elegant and simple. We observe first how it works in the case of ordinary plane curves, where it is simpler than on surfaces. We fix once for all an integer N and decompose the plane curve into N pieces of equal length. We replace each of these pieces by the straight line segment joining its endpoints. We take an additional step with surfaces in mind, although in the plane it's not strictly necessary: we replace the polygon so obtained by the polygon for which the vertices are the midpoints for the prior polygon. Each of these operations strictly reduces the length.

Fig. XII.4.3. Berger (1994) © Università di Parma and A. Concari

Can we do the same on a surface? Here we encounter the problem of the shortest path between two points: it needs to be well defined. On a surface having, for example, a very small "waist" we need to take points very close together. But, the surface being fixed and compact, there exists a strictly positive shortest length I such that each pair of points whose distance apart is less than I is joined by a unique piece of geodesic of length equal to their distance apart. In the lingo of general Riemannian geometry, the constant in question is called the *radius of injectivity* of the manifold considered, and it is a principal invariant; see, for example, Berger (1999) (see the index) or IV.5 of Berger (2003).

We now take any curve on the surface, and divide it into N pieces of length less than I. We replace these pieces by the shortest paths (which are well defined) joining their endpoints; this step generalizes what was done in the plane. Observe that if the curve considered is a geodesic polygon and, if the points of division are precisely the vertices, this operation preserves this polygon and thus does not strictly reduce its length. Now the second operation consists of taking the midpoints of the polygon obtained after the first step and joining these midpoints by the shortest paths that connect them. This time (we need to use the fact that, in the geometry of a surface, a true triangle has sides strictly less than the sum of the other two) the new polygon will have a length strictly smaller than the initial curve, unless all the vertex angles of this polygon are **equal to** π. Now this happens for a periodic geodesic and only in this case!

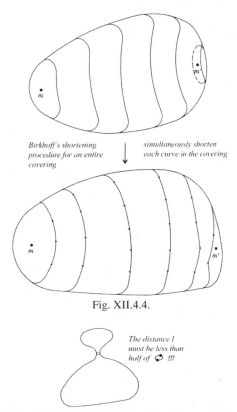

Fig. XII.4.4.

Fig. XII.4.5. Berger (1994) © Università di Parma and A. Concari

The conclusion is, by iterating this shortening procedure simultaneously for the whole covering — which cannot **tear** — there will finally be convergence of a curve situated "somewhere in the middle of the covering" toward a periodic geodesic. If we hadn't replaced all curves by geodesic polygons there would have been a subtle problem of convergence, because the set of curves is of infinite dimension. This is precisely what stopped Poincaré. But the geodesic polygons form a finite dimensional space (of dimension equal to twice the number of vertices). We can thus use the compactness of the set.

Readers familiar with certain parts of analysis will recognize the minimax principle here. We will in fact return to it soon.

Note. The geodesic obtained by Birkhoff's method isn't always **simple**, i.e. it can have self-intersections. For this subject consult Calabi and Cao (1992), or Berger (1993). There it is shown that Birkhoff's procedure always yields a simple geodesic if the surface has nonnegative curvature everywhere.

In fact, there always exist simple periodic geodesics, but they are much more difficult to obtain. It wasn't until 1978 that there was a really solid proof. The most elegant proof today is perhaps that of Grayson (1989). Poincaré would have liked

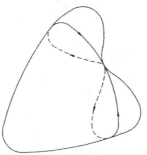

Fig. XII.4.6. A periodic geodesic that isn't simple. Berger (1994) © Università di
Parma and A. Concari

this method very much, which consists of continuously deforming an arbitrary curve,
a deformation that when pursued over time yields a simple geodesic curve. This
deformation is provided by a differential equation in the space of all curves. It is a
partial differential equation of parabolic type (like the heat equation). The derivative
with respect to time is easy to guess: at each point of the curve, it is deformed in
inverse proportion to the **deficiency** of the curve for being geodesic. The whole
difficulty lies in showing that the procedure converges when the time goes to infinity.
It is also necessary to verify that the curve doesn't develop double points.

It's not so astonishing that simple periodic geodesics are difficult to find. There are
two reasons, the first indicated in recent work of Galperin (Galperin, 2003), who has
shown that it is essential that the surfaces considered are differentiable. He proceeds
as follows: we consider a tetrahedron in space, endowed with Euclidean geometry
on its faces. It's a geometry that only has singularities at the four vertices, because
two faces are compatibly joined along an edge. For this geometry, Galperin showed
that, in general, there don't exist any simple periodic geodesics. Here a geodesic
is composed of line segments on the faces in succession, extending over the edges.

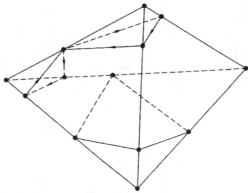

Fig. XII.4.7. Why the case of the tetrahedron is subtle. Berger (1994) © Università di
Parma and A. Concari

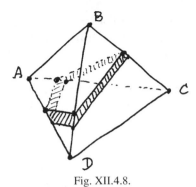

Fig. XII.4.8.

But we want that these segments, four in number if there is simplicity, never go through a vertex, where in fact we wouldn't know how to extend the trajectory. Now if we have such a periodic geodesic (and moreover a whole one-parameter band of them), we see by suitable deformation that this imposes an algebraic relation between the curvatures (Gaussian, but singular) at the vertices of the tetrahedron. A singular curvature is defined as the deficiency from 2π of the sum of the angles of the triangles meeting at the vertex under consideration. Thus for a generic tetrahedron, there is never a simple periodic geodesic.

The second reason is the result of Gruber that says that, in the set of all convex bodies in $\mathbb{R}^3$, there is only a sparse subset (in the sense of Baire; see Sect. VII.13.D) of convex sets that admit at least one periodic geodesic (we don't even ask that it be simple!). See Sect. VII.13.D and p. 1333 of Gruber and Wills (1993).

Poincaré rediscovered. One of the attacks proposed by Poincaré is remarkably simple and elegant: a simple periodic geodesic, according to the Gauss-Bonnet theorem VI.7.3, partitions the surface into two regions for each of which the integral $\int K(m)\,dm$ equals exactly 2π. Poincaré's idea was to look for a simple curve of minimum length among the curves satisfying this integral condition. After numerous partial results over time, we finally have in (Hass & Morgan, 1996) a complete proof.

XII.5. Existence of more than one, of many periodic trajectories; and can we count them?

We look for geodesics with **distinct** supports. In fact, a fundamental remark: for students of mechanics, once a periodic geodesic has been found, then an infinity of them must be considered, because the motion which consists of turning two times, three times, etc. along a periodic geodesic is not the same as that obtained when it is traversed just one time. But for geometers who are investigating all periodic geodesics and whether they more or less fill out the surface, and in what way, it isn't the motion that is interesting, but the trace it leaves behind. It should be remarked that these are periodic geodesics, to be sure, but of lengths L, 2L, 3L, etc. We are thus looking for geodesics that are **geometrically distinct**.

Of course we also eliminate the geodesic traversed in the sense opposite from that of the geodesic considered.

There are many other questions that may be asked concerning the set of periodic trajectories. We will study them as we proceed in our study: their distribution and the counting function of their lengths.

XII.5.A. The case of the torus

In Sect. XII.2.C above we looked at flat structures on a torus; but things are really also easy for every other geometric structure on the torus, whether the torus is embedded in space or has an abstract Riemannian structure. In fact, we will first see how to find two periodic geodesics that are geometrically different: we consider the smallest curve of the "parallel" type and the smallest curve of the "meridian" type. The one type cannot be deformed into the other, nor into the other covered several times. We now need to be more technical. In the space of curves on the torus the equivalence relation of deformation from one to the other by (**free** homotopy — without a fixed base point, in contrast with the ordinary definition of the fundamental group π_1) — yields the equivalence classes of the fundamental group of this torus. In the case of the torus whose fundamental group is $\mathbb{Z}^2$ and thus Abelian, this set is $\mathbb{Z}^2$ itself and we can thus expect many distinct periodic geodesics (i.e. with different

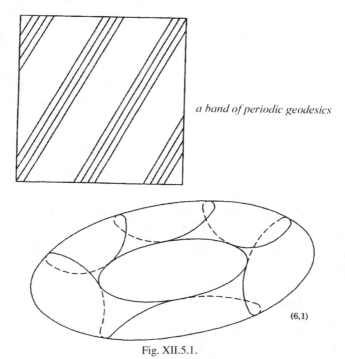

a band of periodic geodesics

(6,1)

Fig. XII.5.1.

supports). It suffices to remark that the supports can coincide, for the classes arising from pairs of positive integers (p, q) and (p', q'), coincide only if these pairs are common multiples of a pair (p'', q''), hence the existence of at least as many distinct periodic geodesics as there are relatively prime pairs of positive integers.

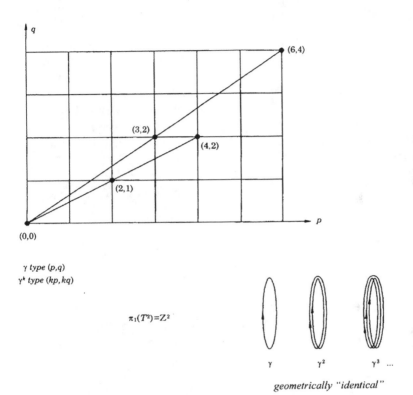

Fig. XII.5.2. A band of periodic geodesics. Berger (1994) © Università di Parma and A. Concari

This launches the question of **counting** the (distinct) periodic geodesics. As in physics for the energy levels, we are led to define the **counting function** of the periodic geodesics of a surface as:

$FC(L) = $ *number of geometrically distinct period geodesics for which the length is less than L.*

As in physics, we are above all interested in an asymptotic estimate of $FC(L)$ when L is very large. An exact value for $FC(L)$ can be realized only for very special surfaces.

Note. In the very special cases of the flat tori (square or not), surfaces of revolution and ellipsoids, the periodic geodesics come in continuous **bands** that are infinite in number and the counting function has no sense for us. In other contexts, billiards for example, counting is effected by treating each continuous band as a single entity (see Chap. XI). This is exactly what we *won't* do here.

An easy geometric estimate of what it costs in length "when we turn around the torus with the meridians or with the parallels" (see the figures) shows that the counting function of an arbitrary (i.e. not necessarily flat) torus grows in at least quadratic fashion:

(XII.5.A.1) $FC(L) \geqslant aL^2$ for each L, where a is a constant.

This comes from the fact that number of pairs of positive integers (p, q) such that $p^2 + q^2 < k^2$ is of order $\pi k^2/4$ and the fact that requiring them to be relatively prime only imposes a subsequent division by the sum of the reciprocals of the squares of the integers, which equals $\pi^2/6$ (see Sect. XI.2.B).

There is much that is still not known about the torus. What is the exact order of growth of FC(L) for tori? Will it be exponential for all tori or, if need be, for **most** tori? And will the merely quadratic case be characteristic of certain tori? It is a sad fact that at the present time the order of magnitude of the counting function isn't known for **any** surface of the type of the torus or the sphere. See however Byalyi and Polterovich (1986) and Bangert (1988). For surfaces of type genus $\geqslant 2$, the situation is different.

XII.5.B. Surfaces of higher genus

We consider a surface having γ holes, the number γ being its **genus**. For $\gamma = 1$ we have a torus. For values of γ greater than 1 and as in the case of tori, the "nice" metrics — the nice structures — can never be obtained by embedding in space; it is necessary to consider abstract surfaces and define suitable geometries on them. Let us recall briefly these results, which are now very classic.

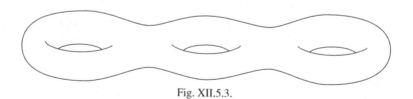

Fig. XII.5.3.

By the two usual methods of construction, we always begin with the hyperbolic plane, with curvature constantly equal to -1. In space we can realize only pieces: the Beltrami surface, etc. What is done next is either that we take the quotient of this plane by a suitable subgroup of the group of its isometries — so that the quotient is a nice compact surface without singularities — or else we fasten together polygons in the hyperbolic plane, with conditions on the angles at the vertices in order to get a good assemblage. The simplest way is to use hexagons for which all the vertex angles equal $\pi/2$. We continue here what was said in VI.XYZ.

By assembling just two, as indicated in the figure below, we obtain a "pair of pants". Observe that the three "circles" that form the boundary of a pair of pants are in fact periodic geodesics. Now, with two pairs of pants we can construct a surface of genus 2, and so forth. We also analyzed this construction in VI.XYZ, in the subsection entitled "space forms".

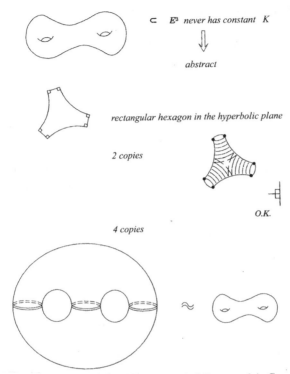

E³ never has constant K

abstract

rectangular hexagon in the hyperbolic plane

2 copies

O.K.

4 copies

Fig. XII.5.4. Berger (1994) © Università di Parma and A. Concari

These hexagons depend on enough parameters so that for genus for genus $\gamma \geqslant 2$ we can place on any surface of genus γ all the structures which are locally hyperbolic, which is the same thing as – this is classic – the Riemannian structures with constant Gaussian curvature equal to -1. An excellent reference for all this is (Buser, 1992). The family of these structures has $6(\gamma - 1)$ parameters (Teichmüller space). Recall only that, according to Blaschke's formula (Gauss-Bonnet), the total area of the surface always equals $4\pi(\gamma - 1)$, since this formula says that we always have $\int_M K(m)\, dm = 2\pi\chi(s) = 2\pi(2 - 2\gamma)$, where K denotes the Gaussian curvature of the surface.

What happens now with the periodic geodesics, both in the case of constant curvature and in the general case, whether embedded or abstract? Here there is a perfect and detailed answer. A rough beginning would consist of proceeding as with the torus, but noting that the fundamental group, along with its conjugate classes, is **enormous**, once the genus is greater than 1. It is of exponential growth; in a sense the simplest thing would be to say that there is an enormity of ways of moving about the holes, by alternating in all possible ways the choice of holes. Whereas for the torus the choice was reduced to that of two integers because the two different ways of turning about the holes **commuted**. By this single consideration of classes of homotopic curves it can easily be shown, by the same technique as for the torus, that the counting function is of exponential growth for each metric on a surface of genus greater than 1.

The detailed theory of surfaces of constant curvature equal to -1 (see (Buser, 1992)) says that the counting function of these periodic geodesics has exactly the asymptotic value

$$\liminf_{L \to \infty}(\log(FC(L))/L) = 1.$$

And for the other geometries on such a surface? We call to the rescue the *fundamental theorem of conformal representation*, which states that each Riemannian metric g on a surface of genus $\gamma \leq 2$ is "proportional" to a metric g_0 with curvature -1. I.e. there exists a numerical function f on the surface and a metric g_0 such that $g = f \cdot g_0$. In passing from g to g_0, distances are in general changed, but not the angles, whence the name conformal. Beginning there, in a very detailed study in (Katok, 1988) there is the following magnificent result:

(XII.5.B.1) *For any metric g of the same volume as for those of curvature -1, the counting function satisfies*

$$\liminf_{L \to \infty}(\log(FC(L))/L) \geq 1.$$

But we have much more: equality occurs only if the metric g has constant curvature. Otherwise expressed: the slightest modification (with constant volume for normalizing, of course) of a metric of curvature -1 results in there being **more periodic geodesics** than in the case of constant curvature! The irregularities, the bumps, create exponential hordes of periodic geodesics.

We can also interpret Katok's result by saying that, on surfaces of genus greater than 1, we have a simple global geometric criterion for expressing that metrics of constant curvature are the "nicest".

Important remark. We might ask whether or not the counting function takes into account whether the geodesics are geometrically different. In fact this is of no importance; there is exponential growth and the iteration in L, 2L, 3L, etc., linear in L, is thus of no consequence.

Notes. Readers may be very astonished: the **genus** does not appear in the above formulas, whereas we might suppose that the number of periodic geodesics would grow with the number of holes. There in fact exists a more refined formula, but where the genus appears only in terms subsidiary to the asymptotic development of FC(L); see 9.6 of (Buser, 1992). The genus also appears in the eigenvalues of the Laplacian on the surface considered, which may seem surprising at first but not after reading Buser's text.

We now recall XII.3.C and Weinstein's counterexample. It has to do with studying the distribution of the periodic geodesics in space or/and in phase space.

Here — no more than in Weinstein's example — can we hope for everywhere density in phase. It suffices to append a "Weinstein fragment" to any surface whatsoever, as in the figure:

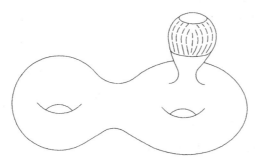

Fig. XII.5.5. A surface about which not much can be said. Berger (1994) © Università di Parma and A. Concari

Topology and the two or more holes don't suffice. An objection might be raised that the drawing isn't **generic** — by which we understand a result valid for almost every surface, or alternatively with probability equal to one. In fact, to be precise this definition requires a measure on the space of surfaces. Since no reasonable measure exists, the word generic here uses a topology on the space, which is easy to define. Generic will then mean with the exception of a sparse subset; see VII.3.D.

However, even generically the periodic geodesics are not dense in phase. The proof is immediate, but modulo the so-called "KAM" theorem of Kolmogorov-Arnold-Moser on the invariant tori; see for example Chap. 6 of (Sinai 1987). We deduce that for each small perturbation of the above figure "a la Weinstein", the phase space still possesses invariant tori (under the geodesic flow), which trap the geodesics enclosed between them.

Now it needs to be remarked that on surfaces of genus greater than 1 there are in fact **three** types of metrics: those with constant curvature, the general metrics, but also the intermediate class of those with strictly negative curvature. Now, for this last class, it has been known since E. Hopf in 1939 that the behavior of the geodesics is excellent — we have full ergodicity as we will see in 5. A below. For the moment we should just retain the idea that the periodic geodesics are everywhere dense in phase.

This distribution isn't in general uniform for the natural measure (called Liouville; see 5. A below) on phase space; more precisely it is uniform if and only if the curvature is constant: this too is a subtle result of Katok (see (Katok, 1988)); we can thus say that for metrics with strictly negative curvature on the surface we have a perfect description of the periodic geodesics. References for the preceding are Ballmann, Gromov and Schroeder (1985), Mané (1987), and Chap. 7 of Sinai (1987). Let us say that the conjunction of several holes (topology) — together with the divergence of the geodesics assured by negative curvature — guarantees a well understood behavior; see also the recent reference (Pollicott, 1994).

There remains a **very simple** question, which applies to surfaces of all topological types: are the periodic geodesics always dense in the space? Otherwise expressed: is it the case that the union of their supports is an everywhere dense subset of the surface? To our knowledge and for all surface types no expert has the least idea whether the answer is yes or no; Weinstein's counterexample was concerned with phase space.

XII.5.C. The sphere: the three Lusternik-Schnirelman geodesics.

In this entire section, we consider only surfaces with the topological type of the sphere, abstract or embedded.

It is very difficult to find or more than one periodic geodesics on a surface. Birkhoff had an idea that works only in certain cases which we will mention below. Meanwhile it will be useful and hopefully agreeable for readers that we reformulate our construction of coverings and the minimax principal above. We consider the space of all curves on the surface and, even though it is of infinite dimension, we visualize it, as an aid to our imagination, as a surface in ordinary space. The vertical direction and the corresponding scale represent the lengths of the curves shown as points of this "surface". The scale reading zero corresponds to curves of zero length, i.e. to the points of the surface studied. This is the shaded region in the figure. It is easy to show that the periodic geodesics are the points of this surface corresponding to curves where the tangent plane is horizontal — this because the periodic geodesics are precisely the curves for which the length of all the curves infinitesimally close doesn't vary, i.e. these are stationary points in the space of curves.

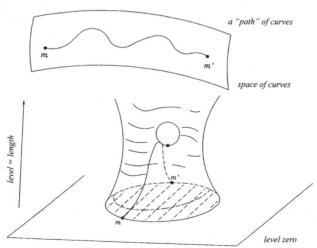

Fig. XII.5.6. Berger (1994) © Università di Parma and A. Concari

We now represent the set of curves of a covering of the surface. In our "surface of curves", it is a **path** joining two points of zero scale reading. The fact that a covering cannot indefinitely have all its curves shortened without **tearing** gets translated to the shape of our space of curves by the fact that this space has at least one hole and that a path of a covering is represented there by a curve leaving from a point of scale reading zero which must then pass through the hole and then return to scale reading zero. The bottom of the hole is a **mountain pass** (saddle point) and the tangent plane there is horizontal; this point then corresponds to a periodic geodesic. The minimax principal is equivalent to saying that there exists a path of this type (a covering) passing exactly through the pass. Birkhoff's shortening technique has the effect of deforming each such path into a path for which the highest point actually loses height. At the end of the day, if we iterate this procedure indefinitely we obtain a path going through the pass. And there we get **stuck**, because there really is a hole.

Otherwise expressed, we use this: to get from one valley to another, we need to traverse a pass. Masochists are allowed to waste their energy by not going straight through the pass, but doubly all the worse for them because, other than suffering useless fatigue, they will never go through the pass at the place where the tangent plane is horizontal and where we can pleasantly picnic (if there isn't too much wind).

Readers have already gleaned what needs to be done for finding other periodic geodesics: we need to show that the space of curves has several holes, even an infinity; and they have likely found the basic idea: we need to consider coverings that yield − in the space of curves − paths passing through different holes.

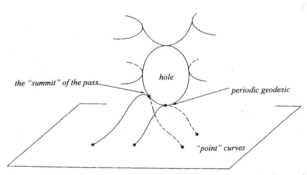

Fig. XII.5.7. Berger (1994) © Università di Parma and A. Concari

Only there is a fundamental difficulty, foreseeable from the beginning of this section: there is a risk that the passes provide only the same periodic geodesic but just traversed several times. One way of being certain that this is not the case is to show that there exist holes for which the passes have non proportional scale readings. But in what follows, we will have to be more subtle.

A very difficult problem is involved. The first result announced was that of Lusternik and Schnirelman in 1927: on each surface there are at least three periodic geodesics, which are moreover simple. Their idea was to consider not only the holes

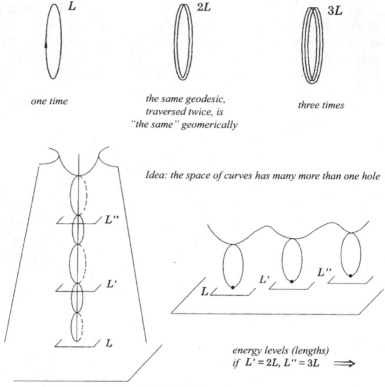

one time

the same geodesic,
traversed twice, is
"the same" geomerically

three times

Idea: the space of curves has many more than one hole

energy levels (lengths)
if $L' = 2L$, $L'' = 3L$ $\implies$

Fig. XII.5.8. Berger (1994) © Università di Parma and A. Concari

through which curves pass that have scale (height) reading zero at their ends, but more generally (don't forget that the space of curves is of infinite dimension and that representing it by a surface in ordinary space is a very crude approximation) "surfaces of curves" and the notion of holes for such objects. The corresponding "surfaces" and "holes" are not difficult to find. We cover the sphere by a **three-**parameter set formed by all the circles on the canonical sphere, including the circles of radius zero (the points) and − at the other extreme − the equators.

But the existence of a corresponding minimax is another story. To our knowledge, the first complete proof is that of (Grayson, 1989). It is very pretty, because it is a sort of heat equation for continuous deformation of the initial "triple" covering of Lusternik-Schnirelman; it uses complicated refined methods of partial differential equations. A proof by "elementary" methods, along with all the necessary geometrical and historical references, can be found in (Taimanov, 1993).

Readers should be aware that the method of passes and holes bears the name **Morse theory**. This very good function method for studying geodesics joining two points and shows immediately that there is an infinity of them for numerous pairs of points. They will only be the same for all points if they are on a surface with only periodic geodesics. Below we will encounter the counting function associated with such geodesics joining two points.

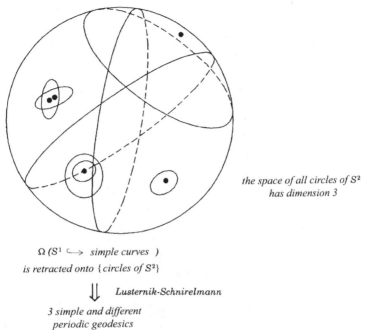

the space of all circles of S²
has dimension 3

$\Omega\ (S^1 \hookrightarrow\ simple\ curves\ )$
is retracted onto {*circles of S²*}

⇓ *Lusternik-Schnirelmann*

3 simple and different
periodic geodesics

Fig. XII.5.9. Berger (1994) © Università di Parma and A. Concari

But for geodesics, the circumstance of having an infinity of holes leads nowhere with respect to different supports, at least for being able to control the heights. There is another difficulty: the space of curves joining two points is a good manifold (of infinite dimension but with good finite dimensional approximations), but in contrast the space of closed curves is not the space of mappings of S^1 into the surface, but a quotient of this space: geometrically identical curves must be identified; the change of orientation and the iteration (which correspond to actions of the groups $\mathbb{Z}_p$) then give singularities in the quotient.

XII.5.D. The sphere: an infinity of periodic geodesics

We mention first that it's only since 1992 that we know that essentially **every** surface — in space or abstract — admits an infinite number of periodic geodesics. We are going to exhibit some of the main points of this recent result, but we will be far from rigorous and precise. There are actually several mutually exclusive cases and, for each case, a proof of a completely different nature. It is in fact a diabolical dichotomy, and even more a trichotomy. If the first method — due to Franks — doesn't work, we will see that it's because the surface has a waist; but then, for surfaces with waist we use a completely different method, due to Bangert.

The initial idea for finding a second periodic geodesic is Birkhoff's; it essentially consists of relying on a **simple** periodic geodesic c (we now know that there is at least one, although Birkhoff himself didn't know this). With each point m of c and

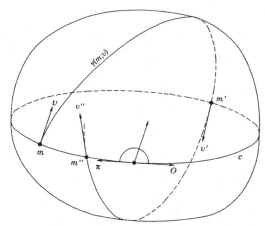

Fig. XII.5.10. The mapping defined by Birkhoff. Berger (1994) © Università di Parma
and A. Concari

with each unit vector v tangent to the surface at m, we associate the geodesic γ
that starts at m with velocity v. Since we are on a topological sphere, c divides
our surface into two "hemispheres". We suppose that the vector v points toward the
first of these. Normally our geodesic $\gamma(m; v)$ moves in the first hemisphere, then
returns to cross c at the pair (m', v'). This last provides a geodesic that moves in the
second hemisphere and returns to cross c at the pair (m'', v''); and so we continue on,
obtaining (m''', v'''), etc. But if (m, v) coincides with (m'', v''), or more generally
with (m'''', v'''') further on, then the geodesic $\gamma(m; v)$ will necessarily be periodic. In
practice it is better to consider the operator f which associates (m'', v'') with (m, v)
in the plane, where (m, v) is described by two coordinates, being respectively the
arc length along c and the angle between 0 and π that defines v. All together these
points form an annulus for which the two boundary circles correspond to the values
0 and π. We want to know if this operation has periodic points. Birkhoff stayed with
the general case and couldn't complete the argument. This annulus reminds us of the
annulus introduced by Birkhoff for billiards in XI.10.B and used by Mather for the
existence of "reversing" trajectories. Here too, in the abstract Riemannian case, it is
fundamental for obtaining the result that we have a dynamical system on this ring,
thus that we have a measure that is preserved by f. Here too this measure is the
product $ds \cdot du$, where ds is the differential of arc length of the geodesic and du that
of the cosine of the angle of departure at the corresponding point. The fact that our
mapping f preserves this measure is a particular case of the fact that geodesic flow
preserves Liouville measure; see Section 5.A below.

Even assuming the existence of a simple geodesic, Birkhoff was only able to
complete the argument in the case called **rotating**, which we will explain. One of
the difficulties is what can happen on the two boundary circles of the annulus. There
are two cases to distinguish, the easy case being where these boundaries rotate
under the Birkhoff mapping. Being on the boundary is precisely the same as considering geodesics that are immediate neighbors of the geodesic c. We thus see the
notion of Riemannian geometry of conjugate points on c arising naturally. If the next

conjugate point doesn't coincide with the starting point, then we actually rotate — this for each starting point according to the classic theory of Sturm; and, moreover, in one sense on one boundary and in the other sense for the other. Birkhoff concluded then the existence of at least two other geodesics, thanks to his proof of the famous posthumous theorem of Poincaré.

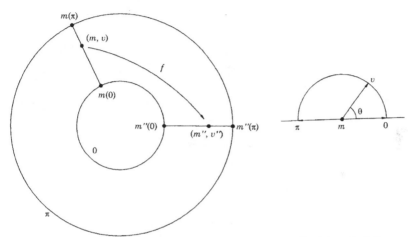

Fig. XII.5.11. The Birkhoff-Poincaré annulus. Berger (1994) © Università di Parma and A. Concari

That in this case there is moreover an infinity of periodic geodesics, i.e. an infinity of fixed points under the Birkhoff mapping, is a classic result of W. Neumann; for the right context see Appendix 9 of (Arnold, 1978).

Now Franks was able to treat the **non rotating** case. He showed this: the Birkhoff mapping in this case either has no fixed point or has an infinity of them. But here there are at least two, thanks to the Lusternik-Schnirelman theorem; there is thus indeed an infinity. Frank's proof cannot be explained in few words; it also uses much prior work on the dynamics of the annulus, i.e. refinements of the work of Birkhoff.

It was later simplified considerably by N. Hingston; see (Hingston, 1993). Why isn't everything now complete? Because we have supposed that the geodesic $\gamma(m; v)$ had to return to cross c. Now some sketches show that this isn't true in all generality, typically if the surface isn't convex. What can happen is that this geodesic accumulates about a periodic geodesic, for example that of a waist; see Fig. XII.5.11.

For simplicity, we suppose that the surface has a waist (the complete proof of Bangert is a little more refined). Here is how Bangert finds an infinity of geodesics. The idea is that of paths of curves passing through holes for which it will be possible to control the heights, i.e. lengths.

The first way is practically the same as for finding — as above — a first periodic geodesic: we start from the geodesic with waist and reduce it to a point by a covering

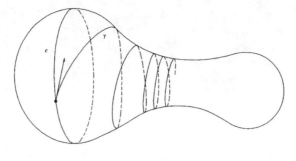

γ won't return to cross c

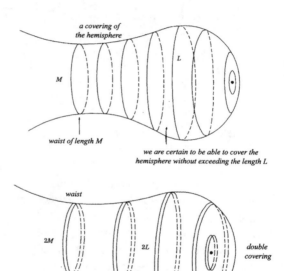

Fig. XII.5.12. Berger (1994) © Università di Parma and A. Concari

of a single hemisphere. Everything moreover happens in this single hemisphere. One of Bangert's ideas is to see what happens on a single side of the waist, by retracting the iterates *n* times from the waist. If this is done carefully, in the manner of the spool method (see the figure), the lengths will be sufficiently controlled to permit the conclusion, after some subtle work with Morse theory.

The counting function for the sphere. If we have an infinity of periodic geodesics, we want to know the asymptotic behavior of FC(L). By refining the methods of Bangert and Franks and by using in particular refinements of Morse theory for the space of closed curves, N. Hingston ((Hingston, 1993)) — by studying each case of the trichotomy — succeeded in showing that we always have:

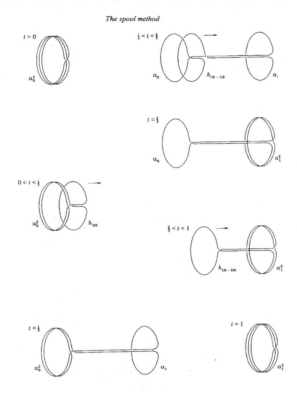

The spool method

γ *will never return to cross* c

Fig. XII.5.13. The figure shows what must be done (as opposed to Fig. XII.5.12) for controlling the length permanently. Bangert (1980) © Springer and V. Bangert

(XII.5.D.1)
$$FC(L) > a\frac{L}{\log L}.$$

The appearance here of $\frac{L}{\log L}$ isn't surprising. In fact, existence is established in all cases by looking for prime numbers k associated with the iterates of the periodic geodesics that satisfy some supplementary properties — so as not to cover geodesics already counted — which are always of the form $k = na + b$, where a, b are relatively prime and n is some integer. A famous theorem of Dirichlet then states that there exist infinitely many such prime numbers and that their density is in $a\frac{L}{\log L}$. The direct connection between k and the lengths allows the conclusion.

Moreover, she showed that, in the space of metrics with positive curvature on the sphere, there exists an open dense set for the C^2 topology for which the counting function is at least quadratic. The reason for quadratic is, with some refinements surely, the same as why the number of pairs (p, q) of relatively prime integers such that $p^2 + q^2 \leq L^2$ is quadratic in L.

We repeat once more, however, that the exact order of magnitude of FC(L) is not known for any surface of toroidal or spherical type. It seems that it isn't even possible at present to conjecture whether, for example, the growth is always at least quadratic or greater; or again to relate this growth to the geometry of the surface.

XII.6. What behavior can be expected for other trajectories? Ergodicity, entropies

XII.6.A. Surfaces of higher genus

Like the problems of periodic geodesics, those with ergodicity are well understood in the case of metrics with negative curvature. This thanks to G. Hedlund in 1935 (Hedlund, 1935) for constant curvature and to E. Hopf in 1939 (Hopf, 1939) for strictly negative, but not necessarily constant, curvature. Hopf's result is difficult, but has the consequence that for surfaces of negative curvature there is ergodicity in the strongest possible sense. In particular almost all the geodesics are everywhere dense, uniformly distributed in phase space.

In order to explain all this, we need to introduce several objects. Readers can also refer to XI.XYZ. The *phase space* of a Riemannian surface (M, g) is the *unitary tangent fiber*, denoted UM; recall that when we speak of *vector velocity* it has to do with (position, velocity), i.e. the mechanical state of the particle. It is the three dimensional manifold composed of the collection of tangent vectors of length 1 at the various points of the surface. The canonical projection is $p : UM \to M$ and the **geodesic flow** is the one-parameter group G^t composed of mappings $G^t : UM \to UM$ defined by:

(XII.6.A.1) $G^t(x)$ *is the vector velocity at time t of the geodesic for which the initial velocity at time zero is the vector x.*

Finally, UM possesses a canonical measure, defined essentially as the "product" of the basic measure for (M, g) by the length (of total amount 2π) of unit circles composed of the fibers $U_m M$ of the unit vectors tangent to M at the point m. An essential point is that this measure, called **Liouville's**, is invariant under geodesic flow. Recall XI.XYZ: ergodic theory in fact has, as its primary context, compact sets endowed with a measure and a bijection f preserving this measure. This theory then studies the behavior of the iterates f^k of f when k becomes very large. This is the discrete context . We are in fact in the continuous context, but there is practically no difference here between the discrete and the continuous: take $f = G^1$.

Hopf's result then says that, in UM, the velocity vectors that generate a geodesic for which the set of velocity vectors is dense in UM form a subset of UM of total Liouville measure, a hypothesis which is clearly not satisfied, for example, by the tangent vectors to the periodic geodesics. But these aren't the only exceptions: there are the geodesics that are asymptotic to periodic geodesics. In fact a lot more is known, practically an ideal amount: apart from a set of measure zero, each geodesic is everywhere dense, and **uniformly** so. This means that the proportion of time spent in any fixed open set U, however small, equals area(U)/ area(M). The "darkening" of the surface by this geodesic is thus a very uniform gray, increasingly dark with increasing time.

But nothing is known at the present moment in the case of a metric of arbitrary curvature. In any event, the geodesic flow is certainly not ergodic in general, even for a generic metric. This can be seen by the same argument using the KAM theorem as in Section 4.B. For any surface it suffices to adjoin a piece of a surface of revolution; a positive measure of geodesics that stay in this piece will be trapped, and their tangent vectors will never be able "to go somewhere else".

Very, very, briefly, here is the idea of E. Hopf's proof. Being given a unit vector v in UM, the fact that the curvature is negative allows us to show that, through v, there passes a well defined curve (locally at least) A of UM, which has the property that all the geodesics generated by the vectors of A converge, or rather approach one another with exponential speed as $t \to \infty$. Likewise, there exists a second curve B (always with $B(0) = v$), but this time the convergence is for $t \to -\infty$. This also tells us that the geodesics starting from B diverge exponentially as $t \to \infty$ but converge exponentially for $t \to -\infty$. If the desired property were false, we would have a subset of positive measure in UM that is invariant under the geodesic flow. But the effects of the A's and the B's, applied to this subset, easily lead to a contradiction: within a set of measure zero, this set must then fill the whole space.

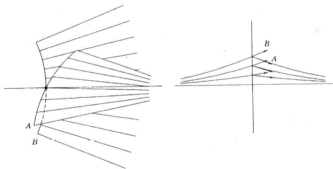

Fig. XII.6.1. In UM; in M. Berger (1994) © Università di Parma and A. Concari

More precisely, we need to show that each continuous (in fact, measurable suffices) function that is invariant under the (geodesic) flow is almost certainly constant. See the precise statement of Birkhoff's theorem in XI.XYZ.1. Technically, Birkhoff's ergodic theorem is applied, which guarantees a limit almost always for the respective means of this function on a geodesic as infinity is approached either positively or negatively. Moreover, these two means are equal almost everywhere. Now the values close to infinity will certainly coincide because of the exponential decrease . Thus these means are constant along the curves A and also along the curves B, as well as along geodesics. Note that the curves, in the classical language of dynamical systems, are called **stable** and **unstable** curves in phase space. We will encounter them in Part C below. In phase space, which is three dimensional, we have a foliation by three families of curves, defining at each point directions on which the function is constant, generating the whole tangent space. We indeed sense that the function will be constant everywhere locally. Next, we assemble the pieces.

There is, however, a substantial technical difficulty, because the means merely exist almost everywhere, for $t \to \infty$ and for $t \to -\infty$. Moreover, if the curves A and B are themselves differentiable, in general they only depend continuously on their point of origin. We need to use a more refined notion, that of *absolute continuity*, which needs to be proved to finally guarantee local constancy. See a very clear exposition of this in the appendix by Misha Brin in (Ballmann, 1995).

Is ergodicity possible on the sphere, specifically on a surface having the topology of the sphere? The dynamicists have asked this question for a long time. The common opinion is that it is impossible. Why? Because surfaces with negative curvature, contrary to the case of the sphere, have both a nontrivial topology which easily yields, as we have seen, lots of periodic geodesics, but mainly a negative curvature. Very recently it has been shown two works have shown that this intuition is false, at least in parts; see (Donnay, 1988), and also (Burns & Gerber, 1989) and (Knieper & Weiss, 1994a). As it is difficult to state these results in the language of dynamical systems — likewise for the result on the counting function of lengths of geodesics joining two points (Section 5. D) — we will briefly recall the various notions of entropy.

XII.6.B. The entropies

We introduced various notions of entropy in XI.XYZ. Here we will give several supplementary explanations adapted to the case of geodesic flow on surfaces, as well as the results themselves for surfaces. In particular we will see in 5.D that for Riemannian geometry there exists a "trivial" definition of topological entropy, but which nonetheless remains difficult to reconcile with the basic definitions.

We need to explain in several ways and in everyday language what it is that these entropies measure. First version: entropy measures the way in which the trajectories of f or of G^t — for us geodesics — diverge exponentially when time (for us the length L) becomes very large. More precisely, the entropy will be positive if there is sufficient exponential divergence with time, and the value of the entropy will be the factor k in e^{kL}. We can also say that, an $\varepsilon > 0$ being fixed, we seek the maximum number of segments of distinct trajectories $t \mapsto G^t(x_i)$ $(0 \leq t \leq L)$, it being understood that two trajectory segments are indistinguishable if they never deviate from one another by more than ε. We study the growth type (exponential) of this number with L, and we take the limit when ε goes to zero.

Another interpretation is that the entropy measures — always in the sense of the exponential factor — the rapidity of loss of information, starting with initial data. There is finally the point of view of Kolmogorov, which consists of observing the rapidity (as k increases) or the precision that approximate information about

$f(x), \ldots, f^k(x)$ (the place of these points in a finite partition) allows retrieval of initial information about x.

It is important to realize that positive entropy does not ensure exponential divergence — for our purposes say of the geodesics starting from all parts of the surface — but only ensures this exponential divergence starting from a certain number of points of the surface. In fashionable language, the positivity of the topological entropy ensures a **chaotic** situation, but only starting from a set of Cantor type (thus nondenumerable). Metric entropy says more: it ensures this chaos starting from a set of positive measure; but this is saying infinitely less than ergodicity, for example.

More precisely, if the metric entropy is strictly positive, then there will be a subset of positive measure on which the restriction of f is ergodic, which ensures a set of positive measure of trajectories everywhere dense in an open set of the space. But this doesn't provide, in general, even a single everywhere dense trajectory. As for the topological entropy, the fact that it is positive says absolutely nothing about the density of trajectories.

On the other hand, for the periodic geodesics, which interest us a lot, it is the result of (Katok, 1980) that ensures the exponential growth of the counting function of periodic geodesics:

(XII.6.B.1) $\liminf_{L \to \infty} \log(FC(L))/L \geqslant h_{top}(G).$

With the notion of topological entropy at hand, we can also now state the result of Katok on the surfaces of genus greater than one mentioned in 4. B. What the text (Katok, 1982) proves is the double inequality

$$\liminf_{L \to \infty} \log(FC(L))/L \geqslant h_{top}(G) \geqslant 1,$$

along with the fact that equality is attained on the right if and only if the curvature is constant (and equal to -1; we recall that we have normalized the volume). Then:

$$h_{top}(G) - \lim_{\varepsilon \to 0} \limsup_{L \to \infty} \frac{1}{L} \log(r(L; \varepsilon)).$$

Of these two definitions, we see why the positivity of Liouville entropy is stronger than that of topological entropy. If the integral is positive, there must exist a set of positive measure of points x, giving an exponential divergence to the geodesic flow G as L goes to infinity.

We recall that there exists (at least) a third entropy, the volumic entropy. It is of interest only when the fundamental group of the surface is infinite, here then for surfaces other than the sphere and real projective space. There are numerous relations between these three entropies. We refer readers to Besson et al. (1995).

XII.6.C. The case of the sphere. The example of Osserman-Donnay

In the case of billiards of XI.10.B, we were very surprised to discover — with Bunimovich — **ergodic** convex billiards, the intuitive idea being that the ergodicity is related to the dispersion of the trajectories, and to the contrary that convexity makes for a concentration of these. We could likewise think that a Riemannian metric on a sphere could never be ergodic. Now in (Donnay, 1988) there was success in constructing an ergodic Riemannian metric on the sphere by improving on an idea of Osserman. Osserman started with a pair of pants – see Fig. XII.5.4 – and closes the pair of pants into a topological sphere by attaching three hemispheres to the three periodic geodesics that make up the boundary. We may suppose that the length of each of the three geodesics is equal to 2π. The hemispheres are then of curvature $+1$. The resulting surface is *denoted by* S.

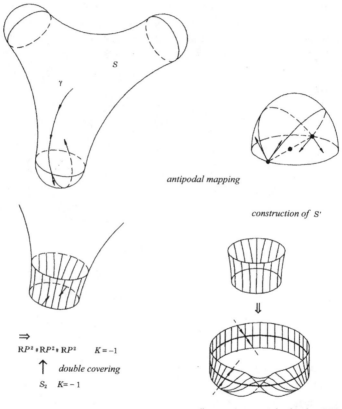

antipodal mapping

construction of S'

$RP^2 \ast RP^2 \ast RP^2$ $K = -1$

↑   *double covering*

S_2 $K = -1$

"we continue straight ahead on Möbius"

Fig. XII.6.2. Antipodal mapping, construction of S', we continue to move "straight ahead" on the Möbius band. Berger (1994) © Università di Parma and A. Concari

What is the behavior of the geodesics on this surface? We start from an arbitrary point of the pair of pants and follow the geodesic. When it comes to the boundary it enters the hemisphere and leaves it at the antipodal point with the vector velocity shown in Fig. XII.6.2. It then enters the pair of pants – and this is the key idea – in the same way as if, instead of having completed the surface at this boundary with a hemisphere, we had identified the antipodes pairwise. In order to better see this identification, we can remark that the effect on a small collar about the boundary yields a Möbius strip. By doing this three times we obtain a surface S' that is compact, smooth, **non orientable** and with constant curvature -1. Topologically, S' is the connected sum of three real projective spaces. Such a surface is just as ergodic as those (of genus greater than 1) that are orientable; see for example Ballmann et al. (1985). If we return to the original surface (topologically a sphere), the ergodic behavior of the geodesic flow on the surface S' induces an ergodic behavior on S.

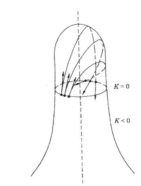

Fig. XII.6.3. Divergent hemisphere

Fig. XII.6.4. Divergence upon entering implies divergence also upon leaving; condition on the meridian: $\frac{dK(l)}{dl} \leq 0$, where K denotes the curvature of the meridian curve and l is the position along the meridian. Berger (1994) © Università di Parma and A. Concari

Now this example of Osserman is smooth, but not enough for us if we are demanding, since the curvature jumps from -1 on the pair of pants to $+1$ on the hemispheres. It is here that the work of Donnay intervenes. We can easily make this construction indefinitely differentiable by smoothing the connections on both sides of the boundaries. The hemispheres will be replaced by surfaces of revolution $\sum', \sum', \sum''$. But we must preserve the condition that the geodesics leaving the pair of pants and entering such a surface satisfy the property that is necessary for ergodicity, which is simply that they diverge upon entering and diverge again upon leaving.

Donnay showed, by studying the infinitesimal variations of the geodesics (what is called a Jacobi field in the jargon of Riemannian geometry), that things go well if the meridian of one of the $\sum$'s has a curvature that satisfies a convexity condition.

Thus ergodic metrics exist on the sphere S^2; but notice that they have, by construction, lots of negative curvature. The problem then arises of finding on S^2 a metric with positive curvature throughout that is ergodic. This problem is presently open.

XII.6.D. Entropy and the length of geodesics joining two given points

We encountered in 4. D above the problem: given two points p and q of the surface, find geodesics (here not necessarily shortest paths) that join these two points. The topology of the space of closed curves on the sphere being infinite, classical Morse theory (see for example (Serre, 1951)) applies, with the restriction that q must never be the conjugate of p on a geodesic going from p to q (this relation is in fact symmetric). It is easy to see that, for any p, the bad q form a set of measure zero. Thus we can say, for almost every pair (p, q) of points of M, there is an infinity of geodesics joining p to q. The case where there would be an infinity of geodesics of the same length isn't considered, since it implies the conjugation of points on a limit geodesic. Thus the *counting function* $FC(p, q; L) = \{$number of geodesics with extreme points p and q, and of length less than L$\}$, is well defined for almost all p and q.

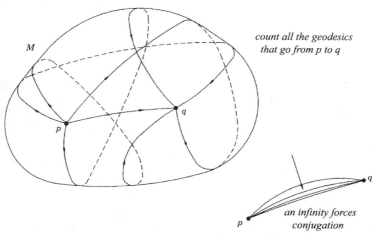

Fig. XII.6.5. Berger (1994) © Università di Parma and A. Concari

Presently no one can say anything about any individual counting function (p, q given). On the other hand Mañé and Paternain, in part independently, have just shown (Paternain & Paternain; 1994), (Mañé, 1997) that the double mean, when p and q range over the surface, yields the topological entropy of the surface by the formula (valid moreover for every dimension and every Riemannian manifold):

$$h_{\text{top}}(G) = \lim_{L \to \infty} \frac{1}{L} \left(\int_{M \times M} \log(FC(p, q; L) \, dp dq \right).$$

For those who like terminology, we could even define a *geodesic entropy* by a formula of this type. The proof unites a geometric part using the exponential mapping of the product M × M with a specialized theorem of Yomdin that allows calculation of an entropy by a formula valid only in the differentiable case (see Gromov, 1987b; we point out that it is rare for dynamical systems formulas to use differentiability.)

It is known that if the topological entropy of the geodesic flow — see for example the preceding section — is nonzero, then the counting function $FC(p, q; L)$ grows exponentially for many pairs (p, q) of points of M.

We can be optimistic and require that we always have:

$$h_{top}(G) = \lim_{L \to \infty} \frac{1}{L} \log(FC(p, q; L)) \text{ for almost all } (p, q).$$

Otherwise we can only require that this is true for generic surfaces.

There are only partial answers, but which justify the above optimism. Specifically:

- (Ma-é, 1997): *if the surface is free of conjugate points, in particular if it is of negative or zero curvature, then*

$$h_{top}(G) = \lim_{L \to \infty} \frac{1}{L} \log(FC(p, q; L)) \text{ for all } (p, q).$$

- (Paternain & Paternain, 1994): *we always have for all p and almost all q*:

$$h_{top}(G) \geqslant \limsup_{L \to \infty} \frac{1}{L} \log(FC(p, q; L)).$$

In contrast, this isn't true in all generality, because (Burns & Paternain, 1997) exhibit a large set of metrics on S^2 where the strict inequality

$$\limsup_{L \to \infty} \frac{1}{L} \log(FC(p, q; L)) < h_{top}$$

is satisfied for an open set of pairs (p, q).

The book (Paternain, 1999) is entirely devoted to geodesic flow.

XII.7. Do the mechanics determine the metric?

Question: suppose that two surfaces (embedded in space or abstract) have the same mechanics, i.e. the same behavior for their trajectories or — to be still more precise — that there is an isomorphism between all their trajectories. Are they then necessarily the same, i.e. isometric? That is, do they have the same geometry?

It is easy to see that the answer is "no" in general; it suffices to think of the Zoll surfaces (other than the sphere, see 2. B). They have their periodic geodesics, of the same length. It is impossible to recover their metric with only their "mechanics".

However, we need to be still more precise. The right notion is **conjugate geodesic flow**. We will say that two Riemannian surfaces (M, g) and (N, h) have geodesic flows that are conjugate if there exists a diffeomorphism $\phi : UM \to UN$ between their unitary tangent fibers which commutes with their geodesic flows, i.e if the diagram below

$$
\begin{array}{ccc}
UM & \xrightarrow{\phi} & UN \\
{\scriptstyle G_M^t}\downarrow & & \downarrow{\scriptstyle G_N^t} \\
UM & \xrightarrow[\phi]{} & UN
\end{array}
$$

is commutative for each time t.

Remarks. The first remark is that we don't specify the degree of differentiability. It is, however, a very important matter; some very subtle phenomena can be produced under low differentiability (see the various references mentioned). To avoid being too technical we will simply assume that we have sufficient differentiability.

The second is that it is important to observe that we don't require that ϕ commute with the natural projections $UM \to M$ and $UN \to N$, i.e. that it be an isomorphism of the fibers, which is too strong a condition. We only want to use the trajectories by themselves, without in any way knowing where they go. You travel in an airplane, but aren't allowed to look out the window!

Some examples. First for the Zoll surface, we need to show the conjugation. It doesn't suffice to know that all the geodesics are periodic or even to know their lengths; we still need to display an isomorphism ϕ. This was done by Weinstein; see Section 2, §4. F of (Besse, 1978).

After Anosov in 1967 we know that the answer is very good in the case of manifolds with strictly negative curvature. Once the geodesic flows are conjugate for two surfaces of this type, they are then necessarily the same (isometric). This theorem was refined in (Otal, 1990a), then in (Croke, Fathi & Feldman, 1992).

In (Croke, 1992) it was proved that if a surface has geodesic flow conjugate to that of a flat torus, then the two are isometric. Geodesic flow is studied in a very detailed way in the remarkable text (Bangert, 1988).

In (Croke & Kleiner, 1994) it was proved that the conjugacy of geodesic flows implies equality of areas. This apparently simple result, is actually not so difficult to to prove for a C^1 conjugation, but much more difficult when it is only known that the conjugation is C^0.

A series of rather obvious results concern the length spectrum of a surface. This is simply the set of lengths of the periodic geodesics. Knowing this set is in general insufficient for recovering the surface, i.e. there exist nonisometric surfaces having the same length spectrum; see for example (Buser, 1992). We mention only briefly that the problem pointed out for billiards at the end of XI.10 on connections between the length spectrum and the spectrum of the Laplacian is a fundamental subject for surfaces and more generally for all Riemannian manifolds; for this see Chaps. IV, V of (Berger, 1999) and VI.7 of (Berger, 2003).

In contrast, the **marked** length spectrumspectrum!marked length|tbf is a much stronger invariant. It involves pairs $(L, [\gamma])$, where L is a length of a periodic geodesic and $[\gamma]$ is its free homotopy class. Then the marked spectrum uniquely determines the metric of the surface: (Otal , 1990a), then Croke et al. (1992).

Note. We need to be aware, in all these results, that the differentiability hypotheses are important. The lower the differentiability, the harder the result. A typical example is that of (Ghys, 1987).

XII.8. Recapitulation and open questions

Are the periodic geodesics dense in space for each surface? It is very interesting to observe the examples given in (Bangert, 1988), for which the periodic geodesics which are the shortest in their free homotopy class aren't dense in space at all — in fact are far from it.

To estimate the order of growth of the counting function of periodic geodesics for at least one surface of the type of the torus or or the sphere.

Is the counting function always at least quadratic on each surface? See N. Hingston's result above. Is it always exponential "in general", i.e. even for a torus with any non flat metric? If we think of an analogy with the billiards, exponential growth doesn't seem likely in general. In fact, for polygonal billiards, we know two results on periodic trajectories: that of Katok (Katok, 1987) which says that, for arbitrary polygonal billiards, the counting function is never exponential. That of Masur (Masur, 1990) says that the counting function is always sandwiched by two quadratic estimates, this for polygonal billiards where the vertex angles are rational multiples of π.

Does there exist an ergodic metric on the sphere with everywhere positive curvature? If not, at least strictly positive Liouville entropy? And if not, having at least one everywhere dense geodesic?

Is it the case that a generic metric on S^2 always has positive topological entropy?

It remains to study the case of the torus, which seems intermediate between that of the sphere and the surfaces of higher genus.

XII.9. Higher dimensions

We have only considered curves and surfaces in this book, being compelled to limit its length in one way or another. We have encountered higher dimensions for convex sets and polytopes, and dimension three for packings and lattices. Of course there exist objects that generalize curves and surfaces in all dimensions. For these objects — the most important being the Riemannian manifolds — there is a geometry, a metric, geodesics and their flow, a Laplacian spectrum and vibrations, heat and wave equations etc., and various notions of curvature. In (Berger, 1999) there is a synthetic exposition, and in the book (Berger, 2003) roughly the same content, but with more detail and with the ideas of the proofs. Riemannian geometry is for everyone; its importance is at least double: on the one hand, it is the model for naturally generalizing Euclidean geometry, since by definition a Riemannian manifold is infinitesimally a Euclidean space throughout. On the other hand, Riemannian manifolds form a natural context for treating Hamiltonian (rational) mechanics, for the solar system in particular.

Bibliography

[B] Berger, M. (1987, 2009) *Geometry I,II*. Berlin/Heidelberg/New York: Springer

[BG] Berger, M., & Gostiaux, B. (1987) *Differential Geometry: Manifolds, Curves and Surfaces*. Berlin/Heidelberg/New York: Springer

Arnold, V. (1978). *Mathematical methods of classical mechanics*. Berlin/Heidelberg/New York: Springer

Ballmann, W. (1995). *Lectures on spaces of nonpositive curvature*. Basel: Birkhäuser

Ballmann,W., Gromov, M., & Schroeder, V. (1985). *Manifolds of nonpositive curvature*. Basel: Birkhäuser

Bangert, V. (1980). Closed Geodesics on complete surfaces. *Annals of Mathematics, 251*, 83–96

Bangert, V. (1988). Mather sets for twist maps and geodesics on tori. In U. Kirchgraber & H. Walther (Eds.), *Dynamics reported* (pp. 1–56). Chichester: Teubner, Wiley

Berger, M. (1993). Encounter with a geometer: Eugenio Calabi. In P. de Bartolomeis, F. Tricerri, & E. Vesentini (Eds.), *Conference in honour of Eugenio Calabi, Manifolds and geometry* (Pisa). Cambridge University Press, 20–60

Berger, M. (1994). Géométrie et dynamique sur une surface. *Rivista di Matematica della Università di Parma, 3*, 3–65

Berger, M. (1999). *Riemannian geometry during the second half of the twentieth century*. Providence: American Mathematical Society

Berger, M. (2003). *A panoramic introduction to Riemannian geometry*. Berlin/Heidelberg/New York: Springer

Besse, A. (1978). *Manifolds all of whose geodesics are closed*. Berlin/Heidelberg/New York: Springer

Besson, G., Courtois, G., & Gallot, S. (1995). Entropies et rigidités des espaces localement symétriques de courbure strictement négative. *Geometric and Functional Analysis, 5*, 731–799

Bliss, G. (1902–1903). The geodesic lines on an anchor ring. *Annals of Mathematics, 4*, 1–20

Burns, K., & Gerber, M. (1989). Real analytic Bernouilli geodesic flows on S2. *Ergodic Theory and Dynamical Systems, 8*, 531–553

Burns, K., & Paternain, G. (1997). Counting geodesics on a Riemannian manifold and topological entropy of geodesic flows. *Ergodic Theory and Dynamical Systems, 17*, 1043–1059

Buser, P. (1992). *Geometry and spectra of compact Riemann surfaces*. Basel: Birkhäuser

Byalyi, M., & Polterovich, L. (1986). Geodesic flows on the two-dimensional torus and phase transition 'commensurability – noncommensurability'. *Functional Analysis and its Applications (Translation of Funktsional.Anal. i Prilozhen.), 20*, 260–266

Calabi, E., & Cao, J. (1992). Simple closed geodesics on convex surfaces. *Journal of Differential Geometry, 36*, 517–549

Chavel, I. (1993). *Riemannian geometry: A modern introduction*. Cambridge: Cambridge University Press

Croke, C. (1992). Volume of balls in manifolds without conjugate points. *International Journal of Mathematics, 3*, 455–467

Croke, C., Fathi, A., & Feldman, J. (1992). The marked length spectrum of a surface of non positive curvature. *Topology, 31*, 847–855

Croke, C., & Kleiner, B. (1994). Conjugacy rigidity for manifolds with a parallel vector field. *Journal of Differential Geometry, 39*, 659–680

Donnay, V. (1988). Geodesics flow on the two-sphere II: ergodicity. *Lecture Notes in Math. n° 1342: Dynamical systems*, Springer, 112–153

Funk, P. (1913). Über Flächen mit lauter gescholssenen geodätischen Linien. *Mathematische Annalen, 74*, 278–300

Galperin, G. (2003). Convex polyhedra without simple closed geodesics. *Regular and Chaotic Dynamics, 8*, 45–58

Ghys, E. (1987). Flots d'Anosov dont les feuilletages stables sont différentiables. *Annales Scientifiques de l'Ecole Normale Supérieure, 20*, 251–270

Grayson, M. (1989). Shortening imbedded curves. *Annals of Mathematics, 129*, 71–111

Gromoll, D., & Grove, K. (1981). On metrics on S2 all of whose geodesics are closed. *Inventiones Mathematicae, 65*, 175–177

Gromov, M. (1987b). Entropy, homology and semi-algebraic geometry. In *Séminaire Bourbaki* (pp. 145–146). Paris: Société mathématique de France, 225–240

Gruber, P., & Wills, J. (Eds.). (1993). *Handbook of convex geometry.* Amsterdam: North-Holland

Hadamard, J. (1898). Les surfaces à courbure opposées et leurs lignes géodésiques. *Journal de Mathématiques Pures et Appliquées, 4,* 27–73

Hass, J., & Morgan, F. (1996). Geodesics and soap bubbles on surfaces. *Mathematische Zeitschrift, 223,* 185–196

Hedlund, G. (1935). On the metric transitivity of the geodesics on closed surfaces of constant negative curvature. *Annals of Mathematics, 35,* 787–808

Hingston, N. (1993). On the growth of the number of closed geodesics on the two-sphere. *International Mathematics Research Notices, 9,* 253–262

Hopf, E. (1939). Statistik des geodätischen Linien in Mannigfaltigkeiten negativer Krümmung. Ber. Vehr. Sächs. Akad. Wiss. Leipzig, 91, 261–304

Katok, A. (1980). Lyapunov exponents, entropy and periodic orbits for diffeomorphisms. *Publications mathm atiques de lílnstitut des hautes études scientifiques, 51,* 137–173

Katok, A. (1982). Entropy and closed geodesics. *Ergodic Theory Dynam. Systems, 2,* 339–365

Katok, A. (1987). The growth rate of singular and periodic orbits for a polygonal billiard. *Communications in Mathematical Physics, 111,* 151–160

Katok, A. (1988). Four applications of conformal equivalence to geometry and dynamics. *Ergodic Theory Dynam. Systems, 8,* 139–152

Klingenberg, W. (1982). *Riemannian geometry* (2nd ed.). Berlin: De Gruyter, 1995

Knieper, G., & Weiss, H. (1994a). A surface with positive curvature and positive topological entropy. *Journal of Differential Geometry, 39,* 229–249

Le Brun, C., & Mason, L. (2002). Zoll Manifolds and complex Surfaces, *Journal of Differential Geometry, 61,* 453–535

Mané, R. (1987). *Ergodic theory and differentiable dynamics.* Berlin/Heidelberg/New York: Springer

Mané, R. (1997). On the topological entropy of geodesic flows. *Journal of Differential Geometry, 45,* 74–93

Masur, H. (1990). The growth rate of trajectories of a quadratic differential. *Ergodic Theory Dynam. Systems, 10,* 151–176

Otal, J.-P. (1990a). Le spectre marqué des surfaces à courbure négative. *Annals of Mathematics, 131,* 151–162

Paternain, G. (1999). *Geodesic flows.* Basel: Birkhäuser

Paternain, G., & Paternain, M. (1994). Topological entropy versus geodesic entropy. *International Journal of Mathematics, 5,* 213–218

Poincaré, H. (1905). Sur les lignes géodésiques des surface convexes. *Transactions of the American Mathematical, 6,* 237–274

Pollicott, M. (1994). Closed geodesic distribution for manifolds of non positive curvature. Coventry: Warwick University

Rademacher, H. (1994). On a generic property of geodesic flows. *Mathematische Annalen, 298,* 101–116

Serre, J.-P. (1951). Homologie singulière des espaces fibrés. *Annals of Mathematics, 54,* 425–505

Sinai, Y. (Ed.). (1987). *Dynamicals systems II, encyclopaedia of mathematical sciences* (Vol. 2). Berlin/Heidelberg/New York: Springer

Smale, S. (1998). Mathematical problems for the next century, *The Mathematical Intelligencer, 20*(2), 11–27

Taimanov, I. (1993). On the existence of three non self-intersecting closed geodesics on manifolds homeomorphic to the 2-sphere. *Russian Academy of Science. Izvestiya: Mathematics, 40,* 565–590

Selected Abbreviations for Journal Titles

Acta Arith. = Acta Arithmetica

Acta Math. = Acta Mathematica

Adv. Math. = Advances in Mathematics

Amer. J. Math. = American Journal of Mathematics

Amer. Math. Monthly = The American Mathematical Monthly

Ann. Mat. Pura Appl. = Annali di Matematica Pura ed Applicata

Annals of Math. = Annals of Mathematics

Annals Prob. = The Annals of Probability

Arch. Math. (Basel) = Archiv der Mathematik

Atti Accad. Naz. Lincei Cl. Sci. Fis. Mat. Natur. Rend. Lincei (9) Mat. Appl. = Atti della Accademia Nazionale dei Lincei. Classe di Scienze Fisiche, Matematiche e Naturali.

Rendiconti Lincei. Serie IX. Matematica e Applicazioni. Accad. Naz. Lincei, Rome.

Beiträge Algebra Geom. = Beiträge zur Algebra und Geometrie

Bull. Amer. Math. Soc. = Bulletin of the American Mathematical Society

Bull. London Math. Soc. = The Bulletin of the London Mathematical Society

Bull. Sci. Math. = Bulletin des sciences mathématiques

Bull. Soc. Math. France = Bulletin de la Société MathÈmatique de France

Comment. Math. Helv. = Commentarii Mathematici Helvetici

Comm. Math. Phys. = Communications in Mathematical Physics

Comm. Pure Appl. Math. = Communications on Pure and Applied Mathematics

Comment. Phys.-Math. = Commentationes Physico-Mathematicae

Compos. Math. = Compositio Mathematica

Contemp. Math. = Contemporary Mathematics

C. R. Math. Acad. Sci. Paris = Comptes Rendus, AcadÈmie des sciences de Paris

Discrete Comput. Geom. = Discrete & Computational Geometry

Duke Math J. = Duke Mathematical Journal

Elem. Math. = Elemente der Mathematik

Enseign. Math. (2) = Líenseignement mathématique

Ergodic Theory Dynam. Systems = Ergodic Theory and Dynamical Systems

Experiment. Math. = Experimental Mathematics

M. Berger, *Geometry Revealed*, DOI 10.1007/978-3-540-70997-8,
© Springer-Verlag Berlin Heidelberg 2010

Expo. Math. = Expositiones Mathematicae

Funct. Anal. Appl. = Functional Analysis and its Applications (Translation of Funktsional. Anal. i Prilozhen.)

Fund. Math. = Fundamenta Mathematicae

Geom. Dedicata = Geometriae Dedicata

Geom. Funct. Anal. = Geometric and Functional Analysis

Gaz. Math. = Gazette des mathÈmaticiens (Soc. Math. France)

Helv. Phys. Acta = Helvetica Physica Acta

Indag. Math. = Indagationes Mathematicae

Int. Math. Res. Notices = International Mathematics Research Notices

Internat. J. Math. = International Journal of Mathematics

Inventiones Math. = Inventiones Mathematicae

Israel J. Math. = Israel Journal of Mathematics

Jahresber. Deutsch. Math. Verein. = Jahresbericht der Deutschen Mathematiker-Vereinigung (DMV)

J. Amer. Math. Soc. = Journal of the American Mathematical Society

J. Combin. Theory Ser. A = Journal of Combinatorial Theory. Series A

J. Complexity = Journal of Complexity

J. Differential Geom. = Journal of Differential Geometry

J. Funct. Anal. = Journal of Functional Analysis

J. Geom. Anal. = Journal for Geometric Analysis

J. Geom. Phys. = Journal of Geometry and Physics

J. London Math. Soc. = Journal of the London Mathematical Society, Second Series

J. Math. Pures Appl. = Journal de Mathématiques Pures et Appliquée

J. Reine Angew. Math. = Journal f,r die Reine und Angewandte Mathematik

J. Stat. Phys. = Journal of Statistical Physics

Math. Ann. = Mathematische Annalen

Math. Intelligencer = The Mathematical Intelligencer

Math. Mag. = Mathematics Magazine

Math. Res. Lett. = Mathematical Research Letters

Mat. Sb. [ou Mat. USSR Sb.] (Translated in Sb. Math.) = Rossiĭskaya Akademiya Nauka

Matematicheskiĭ Sbornik. ìNaukaî, Moscow. (Translated in Sb. Math.)

Math. Nachr. = Mathematische Nachrichten

Math. Scand. = Mathematica Scandinavica

Math. Z. = Mathematische Zeitschrift

Mathematika = Mathematika. A Journal of Pure and Applied Mathematics

Michigan Math. J. = Michigan Mathematical Journal

Mitt. Dtsch Math. Ver. = Mitteilungen der Deutschen Mathematiker Vereinigung

Mon. Not. Roy. Astron. Soc. = Monthly Notices of the Royal Astronomical Society

Monatsh. Math. = Monatshefte f„r Mathematik

Notices Amer. Math. Soc. = Notices of the American Mathematical Society

Pacific J. Math = Pacific Journal of Mathematics

Proc. Amer. Math. Soc. = Proceedings of the American Mathematical Society

Proc. Natl. Acad. Sci. USA = Proceedings of the National Academy of Sciences of the USA

Progr. Math. = Progress in Mathematics

Publ. Math. Inst. Hautes . . . tudes Sci. = Publications mathmatiques de lílnstitut des hautes études scientifiques

Q. J. Math. = The Quarterly Journal of Mathematics

Regul. Chaotic Dyn. = Regular and Chaotic Dynamics

Rend. Mat. Appl., VII = Atti della Accademia Nazionale dei Lincei. Classe di Scienze Fisiche, Matematiche e Naturali. Rendiconti Lincei. Serie IX. Matematica e Applicazioni. Accad. Naz. Lincei, Rome.

Resenhas = Resenhas do Instituto de Matem·tica e Estatìstica da Universidade de S„o Paulo

Results Math. = Results in Mathematics

Rev. Mat. Complut. = Revista Matemática Complutense. Serv. Publ. Univ. Complutense Madrid, Madrid.

Russian Math. Surveys = Russian Mathematical Surveys

Sb. Math. = Sbornik Mathematics. Russ. Acad. Sci., Moscow. (Translation of Mat. Sb.)

Studia Math. = Studia Mathematica

Tohoku Math. J. = The Tohoku Mathematical Journal

Trans. Cambridge Phil. Soc. = Transactions of the Cambridge Philosophical Society

Trans. Amer. Math. Soc. = Transactions of the American Mathematical Society

Name Index

Subject Index

Symbol Index

827

Printed in the United States
By Bookmasters